# ANIMAL BIOLOGY

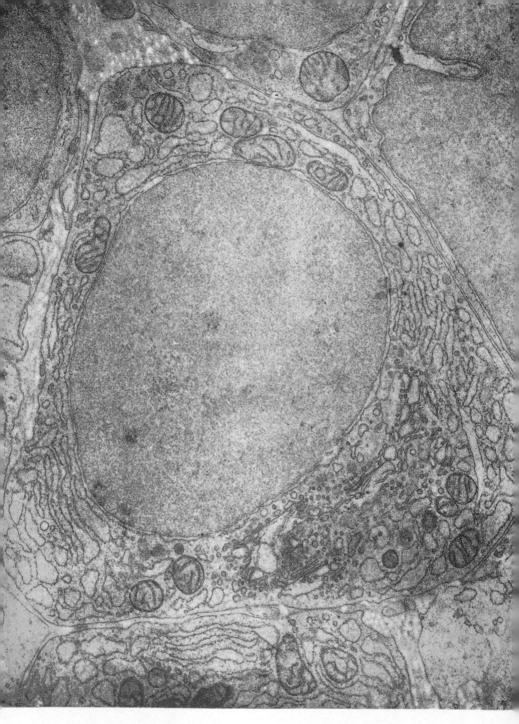

Electronmicrograph of a lymphocyte from a sheep lymph node (compare with Fig. 18) ($\times$ 27,000).  (*E. H. Mercer.*)

# ANIMAL BIOLOGY

*By*

## A. J. GROVE, M.A., D.Sc.
LATE PROFESSOR OF ZOOLOGY IN THE UNIVERSITY OF LONDON

*and*

## G. E. NEWELL, D.Sc., Ph.D.
LATE UNIVERSITY PROFESSOR OF ZOOLOGY, QUEEN MARY COLLEGE
UNIVERSITY OF LONDON

*in collaboration with*

## J. D. CARTHY, M.A., Ph.D.
LATE SCIENTIFIC DIRECTOR, FIELD STUDIES COUNCIL, LONDON
SOMETIME LECTURER IN ZOOLOGY, QUEEN MARY COLLEGE

## E. H. MERCER, D.Sc., Ph.D.
ELECTRON MICROSCOPE UNIT, THE JOHN CURTIN SCHOOL OF MEDICAL
RESEARCH, THE AUSTRALIAN NATIONAL UNIVERSITY, CANBERRA

*and*

## P. F. NEWELL, Ph.D.
LECTURER IN ZOOLOGY, WESTFIELD COLLEGE
UNIVERSITY OF LONDON

## UNIVERSITY TUTORIAL PRESS LTD
842 YEOVIL ROAD, SLOUGH, SL1 4JQ

*Published* 1942
*Second Edition* 1944.   *Reprinted* 1945, 1946, 1947, 1948
*Third Edition* 1950
*Fourth Edition* 1953.   *Reprinted* 1955
*Fifth Edition* 1957.   *Reprinted, with minor alterations,* 1958
*Sixth Edition, revised and reset,* 1961
*Reprinted, with minor alterations,* 1962, 1964
*Reprinted* 1965
*Seventh Edition, revised and reset,* 1966.   *Reprinted* 1968
*Eighth Edition,* 1969.   *Reprinted* 1971
*Ninth Edition* 1974.   *Reprinted* 1977, 1979

ISBN: 0 7231 0641 X

REPRODUCED, PRINTED AND BOUND IN GREAT BRITAIN BY
FAKENHAM PRESS LIMITED, FAKENHAM, NORFOLK

# FOREWORD

By the late C. F. A. PANTIN, M.A., Sc.D.(Cantab.), F.R.S.
Fellow of Trinity College, Cambridge, University Professor of Zoology.

THE old method of approaching Zoology by the detailed study of the anatomy of a few types has much to recommend it. It suffers, however, from two disadvantages. First, a few types chosen from the vast array of different kinds of animals are necessarily so different from each other that it is not easy for the student to compare them in more than a superficial way. Second, it leaves out that aspect which has received more attention than any other in recent years: the relation of structure to the functional requirements of the environment.

In this book the authors have overcome these difficulties with success. They have put the comparative method before the student from the very beginning, and they have made clear throughout the book the functional significance of structure and its relation to the circumstances of the animal's existence.

C. F. A. PANTIN.

CAMBRIDGE.


# PREFACE

IN this book the method of presentation adopted has been to emphasise the broad general principles of Animal Biology with a leaning towards the physiological aspect. This has involved the utilisation of the "types" included in most syllabuses more as "pegs on which to hang a tale" than as objects for intensive and isolated study and description. As a natural corollary of such treatment it is often the comparative aspect which is important— particularly in the vertebrates—though the arrangement is such that individual types can easily be studied separately. Equally frequently, however, a detailed account of one example has served to illustrate the salient features of a particular grade of organisation. Even so, an attempt has been made to give a picture of the organism as a whole and an explanation of "the way it works". In this, we have endeavoured to include the most recent work available to us, even at times at the risk of appearing to go beyond

v

the bounds of an elementary treatment of the subject. However, in order not to distract from the general sequence of presentation, such material has sometimes been relegated to small type. Also, we have not been deterred from repetition where adequate emphasis required it.

Sufficient, and in some cases more than sufficient, has been provided for the requirements of candidates for the General Certificate of Education at Advanced Level, University Scholarship papers, and for Medical students. The discerning student can select those portions appropriate to his needs, though it is hoped that his interest in the subject may have been sufficiently aroused to induce him to add to his essential examinational knowledge by exploring regions outside his syllabus.

It will be noticed that the chapter dealing with the classification of the animal kingdom is fuller than is usual in books of this standard. This has been done in the hope that the student will be led to appreciate the enormous diversity of animal forms. Our reasons for the particularly extensive treatment of the Chordata are, firstly, to show that a true presentation of any group is impossible without reference to fossil forms, and, secondly, because the chordates afford a very good example of evolution within a phylum. Also, the treatment of histology is unusual, for, in addition to the usual description of the tissues, there have been included additional physiological topics which are not appropriate in other sections.

In general, a textbook for elementary students serves two main purposes: it is a supplement to instruction by lesson or lecture; and a source of information for students not attending a formal course of study. It may be said with some truth that many textbooks do not fulfil the latter purpose because of their neglect of those broad generalisations which are common alike to elementary and advanced work. For the young student such generalisations present considerable difficulty (because of the restricted number of examples which can be studied in the time available to him), albeit they are rightly required in answering examinational questions. We have, therefore, tried to keep this need ever before us and it has largely determined the arrangement of the book. The student is recommended to use these "general" portions both as introductions to, and summaries of, the subject-matter dealt with in the succeeding sections.

Of the illustrations, most have been drawn by us especially for this book: a few are reproduced by permission from the *School*

*Science Review* and original papers, whilst in Part IV a few are included from *Textbook of Zoology* by H. G. Wells and A. M. Davies. It must be emphasised that in most cases the figures are entirely diagrammatic; their purpose is to bring out one or two particular points and not to provide a realistic "picture" of the animal or a part of it.

No study of Animal Biology can be complete which ignores practical acquaintance with the animals described, for, as has been rightly said, "Zoology is the study of animals and not merely the study of books about animals". We cannot too strongly urge that, for true appreciation, a living animal is preferable to a dead one, and a dead one is better than any drawing or written description. A list of useful practical manuals is given at the front of the book.

We are indebted for wise criticism and for the reading of the proofs to Dr D. G. Catcheside, Mr A. M. Clark, Dr N. H. Howes, Dr C. F. A. Pantin, Mr B. J. Rickard, Dr W. A. M Smart, Dr J. E. Smith, Miss E. V. Southall, and Dr R. H. Whitehouse.


# PREFACE TO THE NINTH EDITION

Several of the original sections of this book have been extensively rewritten and new illustrations have been provided. It remains true, in this, as in previous editions, that more detail than is required by syllabuses for school examinations is given in most sections and some guidance may be helpful to enable a student to read the various chapters in the most profitable sequence. For example, Chapter I may well be left on one side until rather late in a VIth form course.

# CONTENTS

CHAPTER                                                                      PAGE

I. INTRODUCTION .. .. .. .. .. .. .. 1
   Physics and Biology .. .. .. .. .. .. 9
   Molecular Genetics .. .. .. .. .. .. 14

II. THE CELL AND CELL CYCLE .. .. .. .. 28
   Development, Differentiation and Morphogenesis .. .. 57

III. THE PRINCIPLES UPON WHICH ANIMALS ARE
   CLASSIFIED—TAXONOMY .. .. .. .. 66

IV. PROTOZOA .. .. .. .. .. .. .. 71
   *Amoeba* .. .. .. .. .. .. .. .. 71
   Differences between Animals and Plants .. ,. .. 77
   *Paramecium* .. .. .. .. .. .. .. 79
   *Monocystis* .. .. .. .. .. .. .. 92
   Malaria Parasite .. .. .. .. .. .. 96
   Sexual Processes in the Protozoa .. .. .. .. 103
   *Entamoeba histolytica* .. .. .. .. .. .. 104
   Trypanosomes .. .. .. .. .. .. .. 105
   *Euglena* .. .. .. .. .. .. .. .. 110
   Behaviour of the Protozoa .. .. .. .. .. 116

V. METAZOA .. .. .. .. .. .. .. .. 121

VI. DIPLOBLASTICA (COELENTERATA) .. .. .. .. 127
   Introduction .. .. .. .. .. .. .. 127
   *Hydra* .. .. .. .. .. .. .. .. 130
   *Obelia* .. .. .. .. .. .. .. .. 148

VII. TRIPLOBLASTICA .. .. .. .. .. 157

VIII. PLATYHELMINTHES .. .. .. .. .. .. 162
   *Turbellaria* .. .. .. .. .. .. .. 162
   Parasitic Flatworms .. .. .. .. .. .. 169
   Ectoparasitic Flukes—Monogenoidea .. .. .. 170
   Cestoda .. .. .. .. .. .. .. .. 172
   *Taenia solium* .. .. .. .. .. .. .. 173
   Other Tapeworms .. .. .. .. .. .. 178
   Endoparasitic Flukes—Digenea .. .. .. .. 181
   *Fasciola hepatica* .. .. .. .. .. .. 181

IX. NEMATODA .. .. .. .. .. .. .. 193
   Introduction .. .. .. .. .. .. .. 193
   *Ascaris* .. .. .. .. .. .. .. .. 195
   Other Nematodes .. .. .. .. .. .. 204

| CHAPTER | | PAGE |
|---|---|---|
| X. | PARASITISM .. .. .. .. .. .. .. | 208 |
| XI. | THE COELOMATE ANIMALS .. .. .. .. .. | 214 |
| XII. | ANNELIDA.. .. .. .. .. .. .. .. | 223 |
| | Polychaeta—*Nereis* .. .. .. .. .. .. | 223 |
| | Oligochaeta—*Lumbricus* .. .. .. .. .. | 236 |
| | Hirudinia—*Hirudo* .. .. .. .. .. .. | 267 |
| XIII. | ARTHROPODA .. .. .. .. .. .. .. | 268 |
| | Introduction .. .. .. .. .. .. .. | 268 |
| | Crustacea—the Crayfish .. .. .. .. .. | 272 |
| | Other Crustacea .. .. .. .. .. .. | 308 |
| | The Cockroach (*Periplaneta*) .. .. .. .. .. | 314 |
| | Some other Insects .. .. .. .. .. .. | 334 |
| XIV. | MOLLUSCA .. .. .. .. .. .. .. | 345 |
| | Introduction .. .. .. .. .. .. .. | 345 |
| | *Helix aspersa* .. .. .. .. .. .. .. | 348 |
| | Classes of Molluscs .. .. .. .. .. .. | 357 |
| XV. | CHORDATA .. .. .. .. .. .. .. | 362 |
| | Introduction .. .. .. .. .. .. .. | 362 |
| | Differences between Chordate and Non-chordate Animals | 365 |
| XVI. | ACRANIA .. .. .. .. .. .. .. .. | 366 |
| | Introduction .. .. .. .. .. .. .. | 366 |
| | Cephalochordata—Amphioxus .. .. .. .. | 366 |
| | The Affinities of the Cephalochordates .. .. .. | 387 |
| XVII. | CRANIATA (VERTEBRATES) .. .. .. .. .. | 390 |
| | Introduction .. .. .. .. .. .. .. | 390 |
| | Histology .. .. .. .. .. .. .. | 391 |
| | External Features .. .. .. .. .. .. | 400 |
| | The Skin .. .. .. .. .. .. .. | 407 |
| | Metamerism .. .. .. .. .. .. .. | 415 |
| | The Skeleton .. .. .. .. .. .. .. | 417 |
| | Muscles .. .. .. .. .. .. .. .. | 465 |
| | The Body Cavity .. .. .. .. .. .. | 481 |
| | The Alimentary System .. .. .. .. .. | 482 |
| | Histology of the Alimentary Canal and Associated Glands | 491 |
| | Nutrition .. .. .. .. .. .. .. | 499 |
| | The Blood Vascular System .. .. .. .. .. | 510 |
| | Histology and Physiology of Blood .. .. .. .. | 533 |
| | The Respiratory System .. .. .. .. .. | 540 |
| | The Neuro-sensory System .. .. .. .. .. | 559 |
| | Histology and Physiology of Nervous Tissue .. .. | 590 |
| | The Sensory System .. .. .. .. .. .. | 603 |
| | The Renal and Reproductive Systems .. .. .. | 633 |
| | Histology and Physiology of Kidneys .. .. .. | 648 |
| | Histology of Gonads .. .. .. .. .. .. | 653 |
| | The Endocrine System—the Ductless Glands .. .. | 655 |

| CHAPTER | | PAGE |
|---|---|---|
| XVIII. | THE GERM-CELLS—GAMETOGENESIS .. .. .. | 673 |
| | Introduction .. .. .. .. .. .. .. | 673 |
| | Gametogenesis .. .. .. .. .. .. .. | 674 |
| | Meiosis .. .. .. .. .. .. .. .. | 677 |
| | The Significance of Maturation and Fertilisation .. .. | 681 |

| XIX. | THE DEVELOPMENT OF THE VERTEBRATES— | |
| | CHORDATE EMBRYOLOGY .. .. .. .. | 682 |
| | Introduction .. .. .. .. .. .. .. | 682 |
| | The Development of Amphioxus .. .. .. .. | 691 |
| | The Development of the Frog .. .. .. .. | 704 |
| | The Development of the Chick .. .. .. .. | 731 |
| | The Development of Mammals .. .. .. .. | 765 |
| | The Development of the Rabbit .. .. .. .. | 768 |

| XX. | GENETICS .. .. .. .. .. .. .. .. | 779 |
| | Introduction .. .. .. .. .. .. .. | 779 |
| | Heritable Characters and their Behaviour .. .. .. | 780 |
| | Mutations .. .. .. .. .. .. | 793 |

| XXI. | THE INTERRELATIONS AND ORIGINS OF ANIMALS— | |
| | ORGANIC EVOLUTION .. .. .. .. .. | 801 |
| | Introduction .. .. .. .. .. .. | 801 |
| | The Fact of Change .. .. .. .. .. .. | 802 |
| | The Extent of Change .. .. .. .. .. .. | 804 |
| | Evidence from Classification .. .. .. .. | 804 |
| | Evidence from Geology and Palaeontology .. .. | 805 |
| | Evidences from Comparative Morphology and Anatomy | 814 |
| | Evidences from Embryology... .. .. .. .. | 815 |
| | Evidences from Geographical Distribution .. .. | 817 |
| | Evidence from Comparative Physiology and Genetics .. | 820 |
| | The Way in which Evolution has Come About .. .. | 824 |

| XXII. | ECOLOGY .. .. .. .. .. .. .. .. | 830 |

| XXIII. | THE MAIN GROUPS OF ANIMALS.. .. .. .. | 847 |

| | SELECTED FURTHER READING .. .. .. | 883 |

| | INDEX .. .. .. .. .. .. .. .. | 887 |

# PLATES

FRONTISPIECE *facing Title*
Electronmicrograph of a lymphocyte.

*facing page*

PLATE I        ..    ..    ..    ..    ..    ..    ..    ..    ..    38
Model of a haemoglobin molecule.

PLATE II       ..    ..    ..    ..    ..    ..    ..    ..    ..    39
Photomicrographs of stages in mitosis in a fertilised egg of *Ascaris*.

PLATE III      ..    ..    ..    ..    ..    ..    ..    ..    ..    314
Electronmicrographs of flagella of *Deltatrichonympha*.

PLATE IV       ..    ..    ..    ..    ..    ..    ..    ..    ..    315
A Cockroach and a Locust.

PLATE V        ..    ..    ..    ..    ..    ..    ..    ..    ..    518
The chief features of a generalised craniate.

PLATE VI       ..    ..    ..    ..    ..    ..    ..    ..    ..    519
DOGFISH—Vascular System.

PLATE VII      ..    ..    ..    ..    ..    ..    ..    ..    ..    524
FROG—Ventral dissection of the heart.

PLATE VIII     ..    ..    ..    ..    ..    ..    ..    ..    ..    525
FROG—Vascular system.

PLATE IX       ..    ..    ..    ..    ..    ..    ..    ..    ..    530
MAMMAL—Ventral dissection of the heart (pericardium removed).

PLATE X        ..    ..    ..    ..    ..    ..    ..    ..    ..    531
RABBIT—Vascular system.

# ANIMAL BIOLOGY

## CHAPTER I

### INTRODUCTION

Biology is the science which deals with the study of all the varied aspects of life or living organisms but is usually, because of its vastness, treated as two subjects, animal biology (zoology) and plant biology (botany). In practice both have undergone considerable fragmentation into many subsidiary branches, some of which are often considered to be sciences in their own right. This seems to be the inevitable result of increasing knowledge and specialisation and biology is not unique in this respect. Practically every other science has had a similar history. But fortunately it is still possible, at any rate up to a certain level, to maintain an essential unity in the study of animals whilst at the same time drawing information from all the biological sciences as well as from chemistry, physics, and geology. The same is true for botany. Animal biology takes account of the form and structure (morphology and anatomy) of the parts of animals; of microscopic anatomy (histology); of the mode of functioning of the parts (physiology); of the manner in which animals develop (embryology); and of links between one generation and the next (genetics). Of increasing importance is the study of animal behaviour and of the relations between animals and their environment (ecology), whilst new discoveries continue to be made in the study of inhabitants of the earth in past geological time (palaeontology). None of these categories are hard and fast ones and may best be thought of as pointing to differences of emphasis in the way of studying animals. All of them contribute information which is used in the naming and classifying of animals (taxonomy). Moreover, the list is not a complete one and could be greatly extended by the inclusion of less well-known but nevertheless important branches of the subject, some of which will be mentioned later.

Living organisms, both plants and animals, are distinguished from dead material by their activities, so that it is essential from the beginning to have a general idea of what these vital activities are. They can be described under the headings of **locomotion, nutrition, growth, respiration, excretion, sensitivity**, and **reproduction**. All of these activities involve chemical changes, the sum total of which is termed **metabolism**. Many of these names are familiar but in biology some of them have a special meaning and require definition and explanation.

**Locomotion or Movement.**—The idea of movement from one place to another is familiar to every one and the majority of animals are able to move freely in the environment in which they live. All movements, however, do not result in progression from place to place but may be concerned with feeding and other activities, many of them internal.

**Nutrition.**—This is naturally associated with the process of feeding, but really involves much more than the mere taking in of food. Food is used by animals as a means of obtaining energy for their vital activities and the materials necessary for body building. This energy is locked up in various materials available to them for food. Often these substances are not in a suitable form for the energy to be immediately available and consequently the food has to be altered. It will be seen, then, that the process of nutrition includes several sub-processes. Firstly, the food must be taken into the body. This is called **ingestion**. Secondly, the food is frequently not soluble or diffusible and must be made so in order that it may be absorbed into, and distributed about, the body. This is **digestion** and **absorption**. But there is another and equally important reason why the complex organic substances comprising the food must be broken down into simpler molecules. It is this: only a few organic compounds are suitable for acceptance into the body of an animal which is composed chiefly of compounds that are highly characteristic for any given species. Therefore the food has to be broken down into smaller molecules which can then be synthesised into those more complex ones required for the maintenance of particular cells and tissues, as well as for the formation of high-energy compounds, whose breakdown results in the release of the free energy necessary for the continuation of life. These synthetic processes are sometimes termed **assimilation**.

Not all the food taken in by an animal can be, or is, digested and absorbed. That which remains is eliminated from the body by the process of **egestion** and forms the **faeces**.

**Growth.**—It has been seen that food material absorbed into the animal body may be used either as a source of energy or for building up new body substance. If all the absorbed food is used to supply energy and is consequently broken down, then there is nothing available for adding to the body substance and the animal cannot grow. If, however, the intake is greater than immediate demands, then the body material can be added to. Under these conditions the body may increase in size, in weight, in complexity, may change in form and in a variety of other ways, and may be said to grow. However, growth in a living organism is different from mere increase in size (cf. "growth" of crystals) which may be nothing more than accretion of new material around pre-existing material or even swelling by the absorption of water. Growth is the interposition or gradual wedging of new material *within* that already in existence and is invariably accompanied by differentiation and organisation. But so stated it might appear that new substances are *permanently* added to the body. This is not so. On the contrary, there is throughout life a continual breaking down of all the cells and tissues of the body and as constant a renewal.

This has been proved by the substitution of isotopes for some of the atoms normally composing the food and substances in the body, as, for example, $^{14}C$ for $^{12}C$, $^{2}H$ (deuterium) for H, $^{15}N$ for $^{14}N$, and so on. This makes it possible for the atoms to be followed as long as they remain in the body. They are "labelled", so to speak, and their rates of accumulation and disappearance can be measured. Very often the exchange is rapid, the whole of the liver fat in fully-fed rabbits being broken down and renewed in a matter of hours. More surprisingly, even such apparently permanent structures as bones are no exception to the rule, for their component atoms are continually being discarded and replaced by new ones entering the body with the food. It will be understood, therefore, that growth is certainly not the mere addition of new substances to the body. It is an expression of the excess of assimilative and synthetic processes over those of degradation into simpler compounds, but such an excess can be brought about either by an increased rate of synthesis or by a decreased rate of degradation. In this instance again, the use of radio active isotopes enables the answer to be given that in many (perhaps all) instances rapid growth is due to the slowing down of the rate at which body substances are broken down.

**Respiration.**—In everyday language respiration simply means breathing but in biology it concludes all the complex processes concerned with the liberation of energy from the food (**internal** or **tissue respiration**). Gaseous interchange with the environment (**breathing** or **external respiration**) is merely one section of the energy cycle during which the waste products, carbon dioxide and water, are got rid of and oxygen is renewed to the system. From this bald statement it might be inferred that in some way, perhaps nearly comparable with combustion, an animal obtains its energy by oxidation of the food stuffs, as for example might be represented by the equation $C_6H_{12}O_6 \rightarrow 6\ CO_2 + 6\ H_2O +$ Energy, which shows how glucose can be oxidised to give carbon dioxide, water, and energy. But this would give a totally wrong impression of the energy cycle which, in short, consists of a complex series of reactions each catalysed by a particular enzyme, the bulk of them taking place anaerobically. Oxygen enters into the cycle only at the end of the breakdown processes and can be thought of as oxidising certain of the enzymes—not the food directly ( Chapter XVII).

**Excretion.**—Some of the vital activities which involve the breaking down of materials within the body result in the formation of substances which are of no further use to the animal and may even be harmful and consequently have to be got rid of. In the process of respiration, for example, carbon dioxide and water are produced, and these substances are passed out of the body. This is sometimes spoken of as carbonaceous excretion. When proteins are broken down **nitrogenous** substances are formed and then eliminated. This is called **nitrogenous excretion.** Excretion, then, is the elimination from the body of unwanted materials (chiefly water, carbon dioxide, and nitrogenous substances) produced as the result of vital activities.

**Reproduction.**—This is a unique property possessed by living organisms alone: the capacity to produce new individuals resembling the original animal in all essential respects. Its manifestations are many and varied, from the simple division of the parental body into two or more parts (each part growing into a new individual similar to the parent) to the extremely complicated reproductive processes of higher animals. In this way continuity of the race is maintained.

**Sensitivity.**—It is a matter of common experience that organisms, not only highly complicated beings like ourselves, but other animals and plants also, continually throughout their lives react to changes in their environment which, it must be borne in mind, is made up not only of the surroundings in which they live but also of factors originating within their bodies (internal environment). These changes in the environment are detected in higher animals by special parts of the body termed receptor or sense organs of which well-known examples are the eyes, ears, nose, tactile organs, and the like, and by less obvious structures in the muscles, gut, and other parts of the body. Each of these receptors is particularly sensitive to one kind of change in the environment which is then said to stimulate the organ since it causes it to initiate nerve impulses which may, after being relayed through the nervous system, cause other impulses to reach the effector organs of the body (muscles, glands, etc.). These then respond by an appropriate change involving the release of energy; a muscle, for example, responds by a contraction. It must not, however, for one moment be imagined that this ability to react to changes in the environment of necessity requires the development of special structures for the reception of stimuli, for the conduction of messages, or for the carrying out of responses. Sensitivity and the ability to respond to stimuli are properties common to all living organisms, even the very simplest, where structural differentiation into receptors and effectors is wholly lacking. Thus, it may be said of all living organisms that, in the customary language of the physiologist, a stimulus evokes or causes a response. An analysis of this statement and a consideration of all its implications would require a lengthy discussion, but it is not, perhaps, out of place here to point out that it is the exception rather than the rule for higher animals living under natural conditions to respond in a direct way to simple changes in the external environment (such as changes in light intensity, etc.). More usually they react to complex situations which have some significance in the life of the animal. It is true that some simple stimuli, even in higher animals, may, through the intermediary of the nervous system, elicit a simple or "reflex" response, but this type of behaviour seems to become more and more important with increasing simplicity of structure. Yet, even in the simplest of animals, such reflex responses probably do not account for the total behaviour pattern.

It is this ability to react in such a way that the integrity of the organism is preserved that is the very essence of life and it will be readily understood that the vital activities, which have been given and dealt with separately, are really just particular aspects of the reactions of an organism with its environment. This splitting or analysis of complicated events into their simpler constituents is the method of the physiologist. It is sometimes termed the "causal-analytic" method and has yielded, and continues to yield, valuable and far-reaching results. But, because many activities can receive an explanation in physico-chemical terms—which is the chief aim of physiology—this does not preclude other ways of studying living organisms. Different, but equally valuable results, as for example, Darwin's Theory of Natural Selection, may be obtained by studying organisms as harmoniously working wholes and not as collections of separable activities. When this is done it becomes apparent that organisms are rather different from mere machines controlled by physico-chemical laws. A machine uses energy to carry out the purpose for which it was made and, like a living organism, wears out in the process. But, unlike any machine, the living organism rebuilds, repairs, refashions, and alters its parts, without ceasing to function or to maintain its essential organisation. It is only when death puts an end to its activities that this ability is lost. One essential difference between a machine and a living organism is that, whereas a machine is controlled from outside (by man), the organism is controlled from within itself, and since all its activities are directed to the maintenance of life and the perpetuation of the race its behaviour may be said to be purposive. It must, however, be emphasised that the use of the term "purposive" is in no way intended to imply that the organism consciously directs its activities, although conscious actions play a large part in the lives of men and also of other higher animals.

All existing animals must be efficient since they have survived; the terms "higher" and "lower" animals have no real meaning except in relation to the now widely accepted idea that there has been in operation from the time life first appeared on this earth, a process called "organic evolution" (Chapter XXI), by which the complex have been derived from the simpler animals. Hence, the terms "higher" and "lower" animals are really contractions for animals higher or lower in the evolutionary scale of complexity.

One consequence of an increase in complexity is the capacity for a greater range of response to environmental conditions leading to a greater independence of the immediate surroundings. Thus, though the ancestors of all living organisms were originally aquatic, indeed, most of them marine, many lines have, by developing complicated devices for preserving the constancy of their internal environment, been able to adapt themselves to life on dry land. Paramount among these are such groups as the insects, birds, and mammals, each representing the climax of a particular evolutionary line.

Living organisms* are, then, essentially dynamic and self-maintaining systems perpetually engaged in varied activities and these are always of a "directive" nature in the sense that they appear to be directed to attain a particular goal or end result, which

---

* As is well known, certain animal and plant diseases are caused by "organisms" which cannot be satisfactorily described as either simple plants or animals. In size they are much more minute than any bacteria and will pass through all but the finest filters. They are therefore called "filterable viruses" or simply viruses. Of their many remarkable properties, there may be singled out the facts that they can be crystallised from solutions, and that the crystals when redissolved will infect living tissues.

At one time these were regarded as submicroscopic forms of life, providing as it were a link between non-living and living matter. It is now known that

is of advantage to the organism. Seen as a whole, these vital activities have a cyclic nature since all organisms undergo development, reproduction, and, as individuals, death. This life-cycle, relatively simple in some, exceedingly complex in others, is perhaps the most characteristic feature of living organisms.

**Cells and Protoplasm.**—When an animal of any complexity is examined it is very soon found that some of its constituent parts are dead, as for example, the jelly of jellyfishes, the shell of a snail, the bristles of a worm, the hooves of a horse—and so on. All of these play an essential part in the lives of the animals concerned, but it is instructive to contrast them with the living material which in most animals forms the bulk of their bodies and which is responsible for carrying out all the vital activities. To this living material, no matter in what part of the animal nor in what animal it occurs, the name of " **protoplasm** " was given. This is not to say that the protoplasm of all animals was believed to be identical.

Living protoplasm was at one time thought of as being a greyish jelly-like substance and as an aqueous solution of a mixture of substances. Water comprises a very large part of it—perhaps 70-80 per cent. Of the remaining 20-30 per cent. proteins form the largest part but their large molecules are not in true solution. Nor, of course, are the fats or lipoids which often occur. They and the proteins are colloids which can, with water, form a fluid (sol) or a jelly (gel). Moreover, living protoplasm can change from gel to sol and *vice versa*, often with great rapidity. Other constituents, such as salts and some carbohydrates, are in true solution.

One of the important properties of protoplasm is its ability to form a thin membrane at its surface. This is probably composed of a double layer of lipoid molecules with a layer of protein molecules sandwiched between them. Such a membrane is selectively permeable, allowing water molecules to pass through it but offering a considerable barrier to salts and larger molecules.

Thanks to studies with the electron microscope we now know that the apparent formlessness of protoplasm is an illusion. It has, in fact, an intricate architecture but with, nevertheless, many features in common in all living organisms. Perhaps the time has come for

most viruses are particles consisting of a core of nucleic acid (either RNA or DNA) within a protein shell and in this respect they resemble parts of the cells of which living organisms are composed. No virus is capable of independent life. They can multiply only within the cells of an organism. For these and many other reasons, therefore, viruses are now regarded as having been derived from the cells of living organisms rather than representing simpler and more archaic forms of life.

the term "protoplasm" to be dropped, carrying as it does overtones of meaning not in keeping with modern thought.

Moreover, since the comparatively early days of microscopy it has been known that the substance of most living organisms occurs in minute, discrete units. It was this observation which led Schleiden and Schwann in 1838 to put forward the now universally accepted theory, usually called the "cell-doctrine". This may be briefly summarised by the statement that living organisms are built of units called **cells**, together with materials produced by these cells. The term "cell" is far from being self-explanatory and was originally applied by Robert Hooke (1667) to the cells of which plant tissues are composed. As is well known, in these units, it is the cellulose cell wall that is the most obvious feature and in the less specialised plant tissues (*e.g.* parenchyma) the hexagonal cell walls give to the tissues an appearance somewhat like a honeycomb. Animal cells, none of which has a cellulose cell wall, never present such an appearance and it is now recognised that it is the *contents* of the cells which are of paramount importance. A very brief definition of a cell is "the nucleus and cytoplasm within its sphere of action". These terms receive an explanation in Chapter II.

## Molecular Biology

In the past the unity of biology has found its most satisfying expression in two generalisations: the all-embracing theory of evolution, which unites living organisms in terms of their descent from common ancestral forms, and the universality of the cell in one form or another as the structural and functional unit of living matter. While resting secure on these two very different but equally universal concepts, it has also been felt that a more fundamental unity would be achieved ultimately in terms of the physics and chemistry of the molecules composing living matter.

This stage is coming into being to-day owing to the extraordinarily rapid advance of biochemistry; indeed, while a detailed deductive biology is still far off it is felt by many that its broad outlines are revealed and that it is already possible and desirable to introduce the study of biology along the lines of its probable ultimate form. A coherent unified theory of biochemistry, genetics, and morphology is being developed in terms of the concept of self-replicating genetic molecules with the function of guiding the cellular mechanism for protein synthesis to synthesise specific enzymes and structural proteins. For this reason the movement is called molecular biology.

Comparative biochemistry has yielded abundant proof of the identity of the chemical basis of all types of organisms. Information concerning the chemical basis of inheritance, for example, may be sought in bacteria, protozoa, plants, or animals, as is convenient on technical grounds. The molecular biological viewpoint thus emphasises the arbitrary character of the subdivision of biology into botany and zoology and makes it seem undesirable that, in an introductory text, the discussion of the science of life should be separated into two disciplines. While admitting that for practical reasons and for more detailed study, the traditional division will continue, an integrated attitude will be adopted in the remainder of this chapter. Assuming then that the future description of biological phenomena will only be considered valid when couched in physical and chemical language, or in a form translatable into such terms, we shall outline, in molecular terms, some generalised problems posed by the existence and behaviour of living matter.

## PHYSICS AND BIOLOGY

**A Modern Definition of Life.**—Attempts to define living matter inevitably do so by contrasting it with inorganic or lifeless matter: living matter exhibits such and such a property (movement, sensitivity, reproduction, etc.) which inorganic matter does not (p. 1). This is a common-sense definition with which we all agree. It can be given a more generalised aspect by noting that living organisms are characterised by molecules arranged in ways that are extremely unlikely to occur by chance and are therefore immensely improbable but which, in the face of the degrading disorganising influences everywhere apparent in the inorganic world, have persisted for enormously long periods of time (of the order of $10^9$ years) and, moreover, in that long time have constantly tended to generate more and more improbable configurations. The various attributes singled out in the common-sense definition are seen in this more general light to be usually the expression of one or other of the special devices employed to ensure the preservation of the organisation.

The generalised view contrasts the increasing complexity of biological organisation with the tendency elsewhere in the inorganic universe to run downhill towards a state of increased disorder or, in thermodynamical language, to change in a direction of increasing entropy. This characteristic of the non-living world is the **Second Law of Thermodynamics** and it is precisely by its affront to our expectations based on this law that we recognise living matter.

The laws of physics are of two kinds: the first is exemplified by the laws of conservation of matter and energy. These laws have a universal validity and living organisms are no exception: experiment shows that in all vital exchanges matter and energy are always conserved. The second type of law is statistical, a statement of probable behaviour, and it is this kind of law that living processes seem to circumvent by maintaining and increasing their highly improbable, ordered states. However, the consequences of the second law are also inescapable and its violation is only apparent. Stated more fully the law predicts that any spontaneous change in a closed system, *i.e.* one in which energy and mass neither enter nor leave, will be towards a state of greater probability. The law applies to *living matter and its environment* which together constitute a closed system in the thermodynamical sense. The ordered structures are in fact built up and maintained at the expense of the environment; the local decrease in entropy they effect is more than compensated for by the increase in entropy of the environment (Fig. 1).

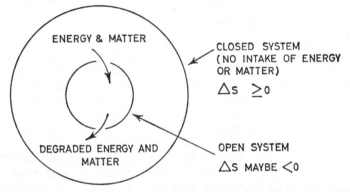

Fig. 1. The relationship of open and closed thermodynamic systems. The entropy change ($\triangle S$) is always positive for the closed system. In the open system matter and energy may be taken from the larger closed system and $\triangle S$ may be negative, *i.e.* the degree of order in the system may be increased.

Limited systems of this kind, which form part of a larger whole itself thermodynamically closed and to which the second law applies, are called **open** systems. Biological systems are thus open and some of their important characteristics follow from this. Open systems take in energy and mass from the larger systems and use these to decrease their entropy or increase their organisation. At the same time there is an equal or greater gain in the entropy of their surroundings. This statement may be put in an alternative form by speaking in terms of "negative entropy" or **negentropy,** which is then a positive measure of the amount of order in a system. Open systems may increase internal order by extracting negentropy from their surrounds or, in E. Schrödinger's aphorism: "organisms feed on negentropy".

The degrading influences of the environment are countered by various expedients which may be classified according to whether they act to circumvent short or longer term effects of the environment. Short term homeostatic devices act simply to maintain the *status quo*. They may be biochemical or physiological in nature and are concerned with keeping the organism in one piece and its internal organisation intact by obtaining sufficient energy and matter, by controlling certain internal parameters, and by avoiding dangerous situations. Examples of these devices will be met with in subsequent chapters. Long term devices act to secure an organism against total loss by increasing the number of examples of that organism and this involves the exact reproduction of its parts and whole, a property always insisted on as a characteristic of living matter.

It is possible to conceive of populations of organisms which persist in time simply by the piecemeal duplication by each member of its individual parts followed by its separation into two parts leading to an increase in numbers sufficient to offset the inevitable losses. It is thought by many that in an earlier period of chemical evolution, which preceded the appearance of present day cellular organisms, reproduction at a molecular level was effected in this way. However, in all biological organisms of which we have knowledge to-day, reproduction is a more complex process having as its basis the preservation and replication of a plan or record of the entire organism in the form of a **genetic substance**. Only the genetic substance replicates in the piecemeal molecular fashion of the hypothetical forms of the precursors of life and has the additional essential property that, in an appropriate medium, it is able to guide the synthesis of all the materials required to reconstruct a new organism. The advantages of this method of recording and con serving an organisation seem to lie mainly in the ability of the genetic record in some simple, stabilised, and protected form to survive adverse conditions which would certainly destroy the more sensitive fully developed organism. In the sequel there will be described numerous examples of specialised devices which preserve the genetic material and hand it on intact. As will be seen later it also offers a simple means for generating new types of organisms by effecting in the first place the changes in the plans alone.

Certain *à priori* properties can be predicted of a genetic substance. In the first place, as Schrödinger points out, to obtain the required stability and permanence in a genetic record we are virtually forced to suppose that it exists in the form of covalently bonded molecules. No other material association except one bonded by covalent chemical linkages could persist for millions of years. Secondly, it must be sufficiently complex and large enough to store the information needed to reconstruct some thousands of other complex compounds in an organism. We are thus led to suppose some very high molecular weight complex polymers as genetic substances.

Since organisms are composed of molecules, many of them of high molecular weight, of quite definite chemical composition, and not found in the inorganic world, it is likely that these instructions to form them will be divisible into units at least equal in number to that of the compounds required to ensure reconstruction. It is these units which are likely to correspond to the units of inheritance, the **genes**, discovered by geneticists in actual organisms (Chapter XX). Each of these units must be duplicated to provide a complete copy of instruction to form a new organism. Following the usage of genetics the complete genetic apparatus may be referred to as the **genome** and the rest of the organism the **phenotype**.

There are two ways in which the mechanical problem of ensuring that descendant organisms will receive complete copies of every single unit of inheritance, every gene, can be achieved. The first is to make so many copies of each that on statistical grounds alone a copy of each is certain to be present in all parts should the present organisms divide; alternatively it is less wasteful and more efficient to have all the units linked together in a string and be transported as a whole, when it would be necessary to make only one copy of the entire instructions. Should the actual physical dimensions of the entire copy become too great for convenient handling, we may anticipate that it will be subdivided into a small number of groups of a more convenient size for subdivision and transport. These we shall recognise later as the **chromosomes**.

We may speculate that the first and simpler method probably prevailed in an early stage of chemical evolution and was later replaced by the second since this proved more efficient. However it may have occurred, the greater part of the inheritance in actual organisms is carried by the second method. Nevertheless, an as yet undetermined amount of instruction is inevitably handed on, when two organisms are produced by division, through the inevitable carryover of fragments of the parent organism itself.

One other property needs to be possessed by the genetic material if it is to record the construction of organisms with an indefinitely prolonged expectation of life. It must possess a sufficient degree of instability to allow of the formation of new varieties of organisms with new properties. Without this degree of flexibility organisms would be at the mercy of environmental changes, conservatives perishing in a changing world; with a sufficient supply of such variation new organisms adapted to new conditions may emerge and better adaption to existing conditions be arrived at. This is the basis of the evolutionary movement towards complexity which was noted as a further characteristic of living matter. The environment itself ensures the evolution of a population of organisms by favouring the survival of certain types of variant, and the process is for this reason referred to as **natural selection** (p. 825).

In a chemical sense the origin of variation must be sought in the actual molecules which are the material basis of inheritance. Such changes are referred to as **mutations**, and are thought to have a variety of origins: chemical accidents affecting the structure of the constituents of the molecules, the chemical effects of high energy radiation (cosmic rays) or, even, by errors occuring during copying of the genetic material (p. 827).

All of these are chance happenings, most will be deleterious, rarely will advantage accrue from them, but it is precisely these advantageous changes that natural selection will conserve (Chapter XXI).

It is worthwhile contrasting here the different role of chance in generating the spontaneous movement of open and closed systems. The closed system moves inexorably towards a state of uniformity and equilibrium; each spontaneous change is in a direction of greater probability or of increasing entropy. An open system, capable of self-reproduction with chance variation, may produce a rare and occasional variant, which is favoured by the selective action of the environment. Such a variant is then preserved by self-reproduction. Since an increase in complexity is likely to confer some advantages on an organism, a mechanism of this sort will be likely to generate complexity spontaneously, an exactly opposed movement to that occurring in the closed system. Organisms thus evolve inevitably and no physical law is evaded in the process; the increased order (Fig. 1) is paid for by the disorder produced in the organism's environment.

Having made these more or less à priori statements the deductive approach runs into difficulties through lack of information concerning the potentialities of the many polymers available. Are the actual polymers which, as we shall see, play the central roles in the living systems found on this planet, the only possible candidates for these roles? Is there an element of chance in the evolution of these systems leaving open to us no other way than actual observations to discover what polymers are in use?

Whatever the answers an advanced chemistry will make to these questions, an analysis of all existing forms of life has revealed only one form of genetic substance on this planet: the polynucleotide, **nucleic acid**. No general principle seems to exclude another polymer playing the part elsewhere, although the peculiar appropriateness of the polynucleotides will become evident later. Similarly it has been found that the necessary working parts of organisms, whose structure alone is recorded in the genetic material, are based on one other class of high molecular weight polymers, **the proteins.** The universality of this partnership between the nucleic acids and the proteins suggests that all existing forms of life on this planet have a common origin and represent a single form of living matter.

**Other Forms of Life. Artefacts.**—With the more detailed appreciation of what features characterise living matter, it becomes practical to consider the possibility of constructing living artefacts and to speculate more fruitfully about the existence of forms of life differing from those found on this planet. Defining living matter as an improbable system of molecules, which ensures its persistence in time by obtaining negentropy from its environment, we can readily admit of the existence of other systems based on different biochemical foundations. Further, it becomes difficult to frame a definition which will deny the title of "living organisms" to the several mechanical devices recently constructed with the object of illustrating homeostasis, goal seeking behaviour, and reproduction. At present the value of these hardware systems (mechanical and electrical artefacts) is largely heuristic, worthwhile for the insight they bring on the behaviour of protoplasmic or wet-chemical analogues; but it seems unquestionable that these "organisms" are about to embark on an elaborate evolution in which their likeness to biological organisms will increase.

## MOLECULAR GENETICS

**Nucleic Acids, DNA.**—The proof that the genetic material in living organisms is nucleic acid is based on the circumstantial evidence that it is a constituent of all cells and in particular of such specialised cellular products as spores, spermatozoa, and viruses, which obviously convey genetic information, and on a few more definite situations (which will be discussed later) where it is established that the transfer of nucleic acid (and nucleic acid alone in sufficient quantity) has altered the genetic character of receptor organisms. The best proof, however, is the mass of steadily accumulating biochemical evidence on the exact manner in which nucleic acid actually stores the genetic information and makes it available to the organism. This we shall now discuss.

When extracted from cells nucleic acid forms exceedingly viscous solutions in water, an indication that it consists of very long molecules. Accurate estimates from physico-chemical data indicate that the molecules may have a width of 30 Å* and lengths greater than a millimeter! The corresponding molecular weights are tens of millions. A similar order of size is obtained from electron

Fig. 2. An autoradiograph of bacterial DNA, radioactive atoms incorporated in the molecule are permitted to affect photographic emulsion to produce the radiograph, scale equals 0·1 mm (from Cairns).

micrographs of single molecules and of autoradiographs of single radioactive molecules (Fig. 2).

The analysis of the substances, which are produced by hydrolysing DNA, reveals that the major components of the molecule are phosphoric acid, a pentose sugar, and several complex heterocyclic bases. The 5-carbon sugar is **desoxyribose** and hence this

* An Angstrom unit, Å, is a ten thousandth part of 1 μm, which in turn equals a thousandth of a millimetre, therefore 1 mm = 10,000,000 Å.

acid is called desoxyribonucleic acid, or DNA for short.   The sugar
is **ribose**, in the related polynucleotide, ribose nucleic acid (RNA).

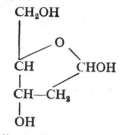

Desoxyribose, the sugar of DNA.

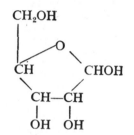

Ribose, the sugar of RNA.

The sugars are made the basis of classification of nucleic acids into
DNA and RNA, since the two kinds of nucleic acid are found to
have very distinct functions.   In most organisms DNA has been
shown to be the primary genetic substance and RNA plays an
intermediate role between DNA and the protein produced in the
cell.   In some instances RNA is, however, the primary genetic
material.

Of the four heterocyclic bases found in DNA,

two are purines:

**1.** Adenine.

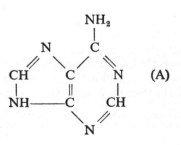

2. Guanine,

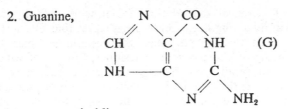

(G)

and two are pyrimidines:

3. Thymine,

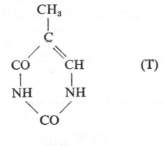

(T)

4. Cytosine,

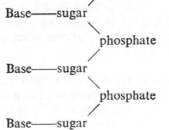

(C)

The relative amounts of these four bases are different in the DNA from various organisms: the sum of the purines taken together always equals the sum of the two pyrimidines, a feature which has important consequences for the structure of the whole molecule. Since in other respects all DNA's are alike, it is clear that the ability of the molecule to convey genetic information must depend on the number and arrangement of these bases. Partial hydrolysis of DNA yields **nucleotides** in which sugar, phosphate, and base are linked:

Base——sugar

phosphate

Base——sugar

phosphate

Base——sugar

These results of chemical degradation, taken with the physical properties indicating a long linear molecule, suggest that DNA is a linear polymer of nucleotides in which the continuous chain or backbone is a string of phosphoric acid residues, to each of which is attached a sugar residue and a base. Since the base may be distributed in any order the total molecule has an enormous number of possible internal arrangements as is necessary for a genetic material.

Further elucidation of the stereochemical structure of DNA followed an X-ray crystallographic analysis by Watson, Crick, and Wilkins of fibres drawn from solutions of the material. This work showed that the unit in the crystal consisted of two chains twisted together in opposite senses to form a double helix (Fig. 3). To obtain the close packing required for this structure Watson and

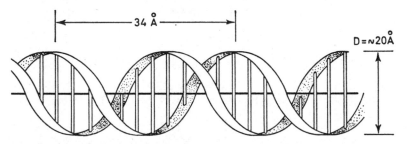

Fig. 3. Diagrammatic representation of a short length of the DNA molecule according to Watson and Crick. The two ribbons symbolise the sugar-phosphate backbones of the two chains and the horizontal rods the pairs of bases holding the backbones together. The horizontal line marks the long axis.

Crick found it necessary to pair each large pyrimidine base with a smaller purine (Fig. 4). This arrangement automatically accounts for the equality in base ratio mentioned above; the requirement of close packing yields the pairing rules:

A pairs with T
G pairs with C.

As a consequence of these pairing rules, in the double helix each chain is the complement of the other and each is uniquely determined by its partner.

**The Self-replication of DNA.**—It is this feature which led Watson and Crick to their dramatic proposal for a molecular mechanism for the replication of the double helix, which to-day forms the starting point for all speculation on this theme. They suggested that

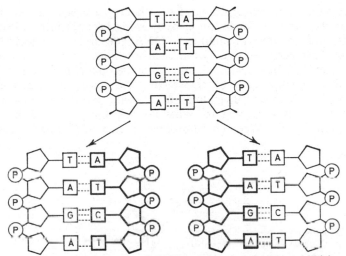

Fig. 4.  Scheme of replication of DNA according to the Watson-Crick model.
H-bonding between base pairs is indicated by broken lines.   The two newly
added strands in the daughter helices are shown in bold lines.

the two chains separate and that each then forms the template for the
assembly of the units for the build-up of its complement, thus leading
to the formation of two identical double helixes (Fig. 5).

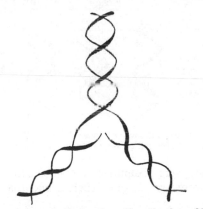

Fig. 5.  A hypothesis for the self-replication of DNA.

The exact mechanism of replication and the accessory biochemical
reactions are not yet fully clear.   A useful picture is to suppose that
the chains begin to unwind and that new units to form the complement
chains are added as each base pair in the parent chains separate.

A number of consequences of this theory have been tested and are in its favour. For instance, by growing bacteria in a medium containing tritiated thymidine, a DNA precursor in which $^1H$ atoms, have been replaced by radioactive tritium $^3H$ atoms, all the DNA can be labelled with tritium. It is then possible to follow its distribution in progeny by transferring the colony to a medium containing unlabelled thymidine and allowing growth to continue. A distribution according to Watson-Crick theory would be as shown in Fig. 6. When chromosomes of the first ($F_1$) and second

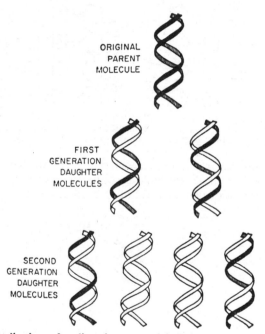

ORIGINAL
PARENT
MOLECULE

FIRST
GENERATION
DAUGHTER
MOLECULES

SECOND
GENERATION
DAUGHTER
MOLECULES

Fig. 6. Distribution of radioactive tracer labelled DNA strands (in white) during successive cell divisions (from Meselson and Stahl).

($F_2$) generations were examined by autoradiography, pairs of homologous chromosomes, one labelled and one unlabelled were noted.

An even more vivid visual demonstration was obtained by Cairns using bacterial DNA. He broke up bacterial cells carefully and allowed the unbroken radioactive DNA to settle on to photographic emulsion to register its autoradiograph. The DNA of *Escherichia coli* (as was expected from genetical analysis) proved to be a single closed strand. Double labelled helices were recognised by their dense autoradiograph and helices containing only one labelled strand by a print half as dense. The actual separation in replication is also demonstrated in Fig. 2.

Chemical studies show that polynucleotides, being linked covalently, have a high stability within the physico-chemical range encountered near the surface of the earth, which is the first requirement of a genetic record. DNA-type polynucleotides have also the required degree of complexity, produced by permutation of the bases. The production of differences among organisms can be understood in terms of alterations in the base sequence, which may take the form of the insertion of the wrong base during replication, the loss of a base, or the addition of too many or other unknown changes. Such alterations could lead to a change in the genetic instructions and thus to the formation of a protein with different properties. Such changes are called **mutations**; they are the raw material on which evolutionary change depends. The new protein may be a new enzyme capable of handling a reaction not possible in the parent organism, and the new organisms may thus be able to utilise this reaction, *e.g.* hydrolysis of a new substrate, to survive in conditions fatal to its precursor. By the slow, accumulation of many such genetic changes the somatic character of an organism (its phenotype) may change indefinitely.

**Proteins.**—The key role of proteins in the cellular economy was recognised much earlier than that of nucleic acid and their function as enzymes or organic catalysts was established almost a hundred years ago. Their structure as polypeptides of high molecular weight formed by the condensation of amino acids with the elimination of water:

$$CH_3$$
$$|$$
$$NH_2.CH_2.COOH + NH_2.CH.COOH$$

glycine          alanine          *amino acids*

$$\downarrow$$

$$NH_2.CH_2.CO.NH.CH.COOH \quad + \quad H_2O$$
$$|$$
$$CH_3$$

glycylalanine                    *a peptide*

—CO.NH.CH—  =  *the peptide bond*

was proved by hydrolysis and by synthesis of artificial peptides. The long chains are usually folded to yield a secondary structure in actual proteins (Plate I.).

There are about twenty amino acids each having a different side group attached to the carbon atom. The molecular weights of proteins run from a few thousand to several million; thus there may be many hundreds of amino acid residues in any protein. While the amino acid composition of many proteins is known, the exact order of the residues in the polypeptide (a much more difficult problem) is known for only a few proteins.

The properties of a protein and the secondary folding assumed by the main chain is determined by its amino acid sequence. In many cases, when conditions permit, segments of polypeptide chain assume the form of a tightly bonded single chain spiral called an α-helix. Further folding, controlled largely by close packing and side chain interaction, may cause the molecule to assume a compact rounded form, the case with most soluble enzymes. On the other hand, fibrous forms arise when the attraction between different chains predominates over the attraction between parts of the same chain.

Proteins are fundamental to the functioning of all cells, without exception; they may be divided on the grounds of function into two classes:

(a) enzymes or catalysts of biological reactions;
(b) structural proteins (often fibres) whose predominant role is mechanical (reinforcement, contractibility, etc.). Structural proteins may also have enzyme properties.

The cell in structure and function is literally the creation of its population of enzymes: their type, concentration, location, and activity. By means of enzymes the cell handles the processing of the raw materials and synthesises multitudes of biological components with the notable exception of proteins themselves. The synthesis of proteins (with the exception of some minor and group modification which may be enzyme-effected) requires the participation of nucleic acids.

It is indirectly through its control over the synthesis of enzymes that the nucleic acid of the genetic apparatus or genome controls the composition of the phenotype. Thus in formal terms the problem of the replication of a biological organism is resolved into the problem of how each of the proteins of the organism is recorded or coded in the genome and how these coded instructions in turn are used to guide the ordered synthesis of the proteins, each of which has its specified role in the construction and maintenance of the organism.

**Enzymes as Catalysts.**—Enzymes are catalysts in the same sense as this word is used in inorganic chemistry. That is, if a reaction between two molecules is possible, this may be promoted (or catalysed) by adsorbing the two together in close proximity on the surface of a catalyst.

The special feature of organic catalysts is that they catalyse only specific reactions. This specificity arises from the fact that the surfaces of large protein molecules, constructed from intricately folded and shaped segments of polypeptide chains, contain grooves or moulds shaped to fit exactly parts of other molecules. Those molecules which fit (*i.e.* have complementary parts) may be adsorbed and not others. When two molecules are adsorbed side by side in a very close complementary fit with atoms of the surface, the primary bonds holding the two molecules together are so offset by the numerous weaker secondary bonds formed with the surrounding atoms, that they may exchange parts on separating from the protein surface. The reaction is thus effected without raising the reactants to a high temperature and it is a marked characteristic of biological reactions that an enormous variety of reactions can be effected with great rapidity at low temperatures (37° C or less).

It commonly happens in cellular reaction that the products of one enzymatic reaction forms the substances for another enzymatic reaction and so on:

$$A \rightleftharpoons B \rightleftharpoons C \rightleftharpoons D \rightleftharpoons X$$
$$\quad E_1 \quad\quad E_2 \quad\quad E_3 \quad\quad E_n$$

Chain reactions of this type may be of several kinds, in each case their functioning depends upon the presence of the proper enzymes ($E_1$, $E_n$, etc.) in close proximity. This is often achieved by adsorbing the enzymes in the appropriate order on the membranous surfaces which are common constituents of biological systems (p. 44).

Relation between DNA and proteins. The amino acid code. Since both DNA and protein are long linear polymers it might seem possible that a sequence of amino acids could be represented or coded by a simple sequence of nucleotides, each one corresponding to an amino acid; with only four nucleotides to choose from, it is at once obvious that this simple code could not specify as many as twenty amino acids.

Two different nucleotides can be arranged in sixteen ($4^2$) different ways and thus could describe sixteen acids; three can code sixty-four ($4^3$). At least three are then needed to code the twenty acids, but in this case the code is likely to be "degenerate" meaning that, since there are sixty-four possibilities, some amino acids may be coded by more than one triplet of bases or some combinations are meaningless.

The "cracking" of the amino acid code, *i.e.* the finding of which group of three nucleotides represents which amino acid, is engaging much attention at the present time. The biochemical techniques used to deduce the coding of amino acids are very ingenious and often subtle.

The most successful present method is to use synthetic poly-nucleotides of known composition as primers or templates in cell-free systems capable of synthesising proteins to find what amino acids they can add to polypeptides.

Table 1 is a list of the probable code assignments at present available (1965).

## TABLE 1
### GENETIC CODE FOR POLYMERISATION OF THE AMINO ACIDS

| AMINO ACID | TRIPLET CODE LETTER* |
|---|---|
| Alanine | GCU, GCA§ |
| Arginine | CGU, CGA§ |
| Asparagine | AAU |
| Aspartic acid | GAU |
| Cysteine | UGA |
| Glutamic acid | GAA |
| Glutamine | CAA |
| Glycine | GGU, GGA§ |
| Histidine | CAU |
| Isoleucine | AUU |
| Leucine | UUG, CUU, CUG§ |
| Lysine | AAA |
| Methionine | AUG |
| Phenylalanine | UUU |
| Proline | CCU, CCA§ |
| Serine | UCU, UCG§ |
| Threonine | ACU, ACA§ |
| Tryptophan | UGG |
| Tyrosine | UAU |
| Valine | GUU, GUA§ |

* These are the codes as represented in RNA, where uracil (U) takes the place of thymine (T).    Code is read from left to right.
§ Possible alternatives.   In the terminal position U may replace C and A replace G.

By means of a code of this kind each sequence of amino acids in a protein is formally related to a sequence of nucleotides in the genome.

The study of inheritance (genetics), which forms the subject matter of Chapter XX, has shown that there are recognisable units of inheritance called genes which are distributed among an organism's progeny according to simple rules.    Genetical analysis has shown that genes form linear sequences which can be depicted in a genetic map.   With the recognition of DNA as the genetic substance and of the codes for protein in the form of poly-nucleotide sequences of the DNA molecule, it seems plausible to identify a polynucleotide sequence coding a single polypeptide chain with at least one level of the classical gene.   This identification is a restatement of the earlier "one gene, one enzyme" concept broadened to cover all types of protein (Fig. 7).

In bacteria it was established genetically that there is only one linear sequence of genes (or one chromosome) and it is now known that there is only one DNA molecule (Fig. 2). The many nucleotide sequences are therefore linked end to end in the long molecule and the identification is complete. In complex organisms there are probably numerous separate DNA molecules; associated with them are many proteins some of which may be linking molecules yielding structures such as seen in Fig. 8, but the actual structure remains obscure.

Before turning to consider the problem of detailed biochemical mechanisms by means of which the molecular blue prints of the

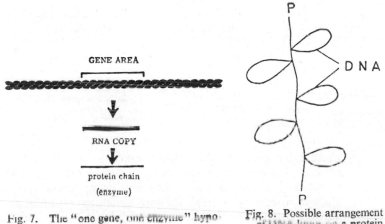

GENE AREA

RNA COPY

protein chain
(enzyme)

Fig. 7. The "one gene, one enzyme" hypothesis.

P

DNA

P

Fig. 8. Possible arrangement of DNA loops on a protein backbone to form a chromosome.

genome are translated into proteins, it is desirable to restate in summary form the ideas just developed which seem likely to become the fundamental axioms of the theoretical biology of the future:

(a) biological inheritance is a property based on the precise self-replication of large linear genetic molecules, the polynucleotides;

(b) the genetic molecules are able to guide the synthesis of other molecules, in the first instance proteins, which as enzymes catalyse the construction of other molecules, which form along with the structural proteins the material basis of organisms. These relationships are depicted overleaf—

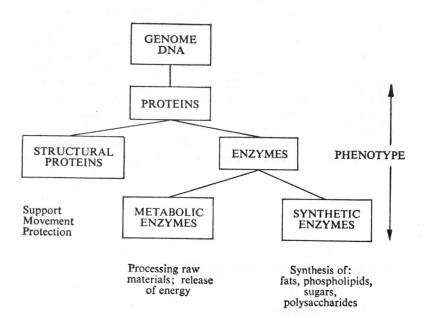

(c) structural changes in the genetic molecules are the origin
of changes in other molecules and therefore of the organisms
of which they are a part. Selection from among these
changed forms by natural agencies generates evolutionary
change.

One of the few examples worked out in any detail concerns
modified haemoglobin, the oxygen transporting protein of the red
blood cell. A considerable number of abnormal haemoglobins
are known from their association with various abnormal blood
conditions but we need only consider one as an example. The
haemoglobin of sickle cell anaemia (HbS) (so called on account of
the shape of the erythrocytes) differs from normal haemoglobin
(HbA) in only one amino acid residue out of some three hundred.
The affected polypeptide sequences are:

HbA (normal)... His. Val. Leu. Thr. Pro. *Glu.* Glu. Lys. ...

                                             ↓
HbS            ... His. Val. Leu. Thr. Pro. *Val.* Glu. Lys. ...

A glutamic acid residue is replaced by a valine.

The triplet codes for these residues are:

glutamic acid        GAA

valine               GUU or GUA

from which it is seen that a single change of a base from adenine to uracil is sufficient to change the code from glutamic acid to valine.

A description of the biochemical mechanisms, which read off the genetic codes and translate them into actual proteins requires first a discussion of cellular structure. This will form the subject matter of the following chapter.

# CHAPTER II

## THE CELL AND CELL CYCLE

The separation of an organism into two parts, genetic records and working parts, necessitates a distinct life-cycle in which several events—duplication of records, separation of the new copy from the old, and the rebuilding of a new organism from the new copy—may be distinguished. A phase of synthesis occurs in all such life-cycles in which the genetic record (genome) is replicated and also directs the construction of a phenotype, whose function it is then to gather and process the energy and raw materials needed to reduplicate

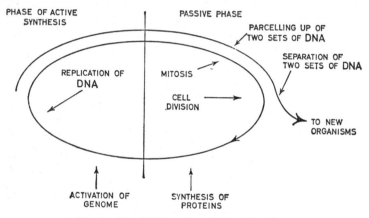

Fig. 9. The DNA cycle and the cell cycle.

the genome (Fig. 9). As a consequence of mutation or a new combination of genes, which becomes possible in situations where the genetic material from two organisms is pooled (see later), the new organism may not be an exact copy of the old. It then represents a further experiment subjected to trial by natural selection and may or may not prove to be more viable than its parent. As mentioned earlier, it is this constant generation of variants and the selection from among them by the environment, that maintains the drift towards greater and greater complexity.

The viral type of organisation is of interest here since it has some exceptional features which help to illuminate the relationship between a genetic molecule and the biological organisation it controls. The particle usually described as a "virus" (more correctly virion) is no more than a genetic molecule (DNA or RNA) associated with a few accessory, largely protective molecules and is the passive phase of the viral system. Its active phase is passed in another organism, to which it has gained entry. Here it disorganises its host's DNA and borrows the use of its synthetic mechanism to reproduce itself. The viral genome is reduced in size since it need only code for a few special specific proteins it cannot borrow from its host.

**The Cell.**—The primordial structural feature of biological organisms is the enclosing or limiting membrane, the plasma membrane, which surrounds the system forming a cell. Larger organisms consist of many cells limited by such membranes. The existence of a plasma membrane might be regarded as definitive of the cell since without it, it is difficult to conceive of the elements remaining together to form an organised unit. It is speculated that the appearance of the membrane-enclosed cell marked the end of the stage of chemical evolution and the ushering-in of a stage of biological evolution by discrete organisms which we would unhesitatingly describe as living.

Within the cell a further centrally placed region, the nucleus, is delimited by a second membrane, the nuclear membrane, and it is here that the genetic apparatus is located during the phase in which it is directing cellular activity. The cellular region outside the nucleus is the cytoplasm—and most other cellular activity other than genetic is located here. This primary structural subdivision of the cell into nucleus and cytoplasm, thus parallels the functional division into genetical and maintenance activities (Fig. 10).

Most cells are very small, no more than 10-30 $\mu m$ in diameter; certain physical considerations appear to govern this size and also the ratio of the volume of the nucleus to the cytoplasm. The dominating physical factor is the rate of diffusion of the many kinds of molecules on whose exchange between the various parts of a cell its function depends. Cells must obtain oxygen by diffusion inwards from their surroundings and they must excrete their waste products by the reverse route. Both the demands for oxygen and the production of excreta are proportional to cell volume and this increases as the radius cubed. The surface through which molecules enter or leave increases only as $r^2$. Clearly there is a radius at which, given the type of chemical reactions involved, an optimal condition is reached beyond which a cell begins to choke.

Similarly, as we shall see more clearly shortly, there are important molecular exchanges between nucleus and cytoplasm by which a certain amount of genetic material can support a given amount of cytoplasmic apparatus, but only that amount.

Many cells do in fact exceed the optimal size expected on these simple considerations, but there are always special reasons why this can be so. The more general rule is that on achieving a certain size by growth, the cell divides into two and restores the more optimal size. Larger cells may have shapes which depart from the spherical thus bringing all inner parts nearer to the surface; their nuclei similarly may be non-spherical. The nuclei may be polyploid, *i.e.* have the genetic material duplicated many times without division, or the cell may be multinucleate. In ciliates a special secondary nucleus much larger than the primary and containing a large number

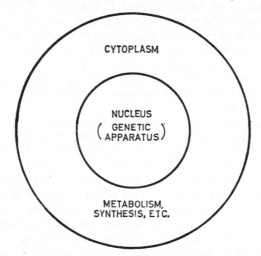

Fig. 10. Subdivision of the cell structurally into nucleus and cytoplasm reflects its two functional divisions.

of copies of the genome is developed to handle the demands of the large volume of cytoplasm (pp. 39, 91).

**The Plasma Membrane.**—The mechanical function of the plasma membrane has been mentioned. In many cells its strength is supplemented by extracellular deposits forming cuticles, capsules, cell walls, etc. Its other important property, which may be considered in a little detail since it affords an example of the homeostatic properties of biological organisms mentioned earlier, is to create and maintain a special micro-environment for the intracellular organelles. Analysis of intracellular fluids shows that the amounts of many inorganic ions, $K^+$, $Na^+$, $Mg^{++}$, $Ca^{++}$, $Cl^-$, and $SO_4^-$, etc., differ from that found in extracellular fluids and that, further, alteration of the extracellular concentrations does not (within limits) alter the intracellular composition. The membrane acts homeostatically, *i.e.* to preserve the constancy of the internal milieu. Since many enzymes function optimally at definite pH values and in the presence of certain ions (co-enzymes), the point of maintaining constancy is evident. There is in all cells a relatively higher level of potassium and a lower level of sodium than usual extracellularly. Diffusion would act to even out these differences (*i.e.* to increase the entropy according to the Second Law of Thermodynamics); it is clear then that some agency located in or near the membrane is acting in opposition to the tendency of diffusion. That this agency consumes energy (the open system) to bring about this imbalance, is shown by equality being quickly established when the cell's metabolism is poisoned. Experiment shows that both $Na^+$ and $K^+$ ions enter the cell through the membrane, but that the sodium is "pumped" out again. The energy for this purpose ultimately derives from the food intake. It may also be necessary to pump out the water that has entered the cell. This again is effected by mechanisms which use metabolic energy.

All membranes are found to be more permeable to fatty type substances (lipids) than to polar (molecules one end of which is electrically charged differently from the other) which suggests that the membrane itself has a lipid character. Actual analyses of membranes, when these are separated from cells, show they consist of about equal parts of a mixture of phospholipids and proteins.

Such findings suggest that cells are covered by a continuous membrane of phospholipid and that this sheet is responsible for their permeability and for their low electrical conductivity. The protein component is less easy to localise. The view of Danielli and Harvey, that the protein is disposed in the form of layers on either side of the lipid, is widely accepted (Fig. 11). The image of the plasma membrane, when this is sectioned at right angles and viewed in the electron microscope (Frontispiece), has been interpreted in terms of a protein-phospholipid-protein sandwich. In such images the membrane appears to be of the order of 70 Å thick and to consist of two dense regions and a less dense central region, each about 20 Å thick. It is not easy to interpret such images of fixed and stained material in molecular terms, but its likeness to the model structure of Fig. 11 is at once evident.

On account of the constant thermal agitation to which all cellular structures are subjected it is probable that transient ("statistical") breaks or pores will occur in membranes of this thickness and these could allow the leakage of water and other small molecules.

It has also been postulated that structural pores of a more definite and permanent character exist, consisting of proteins intercalated in the lipid layers perhaps in a regular mosaic pattern. Such proteins could be the carrier molecules involved in the workings of a specific transport route.

As will be seen later, in most cells there are internal membrane systems, or cytomembranes, which have essentially the same basic composition and structure as the plasma membrane. Such mem-

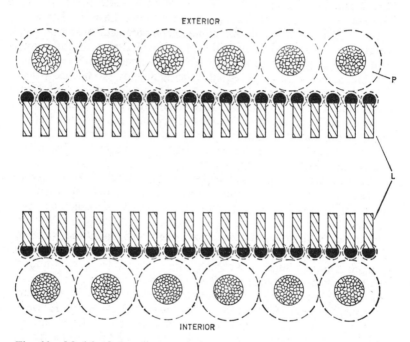

Fig. 11. Model of the plasma membrane proposed by Danielli showing a phospholipid layer covered on each side by protein layers, compare with the Frontispiece (from Davson and Danielli). L, lipid; P, protein.

branes serve to subdivide the internal space in the same way as the plasma membrane marks off the entire cellular space from the environment. They separate various functions (by separating enzymes and substrates) and produce a channelled flow of metabolites from site to site. The most important of these is the nuclear membrane which segregates the genetic apparatus.

**Bacterial Cells.**—Bacteria are exceptionally small cells of the order of 1 μm in diameter and are atypical in other respects. They are, however, so conveniently cultivated and have so many other advantages biochemically that they are to-day the mainstay of the biochemist working on problems of cellular inheritance and protein synthesis. They are enclosed by a plasma membrane and by a more or less thick cell wall which helps to counteract the strong osmotic pressure generated by the highly concentrated cell contents (Fig. 12). There is little formed cytoplasmic structure, and only a small cluster of internal membranes (mesosome).

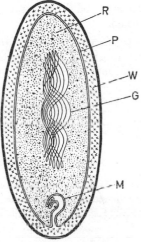

For all their structural simplicity bacteria are sufficiently complex biochemically and contain perhaps 500-1,000 proteins, many lipids and phospholipids. They are able to utilise numerous substrates as source of food and energy, evidence for the possession of many enzymes, and can synthesise numerous other compounds including, of course, DNA. The intracellular enzymes are either simply dissolved in the cell water or associated with the plasma membrane and its involutions.

The DNA lies centrally within the cell and is seen clearly (in electron micrographs) as a skein of loosely bundled threads with no retaining nuclear membrane. The absence of a nuclear membrane in bacteria could be interpreted as a primitive feature. Their smallness and the elaborate retaining wall, which relates them to plants, are likely to be later refinements. In some larger bacteria there is a considerable elaboration of cyto-

Fig. 12. The structure of a bacterium. G, DNA molecule rolled into a loose skein; P, plasma membrane; W, cell wall; R, ribosomes. The fine stipple indicates protein. There is no nuclear membrane. M, is a mesosome, a whorl of membranes. Based on electron micrographs of sectioned cells.

plasmic membranes, apparently involutions of the plasma membrane; these may be sites for specialised metabolic enzyme systems as is the case in more elaborate cells (see later). Endospores, in which a sample of DNA is wrapped in many protective coats of resistant membrane, arise in this way.

The bacterial genome consists of a single closed loop of DNA about 1·5 mm long, tightly wound into a bundle less than 1 μm in its largest dimension. The peripheral cytoplasm contains numerous proteins (enzymes) and ribonucleic acid. A large amount of the ribonucleic acid (RNA) is in the form of macromolecules of diameter about 80 Å which are seen as dense particles in electron micrographs. These are called **ribosomes**.

Several fundamentally important experiments to prove that DNA is a genetic substance have been carried out using bacteria:

(a) **Cellular transformation** using extracted DNA. Highly purified DNA extracted from some types of bacteria when added to cultures of other bacteria can be taken up by some of these and incorporated into their genome when they acquire properties characteristic of the donor strain. Such cells are said to have been transformed.

(b) **Sexual transformation.**—Bacteria normally reproduce by simple subdivision; they can however also interact in an essentially sexual manner in that one donor cell (male) contributes DNA to a

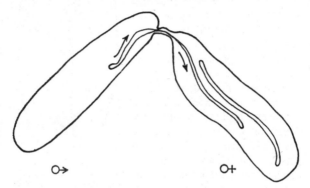

Fig. 13. Sexual transformation in bacteria. A connecting tubule has formed between the mating pair and the donor (♂) is passing a thread of DNA into the acceptor cell (♀). Should the connection be interrupted the thread may break before the entire DNA thread has been transferred.

second acceptor cell (female). The cells meet and form a narrow connecting channel through which the thread of DNA is transferred slowly—the operation requiring many minutes (Fig. 13). If the conjugating pair is disturbed in this time they separate, breaking the thread. The results then show that the number of transferred genes (*i.e.* length of DNA) is proportional to the time the two were joined. The transferred genes are recognised by the altered character of cells resulting from the subsequent divisions of the acceptor.

(c) **Viral infection.**—A special class of virus is associated with bacteria, the bacteriophages (phages), whose study has cast much light on the fundamental problems of genetics. The phages contain a single strand of DNA enclosed in a capsule and equipped with a special attachment device for gaining entry to bacterial cells (Fig. 14).

Once the phage DNA enters, it dissociates the bacterial DNA and uses the cellular equipment to manufacture more phage particles (see below). Experiments using phage with labelled DNA show beyond question that transfer of the DNA does occur and that this alone accomplishes the transformation.

**The Role of Ribonucleic Acid in Protein Synthesis.**—The extreme convenience of bacteria for biochemical studies has greatly assisted the study of the relationship between DNA, RNA, and proteins. In the previous chapter some purely formal aspects of the recording of the structure of proteins in a genetic apparatus were discussed. Reasons for believing that in bacteria and probably also in higher

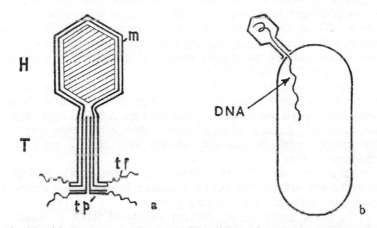

Fig. 14. (a) Fine structure of a T₂ type bacteriophage. H, head; T, tail; m, head membrane; tf, tail fibres; tp, tail plug. The DNA is contained in the head (shown shaded). The phage is about 2,000 Å in length; (b) illustrating how the bacteriophage attaches to a bacterium and passes its DNA into the cell.

organisms the genetic record is a long thread of polynucleotide nucleic acid have been given. We shall now discuss in more detail how the information stored in the DNA in the form of a sequence of nucleotides is actually effective in controlling the type of polypeptide produced by the machinery of protein synthesis. It is remarkable that in all organisms studied this process has proved to be the same in essentials: biochemical experiments on plants, animals, and bacteria have all contributed to establish the present theory.

In the first place a number of experiments carried out on cells larger than those of bacteria have established that protein synthesis is mainly a cytoplasmic function. For example, it is possible in a special case, the larger single celled alga *Acetabularia* (which may be 5 cm long), to remove the single nucleus which is located in the rhizoid and to observe that protein synthesis continues for some time though eventually ceasing.

The most characteristic cytochemical feature of all cells engaged in protein synthesis (bacteria and larger cells) is the attraction of their cytoplasm for basic dyes (basophilia) due to the presence of RNA. Electron microscopically, such cytoplasms are richly populated with the small dense particles, the ribosomes. In bacteria these are about 80 Å in diameter, in all larger cells they are larger, about 150-200 Å across.

By homogenising cells mechanically and subjecting the homogenate to fractional centrifugation ribosomes may be obtained among the finer fragments of the disintegrated cell, called microsomes. These contain many diverse fragments, principally membranes and ribosomes. Treatment with detergents can remove the membranes yielding purer ribosomes. On analysis they prove to contain both RNA and protein.

The invariable presence of ribosomes in cells making proteins suggests that they are the actual sites of synthesis and this is confirmed by many experiments. When cells are supplied with amino acids labelled by means of a radioactive atom, which they may use for protein synthesis, the radioactivity very rapidly enters the ribosome fraction.

Ribosomes are also an essential ingredient of the artificial reconstituted cell systems (to be described below) which are able to incorporate labelled amino acids *in vitro*.

The participation of yet another variety of RNA is revealed when the problem of how the amino acids are induced to attach themselves to ribosomes is considered. A more detailed analysis of all cell fractions reveals the presence of RNA other than that in the ribosomes. When a small amount of labelled amino acid is supplied to a bacterial colony and the label looked for immediately after it has entered the cells, it is found in close association with a soluble lower molecular weight RNA (s-RNA). At a later stage it moves to the ribosomes and finally to formed protein.

In experiments *in vitro* in which cell systems are reconstituted it is found that labelled amino acids are only incorporated into proteins in the presence of both an energy generating source (in cells this is the mitochondrion) and of activating enzymes. Under these conditions the amino acid is linked to soluble RNA, also called transfer-RNA (or s-RNA), whose function seems to be to hold the amino acid to the ribosomal surface. The properties of the several RNAs is summarised in Table 2.

TABLE 2. BACTERIAL RNA

| TYPE | PER CENT. OF RNA OF CELL | VALUE | FUNCTION |
|---|---|---|---|
| Ribosomal | 80 | 70 (30 and 50) | Site of protein synthesis. |
| Transfer (RNA) | 15 | 4 | Attachment of amino acid to ribosome. |
| Messenger (RNA) | 5 | 8-30 | Information transfer. |

These experiments show that the DNA molecule of the genome can not itself directly act as a template to guide the assembly of

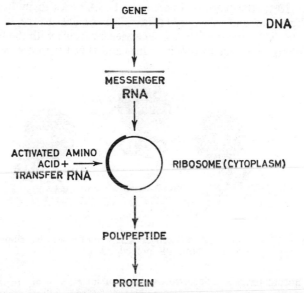

Fig. 15. The synthesis of protein under the control of a nucleotide sequence (gene) of the DNA. The messenger RNA is a working copy of the gene sequence and conveys the information to the site of assembly, the ribosome, in the cytoplasm. The amino acids are transported and held to the ribosome by the transfer RNA.

amino acids in the cytoplasm but that it does this indirectly through the intermediary of a ribonucleic acid. It could be supposed that a DNA sequence (master copy) formed an RNA working copy by base pairing as in DNA replication and that this RNA moved off into the cytoplasm to act as the assembly template. This theory at first met the objection that the base composition of the ribosomes (about 80 per cent. of the cell's RNA, see Table 2) did not resemble that of the DNA. This led to the discovery of a further type of soluble RNA (m-RNA) having the required base composition and a higher molecular weight than s-RNA The present view due to

Jacob and Monod is that this RNA is the direct carrier of the genetic information from the DNA to the cytoplasmic sites of synthesis. For obvious reasons this variety of RNA (see Table 2) is called messenger RNA (m-RNA).

Enzymes of a type known as RNA-polymerases are universally distributed which, in the presence of DNA as a primer and the appropriate nucleotide triphosphates, can polymerise these *in vitro* to an RNA having a base composition reflecting that of the primer.

Messenger RNA is single stranded and of high molecular weight ($5\text{-}30 \times 10^6$); the place of one of the DNA bases thymidine (T) is taken by the related base uracil (U). The current theory, illustrated in Fig. 16, is then that s-RNA is a carrier molecule with the function of attaching amino acids to ribosomes and that the messenger RNA

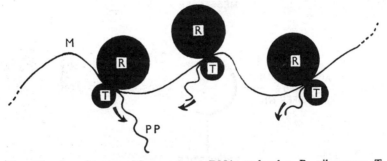

Fig. 16. A polysome. M, messenger RNA molecule; R, ribosome; T, transfer RNA; PP, polypeptide produced.

is the actual template determining the order of the amino acids in the resulting polypeptide. When isolated from cells ribosomes are usually found in small clusters (3-7) strung along a long thread of RNA, the messenger, the whole being referred to as a **polysome**. It is supposed that these ribosomes are moving along the messenger thread "reading-off the code" by assembling the amino acids to form the polypeptide as they move. This hypothesis is illustrated in Fig. 16. In bacteria the messenger has a short life probably only assembling about twenty protein molecules.

The primary structure of a protein (sequence of amino acid residues) is thus determined ultimately by a nucleotide sequence in the genome. The secondary structure, the folded form assumed by the polypeptide on being released from its site of formation, is thought to be assumed spontaneously by the free chain and to be a consequence of a balance struck between intramolecular and inter-molecular interaction of side chains. These interactions are

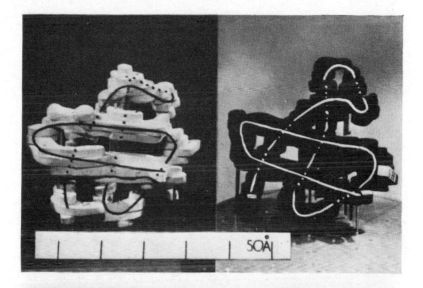

The haemoglobin molecule.   By building up segments which represent regions of high density detected by X-ray analysis, it is possible to construct a three-dimensional model of the molecular configuration.   *Above:* The two kinds of protein chain (α in white, β in black) found in haemoglobin; the discs represent the haem group.   The lines show the "backbones" of the chains.   *Below:* The whole molecule assembled from two of each of the constituent chains shown above.   (*By kind permission of M. F. Perutz.*)

## PLATE I.

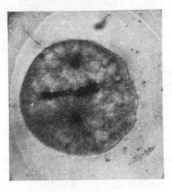

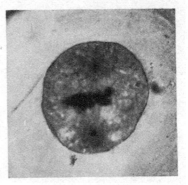

METAPHASE OF 1ST DIVISION         ANAPHASE OF 1ST DIVISION

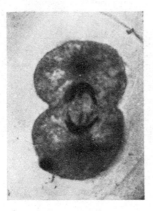

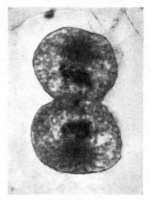

EARLY TELOPHASE OF 1ST DIVISION         LATE TELOPHASE OF 1ST DIVISION

Photomicrographs of stages in mitosis in a fertilised egg of *Ascaris*.

PLATE II.

determined in turn by the original specification of the amino acid sequence by the DNA.

The secondary folding of the polypeptide will be responsible for the specificity of the molecule should it be an enzyme. In structure-forming proteins this shape will control further steps in a spontaneous epigenetic build-up into formations of larger dimensions. The importance of such processes in the self-construction of organisms will be referred to later (p. 47).

**More Complex Cells.**—NUCLEUS AND CYTOPLASM.—In all cells, other than bacteria and a few primitive algae, the genetic apparatus is separated from the rest of the cellular apparatus by a nuclear membrane. This segregation is maintained during the time the

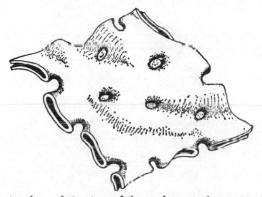

Fig. 17. The two layered structure of the nuclear membrane showing "pores" which are in a hexagonal array.

cell is in a phase of assimilation and synthesis; the nuclear membrane, however, usually breaks down immediately preceding the division of the genetic material and cell division. This membrane consists of two distinct sheets pierced by a regularly arranged system of "pores" having diameters of about 500 Å. The so-called pores, partly closed by a spongy material, are thought to be the principal avenues of communication between the nuclear and cytoplasmic compartments (Fig. 17). Through them the precursors required for DNA and RNA synthesis must pass inwards from the cytoplasm and messenger RNA and ribosomes outwards to the cytoplasm.

The single sheets of the nuclear membrane appear less thick and dense in electron micrographs than the plasma membrane but display after suitable staining the same triple layered structure and it is supposed that their fine structure is essentially like that of the plasma membrane.

The cytoplasm contains numerous particles and those structural features which distinguish one differentiated cell from another are developed here. The nucleus is relatively featureless except when the cell is dividing (see under). One or more densely staining

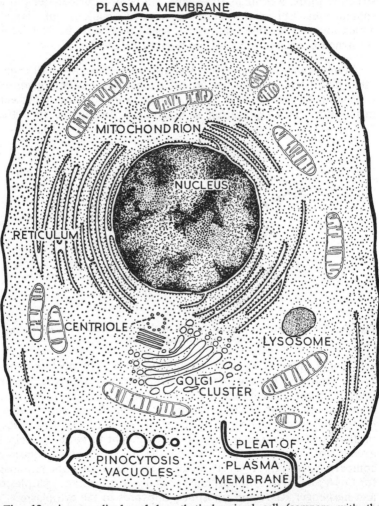

Fig. 18. A generalised and hypothetical animal cell (compare with the Frontispiece).

irregular particles referred to as **nucleoli** are usually visible. It is assumed from this that the DNA and associated proteins which form chromosomes are dispersed to a molecular level during the active phase of cell synthesis. This would seem to be necessary to allow the copying of the DNA sequences for the formation of m-RNA or for self-replication.

We describe the cytoplasm first, and in order not to become involved in the first place with the specialised features of a highly differentiated cell, we shall describe a hypothetical cell which is approximated to (in the structural sense) by some germinal cells, by certain tumour cells, and by liver cells (Fig. 18).

In the description which follows information derived both from light microscopy and electron microscopy is combined. The light microscope may be used with living subjects or on material fixed and stained in various ways to indicate the chemical nature of the constituents. The resolving power of instruments using light is of the order of 0·3μm; for structures having smaller dimensions than this, it is necessary to have recourse to electron microscopes with resolving power better than 0·001μm (or 10 Å). Very regular structures, such as fibres and crystals, may be studied by means of X-ray diffraction, which permits the actual atoms composing the molecules to be located.

CYTOPLASM.—In the cytoplasm of most cells it is possible to distinguish numerous small particles. Some of these are fat and are not permanent features; others appear to be present in virtually all cells indicating that they are necessary for cellular function. Many permanent and universally occurring cellular organelles have a closed membranous framework which serves both to separate one liquid phase from another and as a scaffolding to support enzyme systems. Many of these are vascular or sac-like and are literally reaction vessels in which metabolites are concentrated (perhaps by the action of transferring enzymes embedded in the membranes) along with enzymes to facilitate their interaction (Fig. 19).

The largest organelle is, of course, the nucleus already mentioned. Among the smaller organelles are the following:

*Mitochondria.*—Mitochondria are elongate objects with a diameter 0·5-1μm; they are very numerous (100s-1,000s in most cells) and are characterised histochemically by staining in life with the dye *Janus green* which indicates that they are the seat of sub-stances capable of oxidising this dye.

Mitochondria can be separated from many tissues by homogenation of a mass of cells followed by fractional centrifugation and may then be analysed chemically. They contain about equal parts of phospholipid (membranes) and proteins. Among the proteins are typically many enzymes capable of oxidation and dehydrogenation (oxidases and dehydrogenases), these being responsible for the *Janus green* reaction.

In electron micrographs of cells or of separated mitochondria, these present a characteristic appearance which is as definitive as their enzyme content (Fig. 20). Every mitochondrion has two enclosing membranes: an external one of simple form and an internal one which is thrown into folds, pleats or tubules forming patterns often characteristic of the organism or organ from which the organelles were obtained.

When the mitochondrion itself is ruptured some enzymes are released suggesting that these were in solution in the inner chamber, others remain associated with the membrane fragments. These latter may be seen under the electron microscope in ordered patterns on the inner surfaces of the membranes. It is thought that the efficiency of the multienzyme systems of these organelles,

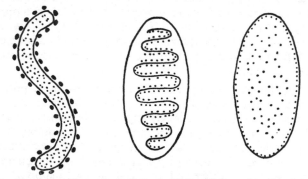

Fig. 19.    Examples of membrane bounded intracellular organelles.
*Left:* particle covered endoplasmic reticulum (or ergastoplasm). The particles on the surface of the membrane are *ribosomes.*
*Centre:* a melanosome, the organelle for the synthesis of melanin in vertebrate melanocytes. The enzyme tyrosinase forms a framework within the vesicle.
*Right:* a lysosome, a vesicle containing hydrolytic enzymes responsible for intracellular digestion.

in which the product of one reaction normally forms the substrate for the next, is dependent on these close-packed ordered arrays on the inner surfaces.

One of the main products of mitochondrial activity is the formation of adenosine triphosphate (ATP), a molecule from which the phosphate groups may be easily detached yielding energy for chemical reactions. Many enzymatically catalysed reactions take place between molecules to which ATP has been attached. Because of its role as a generator of molecules charged with energy the mitochondrion is appropriately referred to as the "powerhouse of the cell".

Mitochondria are found in all cells capable of respiration except in such small forms as bacteria (which in any event are the same size as mitochondria) where it is possible that the respiratory enzymes are located on involutions of the plasma membrane forming a system itself not unlike a mitochondrion. Circulating mammalian erythrocytes lack a nucleus and have no more than traces of mitochondria, although they possessed both during their developmental phases.

In differentiated cells the number of mitochondria is an indication of the respiratory demands of the cell. They are common and large with well developed internal membranes in cells performing

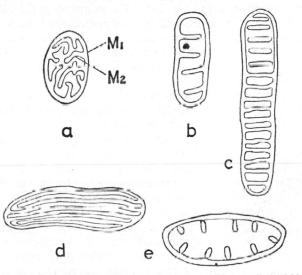

Fig. 20. Varieties of mitochondria (a) protozoon, (b) liver cell, (c) kidney cell, (d) snail ovotestis, (e) plant cell.

much mechanical work (e.g. myocytes) or molecular transport (kidney glomeruli).

The reproduction of mitochondria poses several issues. During cell division they are ordinarily shared approximately equally between the daughter cells as are all numerous cytoplasmic particles. They appear also to be able to increase in length and in numbers during cellular activity or when special demands are made upon a cell in differentiation. This appears to involve both an increase in mitochondrial mass and some process of fission by lateral constriction when the organelle becomes too large. The phospholipid is probably derived from a generalised cell source, since membranes of a similar type are the bases of several intracellular organelles as well as of the plasma membrane. The enzymes are thought to be coded in the

DNA, although there is some evidence to suggest that additional strands of nucleic acid are located within the mitochondrion itself.

*Phospholipid Inclusions, the Golgi Body, Smooth Membranes.*— When fixed cells are stained for phospholipids or treated with solutions containing silver salts or osmium compounds various densely staining cytoplasmic particulates are revealed.   In particular in one region of the cytoplasm usually close to the nuclear membrane an irregularly shaped body can be demonstrated in most cells and is called the Golgi body (or apparatus) (Fig. 18).   It is particularly evident in secretory cells (Fig. 23).   The morphology (and even the existence) of this structure was much disputed until the advent of electron microscopy which revealed in the Golgi region a characteristic array of paired membranes and vacuoles (Fig. 21).   Although

the details and degree of development is extremely variable the *pattern* of the membranes is unmistakable and the Golgi is now recognised as an ubiquitous cellular organelle.

Clusters of Golgi membranes have been separated from various germinal cells (where they are common) and examined biochemically, thus establishing their existence prior to fixation for microscopy.

Fig. 21.  Paired membranes found in the Golgi region.

The function of these inclusions remains obscure.  As mentioned they are elaborately developed in secretory cells where vacuoles containing the secretion apparently arise in the neighbourhood of the Golgi cluster.

This association of membranes and secretion granules has given rise to the theory that the vacuoles serve to accumulate the secretion which is elaborated elsewhere.   Cells, which appear to have little need for membranous organelles in their functioning, have poorly developed Golgi clusters and this feature taken along with their peculiar morphology (Fig. 23) suggests again that the cluster may be the cell centre for the elaboration and dissemination of phospholipid membranes.

Most cells have numbers of other pairs of smooth membranes and vacuoles or vesicles whose function is unknown and which are often lumped together and regarded as belonging to a cytoplasmic network (reticulum) to be discussed below.

In certain cells, such as striated muscle cells, where there is good reason, on the grounds of their great metabolic activity associated with the conversion of chemical energy into mechanical work, to suppose that a channelled transport of metabolites would be required, these smooth membranes form a veritable interconnected system of pathways enveloping the bundles of myofilaments and mitochondria.

*β-cytomembranes.*—A further type of paired cytoplasmic membrane which is really a pleat formed by the invagination of the plasma membrane penetrating sometimes deeply into a cell are referred to as β-cytomembranes (Fig. 22) and are common in cells responsible for the transport of water or of ions. The compartments formed by

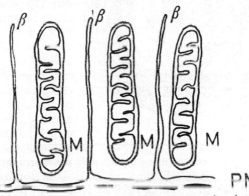

Fig. 22. Invaginating pleats of plasma membrane penetrating deeply into the cytoplasm (β-cytomembranes). PM, plasma membrane; β, the pleats of β-cytomembranes; M, mitochondria occupying the spaces partitioned off by the invaginating pleats.

the deeply penetrating pleats are often occupied by large mitochondria which provide the energy in the form of ATP required by the transporting mechanisms.

*Centrioles and basal bodies.*—These are dealt with on p. 50.

**Protein Synthesis in Complex Cells.** The synthesis of specific proteins by complex cells takes place in steps essentially similar to those in bacteria. The presence of the nuclear membrane is not known to introduce any biochemical complication. Ribosomes are large (150-200 Å diameter) and conspicuous and various soluble RNAs are known.

Protein synthesis in the cells of multicellular organisms is inevitably associated with the process of differentiation (to be discussed later) which leads to the appearance of distinct cytological differences between cells of different functions; nowhere is this more apparent than in the several classes of cells forming proteins.

In early embryonic cells and in many germinal cells, which preserve an embryonic character while supplying cells to a differentiating tissue, after allowance is made for the presence of the large particulate organelles described above, the cytoplasm resembles that of bacteria in being characterised by many *free* ribosomes.

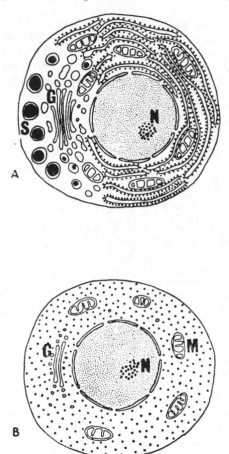

The proteins assembled on these ribosomes are released directly into the cell sap where they may either remain in solution or proceed to aggregate and form an inclusion characteristic of that type of cell (Fig. 23b). In the reticulocyte, the precursor of the erythrocyte, for example, the respiratory protein haemoglobin simply accumulates in solution in the cytoplasm. In myocytes (muscle cells) the protein forms fibrils which accumulate and produce the characteristic appearance of a muscle cell. Such cells may be called **retaining cells** to distinguish them from secretory cells where the protein synthesised is secreted from the cell after a temporary accumulation. Retaining cells become differentiated by the protein they synthesise and retain; as the cell enclosure fills, some sort of mass action effect inhibits synthesis; the

Fig. 23.    Cytology of A, secreting cell and B, protein retaining cell.

ribosomes and kinds of RNA diminish in amount, the nucleus becomes condensed and may even disappear.

On the other hand **secretory cells** are characterised by the development of a specialised internal membrane system whose function appears to be to collect and transport the secretion and to guide raw materials to the sites of synthesis. Characteristically the ribosomes in such cells are closely associated with the membranes which are therefore referred to as **particle-covered membranes** (Fig. 23*a*). The ensemble of cytomembranes is often referred to as the **reticulum** (endoplasmic reticulum and ER are synonyms).

Retaining cells have usually one phase of synthesis when the cell is stocked with protein after which the rate of synthesis declines and the cell enters upon a prolonged life in its differentiated state when the accumulated protein enables it to perform some definite function (oxygen transport in red cells, contraction in muscle cells, protection by keratinocytes) on behalf of the entire organism.

Secreting cells, however, continue their synthetic activities indefinitely synthesising and secreting either in a cyclic fashion or on demand. The secretions may be enzymes effecting extracellular digestion of food stuffs in the alimentary canal, they may be signal materials exchanged by neural cells of the nervous system, or external deposits, membranes and fibres, put to a variety of uses.

Messenger RNA in bacterial cells is known to have a rather short life and to be capable of handling only a small number of protein molecules. Such cells are thus able to switch rapidly from synthesising one protein to another (see below). In more complex cells such lability would be undesirable during differentiation and it is found that the messenger has in fact a much longer life and is able to assemble a larger number of protein molecules than the bacterial m-RNA.

**The Formation of Structures of Larger Dimensions.**—We have seen that the secondary folding of a polypeptide chain is determined by its amino acid sequence and therefore ultimately by a polynucleotide sequence in the genome. In the case of some structural proteins responsible for the larger elements of anatomy visible to the unaided eye, the further aggregation leading to these larger dimensions occurs spontaneously and depends on the shape and distribution of electrical charges over the surface of the molecule (its "reaction profile"). This again is a function of the primary amino acid sequence and the secondary folding.

The best known example of this type of protein is **collagen**, the principal bonding fibre of the mesodermal tissues of vertebrates. Collagen is secreted as a long thin molecule (3,000 Å long and 13·6 Å wide) by the wandering fibrocytes of these tissues. This soluble precursor molecule is called tropocollagen and consists of three polypeptide chains twisted together helically at a steep angle. The structure is compact and stable and is made possible by the frequent occurrence of the sequence: glycine-proline-hydroxyproline; the sequences of the entire molecule we must suppose are coded in the collagen genes of the genome.

A variety of fibrillar structures will separate from solutions of tropocollagen (Fig. 24), each representing a different mode of aggregation of the long molecules. Only one form, that having a periodicity of the order of 640-700 Å, occurs normally in tissues.

Collagen is an extracellular fibre, a secreted product, but similar principles of aggregation apply to intracellular fibrils of which two

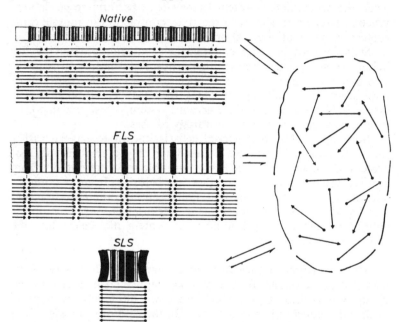

Fig. 24. Diagram to illustrate the reversible formation of different ordered arrangements of collagen macromolecules from solution. The packing arrangements are indicated under the band patterns observed electron-microscopically for each type of aggregate when stained with phosphotungstic acid. Native, arrangement found in living tissue; FLS and SLS, two other arrangements which may occur artificially. (After Schmitt.)

important examples in vertebrates are muscle fibrils and the keratin fibrils of the epidermis.

Not all structure-forming materials are proteins coded directly in the genome. Other important structural molecules are the phospholipids, the basis of all cellular membranes, which are the end products of a chain of enzymatically controlled syntheses. In this case the genetic influence is brought to bear indirectly through its control over the synthesis of these enzymes. Phospholipids once synthesised form membranes spontaneously in aqueous solutions. Each membrane is a bimolecular leaflet, a sort of 2-dimensional crystal, in which the lipid molecules are oriented at right angles to the surface of the membrane. Such membranes may readily absorb layers of protein yielding structures like that of Fig. 11.

**Genetic Units Outside of the Nucleus. Cytoplasmic Inheritance.** —While many structural materials produced by cells can be regarded as spontaneously self-organising, there are hints to suggest that in some instances the construction of systems at a higher level of organisation may require a priming agent, some independent template, to initiate the aggregation of the monomeric units. The existence of such agents would mean that not all inherited information is channelled through the nuclear genes. The loss of these templates would mean to the cell the loss of the know-how of construction of that particular system.

It is possible that examples of such priming agents may be numerous in the cytoplasm and be divided equally between daughter cells at cell division ensuring their inheritance. Only in the event of there being few of them does the probability of one daughter cell not inheriting one of them become significantly large. This situation does arise in the case of the chloroplast, the mitochondrial-like body in plant cells which contains chlorophyll and various enzymes and which is responsible for photosynthesis. In unicellular plants it may happen, when division is rapid, that cell division outstrips plastid replication and a clone of colourless chlorophyll-free cells incapable of photosynthesis results. No doubt the many enzymes involved are coded in the genome but some element of chloroplast structure is carried by the plastid itself and cannot be regenerated by a cell which has lost it. Colourless patches on leaves are due to a similar loss.

In bacteria a class of unattached genetic units the **episomes**, consisting of DNA not physically united with the main bacterial chromosome, are known. These may move independently of the main strand of DNA in various experiments in which bacteria conjugate and confer properties independently of it. We meet here again a reminder of the independence of genetic materials; these molecules need only to find a suitable cytoplasm to ensure their self-replication. Their effect on the host cell may be beneficent or deleterious or more or less neutral. They are genetic analogues of the symbionts, parasites, and pathogens recognised among whole organisms.

A particularly interesting particle of the template class is the basal body (or kinetosome) which is universally associated with the whip-like organs of motility found on the surfaces of many cells, cilia and flagella (p. 81). These organelles are found in all types of organisms except arthropods, nematodes, higher plants, and some groups of protozoa. Everywhere they have the same fine structure. The surface of the cilium, a thimble-like extension of the plasma membrane, sheathes a bundle of nine, peripheral, paired fibres and a central pair of different character. This invariable pattern is reproduced with the absence of the central paired fibril in the basal body. Basal bodies always arise from other basal bodies (or centrioles, see below) and the development of surface patterns on fields of basal bodies precedes the development of cilia; the template role of the basal body is obvious but its exact action is obscure.

Centrioles, which form the initiating centres for the asters and spindle of division, have an identical structure (Fig. 18) and an analogous role. The centriole is self-replicating and its duplication normally precedes the division of the chromosomes. Details of this replication in basal bodies has been followed in some detail in certain spermatozoa. The new body begins to form at the side of the parent with its axis at right angles to that of the parent. The meaning of this escapes us; yet there is no doubt that the structural organisation of the parent is being imprinted on the offspring even if the materials used have been presynthesised under nuclear control.

There are microsurgical experiments on protozoa and eggs to suggest that there are structural patterns in the plasma membrane which are necessarily inherited with the surface since this is inevitably handed on at division. Probably the materials of the surface are preformed and a particular organisation is imposed on them as they are added to the existing surface. This is very apparent in the complex organisation of the ciliate surface (p. 82), in which each unit of the surface has an asymmetrical polarity. The basal bodies lie in rows and from each there extends, anteriorly beneath the surface, a tapering fibril and from its opposite face outwards a cilium. The fibrils beneath the surface lie side by side to the left of the line of basal bodies forming the kinetodesma, a line visible in the light microscope (Fig. 41) and recognised as a fundamental morphological feature of the ciliate surface.

**Cell Division and Mitosis.** THE CELL CYCLE.—The normal life-cycle of growing cells must include as a minimum of essentials: the replication of the cell's genetic units, the directed synthesis of numerous proteins required for the maintenance and growth of the cell between divisions, and finally an act of division itself involving the separation of the two copies of the genetic materials and the contents of the cytoplasm (Fig. 9).

The length of a cycle varies greatly: as short as twenty minutes with the small bacteria, it may be of the order of days for larger cells. This of course applies to continuously growing cells; the cleavage divisions of eggs are special cases involving at first no pause for cytoplasmic synthesis and succeeding each other rapidly. In the opposite sense in multicellular organisms as cells differentiate they may, according to the particular function they are coming to assume, divide more or less slowly. In the adult vertebrate, for instance, the germinal cells of certain tissues, whose function demands the continuous production of cells, e.g. the abraded epithelia (internal and external), the seminiferous epithelium of the testis, and the centres of blood cell production, continue to divide rapidly and usually rhythmically.

The cells of the nervous system on the contrary never divide after a certain stage since their function depends upon the maintenance of a more or less permanent system of intercellular bridges in the form of elongated cell processes (axons and dendrites) which could scarcely survive cell division. Other tissues like the liver, maintain a constant cell number, yet can initiate cell division when required to restore a fall in the cell population.

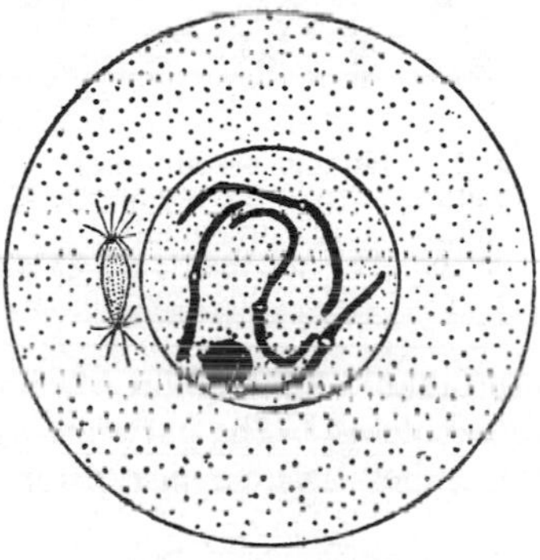

Fig. 25. Early prophase.

MITOSIS.—The replication of the genome may occur at any time in advance of the onset of the next division (prophase). In fact, all the preparations for division (genetic replication, duplication of the centriole, synthesis of the mitotic protein, storage of energy) are made in advance. Cellular activity (synthetic and metabolic) decreases as the genetic material begins to condense and is at a low ebb during division when it is assumed that the condensed genetic material is inert.

The chromosomes, which disappeared from view after the previous division, reappear as the cell approaches division (**prophase**) (Fig. 25). Each is seen as a double strand showing that replication has occurred earlier in the cycle.

The condensation of the genetic material into a number of densely staining rod-shaped particles, the **chromosomes,** is a characteristic preliminary to all forms of cell division. In bacteria (and presumably many viruses) there is one chromosome and one

DNA thread, in this case the two being identical. In more complex organisms, genetic material is divided into several discrete parts,

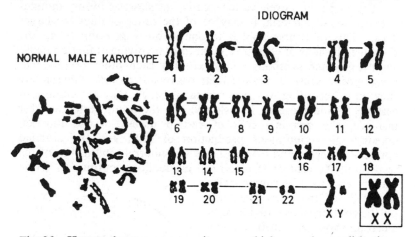

Fig. 26. Human chromosomes seen in a very thinly spread-out cell in tissue culture. Right (idiogram) the 23 chromosomes are arranged in order of size and numbered (the numbering is not final since several of the same size cannot yet be distinguished). Set 23 are the sex chromosomes. Note the smaller Y of the male. Inset and slightly larger is the XX pair taken from a female cell.

presumably on account of the sheer mass of it needed to code all the proteins. The parts are seen as chromosomes in the condensed form they assume just prior to cell division. The chromosomes contain both DNA and several kinds of proteins. The role of these proteins remains a disputed question, the principal point at issue being whether the proteins play a genetic role additional to that of the DNA.

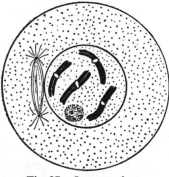

Fig. 27. Late prophase.

The number of chromosomes is different in different organisms, e.g. four in *Drosophila*, the fruit fly used in many genetical experiments, and some hundreds in many plants. The actual number has no particular significance but tends to be larger in more complex organisms (Figs. 26 and 32).

The nuclear envelope usually disperses in prophase or is withdrawn into the endoplasmic reticulum and the cell passes into **metaphase** with the double-stranded chromosomes gathering in a plane at the cell equator to form a metaphase plate (Figs. 27 and 28). Around each centriole in the poles of the elongating cell there now appears a mass of radiating filaments which are said to compose the spindle. The spindle fibres consist of a definite protein (mitotic protein) which can aggregate reversibly to form fibrils. The two centrioles whose structure closely resembles that of the basal body of a cilium (p. 50) appear to function as an instigator of fibril formation, a role recalling that of the basal body in cilium formation.

It is supposed that the mitotic protein is synthesised in the same way as are other proteins and a control exerted over its synthesis would be an effective control over chromosome separation and cell division. In some organisms it is a normal event, in the course

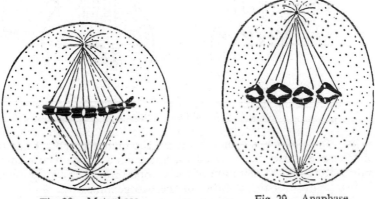

Fig. 28. Metaphase.     Fig. 29. Anaphase.

of developing a differentiated polyploid cell, for chromosome division to occur without subsequent separation or of cell division. In such cells the spindle fails to form; its synthesis has been repressed along with that of other proteins not present in that cell line.

The function of the spindle seems to be to assist in the mechanical act of separating the two sets of chromosomes and perhaps aid the division of the cell (Fig. 31). The actual events take different forms in different cells and have given rise to as many different theories of the mechanism, which it would serve no purpose to describe here. The actual separation of the cell into two also takes various forms which may be said to depend on how much "elbow-room" the cell possesses. Free cells in a liquid medium divide by a simple constriction of the waist of the elongated cell in the plane of the metaphase plate aided by the daughter cells moving apart. When closely confined in a tissue the cell membrane may grow inwards, recruiting vacuoles from the cytoplasmic membrane systems, until a dividing partition is established. Something similar occurs in plants and bacteria whose movements are impeded by their solid cell wall. Here the future division appears transversely between the already divided chromosomes and grows outwards to join the

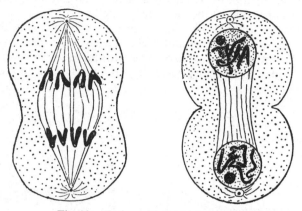

Fig. 30.　Early and late telophase.

walls. In other cases a bud forms on the maternal cell. In spite of their complexity these are all variations on a theme whose common feature is the separation of the cytoplasm into two to provide further centres for the activities of the two genomes.

　　Fragments of the nuclear membrane are detectable microscopically during division and these gather again in the final phase, telophase, with probable additions from other membrane systems to enclose the two bundles of chromosomes and two daughter nuclei (Fig. 30).

CHROMATIDS

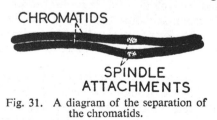

SPINDLE
ATTACHMENTS

Fig. 31.　A diagram of the separation of the chromatids.

　　The cycle of DNA replication, chromosome formation and separation of the two sets of chromosomes (Fig. 9) usually proceeds in phase with the cycle of cell division. The two cycles are, however, under separate control and both in natural and in experimental conditions they may become "uncoupled". Cell division without a doubling of the chromosomes produces a "reduction division" leading to halving of the chromosome number, a normal event in **meiosis** (Chapter XVIII).

A doubling (or further multiplication) without a corresponding cell division leads to polyploidy, a multiplication of the genome. Such polyploid genomes (either as a single nuclear body or many) may support larger cells than normal, the case in many protozoa and plants.

There are certain exceptional cells in which the chromosomes may become enormously enlarged through failure to divide and are visible in the light microscope in such detail as to enable a

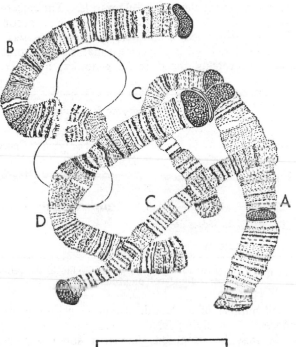

50 μm

Fig. 32. The chromosomes of *Drosophila*. Note the vesicles attached to B.

number of structural conclusions based on genetic analysis to be confirmed visually. For instance, the giant chromosomes in certain organs of dipterous insect larvae are visible as banded threads. The bands are interpreted as loci of genes since they are repeated in the same order in equivalent chromosomes in different organs, but the bands are not equally developed in the different sites (Fig. 32). Such variable appearances are support for the view that different genes are active in different organs.

**Control Systems and Differentiation in Bacteria.**—Cells can synthesise the proteins coded in their genome and these proteins only. Moreover they act in accordance with a programme which ensures the appearance of the appropriate protein at the proper time. Bacterial cells also display adaptability in being able to modify their synthetic activities in response to the state of the environment. These facts clearly indicate the presence of control mechanisms adapting synthesis to internal and external demands. The embryonic development of a complex multicellular organism reveals even more plainly the action of controls. The cells resulting from cleavage and cell division differentiate, *i.e.* become distinguished by the substances accumulating in them or secreted by them. A differentiated cell is a specialist making only some of the total number of proteins coded in its genomes, in other words

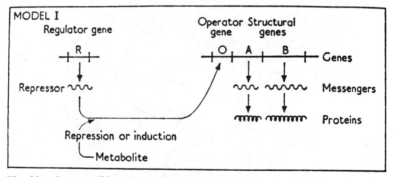

Fig. 33.   One possible scheme for the interaction of genes in the genome of a
bacterium (Jacob and Monod).

only a limited number of nucleotide sequences coding proteins are actively engaged in shedding RNA messengers to control protein synthesis. It may be supposed then that some sequences are repressed and others activated and that some controls act directly on the RNA copying mechanisms.

Very little beyond conjecture can be said of these control mechanisms in complex organisms. In the simpler unicellular bacteria the situation is better understood and we at present are forced to assume that, since so much of the biochemical machinery is identical in all cells, the control systems will be similar too.

The phenomena in bacterial systems, which seems most closely analogous to differentiation in higher forms, are enzyme induction and enzyme repression. A bacterial colony supplied with a substrate it cannot at first utilise, may after a time produce an enzyme capable of attacking the substrate. This is called **enzyme induction**; the bacteria are in a sense now differentiated. We suppose that the induced enzyme was coded in the cells' DNA and that the influence of the substrate was somehow brought to bear on the polynucleotide sequence coding this enzyme to cause it to release the required messenger. Many of the nucleotide sequences of the genome seem therefore to be normally inhibited from producing their RNA complements.

This problem was illuminated by the discovery of Jacob and Monod that the cause of the inhibition was a product of another part of the genome itself, *i.e.* there are **inhibitor genes** whose product can prevent other genes from acting. They proposed that a **repressor gene** *r* acted on an **operator gene** *o* to repress a number of associated genes $g_1$, $g_2$, $g_3$, etc., which constituted an **operand** (Fig. 33). The existence of such genes was established by discovering mutants which affected them. It is the ease with which such mutants can be discovered by bacteriological techniques that distinguishes this work from parallel work on complex cells.

A complementary relation may be supposed between an operator and its repressor. If now a metabolite molecule having a special affinity (based on complementarity) with the repressor molecule enters, the repressor may be itself inactivated and the operator gene released. The associated genes may then release their RNAs and the synthesis of the proteins coded by these sequences follows. In this way an enzyme capable of dealing with the metabolite can be induced. It is assumed that the cell was coded for a suitable enzyme, which of course may not always happen. A particular strain of bacteria cannot be provided in advance with the potential to produce enzymes dealing with every conceivable substrate.

## DEVELOPMENT, DIFFERENTIATION, AND MORPHO-GENESIS

**The Necessity for Development.**—There are advantages in size and in the specialisation of parts which have favoured the appearance of numerous kinds of multicellular organisms. On the other hand, there are advantages in having a small unit for the transmission of inheritance which have led to the retention of the single cell (ovum or spermatozoön) for this purpose. There thus arises in multicellular systems an "alternation of generations", a cycling between a unicellular and a multicellular condition, and the necessity for a development from a germ cell to an adult in each generation. Early in development a distinction can usually be made between **germinal** cells destined to contribute towards the formation of another independent organism and **somatic** cells which differentiate to perform other specialised functions and do not normally contribute to the germ line.

Morphogenesis is of fundamental interest to biologists partly because classification is based largely on morphology and perhaps even more since it is felt that the property of achieving and maintaining a definite size and shape is one of the definitive characteristics of living forms that stands most in need of explanation. The actual course of development in any particular species, which at times seems to follow indirect and makeshift paths, can only achieve a rational explanation in terms of the evolution of development itself. This too has an epigenetic quality, and modifications have to be fitted into an existing development, whose origins must be very remote and evolved to solve constructional problems simpler and quite different from those it solves to-day.

Development involves (*a*) **growth,** *i.e.* an increase in the number of cells (and sometimes their size), (*b*) **differentiation,** the appearance of cells with different functions and cytology, and (*c*) a whole sequence of **morphogenetic events** which shape the final organism and its parts (see also Chapter XIX).

The understanding of all these phenomena remains very imperfect and as yet only a sketch of a molecular biological theory exists. A more or less coherent theory of growth and differentiation can be devised in terms of cellular biochemistry described above but, for technical reasons, experimental support is patchy. It is recognised that a dominating factor in morphogenesis is the overall integration effected by the exchange of numerous chemical substances between the parts of an organism. These exchanges probably control cell division and differentiation through their action on the cells' own control mechanisms. In animals, direct intercellular contact between cells controls cell movement, and perhaps also affects differentiation.

**Control of Cell Division and Size of Organisms.**—The control of cell number both in the whole organism and in its component organs is clearly a primary morphogenetic factor. Even in bacterial and protozoan colonies, in the presence of abundant food, various inhibitory influences accumulate which limit growth. A phase of exponential growth ("log-phase") is followed by a decrease in rate of cell division, so that typically the growth curve is sigmoid in shape and the cell number tends to approach an upper limit. A similar situation seems to prevail in the cell colonies composing multicellular organisms. Probably some substance (or substances) accumulates during growth with an inhibitory action on the synthesis of materials needed for cell division. These inhibitors may be lost by diffusion into the surroundings or consumed in some way, so that their concentration (assuming the concentration is the controlling factor) is a balance between production and loss. The slowing down of growth and the correlation between the growth of different organs, can be understood formally in terms of inhibitors, although these have not yet been isolated.

In the adult there are organs in which the cell number is substantially constant and in which there is little or no cell loss or cell division, e.g. the liver and the brain. In other organs a steady dynamic state prevails between cell production and cell loss, e.g. the mammalian epidermis and blood forming organs, so that production equals those rubbed off or, in the case of corpuscles, destroyed.

In such situations it is usually supposed that information about the state of the cell population (number of cells and their activity) is fed back to the germinal layers to control the rate of production of new cells. Substances known to stimulate erythrocyte production (e.g. erythropoietin) have been isolated. There is little doubt that numerous substances released by cells can act as stimulants and inhibitors of the growth of other cells and organs. These substances are largely responsible for the integration of the entire cellular community and the growth rate of an organ is likely to be a balance struck between these opposing stimulants and inhibitors. The best known examples of such substances are the hormones, products of the various endocrine glands in vertebrates. Such hormone-like compounds are known in most types of multicellular systems including plants and it is likely that many more having less spectacular effects remain to be discovered.

**Control Systems in Complex Cells.**—The notion of the hypothetical inhibitors of cell division is probably (at the biochemical level) similar to that of other cue molecules exchanged by cells and acting as signals which in multicellular organisms are responsible for directing cell development along different paths leading to different differentiated states. The nature of the control mechanisms which these signal molecules activate are almost unknown. However, since the process of protein synthesis is similar in bacteria and in higher organisms, it is plausible to suppose some similarity in the mechanism of control (cf. Fig. 33), although the techniques do not exist to demonstrate them clearly.

From the biochemical viewpoint the problem may be summed up as follows:

With some exceptions all cells in multicellular systems contain the same set of DNA molecules. Development leads to the establishment of a number of specialised environments for cells, each of which constitutes a set of inductive and represssive cues acting on the nuclear genome of each cell and determining which nucleotide sequences will function to produce proteins.

In the differentiated tissues of adult organisms only a limited number of sequences are active and thus only a limited number of the total proteins coded in the genome appear in each cell type. Experiments on differentiated cells separated from adult tissues show that the repression of a large part of the nucleotide complement is more or less permanent.

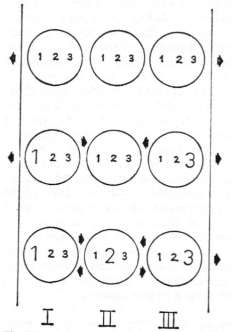

There is an obvious analogy between enzyme induction and repression in bacteria and the production of some proteins with the repression of others in differentiated cells. Direct activation of enzyme systems by metabolites has been demonstrated in a few instances in complex cells. It is usually assumed as a working hypothesis that some system such as that of Fig. 33 will be present in these cells. The activating metabolite molecule is here some product of neighbouring inducing cells, which is able to switch on or off a control system affecting the release of m-RNA by the gene sequences.

Fig. 34. Epigenetic development in a hypothetical tropoblastic organism. Three (1, 2, and 3) linked systems of nucleotide sequences (genes) determining three differentiated states (I, II, and III) are supposed. The size of the numerals indicates the relative activity of the nucleotide sequences.

The epigenetic sequential (one step leading to another) aspect of development can be understood in terms of this type of control system. In Fig. 34 for illustration, a hypothetical three-layered (triploblastic) system is shown. Three polynucleotide groups (genes) whose products each code for a set of enzymes and proteins characteristic of three distinct states of differentiation (I, II, and III) are supposed and indicated by the numbers 1, 2, and 3. Gradients established by exposure to the inside and outside environments bring about the differentiation of inner and outer surfaces (I and III) and the products escaping from these cells influence the development of the enclosed cells bringing about a third state, II. The interchange of signal molecules from all three systems stabilises and enhances the differentiation.

Numerous experiments establish that, once a difference develops in one tissue, its products are able to influence less advanced cell masses. The effect is usually termed **induction** and its obvious epigenic quality can be understood in the above general terms. Biochemically induction probably operates through a genetic control system of the type described for bacteria. We suppose, for example, that some of the products of ectodermal differentiation escape from these cells, penetrating less advanced cells of the mesoderm, and are able to reinforce the action of the genetic inhibitors of the polynucleotide sequences coding the proteins of the ectodermal system. They may also act to induce the formation of the proteins of typical mesodermal cells (fibrocytes, myoblasts, etc.).

Unlike induction in bacteria differentiation in higher organisms is essentially irreversible. When cells are taken from a differentiated tissue and cultivated *in vitro* they usually cease to produce their characteristic products and become visibly de-differentiated. This proves only that continuous production requires the presence of the normal cellular environment. The cells are not de-differentiated in a profound sense, since they will usually differentiate again only in the same sense when transferred to an appropriate environment. The irreversibility is, however, a matter of degree; in embryonic development as divisions continue and cells become separated from the egg by more and more generations, their characters become more irreversibly fixed; earlier traits are more engrained than later. This means that cells early become divided into families of restricted but still broad potentialities, which become more limited later. Epidermal cells, the external epithelial cells, are early marked out in the embryo by their surface location and are differentiated at this stage in the sense of their possibilities of future synthesis. They will not synthesise collagen or muscle proteins, as will members of the mesodermal family, but may form any of a large variety of compounds found in the adult epidermis. In the adult, according to their location in the epidermis, their activities become more narrowly channelled, yet they may be switched over to make any of a number of epidermal components by changing their position. These less stabilised differentiations may be referred to (following Weiss) as **modulations**.

Modulations are more closely analogous in their stability to the induced states in bacteria and a control circuit such as that of

Fig. 33 seems appropriate. The more profound differentiations which distinguish the cells of the three so-called germ layers, have a greater stability. The silent sequences in their genomes are clearly more firmly and finally inhibited. The molecular mechanism of this is unknown. Very likely some of the many proteins (histones) associated with the chromosomes act as repressors in which case the repressor mechanism includes a stage of protein synthesis such as is not represented in bacteria.

This irreversibility of differentiation seems not to occur in the higher plants where it has been shown that almost any cell can give rise to a complete plant when dissociated from other cells in a suitable nutrient medium stimulating cell division. This need to separate the cells again emphasises that it is the position of a cell in the organism as a whole which, by subjecting it to a special environment, controls its activities. During embryogenesis in animals some cell movements, whose causes are still obscure, occurring in its early phases (blastulation and gastrulation) are responsible for establishing these initial positions and it is from this point onwards that differentiation becomes increasingly firmly established. The position that a cell acquires during the early cleavage divisions of an egg normally determines its **cytoplasmic** (as distinct from nuclear) **inheritance**, *i.e.* its initial endowment of factors which influence the activity of the nuclear genes. Two extreme cases are recognised: mosaic development and regulative development. In an egg showing mosaic development a well established cytoplasmic segregation exists before cleavage; the cell boundaries produced by cleavage immediately block out cells having different cytoplasmic endowments which initiate different genic activities. In regulative development the öoplasm is more uniform and the early cleavages lead to equivalent cells. Divergent development seems to be initiated by the position of the cells relative to the environment and to each other (Fig. 34). The development of differences is epigenic and one step leads to the next and so on. Tissue culture studies show that the several cell types which emerge must be present together to obtain further development. Clearly the developing organism is from the very beginning a system co-ordinated by the exchange of signal molecules or other cues between its parts as has been indicated in Fig. 34.

One cell can influence another in two ways: either by physical contact of their surfaces or more remotely by releasing substances which diffuse between them, and these methods of intercellular communication remain the basis of physiological regulation and control in the adult.

These remarks apply equally well to organisms having a predominantly mosaic development. In fact the difference between the two developments lies not in the nature of the final organisation as in the times and means by which differences are established. There is a more massive differential cytoplasmic inheritance in the mosaic egg and differentiation appears early in such eggs and is firmly established; in regulative systems there is a longer period in which cells are "uncommitted", and this may serve as a buffer against disturbances in development.

**The Formation of Organs in Animal Systems.**—Organs are colonies of cells of a few types united to serve a common function. They may be classified from the point of view of construction as being composed of either **close-packed** (usually epithelial) tissues or **open** tissues (composed of few cells). These associations are brought about by secretions from the cells themselves, which are synthesised when the appropriate sequences of the cells' genomes are activated.

In close packed tissues the plasma membranes of the cells are virtually in contact being separated only by thin sheets (about 200 Å thick) of adhesive material which holds them together. Their surfaces are sticky, but this does not impede lateral movement as can be confirmed in time lapse films of such cell aggregates in tissue cultures. The adhesion is specific, like cells tend to adhere more strongly than unlike and thus tend to seek each other out and to clump together. These movements have obvious morphogenetic potential. In tissues subject to mechanical stress the adhesion is reinforced by the formation of thicker patches of dense material called desmosomes on the plasma membranes. These are elaborately developed between muscle cells and the superficial cells of epithelia for obvious reasons.

The open tissues are in effect a meshwork of filaments and gels secreted by cells which wander through them. The most important of these is the **fibrocyte** which secretes the fibrous protein tropocollagen, whose architectural potential is due to its spontaneous aggregation to collagen already referred to and illustrated in Fig. 24. There is a relative abundance of extracellular deposits in this type of tissue and cells do not spontaneously adhere to each other. The secretion of collagen seems to occur as an inductive response to the concentration of products from the epithelia enclosing the mesoderm.

**Molecular Phylogeny.**—When evolution is considered at the molecular level it is clear that its advance depends upon mutations producing new proteins which, either for their structural value or for their effect as enzymes in permitting a new reaction, confer some advantage on an organism. A fully developed molecular biology would be able to correlate the step by step progress of evolution with these mutations and the new proteins which they code.

Most of the fundamental cellular reactions and structures are the same in all organisms and therefore must date back to a primordial cell whose origin and composition is a matter of almost total speculation. Among these reactions are the most fundamental: those linking nucleic acids and proteins. Scarcely less universal are the respiratory enzymes and the energy transfer system based on ATP and other high energy molecules. Mitochondria and chloroplasts (sites of ATP production) are not known in bacteria although their material bases (enzymes and lipoprotein membranes) are, and here perhaps size excludes them. When the higher phyla are examined it will be seen that, although sharing a common cellular equipment and intracellular biochemistry, they are characterised by different structural habits each based on the exploitation of different structure-building macromolecules, themselves either proteins or the product of enzymatic syntheses. Many of these are extracellular materials which form the envelopes protecting cell colonies. Some function as intercellular adhesives without which multicellular forms could not develop or persist. Others, accumulating intracellularly, confer some special property on a cell which can become the basis of a valuable specialised differentiation, e.g. contractile protein systems characteristic of animals and wanting in the higher plants.

When the distribution of these materials is considered phylogenetically it is clear that, although they are widely disseminated, each great subdivision of living organisms is characterised by its spectrum of macromolecules whose properties have influenced decisively its evolution and the way of life of its members. Some few examples may be considered. The rigid **celluloses** and **lignins** of plant cell walls confine each cell in a chamber limiting movement and communication; yet each such cell by forming a constructional brick confers properties equally apparent in a long algal filament as in the giant durable sequoia tree. A related polysaccharide **chitin**, strengthened by association with tanned proteins (arthropodins), is secreted by the epidermal cells of arthropods and forms a rigid, insoluble, protective exoskeleton (p. 268) which commits these organisms to growth by successive moultings and to evolutionary variations within this theme, such as sudden metamorphoses. Both cellulose and chitin are polysaccharides, not directly coded in the genome, but synthesised by enzymes which are so coded. The enzymes are probably structurally related. It is clear that the evolution of the two very different phyla was conditioned by the properties of the two materials and, therefore, by the appearance of the genes coding the needed enzymes in the genomes of the early precursors of the two phyla.

The vertebrates have most of their structural materials in common. Their surfaces are made hard but flexible by the accumulation of the cross-linked protein **keratin** in their epidermal cells; their mesodermal tissues are bonded by the fibrous protein **collagen** which is also the seat for the deposition of the typical bony endoskeleton (p. 398).

None of these materials has preserved an invariable composition throughout evolution. Mutations directly affecting their composition, or that of enzymes involved in fabricating accessory molecules, must have occurred. It is more correct to recognise evolving families of macromolecules. Both the collagens and the keratins have been shown to have lower stability in the tissues of the more primitive, surviving vertebrates; this undoubtedly means that an evolution towards greater stability accompanied the evolution of these animals.

Biochemical evolution has, however, been less spectacular than structural change. The same basic proteins have been used to construct an astonishing variety of morphological forms, e.g. the epidermal keratinised appendages which include hairs, feathers, claws, etc. As yet little progress has been made in these more typically morphological problems. They appear to involve not so much the coding of new proteins as changes in the genes of the control mechanisms which time the morphogenetic events.

# CHAPTER III

## THE PRINCIPLES UPON WHICH ANIMALS ARE CLASSIFIED—TAXONOMY

Up to the present day about a million species of animals have been described and their number is being continually added to as a result of new work. Many of these animals are abundant; others exceedingly rare; whilst the diversity of form displayed is considerable. It is evident, therefore, that some system is required for accurately naming and grouping them to produce order out of what might well be chaos. In some ways the methods used in grouping organisms resemble those which form the logical basis of any system of cataloguing other objects, whether of merchandise in a store or books in a library. Here the articles are carefully named, and then arranged in groups, or sections, so that any particular articles have more features in common with other members of their group than with those in other groups. Small groups can in turn be arranged into larger groups and this placing together of like objects facilitates their identification by the user of the catalogue. It is at once clear that the arrangement of articles in a catalogue will vary with the criteria chosen as a basis of similarity and difference between the articles. Thus books can be (but rarely are) arranged according to size or colour. More usually they are grouped on an author or subject basis. Usually the needs of the user of the catalogue will determine the features which are selected to form the basis of classification. This is true to some extent of the classification of organisms but in this instance there is some division of opinion as to the aims and objects involved. On the one hand there are those who hold that classification (or systematics or taxonomy as it is often called) is entirely utilitarian, a mere convenient filing device or catalogue, and little else. Others, on the other hand, believe that, since it is now conceded that all animals are related to one another by a common descent, a "natural classification" will express the phylogenetic relationships between the groups in the scheme.

Despite these differences of opinion, however, there is, apart from minor details, a very large measure of agreement among all taxonomists as to the way in which the animal kingdom should be classified. Further, in order that the procedure in classification shall remain standardised throughout the world, an International

Committee on Zoological Nomenclature has been established which has published a set of rules for the guidance of taxonomists.

The accepted scheme of classification is a development of that initiated by the Swedish naturalist Linnaeus in his book *Systema Naturae*, whose tenth (and best known) edition was published in 1758. He it was who first devised the binominal system of nomenclature in which every species of animal (or plant) is given two names. One of these, the *specific* or *trivial* name, particularises the species from all others, whilst the other, the *generic* name (which is placed before the specific name), is shared by other related species which are considered to be sufficiently similar to be grouped in the same *genus*. Both specific and generic names are always constructed in latinised form and are always printed in italics, the generic name having a capital and the specific name, a small initial letter. To particularise: one of the most familiar animals to the student is the common frog. Scientifically, this animal is named *Rana temporaria—Rana* being the generic name and *temporaria* the specific name. Such naming is usually sufficient to distinguish an animal from all others, but in works on classification or systematics a refinement is always used to make quite certain that the name is correctly applied. This additional precaution is to add the name, or abbreviated name, of the worker who first applied the specific name to that particular animal. Thus, *Rana temporaria* becomes *Rana temporaria* Linnaeus or, simply *Rana temporaria* L.

So far the term species has not been defined, nor is this an easy matter. Formerly, morphological features were the chief, if not the sole, criteria used in defining species, but nowadays genetics, cytology, physiology, and other branches of zoology contribute their quota of evidence. However, for the present, a species may be looked upon as a group of individual animals which in the sum total of their characters (morphological, physiological, etc.) constantly resemble each other to a greater degree than members of other groups: which form a true interbreeding assemblage, but will not, under natural conditions, produce viable offspring with members of another group. It will be appreciated from this that no single test is sufficient to determine if a group of animals constitutes a species but, nevertheless, the species determinations of museum workers are usually found to be accurate in the field, in the sense that their species usually form groups of individuals which are biologically isolated from each other by differences in habitat, by geographical range, by differences in breeding habits, and so on. It may be said, therefore, that a species has a large measure of biological reality in that it is a term applied to an assemblage of animals which in their *natural state* form an inbreeding group separable by several features from all other groups. The species forms, as it were, the first important break in what, on the evolutionary basis, it must be supposed was once a homogeneous interbreeding population.

Once determined and listed, species can be arranged in larger groups and it is in these that the chief differences in schemes of classification become apparent. Yet, once again, there is a large measure of agreement and the names applied to the main categories are standardised. Thus, as mentioned above, species having many attributes in common are placed in the same genus. So, the common frog, *Rana temporaria*, belongs to the same genus as the larger, continental frog, *Rana esculenta* and with other species of the same genus, is placed in the family **Ranidae.** Families with common characteristics constitute an **order**, and orders in turn are grouped into **classes.** The larger groups of the animal kingdom are the **phyla,** which contain, in some instances, very many classes. The animals in each phylum, although displaying a wide range of form, have their bodies constructed on the same ground-plan, but differences between phyla are very great when compared with those between the other categories.

The full systematic position of the common frog can therefore be given as:

| | | | |
|---|---|---|---|
| KINGDOM: | Animalia. | ORDER: | Anura. |
| SUB-KINGDOM: | Metazoa. | FAMILY: | Ranidae. |
| PHYLUM: | Chordata. | GENUS: | *Rana.* |
| SUB-PHYLUM: | Craniata. | SPECIES: | *temporaria.* |
| CLASS: | Amphibia. | | |

This example will suffice to show the way in which a species is placed in a logical position in the scheme of classification, but to give a broader view a commonly adopted classification of the main phyla is now given:

KINGDOM: Animalia.
    SUB-KINGDOM: PROTOZOA.
        Phylum Protozoa.

    SUB-KINGDOM METAZOA.
        Branch Parazoa.
            Phylum Porifera.
        Branch Eumetazoa.
            Phylum Coelenterata.
              ,, Platyhelminthes.
              ,, Nemertini.

Phylum **Nematoda.**
   ,, **Rotifera.**
   ,, **Chaetognatha.**
   ,, **Polyzoa.**
   ,, **Phoronida.**
   ,, **Brachiopoda.**
   ,, **Annelida.**
   ,, **Echinodermata.**
   ,, **Arthropoda.**
   ,, **Mollusca.**
   ,, **Chordata.**

Other schemes depart from this to a greater or lesser degree but are similar in broad outline. Sometimes groups which are commonly included in a phylum or class as sub-phyla or sub-classes respectively, may be regarded by some systematists as of sufficient distinction to be regarded as phyla or classes, and similar differences of rank are met with right down to genera.

Considered in this way classification would seem to be no more than a convenient filing device in which each type of organism is placed in a particular niche in accordance with its resemblances and differences with others. Yet really it is more than this and in one sense is a convenient form of "shorthand", since, once the systematic position of an animal is stated, a summary of its characters is at once implied. Merely to cite the phylum to which an animal belongs is to state (by implication) a great deal. As an example, the systematic position of the common frog may once again be mentioned. It belongs to the phylum Chordata, and this at once indicates that it is possessed of all those characters contained in the definition of the phylum. It is placed in the sub-phylum Craniata since it has a full complement of craniate features, and is an amphibian for the reasons mentioned in Chapter XXIII. It will be realised that the most useful scheme of classification from this point of view will be one which is based on the *maximum* number of important attributes. Such a scheme is termed a **natural classification** in contrast to one which pays heed to a few characters only and which may be termed an artificial scheme. Thus animals *could* be classified into flying animals, running animals, and swimming animals, but this would result in the bracketing together of types with as little in common as insects, birds, and bats in the first category; as certain insects, non-flying birds, and mammals in the second; and as some protozoa, fishes, and whales in the third.

By many systematists it is held that a natural classification provides a basis whereby the true phylogenetic relationships of the groups is expressed, but it is doubtful if this ideal can, in fact, ever be realised. At best, a natural classification may be said to give an *indication* of the steps taken during evolution. The broad trends will be apparent, but the details are obscured by lack of fossil evidence and by parallel and convergent evolution. Yet it is apparent that the gradually increasing complexity expressed in a taxonomic scheme agrees well with what Darwin termed "descent with modification". This is particularly true of the relationships of the groups within a phylum but the relations between the phyla are more obscure. Yet, although the phyla may have little apparent relation to one another, it is possible to arrange them in order of degree of complexity of organisation and as indicated in the outline classification already given, certain phyla may, in turn, be arranged into still larger groups such as sub-kingdoms and branches which serve to emphasise that the graduations between phyla are not of equal magnitude and that some phyla have features in common indicative of descent from a common stock.

Taken as a whole, therefore, these considerations are evidence for the truth of organic evolution, but the phylogenies suggested by the scheme of classification are not true phylogenies in the sense of pedigrees, since taxonomy is concerned not with individual animals but solely with *groups*.

# CHAPTER IV

## PROTOZOA

With the invention of the microscope in the seventeenth century a whole new realm of animal and plant life was brought to view which was a source of marvel to the naturalists of that time, particularly, perhaps, because of the entertaining descriptions, many of them amazingly accurate, bearing in mind the limitations of early microscopes, given by the Dutch microscopist, Antony van Leeuwenhoek (1632-1723). Many, perhaps most, of the minute animals have bodies composed of a single mass or unit rarely reaching more than 1 mm. across and although this may form quite intricate parts, the body never attains the complexity found in larger animals. Indeed, although many unicells are found only in particular kinds of habitats, they rarely show any obvious structural adaptions to them in the way that larger animals do, so that it is rarely possible to make inferences about their mode of life merely by inspection of their structure. For this reason they are regarded as the simplest of living animals and are called **Protozoa**. Yet in most essential respects they resemble the cells of which the bodies of higher organisms are composed in that they have a nucleus and a full complement of cell organelles, and it is customary nowadays to refer to them as such. Enormous numbers (over 45,000) of different species of protozoa are now known and a few of these, selected to show different grades of structure and different ways of life, will now be described. But no protozoan, even the simplest of them, can be regarded as representing in any direct way forms which originally populated this earth. All are, to a greater or lesser degree, specialised to fill particular present day ecological niches.

### AMOEBA

Various species of *Amoeba* are widely distributed in the sea, in freshwater and in soil whilst some related forms are commensals or parasites. *Amoeba proteus*, an account of which follows, is a large species of *Amoeba* living mainly on the floor of lakes and permanent ponds. It is structurally one of the simplest of animals in that it consists of a single cell just visible to the naked eye.

**Form.**—When examined under the microscope, the first thing that strikes one about *Amoeba* is its irregularity of shape, and

71

careful observation for a few minutes shows that the shape is continually changing. The preservation of its distinct outline is due to a regular orientation of the molecules in the outer layers of the cytoplasm which produces a kind of skin effect to a depth of perhaps one or two microns, the so-called plasmalemma (p. 31) which under the electron microscope is seen to bear a fringe of minute filaments formed mainly of polysaccharides (much smaller than cilia) on its outer surface. They are of unknown function, but may function by sticking the animal to the substratum. The greater part of the cell is made up of cytoplasm (p. 40) in which there is an outer, clear, transparent layer called the ectoplasm and a central mass full of minute granules, the endoplasm. These granules are shown by the electron microscope to be of several different kinds. Firstly, there are numerous small vesicles with ribosomes on their outer surface, which represent the endoplasmic reticulum (p. 47). Then there are minute crystals of carbonyl diurea, an excretory product, and also mitochonitria. There are also several stacks of Golgi bodies. At the hinder end the ectoplasm is lacking. The nucleus is a roughly bun-shaped mass (Fig. 35). It is made still clearer if the animal is subjected to the action of dyes or stains, when it is found that the nucleus retains certain colouring matters to a greater degree than the rest of the protoplasm. The nucleus is a structure of the utmost importance to the animal, for an *Amoeba* from which the nucleus has been removed by microdissection may remain alive for some time but is incapable of carrying out all its vital activities for reasons set out on pp. 39-47.

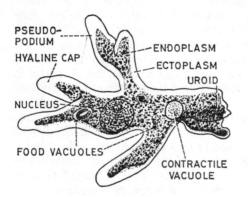

Fig. 35. AMOEBA.—Seen as a transparent object crawling in water on a glass slide.

**Locomotion.**—Change in shape is caused by the apparent flowing of cytoplasm towards certain points where, in consequence, projections called **pseudopodia** appear. Of course, the cytoplasm that enters into them is withdrawn from somewhere else, and therefore, if the formation of pseudopodia is mainly in one direction, the *Amoeba* must be changing its position as well as its shape: in this way it moves about and usually continues in one direction for a considerable time. At one time it was thought that the formation of pseudopodia is brought about by a change in the colloidal nature of the cytoplasm at the point where the pseudopodium appears. The ectoplasm is in the gel condition and, where the protuberance is to be formed, changes to the sol condition. The ectoplasm, particularly at the hinder end, seems to contract and forces the fluid endoplasm outwards. On reaching the surface the endoplasm gels to reform fresh ectoplasm. This process is, of course, a continuous one and proceeds smoothly so that in this way the projection of the cytoplasm increases. The reverse process must, naturally, be taking place near the hinder end of the animal, with the result that here the ectoplasm solates to contribute to the forwardly streaming endoplasm. When floating in water the pseudopodia radiate more or less in all directions giving the animal a star-shaped appearance but when the animal is crawling on a surface the pseudopodia all tend to point in the direction of movement. At the opposite end is a portion of wrinkled cytoplasm, a semi-permanent structure, the uroid.

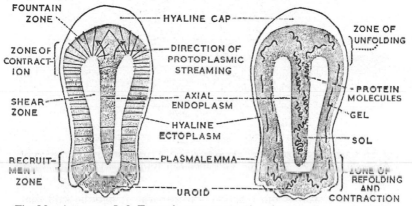

Fig. 36. AMOEBA.—*Left*, Fountain zone contraction theory (Allen *et al.*, 1960); *Right*, Protein molecule contraction theory (Goldacre and Lorch, 1950).

Difficulties remain in explaining, firstly, the mechanism causing contraction of the general ectoplasmic covering and, secondly, that responsible for the change from sol to gel and gel to sol. It may well be that the explanation lies in the arrangement of the large protein molecules of which the cytoplasm is (with water) largely composed. These molecules are possibly joined together to form chains, but in the general ectoplasm they are unravelled and orientated. At the hinder end the chains become folded and contract so forcing forwards the stream of endoplasm. The folding also favours a change to the sol state. At the tip of an advancing pseudopodium the molecules unravel and form ectoplasm. An alternative and more recent theory of amoeboid movement derives from the observation that sometimes endoplasmic granules near the front end begin moving before those at the hinder end, so that progression cannot be due to a squeezing from behind forwards. Instead, it is believed that the region near the tip of the elongating pseudopodium where the everting endoplasm becomes gelled to form its wall (fountain zone) is a region which developes tension. This tension, it is thought, is transmitted to the endoplasm at the hinder end which is pulled along. Both theories require that temporary adhesion between the plasmalemma and the substrate take place for internal streaming alone could not cause locomotion. As it has been shown that some species of *Amoeba* will continue to move in the same direction for long periods, it has been suggested these possess a constant posterior end, responsible perhaps for initiating the changes in molecular configuration. As yet there is no complete or final explanation of amoeboid movement, and some kinds of amoeboid organisms move in quite a different way than *Amoeba*.

**Sensitivity.**—Experiments with *Amoeba* have shown that although it has no definite receptor or effector structures it is sensitive to a variety of stimuli, the whole of the cytoplasm of the

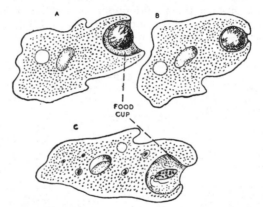

Fig. 37. AMOEBA.—Diagram to show feeding. A, early stage in formation of food-cup; B, later stage; C, with food-cup forming around small living flagellate.

body being sensitive and reactive. In general, it may be said that it reacts to stimuli in such a way as to preserve it in a favourable environment (see p. 116).

**Nutrition.**—In the same mud in which amoebae are found, there are abundant microscopic organisms, usually algae, bacteria, or protozoa, upon which *Amoeba* can feed. When the cytoplasm of *Amoeba* comes into contact with one of these, it flows around it, forming a cup-shaped pseudopodial projection, a food-cup, which completely encloses or ingests the food, together with a drop of water, by the coalescing of the rim of the food-cup. In this way a food-vacuole is formed in the endoplasm. This process is obviously comparable to the phagocytosis carried out by white blood corpuscles (p. 536).

In an *Amoeba* which has been feeding actively, several of these food-vacuoles may be seen and the progress of digestion can be followed. Digestion is effected by processes essentially similar to those in higher animals. Enzymes are secreted into the food-vacuoles by lysosomes fusing with the vacuole (p. 40) along with fluid which can be shown by harmless indicator dyes to be first of all acid, but later, as digestion proceeds, to be distinctly alkaline. The digested material now in solution is absorbed into the surrounding cytoplasm. When all the digestible material has been absorbed the useless remainder is egested by the animal flowing away from it. Sometimes very much more minute vacuoles are formed on the surface at the tips of the pseudopodia and serve to take colloidal material adsorbed on the plasmalemma into the endoplasm. This is known as **pinocytosis**. In this manner amoebae, like white blood corpuscles, can take in food substances and liquids and so some species of *Amoeba* can thrive when denied solid food.

**Respiration and Excretion.**—The whole of the surface of *Amoeba* is in contact with the surrounding water which contains oxygen. Consequently all the oxygen necessary for respiration can pass through the outer surface inwards and the carbon dioxide can pass outwards. This takes place by simple diffusion, for, as the oxygen inside is used up, there will be a lower concentration of oxygen there than in the water outside, and oxygen will tend to pass inwards. Consequently there will be established a gradient of oxygen concentration from the surface into the deeper parts of the cytoplasm. Similar factors will control the passage of the carbon dioxide outwards.

Because of the smallness of the animal, the surface area is large relative to its bulk and no special feature is needed to ensure the diffusion of dissolved substances into and out of the cytoplasm. Similarly, the soluble products of nitrogenous excretion diffuse through the surface.

**Osmoregulation.**—When watching a living *Amoeba*, there will be seen in the cytoplasm a clear circular area which indicates the presence of a spherical cavity containing liquid. This slowly increases in size by the collection of more liquid in it, and then quite suddenly collapses with just the same appearance that is presented by a bladder full of water when it is pricked. Soon it reappears and gradually grows in size as before until it again collapses. It is evident that some liquid is collected in it and periodically expelled from the cytoplasm. This **contractile vacuole** is a mechanism for the control of the water content of the cytoplasm, part of the process of **osmoregulation.** Since the outer layer forms a semi-permeable membrane and the cytoplasm contains osmotically active substances there will be a continual influx of water, which, if it were not expelled, would lead to the rupture of the animal. This excess water collects in the contractile vacuole, being in some way forced out of the cytoplasm into the vacuole—a process which must involve the expenditure of energy since it causes a hypotonic fluid to pass against the osmotic gradient. As would be expected, therefore, substances which "poison" respiratory enzymes and thus prevent the release of energy, stop the contraction of the vacuole and the cell swells and finally bursts. It is possible that the expelled water may contain some excretory products, but there is no reason for supposing that these are more likely to accumulate in the contractile vacuole than they are to diffuse outwards from the surface of the animal. Moreover, the rate at which water is pumped out from the cell is very slow. It only appears fast because the microscope magnifies movement but not time. The absence of a contractile vacuole in most marine amoebae may be due to the fact that the osmotic pressure of the cytoplasm approximates to that of the surrounding medium so that water does not accumulate in the cytoplasm, but many amoebae and other protozoans can withstand wide fluctuations in salinity despite the fact that, because of their small size, they have a large surface/volume ratio. This may be due to some surface effect which slows down the rate of entry of water into the body. Naturally, the rate at which water is discharged from the body depends on the concentration of salts in the outside water, being most rapid in distilled water.

**Growth and Reproduction.**—An actively feeding *Amoeba* assimilates more food than is required for respiratory and other purposes and the excess is used to manufacture new materials which become incorporated in the animal. This leads to growth. Growth, however, does not continue indefinitely; when an optimum is reached, increase in size is replaced by an increase in numbers; reproduction occurs. This process in *Amoeba* is very simple; first, the nucleus divides into two and the two move apart; then gradually, the cytoplasm becomes separated into two approximately equal parts, one around each nucleus. Thus there are two small amoebae instead of one large one. Each one goes on feeding and growing and eventually divides again.

Soil dwelling amoebae belonging to the *limax* group will, under adverse conditions, round off and secrete a protective envelope to form a cyst from which they will emerge when conditions are favourable. When encysted they can withstand desiccation and a wide range of temperatures. Encystment thus affords a means of dispersal and survival. *Amoeba proteus* is believed not to form cysts.

## DIFFERENCES BETWEEN ANIMALS AND PLANTS

As has been mentioned, amoebae will take as food small uni-cellular plants which live in fresh-water ponds. Although of a comparable simplicity of structure these present many points of contrast with *Amoeba*. For example, around the plasmalemma there is a layer of cellulose whilst in the cytoplasm are small bodies, the **chloroplasts**, containing chlorophyll. The presence of the cell wall precludes the putting out of pseudopodia or the development of comparable locomotor structures.

Plant nutrition, too, is different from that of animals. Instead of using an outside source of complex organic food substances, the plant is able to utilise simple inorganic materials and build up (synthesise) the complicated components of its cytoplasm and the compounds it requires for a release of energy. It is able to do this by means of the chloroplasts. A chloroplast, in sunlight, is able to combine carbon dioxide and water from its surroundings to form a simple sugar such as glucose with the evolution of oxygen. To do this, the chlorophyll absorbs the radiant energy of the sun's rays and this energy is then "locked up" in the glucose. This process, **photosynthesis**, is really a very elaborate one and cannot be given in detail here, but for the present purposes the essential fact is that the plant, by means of photosynthesis, is able to build up organic substances (sugars) from inorganic materials. But the nutritive processes of the plant do not stop here. Having elaborated an organic basis to form the starting point, the plant can utilise such inorganic salts as nitrates, phosphates, and sulphates—which again are present in small quantities in the water in which it lives—to build up simple nitrogenous substances (amino acids) from which proteins are finally made. Thus, in addition to synthesising carbo-hydrates, the green plant also carries out **protein synthesis**, using its own, partially elaborated food material and simple inorganic salts and it should be noted that the compounds synthesised are of the same kinds as those which form the food of animals.

It will be seen, therefore, that one of the most important and fundamental differences between animals and plants is their mode of nutrition; the animal feeds on complex organic materials—a type of nutrition which is called **holozoic** (or **heterotrophic**)—while the plant *synthesises its food from inorganic substances*—the **holo-phytic** (or **autotrophic**) type of nutrition. But as has been implied by stating that plants build up their own foodstuffs, plants also carry out degradation processes. Indeed, seeds, which are supplied with reserves of food, are entirely heterotrophic until they have germinated and acquired chlorophyll.

Because of, rather than in spite of, these differences between the two types of organism, there is a close interdependence between them. Since the green plants and certain bacteria are the only organisms which can synthesise carbohydrates and proteins from inorganic materials, animals depend entirely upon them, either immediately or ultimately, for the whole of their food materials and consequently for their continued existence. As a result of this interdependence there are established in nature what are termed "food chains", the starting point of which is always a plant. Thus, many of the common food-fishes live upon small crustacea, such as copepods, which abound in the sea and which feed upon microscopic plants (diatoms) found in the floating life (plankton) of the surface waters. Thus the chain: diatom-copepod-fish-man, is set up. Other examples can be easily worked out, some of which may be very complicated, a single starting point leading to several very different climaxes. The activities of plants and animals are interrelated in yet another important way, namely, they assist in maintaining the constant proportion of carbon dioxide and oxygen in the atmosphere, for, whereas animals are continually using up oxygen in aerobic respiration and giving out carbon dioxide, green plants—when they are actively photosynthesising—are using up carbon dioxide and giving out oxygen, which masks their respiratory activities.

In the simple animals and plants (Protozoa and Protophyta) almost the sole difference between them is that of nutrition, so that it is frequently difficult to say to which category the organisms belong [*e.g.* Mycetozoa (Myxomycetes), and many others—see also Chapter XXIII], but when the animal and plant kingdoms are considered as a whole, the differences are quite obvious. The plants, for example, do not need to search for their food and consequently are fixed, whereas the animals have developed a great variety of devices for moving about in search of, and to capture, their food. A corollary of locomotion is nervous control and the elaboration of sense organs. Further, this difference in habit has led, in the plants, to the development of an almost unlimited capacity for vegetative growth, with the consequent multiplication of parts, resulting in an indefinite size, whereas in animals, size and number of parts is, within certain limits, fixed.

In order to illustrate the vital activities of a simple animal, *Amoeba* was selected because of its simplicity of structure. Many other small animals, however, although having bodies composed of a single unit of protoplasm, and belonging to the Phylum Protozoa, show a much more elaborate organisation and specialisation. A good example among free-living animals is provided by the slipper animalcule, *Paramecium*, a well-known member of the class Ciliophora which has been intensively studied for many years.

## PARAMECIUM

**Form and Habitat.**—*Paramecium* is a genus of minute organisms very common in fresh water where there are decaying vegetable remains and bacteria abound. Several species of *Paramecium* are known but the one most usually studied is *P. caudatum*. This is just visible to the naked eye as an elongated whitish speck. Microscopic examination shows that its body consists of a single cell, as does that of *Amoeba*; but two striking differences from *Amoeba* are seen at once. Firstly, *Paramecium* has a definite and constant shape; and secondly, it swims about through the water (rotating as it does so) or glides over the surface of any solid object with what appears to be great rapidity as compared with the slow crawl of *Amoeba*.

The outline of *Paramecium*, when seen in sharp focus, is not unlike that of the sole of a slipper—which gives it its common name—"slipper animalcule". It is, in reality, shaped rather like a short cigar which has become partially flattened. On one of the flattened surfaces there is a shallow groove, the **oral groove**, from which, extending deeply into the endoplasm in an obliquely backward direction, is a wide tube, the **cytopharynx**, ending in the **cytostome**. The rounded end is usually directed forwards during movement and is therefore termed anterior; the pointed end being posterior, but the animal, on occasion, can reverse the direction of its movement. Although having a definite shape, *Paramecium* is flexible and elastic being able to squeeze itself through narrow openings and bend round obstructions. This constant shape is due to the enclosure of the cytoplasm in a thin but firm membrane formed by its outer surface and which is called a **pellicle**.

The cytoplasm is differentiated into two layers, the ectoplasm and endoplasm. There are two nuclei present, embedded in the endoplasm. They differ in size and appearance, and also in function. The larger is the **meganucleus**, and the smaller is the **micronucleus**, which is lodged in a depression at one side of the meganucleus. The meganucleus seems to control the ordinary "vegetative" activities of the animal while the micronucleus is concerned with reproduction. It has been shown in an allied form that, if the part of the animal containing the micronucleus is cut away by micro-dissection, the part with the meganucleus can restore the form of the body which can feed but cannot reproduce.

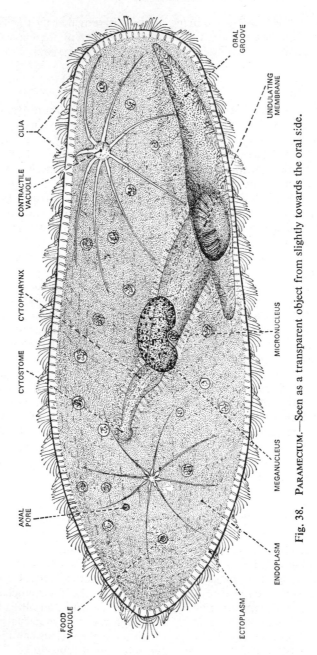

Fig. 38. PARAMECIUM.—Seen as a transparent object from slightly towards the oral side.

**Structure of the Ectoplasm.**—The endoplasm has a more fluid and apparently fairly homogeneous structure, the ectoplasm is firmer and very highly differentiated to form important organelles (see p. 92). It is bounded on the outside by the pellicle which is not smooth but bears regular depressions and ridges which appear as lattice-like markings on its surface. The depressed areas may be rectangular or hexagonal, the latter shape possibly being due to distortion. In the centre of each area protrudes a **cilium**, or in some species, two cilia. Embedded in the ectoplasm at right angles to the surface are numerous small bodies, the **trichocysts**. Pressure or irritant liquids cause each of the trichocysts to eject a long, fine thread,

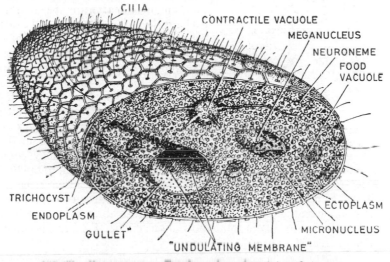

Fig 19 PARAMECIUM.—Cut through to show internal structure.

the precise function of which is uncertain, although it has been suggested that the threads are used for anchoring the animal to the substratum on the rare occasions when it ceases its almost perpetual swimming. In some ciliates the trichocysts almost certainly have a defensive function and even in *Paramecium* they may help to ward off the attacks of predaceous ciliates like *Didinium*. Other predatory ciliates, *e.g. Dileptus*, use their trichocysts to attack and paralyse their victims. The cilia, on the other hand, are rather shorter, very delicate, hair-like prolongations of the ectoplasm. At the base of each is a small swelling or granule which is refringent when viewed in the living condition but stains darkly with haemotoxylin. This is the **basal granule** or **kinetosome**. The kinetosomes (and cilia) are arranged in longitudinal rows each of which is called a **kinety**. Running alongside each kinety is a composite fibre, the **kinetodesma**, which receives contributory fibres from each of the kinetosomes, each fibre ending when it has passed four or five kinetosomes; similar kineties are found in those flagellates whose bodies bear many flagella. Other systems of cortical fibres have been described from ciliates "stained" with silver salts but they have not been identified by electromicrography so that their reality remains in doubt. The function of the kinetodesmata also remains a matter for speculation. Perhaps they are contractile and cause the slight distortions in the shape of the body so often seen when the animal is swimming between obstacles. If so then the kinetodesmata resembles the threads (**myonemes**) known for certainty to be contractile and which occur, for example, in the cortex of certain ciliates like *Stentor* or in the stalk of *Vorticella* where they combine to form a thick thread, the spasmoneme. This is not to say that fibres responsible for co-ordination of the cilia are absent from all ciliates. In some, a system of fibres converges towards a darkly-staining granule on the wall of the gullet and if those fibres are cut the beat of the composite cilia in the hypotrichous ciliate, *Euplotes*, becomes unco-ordinated.

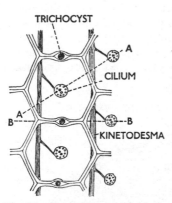

Fig. 40. PARAMECIUM.—Optical section of pellicle (based on Ehret and Powers).

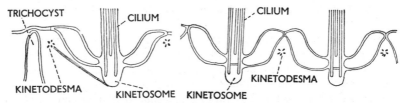
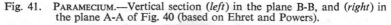

Fig. 41. PARAMECIUM.—Vertical section (*left*) in the plane B-B, and (*right*) in the plane A-A of Fig. 40 (based on Ehret and Powers).

It is by the vibration of the cilia that the animal progresses through the water. In slow movement a cilium stiffens and bends over rapidly to lie nearly parallel with the surface of the body. It then becomes relatively limp and returns more slowly to its original vertical position (Fig. 42). It is the down stroke of the cilium acting against the resistance of the water which provides the impetus. When moving fast, the cilia may come to lie parallel with the body surface at the beginning of the stroke, thus moving through 180° giving a pendular beat. With the introduction of the electron microscope it has been possible to examine in more detail the structure both of cilia and flagella and to show that they both are remarkably similar, not only in structure but presumably in mode of operation, properties which they share with the tails of many kinds of spermatozoa.

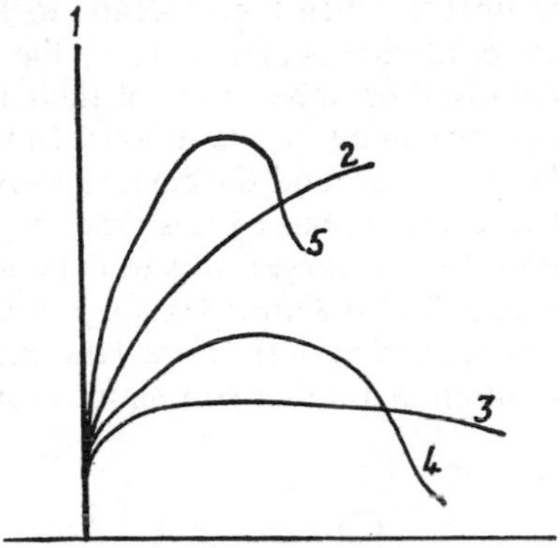

Fig. 42. Diagram to show movements of a cilium, 1-3 effective stroke; 4-5 recovery stroke.

Cilia and flagella (between which no clear distinction can now be made) from many sources, including *Paramecium*, contain within a fine sheath two central fibres and nine outer or peripheral ones (Fig. 43 and Plate III). The central fibres are thinner than the outer ones and are probably formed of a different kind of protein. In some ways they are reminiscent of the filaments within a single muscle fibril. Further details of ciliary structure are given in Fig. 43 which is based on electronmicrographs of cilia from the gill of the mussel, *Mytilus edulis*. The nature of the basal granule and of the ciliary rootlets are not constant features of all cilia. Moreover, in some cilia and flagella there is evidence that each of the nine outer filaments or fibres is paired.

A provisional theory for the way in which ultramicroscopic structure of cilia and flagella is related to their observed manner of movement runs along the following lines (Fig. 44). The basal granules initiate some kind of excitation which causes a contraction

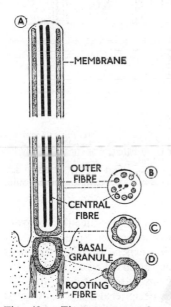

Fig. 43. Fine structure of a cilium. A, longitudinal section of a cilium; B, C, and D, transverse sections at different levels (based on Bradfield, 1955).

in fibre 1 which then spreads to fibres 2 and 9. At the same time
the excitation is picked up by the central fibres and is transmitted to
the distal or upper parts of the cilium by which time fibres 3 and 8
are beginning to contract. Next the excitation spreads circum-
ferentially around the base, successively stimulating fibres 4, 7, 5, and
6 to contract, but by this time the central fibres are in the refractory
state (*i.e.* no longer able to conduct excitation to the tip) and fibres
1, 2, 9, 3, and 8 are relaxing. In this way, by successive contraction
and relaxation of the fibres, the cilium is bent towards the horizontal
position in a rather stiff manner (since its upper regions are contrac-

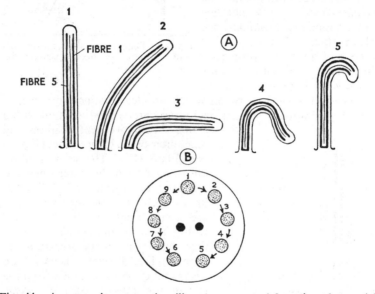

Fig. 44. A, successive stages in ciliary movement, 1-3, active phase; 4-5,
recovery phase. B, diagram to show the direction of excitation of the outer
fibres (after Bradfield).

ting) and is returned to the upright position limply since the recovery
phase is brought about by a wave of contraction spreading only
rather slowly from base to tip. Nothing can be certain at present
and a slightly different mechanism must be invoked to explain the
corkscrew-like waves which spread from base to tip in flagella.
They could, however, be produced if the peripheral fibres contracted
serially in one direction around the circlet provided that the two
central fibres did not form a pathway for rapid conduction of
excitation. As in muscle (p. 476) the energy for contraction is
supplied by the breakdown of ATP.

The cilia do not move independently of one another nor all at the same time but in a characteristic way known as metachronal rhythm, giving an effect rather like that of wind on a field of standing corn. This co-ordination is conceivably brought about by the kinetodesmata but there is little evidence to support this idea.

But, as is well known, isolated cilia continue to beat provided they have a little cytoplasm and basal granule (kinetosome) attached. Their movements are thus intrinsic. Co-ordination of the beat along the ciliary rows is in part due to the excitation handed on step-wise from one cilium to the next, and one ciliate (*Stentor*) there is good evidence that one particular cilium in the gullet acts as a "pacemaker". The spread of excitation is thus rather like that in the vertebrate heart (Chapter XVII).

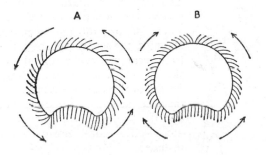

Fig. 45. PARAMECIUM.—Diagram to show action of cilia in producing spiral path as seen from behind. A, diagonally backward beat during forward movement; B, opposing action of cilia of opposite sides during shock reaction.

**Locomotion.**—As already mentioned, *Paramecium* swims through the water comparatively rapidly by means of the beating of its very numerous cilia.* Usually the animal moves with the blunter end of the body leading but it can also swim in the opposite direction by reversal of the ciliary beat. Swimming is never in a straight line, however, for the body traces out a spiral path and at the same time continually rotates about its own axis. This type of motion results from the direction of the ciliary wave or beat which is not directly from front to rear, but diagonally backwards towards the right, so that the wave of ciliary activity takes a spiral course along the body (Fig. 45). The forward component results

* *Paramecium* may swim as fast as 180 mm per minute which, expressed as it is in absolute terms, seems rather slow. If, however, its speed is expressed as the number of times the length of the body is traversed in unit time, it compares favourably with the speed of modern aircraft! The rapidity with which small animals appear to move when under the microscope is, of course, due to the distances, but not the time, being magnified.

from the cilia beating chiefly to the rear, whilst rotation of the body is due to the cilia beating obliquely to the right which causes the animal to roll over on its left side. The spiral nature of the path is caused by the cilia along the oral groove beating more strongly than those on the rest of the body so that if the animal did not rotate on its axis it would swim in circles. As it is, the *mean* path traced out is a straight line passing through the axis of the spiral. When the body is subjected to practically any strong stimulus (pp. 118 and 119) the ciliary beat is reversed so that the animal swims backwards for a short distance before resuming forward locomotion

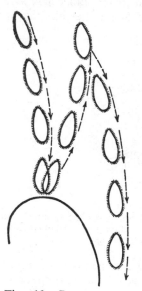

in a different direction. These changes in direction are produced by changes in the beat of the cilia in the following sequence. The cilia momentarily cease beating and reverse the direction of their beat. Then, when forward movement is about to be resumed, the general cilia beat more directly to the rear so that the animal revolves more slowly, whilst at the same time, the cilia on the oral groove beat more strongly and increase the swerve towards the aboral side, and the front end swings round in a wider arc. After this normal swimming is resumed. This type of behaviour is often termed the **shock** or **avoiding reaction** (Fig. 46).

**Nutrition.**—The cilia whose activities have already been described as responsible for locomotion also play an important part

Fig. 46. PARAMECIUM.— Diagram to show avoiding action movements.

in feeding. Not only do they propel the animal through the water, but they cause a current of water to move on to the animal from in front, a cone-shaped volume of water being drawn in towards the oral groove from a considerable distance in advance of the front end. Particles in suspension are thus brought towards the oral groove and cytopharynx whose walls themselves bear cilia which are arranged in longitudinal rows and are sometimes referred to as **undulating membranes**. These serve to waft particles down the cytopharynx towards its narrow basal end, where is found a small area of naked endoplasm. Because of the presence elsewhere of the investing pellicle, this is the only place where the ingestion of food particles can take place and it is therefore frequently called the cell-mouth or **cytostome**. In this respect *Paramecium* differs from *Amoeba* which can take in food at any point on the surface of its naked ectoplasm. When sufficient small particles, which are chiefly bacteria (the main source of food for *Paramecium*), have collected, they are taken into the endoplasm along with a minute drop of water to become enclosed in a food-vacuole. The food-vacuoles formed at intervals travel along in the endoplasm, through which they circulate along a fairly definite course (Fig. 47) (cyclosis of the food-vacuoles), during which the food is subjected to an acid, and later an alkaline, phase of digestion as in *Amoeba*. Eventually all that remains undigested and unabsorbed is cast out at a point near the cytostome, the **cytoproct**. From this description of the feeding mechanism, it will be evident that any suspended particles of suitable size will be swept towards and along the cytopharynx, but much evidence has been brought forward recently to indicate that *Paramecium* has distinct powers of selection over the kind of particles actually engulfed. The discrimination is not an absolute one but it is noticed that more food-vacuoles are formed for certain kinds of particles than for others, the animal even preferring some kinds of bacteria to others. Evidently, even at this level of organisation definite food chains can be established.

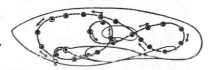

Fig. 47.   PARAMECIUM.—Diagram of path of food-vacuoles.

**Respiration and Excretion.**—The gaseous exchanges associated with respiration take place by diffusion through the pellicle, which also allows the products of nitrogenous excretion to escape.

**Osmoregulation.**—*Paramecium* has two contractile vacuoles—one anterior and one posterior. These have a more complex structure than that of *Amoeba*, for while the vacuole is refilling, a number of very long radiating canals can be seen, along which the water is plainly being conveyed into the vacuole. Otherwise the action of these is similar to that in *Amoeba*, and there is no doubt that they serve the same function, namely the control of the water content of the cell. The posterior vacuole pulsates more rapidly than the anterior one (owing to its proximity to the cytopharynx which provides an additional surface for endosmosis).

**Sensitivity.**—The reactions of *Paramecium* to conditions in its environment and its behaviour are dealt with on pp. 118-19.

**Reproduction.**—The commonest method of multiplication of *Paramecium* is by transverse fission. First the micronucleus and then the meganucleus divides—the former by mitosis (see pp. 51-5) but the latter amitotically—and the animal becomes divided, by a constriction in the transverse plane, into two. But before this can happen certain important organelles, notably the cytopharynx and associated structures, duplicate themselves and draw apart. Each half, by slight changes, soon assumes the form of a normal *Paramecium*. In ciliates, division takes place across the line of the kineties, but in flagellates the division is along the line.

When well fed, *Paramecium* will increase in this way very rapidly, dividing once or twice in a day, producing a series of descendants called a **clone**. But after a time the rate of division slows down and eventually ceases. Also, with an increasing number of successive divisions, cytoplasmic abnormalities become more and more numerous and the stock dies out. Thus a clone shows similarities to the body of a metazoan in that it becomes progressively less viable. In other words, it "ages" or becomes senescent. Yet senescence is not inevitable and a clone can be "rejuvenated" so as to regain its former vigour provided that certain nuclear rearrangements intervene, as they normally do. These nuclear events are of two types, **autogamy** and those which accompany and follow **conjugation**. Of these conjugation will be considered first.

Fig. 48. PARAMECIUM.— Diagram to show a stage in transverse binary fission.

In most cultures of *Paramecium*, the continued multiplication by fission is interrupted by a process termed conjugation, which is of the greatest interest as affording a link between the reproductive processes of higher Protozoa of this type and those of the Metazoa (see Chapter XVIII). Conjugation in *Paramecium* consists of the temporary adhesion by their oral surfaces of two individuals—termed conjugants—followed by certain internal changes which differ according to the species.

The following account refers to *P. caudatum* which has sixteen varieties or strains and in each variety there are two mating types. Conjugation will take place only between opposite mating types belonging to one of the sixteen varieties and can be induced by starving the cultures.

The internal changes concern the nuclei and especially the micronucleus, which becomes very active and separates from the meganucleus (Fig. 49). The meganucleus is at first unaffected, but eventually disintegrates and disappears, being absorbed into the endoplasm. The micronucleus of each individual, on the other hand, increases in size and divides twice, forming four nuclei. Three of these abort, an interesting parallel with certain happenings in the Metazoa, and it is certain that during the divisions of the nuclei there is a reduction in their chromosomes to the haploid number. The remaining one then divides again, forming two (gametic) nuclei. There is now an exchange of nuclei between the two conjugants, one nucleus—the migratory nucleus—from each passing across the region of cytoplasmic continuity between them, into the other. Only rarely is the nucleus accompanied by cytoplasm. This nucleus then fuses with the remaining or stationary nucleus, forming a fusion or zygotic nucleus. The conjugants then separate and conjugation proper is at an end. The subsequent events differ in different species of *Paramecium*, but by a series of divisions of nuclei and cytoplasm, a number of new individuals, each possessing a mega- and a micro-nucleus results. For example, in *P. caudatum* the zygotic nucleus undergoes repeated divisions so that each ex-conjugant contains four pairs of daughter nuclei. The cytoplasm then divides so that there are produced four new individuals, each of which has two nuclei (Fig. 50). Of these, one enlarges to become the meganucleus and the other remains small to form the new micronucleus.

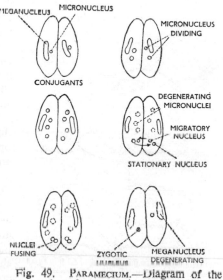

Fig. 49. PARAMECIUM.—Diagram of the nuclear changes during conjugation.

The second type of reproductive process found in *Paramecium* introduces an entirely new factor into the life-history, for in this process of conjugation two co-operating individuals are involved, whereas in binary fission only one participates. Further, the fusion of nuclear material of different origins foreshadows a process associated with reproduction in the Metazoa. A still further parallel may be drawn. After the separation of the conjugants, there is a re-establishing of the meganucleus as well as the micronucleus in the new individuals formed by the subsequent divisions. It will be recalled that at the beginning of conjugation, the old meganucleus disintegrated and disappeared, so that this new one, as well as the micronucleus, is composed, at least in part, of different material. It is possible, therefore, that these new nuclei may contain new combinations of genes (p. 12) which will affect character in the individual and its progeny.

The migratory and stationary nuclei, too, behave in a closely comparable way with the special reproductive bodies—called **gametes**—of the Metazoa and consequently may be called gametic nuclei. The fusion nucleus, again, corresponds to that of the product of the fusion of the two gametes—the **zygote**.

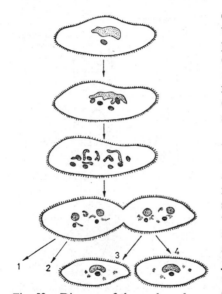

This parallel is still further enhanced by the behaviour of the meganucleus which takes practically no part in the process, and eventually "dies", since it disintegrates. Thus it may be compared to the **soma** (body) of the Metazoa which suffers a like fate.

Any process of reproduction which involves the co-operation of two individuals and the fusion of nuclear material is usually called a **sexual** process, whereas those which do not are termed **asexual**. Fission, therefore, is an asexual process, and conjugation a sexual one. It should be noted, however, that conjugation in itself does not result in reproduction since the number of conjugants is the same at the beginning and at

Fig. 50. Diagram of the nuclear changes leading to the formation of four daughter *Paramecia* from an ex-conjugant.

the end, but it does precede a phase of active asexual multiplication.

It has previously been mentioned that clones of *Paramecium* become senescent in time, provided that neither conjugation nor a second sexual process, **autogamy**, intervenes. Autogamy is essentially similar to conjugation in that the meganucleus degenerates and is replaced by a new one derived by enlargement of one of the products of divisions of the micronuclei but all the events take place in individual animals without the pairing of conjugants so that no exchange of nuclei between partners is possible. It is called autogamy because the micronucleus divides by succesive divisions to give four daughter nuclei each of which has the haploid number of chromosomes. Three of these abort and the remaining one divides once. These final products of nuclear divisions (or, more logically, multiplications) then behave rather like the gametic nuclei formed during conjugation. They fuse to form a zygotic nucleus which then behaves just like the similarly named nucleus of the conjugants. It divides (or multiplies) to form four groups of two nuclei, in each of which one forms a new meganucleus, whilst the other forms a new micronucleus (the original meganucleus having degenerated).

Autogamy is sometimes preceded by a temporary adhesion between two *Paramecia*. This is called cytogamy.

It is now known that the meganucleus is large because in it the chromosome pairs (p. 52) are repeatedly duplicated. It is, in fact, a highly polyploid nucleus (p. 55). Thus, autogamy, like conjugation, has as one of its important consequences, the renewal of the meganucleus. It may now be asked, "What bearing have these events on the phenomena of senescence and rejuvenation of clones?". The answer is that there is now a wealth of experimental data which point to the following conclusions:

(a) As has been mentioned the meganucleus is highly polyploid.

(b) Since it divides amitotically, its chromosomes are distributed *at random* to the daughter *Paramecia*. Therefore, with an increasing number of divisions there will be an increasing disharmony or unbalance and this makes itself felt in a clone by decreased viability and increases structural abnormalities. To put it popularly, successive divisions of the meganucleus result in it becoming more and more worn out.

(c) With autogamy or conjugation a new meganucleus is formed with a correct chromosome balance in each of the daughter individuals so that the stock is renewed in vigour. The event of fundamental importance seems to be the renewal of the meganucleus which then controls the differentiation of the daughter *Paramecia* and continues to control their activities.

(d) Yet, throughout the life-cycle there is a continual interplay between the cytoplasm and the nuclei. For instance, whilst the cytopharynx is a self-duplicating structure, it cannot duplicate itself without the presence of a meganucleus. On the other hand, before division into two daughter *Paramecia* those nuclei nearest the anterior end always form micronuclei and those nearer the posterior end, meganuclei. There is then a regrouping of the nuclei which ensures that each daughter *Paramecium* receives one mega- and one micronucleus. Again, the reduction division of the micronucleus cannot begin if the meganucleus is removed by micro-dissection.

(e) The role of the micronucleus is, then, the handing on of a balanced chromosome and gene complex and breeding experiments show that many characters are controlled in a simple Mendelian fashion (Chapter XX). During autogamy it is clear that, since there is a fusion of identical nuclei, the clone which results from successive divisions must be homozygous for all its genes. It is only as a result of syngamy (p. 105) during conjugation that any gene recombination is possible, so allowing of variations some of which may be favourable ones.

Certain strains of *Paramecium aurelia* will cause clumping and the death of individuals of other strains of the same species. They are called *killer* strains and their victims, *sensitives*. This ability is the result of self-duplicating rod-like particles carried in the cytoplasm of the killers and can thus be inherited by transference through the cytoplasm. These kappa particles, however, although formed of nucleic acid, cannot duplicate themselves unless the necessary gene (p. 24) is present in the micronucleus (this can therefore be transferred in conjugation), so continued inheritance of the killer ability depends on the presence of both cytoplasmic *kappa* particles and nuclear genes. Perhaps these *kappa* particles are a special kind of virus but they behave like genes carried in the cytoplasm. Other particles, termed *pi*, *mu*, and *lambda*, with properties slightly different from *kappa* particles, also occur in some varieties of *P. aurelia*.

**Dispersal.**—The wide distribution of *Paramecium* in fresh-water ponds, particularly those liable to periodic drying, strongly suggests that this animal has some phase during which it is protected from desiccation. It is also known that cultures of *Paramecium* can be obtained from infusions of dry hay. It seems reasonable to assume, therefore, that *Paramecium* can produce cysts although these have not been observed with certainty.

## MONOCYSTIS

*Paramecium*—as has been seen—lives an active free life in an environment inhabited by numerous other organisms, with some of which it must be competing for food, whilst others may prey upon it. There are numbers of lowly animals, however, which live in an entirely different environment where inter-specific competition is reduced to a minimum and many of the conditions of an independent existence are absent. Such an environment is that found within the body of another living organism. Organisms which live in such an environment are called **parasites**, but this term should, strictly, only be applied when the foreign organism lives at the expense of, and is harmful to, the "host". In other instances, the presence of the intruder may not be harmful but may even be advantageous—an alliance which is called **symbiosis** (see p. 144).

The assumption of a parasitic mode of life is reflected in many of the characteristics of structure, and phases in the life-history, of the animals which adopt it and it will be interesting to contrast a simple parasite with an active animal such as *Paramecium*.

*Monocystis\**, an inhabitant of part of the reproductive apparatus of the common earthworm, affords such a striking contrast.

**Form and Habitat.**—Various stages in the life-history of this organism are almost invariably found in the seminal vesicles of the earthworm. A convenient stage at which to begin the description, is when the parasite is a minute nucleated body found in the central cytoplasm of one of the collections of developing spermatozoa called sperm morulae. This is the feeding or trophic stage and the individual is called a **trophozoite**. It feeds and grows at the expense of the protoplasm around it until all the cytoplasm has been consumed, when, for a time, the trophozoite is found surrounded by the tails of the dead spermatozoa of the earthworm so that it

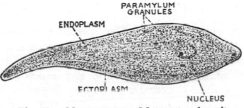

Fig. 51. MONOCYSTIS.—Mature trophozoite.

might be mistaken for a ciliated organism.† Ultimately, these spermatozoa are detached and the trophozoite becomes free. The mature trophozoite has a somewhat elongated cylindrical form, limited externally by a thick, smooth pellicle. Its cytoplasm shows a distinction into an outer ectoplasm and more granular, somewhat opaque, endoplasm. In the ectoplasm are found longitudinally and circularly running tracts of cytoplasm called myonemes, in which the power of contractility is specially developed. Within the endoplasm is the nucleus, which is limited by a delicate membrane, and contains a clear liquid in which are suspended denser bodies called nucleoli.

**Locomotion.**—The trophozoite, surrounded as it is by its food, has little need for movement. It is capable of slow wriggling

* Several species of *Monocystis*, and also related forms now put into separate genera, may be found in the same worm, some of them in the coelom.

† At this stage, *Monocystis* should not be confused with *Rhinchocystis pilosa*, another gregarine similar in form but with hair-like projections of the pellicle, giving it a superficial resemblance to a trophozoite of *Monocystis* surrounded by sperm tails.

movements brought about by rythmical contractions of the myonemes, a type of movement characteristic of the order (Gregarinida) to which *Monocystis* belongs and is consequently called gregarine movement. It can also glide about by wave-like contractions of the myonemes.

**Nutrition.**—At first the trophozoite lives on the cytoplasm of the sperm morula but does not ingest parts of it, for it has no cytostome or other structure for so doing. Presumably, it exudes digestive enzymes which act upon the cytoplasm in the immediate vicinity, rendering it soluble. The digested products are then absorbed through the pellicle. So, unlike *Paramecium* or *Amoeba*, instead of forming food-vacuoles within its cytoplasm, it, like some other parasites, digests its food outside its body. After the sperm tails have been shed, the trophozoite lies freely in the fluid of the vesicula seminalis and doubtless absorbs nutriment from it in the same way as do the developing sperms in the morulae. Reserve food material is stored in the endoplasm in the form of granules of carbohydrate.

**Respiration and Excretion.**—The gaseous exchanges of respiration are carried out by the usual diffusion processes, for the fluid in the seminal vesicle receives supplies of oxygen brought to it by the blood of the earthworm, and any excess carbon dioxide passed into the fluid is conveyed away by the same medium. In like manner the products of nitrogenous excretion are disposed of.

**Sensitivity.**—Little is known of the reactions of the trophozoite to conditions in its environment, but these must be remarkably uniform and call for little active response by the animal. It is, perhaps, worthy of mention that the reproductive cycle is initiated by the pairing of two mature trophozoites. This must necessitate some movement in response to a stimulus to pairing.

**Reproduction.**—When the trophozoite has attained a certain size, it enters upon the reproductive phase and each trophozoite now becomes a **gametocyte** (gamont). The gametocytes come together in pairs and each pair secretes around itself a double-walled cyst. Since the gametocytes do not fuse, but are merely associated together, the cyst is commonly called an **association cyst**, but might equally well be called a **gametocyst**. Within the cyst, each gametocyte undergoes subdivision to produce a number of small nucleated bodies called gametes. In this process the nucleus of the gametocyte divides into two, then four, and so on until a number (frequently sixty-four) of small nuclei are produced. These then migrate to the periphery of the gametocyte where each becomes surrounded by a little cytoplasm and projects from the surface, so that the gametocyte at this stage has rather the appearance of a mulberry. Finally, the several gametes become separated from the rest of the cytoplasm—which remains as the residual cytoplasm—and are set free in the cavity of the cyst. Each gametocyte of the pair usually produces the same number of gametes, which then become active and eventually fuse together in pairs. In one species of *Monocystis* found in the common earthworm, the gametes produced by both gametocytes are exactly similar and are termed isogametes. Consequently, when they are free in the cyst it is impossible to distinguish between them, but it is presumed from other evidence that one of each pair comes from one gametocyte and the other from the other. The fusion of the gametes is complete, involving both cytoplasm and nucleus.

The fusion results in the formation of larger bodies termed **zygotes**, or, because of their later development, **sporoblasts**. Each sporoblast then secretes around itself a tough resistant covering the **sporocyst**. Within the sporocyst, the sporoblast undergoes three successive divisions (one of which is a reduction division), producing eight sickle-shaped or fusiform, nucleated, bodies termed **sporozoites**.

When an infected earthworm dies and decays the sporocysts are released into the soil where, because of their resistant coats, they can remain alive for a considerable time. No further development can take place unless sporocysts are eaten by a worm. If this happens, the coat of the sporocyst is digested in the anterior part of the gut and the released sporocysts enter the cells lining the intestine and pass through them into the body cavity and then into the seminal vesicles. There they enter the sperm mother cells and grow to form trophozoites.

From this account of *Monocystis*, two salient facts emerge. First, the special, though simple characters of the adult trophozoite— when compared with a free living organism such as *Paramecium*— and second, the almost simultaneous production of vast numbers of potential individuals during its reproductive processes. These are related to its environment and way of life. Within the vesiculae seminales of the earthworm, food is abundant, competition, except from other species of *Monocystis*, is absent, and conditions are so uniform that the development of special structures and organelles to carry out the normal "vegetative" activities are unnecessary. But these comfortable conditions have one important drawback; for the continuation of the life-cycle it is essential for the animal to be transferred to another host worm so that it is necessary for the parasite to enter and pass through the outside world where dangers, enemies, and competition abound. To compensate for the almost inevitably high mortality during this dangerous period, the number of potential individuals is vastly increased to ensure that a sufficient number will survive the ordeal and the continuation of the race will be accomplished.

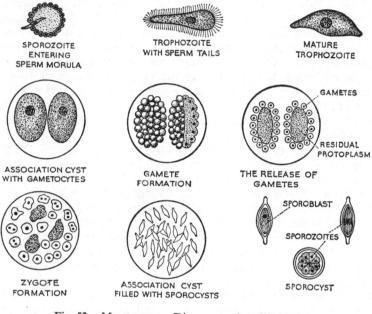

SPOROZOITE ENTERING SPERM MORULA

TROPHOZOITE WITH SPERM TAILS

MATURE TROPHOZOITE

ASSOCIATION CYST WITH GAMETOCYTES

GAMETE FORMATION

GAMETES

RESIDUAL PROTOPLASM

THE RELEASE OF GAMETES

ZYGOTE FORMATION

ASSOCIATION CYST FILLED WITH SPOROCYSTS

SPOROBLAST

SPOROZOITES

SPOROCYST

Fig. 52. MONOCYSTIS.—Diagram to show life-history.

## MALARIA PARASITE

A good example of an animal illustrating in an almost extreme way the parasitic mode of life, is found in *Plasmodium*, the organism responsible for the occurrence of malarial fever in man. Four species of *Plasmodium* are known regularly to attack man (see below) but only *P. vivax* need be described in detail, and the structural details will be given along with the life-history.

**Life-history.**—The life-history of the malaria parasite is a complicated one but may be considered to consist of three main phases or cycles, a **sexual cycle** which begins in man and is continued in the mosquito; an **asexual multiplicative** phase (**sporogony**), also in the mosquito; and, lastly, a **phase** of **growth** and asexual multiplication (**schizogony**) in the liver-cells and blood-corpuscles of man. The last phase or phases have only fairly recently been discovered and will form a convenient point at which to begin a description of the life-cycle.

Infection of man takes place when minute infective stages, the sickle-shaped **sporozoites**, are injected (in a way to be described later) into the blood-stream. There then follows an incubation period of ten days or so, during which the patient shows no symptoms of malaria. Nevertheless, during the first half-hour or so after the sporozoites have been injected into the blood-stream, the blood will produce an infection if introduced into that of another subject. However, after about half an hour the sporozoites disappear completely from the blood, all of them having reached the liver circulation and having entered the parenchymatous cells of the liver. Here each sporozoite grows rapidly to form a **schizont** which then divides **(schizogony)** to produce about 1,000 small **merozoites**. The rupture of the schizont releases the merozoites into the liver sinusoids and the residual cytoplasm of the schizont is then destroyed by the action of phagocytes. This phase of reproduction is termed the **pre-erythrocytic phase** during which the parasites seem to be more or less immune either to the action of quinine which has, therefore, no therapeutic effect, or to resistance **(naturally acquired immunity)** on the part of the host. A similar phase of multiplication (the **exo-erythrocytic phase**) may continue in the liver-cells so forming a reservoir of parasites which can prolong the disease in a latent form indefinitely until the resistance of the host is, for some reason or the other, lowered, when a relapse occurs. The continuance of the exo-erythrocytic parasites is possible only because they are by virtue of their intracellular position, practically unharmed by drugs (see, however, p. 103) or by any acquired immunity on the part of the host which operates successfully only against parasites in the blood.

From the liver sinusoids the merozoites may pass into the general circulation or reinfect liver-cells so as to continue the exo-erythrocytic schizogony. Those which remain in the general blood-stream attack and enter the red blood-corpuscles **(erythrocytes)**. Inside a red corpuscle each merozoite grows to form a **trophozoite**. The elongate shape is lost, the trophozoite becoming disc-like and beginning to feed upon the cytoplasm of the corpuscle in the same kind of way as was described for *Monocystis*. Soon a vacuole appears in the middle of the cytoplasm so that the nucleus is pushed away to one side and the trophozoite assumes the characteristic "signet ring" stage. It continues to grow and living specimens have been observed to thrust out pseudopodia into the cytoplasm of the corpuscle. After about thirty-six hours, growth slows down and the now mature trophozoite—which almost completely fills the enlarged corpuscle and contains numerous light-brown pigment granules (haemozoin, a degeneration product of haemoglobin)—is about to enter upon a phase of multiplication by a process also called **schizogony**. In this condition it is called a **schizont** and its nucleus undergoes repeated divisions, portions of the cytoplasm aggregating around the daughter-nuclei so that a number, up to twenty-four, of small nucleated bodies termed **merozoites (schizozoites)** is formed. A certain amount of residual cytoplasm is left and in this pigment the granules collect. After a lapse of about forty-eight hours, the thin envelope of the exhausted corpuscle bursts and the merozoites are liberated into the plasma. The merozoites,

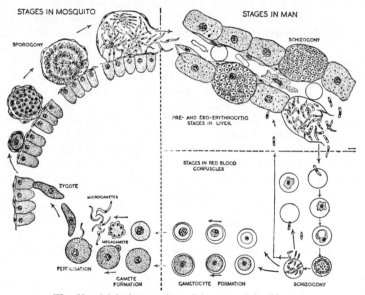

Fig. 53.   Malaria parasite.—Diagram of the life-cycle.

which are shorter and thicker than the sporozoites, then attack fresh corpuscles in exactly the same way and the cycle is repeated.

After a while, this phase of simple asexual multiplication is replaced by the onset of the sexual phase marked by the production of **gametocytes**. The merozoites, instead of becoming trophozoites which develop into schizonts, become rounded compact bodies without a vacuole and seem to grow more slowly. Eventually, two kinds of gametocytes are distinguishable; the so-called female gametocyte (or **megagametocyte**) which has a small nucleus and dense food-laden cytoplasm and the male (or **microgametocyte**) with a large nucleus and clear cytoplasm. Both contain pigment granules, which are more numerous than in the schizont, and are roughly circular in outline. The mature gametocytes undergo no further development in man and in order that the life-history may be continued, it is necessary for the gametocytes to be taken into the body of another host, viz. a mosquito belonging to the genus *Anopheles*, and unless this happens, the gametocytes ultimately degenerate and die.

The females of these insects require blood for food and obtain it by piercing the skin of mammals with their specially modified mouth parts and sucking up the blood liberated from the lacerated capillaries. If such a mosquito obtains, from a human being, blood in which corpuscles containing gametocytes are present, then, on reaching the stomach of the insect, the parasite develops further. Usually, other stages than the gametocyte are also in the meal, but they are digested along with the blood, and degenerate and disappear. The gametocytes, however, can withstand the action

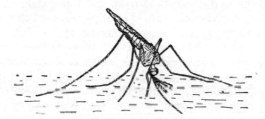

Fig. 54. A female *Anopheles* feeding (drawn from an original photograph).

of the digestive fluid and escape from the corpuscle by rupturing the envelope.

The nucleus in each microgametocyte undergoes repeated division to form either four or eight daughter-nuclei around each of which cytoplasm aggregates to form complete daughter-cells which then break free from the pigmented residual cytoplasm and grow into elongated **microgametes**.

The megagametocyte, after escaping from the corpuscular envelope, shows very little change, though there is good evidence that a reduction of the chromosomes occur which may be considered as comparable to that which takes place in the maturation processes of the gametes of the Metazoa. The nucleus moves towards the periphery of the gametocyte and at this point the cytoplasm projects slightly. The final result of these changes is that the gametocyte has now become a **megagamete**.

The microgametes after their release begin active lashing movements. Some approach the megagamete and one becomes attached to and passes into a cytoplasmic projection, its nucleus fusing with that of the megagamete.   In this way a spherical **zygote** is formed which for a time remains inactive, but eventually becomes elongated or vermiform and moves through the stomach contents until the wall is reached.   Here it enters and passes between cells in the epithelium, finally coming to rest between the epithelium and the sub-epithelial tissues.   In this situation, the zygote becomes enclosed in a **cyst** derived partly from the insect tissues and partly produced by the zygote.   The zygote then increases in size, the cyst enlarging to accommodate it, and its nucleus undergoes repeated divisions producing an enormous number of small nuclei.   At the same time, the cytoplasm of the zygote develops large vacuoles thereby vastly increasing the cytoplasmic surface and the nuclei arrange themselves along the free margins.   Eventually, each nucleus appropriates to itself a small portion of the cytoplasm which becomes elongated and projects from the surface.   Finally, each spindle-shaped nucleated projection is detached from the protoplasm and becomes a **sporozoite**.   A phase of asexual multiplication of this kind interpolated in the sexual cycle is termed **sporogony**.

When the cyst has become filled with sporozoites—some residual cytoplasm is always left behind—it bursts and the sporozoites are discharged into the blood sinus or haemocoel which forms the body cavity of the insect.   In this haemocoel lie the insect's salivary glands, and although the sporozoites may pass to any part of the body of the insect, the majority penetrate into the cells of the salivary glands eventually finding their way into the salivary duct along with the salivary secretion.   It follows then, that when a mosquito in this condition pierces the skin of a human being, some of the sporozoites will pass into the blood-stream, for the insect always injects some of its salivary secretion into the wound before sucking the blood.   The sporozoites are carried to the liver where they begin the pre-erythrocytic phase.

*Plasmodium*, just like *Monocystis*, has in its life-history many of the features associated with a parasitic life.   The parasite has an extremely simple structure correlated with its life within the red blood-corpuscle and liver-cell, and to ensure the continuation of the race there are very active phases of multiplication.   An additional feature is the introduction into the life-history of an alternative host which, because it feeds on the final host, serves as a means of active transference from one human being to another.   Such an intermediary is termed a **vector**, and when the extreme prevalence of malaria in uncontrolled areas is remembered, the success of this feature can be understood.   The vector provides a means whereby a continual stream of reproductive stages can be transferred from an original host to fresh hosts, so that if the former dies or the parasite is killed off by the protective biological reactions of the host itself, or by other means, the continuity of the race is not interrupted. For, of course, when the gametocytes are present in the blood, all the anophelines which "bite" the patient will become reservoirs of sporozoites, and naturally, a mosquito may infect more than one human being.

**The Effect of the Parasite on the Host.**—The disease known as malarial fever or ague is an acute or intermittent fever formerly supposed to be due to harmful vapours given into the atmosphere from marshy land (*malo*, bad; *aria*, air). The feverish condition in man is due to toxins liberated into the blood along with the merozoites when the corpuscle is ruptured at the end of schizogony. Usually an accumulation of toxin is necessary to cause an attack of malaria, so that it only occurs when sufficient cycles of schizogonous multiplication have occurred. There is thus an interval of time between the primary infection and the onset of the fever and this is called the "incubation period".

Three forms of the fever are known and are due to four different species of *Plasmodium*. "**Benign**" tertian fever caused by *P. vivax* is characterised by three-day intervals in the climaxes of fever —the period required for the trophozoite to grow, undergo schizogony, and the merozoites to be liberated. **Ovale tertian** malaria is caused by *P. ovale* which resembles *P. vivax*. **Quartan** fever is due to *P. malariae*—with four-day intervals—and **malignant tertian** or **quotidian** fever to *P. falciparum*, where the fever is irregularly continuous. A patient may be infected by more than one species of *Plasmodium* at the same time, a condition known as multiple or mixed infection.

Although, in general, the plan of the life-histories of the three species is the same, individual differences occur by which they may be distinguished and diagnosis facilitated. For example, the gametocytes of *P. vivax* and *P. malariae* are circular in outline while those of *P. falciparum* are crescent-shaped; the pigment of *P. vivax* is light brown but that of *P. malariae* and *P. falciparum* is dark brown or black, etc. In all three the effect upon the human host is severe. In the mosquito, however, no harmful effects are observed, and, although there must be some drain on the nutritive resources of the insect body, a tolerance seems to have been acquired.

**The Control of Malaria.**—This falls under three headings: (*a*) treatment of the infection in the patient; (*b*) the prevention of infection—prophylaxis; (*c*) control of the vector.

(*a*) At present no drug is known which will kill the sporozoites during their brief sojourn in the blood-stream after injection by the mosquito. Nor is it usually known when this has happened but, fortunately, it is probable that none survives if they do not reach the liver. Primaquin, which is not very toxic to man, and also other drugs, such as paludrine, will prevent the stages in the liver from being completed. When the parasites leave the liver, schizogony continues until the patient acquires some degree of immunity but the schizozoites can also be killed by several drugs, such as quinine and mepacrine, which are called schizonticides. Following a clinical bout of malaria there is a latent phase, for most patients acquire some degree of immunity, and during this period gametocytes are common in the blood. Most schizonticides will not act on these, but quinine and chloroquine are among the more effective gamonticides. Often there is a relapse by the patient. The fever then appears again owing to the re-emergence from the liver of resistant forms of the parasite. Combinations of primaquin, quinine, and chloroquine are used to combat such a relapse.

Certain drugs, *e.g.* paludrine, if taken in small, regular doses, give immunity—not from infection but from the clinical symptoms—by preventing the life-cycle from being completed. Others are, by acting as schizonticides, more effective once the disease has begun. The difficulty of administering suitable drugs in regular doses has been overcome in Brazil by the incorporation of 0.003 per cent. chloroquine in salt used on food.

(b) The prevention of infection can be effected by using protective measures—such as mosquito curtains, etc.—by which the mosquito is prevented from "biting".

(c) It is perfectly clear that if the vector is completely exterminated then infection from one human being to another cannot take place. By draining swampy places and thus reducing the available breeding places of the mosquito (which has an aquatic larva), and by spraying the surface of the water with oil or the introduction of enemies of the larvae, etc., the mosquito population may be completely eliminated. The insecticide D.D.T. has also been used with great effect when sprayed on water which mosquitoes frequent. D.D.T. and other insecticides whose effects last for long periods have been used as well for spraying walls of domestic buildings and this to some extent replaces the necessity of spraying the breeding places of the mosquito which are often difficult to find. This method depends for its success on the habits and life-cycle of the mosquito as well as on the duration of the sexual cycle of the parasite in the mosquito. Spraying the walls on which mosquitoes usually rest after their meal of blood ensures that few survive to bite a second time. Moreover, it is doubtful if the parasites can survive in man for more than about three years, so that a campaign which stops the transmission of the disease for this period will eradicate the disease, even if mosquitoes survive. It follows that a mosquito species with a short average life is a less efficient vector than a long-lived species, since it may not live long enough for the sexual cycle to be completed. Some mosquitoes have a life of less than two weeks but *Anopheles gambiae*, the chief malarial vector of West Africa, is a long-lived species, and this renders control of the disease difficult in that area.

## SEXUAL PROCESSES IN THE PROTOZOA

The three examples which have been dealt with so far, in addition to illustrating the differences in structure, habits, and life-history associated with an independent and a parasitic existence, form an interesting series of increasing complexity of sexual processes.

In *Paramecium* the only permanent fusion is between nuclei, the temporary cytoplasmic continuity between the conjugants merely providing a bridge across which the migratory nuclei can cross, and the capacity to act as gametes belongs to the nuclei alone so that they may be termed gametic nuclei. In *Monocystis*, however, the gametocytes divide into separate bodies, gametes, which fuse completely, both nucleus and cytoplasm; but the gametes are exactly similar to one another and are termed isogametes. In *Plasmodium* also, gametes are produced, but in this instance they are dissimilar; the one, the microgamete, is small and active and the other, the megagamete, large, food-laden, and immobile. Hence they are called anisogametes or heterogametes, and with this anisogamy there is associated a difference in behaviour between the two types. An exactly comparable phenomenon is found in the Metazoa where also the gametes are anisogamous. The smaller, actively motile gamete is termed a spermatozoon and the larger, food-laden inert gamete the ovum or egg.

In the Metazoa this distinction between the two types of gametes introduces a fresh concept, namely that of sex. The spermatozoon is called a male gamete and the ovum, the female gamete. Usually, but by no means always, an individual metazoon produces one kind of gamete only and it is often easy to tell at a glance males from females because of morphological differences. This is spoken of as **sexual dimorphism**. But often adult males and females seem to be identical apart from the different gametes they produce. Also, many animals produce both male and female gametes and are then said to be **hermaphrodites**. The basic fact of sex, then, lies in the different potentialities of the gametes and for this reason it is easy to see how the production and fusion of microgametes and megagametes in *Plasmodium* is exactly in line with that of spermatozoa and ova. But even in the case of the isogametes of *Monocystis* and the gametic nuclei of *Paramecium*, the same fusion (syngamy) occurs so that these, too, are true sexual processes although the difference between the co-operating structures is not evident. In *Paramecium*, however, there is a kind of anisogamy in that the migratory nucleus is active and the stationary nucleus, passive, showing differing behaviour through little outward differences.

It will be understood, however, that the gametes of Protozoa are not necessarily strictly comparable with those of the Metazoa, for these are *always* haploid (the nuclei containing only half the somatic number of chromosomes—see Chapter XVIII). The stage in the life-history at which the reduction division takes place in the Protozoa is *variable*.

A few different kinds of Protozoa which are of interest from other points of view will now be considered.

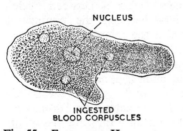

Fig. 55. ENTAMOEBA HISTOLYTICA.

NUCLEUS

INGESTED
BLOOD CORPUSCLES

### ENTAMOEBA HISTOLYTICA

*E. histolytica* is a protozoan very like *Amoeba* in all its essential characteristics but it has a parasitic mode of life and lives in the large intestine of man, where it may (but not always) produce that kind of dysentery known as amoebic dysentery. Under some conditions, particularly in tropical countries, it invades the tissues of the intestinal wall, the penetration into the cells being probably effected by the secretion of proteolytic enzymes. These amoebae ingest solid particles and blood-corpuscles as do other amoebae. There is no contractile vacuole, the fluid in which the animal lives being isotonic with the cytoplasm so that a special mechanism for osmoregulation is not necessary.

The reproductive activities of *Entamoeba* resemble those of other amoebae in that its normal method of multiplication is by binary fission. Probably associated with the method of transference from host to host, however, *Entamoeba* encysts when it passes from the ulcerated tissue in which it normally lives, into the lumen of the intestine. The cyst is secreted by the organism and during the passage of the encysted form through the intestine, the nucleus undergoes two divisions so that when the cyst reaches the outside world it contains a mass of cytoplasm with four nuclei. In this form it is stated to be "infective" and if swallowed by a human being will, when it reaches the intestine, rupture, liberating eight potential *Entamoebae*, though it seems uncertain whether it penetrates the tissue in the quadrinucleate stage or whether it first divides into four small amoebae.

By way of contrast, other Entamoebae, such as *E. coli*, which is found among other harmless organisms in the intestine, are always non-pathogenic. They engulf bacteria, but as the ones consumed are not pathogens, this is hardly of special benefit.

## TRYPANOSOMES

Another interesting example of the parasitic Protozoa is the organism, *Trypanosoma gambiense*, which is responsible for the dread tropical disease, sleeping sickness, which makes regions of West and Central Africa where it occurs practically uninhabitable by man. A similar organism, *T. rhodesiense*, occurs mainly in East Africa. Not all trypanosomes are pathogenic, however, though many occur in the blood of vertebrates, and near relatives of trypanosomes are found in some invertebrates.

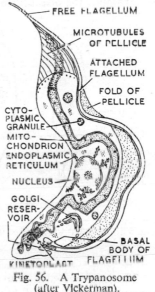

Fig. 56. A Trypanosome (after Vickerman).

One of the important characteristics of the genus *Trypanosoma* is that it is polymorphic, that is, the form of its body and the arrangement of the organelles vary with the different stages in the life-history. These variants are structurally related in a readily understandable way and can thus be regarded as modifications of a plan common to the genus throughout its life-cycle, the simpler forms being easily derivable from the more complex and *vice versa*. Four main forms are distinguished, namely, *trypanosome, crithridia, leptomonas,* and *leishmania* (Fig. 57) and of these the trypanosome can be regarded as the adult and the others as developmental stages some or all of which occur during that part of the life-cycle spent in the secondary host, which is always an invertebrate and commonly an insect. The names of the different stages are derived from their resemblance to the adult of genera related to *Trypanosoma* and the identification of a genus therefore depends on a full knowledge of its life-history and developmental stages.

The adult form or trypanosome proper consists of a fusiform protoplasmic body, pointed at both ends, the shape being maintained by the presence of a firm pellicle. Within the cytoplasm is a

large, single **nucleus**.  Arising from a pocket, representing the gullet
of *Euglena* (p. 111) is the flagellum, which is continued along the
whole length and beyond the body.   Throughout the greater part of
its length it is joined to the cytoplasm of the body which is, so to
speak, pulled out to form the **undulating membrane** along which
waves pass from base to tip.   At the base of the flagellum is a darkly
staining granule called the **basal granule**.   Near the basal granule,
within the single, large mitochondrion, is a small mass of D.N.A.
often called the kinetoplast (p. 44).   So far as many trypanosomes
are concerned, this description applies to what might be regarded
as the "typical" form, though departures from it frequently occur,
for example, the insect form has a larger mitochondrial system.

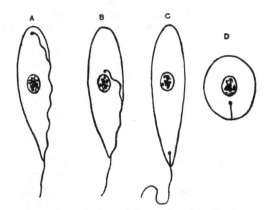

Fig. 57.   Diagram to show the interelations of various forms which may occur
in the life-cycle of a trypanosome.   A, trypanosome; B, crithridia; C, leptomonas; D, leishmania.

The trypanosomes swim freely in the blood-plasma, not only by
vibratile movements of the undulating membrane and flagellum,
the free end of the flagellum pointing forwards, but also by the whole
body being thrown into waves.   The animals feed by absorbing
food dissolved in the blood of their hosts whilst pinocytosis occurs
through the "gullet".   The pellicle also serves as the surface through
which gaseous interchanges occur as well as for the elimination of
nitrogenous excretory products.   As is common in Protozoa living
in body fluids, no special mechanism, such as contracile vacuole,
for osmoregulation, is present.

**Reproduction.**—Longitudinal binary fission is the usual method
of multiplication in the trypanosomes for all stages in the life-history.
First the basal granule and kinetoplast divide, followed by the

nucleus. Then a longitudinal division of the cytoplasm takes place and in some instances the flagellum also may divide, but usually, one-half of the cytoplasm takes the old flagellum and the other forms a new one, this being done before the complete separation of the cytoplasm. This seems to be the only method of reproduction, the evidence for sexual processes being extremely doubtful.

A familiar example of a trypanosome whose life-history is well understood is *Trypanosoma lewisi*, a blood parasite of the rat.

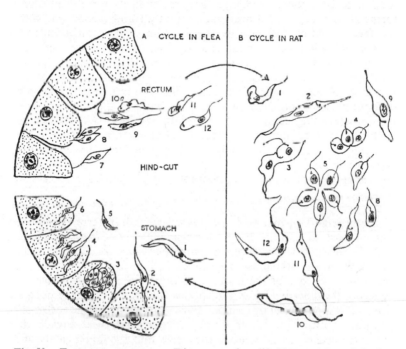

Fig. 58. TRYPANOSOMA LEWISI.—Diagram to show life-history. A, cycle in flea, B, cycle in rat. Numbers are referred to in the text.

The secondary host and vector is the rat-flea whose faeces, containing infective forms (Fig. 58A, 10, 11, 12) of the parasite, are taken in by the rat when it licks its fur. After a lapse of a few days, fully formed trypanosomes (Fig. 58B, 2) appear in the blood-stream of the rat where for a period of a few weeks they multiply asexually by binary and multiple fission (Fig. 58B, 3, 4, 5), giving rise to a great variety of forms all of which, however, are referable to the trypanosome, crithridia (9), leptomonas (6), or leishmania (4), types already described. After some time multiplication ceases (owing to the production in the blood of some anti-body) and the only forms found are typical trypanosomes with a curved body rather pointed at the end from which projects the flagellum (Fig. 58B, 10, 11). Later still, even these forms of parasite disappear from the blood of the rat which is then immune to subsequent infection. For the life-cycle of the parasite to be completed, blood containing the fully-formed trypanosomes, produced during the later stages of infection, must be ingested by a flea in whose body they undergo a series of changes and multiplication until infective forms (Fig. 58A, 11, 12) are again produced and passed out with the faeces. After ingestion by the flea the trypanosomes burrow in the epithelial cells lining the stomach where they undergo multiple fission to form masses of oval forms (Fig. 58A, 3). These may break free and once again enter the stomach as small trypanosomes which may invade other epithelial cells and repeat the process of multiplication or, alternatively, pass back to the hind-gut and rectum where they assume the crithridial form (Fig. 58A, 5). These crithridia in turn multiply by fission, but after a time no more crithridia are produced but only small trypanosomes which are the infective forms.

These changes in form are accompanied by changes in metabolic processes.

**Pathogenic Trypanosomes.**—*T. gambiense* is transferred from one human host to another by a blood-sucking fly, *Glossina palpalis* and *T. rhodesiense* by *Glossina morsitans*, flies known as **tsetse flies**. When the trypanosomes are sucked up by the fly they are retained in the gut of the insect, and those which there survive the action of the digestive juices multiply by binary fission in the normal way. During this multiplication, forms known as "slender forms" from their rather narrower shape, appear, and these invade the anterior part of the gut whence they pass into the labial cavity and part of the mouth parts known as the hypopharynx, and eventually into the salivary glands. Here they become attached to the cells in the glandular part and a further phase of multiplication occurs, during which crithridia and, later on, forms similar to those found in vertebrate blood are produced. Under these conditions the cavity of the gland is filled with trypanosomes and the fly is said to be "infective". A human being "bitten" by an infective fly will receive, along with the salivary secretion, trypanosomes which will continue to live and multiply and the disease will follow.

This example of a pathogenic parasitic protozoan presents a very different picture from that of *Plasmodium*, particularly in the absence of sexual stages in the insect host, yet the fly acts in the same way as the **vector** by which the transmission of the form found in the mammal is successfully accomplished. The significance of the phases in the tsetse fly is, however, difficult to understand.

Other trypanosomes which are economically important are those causing diseases in domestic animals. *Trypanosoma brucei* is a scourge of cattle in South and Central Africa and is transmitted by the bite of a tsetse fly, *Glossina morsitans*, the disease being termed "nagana".

**Control of Trypanosomes.**—Attempts which follow the same general lines already mentioned in the control of malaria have been made, and are being continued, to deal with pathogenic trypanosomes. Men and animals infected are treated with drugs, including

Fig. 59. A Tsetse Fly (after Austen).

compounds of antimony and trypan dyes now largely superseded by antrycide and its successors. Tsetse flies are difficult to wipe out and recent measures have been chiefly aimed at drastic reduction of wild game. Wild game animals which harbour trypanosomes without showing signs of disease form a reservoir from which tsetse flies can pick up trypanosomes and transmit them to domestic animals and their extermination in cattle-rearing areas seems to be an unfortunate but necessary precaution. Moreover, the present view is that the human trypanosomes, *T. gambiense* and *T. rhodesiense*, are form variants of *T. brucei*, from which they differ only in their ability to infect man, and there is some evidence that in some districts game animals form a reservoir for *T. rhodesiense* and, more doubtfully, for *T. gambiense* also. But the real argument in favour of game eradication, which can be combined with partial clearing of the bush, is that game form the essential food of tsetse flies of the *morsitans* group which are vectors of both cattle and human trypanosomes. Thus by denying animals for food to the flies, the flies are controlled in numbers. Though this seems to be official policy, it is not only repellent but is diametrically opposed to recent views on "game cropping" and the preservation of African wild-life. But the tsetse flies of the *palpalis* group are water-side dwellers, feeding on birds and reptiles as well as mammals and so game eradication can play little part in controlling them. The successes for animals reported for antrycide, which kills all stages and confers some immunity have not, unfortunately, been so spectacular as was hoped.

## EUGLENA

This account of the various kinds of specialisation among the Protozoa would be incomplete without reference to examples which show features not usually met with in animals, and which may be on the border line between animals and plants showing some features peculiar to both.

**Form and Habitat.**—A flagellate such as *Euglena viridis* may first be considered. *E. viridis* occurs commonly and often in great abundance in stagnant water containing much nitrogenous organic matter. Pools or ditches which are contaminated with the urine and faeces of farm animals almost invariably contain euglenoids of more than one species and one of the commonest is *E. viridis*. The individuals are minute in size, spindle-shaped, the end which is foremost when the organism moves being blunt, and the opposite end tapering to a point. The cytoplasm forming the body is surrounded by a distinct **pellicle**, which, though firm and strong, is very thin and sufficiently flexible and elastic to permit of considerable and characteristic changes of form which the body occasionally exhibits. These changes consist of the contraction of one part of the body and the dilatation of another, and are spoken of as **euglenoid movement** when they occur here or in other Protozoa. At the blunt end of the body, there is a flask-shaped gullet or cytopharynx which serves for the intake of food particles in certain near relatives which feed holozoically but which in *Euglena* is a gullet in name only. At one side, for convenience termed the dorsal side, is a prominent red pigment spot or **stigma** lying close against the wall of the gullet. From the inner side of the ventral wall of the gullet a long flagellum arises and on its root is a swelling known to be sensitive to light and termed, therefore, the photoreceptor. A second, short flagellum is also present close to it. Electronmicrographs reveal that a flagellum has the same essential structure as a cilium but that of *Euglena* has, in addition, a row of fine threads, called mastigonemes, along one edge. Their function is unknown. A large contractile vacuole lies near to the ventral wall of the gullet into which it discharges when it bursts, but before it does so, smaller

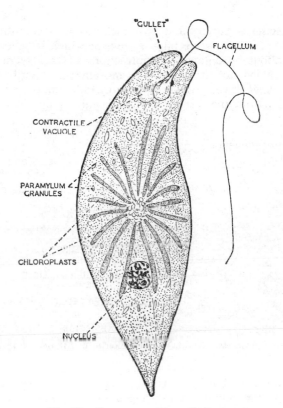

Fig. 60. Euglena viridis.—Seen as a
transparent object.

vacuoles can be seen near to it and these run together to form the next functional contractile vacuole.

The nucleus is large, is usually situated near the pointed end of the organism, and contains a karyosome. The green colour of *E. viridis* is due to the presence in the cytoplasm of chlorophyll-containing chloroplasts which take the form of rod-like structures radiating from a common centre so as to give a stellate appearance which, however, disappears in culture. Also in the cytoplasm are numerous large ellipsoidal granules of a carbohydrate similar to glycogen paramylum. These granules originate around special cytoplasmic centres which have been wrongly called pyrenoids, a term restricted to a structure found in relation with each of the chloroplasts.

**Locomotion.**—*Euglena* has two methods of locomotion. The first of these is a slow creeping movement which is believed to be brought about by differential contractions of the myonemes and hence is allied to the euglenoid movements described above. Secondly, *Euglena* can swim fairly rapidly (3.6 mm. per minute) and this results from the lashing of the flagellum, not in any direct

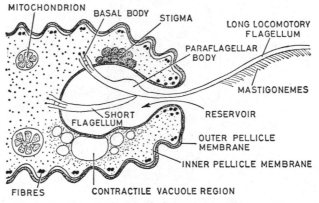

Fig. 61. EUGLENA.—Sagittal half of anterior end (based on Leedale).

manner, however, but by movements of the whole body which are induced by the activity of the flagellum in a manner only fairly recently understood. When *Euglena* is swimming the long flagellum trails obliquely to the rear (Fig. 60) and is thrown into waves which pass along it at about twelve per second and with *increasing velocity and amplitude* from base to tip. This shows that the movement of the flagellum is no mere passive flicking such as is caused in the lash of a whip when the handle is rapidly moved but that the flagellum contributes to its own activity and must be expending energy in the process. Also, a flagellum detached from the body will continue its lashing provided that the blepharoplast remains attached. As the animal moves forward it rotates on its own axis and also gyrates

(Fig. 62). These movements are the result of the reaction of the body to the forces generated by the beating of the flagellum and the whole body then behaves like a rotating inclined plane and therefore moves forward through the water. Mechanical models which can be caused to rotate and gyrate show the adequacy of this theory of flagellar action to account for the swimming of flagellates and there seems little doubt that in flagellates with trailing flagella this indirect effect of the flagellum in locomotion is the main one. In some flagellates, however (e.g. *Peranema*, Fig. 65), one flagellum remains directed forward during swimming and probably contributes directly to locomotion, the flagellum and not the body forming the main tractor acting like a "screw propeller", the trailing flagellum remaining motionless and adherent to the pellicle.

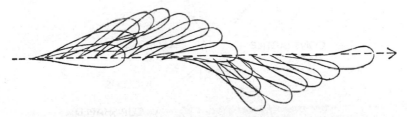

Fig. 62. EUGLENA.—Successive positions taken up during forward locomotion through one complete gyration and rotation (after Lowndes).

**Nutrition.**—Since *E. viridis* contains chloroplasts, it can synthesise carbohydrates from carbon dioxide and water as can any ordinary green plant and consequently its nutrition may be said to be holophytic. Indeed, some flagellates, e.g. *Chlamydomonas* (Fig. 63) are not only plant-like in their nutrition but have a cellulose cell wall, rather like that of a plant cell, secreted by their ectoplasm. *E. viridis*, though commonly considered to be an animal, is really closely related to many simple plants, and in fact is equally generally claimed by botanists to be such. Other closely allied flagellates, such as *Peranema*, are colourless and their nutrition is undoubtedly holozoic, solid food particles being taken into the cytoplasm between three minute rods lying in the wall of the gullet. Again, some species (e.g. *E. gracilis*) can, under certain conditions where photosynthesis becomes impossible, live saprozoically, obtaining their food by absorbing soluble organic material usually of a rather simple nature from the water in which they live. There is evidence that all euglenoids require some simple organic as well as inorganic material in the water in which they live; the species vary in their synthetic abilities so that some can use inorganic substrates, others simple organic compounds like amino acids while others need peptones as their synthetic starting point.

The small, colourless flagellates, e.g. *Polytoma*, belonging to the same order (Phytomonadina) as *Chlamydomonas* are always saprozoic as can be shown in the laboratory by cultures grown in solutions of different pure substances. Most colourless, free-living flagellates require relatively complex organic substances but *Polytoma* can build up food and protoplasm from simple compounds such as acetic and butyric acid plus ammonium salts.

**Respiration and Excretion.**—Oxygen for purposes of respiration diffuses into the cytoplasm through the pellicle and carbon dioxide

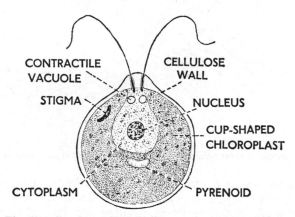

CONTRACTILE VACUOLE

CELLULOSE WALL

STIGMA

NUCLEUS

CUP-SHAPED CHLOROPLAST

CYTOPLASM

PYRENOID

Fig. 63. CHLAMYDOMONAS.—Seen as a transparent object.

outwards, but when the organism is actively photosynthesising, this is masked by the intake of carbon dioxide and the output of oxygen.

Such excretory products as are produced will also diffuse outwards.

**Sensitivity.**—As would be expected in a motile, free-living organism *E. viridis* reacts to a variety of stimuli as do other Protozoa (see p. 119).

**Osmoregulation.**—The contractile vacuole is stated to discharge its contents periodically into the gullet which becomes momentarily enlarged until the additional contents pass out to the exterior.

Subsidiary vacuoles have been seen to form around the contractile vacuole, and later open into it. The apparatus functions, then, in a similar manner to that of other Protozoa.

**Reproduction.**—*Euglena* multiplies by longitudinal fission. The organism ceases to swim about and encloses itself in an envelope of mucilage. The flagellum is withdrawn. The nucleus and chloroplasts divide and this is followed by the division of the cytoplasm.

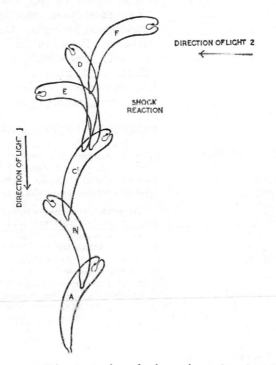

Fig. 64. EUGLENA.—Diagram to show shock reaction and reorientation to light (flagellum not shown). From A to C, the animal is moving forward in a spiral fashion towards the first light source. At C, the first light is switched off and the second light switched on. The shock reaction occurs at D, when the pigment spot shields the photoreceptor from the light. A series of such shock reactions will occur until the axis of the spiral path traced out by the animal is parallel with the beams from the new light source.

Under certain conditions, encystment may occur. The cyst is composed of a special carbohydrate, yellowish-brown in colour, and division may take place within the cyst. So far as is known at present, no sexual process takes place in the life-history.

The consideration of all these examples shows that, simple as Protozoa are in that the body of the organisms consists merely of a single unit of cytoplasm containing a nucleus, yet all kinds of modifications can take place producing organelles designed to carry out particular functions.

It is important to notice that in the Protozoa differentiation is not a result of fission, whereas, as will be seen later, in the Metazoa the differentiation of the developing organism is *preceded by its division into cells* (pp. 121-6), which show progressive modification as development proceeds. On the other hand, each fission in the Protozoa is, as a rule, preceded by a *loss* of its differentiated parts which are only later re-acquired to the same degree by the daughter organisms.

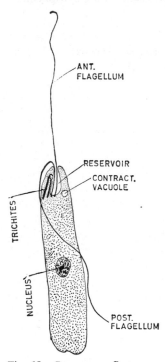

Fig. 65. *Peranema.*—Seen as a transparent object.

### BEHAVIOUR OF THE PROTOZOA

It has been shown that Protozoa are sensitive to changes in the environment in which they live and the responses which they make to such changes usually follow a constant sequence for any particular organism.

**Amoeba.**—Special structures for the reception of stimuli have not been detected in *Amoeba* although, to judge from its responses, it seems sensitive to a wide variety of changes in its environment. Semipermanent changes, for example, are produced by even minute changes in the chemical composition of the surrounding water. A change to distilled water, the addition of various sugars, salts, and acids to the surrounding water cause a rounding off of the body at first, but this is then followed by the protrusion of a single pseudopodium, by which locomotion is effected, in contrast to the normal radiating form of the body with its numerous pseudopodia. Normal body form and movement also seem to depend on the maintenance of a correct $pH$ value in the medium. Local stimulation, by the application of chemicals to small areas of the body, always seems to produce a negative response or withdrawal of the body. The stimulated region first of all contracts away from the stimulus and then pseudopodia become active on the opposite side of the body, so that the animal moves away from the source of the stimulus. Whilst nearly all local changes in the chemical nature of the water initiate this withdrawal response, a few chemicals stimulate the formation of food-cups. The normal reaction to food particles may, itself, be one particular example of a response to chemical stimulation by the formation of a food-cup. It is, however, extremely difficult to distinguish between the effects due to contact with the food and those due to substances diffusing from it. Certain it is, though, that *Amoeba* shows considerable powers of discrimination over the particles which it engulfs, but its selective activities are not always true indicators of the nutritional value of the particles. A useless substance such as carmine is, for example, engulfed more readily than gluten (a protein from cereals), but, again, the carmine is more speedily egested. In general, however, the vast majority of acceptable particles are of nutritional value.

*Amoeba* is sensitive to light gradients as is shown by the fact that its movement is a zig-zag path away from any intense source of light. Constant or uniform illumination is not so effective in producing a response as is a sudden increase in intensity. The animals, however, always seem to react negatively; an increase in illumination apparently inhibiting the formation of pseudopodia on the illuminated side but not on the shaded parts, so that the animals move away from the light source. These reactions, carried out in nature, result in *Amoeba* inhabiting regions of diffuse or dim light. Intense and sudden increases in illumination cause complete cessa-

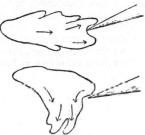

Fig. 66. Diagram to show the avoiding reaction to contact stimulus by *Amoeba*.

tion of movement and rounding off of the body, but if the high level of illumination is maintained, then, after a while, movement is resumed. Thus *Amoeba* adapts itself to new conditions.

The temperature of the surrounding medium has, also, interesting effects. Within certain limits, the speed of movement increases with a rise in temperature in a way which can be roughly expressed by van't Hoff's formula for the relation between temperature and velocity in chemical reactions (*i.e.* rate is doubled for 10° C rise in temperature). Maximum, optimum, and minimum temperatures have been recognised for activity and successful life in *Amoeba* as for other organisms. Local application of heat has similar effects to other stimuli, namely, cessation of movement on the stimulated side followed by movement of the body away from the stimulus.

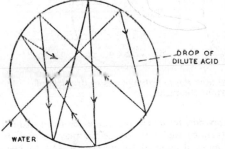

Fig. 67. Diagram to show how shock reactions can produce aggregations of *Paramecium* in a drop of dilute acid.

Local mechanical stimulation has varying effects, but if above a certain intensity, it evokes the withdrawal response as for many other stimuli. Weaker stimuli (*e.g.* vibrations such as those produced by certain ciliates) elicit the formation of pseudopodia on the stimulated side (*i.e.* a positive response) and this is followed by the formation of a food-cup of greater diameter, better adapted to encircle a motile prey.

There is much evidence to suggest that the magnitude of the response to various stimuli varies with the strength of the stimulus, but this relation is often masked by sensory adaptation or acclimatisation to the stimulus. Reactions show certain features which are characteristic of higher animals; a reaction time which varies with the strength of the stimulus; a latent period during which no further response will take place; the summation of inadequate stimuli, and so on. Thus, when subjected to a weak but steady electric current, *Amoeba* moves towards the cathode, apparently because pseudopod formation is inhibited on the side of the anode. Stronger currents produce cessation of movement and ultimate disintegration, beginning on the side of the anode. These effects, however, are most probably not a direct result of the action of the current, but it is likely that the current sets free certain chemical substances within the body and also produces regional changes in acidity and alkalinity. It is these changes which provide the stimuli for the changed behaviour.

Gravitational responses have, apparently, not yet been reported for *Amoeba*.

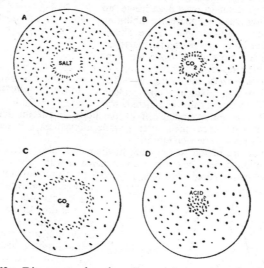

Fig. 68.   Diagram to show how *Paramecium* reacts to droplets of certain chemicals.

**Paramecium.**—*Paramecium* responds to much the same kinds of stimuli as does *Amoeba*, and to some others as well. Its behaviour may be said, in general, to be very stereotyped but well adapted to preserve it and to keep it within the bounds of a normally favourable environment. Special organelles for the reception of stimuli are not evident, although certain regions of the body seem to be more sensitive than others. The difficulties involved in the application of local stimuli to such a small and rapidly moving animal, however, deserve emphasis. The chief noticeable effect of the addition to, and thorough mixing of, various chemical substances with water containing *Paramecium* is on the speed of movement. Some chemicals increase the speed; others slow it down; many are toxic even in very dilute solutions.

There is a good deal of evidence which suggests regional sensitivity, the anterior end being the most sensitive. Despite this, there seems to be no relation between the direction of movement and the point of application of the stimulus; the avoiding reaction is always elicited. Contact with solid objects elicits the avoiding reaction as a rule, but on rarer occasions the animals will come to rest on the object when the cilia which are touching the object will cease to beat, but feeding currents are still maintained by the other cilia. Jarring causes the animals to swim downwards.

*Paramecium* apparently selects certain kinds of food particles in preference to others. This is shown by the formation of a greater or smaller number of food-vacuoles when the organism is fed on different kinds of food materials.

Some chemical substances tend to induce the animals to collect together and this effect is also produced by the avoiding reaction. If, for example, a drop of solution is carefully placed in a *Paramecium* culture, the animals may swim straight into it and, having got there, they show the avoiding reaction whenever they come into contact with the boundary of the drop. This will produce aggregation. On the other hand, they behave in a contrary way with other substances, giving the avoiding reaction whenever they meet the outside boundary of the drop so that they are unable to enter it. Accumulation, therefore, depends upon the animals entering a region at random and, having arrived there, being in a sense imprisoned (see Fig. 67).

Up to a point, a rise in temperature causes a general increase in the speed of all the activities of the animals. If placed in a temperature gradient, they tend to accumulate in regions of moderate temperature, and this is brought about by avoiding reactions towards the regions of higher or lower temperature.

In a constant electric current of moderate strength, the animals swim towards the cathode because the current causes a partial reversal of the ciliary beat, more cilia driving towards than away from the cathode. With stronger currents, more cilia beat backwards than forwards and the animals swim towards the anode.

It has been found that *Paramecium* is negatively geotactic, but only weakly so. The reaction is possibly brought about by heavier particles in the cytoplasm acting as statoliths, but it is also stated that the posterior end is the heavier which would cause it to be directed downwards and hence this promotes upward swimming.

**Euglena.**—Except in its response to changes in light intensity, *Euglena* has been less studied than either *Amoeba* or *Paramecium*, but seems to react in much the same general sort of way as other Protozoa in giving an avoiding reaction to most unpleasant stimuli. Like most organisms which possess chlorophyll, *Euglena* reacts positively towards light and, in fact, it may be described as being positively phototactic since it orientates itself parallel to the light beams and moves towards the source of the illumination though the strength of this reaction is greatest at approximately twenty-four hour intervals. The way in which this reaction is brought about has been described for a fairly sluggishly moving species. Like others of its kind, it rotates and gyrates as it moves, but when light falls upon it from one side only, it reacts by swerving violently to one side as a result of a sudden flexure of the body—the so-called "shock reaction". These shock reactions occur at intervals, in fact, whenever the axis of the spiral path is not parallel with the light beams and it has been noted that, for such sluggishly moving species, the reaction takes place whenever the photoreceptor is shaded by the pigment spot, caused by the normal rotation, the body then bending towards the ventral surface, *i.e.* away from the light. The next rotation of the body will then bring the anterior end pointing towards the light and this rotation is also accompanied by a slight straightening out of the body. In this way, a series of shock reactions results in the path followed by the organism curving gradually towards the source of illumination.   In more active species, a diminution in the light intensity causes a swerve towards the dorsal surface which is, unlike the creeping forms, always maintained towards the outside of the spiral path.   Otherwise, the mechanism of orientation seems to be the same (Fig. 64).

Compared with the higher animals, Protozoa are morphologically and anatomically simple and this conditions the relative simplicity of their behaviour. Nevertheless, their reactions show similarities in many important particulars with those of higher animals and are perfectly adequate to maintain the organisms within their normal environment.   Thus, certain ciliates, especially those which, for extended periods, are commonly attached to the substratum, show a *variety* of responses to unpleasant stimuli, first one and then others being manifested in turn in order to effect an escape from the discomfort.   There is also much evidence to show that ciliates are capable of modifying their behaviour as a result of previous treatment or experiences and, among others, *Paramecium* will "learn" to turn back from a light-dark boundary after the crossing has been found, many times, to be associated with a noxious stimulus. This is, to all intents and purposes, the establishment of a "conditioned reflex"— that is, the animals associate, after a time, a neutral stimulus (the crossing of the boundary from lightness to darkness) with a potent stimulus (some noxious chemical substance, or heat) and react by reversing their direction of movement even when the potent stimulus is withdrawn.

# CHAPTER V

## METAZOA

All the examples of animals so far considered belong to the Protozoa and all have been of small size, a feature characteristic of the Protozoa as a whole. This minuteness of size can be correlated with the simple unicellular condition, for the retention by these organisms of a limiting membrane of a permeable nature precludes the development in them of structures which will give the strength and rigidity necessary in a size beyond a certain upper limit. (Some Protozoa compensate for the delicacy of their outside limiting membranes by the secretion of "shells" and other "skeletal" structures, but this development is also limited.) Strength alone is, however, not the sole, or most important factor, for since many of their activities involve an exchange of substances between the cytoplasm of the organism and the surrounding liquid medium, the efficiency of these processes is governed by the surface/volume ratio; the bigger the animal, the relatively smaller its surface, since the surface increases as the square of its dimensions and the volume as the cube. Yet, on the other hand, a large surface/volume ratio equally results in a greater exposure to adverse factors in the environment.

When large animals are examined it is found that their bodies are composed of a number—often a very large number—of cells. Each cell may be compared to the protozoan body since it is potentially capable of performing all the essential activities of such a body, but specialisation invariably takes place and the cells are not all alike. This compound character of the larger animals opens up vast possibilities of increased complexity of body form and structure. It is this complexity, as well as the formation of the body by many cells, that has led to the separation, in classification, of these animals, termed Metazoa, from the Protozoa.

Metazoa are, then, distinguished, not only by their larger size, but by the high degree of differentiation and specialisation of their various component parts, which is spoken of as "morphological differentiation". This affects not only the larger parts but also the individual cells. Morphological complexity must not, however, be thought of as an end in itself but is better regarded as being to a large extent "forced" upon the animals by mere physical factors as a larger size is attained. This thought is more accurately expressed

121

by the statement that increase in the size of the body is correlated with morphological complexity and it may be noted that within certain limits, the larger the animal, the more complex its structure. Hand in hand with morphological differentiation of structure goes "physiological division of labour" which merely implies that any given part is specialised to perform a definite function for the whole organism and that function alone.   This, of course, results also in a "loss of plasticity" by the specialised part, an effect that is, however, compensated for by an increase in "efficiency".   Differentiation and specialisation are not, however, the sole prerogatives of the Metazoa, for they are also met with in the Protozoa, notably in the Ciliophora where a high degree of complexity is shown by the single unit which forms the body; but such differentiation never attains the complexity met with in the Metazoa where it follows as a result of the formation of different types of cells, and, where also it seems to bear a definite relationship to the type of bodily organisation and to the size of the animal, as will be apparent from a consideration of the various systems and organs met with in the Metazoa.

Thus, in the Metazoa a special region of the body is set aside for dealing with the food.   It may be merely a relatively simple sac or, as in higher animals, it may be a complicated "gut", but it always has the power of enclosing the food within its cavity and of secreting on to the food digestive enzymes which render the major part of it soluble and capable of absorption and assimilation by the rest of the body.   Such a digestive system is essential in larger animals, for they require large quantities of food to maintain their vital activities and the food is usually of such a size that it cannot be taken into food-vacuoles as in the Protozoa and must as a preliminary to actual digestion be broken down into small particles, when digestion may be completed in the cavity of the digestive system, or, alternatively, the particles may be enclosed in vacuoles in the cells and dealt with in much the same way as in *Amoeba*, for example.

In the larger and more complex Metazoa many parts of the body inevitably lie at some considerable distance from the digestive system and so cannot receive their nourishment by mere diffusion. Some sort of transport system is then developed to get over this difficulty and may take the form of tubular prolongations of the digestive sac (*e.g.* in jelly-fish) which pass among the tissues or else of a blood vascular system as in most higher animals.

In larger organisms the outer surface, which alone is in contact with the outside world, and therefore retains its sensitive nature, is at some appreciable distance from the deeper organs and tissues. It is chiefly in these tissues that the effectors, that is, structures which when stimulated enable the animal to respond to changes in the environment, are developed, such as muscles and glands.

In the Protozoa all the effector organelles of the body are in direct cytoplasmic continuity, and only in certain of the higher forms are portions of the cytoplasm set aside for the function of co-ordinating their activities. In general it may be assumed that all the cytoplasm of the protozoan body is of a sufficiently generalised nature to retain its powers of sensitivity and conduction as well as forming the receptors. In the Metazoa, however, as a necessary corollary of increase in size, the effector structures lie at some appreciable distance from the receptors, and so a special conducting and co-ordinating system, the nervous system, is developed. This is built up of cells (**nerve-cells** or **neurons**) which have as their special feature a number of branching cytoplasmic prolongations (**nerve-fibres**) arising from the body of the cell. The nervous system is often regarded as part of a larger system, the neurosensory system which in its simplest conceivable form would consist of a series of cells (receptor cells) lying in the surface layer of the body, and from each of these a nerve-fibre passing to an effector in its neighbourhood (Fig. 69). Such a simple arrangement is never realised, and even in the simplest of neuro-sensory systems there is no *direct* connection between the receptor and effector cells. The nerve impulses are always

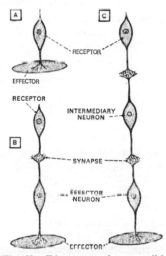

Fig. 69. Diagram to show possible linkages between receptor and effector structures. A hypothetical. B and C possibilities in coelenterates.

conducted to the effectors through a chain consisting of a variable number of neurons. Moreover, the component neurons of the system are not joined together in actual cytoplasmic continuity, for the fibres from neighbouring neurons intermesh so that minute gaps filled with non-nervous material are left between the twigs. These gaps are termed **synapses**. In almost all instances *at least* two neurons (and hence two synapses) are found in the chain between the receptor fibre and the effectors (Fig. 69).

The receptor cell is concerned, as its name implies, with the reception of stimuli, that is, a change in the environment in its proximity causes it to react in some way so that nerve impulses are generated in the nerve-fibre taking origin from its base. A receptor cell can therefore be regarded as a "*transducer*". The fibre (**afferent** or **sensory fibre**) then synapses with one from another neuron (**intermediary, association, internuncial,** or **relay neuron**) from which yet another process hands on impulses, across the second synapse in the chain, to the fibre of the third cell (**efferent** or **motor neuron**). From this neuron a fibre (**efferent fibre**) runs direct to the effector. Such a chain of neurons is said to form a **simple reflex arc**, since stimulation of any particular receptor will cause a particular effector to respond; the stimulus, as it were, is inevitably *reflected* along a certain track, so that a uniform response always recurs after stimulation.

In many of the lower Metazoa the neuro-sensory system does, indeed, show a fairly close approximation to a system of simple reflex arcs linked together in a criss-cross fashion. The receptor cells are hardly, if at all, morphologically specialised, and lie scattered in the outside layer of the body. The intermediary (internuncial) and efferent neurons consist of cells each of which gives origin to a number of nerve-fibres, all short and of more or less the same length. The nerve-fibres intermesh to form a network which is usually fairly superficial in position and is termed a **nerve net**.

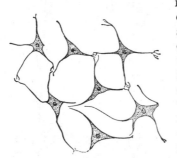

Fig. 70.   Diagram of a nerve net.

In the higher Metazoa the nerve-fibres, as a general rule, are not equally well developed on all sides of the nerve-cell, one or more of them being especially long. Moreover, the fibres become bound together by sheaths to form long conducting tracts, the **nerves**. Also, the cell-bodies, especially of the intermediary neurons tend to become aggregated in certain regions to form a **central nervous system**, when the nerves which connect it to the effectors and receptors are said to form the **peripheral nervous system**. The significance of these departures from the nerve net condition is that the combination and linking up of the various reflex arcs take place in the central nervous system instead of peripherally, and that the elongation of the nerve-fibres means that the number of possible synapses in the arcs is restricted. Any impulses due to the stimulation of a receptor are therefore conducted along definite pathways and not in *any* direction as is possible in an idealised nerve net, and the reflexes become precise or specific instead of diffuse.

Nerve-fibres will respond, that is, transmit impulses (see Chapter XVII) if stimulated in a variety of ways; for example, by injury, or chemically or electrically, although, under natural conditions in the body, nerve impulses are most commonly initiated in a receptor cell when it is stimulated. A stimulus above a certain minimum intensity must be applied for a certain minimum time to any nerve (or receptor) before nerve impulses will be evoked in it (threshold of excitability).

After the passage of an impulse a nerve-fibre becomes, for a brief period, incapable of passing further impulses, this interval being referred to as the refractory period. As a result, continued stimulation calls forth, not one continued response, but a chain or volley of impulses, an effect which may be compared to the sustained pressure on the trigger of a machine gun releasing a stream of bullets from the muzzle. A very weak stimulus may cause but a single impulse to travel, but a strong stimulus results in a succession of impulses, all of which, however, are of the same magnitude. But the stronger the stimulus, the more rapidly do the impulses succeed each other up to a rate which is determined by the refractory period of that particular fibre.

The impulse in a nerve-fibre is of an "all-or-nothing" character. That is to say, a stimulus is either adequate to promote a complete response or none at all. A partial or weak impulse cannot be set up, and once a nerve impulse has started along a nerve it passes (unless artificially interrupted) with undiminished strength to the other end. This aspect of nervous conduction has been likened to the effect of firing a train of gunpowder. It follows that the impulses are not conducted passively like electricity along a wire but that the nerve-fibre itself maintains as well as propagates the disturbance.

Most of the smaller Metazoa are aquatic in habit and their relatively large outer surface provides an area adequate for the gaseous interchanges necessary for respiration and also allows the nitrogenous waste materials to diffuse outwards sufficiently rapidly. The larger Metazoa which have small external surfaces relative to their size and are often enclosed in protective structures, display an astonishing variety of "respiratory organs". All of these, however, have in common the fact that they are devices for increasing the respiratory surface at which the gaseous interchange can take place. They may be branched or laminated structures or they may be formed of very numerous branching air ducts, the finer twigs of which end in minute blind sacs.

Organs of nitrogenous excretion also differ widely in structure, mode of development, and position, but like organs of respiration, they are distinguished by their large excreting surface which usually lines a complicated system of fine, branching tubules (nephridia or kidneys), opening directly or indirectly to the exterior.

It is evident that the specialisation outlined above leads to a corresponding loss of plasticity in the specialised parts of the body and it is not surprising, therefore, that only certain regions of the body have the power of reproduction. In many ways the reproductive cells themselves must be regarded as extremely specialised, but they retain, of course, the power of giving rise (under suitable conditions) to a complete new organism. In this sense they are less specialised than other cells of the body, and it is possible therefore to draw a distinction between the reproductive cells ("germplasm") and the rest of the body ("soma").

Finally, it must be pointed out that the degree of differentiation and specialisation met with in the Metazoa is extremely variable and results in the recognition of different grades of structure which form a basis for the classification of the multi-celled animals into two main groups: the Diploblastica, the two-layered animals comprising the Coelenterata, and the Triploblastica, the three-layered animals. Of these groups, the Diploblastica are without doubt the simpler, but it must not be imagined that the distinction between diploblastic and triploblastic animals is a perfectly sharp one. Strictly speaking, only the simpler coelenterates are truly diploblastic, for the higher forms approach the condition found in the simpler triploblastic animals, not only in having what amounts to a third layer of cells, but also in their large measure of bilateral symmetry. Indeed, by some it is believed that coelenterates have been evolved by secondary simplification from simple flatworms (Rhabdocoelida) which themselves arose from ciliate protozoans. There is little to commend such a theory.

# CHAPTER VI

## DIPLOBLASTICA (COELENTERATA)

### INTRODUCTION

Animals belonging to the Phylum **Coelenterata** are all relatively simple aquatic Metazoa and include such well-known creatures as jelly-fish, sea anemones, and corals as well as common but less conspicuous types simulating plant forms, the "**zoophytes**". The form of the body is extremely variable and may be fairly simple or relatively complex but, nevertheless, it is possible to visualise a schematic coelenterate which incorporates most of the main features and which, by reducing the variation to a common plan, renders more intelligible the anatomy of the types subsequently to be described.

In its simplest conceivable form such a generalised coelenterate has a simple sac-like body whose walls are built up of two layers of cells, an outer one (**ectoderm**) and an inner layer (**endoderm**),* the two cell-layers being separated from each other by a layer of jelly (**mesogloea**),which, however, particularly in sea anemones is permeated by a system of collagenous fibres and may contain cells of several types. In all except the Hydrozoa there is a considerable inwandering of cells from both ectoderm and endoderm into the mesogloea where connective tissue and, more rarely, muscles may be formed from them. In this respect the larger coelenterates differ only in degree from the simpler triploblastic animals. The cavity of the sac is the digestive cavity in enteron and opens to the outside world by a single aperture, the mouth. In practice no *adult* coelenterate is ever found in such a simple form.

Coelenterates are for the most part fairly small animals, but they are often found joined together to form colonies, the units of which are termed individuals or **zooids**. As a rule the zooids are not all alike and among them two main types may usually be recognised. These are the **hydroids**, the feeding zooids or **polyps**, and the **medusae**, which are specialised to carry the reproductive organs. Both hydroids and medusae may be readily derived from the schematic sac-like form described. The hydroid retains the sac-like body whilst the upper end of the sac, just below the mouth, is drawn out to form a circlet of tentacles. In colonial coelenterates

---

* These are the more usual terms for the two main cell-layers but recently the words "epidermis" and "endodermis" are coming into common usage.

the hydroids are joined to the main part of the colony by a tubular prolongation from the lower end of the sac. The typical medusa, on the other hand, may be imagined to be derived from the basic form by a flattening of the sac from end to end so that a somewhat bell-like structure is produced. At the same time there is pronounced thickening of the mesogloea and the enteron becomes restricted to a small cavity into which the mouth opens; to the cavities of a number of **radial canals** which join the central cavity

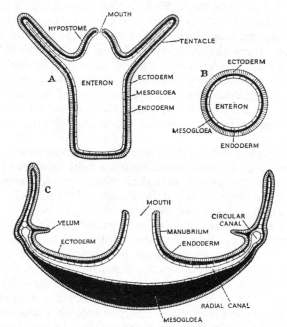

Fig. 71. Generalised coelenterate zooids.—A, vertical section of a hydroid; B, transverse section of a hydroid; C, vertical section of a medusa orientated in the same way as the hydroid.

to a **circular canal**; and to the cavity of the circular canal which runs around the margin of the bell. Tentacles are invariably present and usually are numerous, forming a fringe around the margin of the medusa.

The relations of the medusa and hydroid zooids and their fundamental similarity are shown in Fig. 71.

It must not, however, be assumed that this description is anything more than a convenient generalisation and in some coelenterate colonies a great variety of zooids, each specialised to carry out a particular role in the life of the colony, is met with—a phenomenon called **polymorphism**. But practically all the types of zooids in colonial coelenterates can be regarded as variants of either the medusoid or hydroid form from which they depart to a greater or lesser degree. Not all coelenterates are colonial or polymorphic: some are solitary, but in these also the adult individuals can usually be seen to resemble either the medusoid or hydroid phase.

It will be evident from the above description that a coelenterate is radially symmetrical. That is to say, when viewed from either end it is symmetrical with respect to any plane passing through its longitudinal axis and a plane passing through *any* diameter bisects the body into equal right and left halves. Not all coelenterates are strictly radially symmetrical. Some of the comb-jellies (Ctenophora), for example, are flattened from side to side and have but a single pair of tentacles which arise from the side of the body. The body is therefore symmetrical about a single plane alone and is therefore said to be bilaterally symmetrical (see p. 162).

Coelenterates are for the most part sedentary creatures, attached to the substratum, and those that live freely in the sea have but limited powers of swimming so that they are largely at the mercy of currents. They cannot move about in search of their food, being able merely to entrap such as comes near to them—habits which they share with many other radially symmetrical animals. Radial symmetry, therefore, fits them to receive food from any direction.

Coelenterates are, then, adapted to a passive way of living and locomotor structures and organs of special sense—structures which are only well developed in more active creatures—are not, as a rule, conspicuous. Nevertheless, it is interesting to find that in the free-swimming medusae, notably the jelly-fishes, sense organs are present, although they seem to be concerned merely with detecting differences in the intensity of the light and the orientation of the bell in the water, and so initiating reflexes which help to keep the animal in favourable surroundings.

It is not surprising, therefore, that most of the features in which the adult sedentary phase of the various coelenterate groups depart from the schematic or generalised hydroid are, within the limitations imposed by the diploblastic grade of organisation, in the nature of adaptations to a sedentary way of living. These departures are greater in certain groups than others, and those which have a greater complexity of structure are regarded as more "advanced" when compared with the simpler or more "primitive" members of the group. It does not, however, follow that any particular type will be uniformly advanced in all its features of organisation. In general it seems true that there is a tendency for the hydroid and medusae to become larger—a trend which may be noticed in

many other groups of animals.   This increase in size of the animal
is usually accompanied by an increasing sharpness of cell differentiation and also by a tendency for cells of a particular kind to become
aggregated to form tissues.   Thus in the jelly-fishes and in the sea
anemones definite muscles and nerve tracts are formed, so that in
this respect they approach the Triploblastica.   Also, in the higher
coelenterates the mesogloea is far from being a structureless
jelly, but becomes so invaded by cells (which may lay down
a skeletal system of spicules or fibres) that in many respects it
simulates the mesoderm of the three-layered animals.

## HYDRA

One of the commonest and most readily obtainable of
coelenterates is the small fresh-water polyp, commonly called
hydra, several genera and species of which are found in European
fresh-water ponds.   Not only is it very common but it serves
as a good example of the hydroid structure in a very simple
form.

The following account does not refer to any particular species
but to hydras in general.

**Form.**—The body is cylindrical and, during life, the lower
end (basal disc), perforated by a small pore, is normally attached
to the substratum or to water weeds.   The other end of the
body bears a number (up to ten) of long, hollow tentacles which arise
from around the base of a cone (hypostome or oral cone) at the apex
of which is situated the mouth.

**The Body Wall.**—The wall of the sac-like body is built on the
typical coelenterate plan and consists of an outer layer of cells, the
ectoderm, and an inner layer of cells, the endoderm.   The mesogloea
which separates the two cell-layers remains as a thin layer of structureless jelly, as seen in the light microscope but appears granular
or sometimes fibrous in electronmicrographs, depending on the
degree of contraction of the animal during fixation.   It is not limited
by a distinct membrane and penetrates in between the cells of the
body wall all of which in a sense can be considered as being
embedded in it.

Despite its simple anatomy, considerable differentiation has
taken place in the cells of the ectoderm and, to a lesser degree, in the
endoderm, so that several distinct kinds of cells enter into the
composition of the body.   The ectodermal cells are of seven main

kinds: **musculo-epithelial cells, interstitial cells, nematoblasts, nerve-cells** (including neuro-secretory cells), **sensory (receptor) cells, gland-cells,** and **germ-cells.**

The musculo-epithelial cells form the bulk of the ectoderm and are large cone-shaped cells having their broader bases lying away from

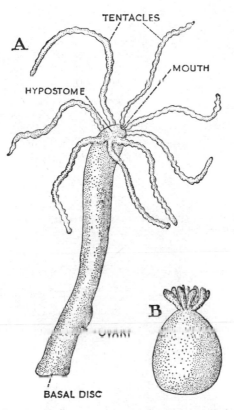

Fig. 72. HYDRA.—A, expanded; B, contracted.

the mesogloea, that is, on the outside of the ectoderm where they link up with one another to form a continuous covering to the body whilst still leaving interstices in the deeper parts of the ectoderm. Overlying the musculo-epithelial cells the electron microscope reveals a thin cuticle continuous over the body except on the basal disc. The ends of the musculo-epithelial cells nearest to the mesogloea are drawn out to form contractile processes which lie along the mesogloea parallel to the longitudinal axis of the body. They overlap and interdigitate with those of adjacent cells and are connected

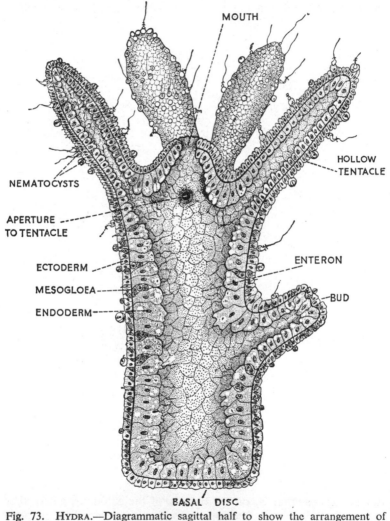

MOUTH

HOLLOW
TENTACLE

NEMATOCYSTS

APERTURE
TO TENTACLE

ECTODERM

MESOGLOEA

ENDODERM

ENTERON

BUD

BASAL DISC

Fig. 73. Hydra.—Diagrammatic sagittal half to show the arrangement of the parts.

by desmosomes as in the epithelia of higher animals. These contractile processes are termed **muscle tails.**

The interstitial cells lie in the interstices between the musculo-epithelial cells. They are small rounded cells which are not specialised in any way and consequently have the power of giving rise to any of the other kinds of ectodermal cells, including the reproductive cells. They have as their main function the replacement of the nematoblasts which are continually being used up.

The nematoblasts are extremely specialised cells usually lying in groups or batteries embedded in the superficial layers of the ectoderm. Each nematoblast secretes a most remarkable structure, the **nematocyst**, which, as will be seen later, is an extremely efficient

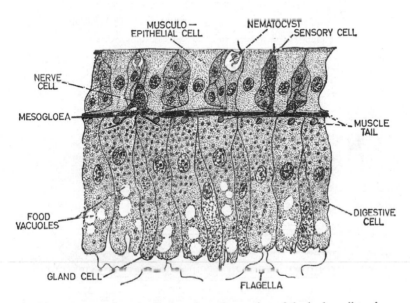

Fig. 74. HYDRA.—Longitudinal section of a portion of the body wall to show the different types of cells.

weapon and is not only used for defence and for immobilising and killing the animals upon which the hydra feeds, but also plays a part in locomotion.

Four kinds of nematocyst are found in a hydra, the largest and most complex being the **penetrants** or **stenoteles**. Each of these (Figs. 75 and 76) has three main parts, namely: (1) the capsule with its lid or operculum; (2) a short, but wide, tube inverted into the capsule, tapering towards its inner end and differentiated into a shaft (the wider part) and a spinneret which bears on its inner surface three pointed stylets or barbs and six spiral ridges beset with minute spines; (3) a coiled filament which lies in the fluid-filled capsule and winds around the shaft and spinneret, to whose inner end it is attached, so plugging its inner aperture. The walls of both the capsule and thread are formed of a collagen-like material. Lying just to one side of the operculum there projects from the surface of the nematoblast a small, sensory bristle, the **cnidocil** or trigger. On the wall of the nematocyst capsule is a number of refractile rods which are attached to contractile threads running to the base of the nematocyst itself. Reference to Fig. 75 will show that, before

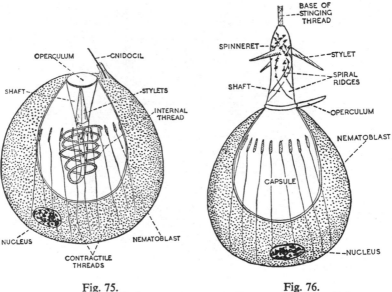

Fig. 75.
PENETRANT.—Before discharge.

Fig. 76.
PENETRANT.—After discharge.

discharge, the points of the stylets converge towards the operculum so as to form a sort of nozzle.

When the nematocyst is suitably stimulated, as for example, by a passing water animal coming into contact with the cnidocil, or even some other part of the nematoblast (the nematocysts of most coelenterates are devoid of a cnidocil), a long thread is shot out which can penetrate the prey and which carries a protein poison and probably also 5-hydroxytryptamine that exerts a paralysing effect, even on such relatively large animals as water-fleas (*Daphnia*) or small worms, for it is about 75 per cent. as toxic as cobra venom.

A series of careful observations on the very large nematocysts of the "anemone", *Corynactis*, make it clear that these nematocysts have before discharge a *hollow*, coiled thread with closely-packed barbs along its inner surface. The base of the thread is attached to a nipple-like prominence at the pole of the capsule (which probably represents the structure termed the operculum of the nematocysts of a hydra), so that an operculum as such is absent. Neither is there a shaft nor a spinneret, all of which may have been reported in error in a hydra.

Observations on threads in the act of discharge leave little doubt that, as believed long ago, the thread turns inside out at the same time as the intra-capsular thread uncoils like a hank of rope. As the thread everts continuously at its tip the barbs flick out and take up their positions in three spiral rows on the outside of the thread, the spiralling being due to the twisting of the thread as it everts.

Most earlier workers placed great emphasis on the contraction of the capsule in causing eversion of the thread, but it has been shown by observations and experiments on dried, undischarged nematocysts that the thread itself contains an essential part of the mechanism for its uncoiling and eversion. The addition of water or water vapour will cause these dried nematocysts to discharge their filaments, lengthening and also swelling in girth. But threads cut off from the capsule whilst lengthening and swelling will not evert so that clearly the capsule also plays an essential part in discharge. What seems to happen is that, when suitably stimulated, the capsular contents swell, and thus rupture the nipple at the base of the thread. This allows water to enter and the thread begins to turn inside out. The pressure of the intracapsular fluid then assists in eversion, which is facilitated by the increase in diameter, so that the part of the thread still undischarged is forced up a tube of wider bore.

To account for the swelling of thread which results in an increase in girth as well as length, it is suggested that its walls are composed of fibrillar aggregates, possibly of a keratin-like protein, orientated at an angle of between 45 and 90 degrees to the longitudinal axis. In the undischarged thread the walls are partially dehydrated and swell when water enters after rupture at the nipple. An indication that there is an orientated fibrillar structure to the walls is that the walls are positively birefringent with respect to their length when viewed with polarised light.

It is found that the nematoblasts arise chiefly from the interstitial cells in the middle third of the body and none forms above the bases of the tentacles or in the lower third of the body. Yet nematocysts are more plentiful on the tentacles than elsewhere. This apparent anomaly is explained by the extraordinary travels undertaken by the nematoblasts before the nematocyst is fully formed. Young nematoblasts are passed from the ectoderm, through the mesogloea, into the cells of the endoderm. The endoderm cells then pass the nematoblasts into the enteron or else extrude them, enclosed in a mass of endodermal cytoplasm. Alternatively, an endoderm cell with a young nematoblast may detach itself and wander, amoeboid fashion, in the enteric cavity. Whichever of the three methods is followed the end result is much the same. The young nematoblasts after journeying in the enteron are picked up and ingested by endodermal cells, which transmit them once again to the ectoderm, as and when they are required. Arrived at their final situation, each nematoblast becomes correctly orientated and completes the differentiation of the nematocyst.

The remaining three types (Fig. 77) of nematocysts of a hydra are termed **volvents (desmonemes)**, **large glutinants (holotrichous**

**isorhizas**), and **small glutinants (atrichous isorhizas).** All are essentially similar in that they consist of a sac containing a coiled thread, that they are formed within a nematoblast, and that this bears a cnidocil, but they differ from one another in the form and functions of the thread. The thread of the volvents when discharged winds into a tight coil around suitably sized objects that it touches, serving to retain prey, particularly small crustacea whose exoskeletons are raised into bristles. The thread of the glutinants, on the other hand, remains straight and seems to have adhesive properties,

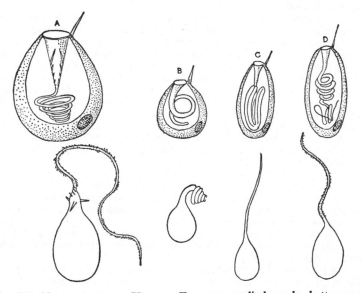

Fig. 77. NEMATOCYSTS OF HYDRA.—Top row, undischarged; bottom row, discharged. A, penetrant (stenotele). B, volvent (desmoneme). C, a small glutinant (atrichous isorhiza). D, a large glutinant (holotrichous isorhiza) (after Ewer).

although it is true that the large glutinants can also penetrate tissues of small animals.

In the adult the nerve-cells, or ganglion cells derived from the ectoderm, lie outside the muscle tails and between the bases of the other cells, and form a network on the endodermal as well as on the ectodermal surface of the mesogloea. Each nerve-cell consists of a cell-body, containing the nucleus, and a number of branching processes (nerve-fibres). The finer twigs of the branches intermesh with those from neighbouring cells to form a nerve net. Certain of the nerve-fibres are connected to the musculo-epithelial cells, whilst others take origin in receptor cells.

The sensory cells or receptors are small columnar cells found more plentifully in the ectoderm than in the endoderm. Each bears at its free surface a minute conical projection whilst its lower end is prolonged into a nerve-fibre.

The neurosecretory cells resemble in most respects the ganglion cells but contain large electron-dense particles. Gland-cells (strictly speaking, musculo-gland-cells) of one kind are found on the basal disc where they produce a sticky secretion by means of which the hydra adheres to the substratum. Other gland-cells are found around the rim of the hypostome where they are of a mucus-secreting type, the secretion facilitating the "swallowing" of large food particles.

During the warmer months of the year the interstitial cells in certain restricted regions of the body multiply by repeated division to form the reproductive cells. These regions of active germ-cell proliferation are termed **gonads**, a general term which can be applied to an organ producing gametes in any animal.

The endoderm is histologically simple when compared with the ectoderm, for it consists chiefly of a single layer of large cells resting on the mesogloea. These are of two types, digestive cells and gland-cells. The digestive cells have their bases prolonged into muscle tails which run at right angles to the main body axis, so that they are musculo-epithelial cells. They have the power of forming pseudopodia, and microvilli, and of engulfing food particles in food-vacuoles. But, instead of extruding pseudopodia from their free surfaces, they may bear two to four flagella. The gland-cells lack muscle tails, do not rest on the mesogloea, and are concerned with secreting digestive juice and mucus.

A few small interstitial cells are found between the large endodermal cells. Scattered receptor cells and nerve-cells have also been reported in this layer.

In one species of hydra (*Chlorohydra viridissima*), small, spherical green algae—**zoochlorellae**—are found living in the large endoderm cells in such numbers that the whole animal is quite green in colour. It must not, however, be thought that *Chlorohydra* is unique in having this curious feature. In point of fact, not only many other coelenterates (*e.g.* corals and sea anemones), but also certain flatworms (e.g. *Convoluta*) and other animals have similar (not necessarily identical) algae living in their tissues. The role of these algae will be discussed later on.

**Movements.**—As a few moments' observation of a living specimen will serve to show, the animal can alter the shape of its body to
a very remarkable degree. It may be so extended that it is a very
long slender cylinder or it may be so contracted as to be like a
small barrel. Contraction is brought about by the contraction of
the muscle tails of the musculo-epithelial cells whilst extension of
the body is effected, partly by the elasticity of the mesogloea, and
partly by the contraction of the muscle tails of the endoderm cells
which causes the body to become thinner. In that the mesogloea
serves for the attachment of the muscle tail system, it may be
regarded as having a skeletal function.

The tentacles are very mobile structures, for they can be bent
in any direction and can be rapidly extended or shortened. Their
movements take place in an apparently "purposive" manner.
For example, if the animal in the extended condition is touched
with a needle, or if the container is suddenly jarred, the tentacles
and the body immediately contract. This withdrawal response
occurs after the application of an "unpleasant" stimulus. A period
of rest is, however, soon followed by a gradual extension of the body
and tentacles and, in the presence of food, feeding is soon begun.

Rhythmic nervous activity can be recorded electrically from
an individual hydra; it is unrelated to locomotory activity but the
rhythm is accelerated when light is shone on the disc. It seems
possible that this activity is taking place in the inner nerve net.
In addition, when the hydra shortens, as it does every eight or
ten minutes, there are bursts of larger potentials which are, however,
inhibited by light. They seem to come from a centre near the
mouth and to be conducted in the outer nerve net. Other potentials
are recorded when the animal is somersaulting or walking.

**Feeding.**—If a small water animal comes into contact with a
tentacle the penetrant and volvent nematocysts of that region are
immediately discharged (p. 134). The penetrant threads enter the
tissues and a poisonous fluid is injected which gradually paralyses
the prey, at the same time releasing tissue fluids which play an
important part in evoking the feeding reactions (see below). The
volvents coil around any projections on the prey, particularly around
the bristles such as are found on small crustaceans, helping to hold
it fast until it is ready to be engulfed. The tentacle holding the
captured animal then shortens and usually curls over towards the
mouth. Gradually the other tentacles bend over to the one holding
the prey and finally, all the tentacles shorten so that the food is
pulled into position just above the mouth. This then opens widely
and engulfs the prey.

But a hydra will normally open its mouth to and swallow only living prey. It will, however, swallow dead animals and even bits of gelatine which have, for example, been soaked in crustacean juice. This suggests that some chemical principle is necessary to evoke the feeding reaction. Certain experiments suggested that this substance is glutathione, a substance universally present in all animal tissue fluids and which, if added to the water in which a hydra is living, evokes very powerful movements of the tentacles and opening of the mouth even if food is absent. It is thought that glutathione is normally released by the puncturing of the prey by the penetrant nematocysts. Other observations suggest that the glutathione (or substance with similar properties) may also come from the discharged nematocysts themselves. In a sense, glutathione acts rather

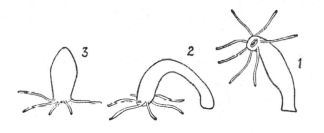

Fig. 78. HYDRA.—Diagram to show method of locomotion (after Ewer).

like a hormone. It stimulates the mouth to open before the prey touches it. Presumably it makes its effects felt by stimulating chemo-receptors near the mouth and these send impulses into the nerve net.

**Locomotion.** Although usually attached by a sticky secretion formed by gland-cells in its basal disc with its main body axis at right angles to the substratum and not orientated with respect to gravity, a hydra can, on occasion, move about from place to place. The body first extends and then bends over so as to bring the tentacles down on to the substratum. By means of the small glutinants the tentacles adhere and then the basal disc is freed. When this has happened the body contracts strongly until it appears as a small knob supported by the tentacles. This phase is followed by an extension of the body, which then bends over and acquires a new attachment to the substratum near to the still attached tentacles (see Fig. 78). The tentacles are then freed and the process repeated. This kind of locomotion may be termed "walking" and is the normal method of progression. More rarely, however, a hydra floats at the surface by means of a bubble of gas secreted by certain cells in the centre of the basal disc and is then carried about passively by means of water currents.

**Behaviour.**—Although the movements connected with feeding have at first sight every appearance of being purposeful, yet it has been shown by experiments that they are largely automatic, that is, governed by events in the external environment. It is found, for example that stroking the tentacle with a thin glass rod causes the tentacle to become shorter and also to bend towards the side which is stimulated in this way. The shortening is due to the contraction of the muscle tails of the musculo-epithelial cells, whilst bending is

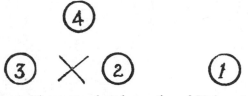

Fig. 79. HYDRA.—Diagram to show locomotion of Hydra. 1, 2, 3, and 4 represent alternative positions of the foot in taking a step after the tentacles have been fixed at X (modified after Ewer).

due to a greater contraction on the side that is being stroked. The contraction and bending is at first strictly a local affair, but later spreads along the tentacle and still later, the other tentacles bend over in the same way. It seems that the gradual spreading of the area of contraction which follows stimulation of any part of the ectoderm is due to the conduction in all directions of impulses in the nerve net. It is also apparent that since the response is greatest near to the point of stimulation and gets progressively less in more distant regions, the nerve net offers some resistance to the passage of nerve impulses. This resistance occurs at the numerous synapses; but it seems that whilst the first impulse may fail to pass across any synapse, its arrival there lowers the resistance to later impulses which are therefore able to reach other neurons —a phenomenon termed inter-neural facilitation. The progressive diminution in the responses away from the stimulated area is due, not to a decrease in the intensity of impulses in the nerve-fibres (which would be a contradiction of the "all or nothing" principle), but to a certain number of impulses failing to pass the synapses.

The distance to which excitation is propagated through the nerve net depends, to a large extent, on the strength of the stimulus applied; the stronger the stimulus, the greater the distance (as well as the intensity) through which the disturbance spreads. This is because a stronger stimulus initiates a larger number of nerve impulses so that more synapses are passed.

During life the tentacles are usually extended and curved away from the mouth and may project downwards at the sides of the body. When a small animal is caught on one of the tentacles it produces a local contact stimulus and the tentacle shortens and bends towards the stimulated point. If the prey is caught on the upper surface of the tentacle then, of course, the bending is towards the mouth but if, as may happen, it is caught on the lower surface

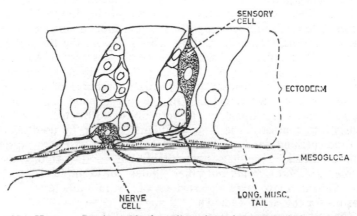

Fig. 80. HYDRA.—Portion of body wall to show the supposed arrangement of sensory cells and nerve-cells (dark shading).

bending is at first away from the mouth. Whichever happens, the other tentacles bend towards the point where the prey is caught and then come into contact with it. These tentacles in turn receive a direct stimulus and all the tentacles gradually shorten, so that the prey is drawn into a position just above the hypostome. The mouth opens and as a conclusion to this co-ordinated set of movements the prey is engulfed (see also p. 139).

It has already been mentioned that hydra, if stimulated violently, has the power of contracting very suddenly, a movement that is brought about by the simultaneous contraction of all the muscle tails of the ectoderm. The impulses are in this case conveyed with rapidity to all parts of the animal by special tracts of nerve-cells which are said to form "through-conduction paths".

The discharge of the nematocysts may be quite independent of nervous control as is shown by the observation that completely anaesthetised coelenterates discharge their nematocysts in the usual way when stimulated, whilst even nematocysts completely isolated from the body will shoot out their threads if an adequate stimulus is applied to them. The small glutinants which, it will be remembered, are used for the attachment of the tentacles to the substratum when the animal is walking, are easily stimulated to discharge by light mechanical stimulation provided it is of sufficient duration such as occurs when a tentacle is placed in position during locomotion. Their discharge is, however, inhibited by the presence of food. The remaining three types of nematocysts are not discharged during walking and this is due to their requiring a much greater mechanical stimulation, except when certain chemical substances are present, when the threshold for discharge is greatly lowered. Thus, the presence of food extracts, particularly of those animals on which a hydra normally feeds, sensitises, that is, lowers the threshold to mechanical stimulation at which the penetrants and volvents discharge, the significance of this being that a light touch by a suitable food animal causes an immediate discharge of the appropriate kinds of nematocysts. On the other hand, nematocysts are not wasted by discharge at every object with which the body comes into contact. The large glutinants seem to be used neither in food-catching nor in walking, but to be reserved for the special task of attacking animals not suitable for food. Among these animals are *Paramecium*, but other ciliates, for example, *Trichodina* and *Kerona*, move over the surface of hydra unmolested. The reasons for this are not known.

Thus, although not under the direct control of the nervous system* and therefore termed independent effectors, the four types of nematocyst, by their discharge in response to differing stimuli, are each enabled to play an effective part in the three separate functions of locomotion, food-capture, and defence.

**Digestion.**—The food which has been transferred to the mouth by the tentacles and ingested into the enteron may remain alive there for some time but, sooner or later, it is killed by juices secreted by the gland cells of the endoderm. This secretion contains digestive enzymes, the chief function of which is to reduce the food to such a state that it easily fragments into small particles which are then taken into food-vacuoles by the pseudopodia of the digestive cells. It is here that digestion is completed, the insoluble remains

* There is some evidence that the threshold to discharge of nematocysts by mechanical or chemical stimuli may be lowered by impulses through the nerve net.

being later returned to the enteron, finally to be wafted out through the mouth by the current of water maintained by the flagella of the flagellated endodermal cells. Distribution of the smaller food particles is also effected by certain endodermal cells which after engulfing the particles lose contact with their neighbours and wander in an amoeboid fashion through the enteron to parts where the food is needed. Even portions of the cytoplasm of the endodermal cells may act as food carriers in a similar way.

Egestion of very large particles may take place by a sudden muscular contraction of the body so that the contents of the enteron are suddenly shot out. Digestion is thus carried out in two phases: an extracellular and an intracellular phase. This is very interesting, for in the former it resembles digestion in most higher animals, whilst the latter is strictly comparable with the process in a protozoan such as *Amoeba* as well as in flatworms, nemertines, and some other animals. Other coelenterates also have similar digestive processes, the probable reason for the retention of the intracellular phase being that a sufficient concentration of digestive juice could not be produced in an enteron, which is in more or less free communication with the surrounding water.

**Nutrition in Coelenterates.**—In a hydra, the soluble products of digestion are distributed of necessity by diffusion from the endodermal cells to other parts of the body. This is a slow process which can be effective only for short distances, *i.e.* through the mesogloea to the ectoderm. Many of the larger coelenterates have various devices for increasing the surface area of the endoderm so that the amount of food capable of being digested is greatly increased. In the sea anemones and corals, for example, the enteron is incompletely divided up by numerous radially arranged endodermal folds termed "mesenteries". In the jelly-fishes there is a complicated system of branched radial canals joining the part of the enteron into which the mouth opens to a canal around the margin of the bell. This arrangement is a device to ensure that the fluid in the enteron is circulated throughout the body. In many ways it is comparable with the water vascular system of echinoderms. Increase of surface in the larger jelly-fishes is brought about by special filaments arising from the corners of the mouth and these filaments may carry out extracellular digestion.

Coelenterates are predominantly carnivorous creatures, *i.e.* they derive their energy from the breakdown of proteins, and much of their excretory products are nitrogenous. Large amounts of inorganic salts, notably phosphates, are also excreted. Now proteins are not capable of complete breakdown to simple inorganic

substances in the animal body, but during their breakdown ammonia
is liberated and, in higher animals, is indirectly converted to urea.
The elimination of substances such as ammonia, carbon dioxide,
and salts from the body may be a matter of some difficulty, and this
may in part explain the very common occurrence of unicellular
algae in the endoderm. These algae (zoochlorellae if green,
zooxanthellae if brown) are **symbionts**. That is, they and the
animal live together in complete harmony to their mutual benefit,
an association spoken of as **symbiosis**. These symbiotic algae, like
all other green plants, build up carbohydrates from carbon dioxide
and water (photosynthesis), whilst proteins are synthesised by a
combination of the elements of the carbohydrates with nitrogen,
taken in as a solution of nitrates from the surrounding water.
It is probably for this reason that *Chlorohydra viridissima* is strongly
positively phototactic, always tending to move towards a source
of light. During photosynthesis, oxygen is given off and is utilised
by the animal symbiont for respiratory purposes, the resulting
carbon dioxide then being available for the photosynthetic processes
of the algae. It has also been shown that the algae absorb and
utilise much of the phosphates and nitrogenous waste excreted by
the animal so that in addition to shelter the algae receive substances
of use in the synthesis of their food, whilst they also "contribute to
their keep" by supplying the animal with a certain amount of the
oxygen that it requires. That the symbiotic algae are of importance
in the nutrition of a hydra was shown by rearing normal animals as
controls and comparing with others from which the algae had been
killed off, under a variety of conditions. When food was plentifully
supplied those without grew as rapidly as those with symbionts.
But when the food supply was cut down the colourless hydras
grew more slowly than the controls. Moreover, when food was
completely withheld the controls lived much longer than the colourless hydras. It has not as yet been proved that hydra uses symbionts directly as food although by the use of [14]C some sea anemones
have been shown to do so. Zooxanthellae are essential to reef-
forming corals for as well as acting like the symbionts of a hydra
their removal of phosphates makes possible calcification of the
skeleton since phosphates act as "crystal poisons", preventing the
laying down of calcium carbonate. It is interesting that, during
the formation of the oocytes, zoochlorellae become more concentrated in the endodermal cells below the ovary than elsewhere
in the endoderm and then, in some way as yet unknown, pass
through the mesogloea into the developing oocytes. Yet they do
not move into a developing testis nor is there any accumulation of
zoochlorellae in the endoderm below a testis so that it seems as if

an ovary exerts some attractive influence in the zoochlorellae in the endoderm below it.

**Gaseous Interchange and Excretion.**—In the absence of definite organs of respiration or of nitrogenous excretion it is concluded that both the gaseous interchange necessary for respiration as well as the excretion of waste nitrogenous matter takes place over the whole surface of the body wherever it is in contact with the water. The various devices for the increase of the surface lining the enteron in other coelenterates no doubt play a large part in ensuring the efficiency of this simple method of getting rid of waste products and of absorbing oxygen, for in the coelenterates, as in many other

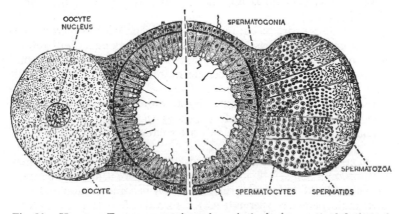

-Fig. 81. HYDRA.—Transverse sections through the body—on the left through an ovary—on the right through a testis.

aquatic animals (echinoderms, annelids, etc.), the endoderm is almost certainly respiratory.

**Osmoregulation.**—As has already been mentioned, the large surface presented to the water by both the ectoderm and endoderm of the body facilitates gaseous interchange and nitrogenous excretion but, on the other hand, these same conditions favour the influx of water into the tissues in fresh-water coelenterates. It is known that the cells of a hydra osmoregulate and actively take up ions from the surrounding water but the mechanism is not known.

**Reproduction.** ASEXUAL.—Hydras and many other coelenterates reproduce asexually by budding, but whereas in hydras the buds break free and become fresh solitary individuals, in many coelenterates the buds do not separate and budding results in the formation of a colony. Budding in a hydra takes place chiefly during the summer months when food is plentiful. Both ectoderm and endoderm grow rapidly at one point, and a protuberance containing a diverticulum from the enteron is thus produced at the surface. This increases in size, tentacles develop at its free end, and a mouth is soon opened in the centre of them. Gradually the bud grows into a new hydra, the attachment to the parent becomes constricted, the enteric cavities separate, and finally the young individual is detached from the parent. The young migrate upwards towards the surface of the water, a response termed negative geotaxis. This type of behaviour lasts, however, for about three days only, its significance being that it aids in the dispersal of the young organisms. Several buds sometimes appear at one and the same time on a single hydra.

Again, if a hydra be cut into two parts each will soon grow so as to restore its lost parts. Even quite small fragments will reform into complete animals provided that they are not below a certain minimal size and that both ectoderm and endoderm are present. This is a fairly extreme example of regeneration, a property which is possessed to a varying degree by all animals, but as a rule, it is only in comparatively simple creatures that it can result in any extensive replacement of the body after damage.

SEXUAL.—Gonads, male (testes) and female (ovaries), may develop on a single individual which is therefore said to be bisexual or hermaphrodite, but as a rule, the testes become mature and shed their male gametes (spermatozoa) before the female gametes (ova) are ripe so that self-fertilisation does not take place. In most species the sexes are separate. In the earlier stages a gonad destined to be a testis closely resembles a young ovary in that it consists of a mass of small rounded cells which have arisen by multiplication of the interstitial cells of that region to form germ mother-cells. During later stages, however, the sperm mother-cells (spermatogonia) in the testis undergo a reduction division (Chapter XVIII) to form large numbers of spermatozoa, but in the ovary only one cell out of all the potential ova attains maturity. This cell, the oocyte, which grows at the expense of all the rest, then undergoes a reduction division to form one enormous ovum laden with yolk granules, and two smaller functionless cells. The two gonads may also be distinguished by their relative positions in the body. The testes lie near the oral end, the ovaries nearer the base.

Sexual reproduction takes place in the autumn and in hydras, as in all other Metazoa, is initiated by the fertilisation of an egg-cell by a spermatozoon. The spermatozoa are released by rupture of the ectoderm over the testis and are carried partly by swimming, partly by water currents, to the egg-cell on another hydra. The fertilised egg then almost immediately begins to divide (**cleavage**) with the result that a hollow ball of cells, all more or less equal in size, is formed. This hollow spherical embryo is termed a **blastula**, whilst its central cavity is the **blastocoel**. It is soon succeeded by a second embryonic stage, the **gastrula**. This is effected by a process termed **gastrulation**, in which the cells from the wall of the blastula

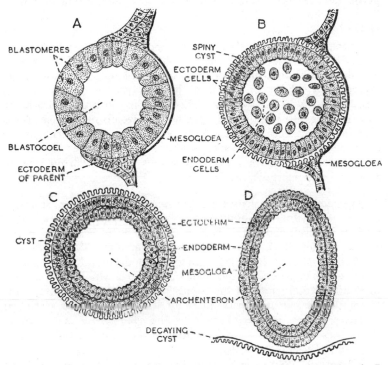

Fig. 82. HYDRA.—Development of the zygote. A, blastula. B and C, gastrulae. D, young larva released from cyst.

migrate into the blastocoel from all directions and gradually obliterate it. Later still, a new cavity, the enteron, appears in the mass of centrally placed cells. Whilst these changes have been taking place the outer layer of cells (ectoderm) have secreted a horny envelope or cyst around the embryo. The embryo is then shed from the ovary and falls to the bottom of the pond, where it remains quiescent until the next spring. With the advent of the warmer weather, the little two-layered embryo emerges from its cyst, its wall becomes perforated at one place to form the mouth, tentacles grow out, and the larva (as it is called) is to all intents and purposes a young hydra.

Hydra is peculiar in its retention of the solitary condition for, although multiplying by budding, the buds soon become separated from the parent to become self-supporting. Budding, however, allows of the formation of a type of colony very common in the group of coelenterates to which a hydra belongs. This is well shown by the next example, *Obelia*, which also has polymorphic zooids.

## OBELIA

**Form.**—*Obelia geniculata* is a marine coelenterate which forms small, branching colonies attached to the surface of sea-weeds, usually to the fronds of *Laminaria*. Each colony consists of a stem **(hydrocaulus)** about one inch in length, bearing zooids, and is attached to the weed by a branched rooting portion **(hydrorhiza)**. The hydrocaulus, together with the hydrorhiza, forms the **coenosarc**. The individual zooids of the colony are attached alternately on opposite sides of the hydrocaulus and are of two types—**hydranths** and **blastostyles**. Both are modifications of the hydroid type of zooid.

**Hydranths.**—Each hydranth has a somewhat similar structure to that of a hydra. Its body is sac-like and is connected by a hollow stalk at its lower end to the hydrocaulus. The body wall is built of ectoderm, mesogloea, and endoderm and encloses a simple enteron. At the free end of the hydranth is an aperture, the mouth, immediately below which is a conical **hypostome** or oral cone which is about one-third of the total length of the zooid. From around the base of the hypostome, where it joins the rest of the body, arises a circlet of tentacles. These are very numerous and *solid* for the endoderm in them does not enclose an extension of the enteron, but forms a solid rod of vacuolated cells.

The whole colony is surrounded by a cuticular secretion of the ectoderm (stated to be a form of chitin), called the **perisarc**. This is not in contact with the ectoderm, but is separated from it by a space, except at occasional places where the ectodermal cells extend to meet it. The tubular perisarc expands at the end of each branch bearing a hydranth to form a conical cup which protects the zooid. This cup is the **hydrotheca** and since it is open at its distal end, the zooid can be protruded from, or retracted into, it. Between each hydrotheca and the stem the perisarc has annular constrictions, and also there is usually one in the main stem itself just below the base of each zooid. In the interior of each hydrotheca,

near its base, there is a ring-like shelf on which rests the base
of the hydranth.

From this description it will be realised that the whole of the
colony including the stem and hydrorhiza is a hollow system
throughout which the enteron is everywhere continuous.

The finer details of the feeding zooid or hydranth are shown in
Fig. 84. It will be seen that its structure is closely similar to that

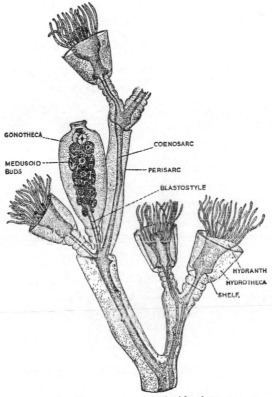

Fig. 83. OBELIA.—Hydroid colony.

of a hydra. The ectoderm is thin and consists chiefly of musculo-
epithelial cells. Interstitial cells are practically absent and the
nematocysts, which are especially abundant on the tentacles where
they are arranged in annular batteries, originate from nematoblasts
in the basal parts of the hydranths and in the coenosarc and reach
their final positions by active migration. A nerve net is present on
both sides of the mesogloea. The endoderm cells are large and have

granular contents: most of them are furnished with flagella.  These may be withdrawn and replaced by pseudopodia.

**Blastostyles.**—When an erect stem has reached its full development it produces special elongated zooids called blastostyles.  They arise as buds in the axils of some of the lower hydranths and have the form of a narrow cylindrical tube closed at the distal end.  They are really simplified zooids without mouth or tentacles.  A

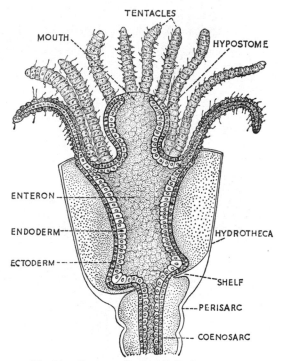

Fig. 84.  OBELIA.—Sagittal half of a hydranth.

blastostyle is enclosed in a special vase-like portion of the perisarc termed the **gonotheca**.  Each blastostyle gives rise to numerous lateral buds which develop into **medusae** which, when fully developed, are set free and swim away from the parent colony, thus helping to disperse the species.

**Medusae.**—The **medusa** has the form of a disc which is concave on one side, convex on the other.  From the middle of the concave surface there hangs a short projection, the **manubrium,** which, together with the disc or bell, as it is often called, gives an umbrella-like appearance to the medusa. The mouth, which is situated at the apex of the manubrium leads into a small central part of the enteron and from here there radiate four radial canals, which in turn open into a circular canal near the margin of the bell. The endoderm is, however, not confined to the central sac and canal system, which together represent the enteron, but is continuous as a fairly thin layer between the canals. This plate of endoderm, the endodermal lamella, may be regarded as being formed by the fusion of an upper and lower sheet, the fusion having taken place at all points except in the region of the central cavity of the enteron and along the canals. Between the ectoderm and endoderm lies the thick mesogloea, forming the bulk of the animal.

Around the margin of the bell are numerous tentacles which, like those of the hydranth, have a solid core of endoderm. Their

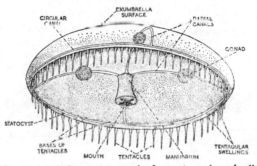

Fig. 85. OBELIA.—Medusa seen partly from the sub-umbrella surface.

number increases with the age of the medusa. For descriptive purpose it is convenient to fix the position of these tentacles by reference to particular radii of the bell. Thus the four radial canals mark out four principal radii termed **per-radii**. The tentacles at the ends of the per-radii are then the **per-radial** tentacles. Bisecting the angles between the per-radii are four **inter-radii** with **inter-radial tentacles**. Finally, the position of eight **ad-radial tentacles** may be fixed by radii drawn so as to bisect the eight angles already mentioned. All of the tentacles have somewhat swollen bases and these swellings are due to the accumulation of interstitial cells which are, to all intents and purposes, absent from other regions of the ecto-derm. The accumulations of interstitial cells provide a reservoir of cells from which nematoblasts are renewed as they are used up by the tentacles. Sometimes pigment granules are found in the ectodermal cells at the bases of the tentacles, but it has been shown that they have no visual function and are in all probability accumulations of excretory matter. The medusa is well equipped with contractile elements and these are almost entirely confined to the lower surface of the bell. Here, the contractile tails of the musculo-epithelial cells are much more prominent than those of a hydra, and in fact, are so large in proportion to the rest of the cell, that they may be said to form definite muscle-cells. The muscle tails are orientated to form fairly regular tracts of which the most important are a circular tract and several radial tracts.

THE NERVOUS SYSTEM.—This is built on essentially the same plan as that described for a hydra and one which is typical of all coelenterates. That is, there is a nerve net on each side of the mesogloea but, in the medusa, nerve-cells and -fibres are especially concentrated around the margin of the bell to form a double nerve ring. The presence of this double nerve ring is associated with the

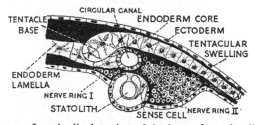

Fig. 86. OBELIA.—Longitudinal section of the base of an ad-radial tentacle.

development of receptor organs—statocysts—structures which would be of little use to the sedentary colonial phase.

SENSE ORGANS (RECEPTORS).—At the base of each of the ad-radial tentacles on the sub-umbrella surface, just inside the margin of the bell, is a small fluid-filled ectodermal sac containing a particle of calcium carbonate secreted by a single large cell. The wall of the sac is composed of an epithelial layer which in one region, on the side nearest to the margin of the bell, is sensory and its cells bear fine protoplasmic processes which arch over the particle of calcium carbonate. The sac is called a **statocyst** or **marginal vesicle** whilst the particle is a **statolith**. The marginal vesicles have been shown to be receptor organs concerned with the detection of changes in the orientation of the medusa. When the medusa is horizontal the statolith hangs vertically downwards and so does not impinge on or stimulate the processes of the sensory cells, but, when the medusa is inclined to the horizontal the statolith falls against the sensory cells which are then stimulated. Nerve impulses are initiated and transmitted by fine fibrils, leaving the bases of the cells to the nerve ring. It follows from the relationship of the sensory processes to the statolith that stimulation takes place only when the medusa tips over to the side bearing those particular statocysts. If the medusa is inclined in the opposite direction, the statocysts of the opposite side are stimulated. The effect of such stimulation is to cause the ectodermal muscle tails to contract more rapidly on the side stimulated so that the bell is brought again into the horizontal position. The statocysts probably have another equally important function, viz. of encouraging the swimming movements which keep the medusae

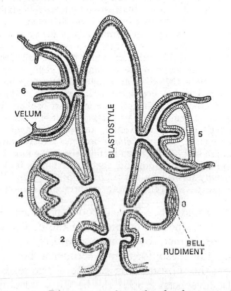

Fig. 87. OBELIA.—Diagram to show the development of medusae from a blastostyle.

in the surface waters, for it is clear that the swimming movements will stimulate the statocysts, which will then initiate reflexes through the circular nerve and nerve net and so stimulate the muscle tails to further contractions. The statocysts may therefore be said to play a part in reinforcing the rhythmical impulses originating in the nervous system and controlling the swimming movements.

SWIMMING OF THE MEDUSAE.—The medusa swims through the water by alternately opening and closing the bell in much the same way that an umbrella is opened and shut (with the important difference that the surface of the medusa remains smooth throughout the movements). Shutting is caused by contraction of the ectodermal muscle tails, the wave of contraction spreading from the sub-umbrella surface, where the contractile elements are best developed, and not occurring at all on the upper surface except at its sides. Opening of the bell is brought about chiefly by the elastic mesogloea regaining its shape and partly by contraction of the muscle tails in the middle of the upper surface.

DEVELOPMENT OF THE MEDUSA.—The first stage in the development of a medusa is seen as a bud which is a simple small diverticulum of the cavity of the blastostyle. This diverticulum grows longer and then enlarges at its extremity so that it becomes a little vesicle enclosed by ectoderm, mesogloea, and endoderm, and connected with the blastostyle by a narrow stalk.

A cavity below the sub-umbrella surface is now developed in a peculiar way. The cavity is not formed by the simple moulding of the vesicle to the shape of a medusa, but the distal ectoderm separates into two layers; the inner layer in turn acquires a cavity, and this little sac, which may be called the bell-rudiment, enlarges and assumes the shape of the sub-umbrella cavity with the manubrium in its centre. The cavity, therefore, although lined throughout by ectoderm, is at first closed by a layer of ectoderm which extends from one margin of the umbrella to the other. This membrane is at last broken through and the remnant of it forms a circular shelf projecting inwards from the margin of the umbrella. This shelf is called the velum, and is permanent and conspicuous in most hydroid medusae, but in *Obelia* it becomes a mere vestige.

When the medusa bud has acquired its final shape the mouth breaks through at the extremity of the manubrium and finally the stalk is broken, the aperture into the enteric cavity formed by the rupture closes up, and the little medusa becomes free and escapes from the gonotheca through the aperture at its end.

REPRODUCTIVE ORGANS.—The reproductive organs of the medusa are four in number, one in the middle of the course of each radial canal. They are formed by ventral diverticula of these canals, which, pushing the mesogloea and ectoderm before them, form projections from the sub-umbrella surface. The sexes are separate. The germ-cells have been traced in the early development of the medusa on the blastostyle as originating in the ectoderm of the manubrium, but they afterwards migrate to the endoderm, and then to the gonads, where before maturity they leave the endoderm and lie between the ectoderm of the sub-umbrella surface, and the mesogloea. When ripe they burst through the ectoderm into the sea, where fertilisation and development take place after which the medusae die.

**Life-history.**—The fertilised ovum segments to form a hollow blastula, which becomes solid by the delamination of cells from the blastula wall. The outer layer of cells becomes uniformly ciliated and the larva is now called a **planula**. A cavity appears, by splitting in the solid endoderm. The planula is easily converted into a single hydroid zooid. It attaches itself by one end to stone or weed, forms a mouth at the other, and then the circlet of tentacles grows out below the mouth. By budding, the colonial hydroid phase is produced and so the life-cycle is completed.

It has been seen that in the life-history of *Obelia* there is a regular alternation of the colonial hydroid and the medusoid phases. In this instance as in many others, the medusae never give rise to other medusae but to eggs which, after fertilisation, develop into hydroids from which other hydroids and medusae are produced by budding. This phenomenon was formerly called an "alternation of generations"—an expression which implies that the organism exists in two distinct forms which alternate regularly in the life-cycle—one "generation" having the power to reproduce the next by asexual means, which then in turn by sexual methods gives rise to the next "generation". A typical and clear-cut example of such a life-history among the plants is that of the fern (*Dryopteris*) where the plant (**sporophyte**) produces spores having the haploid number of chromosomes. The spores develop into small heart-shaped structures (**prothalli** or **gametophytes**) every cell of which is haploid and on the gametophytes are borne the male and female reproductive organs. After fertilisation, the egg-cell gives rise to a new sporophyte which is diploid. Thus a true alternation between an asexual diploid (sporophytic) and a sexual haploid (gametophytic) generation is seen.

Such a life-history forms a marked contrast to that of *Obelia* where, with the exception of the gametes, the cells of both "generations" are all diploid and where, moreover, the "generation" bearing the gametes does *not* arise by the division of a single cell. There is, in fact, no true alternation at all in *Obelia* between a sexual and an asexual "generation". The gametes carried in the gonads of the medusa *do not actually arise there* but may be observed in various stages of maturity in the ectoderm of the blastostyle, and only later do they migrate into the medusa and take up their position on the radial canals so that it is clearly impossible to differentiate between sexual and asexual "generations". The alternation of fixed and sedentary phases is merely a particular example of polymorphism which, as has been stated above, involves the setting aside of some zooids (hydranths) for feeding the colony, some (blastostyles) for budding off a third, motile type (medusae) whose function is usually said to be to carry away the gametes from the parent colony and so effect dispersal and prevent overcrowding of the species. But there is an enormous wastage of medusae and gametes for the chances of the eggs being fertilised are slight. Most planulae also perish before reaching the stage at which they can

settle and metamorphose and those that do so may not find themselves in a situation in which the hydroids can survive. So that the
role of the planktonic stage is difficult to assess. Moreover, many
hydroids produce only incomplete medusae which, although carrying
the gametes, are not released from the parent colony and yet these
species are as widely dispersed as is *Obelia*. The generally accepted
idea that a planktonic phase in the life-history of marine animals is
to ensure dispersion is probably untrue in most instances.

# CHAPTER VII

## THE THREE-LAYERED ANIMALS—TRIPLOBLASTICA

It has been seen that the coelenterates became specialised in different ways but produced two main types of individuals, hydroids and medusae. In both of these the body is built up of different types of cells arranged in two distinct layers, the ectoderm and the endoderm, separated by a secretion of the cells, the mesogloea. Despite the cell-specialisation there is, however, no great histological differentiation; that is to say, the cells although definitely specialised are not aggregated to form tissues. In this, as in other ways, the coelenterates differ from other Metazoa which occupy a place higher in the scheme of animal classification. It appears that the two-layered (diploblastic) plan of the coelenterate body, although showing a great advance in size and complexity when compared to the Protozoa, nevertheless, has led to the establishment of a group in which the upper limit of complexity is definitely fixed. Undoubtedly, one of the causes contributing to this limitation of complexity is the absence of any efficient transporting sytem so that substances passing from the digestive layer (endoderm) to other parts of the body must do so by diffusion alone. Extensive separation of the ectoderm from the endoderm is therefore rarely found (except in medusae) and practically no structures other than those formed by folding of the cell-layers and mesogloea are present. The development of a third layer of cells (**mesoderm**), lying between the ectoderm and the endoderm opens up further possibilities of increase in size and complexity, and animals built on this three-layered plan are called Triploblastica.

In these animals, the three layers of cells (**germ-layers***) from which the body is built are present *as such* only in the embryo, for, as development proceeds, the cells rapidly become specialised and

---

* In most animals the adult organs do not arise in a straightforward way from cells formed by division of the zygote, but instead, the cells at first become arranged to form sheets or layers which are termed **germ-layers.** The outermost layer is called the **ectoderm**; the inner layer, which forms the lining of the gut, is the **endoderm,** whilst the middle layer, lying between the ectoderm and the endoderm is the **mesoderm.**

In the Diploblastica, only the ectoderm and the endoderm are present, and since this is believed to represent the primitive condition in Metazoa, these two layers are often called the **primary germ-layers.** It is further very generally held that homologous structures in different animals always arise from the same germ-layers. In a broad sense this is certainly true. For example, the ectoderm, among other things, always gives rise to the epidermis and to the nervous system; the mesoderm to the muscles, connective tissues, and blood vascular system;

157

grouped *together* to form tissues whilst these in turn become arranged to form the organs of the body. Perhaps the most interesting developments which take place in the three layers of cells are those affecting the mesoderm, which everywhere separates the ectoderm from the endoderm lining the gut. In the simplest Triploblastica, the mesoderm gives rise to the reproductive and general muscular systems and also to a peculiar sort of loose, cellular tissue called **mesenchyme** or **parenchyma**, which forms a packing around all the internal organs. No obvious conducting system is recognisable and this function must be attributed to the parenchyma. Not only must it convey soluble food absorbed by the gut to various parts of the body, but it must also act as a carrier of dissolved gases and of waste nitrogenous matter. It has been shown that conveyance of substances is sometimes accomplished by special wandering cells which occur in it.

The ectoderm and the endoderm retain most of the functions that they have in the coelenterates. The ectoderm usually remains as the outer, protective layer of the body and, as might be expected, retains its sensitive nature and gives rise to the sensory and nervous systems and the excretory system. The endoderm forms the lining of the gut and is digestive and absorptive in function.

Animals built on this relatively simple three-layered plan in which there is no body cavity form a division of the Triploblastica, the **Acoelomata**. All the other three-layered animals are grouped together as the **Coelomata** and are distinguished from the Acoelomata by the fact that the mesoderm becomes split into two layers, a **splanchnic layer** surrounding the endoderm and a **somatic layer** underlying the ectoderm, whilst between the two layers is a more or less extensive fluid-filled cavity, the body cavity or **coelom**.

Even in the simpler three-layered animals, the **Acoelomata**, the organs of the body may attain a considerable degree of complexity and most of the structures present are unrepresented in the Diploblastica, where, as has been pointed out, except in the higher forms, the cells of the body cannot be said to form definite tissues. In

whilst the endoderm forms the lining of the mesenteron and its derivatives. There are, however, many objections to a rigid interpretation of this **germ-layer theory**. In some animals distinct germ-layers are difficult to recognise, and frequently the mesoderm consists of masses of cells (**mesenchyme**) which are not connected together to form sheets or layers. Again, there are no really sure criteria for the identification of the germ-layers in the embryo, for in different types of animals the layers are variable in position, extent, and mode of origin. Lastly, the cells of the germ-layers are to some extent interchangeable and the cells usually designated mesoderm, in many animals receive at a comparatively late stage in development, contributions from both the ectoderm and endoderm.

Nevertheless, the notion that the metazoan body is built up in the embryo from germ-layers is useful for purposes of description.

contrast to the diploblasts the structures which carry out the various physiological processes in the triploblasts are concentrated and localised in definite regions so that definite organs and systems of organs become recognisable. Localisation of function and complication of structure reach their zenith only in the higher coelomates, which are dealt with in more detail later. At this stage it will be profitable to discuss these processes only so far as they affect the organisation of the acoelomate animals.

The muscles of the body instead of being a number of more or less diffuse contractile units are localised in definite layers quite distinct from the ectodermal and endodermal epithelia. Typically the muscles are arranged in two main beds or layers, one just below the epidermis and the other around the endoderm of the gut, each

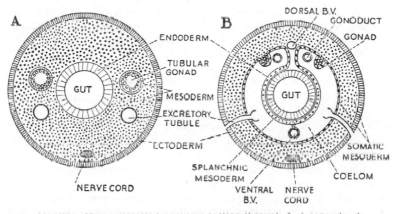

Fig. 88.   Diagrammatic transverse section through A, a generalised
Acoelomate; B, a generalised Coelomate

main layer being built up of longitudinally and circularly running fibres, and each layer is continuous throughout its extent.

The neuro-sensory system is even more clearly subject to the process of concentration, for part of it forms a compact mass of nervous tissue which, instead of occupying a superficial position, lies as a number of longitudinal tracts in deeper positions in the body. This central nervous system remains connected to the receptor organs which lie near the surface of the body and to the effector organs by a series of nerves. This concentration of the nervous tissue has resulted in a change in the form of the nerve-cells compared with those found in the diploblasts. Since the cell-bodies now lie mainly in the central nervous system, in order to maintain connection with the receptor and effector structures, one cell process, at least, must become elongated and pass out of the central nervous system to its appropriate destination. It is these long processes, each of which forms a nerve-fibre, which, when bound together by connective tissue, form the nerves of the peripheral nervous system.

The central nervous system forms a co-ordinating centre in which impulses arriving from the receptor organs by afferent (sensory) nerve-fibres are linked by means of internuncial (relay) nerve-cells to the appropriate efferent (motor) nerve-cells whose fibres transmit motor impulses to the effector organs. The number of cells in the central nervous system is very great indeed and a great variety of linkages between afferent and efferent fibres is made possible. This aspect of the activities of the central nervous system may be likened to those of a telephone exchange. The central nervous system also allows of central control, for it provides a means whereby the various effector organs can be brought into play simultaneously so that a large animal can *react as a whole* to a single stimulus and the behaviour of the animal becomes something more than a series of locally controlled responses. For this some sort of "feed-back" mechanism is required, whereby information about the state of activity of the various effectors (mainly muscles) can be supplied to the central nervous system. A system of small sense organs (proprioceptors) in the muscles and joints with afferent fibres fills this role but it must be admitted that a proprioceptive system has been histologically demonstrated in only a few groups of animals apart from vertebrates.

Usually one end, the front, of the central nervous system is swollen out to form a "brain" which in its simplest condition is merely a correlation centre for impulses entering the body from receptor organs especially well developed and concentrated at the front end. In higher animals the "brain" acquires additional functions. Complete control of all the effector organs of the body is, however, never gained by the central nervous system, even in the highest animals, for some of the organs always remain to a large extent under the local control of a part of the nervous system developed in close proximity to them. These local control centres remain in a diffuse condition and behave in a similar way to the diffuse or nerve net system of a coelenterate so that the behaviour of the organs controlled in this way consists of a series of responses to *local* stimuli. The importance of the nerve nets diminishes as that of the central nervous system increases, but in the acoelomates many of the body movements are controlled by the sub-epidermal nerve plexus (nerve net) and the behaviour of a flatworm, for instance, whose "brain" has been removed does not differ funda-mentally from that of the intact animal.

The gut of the simpler acoelomates is worthy of mention, because in its structure and mode of functioning it has many points of similarity to the enteron of the coelenterates. In the flatworms, for example, it has a single opening, the mouth, which serves for the exit of faeces as well as the entry of food. The absence of a posterior opening (anus) means that food cannot pass in a steady stream through the gut as it does in higher animals, and this may in part explain why the surface of the gut is so vastly increased in many flatworms by the development of lateral pouches. These allow of a fresh meal being taken in before the remains of the last have been passed out.

Definite excretory organs are present. They take the form of much-branched tubules, the ultimate branches of which end blindly in peculiar large cells called flame-cells or flame-bulbs. The larger branches empty into a variable number of drainage tubules, which in turn open to the exterior. Excretory organs of this type are termed nephridia, and it is probable that their primary function is the elimination of excess water from the body and that the ability to excrete nitrogenous waste has been secondarily acquired; indeed the evidence that they really function as excretory organs is meagre.

The greatest complications and variations are seen in the reproductive systems and the description of these and of other points of interest are best left until representative types of acoelomate animals are dealt with as such. But essentially each system consists of gamete-producing cells arranged in various ways to form gonads and in relation with these gonads are ducts to convey the gametes to the exterior.

# CHAPTER VIII

## PLATYHELMINTHES

The flatworms form a large group of relatively simple acoelomate three-layered animals. Many are free-living but others are parasitic on, or actually in, the bodies of larger animals. Yet they all have a common plan of organisation, and an understanding of the structure and habits of those that lead an independent life is a helpful preliminary to an understanding of the others.

### TURBELLARIA

The free-living flatworms are for the most part small creatures and those that are not actually inhabitants of the sea or fresh water

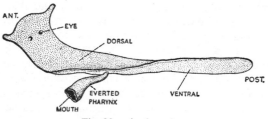

Fig. 89. A planarian.

live in damp situations, for their bodies are in no way protected against desiccation. The planarians are among the commonest examples of turbellarian worms.

**Form.**—In common with other flatworms the Turbellaria have flattened leaf-like bodies and these have a more or less oval outline. There is a definite anterior end, which is always directed forwards as the animal moves along any particular path, but apart from the presence of organs of special sense, this is not very clearly differentiated from the rest of the body and so does not form a definite "head". Not only do turbellarians move with one particular end normally in front, but one particular surface is always uppermost. This is known as the **dorsal surface**, whilst that next to the substratum is called the **ventral surface**. The various organs of the body are symmetrically disposed with regard to the mid-line so that a plane passing through this mid-line at right angles to the dorsal and ventral surfaces would bisect the animal into similar

162

right and left halves. Flatworms and other animals with this type of symmetry are thus said to be bilaterally symmetrical, and this bilaterality, which is in sharp contrast with the radial symmetry of most coelenterates and starfishes, is directly related to progression in which one end is directed forwards.

**The Body Wall.**—The body of a turbellarian is covered externally by a ciliated epithelium which, in addition to ciliated cells, contains mucous and other gland-cells. Some of the cells contain deeply staining rods or granules (rhabdites) which rapidly swell to form a protective slime when discharged. These rhabdites originate in the mesenchyme. Immediately below the epithelium lie the muscular layers of the body wall. These muscles are arranged in a way which is typical of many other invertebrate animals there being two layers of muscle-fibres. The fibres of the outer layer (i.e. the layer next to the epidermis) run in a circular manner around the body, whilst those of the inner layer lie parallel to the longitudinal axis. The two layers are termed respectively the circular and longitudinal muscles (see Fig. 97). Dorso-ventral muscles are also present, forming vertical bundles.

**The Alimentary Canal.**—The position of the mouth, although usually on the ventral surface, is variable in the different species. It opens into the gut or enteron which, in its simplest condition, is merely an endodermal sac, but which in larger flatworms acquires a complex arrangement because of the very numerous branched caeca which arise from its main branches. This complication may be regarded as a means of increasing the surface concerned with digestion, absorption, and distribution of food. In all types,* however, the gut is simple in that it has only one opening to the exterior, namely, the mouth. The first part of the gut, into which the mouth opens, is termed the pharynx. In most free-living flatworms this is a wide, sac-like region, which although normally folded inside the body is capable of eversion by being turned inside out by means of special muscles.

**The Parenchymatous Tissue.**—Lying between the gut and muscular layers of the body wall is the parenchymatous tissue or mesenchyme derived from the mesoderm. Its role is to form a packing around the viscera, and also to effect the transference of soluble substances from one part of the body to another.

**The Reproductive System.**—The reproductive system, which is also derived from the mesoderm, is exceedingly complex, and

* In a few complex forms (polyclads) one or more anal pores are present.

with few exceptions the animals are hermaphrodite; that is to say, both male and female organs are found in one and the same individual. The position and the arrangement of the ducts associated with the gonads is extremely variable in the different types of Turbellaria even in closely related species, so that it must suffice to describe a generalised type. The great diversity of genitalia can be better understood when it is realised that all are devices for ensuring cross-fertilisation between hermaphrodite animals and for the provision of the fertilised eggs with yolk and shells. Moreover, there is a common plan in the structure around the openings of the various ducts since all of them can be thought of as variants of a musculo-glandular organ (Fig. 90). This is easy to see in the

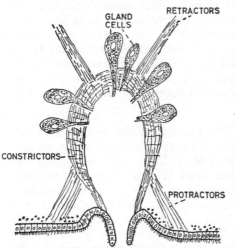

penis, for example, but applies also to parts of the female ducts. The male organs consist of paired **testes** and their ducts to the exterior. The duct (**vas deferens**) from each testis joins its fellow of the opposite side and opens into a protrusible copulatory organ, the **penis**, which in turn opens into an ectodermal pocket, the **genital atrium**. The female organs are more complicated. The paired gonads are as a rule each divided into an egg-producing region, the **germarium**, and a yolk-forming region, the **vitellarium**.

Fig. 90. Musculo-glandular organ. Vertical section.

In some species the vitellarium is completely separated from the egg-forming portion and the two organs are then referred to respectively as **vitelline gland** and **ovary**. A Mehlis' **gland** which secretes a capsule around the fertilised egg is also present. Its function is probably also to provide a substance which lubricates the eggs. The ducts from the germarium and vitellarium may be common to both or may be separate throughout their entire length. They also open into the genital atrium. Finally, it may be said that both male and female ducts may bear dilatations which respectively function as storage organs for eggs or for sperms.

**The Neuro-sensory System.**—The nervous system is greatly different from that of the coelenterates, for in addition to a superficial nerve net just below the epidermis, there is a deeper plexus embedded in the parenchyma below the muscles of the body wall. Moreover, nervous tissue is especially concentrated in a bilobed mass (**cerebral ganglia** or "brain") near the anterior end of the body and in certain longitudinal tracts or nerves in connection with the "brain". These concentrations of nervous material form the central

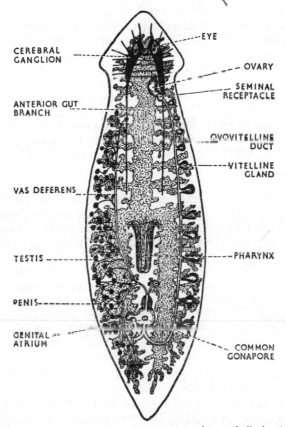

Fig. 91. Diagram to show the main organ systems in a turbellarian (a Triclad).

nervous system and act as a co-ordinating centre for the various nerve impulses. Because of this, the behaviour of the animal as a whole does not merely consist of a series of locally controlled responses and if the brain is removed locomotion, feeding, and breeding is upset. It has been mentioned above that the organs of special sense are situated mainly at the anterior end. They are in fact found just in that region where they are of most use to the animal, for the anterior end is the first part to encounter any change in the environment when the animal is moving. These sense organs are as a rule of two main types, **eyes** and **chemo-receptors**, but in a few species **statocysts** are also present.

**The Excretory System.**—The **excretory organs** consist of a number of branched tubules, the finer twigs of which end in special hollow structures called **flame-bulbs**, whilst one or more of the larger tubes opens to the exterior. These flame-bulbs are worthy of special mention, for not only are they found in all flatworms, but in other types of animals. Each flame-bulb may be a single large cell which as a rule has a number of cytoplasmic prolongations but in many flatworms several flame-bulbs are derived from a single cell to give a more complicated arrangement to the terminal twigs of the excretory tubules.

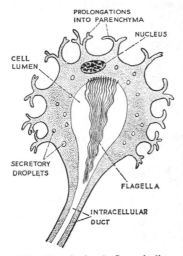

The nucleus is displaced to one side of the cell and the cytoplasm is hollowed out to form a large central cavity which is continuous with that of the finer tubules. A bunch of flagella hangs down into the cavity of the bulb and it is the flickering movement of these flagella that gives the flame-bulb its name. It will be understood from this description that the cavities of the flame-bulbs and of all except the largest tubules are

Fig. 92. A simple flame-bulb.

intra-cellular. It is probable that the soluble excretory products diffuse through from the surrounding tissues into the lumen of the flame-bulb and that the function of the flagella is to keep the fluid circulating, so avoiding any "dead space" in the fluid in the tubule, but the flame may have other uses. On the face of it, it would seem unlikely that the flagella could set up sufficient suction pressure to draw fluid into the lumen of the flame-bulb from the surrounding parenchyma, and yet there is a good deal of evidence that this is their function, or at least that they play some very important part in this process. Thus, it has been found in some land nemertines and planarians that not all the flame-bulbs are active at one and the same time, and that the flames in inactive ones can be stimulated to flicker by the injection of water into the neighbouring parenchyma, after which fluid passes down the tubule. This experiment also indicates that one important function of a flame-bulb nephridial system, perhaps the most important function, is the elimination of excess water from the body, part of the process of osmoregulation. It has also been observed that the flame-bulb system is better developed in fresh-water triclads than in their marine relatives where it is presumed there is less excess water to get rid of from the body.

Fig. 91 shows the arrangement of the main organ system in a planarian, which is one of the commonest types of all the free-living flatworms.

**Movement.**—Planarians usually glide along the substratum or the surface film by the action of the cilia of the epidermis. More active movement, which may cause the animal to swim, sometimes takes place due to the supplementary action of the muscles, which set up a regular series of ripples in the body wall. The ripples start at the anterior and pass towards the posterior end, thus helping to drive the body along when it is in contact with any surface. This muscular type of movement is especially pronounced in the larger free-living flatworms in which ciliary action would not suffice to drive the body along. Planarians do not move in even an approximately straight line, but trace out a tortuous path. This is because the head is continually bending from side to side by the alternate contraction and relaxation of the longitudinal muscles on each side of the body. Now, these muscles are under the control of the bilaterally symmetrical central nervous system which sends excitatory impulses first to one side of the body and then to the other. When a planarian is more strongly stimulated from the outside, then more impulses enter the cerebral ganglia from the sense organs in the head and excitatory impulses to the muscles are sent out in more rapid succession, with a result that the path of movement becomes more irregular.

Such an increase in response may be brought about by increasing the strength of any relevant stimulus. For example, if a piece of food such as a dead shrimp is placed in the water, the chemo-receptors are strongly stimulated, and increasingly so as the food is approached. As a result of this the nearer the animal approaches to the food the more tortuous becomes its path, but when within a few centimetres of the food the behaviour changes, and the animal moves in a practically straight line towards the food and then engulfs it. The significance of this type of behaviour is quite clear, for the apparent random movements of the animal sooner or later bring it near to, and keep it in the proximity of the food. Other stimuli act in much the same way. The slow, wiggly path of a planarian tends to keep it beneath the stones under which it normally spends its time during the day, for planarians hunt for food only by night. If by any chance the animal wanders out from beneath the stone it becomes stimulated by the light and immediately its path becomes more wiggly. As in the previous instance this limits

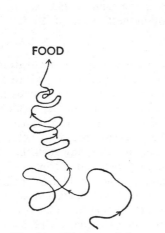

FOOD

Fig. 93. Diagram of the path of a planarian approaching food.

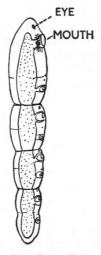

EYE

MOUTH

Fig. 94. Asexual reproduction in turbellarians —a short chain of zooids of *Microstomum*, which are formed in pairs. Each will later separate and form a new individual.

its sphere of movement and increases the chances of its again moving into cover beneath the stone.

**Feeding.**—Most free-living flatworms, including planarians, engulf their food by enclosing it in the everted pharynx, but a few merely suck up small particles or juices by means of a muscular pharynx which is not capable of eversion. All are carnivorous and in all types digestion takes place in much the same way as in coelenterates by two distinct phases, an extra-cellular and an intra-cellular, and in some species which attack large prey digestive juice is poured on to the food before it is engulfed, thus making swallowing easier.

**Gaseous Interchange.**—Free-living flatworms are fairly active animals and therefore require continual supplies of oxygen, and at the same time they have to get rid of carbon dioxide. As in coelenterates the gaseous interchange is carried out over the whole surface. It will be realised that this simple method of respiratory exchange is possible only in creatures whose external surface is large relative to the size of the body. This condition is fulfilled in all flatworms, which in addition to being of a small size (as a rule) are so extremely flattened that their surface is large compared to the body bulk.

**Life-history.**—The method of reproduction is subject to a good deal of variation, but despite the hermaphrodite condition of the gonads, cross-fertilisation is the rule. The fertilised eggs of some species hatch out small free-swimming ciliated larvae which after a time metamorphose into adults. In others the larval stage is omitted and the eggs develop directly into miniature adults.

**Asexual Reproduction and Regeneration.**—Some of the simpler turbellarians (rhabdocoels) reproduce asexually by forming chains of zooids (Fig. 94), each of which breaks free and attains full size. Some planarians can multiply by transverse fission—literally tearing themselves apart. The hinder end fixes itself to the substratum and then the front part jerks forward so strongly that the body breaks. Each part then regenerates the lost parts.

In view of this it is not surprising to find that many, perhaps most, planarians have a great ability to regenerate parts lost by damage. Quite small pieces can regenerate the whole worm, although with progressive reduction below a certain size, first the head and then more posterior parts fail to regenerate.

## PARASITIC FLATWORMS

**Introduction.**—Recent work on larval structure has shown that the traditional divisions of the parasitic platyhelminthes is unacceptable. Three main types are now recognised, **monogenetic flukes (Monogenoidea)**, **tapeworms (Cestoda)**, and the **digenetic flukes (Trematoda)**. The Monogena are separated from the Trematodes by the structure of the larval haptor. They are **ectoparasites** and so both adults and larvae live on, or near, the outsides of their hosts and they have simple life-histories. The tapeworms are specialised **endoparasites** and when adult they usually live in the gut. The life-cycle is complex, requiring an intermediate host in which the larvae often form infective cysts in the muscles. Asexual reproduction often occurs at this stage. The digenetic flukes are endoparasites with a very complex life-cycle which often involves two intermediate hosts, one of which is often a snail. In their adult phase their body form resembles the Monogena. They are of great economic importance.

## ECTOPARASITIC FLUKES—MONOGENOIDEA

**Form and Structure.**—A common ectoparasitic monogenetic fluke is *Polystoma*. It will quite possibly be found by the student during the dissection of the frog, for the parasite is sometimes found in the cloaca or bladder. More rarely, it occurs attached to the gills

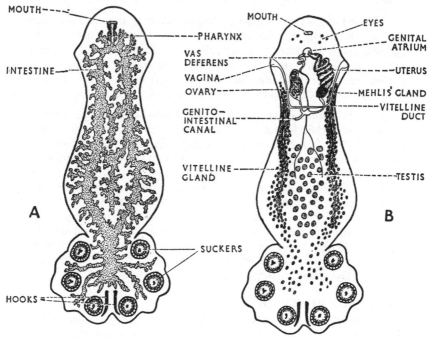

Fig. 95. POLYSTOMA.—A, diagram to show the gut. B, diagram to show the reproductive system.

of tadpoles. The internal structure and systems of organs do not depart very greatly from the plan described for the Turbellaria or from that presently to be described in some detail for the liver fluke. The most noticeable differences from the turbellarian structure, however, are to be seen in the external features. Certain cells of the mesenchyme secrete a thick "cuticle" and the cells then sink inwards into the body and no longer form a continuous layer (see p. 173). Cilia are completely absent. The posterior end of the body bears six suckers as well as a number of curved hooks, so that the parasite is enabled to cling securely to its host. The parasite feeds by sucking juices from the tissues of its host, so that it is not surprising to find that an eversible pharynx is absent and, in fact, the pharynx is represented by a small suctorial bulb. From the pharynx a short oesophagus leads into the intestine, which has two main branches passing to the posterior end of the body. These branches have many lateral caeca and unite at the posterior end of the body.

The **excretory system** calls for little comment, but it may be mentioned that there are two main lateral excretory canals which open outwards towards the anterior end of the body.

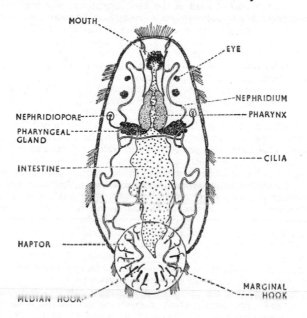

Fig. 96. POLYSTOMA.—The onchomiracidium.

The **nervous system** also resembles that of the Turbellaria and in some ectoparasitic flukes **eyes** are also present, but they are never as well developed as in the Turbellaria. There are two groups of simple eyes near the anterior end in *Polystoma*.

The **reproductive system** is arranged on the same fundamental plan as that outlined for the Turbellaria. There is a single follicular **testis** lying medianly in the body. From it the **vas deferens** passes forward to open by the **penis** at the **genital atrium**. The ducts from the **ovary**, from the **vitelline glands**, and from the **shell gland** converge to a point about one-third from the anterior end of the body, and from this point of fusion of the ducts, two other ducts arise. One, the **uterus**, passes forward to the genital atrium; a second, the **genito-intestinal canal**, opens into the right limb of the intestine. There are, however, two other ducts, the paired **vaginae**, which open right and left of the mid-line on the lateral surface towards the anterior end of the body. They lead into the vitelline ducts and serve as an entrance for sperms from the same or another fluke, self-fertilisation often taking place.

Life-history.—The egg capsules, each containing one egg and a few yolk-cells, are laid in the early spring and from them hatch small larvae with five hoops of cilia, four eye spots, and a large posterior disc bearing sixteen marginal hooks. The rudiments of most of the adult organs are also present. These larvae, **onchomiracidia (gyrodactyloids)**, swim actively by means of cilia but have a life of only forty-eight hours unless they are successful in finding a new host which must be, not a fully grown frog, but a tadpole in the correct stage of development. The future course of development seems to depend on the precise stage of growth reached by the tadpole host. If a larva comes near to a tadpole in which internal gills (p. 730) are already formed it enters the opening into the gill chamber and clings to the gills by means of the hooks. There, feeding on mucus and blood, it lives and grows for about two months until the tadpole is about to metamorphose into a tiny frog. The gill parasites then leave the gills and migrate down the gut into the cloaca and become fully mature in about three years. Now it sometimes happens that the larvae after hatching come into contact with much younger tadpoles in which the external gills are still present, that is, tadpoles which are less than about a fortnight old. When this happens the larvae attach themselves to the external gills, grow extremely rapidly, and attain sexual maturity without developing the full complement of adult characters. This state of affairs, in which a larva becomes precociously sexually mature, is called **neoteny**. The neotenous larvae are capable, after from fifteen to eighteen days, of producing eggs which are self-fertilised and which will develop into a new generation of larvae which seem structurally identical with those hatching from the eggs of truly adult flukes. Yet they are physiologically different, as is shown by their behaviour, for no matter what the age of the tadpoles on which they settle, none is capable of giving rise to neotenous larvae and they can develop into internal gill parasites only.

Thus it appears that the course of development can be influenced by the growth stage of the host and this effect probably makes itself felt by means of the changing hormone balance in the blood of the growing tadpole. On the other hand, the larvae which hatch from the eggs of the neotenous, external gill parasites, have a more inflexible course of development and seem to be uninfluenced by the growth stage of the tadpole.

## CESTODA

**Introduction.**—Most tapeworms are long ribbon-like flatworms in which the body is made up of a number of segments.* Tapeworms are found as gut parasites in a great variety of vertebrate hosts. A few, however, resemble the flukes in being leaf-like and unsegmented, and also in their general anatomy, but tapeworms

* The terms **segments** and **segmentation** have come to have a special meaning to zoologists. They are usually retained to describe **metamerism** (see p. 221), a phenomenon which is not found in tapeworms. It should be appreciated, therefore, that in this description the term **segment** is used solely for convenience and in no way implies metamerism.

have no trace of an alimentary canal. The epidermis is lost in the early stages of development and thus neither ectoderm nor endoderm are represented in the adult which, however, in all other features is clearly at the triploblastic level of organisation, although the whole of the body appears to be built from derivatives of the mesoderm. The great length of the better-known tapeworms is not due to a simple elongation of the body, but to the continual addition of new segments or **proglottides** from a region of proliferation situated near one end of the animal. Each **proglottis** is in many ways comparable to a complete flatworm in that it contains an hermaphrodite set of reproductive organs and portions of the

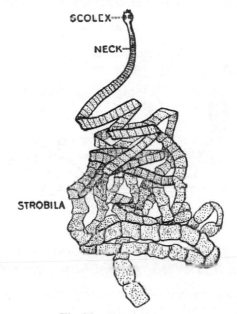

Fig. 97. TAENIA SOLIUM.

excretory and nervous systems which are common to all the segments.

## TAENIA SOLIUM

**Form.**—*Taenia solium*, formerly a common intestinal parasite of Europeans, is a ribbon-like animal which may reach a length of 5 m. One end of the "tape" or **strobila** is attached to the intestinal wall by the "head" or **scolex**. This is a small muscular knob which bears four hollow suckers and a terminal cone, the **rostellum**. On the rostellum is a double row of curved hooks made of chitin plus tanned protein. Just behind the scolex is a narrow region, the "neck" or region of proliferation, in which segmentation is not very obvious. This is because the segments in this region are small and newly formed. The youngest proglottides are therefore found immediately behind the "neck", whilst the oldest are at the other (the posterior) end.

**The Nervous System.**—The nervous system consists of a transverse ganglionated (*i.e.* containing nerve-cells) band in the scolex, from which slender nerves are given off to each of the suckers, and also two stouter nerves which run the whole length of the strobila, one on each side near to the lateral margin of the proglottides.

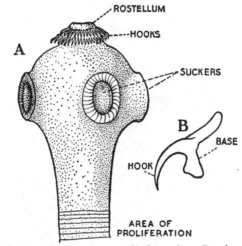

Fig. 98. Taenia solium.—A, the scolex; B, a hook.

**The Excretory System.**—The excretory system consists of many flame-bulbs lying fairly superficially in each proglottis and of a number of drainage tubules. These empty into the main excretory tubules, which run laterally in the proglottides throughout the whole length of the animal. In the anterior region there are usually two canals on each side, an upper and a lower one, but towards the posterior end only one large one on each side is found. The lateral canals join in the mid-line in the scolex and they are also linked together in each proglottis by a transverse canal. The lateral canals open to the exterior by a median pore at the hinder margin of the last segment in the young worm before any proglottides have been detached but, once this has happened, the ends of the lateral canals act as excretory pores.

**The Proglottides.** Each proglottis contains a complete hermaphrodite set of genitalia, but the male organs develop before the female organs so that anterior segments have functional male organs only; those somewhere near the middle of the body have both male and female organs in a state of maturity, whilst in those at the posterior end the male organs have degenerated and the whole of the proglottis is practically filled by the enlarged uterus, which is full of fertilised "eggs" (really embryos inside the egg capsules).

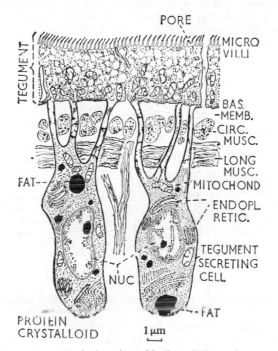

Fig. 99. TAPEWORM.—Vertical section of body wall (from electron micrograph, modified after Threadgold).

A proglottis from the middle region of the strobila is rectangular in outline and covered by a thick layer, usually described as a "cuticle", but which is in reality a layer of living cytoplasm containing mitochondria and bearing on its outer surface a brush-like layer of microvilli. In fact the "cuticle", like that of trematodes, seems to be a superficial extension of certain cells of the mesenchyme and is now called the **tegument**. Immediately below the tegument is a basement membrane and then follows a layer of circular muscles and below this again a much thicker bed of longitudinal muscles.

The "cuticle"-forming cells are found just below the longitudinal muscles embedded in the outer part of the parenchyma outside a second layer of longitudinal muscles. The tegument is perforated at intervals by fine canals at the bottom of some of which are the openings of gland-cells, and of others, free nerve endings. Other minute canals which perforate the tegument branch and open into the parenchyma. They are of two types, (a) those which represent gaps between the tegument-forming cells and (b) those which open into these cells. These canals together with the numerous vacuoles in the outer layers suggest a system of pinocytosis for taking substances into the proglottid of the worm, whilst the microvilli greatly increase the absorptive surface. Yet another layer of muscles,

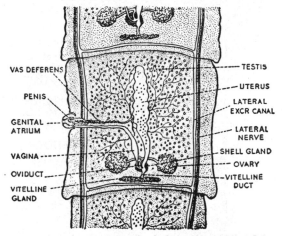

Fig. 100. Taenia solium.—A proglottis from the middle of the strobila.

whose fibres run transversely from one side of the proglottis to the other, is found below the inner longitudinal muscles. Towards each lateral margin is found the longitudinal nerve and just median to this, the longitudinal excretory canal. The transverse excretory canal lies towards the posterior end of the proglottis. Practically all the central part is occupied by the reproductive organs. Large amounts of glycogen are stored in the proglottides.

**The Reproductive System.**—The male organs consist of a very much subdivided testis from which sperm is collected by small efferent ducts, and a coiled vas deferens which terminates in a protrusible penis at the genital atrium. The female organs comprise a bilobed ovary, lying towards the hinder end of the proglottis; a median vitelline gland and a shell gland surrounding the base of the uterus and oviduct. The short oviduct is joined by the vitelline duct and from this point a narrow tube, the vagina, runs obliquely forward to open at the genital atrium, whilst from the base of the oviduct a blind sac, which in more mature proglottides enlarges and becomes branched, passes towards the anterior end. This sac is the uterus, and in some of the more primitive tapeworms it also opens to the exterior.

**The Life-history.**—The sperm travel down the vagina, and it is now reasonably certain that the eggs are fertilised by sperm from the same proglottis. No single host would be able to support two of the larger tapeworms in its gut, so that the possibility of cross-fertilisation between two such worms is precluded. The eggs are fertilised in the lower part of the oviduct and on their passage to the uterus they become enclosed, together with a large yolk-cell, in a thin egg-shell. The so-called shell glands are not directly concerned in shell formation which is due to material exuded from the yolk-cell. In the uterus, development of the egg begins straight away and the first cleavage of the egg separates a larger cell, the megamere from an embryonic cell. The megamere divides a few times and its descendants become filled with yolk from the

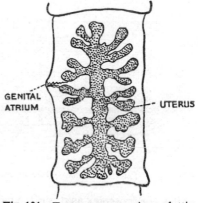

Fig. 101. TAENIA SOLIUM.—A proglottis from near the end of the strobila.

vitelline cell and surround the descendants of the embryonic cell in a kind of envelope. The embryonic cell divides to form a ball of cells the outer ones of which are larger and are termed mesomeres. These in turn secrete a hard case, the **embryophore**, around the smaller embryonic cells which in turn form the embryo proper. This develops six hooks and is known as a **hexacanth embryo**, whilst the whole structure with the egg-shell and embryophore is termed an **onchosphere**. The ripe proglottides at the extreme end of the strobila contain embryos in this stage of development.

The terminal proglottides drop off from the strobila four or five at a time and pass out passively with the faeces of their host. They remain alive for a while on the ground and can perform limited wriggling movements, but eventually they die and disintegrate. The embryos, however, remain infective for some considerable time but do not develop further unless they are eaten by a suitable secondary host, which normally is a pig, but may be another animal or even man. In the stomach of the pig, the egg-shell and embryophore are digested away and the hexacanth is released. By means of its six hooks it rapidly bores its way through the wall of the gut and enters the blood-stream. After passing through the heart it is carried to the tissues, where it becomes encysted. This usually takes place in muscle, very frequently the muscles of the tongue; but other situations may become infected. The parasite then grows to form a fluid-filled bladder, the **cysticercus** or **bladder-worm**. At one point on the surface of the bladder the wall invaginates and at the bottom of the invagination, a small scolex, which resembles that of the adult worm in all essentials, with the exception that its parts are inverted, develops. Later on, when the

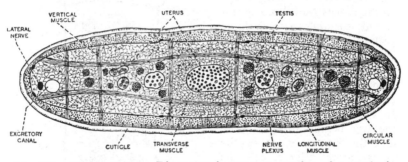

Fig. 102. TAENIA SOLIUM.—Diagrammatic transverse section of a proglottis.

scolex (or **proscolex** as it is often called) is fully formed, the invagination is everted. Pork infected with bladder-worms has a spotted appearance and for this reason is said to be **measly**, but the parasites do not develop further unless measly pork is eaten after being cooked insufficiently to kill them. If, however, pork infected with living parasites is eaten by the final host, a man, the bladder is digested and the young scolex attaches itself to the wall of the intestine, develops a region of proliferation, and begins to bud off proglottides.

### OTHER TAPEWORMS

Although *Taenia solium* is usually known as the common or pork tapeworm, in point of fact it is now very rare. Other tapeworms are, unfortunately, more common, especially on the Continent. All of them have life-histories which are essentially comparable to that of *T. solium*. Thus, the beef tapeworm, *T. saginata*, which may reach a length of twenty feet, and may be distinguished from *T. solium* by the absence of hooks from its club-shaped scolex, is a gut parasite of man, and has as its secondary host, cattle. *T. saginata* is especially abundant in Abyssinia, where it is stated that 90 per cent. of the population

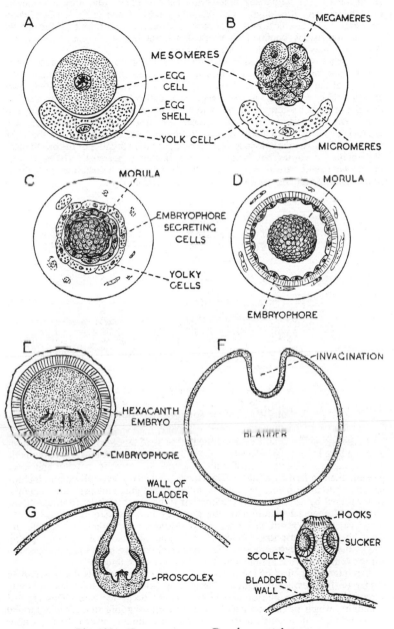

Fig. 103.   TAENIA SOLIUM.—Developmental stages.

are infected.   The commonest tapeworms met with in everyday life are, however, those which parasitise dogs.   The secondary host of one of these (*Dipylidium caninum*) is the dog flea, and of another (*T. serrata*), the rabbit.   Many tapeworms are strictly confined to particular hosts but some, e.g. *Echinococcus granulosus*, may be found in a variety of hosts.

The adult of *Echinococcus granulosus* is a small worm, with only about four proglottides, and is normally found as a relatively harmless parasite in the intestine of the dog.   The bladder-worm, which is an enormous structure known as a **hydatid cyst**, is found in the sheep.   The cyst owes its large size to the fact that the original bladder develops many proscolices and these in turn grow and also form proscolices.   In fact, several generations of bladder-worms are produced, and as the hydatid is often found in vital organs such as the liver and brain, very serious lesions are set up by it, not only because of the mechanical disturbances it causes but because of toxins that it secretes.   Hydatid cysts are by no means rare in man where he lives in close association with dogs, when the onchospheres are easily taken in by accident.

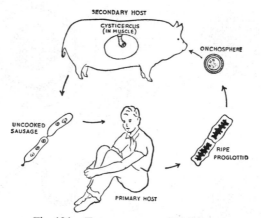

Fig. 104.   TAENIA SOLIUM.—Life-history.

**Pathology.**—The details of the clinical aspects of tapeworm infections cannot be given here but it may be mentioned that, in common with most other parasitic worms, tapeworms give rise to far more severe symptoms in children and weak patients than in healthy adults.   The effects are usually generalised and include nausea, abdominal pains, nervous disorders resembling epilepsy, and anaemia, accompanied by an increase in the number of eosinophil white corpuscles in the blood.   Mechanical irritation of the gut may be so severe as to cause antiperistalsis, with a result that mature proglottides may enter the stomach and there liberate onchospheres, which often invade the tissues of the body, giving rise to the disease, cysticercosis.   The cysticerci seem to have some predilection for the central nervous system, and the disease is often fatal.

An interesting effect of infection by the larger tapeworms is that it seems to confer on the host some degree of immunity from further infection, for only rarely do more than one or two larger taenias occur in man.   This type of immunity, which persists only as long as the primary infection lasts, is termed **premunition**, to distinguish it from the lasting immunity which develops after some bacterial diseases.

## ENDOPARASITIC FLUKES—DIGENEA

Members of the order Digenea, the endoparasitic flukes, are all internal parasites, and some are responsible for serious loss of life in man and domestic animals. The adult flukes are typically parasites of vertebrate animals; but one stage, at least, of their life-history is spent in an invertebrate host, or hosts, of which one is almost invariably a mollusc. This alternation of hosts is responsible for the naming of the order Digenea, members of which are said to have digenetic life-histories. The structure as well as the main outlines of development is fairly constant throughout the whole

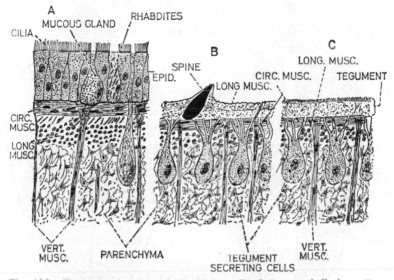

Fig. 105. Transverse section of the body wall of A, a turbellarian, B, a trematode; C, a cestode; as seen in the light microscope.

group, and will be more readily appreciated after a description of one of the best-known examples, *Fasciola hepatica*.

## FASCIOLA HEPATICA

This parasite is an inhabitant of the bile ducts of the sheep and therefore receives the popular name of **liver fluke.**

**Form and Structure.**—Like other flukes it is a leaf-like creature, its size varying from about a half to one inch in length, whilst the width is about half an inch. The anterior end is drawn out to form a conical projection, the head lobe, at the apex of which is situated the oral or anterior sucker perforated by the mouth. Another saucer-shaped sucker—the ventral sucker—is found on the ventral surface, a short distance behind the head lobe. The excretory aperture is single and lies at the extreme posterior end of the body. The body wall is built on the same lines as that of other flatworms, but differs from previous types in that the "cuticle" is very thick.

It appears from recent work with the electron microscope that the "cuticle" of light microscopy is in reality a surface layer of syncytial cytoplasm formed as an extension of "cuticle"-forming cells lying

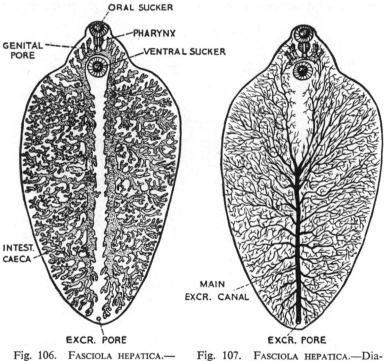

Fig. 106. FASCIOLA HEPATICA.—    Fig. 107. FASCIOLA HEPATICA.—Dia-
Diagram to show the gut.            gram to show the excretory system.

in the parenchyma. This outer layer of cytoplasm contains pinocytotic vesicles, mitochondria, endoplasmic reticulum, vacuoles of various sorts, and at intervals, spines formed of hardened protein. The "cuticle", which is better called by the non-committal word **tegument**, is connected by long necks to the cells which form it. These cells lie below the layer of longitudinal muscles (Fig. 108).

The tegument can absorb glucose and this shows that it is not impermeable. Moreover, experiments with radioactive tracers show that substances can pass from the gut outwards and through the cuticle.

The origin of the "cuticle"-forming cells is still uncertain but they probably belong to the mesenchyme.

As in *Polystoma*, cilia are absent, the epidermis having been finally lost after the last larval (**cercaria**) stage. Below the tegument is a layer of circular muscles and below these again, adjoining the parenchyma, a layer of longitudinal muscles. As in all other flatworms the spaces around the internal organs are filled by the peculiar branching cells of the parenchyma.

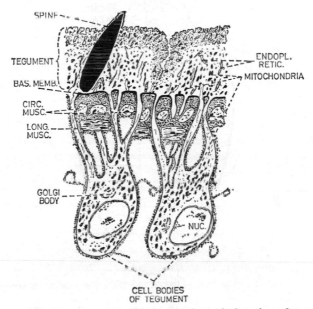

Fig. 108.   FASCIOLA HEPATICA.—Diagrammatic vertical section of outer layers of body from electron micrographs (based on Threadgold, 1963).

The Alimentary Canal.—The mouth, as has already been mentioned, lies in the middle of the anterior sucker. It opens into a short but muscular suctorial pharynx. The oesophagus is very short and almost immediately joins the intestine. This has two main branches and numerous branched caeca which run throughout practically every region of the body, and between which other organs are wedged. The walls of the oesophagus and intestine are composed of a columnar epithelium outside of which is a delicate layer of circular and longitudinal muscles.

**The Excretory System.**—The main excretory canal opens to the exterior at the posterior end of the body. Throughout, it receives many large ducts, which in turn drain smaller tubules whose ultimate branches end in flame-bulbs.

**The Nervous System.**—The nervous system is noteworthy if only for the fact that it is well developed—a condition that is difficult to explain, for sense organs are apparently absent and the sluggish movements which the animal makes would not seem to need such

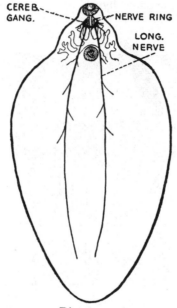

Fig. 109.    FASCIOLA HEPATICA.—Diagram to show the central nervous system.

a large correlation centre. However, there it is! The cerebral ganglia are prominent masses of nervous material and are joined together by a nerve ring around the oesophagus. From the ganglia nerves are given off to the head lobe and to the hinder end of the body, and of these latter, one pair, the lateral nerves, are much larger than any of the others.

**The Reproductive System.**—Like that of most other flatworms the reproductive system is hermaphrodite. The male gonads consist of a pair of very much branched tubules, from the walls of which spermatozoa are shed into the lumen. One **testis** lies just anterior to the second, and the two together occupy a good deal of

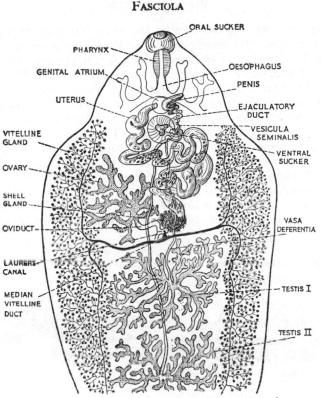

Fig. 110. FASCIOLA HEPATICA.—Diagram of the reproductive system.

the space in the middle of the body. From each testis a duct, the vas deferens, runs forward and then fuses with its fellow to form a wide coiled tube, the vesicular seminalis. This in turn opens by an

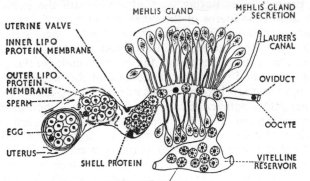

Fig. 111. FASCIOLA HEPATICA.—Diagram to show the relations of the genital ducts.

ejaculatory duct into the duct of the penis which opens into the
genital atrium. The female organs are less complicated than those
of the Monogenea. The ovary is single and lies to one side of the
mid-line, about a third of the way from the anterior end of the body.
From it a short oviduct carries the eggs downwards to a point where
the ducts from the right and left vitelline glands (which are diffuse
glands along the sides of the body) converge to the mid-line and there
fuse to form a median vitelline duct. The vitelline duct and the
oviduct join and from this point a wide coiled tube, the uterus, runs
forward and opens into the genital atrium. A group of "shell
glands" surrounds the base of the uterus. Laurer's canal, which
runs from the point of fusion of the median vitelline duct and oviduct
and opens on the dorsal surface, has as one of its functions the
conveyance to the exterior of excess yolky cells, and possibly eggs.

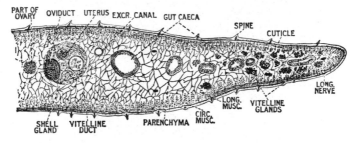

Fig. 112.    FASCIOLA HEPATICA.—Diagrammatic transverse section of part of
the body.

**Life-history.**—It is highly probable that cross-fertilisation is the
rule, pairing flukes having been observed with the penis of one
inserted into the uterus of the other but in most flukes, including
*Fasciola*, copulation takes place by the penis of one partner being
inserted into the Laurer's canal of the other. The eggs, after
fertilisation in the lower part of the oviduct, undergo their maturation divisions and pass towards the base of the uterus; but before
they enter it they become surrounded by yolk-cells. Each egg,
together with a number of yolk-cells (which fairly soon break down
to form a mass of yolk) then becomes enclosed in a shell, which is
formed by droplets of secretion which exude from the yolky cells,
and the shell becomes hardened as the eggs enter the uterus. The
hardening is due to the action of a quinone which tans the protein
(cf. insect cuticle). A small lid or operculum which serves as an
exit for the larva is found at one pole of the egg-shell.

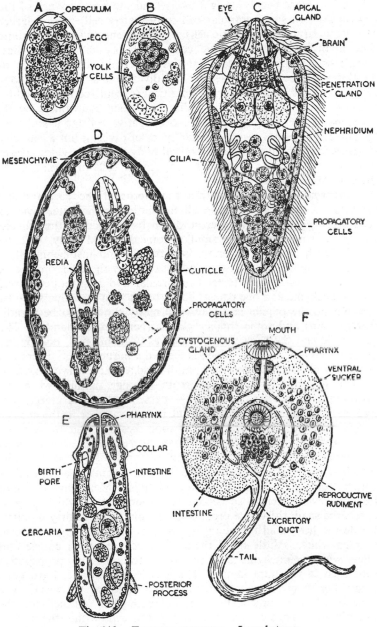

Fig. 113. FASCIOLA HEPATICA.—Larval stages.

Development begins straight away, and the first cleavage of the egg separates a small granular cell (**propagative cell**) from a larger cell (**"ectodermal"** or **somatic cell**). Subsequent divisions for some time involve only the somatic cell, which, as its name implies, gives rise to much of the larval tissues. Only relatively late in development does the propagative cell divide and when it does so it forms two cells, one of which is identical (except for being slightly smaller) with its parent cell, whilst the other contributes by its divisions to the larval body. The propagative cell (as it still may be called) undergoes several similar types of division but eventually it passes to the hinder end of the larval body where it divides several times to form a mass of smaller propagative cells or germ-cells as they are often called.

Meanwhile, whilst still within the uterus of the parent fluke, the body of the larva has been differentiating. The outer layer of cells arranged in five rings around the body become the ciliated epidermis which encloses a smaller conical-shaped body. At the broader end, which is anterior, is a papilla and just behind it, a pair of eye spots. Within the body just below the epidermis lies a delicate layer of muscles formed of circular and longitudinal fibres, the mesenchyme and one pair of flame-bulbs with their ducts, whilst opening on the papilla is a gland (formerly thought to be a rudimentary gut) and some smaller glands (penetration glands). The germ-cells and their derivatives, as already mentioned, occupy the posterior end of the larva which is termed a **miracidium**. There is a larval "brain" at the base of the papilla and a simple nervous system.

The miracidium is released from the egg capsule when the operculum opens after the eggs have passed out of the uterus and, by means of its cilia, it swims about in the water or in the surface film of moisture on vegetation. It soon dies if it fails to reach its secondary host, which is usually *Limnaea truncatula*, a snail which is common in fresh water or damp pastures. If successful in its search it sucks on to its host and by the enzymatic action of its penetration glands digests the host epidermis. Meanwhile the larva loses its own epidermis so becoming a **sporocyst**, in which stage it enters the snail, the whole process taking only about thirty minutes. Inside the host snail the young sporocyst rapidly enlarges to form a stage which is little more than a sac covered by a cuticle and filled with mesenchyme and the derivatives of the propagative cells seen in the miracidium. These cells, meanwhile, have gone some way to giving rise to a third type of larva, the **redia**, by a process which is practically identical with that outlined for the miracidium. Thus each of the propagative cells within the miracidium has divided into a somatic cell and a propagative cell, the latter giving rise to a new generation of propagative cells, to mesenchyme, and to a small gut whilst the somatic cell has formed the epidermis and certain other larval structures. Each redia is an elongated larva with a small mouth, a suctorial pharynx, and a simple intestine. A short way behind the anterior end, the body is produced outwards to form a muscular collar, whilst near to the posterior end is a pair of muscular projections (posterior processes). Just behind the collar is a birth pore through which the next generation of larvae can

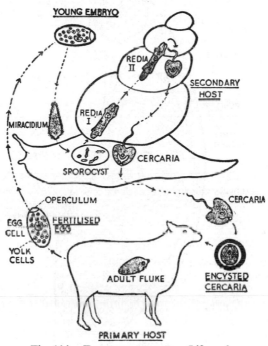

Fig. 114. FASCIOLA HEPATICA.—Life-cycle.

emerge. The propagative cells lie in the cavity of the redia and, in the same way as in earlier types of larvae, they undergo development to form further rediae; but the production of a second generation of rediae takes place in the summer months only, and during the winter the propagative cells of the primary rediae give rise to the final type of larva, the **cercaria**, by cell divisions which follow the usual pattern.

Before this takes place the rediae pass out of the sporocyst, and by means of the collar and posterior lobes of the body they slowly migrate through the tissues of the snail to the digestive gland. It is here that the cercariae are formed inside the rediae. The cercaria, each of which is formed from one of the propagative cells of the redia in the typical manner, is a small heart-shaped larva (just over 0·5 mm long) bearing a long tail. The rudiments of most of the adult organs are present, there being a small anterior sucker perforated by the mouth as well as a ventral sucker. There is a suctorial

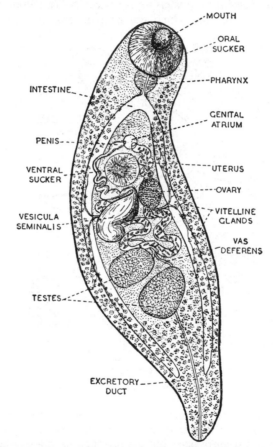

Fig. 115. DOLICHOSACCUS RASTELLUS.—Seen as a transparent object. (Drawn from an actual specimen.)

pharynx and a Ⴖ-shaped intestine, paired excretory tubules and flame-bulbs, as well as the rudiments of the reproductive organs formed by divisions of the propagative cells. The cercariae, when mature, pass out of an aperture in the side of the redia and migrate from the digestive gland to the pulmonary chamber of the host, and finally, to the exterior. After their escape each secretes a cyst around the body by means of special cystogenous cells which lie in groups just below the epidermis. Encystment in the vast majority of instances takes place in water and the encysted larvae can remain quiescent for periods up to twelve months. More rarely, the cercariae encyst on vegetation, but if this happens survival is limited to a few weeks.

Further development only takes place if the cysts are swallowed by an appropriate final host (usually a sheep). The cyst walls are digested and the young flukes which emerge, bore through the wall of the gut and enter the body cavity of the host. After about one or two days they infect the liver by boring through its outer capsule. They feed on the liver cells but cause bleeding, often with serious results to the host. Eventually the flukes enter the bile ducts and grow to maturity, when the life-cycle is complete.

One or two points in this life-history are of theoretical interest. It will be remembered that the rediae arose by the division, differentiation, and subsequent growth of cells which are the direct lineal descendants of the propagative cells, which are set aside at the first division of the fertilised egg. There is thus from the very earliest stage in the development of the zygote a distinction between germinal and somatic cells and the germinal cells alone take part in the production of the subsequent generations of larvae. The formation of the secondary rediae and cercariae follows similar lines, with a result that the germ-cells of the cercaria, and therefore of the adult, are direct descendants of the germ-cells of the previous generation of adults. These facts have given rise to the view that the asexual multiplicative phases of the trematode life-history are to be regarded as an instance of **polyembryony**; part of the substance of the fertilised egg giving rise to successive generations of larvae, whilst part give rise by direct division to the germ-cells of the next generation.

Fig. 116. DOLI-CHOSACCUS RASTELLUS. — A cercaria.

More in line with the majority of other digenetic trematodes, in that the life-history requires two intermediate hosts for its completion, is *Dolichosaccus* (= *Opisthoglyphe*) *rastellus*. This is a rather small and fairly common fluke living in the intestine of the common frog. The eggs are passed out into water where the miracidia hatch and then enter the first intermediate host, a water snail, *Limnaea ovata*. After some time cercariae are released from the snail and after entering the tissues of the larvae of certain

Trichoptera (caddis-flies), Odonata (dragon-flies), or Ephemeroptera (may-flies) they encyst to form metacercariae which infect frogs and grow into adult flukes when the insect larvae containing them are taken as food. It may be mentioned in passing that *Dolichosaccus* is in many ways a more convenient fluke to study than *Fasciola* since it is readily obtainable and its internal organs are far easier to observe (see Fig. 115).

**Pathology and Prevention.**—When sheep are infected with liver flukes the liver becomes seriously affected in structure, and its functions are upset. The sheep is said to be suffering from "liver rot", but the symptoms of the disease are not confined to the liver, and the animals become dropsical and there is great muscular weakness. It is therefore of great importance that such a serious disease should be kept well in check, and this can only be done effectively if the life-history of the parasite is taken into account. If this is borne in mind, several possible methods of preventing the disease from spreading may be devised. First, the parasite may be attacked in the adult stage when it has infected the sheep. This method has not been found to be very practicable. Alternatively, the parasite may be attacked indirectly in its larval stages by the destruction of its secondary host, *Limnaea*. This may be carried out in various ways—by the introduction of ducks to the pastures, when the snail population is soon reduced by being eaten by the ducks; by draining the pastures, when the pond snails cannot survive. This latter method is the one that is usually adopted because of its efficacy.

Neglect of commonsense principles in the selection of grazing grounds for sheep can greatly aggravate outbreaks of fluke disease. For example, during periods of drought shallow ponds and streams dry out and expose formerly inaccessible regions for feeding, regions particularly liable to have a dense population of cercariae. Moreover, cercariae emerge from the snails mainly in July and early August, the months in which droughts are most likely to occur. Obviously sheep should be prevented from grazing on such areas if there is any chance of infected snails having lived there, a particularly important point to bear in mind in view of the practice in some parts of this country of bringing sheep from the hills to graze on the marshes in the summer.

# CHAPTER IX

## ROUNDWORMS—NEMATODA

### INTRODUCTION

To write a short account of the anatomy of the nematodes is in some respects easier than for any other major class or phylum. This is not only because nematodes are among the most structurally simple of all worms but because none of them departs materially from the basic body plan of the phylum. Of no other large group of animals is it possible to speak with such justification of a selected species as a "typical" member. But, because of their vast numbers of species and monotonous structure, classification into categories larger than families is difficult and there is no universally accepted scheme. Of all metazoan classes of animals nematodes are the nearest rivals of insects in absolute numbers and easily outstrip them in the variety of ecological niches they occupy. It is tempting, because of their grimly fascinating life-histories, to think of nematodes mainly as a class of parasitic animals, but such a view is totally misleading. By far the greater number of species are free-living, inhabiting the sea-floor, the bottoms of rivers and lakes and ponds, teeming in soils of all kinds, flourishing in hot springs or icy tundra, and extending geographically from north to south pole. Yet it is equally true that there is a formidable array of parasitic nematodes. Practically every plant or metazoan animal has its quota of nematode parasites. Indeed, in this country the damage done to crops by these worms now exceeds that due to insect pests, whilst all our stock and domestic animals suffer ill-health from their attentions. There can be little doubt that the severity of the effects of nematode infestations are greatly aggravated by monoculture and overcrowding and only a sound ecological approach will help to mitigate them.

Some idea of the numbers in which nematodes can occur is given by the figures which follow:

Inter-tidal marine muddy sands—up to 5 million per square metre.

Arable land—up to 6 billion per acre.

Roots of a single potato plant—up to 40,000.

Despite their abundance, nematodes are never conspicuous and remain largely unnoticed for the very good reasons that the free living species and the plant parasites are nearly all of them of microscopic size, whilst the animal parasites, some of which are large, are hidden from view in internal organs or cavities, the rigid conformity

193

to the nematode body plan precluding the formation of hooks and suckers sufficient to allow them to become ectoparasites of animals. Few students will, however, have missed seeing *Proleptus* sp. in the gut of dogfishes or *Rhabditis maupasi* in the nephridia of earthworms. Most free-living nematodes are saprozoic, feeding on plant or animal remains and perhaps also deriving much of their nutriment from bacteria. Some feed on small plants such as yeasts and small algae. Others again prey on small animals such as protozoans and rotifers. Those which attack larger animals, like worms of various kinds, and which can be termed predators, are provided with spines or teeth around the mouth which are used for piercing the prey. None of them can engulf large particles. They are essentially microphagous or juice feeders. Obviously, by virtue of their structure and habits, it is easy to visualise how saprophagous and herbivorous forms have given rise to plant parasites and how predators and saprozoic types have evolved into animal parasites. Indeed, it is difficult to categorise some species whilst many which are parasitic as adults have free-living juvenile stages.

Most nematodes are physiologically robust. Many can endure the natural extremes of heat, cold, and desiccation. Living nematodes have emerged from moss when it was re-wetted after having been dried for four and a half years. Doubtless this resistance is in part due to the tough cuticle. The vinegar eel, *Turbatrix* (= *Anguillula*) *aceti*, as its name implies, normally lives in vinegar (about 5 per cent. acetic acid) but can withstand for some long time concentrations of up to 14 per cent. acetic acid. Eggs are even more resistant, many kinds remaining viable for years under the most adverse conditions. Some, for example those of *Ascaris*, can withstand prolonged immersion in 12 per cent. formaldehyde, saturated solutions of mercuric chloride, and many other toxic salts. Indeed, 2 per cent. formaldehyde is used as a sterile culture solution for nematode eggs! Juvenile stages are usually less resistant than eggs or adults and some, *e.g.* those of hookworms, survive only within a rather narrow range of temperature and humidity. Nematodes represent one particular branch of the acoelomate grade of Metazoa but many of their fundamental features are unique and attempts to link them closely with groups other than the Nematomorpha are not very convincing. The features common to the two groups are the robust cuticle which necessitates four moults in the life-history, the constancy of number of the large cells of which the body is formed, the peculiar muscle cells, and many negative characters, such as the absence of cilia, nephridia, and segmentation.

It follows therefore that a good idea of nematode anatomy would be given either by describing a hypothetical generalised nematode or by a description of an example chosen as "type". The account which follows is mainly of *Ascaris lumbricoides*, but features of interest shown by different kinds of nematodes will be referred to.

## ASCARIS LUMBRICOIDES

All members of the genus *Ascaris* and of related genera are gut parasites of mammals. *A. lumbricoides* inhabits the small intestine of man, pigs,* and several other mammals; *Parascaris equorum* is common in horses and their allies, whilst the similar, but much smaller ascarids *Toxocara mystax* and *T. canis* parasitise cats and dogs respectively. There are many other well-known related genera and species.

**Form of the Body and External Features.**—By nematode standards, *A. lumbricoides* is a large worm, the males being about 6 in. long and the females half as long again. Moreover, the body is relatively

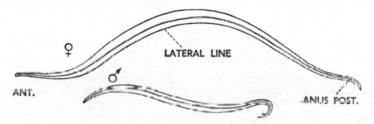

Fig. 117. ASCARIS LUMBRICOIDES.—Side view, male and female, × ⅓.

stout, although it tapers at each end, particularly at the hinder end. Males can be distinguished from females not only by their smaller size but because their tail end curves sharply towards the ventral surface which, on closer inspection, shows a pair of bristles (copulatory or male spicules) protruding from the cloaca, which is situated an inch or so from the tip of the tail. The opening of the female duct is separate from the rectum (so there is no cloaca in this sex) and lies about one-third of the way from the anterior end on the ventral surface. With the aid of a lens the small excretory pore may be seen on the ventral surface, a short way from the anterior end. The three lips which guard the entrance to the mouth are also of use in orientating the animal before dissection, for one of them is dorsal and the other two ventrolateral in position.

The body is encased in a tough but smooth cuticle which is, however, semi-transparent so that some of the internal organs can be seen through it. For example, there are four lines, one dorsal,

* By some these are considered as separate species.

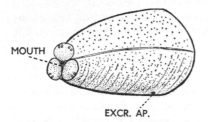

MOUTH

EXCR. AP.

Fig. 118. ASCARIS LUMBRICOIDES.—Anterior end to show mouth guarded by three lips bearing pairs of papillae.

one ventral, and one on each side, the so-called chords, which indicate internal thickenings of the epidermis. Of these the two lateral chords are the most obvious.

**Body Wall.**—As in all other nematodes, the body wall consists of the cuticle, the epidermis, and a layer of longitudinal muscles. The cuticle, secreted by the epidermis, has four main layers; an outer, cortical layer, formed of keratin; a layer of fine fibres; a layer of spongy protein; and, lastly, a layer of interlacing collagen

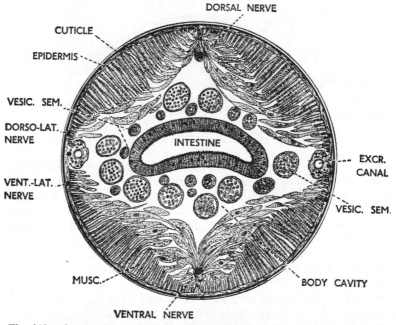

DORSAL NERVE

CUTICLE

EPIDERMIS

VESIC. SEM.

DORSO-LAT. NERVE

VENT.-LAT. NERVE

INTESTINE

EXCR. CANAL

VESIC. SEM.

BODY CAVITY

MUSC.

VENTRAL NERVE

Fig. 119. ASCARIS LUMBRICOIDES.—Diagrammatic transverse section through about the middle of the body.

fibres. It is permeable to metabolites, salts, and water. The epidermis is, like so many other nematode structures, composed of remarkably few cells, so few, in fact, that its cellular nature is not at once apparent in transverse sections when it appears as a finely granular and fibrous layer which is a good deal thicker than elsewhere in the mid-dorsal, mid-ventral, and lateral lines where it forms the chords. The nuclei of the epidermal cells are confined to these chords, along which they lie in longitudinal rows, and tend to occur in clusters. Inside the epidermis lies the layer of longitudinal muscles, each of which is a single remarkable cell with a fibrous contractile part next to the epidermis and a granular cytoplasmic part bulging into the body cavity. The cytoplasmic part of each cell contains the nucleus and is prolonged into a long thin strand. The strands converge mainly towards the dorsal and ventral

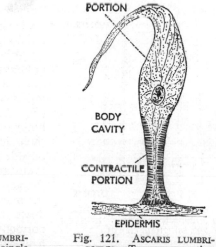

Fig. 120. ASCARIS LUMBRI-COIDES.—Diagram of a single muscle cell dissected out.

Fig. 121. ASCARIS LUMBRI-COIDES.—Transverse section of a single muscle cell.

lines, where they join the epidermis. The epidermal chords thus separate the muscle cells into four columns.

**Body Cavity.**—Between the cytoplasmic parts of the muscle cells and the gut is a fluid-filled cavity which has arisen by excavation of a single huge mesenchyme cell on the dorsal side of the pharynx but some strands of cytoplasm remain, crossing the cavity here and there and forming a thin layer over the gut. Two other large stellate mesenchyme cells (giant cells) are found along the anterior third of each of the lateral chords. Evidently the body cavity is not a coelom (p. 214). It can be thought of as a system of irregular interconnecting spaces within the mesenchyme, perhaps of the same general nature as the smaller, but more numerous cavities in the mesenchyme of Platyhelminthes; it is often termed a pseudocoel. Hydrostatic pressures during life are surprisingly high and vary between 16 and 225 mm of mercury.

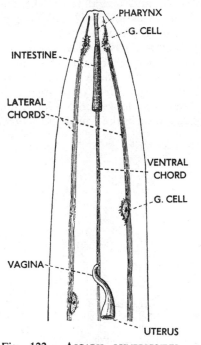

**Alimentary Canal.**—The mouth is a small opening at the extreme anterior end between the three lips. It leads into a long narrow pharynx, tri-radiate in cross-section and lined by a thin cuticle. Its walls are formed by a single layer of epithelial cells but cell outlines are difficult to observe and the cytoplasm is crossed by numerous radial muscle fibres running from the outside bounding membrane to the cuticle (Fig. 124). Connective tissue fibres running from each of the three internal grooves to the outside membrane assist in maintaining the tri-radiate shape of the lumen. Evidently the muscles when they contract dilate the pharynx, which is normally kept shut by the hydrostatic pressure of the body fluid on its walls. The inner end of the pharynx is in many nematodes enlarged to form a muscular bulb and is supplied with

Fig. 122.  ASCARIS LUMBRICOIDES.— Female dissected dorsally to show giant cells on lateral chords.

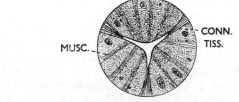

Fig. 123. ASCARIS LUMBRI-COIDES.—Giant cell.

Fig. 124. ASCARIS LUMBRICOIDES.— Transverse section of pharynx.

valves to prevent intestinal contents being forced back into it.  But whether a bulb is present or not, the nematode pharynx acts as a muscular pump forcing food into the intestine. This would seem to be a necessary consequence of the high hydrostatic pressure in the body cavity and it is no mere coincidence that all nematodes are fluid or microfeeders relying on a mechanism to pump food into the intestine.

The intestine is a long tube with walls formed of a single layer of columnar cells bearing a brush-border or layer of microvilli on their inner surfaces—a feature common in cells specialised for rapid secretion or absorption. The intestine joins a narrow flattened rectum which opens by a slit-like anus. Were it not that this slit formed a kind of valve the gut contents would be voided by the pressure of the body fluid, as indeed they are when the anus is opened by a special muscle which raises the dorsal wall of the

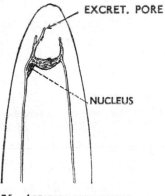

EXCRET. PORE

NUCLEUS

Fig. 125. ASCARIS LUMBRI-
COIDES.—Brush-border cells
from the intestine.

Fig. 126. ASCARIS LUMBRICOIDES.
—Plan of excretory system.

hinder part of the rectum. Strictly speaking, the so-called rectum of the male is a cloaca, for the male duct opens into it.

**Excretory System.**—This consists of a long canal embedded in each lateral line, the two being joined below the pharynx by a network of finer canals. It is from these that a short duct leads to the excretory pore. The whole of this H-shaped canal system is excavated in a single enormous cell, whose nucleus lies on the left-hand lateral canal near its anterior end, whilst two smaller nuclei also occur, one on the bridge and the other on the terminal duct (Fig. 126). The little evidence available suggests that the canal system secretes urea and it has recently been pointed out that the high pressure in the body fluid provides a means whereby excretory products could be eliminated by ultra-filtration.

**Nervous System.**—The plan of the nervous system is similar to that in all other nematodes. It consists of a ganglionated nerve ring around the pharynx from which six nerves run forward to supply sense organs at the anterior end whilst six other main nerves run posteriorly to supply the rest of the body. Of these nerve cords one lies in the dorsal epidermal cord, one in the ventral, and one just above and below each of the lateral cords. Two thin lateral cords run in each lateral line. Ventral and lateral nerve cords are connected at intervals by asymmetrical commissures. The cell bodies of the neurons occur not only in the ganglia joined

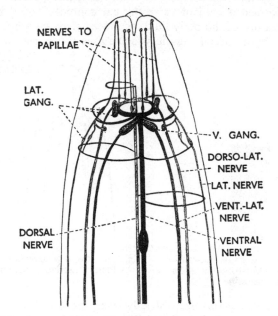

Fig. 127. ASCARIS LUMBRICOIDES.—Anterior end of nervous system, ventral view (based on Goldschmidt).

to the nerve ring but in other smaller ganglia as well as along the ventral nerve cord. The dorsal nerve cord is, however, a motor nerve and has very few cells along its length. As in all other systems, the number of cells is small and constant in number.

**Sense Organs.**—Although always simple in construction, nematodes, particularly free-living ones, are provided with numerous receptors often in the form of innervated bristles. Rings of papillae on the lips are known to be sensory and there are six of these in *Ascaris*, each papilla bearing two receptors which are probably chemo-receptors (Fig. 128 A). In addition, paired head receptors formed of glandular sense organs at the bottom of depressions in the cuticle, termed amphids (Fig. 128 B), are found in all nematodes

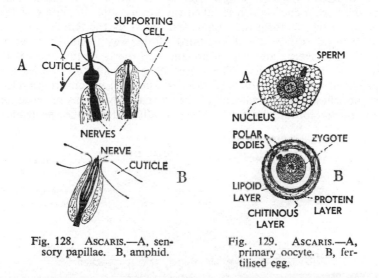

Fig. 128. ASCARIS.—A, sensory papillae. B, amphid.

Fig. 129. ASCARIS.—A, primary oocyte. B, fertilised egg.

but are less obvious in parasitic species. In *Ascaris* they lie on the lateral papillae. Somewhat similar structures, called phasmids, occur on each side of the tail. Both amphids and phasmids are also believed to be chemo-receptors but the sensory physiology of nematodes is obscure.

**Reproductive System.**—The sexes are nearly always separate but just a few nematodes are protandric hermaphrodites (*i.e.* male

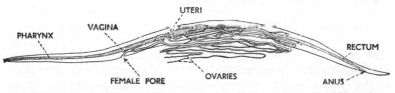

Fig. 130. ASCARIS LUMBRICOIDES.—Dissection of female from the side to show the reproductive system.

gametes mature first). The single testis and paired ovaries are much-coiled, tubular organs ending blindly at their proximal ends but continuing distally to form vesicula seminalis and vas deferens or oviducts and uterus respectively. Germ-cells, budded off in the proximal region, grow and mature in the next part, are then fertilised, and then in many nematodes develop to an advanced stage (but not in *Ascaris*) before being carried to the outside.   The distal part of the female duct, which is called the uterus, joins its fellow of the opposite side to form the vagina and opens by the female pore on the ventral surface about one-third of the way from the anterior end.   The uterus and vagina are the only parts of the ducts with muscles in their walls.   The males of practically all species are provided with a pair of hard bristles (spicules) which are secreted by pouches in the dorsal wall of the cloaca.   By means of retractor and protractor muscles they can be partially withdrawn or protruded

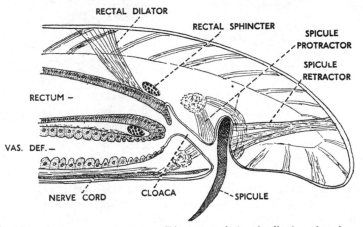

Fig. 131.   ASCARIS LUMBRICOIDES.—Diagrammatic longitudinal section through the hinder end of the male worm.

when the worms are copulating to open the female pore to allow the amoeboid sperms to enter it.

**Life-history.**—The eggs in the secondary oocyte stage are fertilised by the amoeboid spermatozoa (which have travelled up the proximal part of the uterus) and then complete their maturation divisions (see Chapter XVIII).   Almost immediately the egg membrane lifts clear of the cytoplasm and is then called the fertilisation membrane.   This thickens and hardens to form a chitinous shell to which the egg cytoplasm adds a thin inner layer of lipoid material.   As the eggs travel down the uterus, an outside layer of hard, brown protein, a secretion of the uterine wall, is added and the eggs are then laid. They are, as has been mentioned, remarkably resistant structures and protect the eggs or first juvenile stages, which soon form within them once they have experienced a drop in temperature, which is the necessary stimulus for development to begin in *Ascaris*. The eggs are then infective and can under cool, moist conditions, remain so for as long as five years, perhaps more. They do not develop further unless swallowed by a suitable host in whose small intestine the "egg-shell" is dissolved and the juveniles hatch out. Immediately they bore through the intestinal epithelium into the factors of the hepatic portal vein and get carried through the liver to the heart and thence are pumped round the body many times in the blood-stream, during which time they grow. Eventually they become too large to pass through the capillaries of the lung and some stick there and rupture through into the alveoli. The juveniles grow and moult in the lungs and then ascend the bronchi into the pharynx when they are swallowed and eventually reach the small intestine again where they feed, grow, moult, and become adults.

**Nutrition.**—The worms feed on the gut contents of their hosts, much of which by this stage is partially or wholly digested, but, despite this, provision is made for the digestion by the worm of solid materials and diastatic, lipolytic, and proteolytic enzymes occur in its gut. Its intestinal epithelium can engulf small particles and digest them intra-cellularly as well as forming the main absorptive surface. Food absorbed in excess of immediate requirements is stored as glycogen and fats.

**Respiration.**—The intestine of the host is normally (but not always) so poor in oxygen that it is undetectable and it is reasonable to assume that parasites living in it obtain their energy by anaerobic respiration. All the available evidence points to the view that *Ascaris* does, indeed, normally respire anaerobically, breaking down glucose first into lactic acid and, later, forming carbon dioxide and a variety of fatty acids (chiefly valerianic acid) which are excreted through the cuticle. It is these fatty acids, mixed perhaps with traces of other substances, which give the characteristic smell— rather like tinned pineapple—to living ascarids. It seems equally certain that, when oxygen is present, *Ascaris* can respire aerobically (though its oxidative enzyme system is unusual since it cannot be put out of action by cyanides). The haemoglobin present in small amounts in the body wall and in the body cavity fluid has a strong affinity for oxygen and may well play a part in the transport of oxygen even when it is present at low tensions. Not all gut-dwelling nematodes respire in this way. Some which live close to or are embedded in the gut wall may derive a good deal of oxygen from their hosts but are able to respire anaerobically for some time and incur an oxygen debt.

**Relations between Nematode Parasites and their Hosts.**—
Protection from digestion by the enzymes in the small intestine of
the host is chiefly due to the secretion by the parasite of anti-enzymes
but, as has been seen, the cuticle is very resistant, especially when
layered over by host mucus. The presence of a few ascarids seems
to cause but little ill-health, but larger numbers have serious effects,
largely because the worms produce toxins. (If the eyes are rubbed
after handling living ascarids they become very inflamed.) One
human being may house several thousands of ascarids. More
serious harm results if a worm enters and blocks either the pancreatic
duct or the bile duct.

**Locomotion.**—Nematodes, even free-living ones, have only
rather limited powers of locomotion. A few bristly marine species
can elongate and shorten their bodies and so can move in a way
which somewhat resembles that of earthworms (p. 257). Others
can "loop" along, alternately gripping and releasing their hold on
the substratum. But most nematodes, including *Ascaris*, live in
fairly viscous fluids, *e.g.* mud, plant or animal tissues, through which
they move by undulating the body in dorso-ventral waves. These
are caused by differential contractions of the longitudinal muscles
of the body which are antagonised, *i.e.* stretched, by the hydrostatic
pressure of the body fluid (with an average of 100 mm mercury,
a figure much above that in worms with circular and longitudinal
muscles) in conjunction with the cuticle. The collagen fibres of
the inner layers of the cuticle are arranged to form a spiral basket-
work or lattice-work around the body with the fibres at an angle
of about 75° to the longitudinal axis of the body. Any changes in
the shape of the body will tend to distort the angle between the
fibres and so alter the body volume. Shortening will tend to
decrease the volume and lengthening to increase it but, since fluid
is incompressible, the net result of contraction of the muscles will
be to increase the pressure in the body fluid and *vice versa*. Whilst,
therefore, not in themselves elastic, the system of fibres and the
body fluid act as a mechanism which stretches the relaxed muscles.
Other parts of the cuticle are elastic and help to maintain the pressure
in the fluid. The free passage of fluid throughout the body provides,
with the muscles, a means whereby local changes in length (although
of a limited amount when compared with other kinds of worms)
can be brought about.

**Other Nematodes.**—The nematodes which will now be briefly
mentioned, are selected mainly to exemplify different kinds of life-
cycles. All free-living nematodes have a direct type of life-cycle.

From the fertilised eggs hatch out first-stage juveniles which grow, moult, and give rise to second stage juveniles . . . and so on. There are four moults in all. As already mentioned, any sample of soil, or mud taken from fresh waters or the sea, will yield vast numbers of small nematodes, too numerous to mention even by families. More likely to attract attention to themselves are the "vinegar eels" (*Turbatrix aceti*) which are only too often found in old vinegar but can also live on rotting fruit. These small worms feed on yeasts and bacteria. Other small nematodes not very distantly related to vinegar eels are transitional towards a parasitic mode of life as for example, *Rhabditis maupasi*. These small worms, so frequently found as a parasite in the nephridia of earthworms, are juvenile stages. When, and not before, the earthworm dies they feed on its tissues, grow, moult, and become sexually mature (*i.e.* feed saprozoically). Most adults are self-fertilising hermaphrodites but a few are true males and females. From the eggs, whilst they are still in the uterus, hatch juveniles which live and feed in the soil from which they enter and leave the earthworm *via* the nephridio-pores (p. 245). Other kinds feed on plant juices but do not actually enter the plant and these are on the borderline between herbivores and those plant parasites whose adults live in plants but whose juveniles live freely in the soil, as, for example, *Ditylencus dipsaci* which, among other things, attacks and causes great damage to bulbs. *Anguina tritici* causes galls ("cockles") to form instead of normal wheat grains. It is in these galls that the infective juveniles remain dormant until the following spring when they leave them to infect new plants. *Heterodera schactii* attacks sugar-beet and the worms grow to maturity in a gall from which the fertilised females protrude and then become swollen and distorted as they come to contain a vast number of eggs. The eggs can be released in the soil, not singly but enclosed in a gelatinous cyst secreted by the parent, and from this cyst second-stage juveniles are released to infect new plants.

Worms belonging to the order Strongyloidea are gut parasites, mainly of mammals. Some are important scourges of man, as, for example, the hookworms, *Ancylostoma duodenale* and *Necator americana*. They have a very wide distribution in tropical and sub-tropical countries right across the world but are believed to have been introduced to the New World by the slave traffic. When suitable conditions prevail they have occurred outside their normal limits, as for example, in Cornish tin mines and during the building of the St Gothard tunnel through the Alps. The worms live in the small intestine attached by their buccal capsules to its lining through which they suck blood for food. The fertilised eggs pass out with the faeces of the host; if the conditions are moist and the temperature over 18° C (65° F) they hatch. The juveniles feed, grow, and moult twice in the soil but the third stage juveniles retain the cuticle of the previous stage as a sheath. These ensheathed juveniles are the infective stages and can remain infective for long periods. They are strongly negatively geotactic and can work their way to the surface and up grass, etc., even if the host faeces have been buried below several feet of soil. They infect a new host by penetrating the skin (the sheath being left at the surface) and entering the lymphatic system. Eventually they reach the heart, are carried round in the blood-stream, and sooner or later rupture the lung capillaries. From the cavities of the lung they ascend the bronchi and pass to the gut.

Members of the order Filaroidea are blood and connective tissue parasites in a variety of vertebrate hosts. From such deep situations it is impossible for the infective stages to pass to the outside passively and hence reach a new host. In many instances, therefore, transference from host to host is carried out by a vector, frequently a blood-sucking insect. For example, the viviparously produced first-stage juveniles (microfilariae) of *Wucheria* ( = *Filaria*) *bancrofti*, a parasite of the lymphatic system of man in many tropical countries, and which may give rise to serious hypertrophy of the connective tissues—the disease called elephantiasis—are sucked up from the peripheral blood vessels by a night-flying mosquito, *Culex fatigans* (the microfilariae retire to the deeper tissues, particularly to the lungs, during the day). They undergo two moults in the mosquito and the third stage juveniles, which are infective, reach the labium of the mosquito and migrate into the small wound caused by the mandibular and maxillary stylets—they are not, it seems, actively injected. A somewhat similar filarid, *Loa loa*, also a connective tissue parasite, has microfilariae which have a diurnal periodicity opposite to that of *W. bancrofti*, appearing in the superficial blood vessels during the daytime. They are transmitted by a day-flying biting fly (*Chrysops* sp.). Obviously the kind of diurnal migration is related to the habits of the insect vector but how it is controlled is not known. It has, however, been shown that the migrations are influenced by changes in the physiological state of the primary host (Man) and the rhythmic migrations can be reversed if the infected person sleeps by day and remains awake at night.

Some parasitic nematodes are passively transferred from one final host to the next when it eats a secondary host containing the infective stages. *Dracunculus medinensis*, the Guinea worm, is such a one. The adults inhabit the subcutaneous connective tissues of man and other mammals in many tropical countries. The male is small but the female worm may reach a length of three feet or more and when "ripe" travels downwards, usually to the leg. The front end then breaks through the skin of the leg or foot and viviparously produced juveniles are released particularly if the skin blister is immersed in water, as by paddling. The juveniles develop further only if swallowed by fresh-water copepods of the genus *Cyclops* (p. 310) in which after two moults they can infect their final host if the copepods are swallowed in drinking water.

# CHAPTER X

## PARASITISM

In an earlier section (pp. 92-111) an account was given of several protozoan parasites and these, together with the parasitic flatworms and nematodes just described, will serve as useful examples to illustrate some general principles in connection with parasitism.

The parasitic mode of life is to be regarded as a secondary state of affairs, and *when compared with their free-living relatives*, parasites usually show modifications of their bodily structure, which are clearly in the nature of adaptations to a parasitic existence. These modifications may be slight, or may affect all the organs of the body, when at first sight it is often difficult to recognise the group of the animal kingdom to which the parasite belongs. A clue to the relationships is often given only by the mode of development. Thus, for example, the adult *Sacculina*, a parasite related to the barnacles and which lives on crabs, is structurally little more than a bag of eggs plus root-like processes which extend into and draw nourishment from the tissues of the host. Its larva, however, betrays its relationships to other crustaceans (see p. 312).

Endoparasites are more modified than ectoparasites and are usually specialised for life in a restricted region of a particular kind of host. Ectoparasites, which live on the outside of their hosts, are almost invariably provided with organs which enable them to maintain their precarious position. For example, ectoparasites, such as *Polystoma*, have a number of suckers and recurved hooks. This type of adhesive apparatus is found also in ectoparasites belonging to many different groups of animals. Some endoparasites, for instance those which live in the alimentary canal, are also liable to be dislodged and therefore, like ectoparasites, may bear adhesive structures.

It is common to find that the trophic organs (*i.e.* the organs concerned with nutrition) of parasites are reduced when compared with those of their free-living allies, or they may be entirely lacking. This tendency is more pronounced in endoparasites than in ectoparasites, for they live in situations where they are surrounded by food material which is continuously available. Frequently the food is in the fluid state and may be already wholly, or partly digested, which renders an elaborate mechanism for this purpose unnecessary. Thus, most free-living flatworms have an eversible pharynx, which is used to engulf large particles of food; but trematodes have a simple suctorial pharynx, by means of which nourishment is drawn from the tissues of the host. Tapeworms, which lack a gut, absorb through their tegument food digested by the host.

Apart from holophytic and saprophytic forms, most free-living Protozoa are provided with definite organelles for the intake of food, but these are reduced or absent in protozoan parasites. The presence of locomotor organs is also directly associated with the search for food, and they too tend to be reduced in parasitic animals. This is well shown by the absence of locomotor cilia in trematodes and tapeworms, and the reduction or absence of cilia and pseudo-podia in parasitic Protozoa. *Entamoeba*, for example, has fewer pseudopodia than *Amoeba proteus*; none of the Sporozoa has true locomotor structures; cilia are absent from the fully grown Suctoria, which are perhaps the most specialised of parasitic ciliates; whilst the trypanosomes carry out only relatively slower movements by means of their short flagellum and undulating membrane.

Most parasites are relatively sedentary creatures and their environment is rarely subjected to any appreciable variations, so that the importance of the sensory and nervous systems is decreased and they are never developed to any great extent. Thus, whilst turbellarians have well-developed eyes and chemo-receptors, these are greatly reduced in the ectoparasitic trematodes and are lacking altogether in adult endoparasitic flukes and in the tapeworms. It is therefore interesting to find that sense organs (eye spots) are still found in the miracidium larvae of *Fasciola* and other similar flukes, which for a short time lead an independent existence.

The extent to which the trophic, locomotor, and neuro-sensory systems of a parasite are reduced depends to a large extent on the degree of dependence of the parasite on the host. Thus no hard-and-fast line can really be drawn between commensals, which merely live in close association with another animal, and simple ectopara-sites which do but little harm to their hosts. It is apparent, in fact, that the distinctions between parasitism, commensalism, and symbiosis are to some extent subjective. An association between two animals is judged to be parasitic when one partner clearly (according to human opinion) derives some benefit at the expense of the other. Yet this feature is insufficient in itself to distinguish a parasite from a predator, for predators are animals which live by preying on others. Usually, however, predators feed on animals which are smaller and more numerous than themselves, whereas parasites are invariably much smaller than their hosts. But there are exceptions to these general rules and it is difficult to say if such creatures as blood-sucking bats, leeches, and lampreys are

ectoparasites or predators. Although they are usually spoken of
as parasites, several genera of ciliates which are found in
the rectum of the frog cause little inconvenience, so far as is known,
to the host. They move about very actively and some feed by
engulfing particles of faecal matter into a well-formed gullet.
Their trophic and locomotor organelles are as well-developed as in
most free-living forms, which they resemble in powers of movement
and method of taking in food, and they could almost be termed
commensals. Again, the small nematode worm, *Rhabditis maupasi*,
is a facultative parasite (*i.e.* it does not necessarily depend upon a
host for food and shelter) which may be found either free-living in
the soil or as a parasite in the nephridia of the earthworm. It is
apparently not modified in any way by its parasitic life and is almost
identical in structure with other small nematodes which lead an
independent life in the soil.

On the other hand, endoparasites such as *Monocystis*, the
malaria parasite, and the tapeworm, live in situations so rich in
liquid nourishment that an alimentary system is not really necessary
and so is entirely absent. It may be fairly safely assumed that the
absence of a gut in the tapeworm is a secondary feature; that is,
it is due to reduction from the condition in the ancestral group of
flatworms; but both *Monocystis* and the malaria parasite belong
to the class Sporozoa, which includes only parasitic forms and does
not appear to be closely related to any of the free-living Protozoa.
A standard of comparison within the class is therefore lacking,
and it is impossible to say if the absence of trophic and sensory
organelles is a primitive or secondary feature.

So far only the more obvious structural features of parasites
have been mentioned, but it is clear that it is of equal importance
that parasites should be physiologically adapted to their life in the
host. For instance, gut parasites, if they are to survive, must be
able to withstand the digestive juices of their host. Thus, the
tapeworm stimulates the gut of the host to secrete large amounts
of mucus which surround it in an outer protective envelope. It
has also been suggested that added protection from digestion is due
to the secretion by the parasite of a substance ("anti-enzyme") which
neutralises the digestive enzymes of the host, for dead tapeworms are
soon digested in the gut of the hosts. On the other hand, the finding
that the tapeworm's outer layer is, in reality, made of living cytoplasm, makes it likely that this is continually renewed by the cells
which form it. Blood parasites, like trypanosomes and malaria
parasites, are in some way or other also partially able to withstand
the anti-bodies and phagocytes found in the blood; but the number
of parasites that survive is very variable and in many instances the

host is able to get the upper hand. Often the osmotic pressure of the body fluids of the parasite are approximately the same as that of the host rendering osmoregulation unnecessary. Gut parasites are faced with the further problem of obtaining oxygen, for they live under practically anaerobic conditions, and some parasites, like the tapeworm (and *Ascaris*), can obtain their energy by means of anaerobic respiration only. Blood and tissue parasites, on the other hand, receive a plentiful supply of oxygen.

Nematodes, which are perhaps the most successful of all parasites, are found living in practically all types of animals and in a great variety of situations in the body, but they show practically no structural adaptations at all; the parasitic members of the group often being almost identical in structure with the free-living ones. It must be concluded therefore that the parasitic nematodes are very well adapted to the physiological needs demanded by the very diverse situations which they inhabit and that in this particular instance structural adaptation is relatively unimportant.

One of the greatest difficulties that has to be solved by any parasite is that of ensuring that its eggs or offspring shall reach another host. Only in this way can the continuance of the race be maintained after the death of the host. The odds against the offspring reaching another host are always great and many eggs or larvae are therefore wasted. Accordingly, the reproductive powers of parasites are much greater than those of related but non-parasitic forms, so that the likelihood of survival of one of the vast number of offspring produced is increased. The simplest and almost universal way in which an increase in offspring is brought about is by an increased output of fertilised eggs. Thus, both the flukes and the tapeworms produce far more eggs than do the turbellarians. It is, however, by no means uncommon to find that parasites have an additional multiplicative phase at some stage in their life-history. *Entamoeba* forms cysts from which a number of amoebulae emerge when they reach the gut of a new host; the numerous zygotes of *Monocystis* each divide to form eight sporozoites, whilst there are alternating phases of sexual and asexual reproduction in the life-cycle of the malaria parasite. The number of cercariae of a liver fluke released from the snail is much greater than the number of sporocysts because of the multiplication of the rediae, whilst not only do tapeworms increase the number of their proglottides throughout life by a peculiar method of budding (strobilisation), but some, e.g. *Echinococcus*, in addition, produce several generations of bladder-worms.

As has been shown previously, parasites, in particular endo-parasites, have very limited powers of movement and are, as a rule, confined to a small region in a particular host, so that the transference of the eggs or other infective stages to another host is not an easy matter. During such transference the parasite may have to face a totally different set of environmental conditions which may be compounded of the adverse factors of the non-living environment such as temperature, drought, and the like, or may be those experienced within the body of a secondary host, but this will apply mainly to parasites of land animals. Whatever the method of transference, some adjustment to the changed conditions is required. There are two chief ways in which this may be carried out; the eggs or young stages may be passed outside the host and then by chance be picked up by a new host; or the eggs or other immature stages of the parasite in some way or another are taken into a secondary host, where as a rule they develop into a new stage which is then able to infect a new final host. The infective stage of the parasite may be actively transferred to the final host by the secondary host, which is then spoken of as a vector.

Examples of both these methods of dispersal have already been described. Ectoparasites almost invariably lay their eggs on the surface of their host and the young which hatch out are as a rule active and seek out a new host. The larva of *Polystoma*, for example, is ciliated and swims about until it dies or finds a tadpole. The flea, in effect, an active and relatively harmless ectoparasite, lays its eggs in dust and after some time a larva emerges. It feeds on organic particles and changes into a pupa which is a resistant stage and from which in due course a young flea hatches, sometimes after a long period of dormancy, and begins to feed on a new host. The eggs or embryos of gut parasites are as a rule passed out with the faeces of the host and are protected from drying-up by a thick envelope. These stages of the parasite may be infective straight-away when swallowed; or, as described for the tapeworm, they may require to be eaten by a secondary host before they can infect a new final host. The infective stages of many blood parasites, for example malaria parasites and trypanosomes, are transferred to a new host by the bite of a blood-sucking insect.

It may be safely said that for the endurance of any form of parasitic association some degree of adaptation is always required and this involves not only the parasite but the host also. As a corollary to this it follows that a parasite always shows some specificity. That is to say, particular kinds of parasites occur only in or on certain hosts. Further, the specificity extends as a rule to the situation in the body in which the parasite lives. There is nothing in the least surprising about this since many predators (and even herbivores) are restricted to feeding on a few kinds of organisms. Moreover, just as with some predators there is a large amount of tolerance of different kinds of food (a fox will feed on poultry, rabbits, or even mice), so certain parasites seem to be able to invade different host species as has already been mentioned in connection with *Echinococcus granulosus*. With nearly all parasitic relationships there is a high degree of tolerance between host and parasite. *Monocystis* seems to decrease but little the reproductive capacity of its earthworm host and native ungulates act, apparently unharmed, as reservoirs for trypanosomes.

Under natural conditions parasites rarely cause serious disease to their hosts and it is easy to see why this must be so. If a parasite seriously weakened or killed its hosts, the parasitic species itself would, sooner or later, also be wiped out. It is quite otherwise when ecological conditions are altered, as, for example, by over-crowding or the introduction of a species to a new region. Then heavy infections, particularly in young animals, may cause fatalities. The idea is growing that these and other changes brought about by human agency are one of the main causes of the severity of parasitic diseases. Obviously then, there must be some tolerance between parasites and their hosts or both would perish in the "struggle for existence". But the balance is sometimes struck between the species as a whole and not between individual animals, many of which perish—both parasites and hosts. For instance, ichneumon flies lay their eggs in caterpillars. The fly larvae hatch and feed on the caterpillar's tissues but do not always prevent it changing into a chrysalis. But after the ichneumon larvae have become fully grown nothing remains of the chrysalis but an empty shell. Sufficient caterpillars escape infection to prevent elimination of the host insect species, for obviously survival of both host and parasite here depends on a sufficient number of individual hosts remaining uninfected, and the same principle must apply to all instances in which parasites cause severe harm.

# CHAPTER XI

## THE COELOMATE ANIMALS

As their name implies, all coelomate animals are characterised by the possession of a coelom. The group includes all the larger and more highly specialised animals and these are often thought of as comprising four main and several small, but unrelated, stocks. The smaller groups in each of these major divisions can be arranged in a gradually increasing order of complexity so that each group culminates in a highly organised type of animal which is, however, quite unlike that developed along the other three lines.

The first of these four major divisions of the coelomate animals is the platyhelminth-annelid-mollusc stock. It includes the segmented worms and the molluscs, and reaches its climax of complexity in forms like the octopus and its allies. The second branch,

Fig. 132. Diagrammatic longitudinal section through a generalised coelomate.

the echinoderm stock, comprises a variety of animals such as the sea urchins, starfishes, and sea cucumbers. All are extremely specialised, particularly with regard to the condition of the coelom, part of which forms a water vascular system; and yet the neuro-sensory system never reaches a high state of differentiation. The third main coelomate group, itself related to the annelids, is the arthropod stock. It includes the crustaceans (crabs, shrimps, and a host of smaller creatures), the spiders and scorpions, and the insects. In all of them the coelom is greatly reduced, but jointed limbs and neuro-sensory systems are prominently developed. Finally, the chordates, the phylum to which both amphioxus and man himself belong, represent the most advanced of all the coelomate lines of specialisation.

The coelom, although the chief, is not the only feature which distinguishes the coelomate from the acoelomate grade of organisation, for practically all the organ systems show a distinct advance in complexity, whilst a blood vascular system is not represented in acoelomates (with the exception of the nemertine worms).

In fact, it is the recognition of this capacity for the development of greater bodily complexity that enables it to be said that coelomates are at a higher level of organisation than acoelomates, and have gone further in the process of physiological division of labour. Some of the coelomate features seem to be directly correlated with the presence of an extensive body cavity which separates the gut from the body wall. This means that substances absorbed by the gut cannot be carried by *direct* diffusion to other parts of the body, and the transport of dissolved gases is also of an indirect nature. The gut is always provided with both a mouth and an anus and movement of food by contraction of the gut musculature is made easier because the gut wall can contract independently of the body wall, owing to the intervention of a fluid-filled space. The steady passage of the food along the alimentary canal allows various parts of the canal to become specialised. Thus, the anterior end is usually mainly concerned with prehension and ingestion, the middle parts with digestion, and the hinder parts with absorption. The anterior and posterior regions (stomodeum and proctodeum) of the alimentary canal are formed by invaginations of the ectoderm.

The presence of a blood vascular system, whose chief function is to carry dissolved food and gases about the body, is therefore readily understandable in animals built on the coelomate plan. In its simplest form the blood vascular system may be regarded as a system of tubes containing a fluid, the blood, which carries food in solution to all parts of the body and nitrogenous excretory matter to the excretory organs. In the blood may be suspended a respiratory pigment having the power to enter into combination with oxygen where the oxygen tension is high and of releasing it again to the tissues where it is required. In the vertebrates the blood has many other functions and it will suffice at this point to consider the changes in general organisation which are possible once a blood vascular system has been established.

It has been pointed out that an increase in the size of the body is of mathematical necessity correlated with a relative decrease in the outside surface of the body, which is then no longer large enough to carry out the gaseous interchanges necessary for the release of energy or to allow the waste products of metabolism to diffuse away; but with the development of a vascular system, the setting aside of special organs of the body to carry out these functions is at once possible. Thus a special part of the body in contact with the outside world is set aside to carry out gaseous interchange. In this way a respiratory surface is formed and its area is frequently increased by numerous foldings. In some animals, like earth-worms, the respiratory surface is but little differentiated, the general covering of the body remaining thin and overlying a plexus of minute blood vessels, whilst the gut, as in other annelids, may also assist in gaseous interchange. In other more active aquatic animals branched, thin-walled outgrowths from the body wall forming the so-called gills are developed to assist in external respiration, whilst in all animals where most of the surface of the body has become impermeable (often because of the formation of an outer protective layer) the respiratory surfaces are restricted to folded or foliaceous outgrowths from certain regions of the body (gills of crayfish, etc.). True gills are found in fishes and amphibians, whilst in land-living

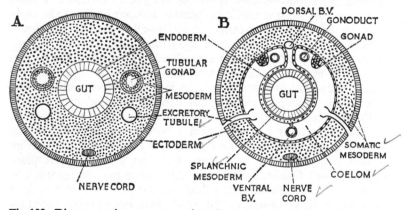

Fig. 133. Diagrammatic transverse sections through A, a generalised acoelomate; B, a generalised coelomate.

animals lungs are often developed. All these respiratory devices have in common two important features: they provide a relatively enormous moist surface which is also always *thin* and usually in close contact with a rich plexus of thin-walled blood vessels, so that diffusion of dissolved oxygen inwards and of dissolved carbon dioxide outwards, is made as easy as possible.

Although the gradual increase in bodily complexity which is evident in the coelomate stocks may not involve all organs of the body to the same extent, other main trends may be briefly noted. A tendency, already well marked in the acoelomates, is an increase in the size and complexity of the central nervous system. This provides for an increase in the number of neurons and possible linkages and interactions between them. This finds expression in the outward behaviour of the animal, for a greater variety of actions are performed, many of which are very intricate. In the higher types of molluscs, in the arthropods, and in the vertebrates not only does there seem to be a "choice" of response to a stimulus, but many nerve impulses originate within the nervous system itself without previous excitation from the receptors, so that some of the behaviour may be motivated from within. The gradual "improvement" of the central nervous system is closely correlated with the development of more conspicuous and sensitive receptor organs which enable the animals to respond to more subtle changes in the environment. The structure and mode of development of the receptors is extremely variable, sometimes even within a group of fairly closely related animals.

The changes which take place in the muscular system are also subject to tremendous variation in the different coelomate groups, but in general it may be said that the body wall muscles, which in the more primitive types form a continuous layer throughout the body, become split up into groups and associated with special locomotor organs, as examples of which may be mentioned the parapodia of annelids, the limbs of arthropods and vertebrates, the "tube-feet" of echinoderms, and the "foot" of molluscs.

The arrangement of the vessels in the blood vascular system has been left for consideration until now, but it is well nigh impossible to give anything but a bare outline. Suffice it to say at this point its general plan is that there is a main dorsal vessel and a main ventral vessel, and that these two are connected by a variable number of commissural vessels which run around the gut. The arrangement of the smaller vessels seems to depend entirely on the arrangement of other organs of the body and each example requires separate treatment.

So far very little has been said about the coelom and its relations to the organs of the body. The term coelom has now come to have a very precise meaning and must not be indiscriminately applied to any cavity within the body. Originally the coelom meant the perivisceral cavity of vertebrate animals, but is now applied to this and to other spaces which are known to be homologous with it. As has been mentioned above, the coelom may be thought of as a space arising as a split in the mesoderm which therefore becomes divided into two layers, a somatic layer next to the skin and a splanchnic layer around the endoderm, and in certain cases, the coelom does indeed arise by this method during development. Its origin is, however, subject to variation, but in all instances the coelom is bounded on all sides by a definite layer of coelomic epithelium, which plays a part in the secretion of the coelomic fluid. Typically, the greater part of the coelom forms the perivisceral cavity or splanchnocoel, a fluid-filled space into which are packed the viscera, which are therefore independent of movements of the body wall muscles. In many animals, however—and this is especially true of the higher animals—certain parts of the coelom become cut off from the general perivisceral cavity to form separate and restricted cavities, whose coelomic nature is only established from a knowledge of their development in the embryo. Again, in some coelomate animals (arthropods and molluscs) the cavities of the blood vascular system become so greatly enlarged that the perivisceral coelom is obliterated and the viscera come to lie in a spacious cavity filled with blood (haemocoel), which thus forms the "body cavity", whilst the true coelom is represented only by the cavity of the excretory organs, of the reproductive system, and in addition, in the molluscs, by the pericardial cavity.

It is highly probable that in the ancestral coelomates the coelom was represented only by a series of segmentally repeated meso-dermal pouches from whose walls the gametes were proliferated, and that only later in phylogeny did these gonadial pouches enlarge and acquire those important relations and functions which are found in most present-day coelomates. These coelomic sacs allowed gametes to accumulate to be released practically simul-taneously during a short breeding season whose onset could be geared to favourable environmental conditions by means of hormones. Each pair of coelomic sacs was, in the primitive coelomates, provided with a pair of ciliated tubes (coelomoducts), developed as outpushings of the coelom to the exterior, which served as a means of exit for the gametes. Other organs to become segmentally arranged were the nephridia, and typically it may be supposed that there was one nephridial tubule to each coelomic sac. The nephridia served as organs by which excess water and nitrogenous waste were eliminated from the body. Primitively, they were somewhat similar to the platyhelminth type; i.e. they consisted of ectodermal tubules projecting into the coelom and ending there in specialised excretory cells. These cells, which are termed solenocytes (tube cells), resemble the flame-bulbs of the flat-worms for they have an intra-cellular lumen into which hang flagella, and the lumen is continuous with that of an intra-cellular duct. They differ from flame-bulbs in that the cell-body is small and rounded and the flagella, which are few in number, are continued into a long slender tubular prolongation of the cell in which the walls are remarkably thin. It has been shown that substances passing into this intra-cellular duct do so through its walls and not through the cytoplasm of the cell-body, as is the case in flame-bulbs. It is probable that in these early forms excretion was also carried out by wandering cells (amoebocytes) which collected up waste products and discharged them directly to the exterior, or else into the coelomic fluid, which was in turn passed to the outside, either directly by way of the coelomoducts, or indirectly by diffusion into the nephridial tubules.

In many coelomates the nephridium may also acquire an opening into the coelom by means of a ciliated funnel **(nephridiostome)** and thus is also able to convey coelomic fluid charged with excretory matter

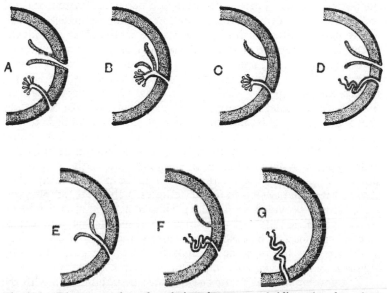

Fig. 134.   Diagram to show the relations between nephridia and coelomoducts (based on Goodrich).
Nephridia and ectoderm, black; coelomoducts and mesoderm, stippled. A, ancestral form.   B-C, possibilities with nephridium bearing solenocytes. D-G, possibilities with nephridium with nephrostome.   F, condition in *Nereis*. D, genital segment and G, other regions of *Lumbricus*.

directly to the outside of the body.   When this happens, however, there is usually a reduction in the coelomoduct system.   In the earthworm, for example, the gonads and coelomoducts are restricted to a few of the anterior segments, so that these segments alone possess a double set of tubules opening to the exterior.   In all other regions of the body segmentally arranged nephridia provide the only means of exit for coelomic fluid.   Economy in the number of tubules per segment is effected by a different process in the polychaetes, for in these worms the coelomoducts and nephridia fuse to form compound organs (**nephromyxia**, see Fig. 134) which carry out the dual functions of excretion (both nitrogenous and aqueous) and of carrying the gametes to the outside of the body.

Yet a different modification is seen in the vertebrates for here the coelom, although the greater part of it forms the continuous perivisceral cavity, opens (in the embryo alone) by a series of coelomoducts to the exterior. These, although segmentally arranged, do not have separate segmental openings, but empty into a common longitudinal duct on each side, which in turn opens externally at the posterior end. There is no trace of a nephridial system and the coelomoducts carry out osmoregulation and nitrogenous excretion whilst parts of the longitudinal ducts are modified as gonoducts. In the adult vertebrate the internal openings of the coelomic tubules have become closed, and excretory substances (including excess water) are extracted from the knot of blood vessels, which become closely applied to the cavity of a small basin-shaped capsule (Bowman's capsule) developed on each tubule. An excretory organ of this type is termed a kidney and is confined to the vertebrates, for it depends for its functioning on the high blood pressure which can be set up only by a muscular heart provided with a special system of valves such as is found in the vertebrates alone. It is not surprising therefore that despite its fairly close phylogenetic relationship to the vertebrates, a chordate (amphioxus) which is not provided with a heart has a nephridial (solenocyte) excretory system and not a kidney. The nephridial and coelomoduct systems are, however, by no means the only ones concerned with excretion, for in some animals the skin, the gut, and other organs may play an equally important or even a greater part.

The development of a true coelom was, then, bound up with the formation of a series of coelomic (gonadial) sacs lying right and left of the gut, a condition which is fairly closely approached by the segmented worms (Annelida) where it is correlated with a wriggling type of progression; but in most other coelomates the coelom has come to form a perivisceral cavity which is continuous throughout the whole length of the body. Nevertheless, in many of the most advanced coelomates (*e.g.* vertebrates) traces of the segmental nature of the coelom are still to be detected and, in addition, it is found that many of the other systems of the body are also segmentally arranged so that the whole body comes to consist of a linear series of segments, each of which is built on the same fundamental plan. Animals whose bodies conform to this arrangement, that is, whose bodies consist of a number of segments, any one of which typically, but not necessarily exactly, resembles all the others, are said to be **metamerically segmented** or, alternatively, to show **metamerism**. This type of organisation must not be confused with that found in the tapeworms, from which it differs in several important respects. Firstly, the body of a metamerically segmented animal consists (within narrow limits) of a number of segments which, for any given species, is constant, and except in certain cases of asexual reproduction, new segments are not added to the body after maturity is reached. From this it follows that all the segments are of the same age and at the same stage of development when the embryonic period is once passed. There is, so to speak, a bit left over at each end and the segmentation of the body may be thought of as involving only that part between the two

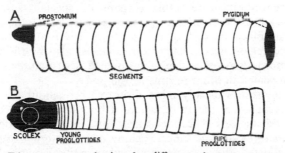

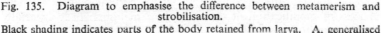

Fig. 135. Diagram to emphasise the difference between metamerism and strobilisation.
Black shading indicates parts of the body retained from larva. A, generalised annelid. B, generalised tapeworm.

ends; and these and the segments retain throughout life a fixed relationship to one another.

Metamerism is, however, far more than the mere division of the body into similar segments. Not only does it mean the serial repetition of homologous parts within the body, but each of these parts is working in co-operation with all the others for the maintenance of the activities of the body as a whole. That is to say, the segmental structures are functionally interdependent and integrated so that the individuality of the body is preserved. In no sense are the segments of the body self-contained units like those of the tapeworms, which in some ways may be regarded as individuals of a sort of colony (strobila). Thus, for example, an annelid worm moves by rhythmical waves of contraction which begin at the posterior end and gradually spread forwards, the muscles of each segment contracting in a regular sequence so that the body as a whole moves forward in an orderly fashion.

In many animals the segmentation is practically complete, that is, most of the main organs of the body are arranged on a metameric plan throughout the length of the body, which is itself incompletely divided into a series of segments by transverse partitions. Each segment is supplied by segmental nerves and blood vessels and typically, has coelomoducts and nephridia. Metamerism is, however, never absolutely complete or uniform throughout the body of any adult animal but is disturbed or obscured by the modification of certain regions of the body. The most important single factor modifying the segmentation is the specialisation of the anterior end of the body to form a "head"—a process spoken of as **cephalisation**. It is often partly due to the development of structures which assist in feeding; but a more universal cause is the concentration of receptors (organs of special sense) at the anterior end. A corresponding concentration of nervous material also takes place in this region to form a "brain", whilst the cephalised region may usually be recognised also by the fusion of two or more segments and by the absence of structures which are serially repeated in other regions of the body.

Metamerism is usually more uniform throughout the body of the embryo than of the adult, where it may become obscured, not only by the process of cephalisation, but by the development of other structures such as limbs. Many of the internal organs may also be affected and become restricted to certain segments. This process of modification of the embryonic metamerism is especially well-shown in the vertebrates.

# CHAPTER XII

## THE SEGMENTED WORMS—ANNELIDA

Some of the most clear-cut and least modified examples of metamerically segmented animals are the annelid worms, and in fact they derive their name, which means "ringed", from the evident segmentation of the body. At the same time they serve to illustrate in a relatively simple way all the main features of a coelomate animal. The phylum Annelida includes such familiar animals as the earthworms and their allies, the leeches, the marine bristle-worms, as well as several less well-known groups.

### POLYCHAETA—NEREIS

In many ways the least specialised annelids are the bristle-worms and one of these, *Nereis*, the rag-worm, several species of which are common in the inter-tidal zone of our coasts, is a convenient example of the Polychaeta for study.

*Nereis diversicolor* (Fig. 136) is certainly the commonest European species, often occurring in many hundreds per square metre on muddy shores. It is particularly abundant at and just above mid-tide level and on muddy shores with a high degree of organic pollution. Like many other species it lives in a U-shaped burrow which it irrigates by waves of muscular contraction in the vertical plane. But the burrow is rarely as simple as this and usually has several galleries and exits. *Nereis virens*, the giant of its kind, may reach a length of over 1 m and lives mainly in mussel banks. *N. diversicolor* is one of the few polychaetes which can tolerate a lowered salinity and a good deal of pollution by organic matter. Indeed, it is commoner in estuaries than on open coasts and can live in practically fresh water, so that it is often common in marshes.

**External Features.**—The body of *Nereis* is long and slender and bears a head at the anterior end, a "tail" at the other. It is roughly oval in cross-section but the dorsal surface is more convex than the ventral one, which is usually flat, but may even become concave when certain of the body muscles contract. A series of ring-like grooves around the body mark off about eighty segments which, apart from those incorporated in the head, are all so alike that a description of one of them would equally well apply to any of the others. This similarity between the segments is clearly indicated

223

from the outside by the **parapodia**, a pair of which is found in every segment. A parapodium is a lateral flap-like projection of the body wall bearing bristles or **chaetae**. It is biramous—consisting of an upper part, the **notopodium**, and a lower one, the **neuropodium**. Both the notopodium and the neuropodium are bilobed, and from each there arises a slender filament, the **cirrus**. The chaetae are grouped into two bundles, one inserted in the neuropodium, the other in the notopodium, and in each bundle two main types of chaetae are found. The vast majority are fine bristles consisting of a long blade jointed to a stouter shaft, which is embedded in an ectodermal pocket, the chaetigerous sac. It is from a single large cell, the formative cell, at the base of the sac that each chaeta takes origin. The slender bristles assist in locomotion by acting as minute paddles, whilst the second type of chaeta (aciculum), which is shorter but very much stouter, has a skeletal function. There is one aciculum in the

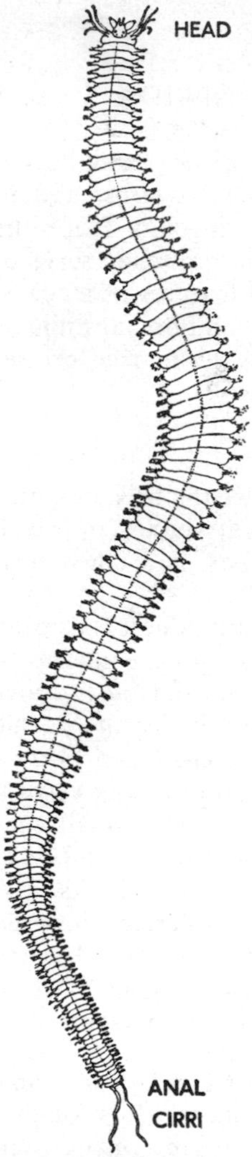

HEAD

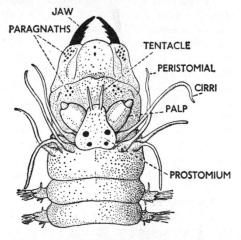

JAW
PARAGNATHS
TENTACLE
PERISTOMIAL
CIRRI
PALP
PROSTOMIUM

ANAL
CIRRI

Fig. 136. NEREIS DIVER-SICOLOR.—Dorsal view.

Fig. 137. NEREIS DIVERSICOLOR.—Dorsal view of anterior end—bucco-pharyngeal region everted.

middle of the notopodial and another in the neuropodial bunch of chaetae. Both types of chaetae appear to consist of chitin and hardened protein (see p. 269).

The head consists of two main parts; a roughly triangular lobe, the **prostomium**, and a ring-like more posterior part, the **peristomium**. The prostomium is not a segment, for it is derived from the preoral lobe of the larva, but the peristomium corresponds to two cephalised segments which have fused together. Various sensory appendages are found on the head. Projecting forwards from the anterior border of the prostomium is a short pair of **tentacles**, and from the ventral side of the prostomium a pair of stout two-jointed **palps**. Both of these types of organ are probably tactile in function. Simple **eyes**, four in number, are found on the dorsal surface of the prostomium. Each eye is formed from an ectodermal vesicle containing a cuticular lens. The visual elements which form the walls of the vesicle are modified ectodermal cells and at the edges are continuous with the epidermis. Other sense organs of doubtful significance are the **nuchal organs**, which are ciliated pits, one on each side of the prostomium. No parapodia are found on the peristomium, but two pairs of **cirri** arise from its anterior part on each side. These cirri are homologous with the neuropodial and notopodial cirri of more posterior segments, and the presence of two pairs on each side of the peristomium is an indication of the fusion which has taken place between two segments during the process of cephalisation. The anal or "tail segment" is distinguished from the true body segments by the absence of parapodia and by the presence of a pair of filamentous processes, the anal cirri. It is frequently termed the pygidium, for it represents the posterior part of the larva so that it does not arise by a process of subdivision or segmentation as do the true body processes, the anal cirri.

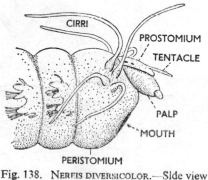

Fig. 138. NEREIS DIVERSICOLOR.—Side view —anterior end.

**The Body Wall.**—The epidermis, which is covered by a thin cuticle, is a single layer of cells of which some are glandular, some are sensory, whilst others are the more or less unspecialised columnar cells or supporting cells. Below the epidermis is a layer of circularly running muscle-fibres and below these, the longitudinal muscles of the body. On the dorsal and ventral parts of the body wall there is a relatively thick layer of circular muscles only in the hinder two-thirds of each segment.   At the sides of each segment, where the parapodia are attached, the circular muscles are displaced to form the anterior and posterior dorso-ventral muscles.   The longitudinal muscles of

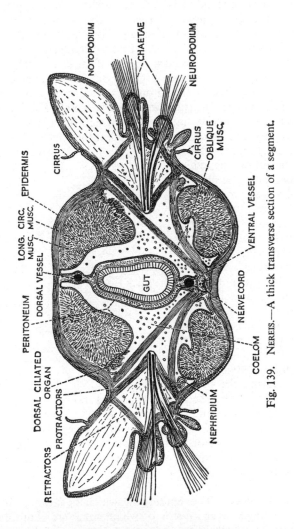

Fig. 139.   NEREIS.—A thick transverse section of a segment.

the body wall do not form a continuous layer around the body nor throughout its length.   They are arranged into four longitudinal bundles, two of which are dorso-lateral and two ventro-lateral in position.   In each of the four bundles the fibres run only the length of one segment, their insertions into the epidermis being largely responsible for the external appearance of segmentation in the worm. The longitudinal muscles are covered on their inner surface by the parietal layer of peritoneum.

**The Coelom.**—The coelom is partially divided up into a linear series of compartments by cross partitions called **septa.** These correspond in position to the external grooves of the body and each septum is formed of a double layer of coelomic epithelium, connective tissue, and muscles. The coelomic compartments, particularly in the anterior part of the body, are, however, not isolated from one another, for the septa are incomplete dorsally and ventrally, so that the coelomic fluid can on occasions circulate more or less freely throughout the body and so is enabled to play a part in the distribution of dissolved gases and in causing changes in turgidity of the body.

The coelom is also partially subdivided by various oblique muscles which run from the dorsal to the ventral body wall. These fall into two main groups according to their insertions, viz., parapodial and interparapodial. There are three parapodial muscles inserted in the anterior surface of each neuropodium and three in the posterior surface. Two slips of interparapodial muscles run from each side of the ventral nerve cord to become inserted in the dorso-lateral body wall just behind each septum. Protrusion of the parapodia is probably very largely due to coelomic fluid being forced into their cavities by the contraction of the body wall muscles, whilst controlled flexures of the parapodia in the vertical as well as in an anterior-posterior direction could be effected by differential contractions of the various oblique muscle-bundles acting against the pressure of the coelomic fluid.

In addition to the oblique muscles, each parapodium has a complicated intrinsic musculature which brings about movements of the various lobes and of the chaetae inserted in them. These muscles appear to be developed as modifications of the circular muscles of the body wall. The largest groups of parapodial muscles are those which take origin from the circular muscles of the body wall around the base of each parapodium and are inserted in the bases of the chaetigerous sacs of the acicula. Simultaneous contraction of all these radiating muscles would cause protrusion of the acicula and central lobes of the parapodium and for this reason the muscles may be termed the protractors of the acicula, which are, therefore in the fullest sense, skeletal structures (pp. 417-65). Other muscles present in each parapodium act as retractors of the acicula, so that it is seen that the parapodia are provided with a complex musculature which, together with the action of the coelomic fluid, is responsible for the varied movements carried out when the worm is swimming or crawling.

**Locomotion.**—Nereids can crawl over the sea-floor and can on occasions swim, although not very rapidly. Probably their swimming merely allows them to exploit water currents and so get carried to more favourable situations. Either when crawling or swimming the longitudinal muscles of the body wall by alternating contraction and relaxation on opposite sides of the body cause a series of side to side flexures of the body to pass, wave-like, from rear to front. These on their own would cause the worm to move backwards. Forward movement is due to the co-ordinated movements of the parapodia and chaetae. These movements also proceed in a wave-like manner which spreads from rear to front, each move involving four to eight segments. Either when crawling or swimming the parapodia on the convex side of a lateral flexure (longitudinal muscles relaxed) are swung forwards whilst their acicula are protruded and dig in. At the same time the parapodial muscles cause the parapodia to swing backwards and so drive the worm forward. Clearly the contraction of the longitudinal muscles of the body wall reinforces this movement. More vigorous movement causes the worm to swim.

**Alimentary Canal.**—The alimentary canal shows a distinct advance on that of the flatworms for it has two openings to the exterior: an anterior one, the mouth, at which food is ingested and a posterior one, the anus, through which undigested food-remains are egested or voided. Food can thus be passed in a continual stream from mouth to anus, whereas in the flatworms, where an anus is not present, food-remains must pass out through the mouth. Moreover, movements of the body wall concerned with progression can take place quite independently of movements of the gut because of the intervention of the coelom. The mouth opens through the peristomium into a wide region termed the buccal cavity and this leads into a very muscular part, the pharynx. Both the buccal cavity and the pharynx are lined by an ectodermal epithelium which is continuous in front with the epidermis of the head, and it is not surprising, therefore, that this lining epithelium bears a cuticle. Local thickenings of the cuticle give rise to paragnaths or teeth (whose number and arrangement is characteristic of the species), but one pair of these cuticular structures near the posterior end of the pharynx is especially large and is termed the "jaws". Regions of the gut lying posteriorly to the pharynx are lined with endoderm and form the mesenteron in contrast to the buccal and pharyngeal regions, which because of their ectodermal origin constitute the stomodeum. The short posterior part of the gut which opens on the pygidium by the anus is also lined by ectoderm and constitutes the proctodeum. The mesenteron is divisible into two main sub-regions, the oesophagus and intestine, and in both of these the epithelium is responsible for the secretion of digestive juices and for the absorption of digested food; but digestive juices are also formed by a pair of laterally placed oesophageal glands.

**Feeding.**—Species of *Nereis* are usually stated to have an entirely carnivorous diet and to feed on small animals which are seized by means of the "jaws" and teeth lying in the wall of the stomodeum. This is made possible by the fact that the whole of the buccal and pharyngeal region can be everted. When this happens the buccal cavity becomes turned inside out, whilst the pharynx is thrust forward so that the "jaws" are opened and come to lie in front of the head. Eversion of the buccal cavity is brought about chiefly by pressure of the coelomic fluid which is forced into the buccal region by the contraction of the body wall muscles, whilst the forward movement of the pharynx is mainly effected by the contraction of the protractor muscles of the pharynx. The folding in of the buccal cavity and retraction of the pharynx is due to the action of retractor muscles accompanied by a relaxation of muscles of the body wall, which allows a redistribution of the coelomic fluid.

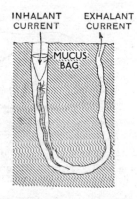

Fig. 140. NEREIS DIVERSI-COLOR.—Filter-feeding in its burrow.

A muscular partition stretching from the gut to the body wall between segments two and three also helps in retraction and eversion of the bucco-pharyngeal regions.

It is undoubtedly true that most nereids are predators but *N. diversicolor* can and does on occasions feed in other ways. Sometimes it will protrude from its burrow and collect small nutritious particles from the surface of the mud. At other times it will secrete a cone of mucus within its burrow and draw a current of water through it by dorso-ventral undulations of the body (Fig. 140). The mucous cone acts like a net and fine particles in the water are strained off. Then, at intervals, the worm engulfs the "net" together with food it has collected. *Platynereis dumerilii* is a vegetarian feeding mainly on *Ulva* which it slices up with its jaws.

**The Blood Vascular System.**—The blood vascular system consists of a system of vessels containing blood in which is dissolved a red pigment (haemoglobin) and which contains cells (corpuscles). Most annelid worms have haemoglobin as a respiratory pigment but some polychaetes have a green one (chlorocruorin) and others again have red and green pigment mixed in their blood. Haemoglobin occurs also in the blood and tissues of many members of invertebrate phyla other than annelida. The pigment is concerned with the conveyance of oxygen from the respiratory surfaces to the tissues (see Chapter XVII); but carbon dioxide, disolved food substances, and excretory products, as well as other substances, are also distributed about the body by the blood and reach the tissues by diffusion through the walls of the vessels. The greater part of the exchanges between the tissues and the blood takes place through the smallest vessels which are termed capillaries, for these alone have thin walls and allow of rapid diffusion. In brief, it may be said that the vascular system of *Nereis* like that of most other animals consists of a distributing system of vessels which communicates with a collecting system by means of networks of capillaries, and that by means of certain contractile vessels the blood is maintained in constant circulation. In *Nereis* the main collecting vessel (dorsal vessel) lies in the mesentery just dorsal to the gut, and is linked to a main ventral vessel—which is the main distributing vessel —just below the gut by a series of segmental commissural vessels. There are two of these in every segment, but they do not pass direct from the ventral to the dorsal vessel, but first supply the body wall and parapodia where they join an extensive capillary network. The gut also receives its blood-supply from segmental vessels arising from the ventral vessel and after circulating through a capillary network in the gut wall the blood is returned to the dorsal vessel by two pairs of vessels in every segment. Most of the larger vessels are contractile.

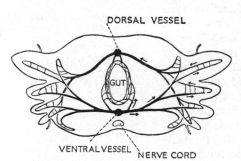

DORSAL VESSEL

GUT

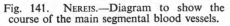

VENTRAL VESSEL     NERVE CORD

Fig. 141.  NEREIS.—Diagram to show the course of the main segmental blood vessels.

**The Excretory System.**—The excretory system consists of segmentally arranged nephridia which are coiled ciliated tubules opening from the coelom by a ciliated funnel, the nephrostome (nephridiostome), and to the exterior by a small pore, the nephridiopore, at the base of the parapodium. This type of nephridium closely resembles in general plan that to be described in more detail for the earthworm, and it is probable that its excretory role is carried out in much the same way in both animals (pp. 260, 261).

**The Neuro-sensory System.**—The central nervous system conforms to a plan which is very common in invertebrate animals in that the central nervous system consists of a mass of nervous tissue, the cerebral or suprapharyngeal ganglia ("brain"), lying in the prostomium, and this is connected by a nerve collar to a ventral nerve cord situated in the mid-ventral line throughout the length of the body.

The "brain", like other parts of the central nervous system, consists of nerve-cells and nerve-fibres, and the bulk of these are concerned with receiving and relaying efferent (sensory) impulses from the organs of special sense situated on the head.

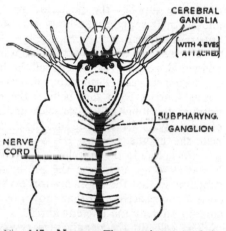

Fig. 142. NEREIS.—The anterior part of the nervous system.

Three main centres are recognised in the "brain"—an anterior one from which a pair of short nerves innervates the prostomial palps; a middle region which receives a stout nerve from each of the four eyes and a pair of short nerves from the prostomial tentacles; and a posterior centre into which run the nerve-fibres from the sensory cells lining the nuchal organs. In all three centres the nerve-cells surround a centrally placed mass of nerve-fibres. The middle region also bears a small pair of lobes (corpora pedunculata), which are association centres which co-ordinate all the impulses entering the "brain". It is most unlikely that the eyes are image-forming and their main function seems to be to mediate the phototactic and photokinetic responses of the worm. Usually nereids respond to increased light intensity by a sudden contraction when they are in a burrow.

If out of a burrow increased light stimulates them to swim or crawl more rapidly (positive orthokinesis) and, later; to burrow again (negative phototaxis), but the sign of this response is reversed during the heteronereid phase.

The anterior pairs of peristomial cirri are supplied by nerves joining the ganglia on the nerve collar. These ganglia may represent nerve centres originally present in the ventral nerve cord which, owing to cephalisation, have shifted dorsalwards and become incorporated in the nerve collar. The two halves of the collar unite in the ventral part of segment three, where they join the "sub-pharyngeal ganglion" which is formed by the fusion of the two anterior ganglia of the ventral nerve cord, as is shown by the nerves arising from it which supply two segments. Thus, one pair of nerves supplies the posterior pair of peristomial cirri whilst another pair supplies the body wall and parapodia of the second apparent segment.

The rest of the nerve cord consists of a chain of elongated segmental ganglia (to which cell bodies are confined) united by short connectives. Four pairs of nerves arise from each ganglion. Nerve I arises from just behind the septum and with Nerve IV supplies the longitudinal muscles and dorsal and ventral surfaces of the body. Nerve II supplies the parapodia and joins the ganglion near the middle. Nerve III is mostly composed of fibres from proprioceptors in the muscles. It lies a short way behind Nerve II. Nerve IV originates from the preseptal part of the next posterior ganglion so that the segmentation evidenced by the septa does not correspond exactly with that shown by the ganglia. All the nerves contain afferent and efferent fibres.

The cells in the ganglia are two types, relay or internuncial neurons and motor neurons; for unlike the central nervous system of vertebrates the afferent cells lie outside the central nervous system. Most of the sensory cell-bodies lie in the epidermis, and their afferent (sensory) fibres pass inwards in the segmental nerves and then synapse with the internuncial cells of the nerve cord; but some of the sensory fibres end in a sub-epidermal nerve plexus of multipolar association cells and their fibres.

The nervous elements which co-ordinate the movement of the gut consist of nerve-cells and nerve-fibres, forming a sub-epithelial plexus in the gut wall connected at the anterior end by a few small nerves to the posterior end of the cerebral ganglia. The internuncial cells are of two main types (a) those with giant axons and (b) those with fine axons which play a similar part in co-ordinating the movements of the body to those in earthworms, linking afferent with efferent fibres within the cord. Motor neurons in the ganglia send fibres in the segmental nerves to the muscles but some of them synapse with relay neurons whose cell-bodies lie along the course of the nerve.

From the brief account given it is plain that the central nervous system of *Nereis* is a mechanism for the co-ordination of the various receptor and effector units of the body, which are linked up in an appropriate way by a very complicated system of nerve-cells and fibres. The animal is thus able to react immediately to any stimulus, no matter where it is received, by suitable movements of the *whole body*. Thus, the central nervous system has as one of its most important properties an integrating effect (unification) on the responses of the separate effector organs so that each one assists in building up a "pattern of behaviour", which although characteristic for any given type of animal can be varied within limits as occasion demands.

Nereids can learn to find the way through simple mazes and to disregard mildly unpleasant stimuli but the ability to retain learned behaviour is greatly diminished if the brain is removed.

**The Reproductive System.**—Coelomoducts as such are absent and are represented by the so-called dorsal ciliated or-

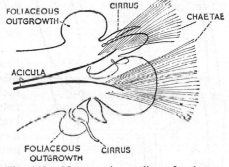

Fig. 143. NEREIS.—An outline of a heteronereid parapodium.

gans. These are segmental patches of ciliated epithelium on the peritoneum just below the dorsal column of longitudinal muscles. The dorsal ciliated organs closely resemble the internal ciliated funnels of the coelomoducts of other polychaetes, and the condition in *Nereis* has probably been secondarily arrived at by the failure of the funnels to acquire a duct to the exterior.

The sexes are separate in *Nereis*, as in most other polychaetes, and the gonads (testes or ovaries) develop by the proliferation of coelomic epithelial cells where they overlie the principal blood vessels of the body. The proliferated cells are spermatogonia or oogonia and break free to undergo further development in the coelom. The gonads are temporary structures and cannot be detected as distinct organs at certain seasons of the year after the gametes have been shed. The development and proliferation of the gametes is, in some species of *Nereis* but not *N. diversicolor*, accompanied by changes in the structure of the posterior half of the body in which the gametes are especially abundant. These changes especially affect the parapodia and the chaetae, but the prostomial eyes become enlarged and the pygidium develops sensory papillae. The mature worms which have undergone these modifications are termed **heteronereids,** and these, instead of living in burrows or in crevices or even creeping about on the sea-floor, then swim actively in the surface waters. Swimming is facilitated by the changes in the structure of the parapodia of the posterior half of the body, for they increase in size and develop flattened foliaceous outgrowths on each of the lobes. The nor-

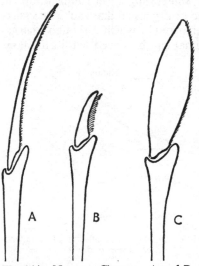

A    B    C

Fig. 144. NEREIS.—Chaetae. A and B, from ordinary parapodia. C, oar-like chaeta of heteronereid.

mal chaetae become replaced by flattened oar-like bristles which are inserted into the neuropodia and notopodia in a fan-like manner, so offering a large surface which renders swimming more efficient.

Metamorphosis to the heteronereid condition is inhibited by "juvenile" hormones (p. 331) formed by neurosecretory cells in the brain. Seasonally these hormones are withdrawn (or inhibited) from the circulation and the worms change to heteronereids. Maturation of the gonads is controlled by the same means. Nereids can regenerate segments lost at the hinder end and the ability to do this is controlled by a hormone formed in the brain.

**Life-history.**—Release of the gametes takes place by rupture of the body wall, a process which is rendered necessary by the absence of coelomoducts, and after the eggs are fertilised in the surface layers of the sea they develop into small ciliated larvae called **nectochaetes,** from which the new adults develop after a gradual metamorphosis, in which most of the larval characters are lost. The larvae which hatch from the eggs of some polychaetes are termed **trochophores.** They are small, transparent organisms having the general appearance of a spinning top. Two main bands of cilia encircle the body whilst in some trochophores there may be a third girdle of cilia. No traces whatsoever of metamerism are present and, in fact, the rudiment of the adult trunk is represented

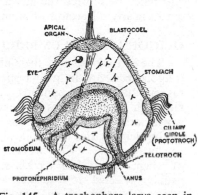

Fig. 145. A trochophore larva seen in optical section.

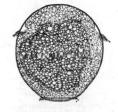

Fig. 146. NEREIS.— Young larva (after Dales).

only by a small region at the lower pole of the larva. There is no coelom at this stage but a spacious blastocoel surrounds the larval gut. Also in the blastocoel is a pair of protonephridia, a certain amount of mesenchyme, and larval muscles. At the upper pole lies an apical sense organ bearing sensory "cilia" and sometimes eyes, whilst below these are the cerebral ganglia. A larva of a very similar nature is also found in the life-history of many molluscs. It is adapted to swimming and feeding on microscopic organisms in the plankton and it is interesting to notice that those

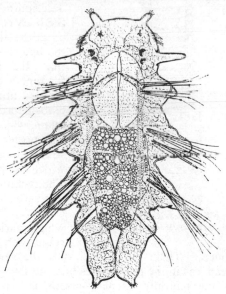

Fig. 147. NEREIS.—Larva with four chaetigerous segments (after Dales)

polychaetes, the vast majority of which have yolky eggs, have larvae which hatch at a later stage of development and which frequently take to life on the bottom directly after hatching. Thus, the provision of food in the form of yolk tends to shorten the larval life and this tendency can be observed in nearly all groups of animals.

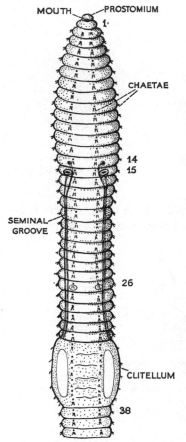

Fig. 148. LUMBRICUS.—Ventral view of the anterior third of the body.

## OLIGOCHAETA—LUMBRICUS

The earthworms and their allies (oligochaetes) are, in many respects, rather specialised annelids, and although many species of earthworms are to be found in the soil of the British Isles and, indeed, in nearly every part of the world, one of them, *Lumbricus terrestris*, is usually selected for detailed description because its anatomy and habits have been studied so fully.

**External Features.**—The general appearance of an earthworm is so well known that it scarcely needs description. It will be recalled that the body is very long compared with its girth; that the anterior third of the body is roughly cylindrical; and that the posterior two-thirds is flattened dorso-ventrally. The pointed anterior end bears a small club-shaped upper lip, the **prostomium**, which overhangs the mouth. The posterior end of the body ends somewhat abruptly in the pygidium, which is perforated by the anus. The metameric nature of the body is clearly indicated on the outside, as in *Nereis*, by the circular grooves which mark off about 150 annuli or rings, each corresponding to a body segment. There is, however, not the same clear indication of a "head" as in the polychaetes but the first body segment, in which lies the ventrally situated mouth, is the peristomium and from its dorsal side there projects forward the prostomium. The ventral surface of the body is easily distinguished from the dorsal surface by its lighter colour and also by the presence of small bristles, the chaetae, which, however, are not so numerous as those of the polychaetes. There are four pairs of chaetae in each segment except the first and last, and of these pairs two are ventral and two are ventrolateral in position. Each chaeta is a curved rod composed of chitin, hardened and strengthened by the addition of sclerotised protein (cf. arthropod cuticle, p. 268), and is embedded in an ectodermal pit, the chaetigerous sac. The two sacs of adjacent members of a pair of chaetae fuse at their inner (proximal) ends and project into the body cavity in the form of a cone to the apex of which are attached the muscles responsible for moving the chaetae. Each chaeta is produced as the secretion of a

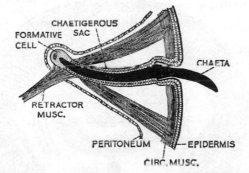

Fig 149. LUMBRICUS.—A chaeta in its sac (after Sajovic—redrawn from Stephenson).

single large ectodermal cell which lies at the bottom of the chaetigerous sac.

Typically, the body wall in each segment is perforated by three apertures, viz. the dorsal pores, which lie in the mid-dorsal line in the anterior part of the segments, and the paired **nephridiopores** which are just in front of the ventral rows of chaetae. Dorsal pores are, however, absent on the first nine segments and from the last, whilst there are no nephridiopores in the first three segments or, again, in the last.

Although, as has already been implied, all the body segments are fundamentally alike in structure, certain of them depart from the typical condition in one or more important respects. Thus, the epidermis in segments thirty-two to thirty-six or thirty-seven is thickened and highly glandular, to form the so-called saddle or **clitellum**, which plays an important part in the reproductive processes. The thickening is most pronounced on the dorsal and dorsolateral surfaces and becomes especially prominent when the worm is sexually mature. The clitellar segments are also distinguished by having very long slender chaetae. Other segments deserving special mention are those in the neighbourhood of the reproductive organs. The male ducts (**vasa deferentia**) open by a pair of slit-like apertures, guarded by lips, on the ventral surface of segment fifteen, whilst the apertures of the **oviducts** are similar but very much smaller, and lie on the ventral surface of the fourteenth segment. Two pairs of minute pores are found on the ventrolateral surface in the grooves between segments nine and ten, and ten and eleven. They are the openings of the **spermathecae**.

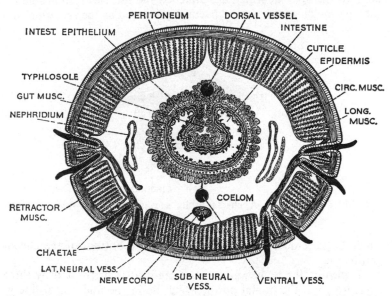

Fig. 150. LUMBRICUS.—Diagrammatic transverse section through the intestinal region of the body.

Extending from the apertures of the vasa deferentia on each side to the clitellum are pigmented lines which mark the position of grooves (**seminal grooves**) in the epidermis. In segments ten, eleven, and twenty-six there are paired masses of glandular tissue which are visible externally as light-coloured swellings around the pairs of ventral chaetae.

**The Body Wall.**—The body wall is covered externally by a thin outer **cuticle** composed of layers of parallel collagenous fibres running at right angles in an upper and lower set and perforated by numerous minute pores through which open the epidermal glands. The cuticle is secreted by the supporting cells of the underlying epidermis. The epidermis is a single layer of cells in which are found mucus-secreting (goblet) cells, albuminous gland cells, sensory cells, and supporting cells. Most of the cells are of the columnar type. Below the epidermis lies a thin layer of nervous tissue (subepidermal plexus) and below this again, except in the intersegmental positions, a layer of circular muscles. The longitudinal muscles are continuous throughout the length of the worm but are arranged in nine blocks or columns. Two of these are dorsolateral, six are ventrolateral, and one is ventral in position. Each column is composed of numerous ribbon-like fibres which are arranged in bundles, the bundles being separated by a connective tissue lamella in which run blood vessels. The muscle-fibres are composed of single elongated cells each with a nucleus, the outer (cortical) part of the cyto plasm being the contractile substance of the fibre. A lamella together with the fibres which lie on each side of it therefore has, in transverse section, a feather-like appearance. The body wall is completed by the somatic layer of peritoneum which thus separates the muscles from the coelom.

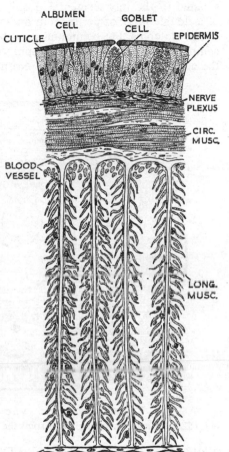

Fig. 151. LUMBRICUS.—Diagrammatic transverse section of a portion of the body wall.

**The Coelom.**—The coelom is a spacious cavity filled with coelomic fluid in which are suspended colourless coelomic corpuscles. Some of these are amoeboid and phagocytic. The coelom is bounded on all sides by a layer of peritoneum (coelomic epithelium) which, where it covers the intestine (splanchnic layer), gives rise to masses of large yellow cells, the chloragogenous tissue. The coelom is subdivided by septa, which are vertical cross-partitions situated in the intersegmental positions. Each septum is formed of a double layer of peritoneum and numerous interlacing bundles of muscle-fibres, but communication between the various coelomic compartments is provided by a sphinctered aperture just dorsal to the nerve cord in each septum. Normally, the sphincter muscle is

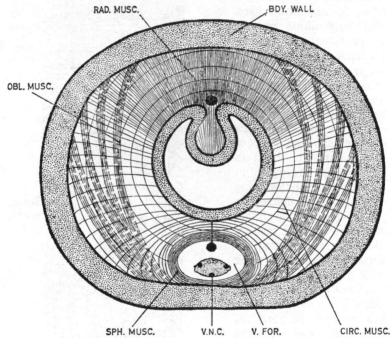

Fig. 152. LUMBRICUS.—Diagram to show the arrangement of the septal muscles.

contracted so that each coelomic compartment is isolated from its neighbours. The coelom opens to the exterior by way of dorsal pores, nephridia, and gonoducts.

**The Alimentary Canal.**—The alimentary canal, although not uniform in diameter or in the nature of its walls, takes the form, when the worm is fully extended, of a straight tube running from the anterior mouth to the terminal anus. When, however, the worm contracts the gut, and in particular the intestine, becomes folded so that it has a sinuous appearance. The mouth, which is

overhung by the prostomium and perforates the first true segment
(peristomium), opens into a short wide, thin-walled, buccal cavity.
This has a small ciliated dorsal diverticulum and also a small ventral
pocket which, however, is not ciliated. The buccal cavity leads
into the pharynx, which has thick muscular walls and extends
back as far as the sixth segment. Like the buccal cavity, the pharynx
has a ciliated dorsal diverticulum and cilia also occur in patches in

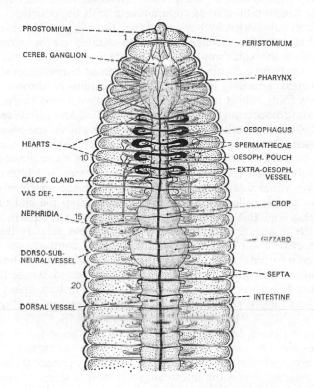

Fig. 153. LUMBRICUS.—The main organs as seen when dissected from the dorsal
surface. (Vesiculae seminales removed.)

other regions. In addition to the mucous goblet cells of the pharyngeal epithelium, there are also glandular masses in clusters on the
dorsal and lateral walls of the pharynx. These glands (pharyngeal
glands) are whitish in colour and secrete mucus and a proteolytic
enzyme into the pharyngeal cavity. The oesophagus which follows
the pharynx and ends in segment fourteen, is a thin-walled tube
slightly constricted in each intersegmental position, but it is not

uniform in structure along its length in which three parts can be recognised: an anterior one with broad longitudinal folds, a middle one (into which open the longitudinal pouches), and a short hind region. Between the lining epithelium (ciliated on its lateral walls) of the middle part of the oesophagus and its circular muscles is on each side a series of lamellae which are continuous with similar ones projecting into the cavities of the oesophageal glands and pouches. The spaces between these lamellae form longitudinal channels whereby the glands communicate with the pouches. It is as if a duct subdivided by horizontal partitions is embedded in the gut wall to provide a storage region and ciliated pathway for the secretion of the calciferous glands before it enters the gut. In segment ten there is a pair of forwardly directed lateral outpushings (oesophageal pouches) in which the epithelium is thrown into numerous folds, whilst in each of segments eleven and twelve is a pair of white calciferous (oesophageal) glands. These glands secrete crystals of calcium carbonate and open into the gut via the oesophageal pouches, their openings being guarded by sphincter muscles and backwardly directed valves. The function of these glands will be referred to later (p. 260). In segments fourteen and fifteen the alimentary canal dilates to form a thin-walled sac which is dark in colour and is called the crop. Its walls overlap the end of the oesophagus and this prevents regurgitation of the food when the crop walls contract. The interior surface of the crop is thrown into numerous folds so that during the frequent contractions of its walls food is forced through a veritable maze of epithelium. The crop opens into a thick-walled muscular region, the gizzard, whose lining epithelium bears a distinct cuticle. The gizzard in turn opens into the last and longest part of the gut, the intestine, which is yellowish in colour because of its covering of chloragogen cells which is much denser here than in parts of the gut lying more anteriorly. For the first two-thirds its dorsal wall is infolded to form the typhlosole, so that in cross-section the lumen of the intestine appears U-shaped. In this anterior region the intestinal walls bear a series of sac-like dilations which regularly expand and contract during life. The rest of the intestine has smooth walls. The typhlosolar epithelium is composed of glandular and ciliated cells. Certain of these gland cells secrete a mucopolysaccharide material which after being carried ventralwards by cilia, solidifies to form a peritrophic membrane which lines the intestine. The muscular coats consist of outer longitudinal fibres and inner circular fibres but these are not continuous across the typhlosole. Like other regions of the gut, the intestine is invested by the splanchnic layer of peritoneum which gives rise to the chloragogen cells.

**The Blood Vascular System.**—The blood has much the same composition as in *Nereis*.

The arrangement of the vessels is very complicated but the general plan, however, is that longitudinally running vessels act as collecting and distributing channels and these are linked up with each other by circularly running vessels in the individual segments. Thus, there is a large longitudinally running **dorsal vessel** which for the greater part of its length is in close contact with the gut. It forms the main collecting vessel of the body and is connected in segments seven to eleven inclusive by five pairs of

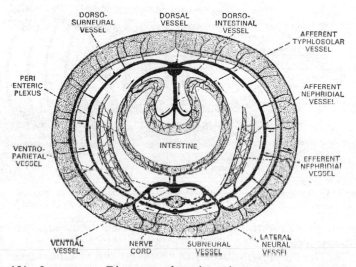

Fig. 154. LUMBRICUS.—Diagram to show the main segmental blood vessels in the intestinal region.

large commissural vessels (**hearts** or **pseudohearts**) to the longitudinal **ventral vessel**, which is suspended in a mesentery just below the gut and forms the main distributing channel. Other important longitudinally running vessels are the **subneural** and the paired **lateral neural** vessels. The lateral neural vessels supply blood to the nerve cord, having received this blood from the ventro-parietal vessels. After passing through the capillaries of the nerve cord, the blood drains into the subneural vessel. On each side of the oesophagus are the paired **extra-oesophageal** vessels which collect blood from the pharynx and oesophagus and body wall and empty it into the

**dorso-subneural** vessels (see below) of segment twelve. The extra-oesophageals are also connected by segmental commissures to the subneural vessel. In every segment behind the eleventh the dorsal and subneural vessels are connected and thus placed in direct communication by a pair of **dorso-subneural** vessels which run in the septa and receive blood from the body wall. The dorsal vessel also receives two pairs of vessels **(dorso-intestinals)** per segment which drain the **peri-enteric plexus** and, in the region of the intestine

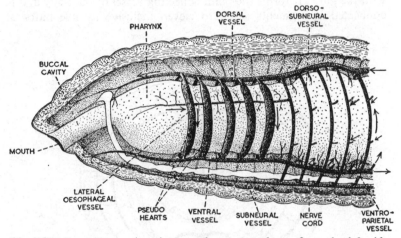

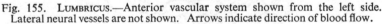

Fig. 155. LUMBRICUS.—Anterior vascular system shown from the left side. Lateral neural vessels are not shown. Arrows indicate direction of blood flow.

proper, gives off three small vessels per segment which supply blood to the typhlosole.

The ventral vessel distributes blood to the various organs of the body by the following branches: **ventro-parietals**, of which there is one pair per segment running to the body wall; and three **ventro-intestinals** which in each segment supply blood to a plexus of vessels (peri-enteric plexus) lying in the wall of the gut between the circular muscles and the gut epithelium. The excretory organs **(nephridia)** receive blood by a number of small vessels **(afferent nephridial vessels)** from the ventro-parietals and return it by three main **efferent nephridial vessels** and a number of smaller ones to a vessel which empties into the dorso-subneural vessel of that segment on each side. The reproductive organs receive most of their blood from branches of the ventro-parietals.

In all instances the vessels which supply blood to the various parts of the body are connected to the vessels which drain these parts by a network of minute vessels termed capillaries.

It will have been gathered from this account that the blood in the living animal is in continual circulation through the blood vessels along which it is driven by the rhythmical contraction of certain of the main vessels. The main dorsal vessel contracts rhythmically in a wave-like fashion from behind forwards and the hearts contract in a similar way downwards in unison with the dorsal vessel and drive the blood towards the ventral vessel. Blood flows forwards in the dorsal vessel, and, behind the hearts, backwards in the ventral vessel. But in front of the hearts, blood has a forward flow in the ventral vessel also. Other vessels are also contractile and muscle-fibres enter into the composition of their walls. The direction of the circulation is aided by a system of valves, which are developments of the epithelium lining the vessels. They occur in the dorsal vessel, in the hearts, and at the points of junction of the chief vessels with the dorsal vessel.

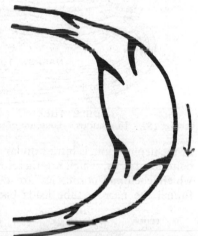

Fig. 156. LUMBRICUS.—Diagram of a pseudoheart to show the valves.

**The Excretory System.**—The organs of nitrogenous excretion consist of segmentally arranged tubules called **nephridia**, which are present in all segments except the first three and the last one. Each nephridium opens from the coelom by a ciliated funnel, the nephridiostome, into a tube which loops about in a complicated way before opening to the exterior by the nephridiopore. Five main portions are recognisable in each nephridium: the nephrostome, the narrow tube, the middle tube, the wide tube, and the muscular tube. The nephrostome (nephridiostome) takes the form of a flattened funnel whose upper wall (upper lip) is much larger than its lower wall (lower lip), whilst the side walls are completed by ciliated "gutter cells" which are similar in nature to, and continuous with, the cells of the narrow tube. The upper lip is built up of a horseshoe-shaped row of marginal cells surrounding a large crescentic central cell and is covered on its upper surface by a layer of coelomic epithelium. The marginal cells are richly ciliated, both on their outer boundaries and on their lower surfaces, whilst the central cell also is ciliated on its lower side. All the cilia beat towards the bottom of the funnel. The lower lip consists merely of a few non-ciliated cells covered on their lower surface by peritoneum. The funnel leads into the narrow ciliated tube into the lumen of which project two rows of cilia. The lumen, like that of the greater part

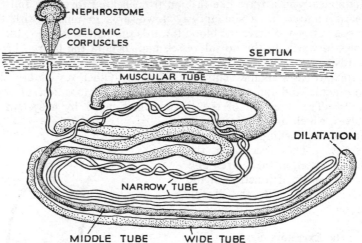

Fig. 157.   LUMBRICUS.—Diagram of a nephridium (altered after Benham).

of the nephridium, is intra-cellular; that is, it perforates a long single column of cells, which are therefore called drain-pipe cells, around which coelomic corpuscles are clustered. From the base of the funnel the narrow tube leads backwards and passes through the

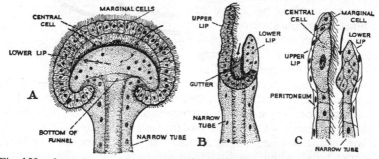

Fig. 158.   LUMBRICUS.—Nephrostome (based on Goodrich).    A, ventral view. B, lateral view.   C, longitudinal section.

septum to form a series of loops in the segment immediately posterior to that in which the nephrostome is situated. The middle tube which follows it is ciliated throughout, the cilia lying in three tracts. It is brown in colour and leads into the wide tube, which like the

**muscular tube** into which it empties, is non-ciliated. This last region, which is the only portion of the nephridium whose lumen is extra-cellular, opens by way of the nephridiopore which is placed near the anterior border of the segment.

The tubular portions of the nephridium are in intimate contact with a plexus of small blood vessels.

**The Nervous System.**—The nervous system is built on the same general plan that is found in *Nereis*. A bilobed ganglionic mass, the cerebral ganglia ("brain"), lies just above the pharynx in segment three and is connected by a nerve collar (circum-pharyngeal connectives) to the ventral nerve cord which runs backwards in the body in the mid-ventral line, just inside the coelom, to the extreme posterior end. The nerve cord is surrounded by a layer of longitudinal muscle-fibres, outside which is an investing layer of peritoneum. The cerebral ganglia, circum-pharyngeal connectives, and ventral

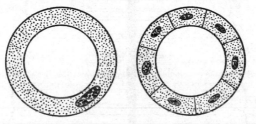

Fig. 159. LUMBRICUS.—Transverse section of A, an intra-cellular duct. B, an extra-cellular duct.

nerve cord together constitute the central nervous system from which a series of nerves (peripheral nervous system) arises and is distributed to the various parts of the body. The prostomium is innervated by a pair of stout nerves from the cerebral ganglia, whilst the paired peristomial nerves arise from the nerve collar. Segment two, also, is innervated by a pair of nerves from the lower ends of the nerve collar. The anterior end of the ventral nerve cord in segment four is swollen out to form the sub-pharyngeal (sub-oesophageal) ganglion and from it arises a pair of nerves to segment two, three pairs of nerves supplying segment three as well as three other pairs to segment four. This state of affairs indicates that the sub-pharyngeal ganglion is a compound structure formed by the fusion of the segmental ganglia of segments two, three, and four. In every segment behind the sub-pharyngeal ganglion the ventral nerve cord gives origin to three pairs of nerves, one of which arises near the anterior end of the segment and the other two pairs close together from the swelling (segmental ganglion) into which the nerve cord enlarges in each segment. Each of the segmental nerves passes out laterally and then divides into two branches, one of which passes dorsalwards in the lateral body wall between the circular and longitudinal muscles, whilst the other branch runs in a similar

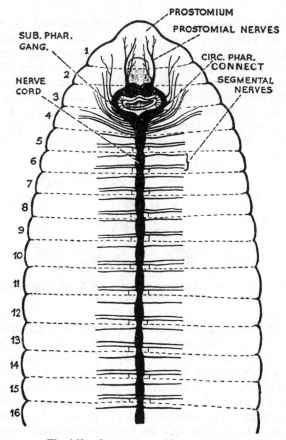

Fig. 160. LUMBRICUS.—Nervous system.

relative position in the ventral body wall. Thus the right and left members of a pair of nerves together form an incomplete **nerve ring** in the body wall. From the nerve rings branches are given off to the muscles where they contribute to the formation of an inter-muscular plexus, and also pass to the epidermis where some of them link up with a sub-epidermal plexus of nerve-cells and fibres.

The gut also is innervated by a plexus of nerves lying just below the lining epithelium, between it and the circular muscles of the gut. This enteric plexus ("sympathetic system") is connected to the central nervous system by six nerves from each side of the nerve collar.

The intermuscular and sub-epithelial plexuses of the gut and skin are comparable to the nerve net of coelenterates.

Like the gross anatomy, the finer structure of the nervous system resembles in essentials that of many other invertebrates. The cerebral ganglia are composed of an outer layer of small ganglion cells which send their processes to a centrally placed mass of nerve-fibres, which are prolongations of sensory cells situated on the anterior end of the body, and also of fibres from internuncial neurons of the ventral nerve cord. The cerebral ganglia seem to be concerned chiefly with relaying impulses from sense organs on the skin at the anterior end. The ventral nerve cord, on the other hand, is built up of motor cells, internuncial cells, and the fibres of these cells, together with the terminations of

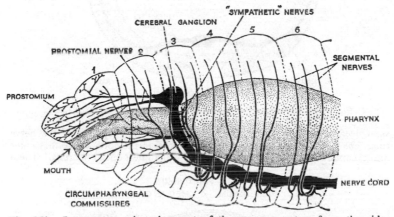

Fig. 161. LUMBRICUS.—Anterior part of the nervous system from the side (after Hess).

fibres from sensory cells. The cells, which are chiefly grouped in the ventral and lateral parts of the cord, are especially abundant in the segmental swellings (ganglia).

The nerve cord itself is surrounded by a connective tissue sheath (epineurium), which is continuous with a median vertical septum, which partially divides the cord into right and left halves. For this reason the nerve cord is frequently spoken of as being a double structure, but throughout its length the right and left halves are connected by transversely running nerve-fibres. The epineurium also continues into a horizontal partition of connective tissue which separates off a more dorsal compartment, through which run the three chief giant fibres. Additional support is lent to the cord by a system of cells and fibres (neuroglia) which are non-nervous and penetrate everywhere between the nerve substance.

The distribution of the various physiological types of cells and their fibres (components) of the nervous system is not fully understood, but it seems to be fairly clearly established that the afferent (sensory) fibres are direct prolongations of sensory cells situated outside the central nervous system. Those concerned with receiving impressions from the outside world lie, for the most part, in the epidermis, but others, which respond to changes in the muscles, lie in between the muscle-fibres and in the nerve rings. Some of the sensory fibres enter the

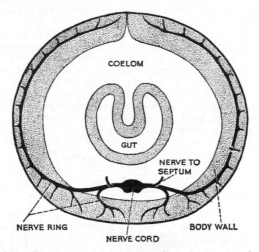

Fig. 162.   LUMBRICUS.—Diagram to show the segmental nerves.

sub-epidermal plexus, whilst others pass without interruption *via* the nerve rings (chiefly in the second pair in each segment) to the cord, where they enter into synaptic connection with processes from the internuncial neurons.   These

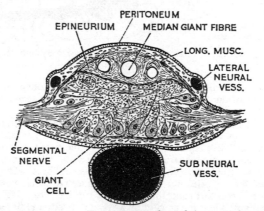

Fig. 163.   LUMBRICUS.—Transverse section of the ventral nerve cord.

link up different segments of the body and also opposite sides of the cord.   The motor cells send efferent fibres to the muscles and other effector organs by way of the nerve rings, but some of the fibres, before leaving the cord, cross over to the opposite side.

In every segment of the body, situated in the ventral part of the nerve cord is a group of especially large cells (**giant cells**) which send segmental processes to a group of three very stout fibres (**giant axons**) lying in the dorsal part of the cord. These giant fibres begin in the sub-pharyngeal ganglion and traverse the whole length of the body to the posterior end, but they are not continuous, for in each segment there is an oblique partition across each fibre (which is probably a sort of synapse). The lateral giant fibres are linked together by cross connections in each segment, and also send small branches both to the fibrous part of the cord and to the middle pair of nerves in each segment. The latter fibres pass out of the cord and enter into synapse with a motor cell, which in turn sends a motor fibre to a muscle. Two smaller giant fibres situated in the ventral part of the cord have also been described. The giant fibre system is perhaps best regarded as a part of the system of internuncial neurons, which has become enlarged and specialised for the conduction of motor impulses at a

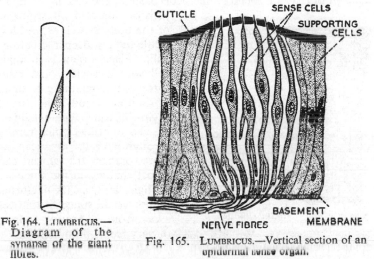

Fig. 164. LUMBRICUS.—
Diagram of the
synapse of the giant
fibres.

Fig. 165. LUMBRICUS.—Vertical section of an
epidermal sense organ.

rapid rate in a system which links up all the body segments. This rapid conduction has been clearly shown by numerous experiments, and it has also been demonstrated that in the living animal the middle fibre conducts from the anterior towards the tail-end, whilst the lateral fibres conduct in an opposite direction. The advantages of such a system are fairly obvious, for it enables all the body segments to contract practically simultaneously, so effecting a very rapid withdrawal of the body from any noxious stimulus and enabling the worm to disappear very quickly into its burrow.

**Sense Organs.**—As in other Metazoa, the organs concerned with the reception of stimuli from the outside world are situated on, or near, the outside of the body. In the earthworm they are relatively simple in structure and consist either of single or of small groups of specialised ectodermal cells. From the base of each cell a sensory (afferent) fibre continues without interruption into the central nervous system, where it synapses with a fibre from an internuncial neuron.

Two main types of receptor organs are found: **epidermal sense organs** and **photosensitive organs (photoreceptors)**. The former are found in the epidermis and each consists of a group of cells each of which terminates in a small hair-like process which perforates, and projects beyond, the cuticle. Epidermal sense organs are particularly abundant at the anterior end, but are also found in the buccal cavity and all the body segments except those of the clitellum. They respond to the tactile stimuli and, in all probability, also to chemical stimuli and changes in temperature. The photosensitive organs are found in the deep parts of the epidermis of the prostomium and all segments of the body as well as, but less frequently, in deeper situations still. Each organ is a single cell with a nucleus and clear cytoplasm containing a small transparent rod, the optic organelle, which focuses light on the neurofibrillae which ramify through the cell. By means of these organs the worm can detect changes in the intensity of light and, as its nocturnal habits would suggest, it moves away from all except sources of very feeble illumination.

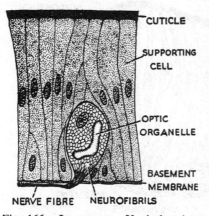

Fig. 166. LUMBRICUS.—Vertical section of the epidermis with a photoreceptor.

Movements and other changes taking place *within* the worm are detected by minute receptor structures (proprioceptors).

**Reproductive System.**—The earthworm is hermaphrodite and in this, and in the restriction of the reproductive organs to a few anterior segments, differs from those less specialised annelids, the Polychaeta. Earthworms are not, however, self-fertilising, but pair together (copulate) and later deposit their eggs in cocoons, and the reproductive processes are determined by two conditions: the ensuring of cross-fertilisation and the deposition of the cocoon. The complicated reproductive apparatus may therefore be conveniently thought of as consisting of three complemental systems which work in harmony to ensure these ends. These parts of the reproductive apparatus are the **gonads** and **gonoducts** (reproductive system proper); the **spermathecae**; and the **clitellum,** together with other regions of the body wall specialised to play a part in the reproductive activities.

The gonads, as in all other coelomate animals, are proliferations of the coelomic epithelium, and the gametes are shed into the coelom and thence pass to the exterior by way of coelomoducts, which therefore function as gonoducts.

There are two pairs of **testes** which are attached near to the mid-ventral line on the posterior faces of the septa separating segments nine from ten and ten from eleven. In other words, in

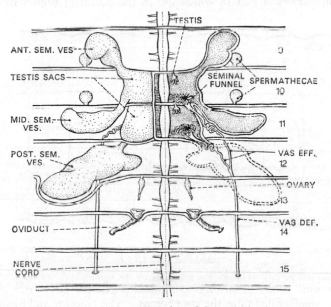

Fig. 167. LUMBRICUS.—Dissection of the reproductive system.

the anterior parts of both segments ten and eleven there is a pair of testes. Each of these is a flat, lobed structure which sheds male germ-cells throughout life but grows continuously and so does not diminish in size. The testes are shut off from the general body cavity by partitions developed from the coelomic epithelium and each pair comes to lie in a **testis sac** (median seminal vesicle). Each testis sac, one in segment ten and the other in segment eleven, is in effect a wide flattened tubular compartment of the coelom extending longitudinally between the septa. It lies just below the ventral vessel and thus encloses the ventral nerve cord. The anterior testis sac communicates anteriorly and posteriorly with a pair of **vesiculae seminales (lateral seminal vesicles)**, which are really nothing more than sac-like outgrowths of the septa at the anterior and posterior ends respectively. The posterior testis sac opens into a single pair of seminal vesicles which lie in the segment immediately behind it. There are thus three pairs of seminal vesicles (lateral seminal vesicles) situated respectively in segments nine, eleven, and twelve. Each seminal vesicle is filled with a nutrient fluid in which are suspended masses of male germ-cells in all stages of development from spermatogonia to spermatozoa. The mature sperms pass back into the testis sacs and are led to the exterior by way of **sperm funnels**, a pair of which lies in the posterior region of each

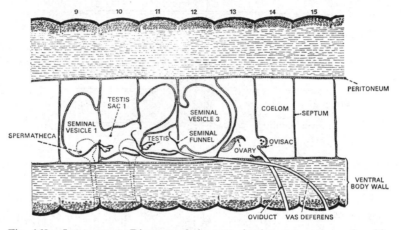

Fig. 168. LUMBRICUS.—Diagram of the reproductive system from the side (modified after Hesse).

testis sac. Each funnel has complicated ciliated folds and opens into a small coiled tube, the **vas efferens**. The anterior and posterior vasa efferentia of each side pass obliquely backwards and join to form a **vas deferens** which gradually dips down into the body wall to open on the ventral surface of segment fifteen.

The **ovaries** are a pair of small carrot-shaped organs attached near the mid-line to the posterior face of the septum separating segment twelve from thirteen. They hang down quite freely in the ventral part of the coelom of segment thirteen, and the ripe **oocytes** which lie at the free narrow end float in the coelomic fluid before passing into the **ovisacs** where the first maturation divisions of the nucleus preparatory to the formation of the polar bodies begin. Each ovisac is a small diverticulum of the septum and opens into the dorsal wall of the funnel to serve as a temporary storage organ for the released oocytes. There is a single pair of **ovarian funnels** situated on the anterior face of the septum, dividing segment thirteen from segment fourteen. Each funnel opens into a short **oviduct** which pierces the body wall and opens on the ventral surface of segment fourteen.

The **spermathecae**, of which there are two pairs, are to be reckoned as accessory structures of the female system, for they serve as reservoirs for sperm received from another worm during copulation. They are small spherical sacs lying in the ventrolateral parts of segments nine and ten and each is attached to the body wall by a short hollow stalk which leads to the external aperture.

The **clitellum**, as has been seen, is a saddle-shaped structure formed of thickened glandular epidermis (particularly on the dorsal and lateral regions) of segments thirty-two to thirty-six or seven. The dorsal and lateral regions of the clitellar segments are rich in groups of gland cells of three types: goblet (mucous) cells; cells with coarse granules which secrete the cocoon membrane; and cells with finer granules which secrete the **albumen** with which the cocoon is filled and which serves to keep the fertilised eggs moist.

Other body wall structures which play a part in the reproductive processes are glandular diverticula from the sacs of the ventral chaetae in segments ten, eleven, and twenty-six, as well as in those of the segments in the region of the clitellum. These glands secrete a substance which enters the body wall of the co-copulant by way of apertures made in it by the long pointed chaetae of the clitellar segments, and so in all probability aids in holding the two partners together whilst they are pairing.

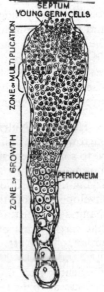

Fig. 169. LUMBRICUS.
—Ovary.

SPERMATOGENESIS.—The testes consist of numerous groups of male germ-cells **(spermatogonia)** in early stages of development. The youngest cells of all lie near the point of attachment of the testis and in them cell boundaries are not visible. Multiplication of the spermatogonia takes place in a band across the middle of the testis and cells proliferated from this region are eventually set free in groups, numbering from sixteen to thirty-two cells, into the seminal vesicles. These cells are **primary spermatocytes**. They divide to form **morulae** or balls of cells composed of sixty-four

**secondary spermatocytes,** which by a further division form a morula of one hundred and twenty-eight **spermatids.** These without further division become transformed into **spermatozoa.** Throughout the numerous divisions the germ-cells get progressively smaller and are attached to a central mass of residual cytoplasm which supplies them with nutriment. The spermatozoa have long rod-shaped heads consisting almost entirely of nuclear material and a short middle piece to which is attached a very long tail. For a time they stick by their heads to the folded walls of the sperm funnels.

OOGENESIS.—The processes resulting in the production of ripe ova are comparable to those concerned in the formation of spermatozoa. The youngest cells (**oogonia**) lie at the base of the ovary and the multiplicative zone is a narrow transverse region also near the base. The bulk of the ovary is occupied by **oocytes,** which do not divide but gradually grow in size as yolk is deposited in their cytoplasm. The actual reduction divisions take place only after the oocytes have been shed from the ovary by rupture of the layer of peritoneum that invests it.

**Habitat. Nutrition.**—Earthworms, as is well known, live in burrows in the soil which, if the worms are to remain active, must be to some extent moist and contain a percentage of organic matter (humus). In dry soil the worms may protect themselves against drying-up by rolling into a ball and secreting a hard mucous envelope around the body. They are, however, tolerant of wide changes in moisture and other soil constituents and it may be said that earthworms of one sort or another are world-wide in distribution. They are, however, rare or absent in acid soils, very heavy clay, and sand. Not only do earthworms swallow soil when it is sufficiently rich in organic particles, but they emerge from their burrows at night to breed or to feed on leaves and other vegetable matter. The leaves are almost invariably seized by their pointed ends and held against the mouth chiefly by suction exerted by the pharynx. Before the leaves are eaten, however, they are moistened by juices containing enzymes secreted by the pharynx. In the oesophagus other enzymes are added, so that before the food reaches the crop it has already undergone a certain amount of digestion. This renders it more easily broken up by the churning action of the muscles of the gut wall, notably those of the crop and gizzard both of which contain small stones. From the gizzard, the food enters the intestine, which is the main digestive and absorptive region, its effective area being greatly increased by the typhlosole. The soluble products of digestion are absorbed into the blood-stream and thus distributed throughout the body. The insoluble remains, together with any soil that has been ingested, are, after passing along the intestine, pushed out at the anus.

**Locomotion.**—The locomotion of earthworms is of interest since it provides a clear-cut and relatively simple example of the results of co-ordinated movements in an animal with a body built on the segmental (metameric) plan.

As may readily be observed when an earthworm starts to crawl, the first few segments become thinner but longer. This is due to a contraction of the circular muscles and a relaxation of the longitudinal muscles in that region, the opposing sets of muscles being

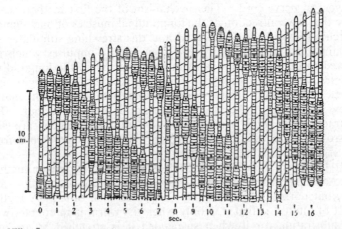

Fig. 170. LUMBRICUS.—Diagram to illustrate the way in which an earthworm moves. The wavy lines pass through equidistant points and show the extent of movement at one-second intervals as given by the figures (after Gray).

antagonised by an increase in the pressure of the coelomic fluid. The thinning and extension of the body gradually spreads to more posterior segments, but before this effect has passed far along the body the chaetae of the anterior segments are protruded so as to grip the substratum. The longitudinal muscles of the anterior segments then contract so that here the body becomes shorter and fatter and the hinder regions of the body are pulled forwards. Successive waves of contraction of the circular muscles are followed up by waves of contraction of the longitudinal muscles and this results in the appearance of waves of expansion following up waves of contraction from the anterior to about the middle of the body, the hinder end being pulled along passively. The septa are important for they normally act as water-tight bulk-heads which transmit pressure changes from one segment to the next by bulging. Their muscles resist pressure changes so that the septa also act as dampers and confine the pressure changes to a few segments. Were it not so, it is difficult to see how the anterior end of a worm could elongate at the same time as the hinder end was shortening. There is, however, a perforation surrounded by a sphincter muscle just above the nerve cord in each septum (Fig. 152). Relaxation of the sphincters allows of free passage of coelomic fluid up and down the length of the worm.

Co-ordination of the waves of muscular contraction is maintained in two distinct ways; by reflexes handed on from one segment to the next by mechanical stimulation and by impulses passing along the ventral nerve cord. The mechanism of the first method seems to be that the contraction of the longitudinal muscles of one segment stretches those in the one behind and this stretching stimulates the contained proprioceptors so that they initiate impulses which are relayed through the nerve cord in that segment to the longitudinal muscles which then contract. The contraction of the longitudinal muscles stimulates other proprioceptors with the result that a reflex inhibition of the circular muscles is brought about. The contractions of both circular and longitudinal muscles are also controlled by reflexes passing through the ventral nerve cord from one segment to the next behind it.

It has been shown by surgical experiments that either method of co-ordination will succeed singly in promoting orderly locomotion, but normally both methods reinforce each other. A further point of interest is that it is only when the worm is in contact with the substratum (or if when suspended it is stretched by its own weight or other means) that the locomotory waves of contraction pass along the body. This means that the tactile receptors in the skin require to be stimulated to initiate the contraction waves which, when once started, are self-propagating.

The fact that the controlling centre for locomotion is the ventral nerve cord is shown by experiments when the cerebral ganglia and sub-pharyngeal ganglia are removed. The worm can still move almost normally, whereas in most higher animals damage to any part of the nervous system which corresponds to a "brain" cause severe disturbances in the power of locomotion.

Earthworms can, when strongly stimulated, contract suddenly throughout their entire length, the practically simultaneous contraction of the longitudinal muscles of all segments being brought about by impulses conducted by the giant fibres. Similar sudden movements in other animals also mediated by a giant fibre system are called escape movements, which describes their purpose very well. So fast can a worm withdraw into its burrow that it can often escape the peck of a bird.

Earthworms also show (bearing in mind their usual method of locomotion) a surprising ability to move over smooth surfaces like glass on which their chaetae can be of little use. They can even climb vertical surfaces and are well able to escape from a glass jar. Faced with such conditions a worm moves partly by "getting a grip" with its sticky mucus and partly by using its mouth as a sucker, the body muscles acting in the usual way.

**Burrowing.**—Earthworms pass the greater part of their time in burrows below the surface of the soil. If, however, they are placed on the surface they can burrow downwards at a speed which varies with the nature of the soil. The worm explores the surface with the prostomium and finding a small crack or fissure, the anterior end is elongated by the contraction of the circular muscles and inserted between the soil particles. The longitudinal muscles of this region then contract causing a thickening of the body, thus forcing the soil particles apart and enlarging the fissure. This process continues until sufficient of the body has entered the soil for the chaetae of the swollen portion to obtain a firm grip. Then, by the contraction of the longitudinal muscles the hinder part of the body can be drawn inwards. Once below the surface a greater portion of the body can be employed in obtaining a firm grip on the sides of the burrow, and, consequently the full effect of the contraction of the circular muscles anterior to the gripping region can be employed in the forward thrust. This is the normal method of burrowing. On occasions, however, earthworms literally eat their way through the soil, especially when the soil is intractable. In soil which is rich in organic matter the same procedure is adopted, so that feeding and burrowing are effected by the same process so that usually much soil is found in the gut.

**Gaseous Interchange.**—The gaseous interchange necessary for respiration takes place mainly through the skin, which is plentifully supplied by looped blood-capillaries (branches of the ventro-parietal vessels), and is kept moist by the secretion of the epidermal mucous gland-cells and also by coelomic fluid passing out through the dorsal pores. Oxygen in solution passes through the cuticle and epidermis into the blood where it enters into chemical combination with the haemoglobin, "dissolved" in the blood, to form oxyhaemoglobin, which gives up the oxygen once again to tissues which have a low oxygen tension. In no instance, however, does the blood come into *direct* contact with the tissue cells so that the oxygen must be handed on through the intermediary of tissue fluids (cf. lymph in vertebrates), and it is probable that the coelomic fluid plays a part in the distribution of dissolved gases about the body. Carbon dioxide is removed from the tissues by the blood and carried in solution in the plasma to the skin, where it is excreted.

**Function of the Chloragogenous Cells.**—Fat is taken up from the gut and stored in these cells, in which glycogen is also deposited and apparently mobilised when required by the worm. Their likeness to the liver of vertebrates is increased by the fact that they also are the site of deamination and formation of ammonia and urea as excretory substances. These pass into the coelomic fluid. It is possible that the cells may collect silica from the soil passing through

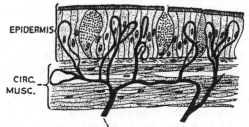

EPIDERMIS

CIRC.
MUSC.

BLOOD VESS.
Fig. 171.    LUMBRICUS.—The superficial looped blood vessels.

the gut as concretions of silica are occasionally found in them. The pigment contained in the cells is probably lipoidal and plays no part in nitrogenous excretion. The chloragogen cells fall into the coelom and after beginning to disintegrate are finally engulfed by phagocytic coelomic corpuscles which may then clump together at the posterior end of the worm to form "brown bodies", or else deposit pigment, derived from chloragogen cells, in various parts of the body.

**Nitrogenous Excretion.**—The organs of nitrogenous excretion are the nephridia which also act as osmoregulators. Coelomic fluid containing ammonia and urea produced by the chloragogenous cells is directed into the nephrostome by ciliary action and passes down the tube of the nephridium. Salts and other substances useful to the body are re-absorbed through the walls of the wide tube and possibly through part of the middle tube. The walls of the brown tube are pigmented by breakdown products of haemoglobin and not purines, as was once thought. The urine which passes out of the worm is hypotonic to the body fluids and thus provides a mechanism whereby excess water can be eliminated. Not all earthworms excrete in the same way. About half the waste nitrogen in *L. terrestris* is got rid of as muco-proteins and most of the other half is urea and ammonia. But this observation refers to the total output and not just that eliminated by the nephridia. In short, the knowledge of execretion in the earthworm is very incomplete and it is not known whether or not ammonia and urea can pass from the blood or coelomic fluid through the walls into the lumen of the nephridium.

To all intents and purposes, osmoregulation in earthworms is similar to that in fresh-water animals. It must not be forgotten, however, that the earthworm is a terrestrial animal and, therefore, on occasions will meet with dry conditions when water conservation will become of paramount importance. This is particularly true of such soft-bodied animals as earthworms with no impermeable outer covering and where a moist outer surface is the site of gaseous interchange. It may well be, therefore, that under arid conditions the urine becomes scanty and hypertonic. It is known that the nephridiopores have sphincters and open only occasionally to allow the urine to escape so that the nephridia must not be visualised as ducts through which coelomic fluid is continuously passing outwards.

**Function of the Calciferous Glands.**—The calciferous glands produce relatively large amounts of calcium carbonate which pass into the lumen of the oesophagus *via* the oesophageal pouches. The chief function of the glands is the excretion of excess calcium absorbed with the food but also carbon dioxide (in combination with calcium) is eliminated from the tissues in this way. The calcium carbonate may help to neutralise organic acids in the food, rendering it more easily dealt with by the alkaline digestive juices of the intestine and before this to flocculate the soil when it is in the crop.

**Life-history.**—Reproduction in the earthworm is usually the result of sexual processes, but many of its aquatic relatives multiply by a type of fission as well as employing normal sexual methods. Sexual reproduction is initiated by a preliminary set of processes (copulation) designed to effect an *exchange* of spermatozoa between two animals. These processes are complicated, both in order to avoid self-fertilisation and because of the difficulties in transference of sperm from one land-living animal to another.

Pairing takes place at night on the surface of the ground. Two worms which have issued from adjacent burrows come to lie

together, "head to tail", with their ventral surfaces in close contact
and in such a position that segments nine to eleven of one worm

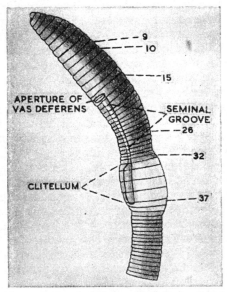

Fig. 172. LUMBRICUS.—Anterior portion of a hardened specimen.

are opposite to the clitellum of the other.   A copious exudation of
mucus from the epidermis then takes place and the body of *each*
worm from segment eight to just in front of the clitellum becomes
invested in a separate
mucous envelope, whilst
in the clitellar regions and
the parts just behind them
the mucus forms a tube
common to *both* partners
which helps to bind them
together.   The secretion
of the diverticula of certain chaetigerous sacs and
the penetration of the
chaetae of the clitellar
segments of one worm

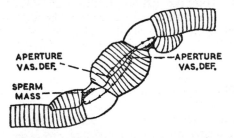

Fig. 173.   EISENIA FOETIDA.—Diagram of two
worms pairing, to show the course of the
seminal fluid.

into the body of the other also help to hold the worms together.
Discharge of the seminal fluid then begins in each worm from the
apertures of the vasa deferentia and the droplets are carried back

*beneath the mucous envelope* in small pits along the line of the seminal grooves to the clitellum. Here the fluid accumulates in the depression between the bodies of the two worms before entering the spermothecae of the opposite partner. The passage of the sperm and fluid along the seminal groove is effected by the action of special muscles situated in the longitudinal muscles of

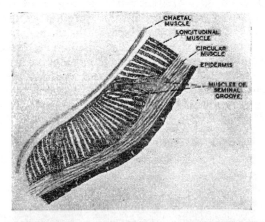

Fig. 174. LUMBRICUS.—Transverse section of the ventrolateral body wall between segments 15-32 to show the arciform muscles.

the body wall immediately below the line of the seminal grooves. In transverse section these muscles are seen to be in the form of an arc attached to the epidermis at two points, so that by their contraction the body wall is pulled inwards at the two points of attachment and a simultaneous contraction, such as occurs when a worm is suddenly killed, of the arciform muscles in all these segments

Fig. 175. Diagram to show the transference of the seminal fluid along the seminal groves.

results in the formation of two grooves separated by a ridge. The more dorsally situated groove is the seminal groove. But, during copulation, the arciform muscles do not contract simultaneously but in repeated succession, beginning at the anterior end so that a series of pits are formed which travel from segment fifteen back to the clitellum. The pits become filled with seminal fluid which is prevented from escaping or drying during its passage to the clitellar region by the overlying mucous tube.

The pairing worms separate soon after the exchange of seminal fluid has been completed.

The **cocoons** into which the eggs (oocytes) are later to be deposited are formed at intervals until all the seminal fluid stored in the spermothecae is used up. The cocoon is a product of the clitellum, and its formation is preceded by the secretion by the epidermal mucous cells of a slime tube which ensheaths the body from segment six to just behind the clitellum. The clitellum secretes the cocoon membrane and then albumen between the membrane and the surface of the clitellum itself. From eight to sixteen eggs then pass back from the oviducal apertures in segment fourteen, probably along the lower of the grooves produced by the action of the arciform muscles, into the cocoon *whilst it is still surrounding the clitellum.* Soon after

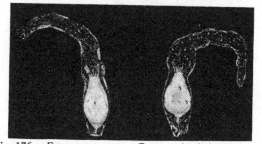

Fig. 176.    EISENIA FOETIDA.—Cocoons in their slime tubes.

this, a wave of expansion of the body of the worm takes place from rear to front so that the cocoon is forced off over the anterior end of the worm. As the cocoon passes over segments nine and ten, seminal fluid is discharged into it from the spermothecae and the eggs are fertilised after the cocoon has been deposited.

The cocoon of *L. terrestris* is a small dark-brown ovoid object about the size of a pea. Cocoons may be found in the earth or under stones. In captivity *Lumbricus* does not apparently produce cocoons readily, but an allied species, the brandling worm (*Eisenia foetida*) will do so and the details of cocoon formation and deposition in this case have been observed. The freshly deposited cocoons of this worm are a pale straw colour and still have the slime tube attached (see Fig. 176). Soon, however, the slime tube dries up and the cocoon becomes progressively darker in colour. The cocoon membrane is tough and elastic and is made up of many layers of the secretion of the large granule-containing gland-cells of the clitellum. Within, the whole of the space is occupied by a thickish albumen (the product of the fine granule-containing cells of the clitellum) in which float the fertilised eggs.

Since the eggs develop within a cocoon, a larval stage such as is found in the Polychaeta is suppressed, and, in the case of *Lumbricus*, only one embryo survives the early stages of development.

These specialised processes for the mutual *exchange* of spermatozoa during pairing, for the formation of albumen-containing cocoons and their deposition, are obviously adaptations to reproduction on dry land but some earthworms reproduce parthenogenetically.

**Earthworms in Relation to Agriculture.**—In most soils except the heaviest clayey and acid kinds, earthworms are abundant. It has been estimated that there may be as many as three million to the acre, equivalent in "biomass" to about six sheep per acre. Their continually burrowing in search of food results in the loosening of the earth particles and the formation of innumerable channels permeating the superficial layers. Such channels ensure the free entry of air into the soil around the roots of growing plants and form ready-made conduits, to conduct away rain water which falls on the surface, at a density of up to 300 tunnels per square metre. Thus the earthworm's activities both drain and aerate the soil. But more than this, some earthworms (but not *L. terrestris*) are continually bringing up fresh soil from below to the surface at a rate in West Europe of about 6 kg per sq. metre per year. This takes the form of the familiar wormcasts which consist of the finest soil particles—for they have passed through the length of the alimentary canal mixed with organic matter in the most suitable form, viz. vegetable tissues broken down and disintegrating, forming a high-quality humus. Such soil has the additional property of having a water-stable, crumby nature which favours plant growth and also returns plant nutrients in a readily available form, such as compounds of nitrogen and phosphorus. Moreover, when earthworms die their decomposing bodies add significant amounts of plant nutrients to the soil and it has been estimated that about four-fifths of the worm population die each year. Various species of *Allolobophora* are chiefly responsible for the formation of wormcasts on the surface but *Lumbricus* deposits below the surface. As the wet wormcasts dry, the wind scatters the fine powder over the surface of the soil and a new layer of tilled earth is in process of formation. The amount of fresh earth brought to the surface annually is considerable.

It is then to the earthworms that the productivity and fertility of land, even that uncultivated by man, may in large measure be due.

There are many species of British earthworms and whilst several kinds may be found living together in one and the same field or garden, on the whole a particular species tends to flourish best in a particular kind of soil. *L. terrestris,* for example, seems to prefer drier kinds of soil. *Allolobophora chlorotica,* although able to withstand a good deal of desiccation, is commonest in wetter soils.

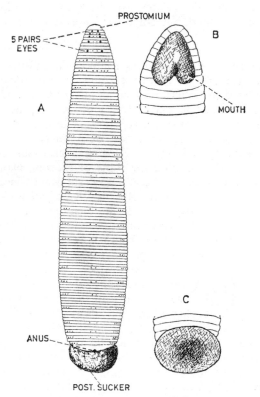

Fig. 177. HIRUDO.—A, dorsal view. The rows of eight dots on certain annuli are sensory papillae and correspond with the segments. Each segment is typically marked with five annuli. B, anterior sucker, ventral view. C, posterior sucker, ventral view.

Most species can live for many months totally submerged in water, yet most will avoid water if given the chance. This explains why earthworms come to the surface when the ground is waterlogged but the many dead worms seen after heavy rain evidently have not been drowned. They are killed by exposure to the ultra-violet light of the sun.

## HIRUDINEA—HIRUDO

The third important class of Annelida is Hirudinea, the leeches. These resemble in so many respects the Oligochaeta that they are often bracketed with them as Clitellata since in both a clitellum is an essential part of their reproductive apparatus. Yet leeches differ from oligochaetes in several important respects; for example, they lack chaetae and their coelom is largely occluded by mesenchyme. The body of a leech is rather thick and is always composed of thirty-three true segments but these

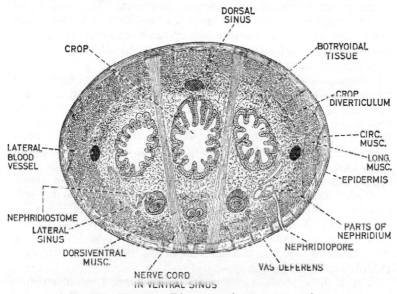

Fig. 178. HIRUDO.—Diagrammatic transverse section.

are difficult to make out externally since each is subdivided by a number of grooves into annuli. All leeches have a strong posterior sucker formed by the last six segments of the body. The sucker around the mouth is smaller and that of *H. medicinalis* is formed from the first four body segments.

Figs. 177, 178 show the main features of the medicinal leech.

Leeches live in the sea and fresh water. A few species occur in damp situations on land. All are predators mainly of invertebrates, although the best known examples are blood-suckers and hence could be called parasites.

# CHAPTER XIII

## THE JOINTED LIMBED ANIMALS (ARTHROPODA)

**INTRODUCTION**

In the introduction to the coelomate animals (p. 157) it was mentioned that evolution along several quite different lines has resulted in the production of forms as divergent from one another as an insect, a starfish, an octopus, and man. It would be of extreme interest to explore all these various lines, but that is not practicable here. To illustrate one of them, however, the arthropods have been selected. These have distinct affinities with the annelid stock but also show important differences.

Characters which may be considered as distinctive of arthropods include the ability of the epidermis to secrete a tough cuticle which functions as an exoskeleton; the presence of jointed limbs or appendages which are serially repeated along the body, some of them being modified for dealing with the food; the main body cavity is a **haemocoel**—that is to say, the true coelom is reduced in size whilst the blood vascular system is so expanded that the main organs of the body lie in blood-filled cavities; cilia are never found.

Characters common to both arthropods and annelids (as well as to some other invertebrates) are the metamerism of the body, the presence of coelomoducts, and the basic plan of the nervous system which consists of a dorsally situated "brain" connected by a collar around the anterior end of the gut to a ventral cord bearing segmental ganglia.

For a fuller understanding of the arthropod characteristics, some of these features may now be considered in more detail.

**The Cuticle.**—Many of the main arthropod features seem to be correlated with the presence of a thick and semi-rigid integument or cuticle which not only invests the whole of the body but forms also the lining of the stomodeum and proctodeum. The cuticle is also continued over the appendages and renders necessary the jointing which is so characteristic of them. This protective armour is not laid down uniformly but as a system of hardened plates, the **sclerites**, joined with the other by more flexible areas, the **articular** or **arthrodial** membranes, so that movement of the sclerites relative to one another is made possible. The various sclerites serve as places for the attachment of the muscles of the body so that, in addition to its more obvious role of protection for

268

the body, the cuticle is, in the narrower sense of the term, a true skeleton but of a particular kind, namely, an **exoskeleton** since it lies mainly on the outside of the body. The cuticle, although secreted by the epidermis, does not lie entirely outside the body, for in certain regions the epidermis sinks inwards beneath the superficial musculature and there lays down a system of rods or plates, the **apodemes,** which serves for the origin of various important muscles concerned with flexion of the abdomen and with movement of the basal parts of the head and thoracic appendages. In some instances the apodemes are so extensive (as, for example, in the thoracic region of many Crustacea) as to form a definite arrangement termed the **endophragmal skeleton.**

It is highly probable that the chemical nature and finer structure of the cuticle is fundamentally similar in all arthropods for it seems always to consist basically of chitin and this is overlain by impermeable non-chitinous layers, while the chitin itself may be impregnated with other substances which alter its properties in important respects. Chitin is a polyglucosamine, that is partly carbohydrate in nature, has a tough but flexible nature, and is freely permeable to gases and to many substances in watery solution. Both in crustaceans and insects, however, the cuticle is seen to be composed of two

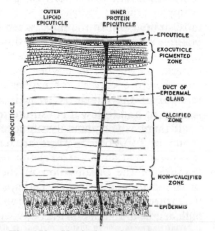

Fig. 179. Vertical section of crustacean cuticle (based on Dennell).

main layers, an inner **endocuticle** immediately over the epidermis and an outer, **epicuticle.** The endocuticle consists of successive layers of chitin which gives it a laminated appearance in vertical section. Its outer layers, often termed the **exocuticle,** are darker in colour owing to the deposition in them of proteins which have been acted on by oxidised polyphenols taking the place of chitin, the layers of exocuticle having undergone a process essentially resembling "tanning" or sclerotisation. This tanning process is carried much further in the insect than in the crustacean cuticle, where, particularly in the larger ones (Malacostraca), much of the hardness results from the laying down of calcium salts in the middle layers of the endocuticle. The epicuticle also is not a single layer but is composed of a layer of hardened protein overlain by an even thinner layer of lipoid substance. It is this lipoid layer which confers on the cuticle its large measure of impermeability to water and its unwettability, though still, in many instances, leaving it permeable to gases. There is often yet a third layer, a cement layer, which protects the waterproofing lipoid layer.

Most worm-like creatures can live only in water or by burrowing in moist soil, the surrounding medium affording some measure of protection as well as buoyancy. Arthropods, on the other hand, are to be found in every conceivable kind of environment deriving considerable mechanical and physiological protection, particularly from desiccation, from their hard and relatively impermeable cuticle. Yet this cuticle by its own intrinsic properties imposes some important restrictions on the activities of arthropods. For example, gradual increase in size with age, such as is usual with most animals up to the point when they stop growing, is no longer possible and growth is limited to short periods of enforced quiescence when the cuticle is cast off, growth occurring as it were in a series of jerks during the time when the new cuticle is still soft, pliable, and therefore capable of expansion. This periodic casting and renewal of the cuticle is termed **ecdysis** whilst the events which lead up to and follow it constitute the moulting cycle (p. 289).

If then, as is commonly supposed, arthropods arose from some group of soft-bodied, worm-like creatures allied to the annelids, it is clear that as the exoskeleton became emphasised certain other features of the organisation of their bodies inevitably underwent progressive modification. Perhaps the most obvious feature of the arthropod cuticle or exoskeleton is its robustness which confers a high degree of rigidity on the body, enclosing it in a protective, jointed suit of armour. This is in complete contrast to the condition prevailing in the annelids where any stiffness is due to turgor set up and maintained by the contraction of the body wall muscles against the coelomic fluid which, like any other fluid, is incompressible. In these animals, then, the coelomic fluid fills the role of a **hydrostatic skeleton**. Progression is either by alternating contraction of circular and longitudinal muscles of the body which bring about waves of contraction and elongation, or else by crawling or swimming during which the body is thrown into a series of waves, usually in a side to side manner. Such a method of locomotion is obviously quite impossible in any armoured creature and arthropods rely, therefore, on the movements of their serially repeated appendages for walking and swimming. In many simpler crustacea (Branchiopoda) the cuticle remains soft and flexible so that limbs and body retain their shape by virtue of the hydrostatic pressure in the contained blood.

**Appendages.**—The appendages are always hollow structures jointed with the body and composed of a number of tubular sclerites, termed **podomeres**, which can also be moved relative to one another. Movement at one joint is, however, usually limited to one plane by peg-and-socket joints, as well as by the arrangement of the muscles. Movement of the limb as a whole in many different planes is due to variation of the plane of movement of the successive joints. The muscles responsible for the movements of the podomeres are arranged in opposing or antagonistic sets, the **flexors** and **extensors**. These muscles take origin from the proximal end of one podomere and are inserted in the proximal part of the next more distal podomere (Fig. 180).

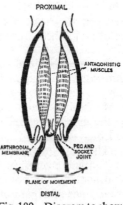

**Haemocoel.**—Among other important features which are correlated with the presence of an exoskeleton may be mentioned the haemocoelic body cavity. The true coelom, which as in other coelomates, develops as a cavity in the embryonic mesoderm, never attains more than insignificant proportions. In the Crustacea the coelom forms part of the cavity of the excretory organs, whilst the cavities of the gonads are also to be regarded as coelomic. In the insects, however, it is virtually absent before the adult stage is reached. The reduction in the size of the coelom is accompanied by a great expansion of the cavities of the blood vascular system, whose walls largely disappear so that the circulating fluid (haemolymph) bathes all the important organs of the body. The body cavity thus comes to consist of a series of interconnecting cavities filled with haemolymph and is called a haemocoel.

Fig. 180. Diagram to show a joint of a crustacean limb.

In the larger arthropods provision is made to keep the blood circulating by means of an enlarged and contractile portion of the dorsal blood vessel which thus forms a pulsating heart. This lies in a special haemocoelic space, called the pericardial cavity, into which open the main blood spaces. Communication between the pericardial cavity and the cavity of the heart is by apertures, termed **ostia**, which are guarded by valves arranged so as only to permit the entry of the blood inwards into the heart.

**Metamerism.**—The chief external signs of metamerism in arthropods are the serially repeated appendages of which there is typically one pair for every segment of the body. As in other animals, arthropod segmentation primarily resides in the meso-derm of the embryo, the paired mesodermal bands being the first structures to become metamerically segmented. Later, other structures follow this primary metamerism but, as might be expected in such a highly organised phylum, metamerism becomes more modified than in the annelids. This is shown partly by a reduction in the number and extent of certain internal organs (coelomoducts, for example) and partly by cephalisation (p. 222). The number of metameres entering into the composition of the body varies con-siderably in the different arthropod groups, but in all a definite head region is recognisable which is built of a pre-segmental part to which is welded a series of segments, commonly six in number. Between the head and the pygidium lie body segments, most or all of which bear appendages.

**Nervous System.**—As in annelids the ventral nerve cord is composed of ganglia linked by nerve-fibres only and from these ganglia arise the segmental nerves which innervate the structures of the body. The anterior end of the ventral nerve cord is joined to the cerebral ganglia by connectives which encircle the anterior end of the gut.

The phylum Arthropoda is certainly one of the most successful and varied groups in the animal kingdom and in an elementary treatment it is impossible to do justice to it; but to illustrate the broad lines of organisation found in aquatic forms a fairly advanced crustacean—the **crayfish**—is selected, and for that of a thorough-going terrestrial form, an insect—the **cockroach**. These will be described first, in some detail, in the pages which follow, but later some mention will also be made of other crustaceans and insects in order to place these two animals in their proper perspective against the background of their respective classes.

**CRUSTACEA—ASTACUS** (*The Crayfish*)

**Habitat.**—The crayfish is fairly common in those rivers in this country which, flowing over chalk or limestone, contain considerable amounts of calcium salts in solution. It lives in holes in banks when it is not feeding. It is omnivorous, feeding on practically any suitable material, alive or dead, which it can obtain.

**The Form of the Body.**—The body is elongated and divisible into three main regions, the **head**, the **thorax**, and the **abdomen**. Between the head and thorax, however, there has been almost complete fusion to form a **cephalothorax** and in this region the segmentation, clearly marked in the abdomen, is not evident externally. The line of demarcation between the head and the thorax is possibly indicated by the **cervical groove**. The abdomen has preserved its original segmentation and is made up of six movable segments and a terminal structure, the **telson**, which is of post-segmental origin. To the ventrolateral margins of the body are articulated the appendages.

**The Cuticle.**—As mentioned previously, the hardened cuticle on the outside of the body functions as an exoskeleton the arrangement of whose parts is best seen in the abdomen where the relations of the sclerites are less obscured by fusion between the segments. In each abdominal segment the exoskeleton consists of a dorsal sclerite,

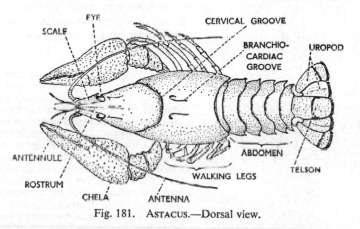

Fig. 181. ASTACUS.—Dorsal view.

the **tergum**, which on each side joins a <-shaped prolongation, the **pleuron**, and of a ventral sclerite, the **sternum**. The portion of the sternum lying between a limb base and the lower part of the pleuron is termed the **epimeron**. All the sclerites are, however, welded together to form a ring around the body. These rings are connected between neighbouring segments by unhardened cuticle which forms articular or arthrodial membranes, allowing of movement between the successive segments. Flexion of the abdomen is, however, limited by the overlapping of the successive terga and, by the presence of definite peg-and-socket joints, to only slight movements laterally. A still greater degree of overlapping dorsally prevents the abdomen from bending past the straight in a dorsal direction but free flexion is possible ventrally. The appendages are joined to the ventrolateral parts of the body by arthrodial membranes inserted in the more lateral parts of each sternum.

In the cephalo thoracic region the sclerites have all fused together to form a continuous shield or **carapace**, which is further modified to form a **branchial chamber** on each side of the body to house the gills which arise from, or near to, the bases of the thoracic limbs. From the dorsolateral margin, as indicated by the branchiocardiac groove on each side, the cuticle has been folded to form the **branchiostegite**, a part of the carapace which extends downwards on each side as far as the ventral surface of the body so forming the outer wall of the gill chamber, its inner wall being formed by the lateral wall of the thorax (Fig. 209). Anteriorly, the carapace is prolonged into a pointed **rostrum**, on each side of which lies an eye. The ventral surface of the cephalothoracic region still retains clear traces of segmentation, for between the bases of the limbs are small **sternal plates**.

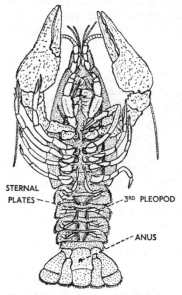

STERNAL PLATES

3RD PLEOPOD

ANUS

Fig. 182. ASTACUS.—Ventral view.

**The Appendages and Metamerism of the Body.**—It is usually assumed that the limbs of the primitive Crustacea were attached to the ventrolateral margins of the body and projected more or less vertically downwards. However, in most living Crustacea, particularly in the larger and more highly organised members like the crayfish and its allies, the original orientation of the limbs may be much altered. Thus, the appendages concerned with mastication and in passing the food into the mouth bend inwards to the mid-line below the mouth; those used for walking can be flexed in various ways and it is only the pleopods which hang vertically downwards. The terms upper, lower, inner, and outer which appear in the figures refer to the limbs as they are normally held in the body.

In the Crustacea, as in practically all large groups of the animal kingdom, are to be found examples of animals which feed on minute particles of food which they sieve off from the surrounding water (**microphagous** or **filter feeders**); those which feed on larger particles or sizeable living animals (**macrophagous feeders** and **predators**); and, again, those which live on or in the bodies of other animals (**parasites**).

The appendages of the Crustacea, in addition to their primary role of locomotion, are used in gathering the food and in passing it to the mouth so that the design of the limbs is always, partially at any rate, correlated with the feeding method adopted. It is clear that since cilia are completely absent, filter feeders will rely on their appendages to promote the required water currents; that the limbs of scavengers and predators will be chiefly adapted for rapid locomotion and prehension, whilst the limbs of the parasitic Crustacea may be expected to undergo some reduction, perhaps to be modified as organs of attachment to the host.   All these trends can, in fact, be noticed in a study of crustacean appendages but two other important factors must be taken into account in their functional morphology, namely, the formation of respiratory surfaces on the limbs and the modification of certain limbs to play their part in the processes of reproduction.

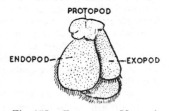

Fig. 183. CRAYFISH.—Uropod.

A very great range of limb form might be expected and, indeed, is to be found but, nevertheless, it is possible both in extinct and living crustaceans to recognise two main types of appendage. These are known as the **phyllopodium (foliaceous limb)** and **stenopodium (biramous limb)**. The phyllopodium is somewhat flattened and its axis bears lobes both on its inner and outer sides, whilst the stenopodium is sub-cylindrical in cross-section and bears two main rami or branches on its basal podomere. A further point of contrast is that the two rami of the stenopodium, the endopodite and exopodite, are very clearly divided by definite joints into numerous **podomeres**, whereas the lobes of the phyllopodium and the main axis are only rarely clearly jointed, flexibility being

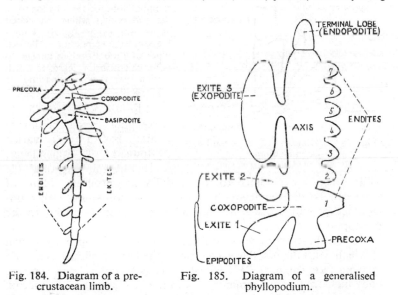

Fig. 184.   Diagram of a pre-crustacean limb.

Fig. 185.   Diagram of a generalised phyllopodium.

due to the retention throughout all the parts of the limb of chitin which is but little hardened.   Both types of limbs are, however, despite differences, clearly related one to the other and possibly also to a common ancestral type about which, unfortunately, little is known.   It seems fairly clear that the pre-crustacean limb had more podomeres than are found in the limbs of any known crustacean and each probably bore an out-growth or lobe from its inner and its outer side.

These outgrowths are known respectively as **endites** and **exites**. Many phyllopodia of modern Crustacea approximate to this condition when it is also common to find that the lobes are fringed with bristles (**setae**), which function as a filter to remove particles of food from the water. Also, the basal endite or endites may be thickened for masticatory purposes, whilst one or more of the exites (usually termed **epipodites**) remain thin-walled to function as a gill. The stenopodium in its less modified form consists of a limb base, the **protopodite** (often formed of three fused podomeres, the **precoxa, coxopodite,** and **basipodite**), from which arise the many jointed rami—**endopodite** nearer to the mid-line of the body and **exopodite** on the outer side. Frequently the precoxa is not evident, having been absorbed into the pleuron. Commonly, two epipodites arise from the protopodite a little nearer the base than the origin of the exopodite (Fig. 186).

The discovery of an extremely interesting fossil crustacean, *Lepidocaris rhyniensis*, whose anterior trunk limbs are typical phyllopodia and whose more posterior limbs are biramous and approach true stenopodia in general form, has done much to confirm the relationships and homologies of the two main types of crustacean limb. These homologies are that the distal exite of the phyllopod is the equivalent of the exopodite; the terminal lobe of the phyllopod is the endopodite, whilst, as stated above, the exites are the homologues of the epipodites. The two types of limb therefore appear to differ chiefly in the degree to which certain parts are emphasised or suppressed.

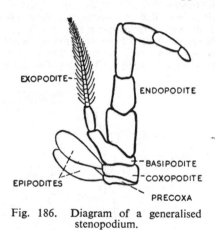

Fig. 186. Diagram of a generalised stenopodium.

The crayfish, with its scavenging and predatory habits, has a very varied equipment of appendages, some carrying out one role while others are modified for quite different purposes. Many of them can be derived from the biramous stenopodium, either unmodified or reduced, whilst a few approximate more closely to the phyllopod form.

In all, the crayfish possesses nineteen pairs of appendages each of which, with the possible exception of the first, is related to a body segment (see Table, p. 281) whilst there is some evidence that the body includes twenty segments—six in the head, eight in the thorax, and six in the abdomen, not including the telson which is not a segment but a post-segmental structure. This 6 + 8 + 6 segmentation is typical of the sub-class Malacostraca to which the crayfish belongs. There are five pairs of head appendages: the **antennules,** the **antennae,** the **mandibles,** the **maxillules,** and the **maxillae.** The first two pairs have a sensory function, whilst the remaining three pairs are **mouth parts** or **trophi**. They surround the mouth and are used in dealing with the food. The first three pairs of thoracic limbs have also been pressed into service for feeding and are called **maxillipeds**. The fourth pair of thoracic appendages are prehensile and form the formidable pincers or **chelae**. The fifth to eighth thoracic appendages are **walking legs**, whilst in the abdominal region the appendages are small but, since in more active relatives of the crayfish they are used for swimming, they are termed **pleopods** or **swimmerets**. In the female the first pair is vestigial, whilst in the male the first two pairs are used for the transference of sperms to the female and are tubular for this purpose. The last pair of

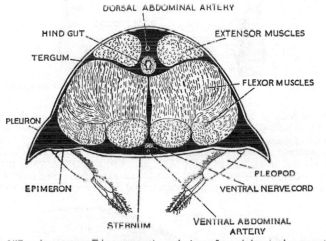

Fig. 187. ASTACUS.—Diagrammatic end view of an abdominal segment.

abdominal appendages are enormously expanded and together with the **telson** form the **tail fan** (Fig. 183).

Each of these appendages may now be considered in more detail. The **antennule** is a biramous structure consisting of two many-jointed rami arising from a limb base formed of three podomeres. In primitive crustaceans the antennule is uniramous and it seems certain that the biramous nature of the crayfish antennule is a secondary state of affairs and that its branches are not the equivalents of endopodite and exopodite. They may be referred to merely as the inner and outer branches. Each is composed of numerous small podomeres but the outer branch is longer than the inner one and bears on its lower side numerous hair-like bristles to which are ascribed the function of **chemo-receptors**. In the proximal podomere of the base of the appendage is a cavity housing the **statocyst** (p. 303).

The **antenna** conforms essentially to the plan of a typical biramous appendage in that it consists of a protopodite of two podomeres (coxopodite and basipodite) from which arise two rami. The outer one of these, which is the exopodite, is flat and triangular so that it is often termed the **scale**. The inner ramus, the endopodite, is long like the lash of a whip and is composed of very numerous podomeres. It is, in fact, so long that it can reach back more than half-way along the body. The coxopodite of the antenna bears a small projection on which opens the duct of the excretory organ (green gland).

Unlike the antenna, in which the endopodite forms the largest part, the **mandible** consists chiefly of the basal parts of the appendage. Its main part,

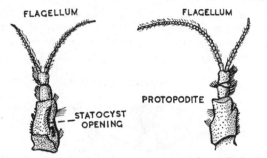

Fig. 188.    CRAYFISH.—Antennule.    A, upper view.    B, lower view.

termed the **body** of the mandible, is a massive sclerite bearing a toothed **incisor process** ventrally on its median surface and dorsally, a blunt **molar process**. The body of the mandible is believed to be the equivalent of an enlarged **precoxa**. A coxopodite is apparently lacking, but jointed with the body is a small **basipodite** and from this arises a small endopodite, or, as it is usually termed, the **palp**, consisting of two podomeres only. There is no exopodite. The mandibles lie right and left of the mouth and the incisor and molar process of one mandible

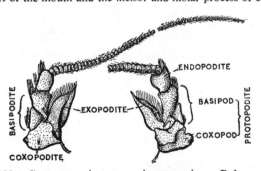

Fig. 189.    CRAYFISH.—Antenna.    A, upper view.    B, lower view.

bite against those of the other and help to push food into the mouth. They have little or no effect in cutting up the food.

The **maxillules**, concerned with guiding the food, also have the limb base and its endites emphasised, lack an exopodite, and have only a small endopodite. The limb base is composed of two flattened sclerites, precoxa-plus-coxopodite and basipodite. Each plate bears a flattened endite fringed with stout bristles projecting inwards towards the middle of the body. Thus it would appear that the maxillule partakes of the nature of a phyllopodium, but in more primitive crustaceans both exopodite and endopodite are represented in the maxillule, so that the small endopodite of the crayfish maxillule is a clear indication that reduction from a biramous condition has taken place.

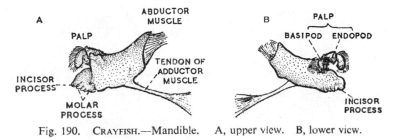

Fig. 190. CRAYFISH.—Mandible. A, upper view. B, lower view.

The maxillae, which lie close behind the maxillules, are also flat structures bearing lobes or endites on their median edge. The two endites are each subdivided into smaller lobes and represent the endites of the precoxa and coxopodite. The small coxopodite bears a small endopodite and a large,

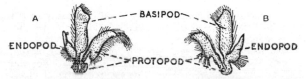

Fig. 191. Crayfish. Maxillule. A, lower view. B, upper view

plate-like exopodite extending both forwards and backwards along the outer surface of the appendage. This exopodite has the important function of maintaining by its to-and-fro movements the current of water which flows over the gills. It is therefore termed the **baler** or **scaphognathite**.

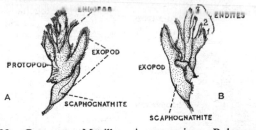

Fig. 192. CRAYFISH.—Maxilla. A, upper view. B, lower view.

Similar in general plan to the maxilla, the first thoracic appendage assists in feeding and is the **first maxilliped**. Its protopodite (coxopodite plus basipodite) has two endites and bears in addition to the small endopodite an epipodite. The **second maxilliped** is similar to the first but is less flattened and lacks endites, thus showing a closer approach to the typical biramous appendage. The **third maxilliped** differs from the second chiefly in being larger and in having a better-developed epipodite. By the possession of endites and epipodites, and in being clearly biramous the maxillipeds are perhaps the most complete of all of the appendages of the crayfish in the sense that they most closely

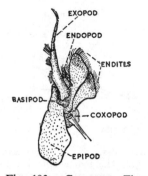

Fig. 193. CRAYFISH.—First maxilliped, lower surface.

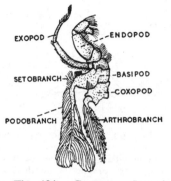

Fig. 194. CRAYFISH.—Second maxilliped, lower surface.

approach the generalised condition. They are also in many ways transitional between the last head appendage, the maxilla, and the hinder thoracic limbs which are the walking legs.

The **fourth thoracic appendage**, which follows the third maxilliped, is the

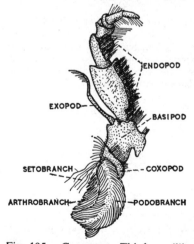

Fig. 195. CRAYFISH.—Third maxilliped, lower surface.

largest of all the limbs and is termed the **cheliped** or great pincers. It and the four succeeding thoracic limbs are of the stenopodial type but differ from a complete stenopodium in that the exopodite is completely suppressed. The limb base is formed by coxopodite and basipodite, the precoxal sclerite having probably been incorporated in the body wall together with its epipodite (p. 276). To the basipodite is attached the five-jointed endopodite, the last two podomeres of which form the pincers. This gripping device is effected by the last but one podomere growing outwards to form a projection as far as the tip of the last podomere which can then bite against this projection and seize the food. An epipodite is attached to the outer edge of the protopodite. The **fifth** and **sixth pairs** of **thoracic limbs** form the first two pairs of **walking legs**. In all respects except that they are more slender, they resemble the chelipeds, whilst the next two pairs of walking legs, the **seventh and eighth thoracic appendages**, differ from the previous legs chiefly in that they **are** not chelate. In the male crayfish the coxopodite of the last walking leg

## Table Summarising Types of Appendages and their Relations to the Segments in the Crayfish.

| Body Region | Segment | Proto-podite | Endo-podite | Exo-podite | Epi-podite | Function |
|---|---|---|---|---|---|---|
| Head | 1 (Preantennal) | ? | ? | — | — | Sensory (olfactory, balance). |
| | 2 Antennule | × | × | ? | — | Sensory (tactile). |
| | 3 Antenna | × | × | × | — | |
| | 4 Mandible | × | × | — | — | } Mouth parts. |
| | 5 Maxillule | × | × | × | — | |
| | 6 Maxilla | × | × | × | — | |
| Thorax | 7 Maxilliped (I) | × | × | × | × | |
| | 8 „ (II) | × | × | × | × | |
| | 9 „ (III) | × | × | — | × | |
| | 10 Cheliped | × | × | — | × | Prehension, offence, and defence. |
| | 11 Walking leg (I) | × | × | — | × | |
| | 12 „ „ (II) | × | × | — | × | (Genital aperture in ♀.) |
| | 13 „ „ (III) | × | × | — | × | |
| | 14 „ „ (IV) | × | × | — | × | (Genital aperture in ♂.) |
| Abdomen | 15 Pleopod (I) | × | × | — | — | Sperm channel in ♂; vestigial in ♀. |
| | 16 „ (II) | × | × | × | — | Sperm channel in ♂; swimmeret in ♀. |
| | 17 „ (III) | × | × | × | — | Swimmerets. |
| | 18 „ (IV) | × | × | × | — | |
| | 19 „ (V) | × | × | × | — | |
| | 20 Uropod | × | × | × | — | Together with telson form tail fan. |

bears the external **aperture** of the **vas deferens**.    In the female the **oviducal aperture** is found on the coxopodite of the second walking leg.

Clearly biramous, the **abdominal appendages** lack both endites and epipodites. Each of the five pairs of **pleopods** or **swimmerets** typically consists of a protopodite (coxopodite plus basipodite) bearing an exopodite and an endopodite, both of

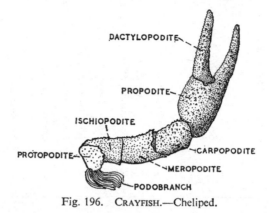

Fig. 196.   CRAYFISH.—Cheliped.

which are fringed by plumose bristles.    The swimmerets of the female crayfish are larger than those of the male and to them are attached the eggs after they have been laid.    The first abdominal appendage in both male and female has no exopodite, whilst that of the male is further modified from the typical condition in that the endopodite has become welded on to the basipodite, these two podomeres also becoming flattened and rolled up lengthwise to form a

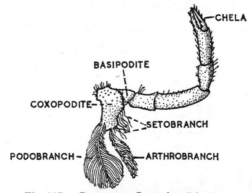

Fig. 197.   CRAYFISH.—Second walking leg.

tubular structure which receives sperms from the aperture of the vas deferens on the last walking leg and transfers them to the female during pairing.

The last pair of abdominal appendages, the **uropods**, biramous like the preceding ones, have undivided protopodites and very broad, flat endopodites and exopodites which, with the telson, form the tail fan (Fig. 181).

**Locomotion and Movement.**—The crayfish has two types of locomotion; **walking** and **darting**. When walking the body is held with the abdomen extended with the four pairs of walking legs in contact with the substratum. The chelipeds are held out in front and serve to counterpoise the weight of the abdomen whilst the antennules and antennae are in continual motion exploring the surroundings. When darting, the animal violently flexes its abdomen with the tail fan fully extended, causing—by the pressure of the abdomen and tail fan against the water—the body to jerk backwards through the water. During this operation, which may be repeated several times, the antennae are directed over the back of the body to detect the presence of obstacles in the rear. The whole movement is very rapid and is resorted to whenever danger threatens or the animal is interfered with. It is therefore called an escape movement and the rapid co-ordination necessary is effected by a giant fibre system in the central nervous system. Unlike its more primitive relatives, the shrimps, prawns, and their

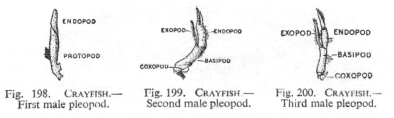

Fig. 198. CRAYFISH.—          Fig. 199. CRAYFISH.—          Fig. 200. CRAYFISH.—
   First male pleopod.           Second male pleopod.          Third male pleopod.

like, the crayfish cannot swim forwards by means of its pleopods which are disproportionately small when compared with the massive body. The arrangement of the muscles responsible for flexion and extension of the abdomen and an example of those in a limb is shown in Figs. 201, 180. In the terminal podomeres of a chelate limb, the flexor muscles become adductors, closing the pincers, and the extensors become abductors, opening the pincers. In many instances the articulation between the podomeres is such as to permit movement in one plane only, the joint acting like a hinge. But where the limb is made up of several podomeres, the planes of possible movement of successive podomeres varies so that the limb as a whole can perform complicated actions in several directions.

**The Endophragmal Skeleton.**—In the thorax and abdomen where the large muscles controlling the movements of the basal podomeres of the limbs and segments of the abdomen are housed, the intrusions of the cuticle (**apodemes**) to provide attachments for the muscles are correspondingly robust and form what is called the endophragmal skeleton. This, in the thorax, consists of a series of **Y**-shaped ingrowths (**endopleurites**) from the lateral wall and upgrowths (**endosternites**) from the sterna. In the abdomen the apodemes are not so prominent.

The Alimentary Canal.—The mouth is an elongated aperture situated on the under side of the head, bordered anteriorly by the shield-like **labrum** and posteriorly by a bilobed lower lip. Originally, in development, the mouth occupied a subterminal position but owing to differential growth of the parts of the head region it comes to lie relatively further back from the anterior end when the adult stage is reached. It is then surrounded by the mouth parts. The stomodeum or fore-gut has three distinct parts, the oesophagus, the gizzard, and the filter chamber. The short oesophagus leads almost

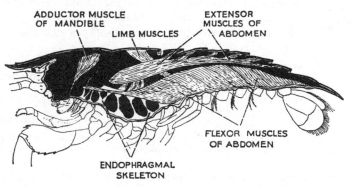

Fig. 201. ASTACUS.—The main muscles of the body (after Huxley).

vertically from the mouth and its opening into the gizzard is guarded by folds of its lining which function as valves, preventing the regurgitation of the food once it is eaten. The gizzard is large and globular and across its roof is a fold. The cuticle lining the gizzard is thickened and hardened (chiefly by calcium salts) in patches dorsally and dorsolaterally to form the "ossicles" which work together to form the crushing mechanism known as the **gastric mill**.

These ossicles, six in number, receive names which really relate them to the terms used in the older description of the crayfish stomodeum. Thus, lying dorsally in the roof of the gizzard is a large, plate-like **cardiac ossicle** and on each side of it, a **pterocardiac ossicle**. Immediately behind the cardiac ossicle lies the **urocardiac ossicle** and in turn behind this a **prepyloric** and a **pyloric ossicle**. At each side of the pyloric ossicle is a **zygocardiac ossicle** which projects forwards along the sides of the gizzard and bears a number of prominences, the **lateral teeth**. The prepyloric ossicle is also prolonged forwards into a long, pointed **median tooth**. The various parts of the gastric mill are shed at each ecdysis.

All the ossicles can be moved relative to one another by means of the **extrinsic muscles** of the gastric mill of which there are two pairs. One pair takes origin from the anterior part of the roof of the carapace and is inserted in the cuticle above the cardiac ossicle, whilst the other pair arises from the posterior part of the carapace and is inserted in the wall of the gizzard above the pyloric ossicle. When these muscles contract they distort the gizzard, drawing the cardiac ossicle forwards and the pyloric ossicle to the rear, whilst the transverse fold across the roof of the gizzard becomes smoothed out and the side of the gizzard forced inwards. This distortion of the gizzard walls causes the median tooth to move forwards between the lateral teeth which become opposed so as to bite together. It follows that any food caught between these moving structures will be crushed and shredded. With the relaxation of the extrinsic

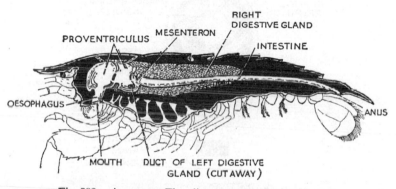

Fig. 202. ASTACUS.—The alimentary canal (after Huxley).

muscles of the mill, the gizzard regains its original shape and the ossicles revert to their former positions largely as a result of the elasticity of the gizzard walls but partly by contraction of the small cardio-pyloric muscles which link the two median ossicles. The rhythmical repetition of these movements subjects the contents of the gizzard to a thorough stirring and crushing so that the size of the food particles gets progressively smaller.

Towards the anterior end of the gizzard there may often be seen, on each side, a mass of calcareous material, each of which is called a **gastrolith**.

Along the mid-ventral line of the gizzard is a deep groove which is, however, less readily discernible because it is arched over by a dense felting of bristles which effectively prevent any solid particles from entering it.

The line of demarcation between the gizzard and the next part of the stomodeum, the filter chamber, is shown on the dorsal side by the points of the three teeth of the gastric mill.   On the ventral wall is a prominent fold, the **cardio-pyloric valve**, which is continued up on each side of the stomodeum so as to make the entrance from the gizzard into the filter chamber quite small.   The filter chamber is distinctly smaller than the gizzard, and moreover, its lumen is reduced by folds and thickenings of its walls which bear bristles in rows and which play an important part in directing the food and digestive juices into definite streams.   The filter chamber bears a ventrolateral pouch on each side into which projects the sides of

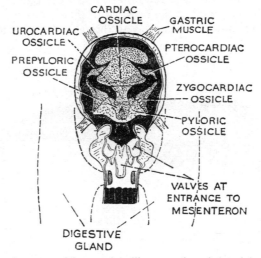

CARDIAC
OSSICLE    GASTRIC
UROCARDIAC           MUSCLE
OSSICLE
PTEROCARDIAC
PREPYLORIC          OSSICLE
OSSICLE

ZYGOCARDIAC
OSSICLE

PYLORIC
OSSICLE

VALVES AT
ENTRANCE TO
MESENTERON

DIGESTIVE
GLAND

Fig. 203. Astacus.—The gastric mill as seen from below (after Huxley).

the **filter** itself.   This is a complex structure made up of two concave curved chitinous plates arising from the cuticle of the ventral wall of the filter chamber.   Each plate rises up in the mid-line to join a prominent ridge and is beset with numerous fine chitinous rods fringed with fine bristles, thus forming an efficient filter and almost completely separating the ventral part of the filter chamber into two ventrolateral channels.   Above the filter, on each side, the walls of the chamber are much thickened by muscle to form the **press,** and the lumen here becomes so small that the filter is almost completely cut off from the dorsal part of the chamber.   This dorsal portion is itself partially separated into a median and two lateral channels by two dorsolateral folds running the length of the chamber.

The very short **mid-gut**, the lining of which arises, embryologically, from the endoderm of the mesenteron, immediately
follows the filter chamber and is not lined by chitin. From its
dorsal wall a blind, forwardly directed pouch, the mid-gut caecum,
projects forwards above the roof of the filter chamber. From
the sides of the mid-gut arise the digestive diverticula. These are
two large yellowish structures very much subdivided into small
lobules which occupy much of the space between the other viscera.

The **intestine**, derived from the embryonic proctodeum, extends
from the end of the mid-gut to the anus on the under side of the
telson. It is lined by unhardened cuticle continuous with the
exoskeleton. At the beginning of the intestine the cuticle is raised
into six squarish patches from each of which a ridge or fold follows
a spiral path along the whole length of the intestine.

**Feeding.**—The food of the crayfish consists of practically any
kind of organic matter, plant or animal, alive or dead. It is therefore said to be omnivorous or a rather indiscriminate scavenger.
Food is seized by the great chelae of the fourth thoracic appendages
or by the smaller pincers of the walking legs and then passed to the
mouth parts. Firstly, the food is gripped by the ischiopodites (see
Fig. 195) of the third maxillipeds posteriorly, and by the mandibles
anteriorly. Held in this position it is gradually shredded by the
movements of the first and second maxillipeds, the maxillae and
maxillules. Both the maxillules and mandibles assist in forcing the
partially shredded food into the mouth, after which its mechanical
breakdown is continued by the gastric mill.

**Digestion and Absorption.**—The fluid in the gizzard contains the
usual three types of digestive enzymes, namely, those which change
insoluble carbohydrates into sugars (amylases); those which split
proteins into amino acids (proteases); and those which split fats
(lipases). The digestive juice is not, however, produced in the
gizzard but travels there from the digestive diverticula, whose cells
are responsible for its secretion. The digestive juice is poured out
into the mid-gut and then, after passing through the filter, flows
along narrow channels on each side of the cardio-pyloric valve to
enter the median ventral groove of the gizzard. Arrived there it is
poured out and is mixed with the food by the churning action of the
gastric mill. Digestion of the food begins straight away in the
gizzard and any of the soluble products pass through the bristles
over the groove and so back to the mid-gut, passing through the
filter *en route*. The continued action of the gastric mill breaks
down the food into very fine particles and these stream back to

the mid-gut by three well-defined routes.   These are along the two dorsolateral channels of the filter chamber and along the median passage immediately ventral to them.   The two dorsolateral channels empty out at the base of the mid-gut caecum where they meet the particles which have travelled along the median channel. These last named have, however, been squeezed by the action of the press so that most of the fluid (containing soluble food) with which they were mixed has been removed and has passed through the filter into the ventrolateral channels of the filter chamber.

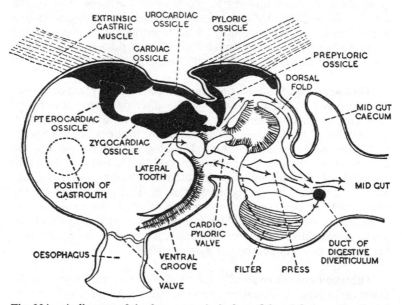

Fig. 204.   A diagram of the fore-gut typical of crayfishes, lobsters, etc.   Large arrows show the route taken by solid particles back to the mid-gut.   Broken lines and smaller arrows indicate the route taken by fluids to and from the digestive diverticula (based on Yonge).

The solid particles which enter the mid-gut are mostly indigestible and are handed on to the intestine to be voided as faeces.

The fluid returned from the gizzard via the ventral groove and filter, or through the filter alone, also enters the mid-gut but close to the entrances to the digestive diverticula into which it flows. Here certain of the cells of the walls of the tubules are absorptive in function and the soluble food is absorbed by them.   The passage of fluid from and into the digestive diverticula is controlled by the pumping action of the walls of the tubules of the gland which are provided with a delicate musculature.

**Growth and Moulting.**—As has been mentioned, arthropods periodically shed their cuticle, or moult, a process which is essential for growth and which is the culmination of a series of events which are repeated cyclically throughout the life of the animal although with decreasing frequency as it gets older. Moulting must not be regarded as a sudden process interrupting at intervals the "normal" life of the animal but as a major activity to which most of the rest are geared. For descriptive purposes the moulting can be considered to begin with the most obvious event, namely, shedding or **ecdysis**. This falls into four main stages: (1) the softening and resorption of the inner layers of the old cuticle; (2) the formation of

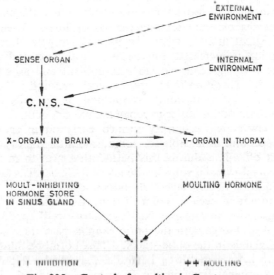

Fig. 205. Control of moulting in Crustacea.

the new cuticle by the epidermis so that the new cuticle lies beneath the old; (3) the splitting of the old cuticle and the withdrawal of the animal from it; (4) growth in size followed by hardening of the new cuticle. Then follows the intermoult period during a brief part of which new tissues are formed but whose chief activities are concerned with accumulating materials for the next ecdysis.

At the beginning of ecdysis the epidermal cells elongate and secrete enzymes which dissolve the inner layers of the endocuticle and "moulting fluid" accumulates in the space so formed. The remains of the cuticle becomes invaded by cells and softened by the withdrawal of calcium salts which are stored for a while in the gastroliths until required for hardening of the new cuticle after which they break up and pass out through the gut.  Resorption of the old cuticle is particularly active along "suture lines" in the carapace above the limb bases and here the old cuticle splits and allows the limbs to be withdrawn.  Certain arthrodial membranes are also broken to facilitate withdrawal of the body from its discarded armour.  It is during the phase of resorption of parts of the old cuticle that muscles are detached and shortly afterwards are rejoined to the new cuticle.

When ecdysis begins the animal becomes sluggish and seeks safety in a place of shelter.  Just before the old cuticle is shed there is an abrupt rise in the glycogen and lipids in the haemolymph whose osmotic pressure also rises.  This is quickly followed by a rapid absorption of water through the gut and so the animal swells and splits the old cuticle along the suture lines.  Then follows a phase of great activity when the animal struggles free.  Oxygen consumption is increased, but because both new and old cuticle cover the body, gaseous exchange is difficult and an oxygen debt is incurred.  Therefore if by any chance shedding is delayed, the animal dies.  After shedding, ventilation of the gills takes place at a maximum rate in order to repay the oxygen debt.

The new cuticle is largely formed early on in ecdysis but it remains soft and distensible for some time after the old cuticle has been cast off.  It is during this period that growth in size occurs and it slows down and stops as the new cuticle hardens and thickens by the addition of new layers of endocuticle.

The moulting cycle is controlled by the balance between two opposing sets of hormones.  Neurosecretory cells (p. 331) forming the so-called X-organs in the optic ganglia pass their secretion to the sinus glands in the optic stalks.  This hormone inhibits ecdysis and antagonises a moulting hormone secreted by a pair of small glands (Y-organs) lying in the front of the thorax.  If the eyestalks are removed growth and ecdysis are precocious.  The rate of production of either hormone and so the balance between is in turn altered by changes in the external and internal environment. For example, conditions of constant light intensity, low temperatures, or the carrying of eggs on the pleopods of the female, inhibit moulting. Plentiful food increases the rate of moulting.  The normal spring moult is initiated by light falling on the animal when it emerges from hibernation.

As will be seen later (pp. 331-2) the moulting cycle of crustaceans has many resemblances to that of insects.

**Autotomy.**—Although during ecdysis and in the normal events of life a limb may occasionally be broken off, the crayfish and its relatives will, under stress of circumstances, break off a limb by reflex action, a phenomenon known as **autotomy**. The breaking off of the limb is brought about by the reflex contraction of a special muscle which flexes the limb until the ischiopodite presses hard against the coxopodite and the appendage snaps at a special plane of weakness, the breaking plane, which is visible externally as a groove around the ischiopodite. Inside this groove is a membrane stretching across the podomere and perforated only by a small hole to let blood into the more distal podomeres. When the limb is cast off the blood clots very rapidly and effectively seals the hole in the membrane so that very little blood is lost. The breaking plane lies between the origins and insertions of the muscles responsible for the movements of the intact limb so that autotomy causes the minimum of damage to the animal commensurate with the loss of an almost complete limb. Moreover, the limb is cast off in a clean, surgical manner. Any intense stimulation of the parts of a limb distal to the breaking plane will provoke autotomy and the adaptive significance of this action is that if a limb is seized then the animal can escape with the loss only of a limb which can be replaced at the next moult. Many other animals autotomise their parts (lizards can lose their tails) and have correspondingly the ability to regenerate them.

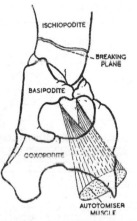

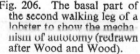

Fig. 206. The basal part of the second walking leg of a lobster to show the mechanism of autotomy (redrawn after Wood and Wood).

**The Blood Vascular System.**—In broad outline the vascular system resembles that of other large crustaceans. It is often spoken of as an "open" one since many of the main organs of the body lie in spaces (haemocoels), others receiving blood from capillaries which discharge into spaces around their cells which therefore are in contact with the circulating fluid (haemolymph) and so can exchange substances with it in a direct manner instead of as in most animals *via* lymph and tissue fluid.

A compact heart lies in a dorsal haemocoel, the pericardial cavity, whose lateral limits are marked externally by the branchio-cardiac grooves in the carapace (Fig. 181). The heart is suspended by six strands of elastic fibrous tissue, the **alae cordis**, originating from the wall of the pericardium. These serve also to distend the heart, that is to stretch the striped muscles in its walls after they have contracted during systole. Opening through the heart walls are three pairs of valved apertures, the **ostia** (one pair dorsal, one pair lateral, and one pair ventral) through which blood enters the heart when it dilates.

From the heart arises a series of seven of arteries which, after branching, convey haemolymph either to haemocoels or to capillary networks related to the various organs. At the origin of each of the seven arteries is a pair of semi-lunar valves which ensure uni-directional flow of the haemolymph. Similar valves occur also along the course of the dorsal abdominal artery. Arising from the anterior end of the heart is the large **anterior aorta** which runs forwards and branches to supply the brain (with capillaries), the antennules, and the eyes. A contractile dilatation near the front end of the anterior aorta acts as an accessory heart. Also from the

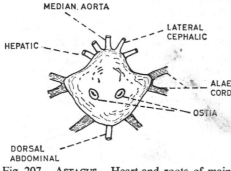

front of the heart springs on each side a **lateral cephalic** artery which has many branches including ones to the antennules, antennae, green glands, anterior parts of the gut, and hepato-pancreas. The paired **hepatic arteries** leave the heart just behind the lateral cephalics and supply the hepato-pancreas and gonads. A median **dorsal abdominal**

Fig. 207. ASTACUS.—Heart and roots of main arteries—dorsal view.

artery runs from the posterior end of the heart to the hinder end of the body, lying above the abdominal muscles (to which it gives branches) and supplying also the intestine. Organs lying below the gut receive blood from the **descending** artery which leaves the posterior end of the heart and runs vertically downwards to the left of the gut and then passes through a gap between the connectives linking the fourth and fifth thoracic ganglia. Then it divides into a forwardly running **rostral subneural** and a backwardly running **caudal subneural** artery which are mainly concerned with supplying blood to the haemocoels in the limbs (including the mouth parts). After circulating in lacunae in the tissues the haemolymph collects in larger haemocoelic spaces called sinuses and these eventually communicate with a large median **ventral thoracic** or **sternal** sinus from which blood passes to the gills. It is then returned from the gills by smaller intersegmental channels which in turn open into the pericardial cavity. Thus, the circulation of the blood is from the heart to the main haemocoels; thence to the sternal sinus; thence through the gills to the pericardial cavity and once more into the heart.

The volume of the haemolymph varies considerably (10-50 per cent. of the body weight) during the moulting cycle and, compared with a vertebrate, circulates at low hydrostatic pressure. When the heart contracts the pressure in the main arteries of a lobster is about 12 mm mercury and is only about 4 mm mercury in the ventral thoracic sinus. A further drop in pressure occurs as the blood passes through the fine channels in the gills yet when the heart

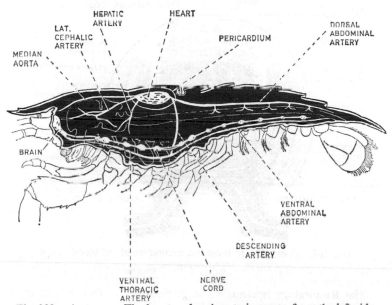

Fig. 208.  ASTACUS.—The heart and main arteries—seen from the left side.

expands the pressure inside it falls 1-2 mm mercury below that in the ventral sinus so that haemolymph is sucked up through the gills into the pericardial cavity and then into the heart through the ostia. The heart thus acts as a force pump during systole and as a suction pump during diastole. Filling of the heart is, then, independent of blood pressure or volume and the whole method of circulation is quite different from that of vertebrates.

In the heart wall there is a complex of nerve-cells and fibres comprising a ganglion which acts as a pacemaker (p. 529) and a system of inhibitory and accelerator fibres linked with the ventral nerve cord, which control the rate and amplitude of the heart beat.

The haemolymph contains the respiratory pigment, haemocyanin, as well as a range of other proteins and inorganic ions in the plasma. Colourless ("white") corpuscles of two main types float in the haemolymph: small ones with clear cytoplasm and large ones with granular cytoplasm. Most corpuscles are phagocytic (p. 540) and all can clump together so helping to form a clot when any part of the animal is injured. One type of small corpuscle plays an important part in clotting for it swells and "explodes" so releasing enzymes which cause the fibrinogen in the plasma to coagulate.

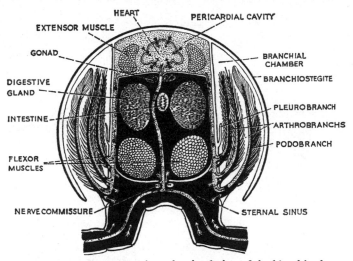

Fig. 209.   A diagram to show the circulation of the blood in the thoracic region.

**The Respiratory System.**—The respiratory surface of the crayfish is provided by the series of gills developed in relation with the thoracic appendages and lying in the branchial chambers enclosed by the branchiostegites at the sides of the body. Each gill is a thin-walled, vascular, feather-like outgrowth of the integument. The gills are arranged in three longitudinal series, the type of gill receiving its name according to the position on the appendage or lateral body wall from which it arises. Thus, those of the outermost series are termed **podobranchs** since they are attached to the bases of the appendages. The next are the **arthrobranchs**, since they take origin from the arthrodial membranes between the bases of the appendages and the body. There may be two arthrobranchs in relation with one appendage, an anterior and a posterior one. The third series, of which a few only are represented in the crayfish, are called **pleurobranchs** because they arise from the sides (pleural region) of the body above the bases of the limbs.

Not all three types of gill are present in relation with all the thoracic appendages and their arrangement may be summarised as follows:

| APPENDAGE | EPI-PODITE | PODO-BRANCH | ARTHROBRANCH | | PLEURO-BRANCH |
|---|---|---|---|---|---|
| | | | ANTERIOR | POSTERIOR | |
| Maxilliped I | X | 0 | 0 | 0 | 0 |
| ,, II | X | X | X | 0 | 0 |
| ,, III | X | X | X | X | 0 |
| Cheliped | X | X | X | X | 0 |
| Walking leg I | X | X | X | X | R |
| ,, ,, II | X | X | X | X | R |
| ,, ,, III | X | X | X | X | R |
| ,, ,, IV | 0 | 0 | 0 | 0 | X |
| TOTAL | 7 | 6 | 6 | 5 | 1 + 3R |

X = present; 0 = absent; R = rudiment.

This somewhat complicated arrangement of the gills may, perhaps, be made easier to understand by reference to the description of the generalised crustacean limbs (p. 276). It will be recalled that in these, both in the phyllopodium and stenopodium, there is a series of exites, commonly three in number, attached to the inner side of the proximal podomeres. The two exites in relation with the precoxa and coxopodite are known as epipodites and function as gills. The precoxal sclerites of the thoracic limbs of the crayfish have become incorporated in the lateral body wall and the epipodites which they bear are the pleurobranchs. The two other main types of gills, the podobranchs and arthrobranchs, are formed by subdivision of the epipodites belonging to the coxopodites of the thoracic limbs. Each of the gills itself has undergone subdivision to form numerous gill filaments which increase the effective respiratory surface, but there is always a small portion of each gill which remains undivided and plate-like, and it is this small portion which is often alone termed an epipodite. It will also be recalled that the three pairs of maxillipeds also bear epipodites, but these are only slightly subdivided and cannot be said to form fully functional gills. From the bases of the maxillipeds and other thoracic appendages arise also a bunch of fine separate filaments which, again, are to be regarded as having arisen from the epipodites and are termed setobranchs.

The gaseous interchange between the blood and the water takes place during the circulation of the blood through the filaments of the gills. The supply of oxygenated water through each branchial chamber is maintained by the paddle-like action of the scaphognathite of the maxilla of its side. These movements cause water to be drawn in under the ventral edges of the branchiostegites into the chamber to be carried forwards and over and between the gills, and thence to be expelled from the front end beneath the head, the exits being situated near the base of the antenna on each side.

When the animal is walking the limb movements themselves may suffice to ventilate the gills and the scaphognathite may slow down or stop beating. Judged by results on other large decapods (*e.g.* lobsters) it appears that the gills can extract about 50 per cent. of the oxygen from the water passing over them yet, thin though the cuticle on the filaments is, it offers an appreciable barrier to diffusion and the haemolymph in the pericardial cavity has a lower oxygen but higher carbon dioxide content than the surrounding water. Despite the fact that the haemocyanin has a low loading tension, it is never fully saturated with oxygen. Nevertheless, most of the oxygen carried by the haemolymph is in combination with haemocyanin.

**Excretion and Osmoregulation.**—The so-called **green glands** or **antennary glands** are usually stated to be the organs of nitrogenous excretion in the crayfish but, as will be seen later, this needs qualification. They are situated one on each side of the anterior end of the cephalothorax and their ducts open out on the bases of the antennae. Each green gland consists of a small **end sac** within which is enclosed

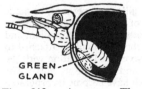

GREEN GLAND

Fig. 210. ASTACUS.—The green gland (after Huxley).

a portion of the coelom, the only part of the coelom (with the possible exception of the cavities of the gonads) which persists into the adult stage. From the end sac arises a duct which passes to the exterior. This duct is not, however, uniform in structure or diameter throughout its length. The part which immediately joins the end sac is very wide and spongy in texture, being permeated by numerous anastomosing canals which form a kind of **labyrinth**. Then follows a narrow tubular portion, the **white tube**, throughout the major portion of which the lumen is broken up by a series of ridges on the walls. The white tube empties into a large, thin-walled sac, sometimes termed the **bladder** and this, by a very short duct, opens out on the protopodite of the antenna.

But the green glands of decapods are not the only, nor indeed, the main organs for the elimination of the end products of nitrogen metabolism, the principal one of which is, as in so many aquatic animals, ionised ammonium carbonate. Ammonia in relatively high concentrations occurs in the haemolymph but only in much smaller amounts in the urine. Presumably (as in fishes) it is largely eliminated through the more permeable parts of the body surface, particularly the gills. Urea, uric acid, and amines occur in small amounts in the haemolymph and are got rid of through the surface and through the green glands.

The green glands produce a urine which is much more dilute than the haemolymph. *i.e.* the urine is hyposmotic to the body fluid. In this respect the crayfish resembles many other fresh water animals in using its excretory organs to eliminate water from the body. But the crayfish is unusual among decapods, even

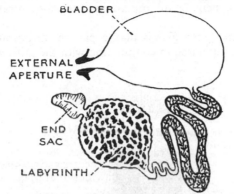

Fig. 211. ASTACUS.—The green gland unravelled (after Marchal).

fresh-water ones, most of which produce a urine isosmotic with the haemolymph. Not only do the green glands get rid of excess water which enters the body (mainly through the gills) because the haemolymph has a higher osmotic pressure than the water in which the animal lives, but they also serve to re-sorb all the ions which occur in the blood. The urine is thought to be formed in the following way. The hydrostatic pressure of the haemolymph is of the order of 27 mm mercury and its colloid osmotic pressure only about 20 mm mercury. It follows that water and small molecules and ions will pass into the excretory organs by a process of ultra-filtration (cf. verte-brate kidney). This occurs mainly in the end sac and labyrinth.

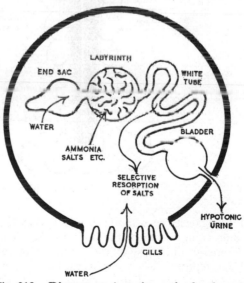

Fig. 212. Diagram to show the mode of action of the green gland.

The rest of each gland is concerned with resorbing salts and glucose so that the urine becomes more dilute.

The gills are also the site of absorption of salts from the surrounding water so helping to maintain the concentration of the haemolymph.

In short, the green glands play a part in nitrogenous excretion and osmoregulation and in conjunction with the gills, which absorb

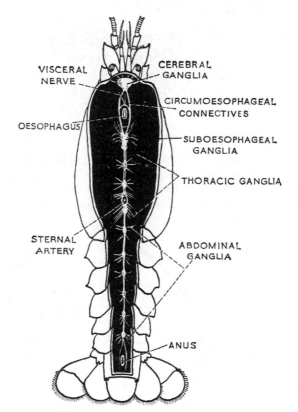

Fig. 213. ASTACUS.—Nervous system.

ions against the osmotic gradient, they help to maintain a correct ionic composition of the body fluids. This entails an expenditure of energy but most of the increases in respiratory rate that have been reported when crustacea have been placed in more dilute media are due not to increases in osmotic work but to increased oxygen consumption of the hydrated muscles.

**The Nervous System.**—In gross form the central nervous system may be seen to be made up of a series of ganglia joined by connectives and cross-connected by commissures so that the whole arrangement can be compared with a ladder (as in worms), although the right and left members of each pair of ganglia are usually so closely approximated that this is not immediately obvious. Typically, it would seem that there should be a pair of ganglia for each body segment, but frequently fusion of ganglia has taken place so that the number of ganglia evident may not be the same as the number of segments. In the front of the head region, lying just behind and beneath the junction of the eye-stalks with the carapace, are found the cerebral ganglia or the "brain". These are believed to represent a pre-segmental portion to which, at the hinder end, have been welded the ganglia of the cephalised segments as far back as, and including, the antennary segment, but no trace of separate ganglia is found externally. As in other invertebrates, the "brain" is morphologically dorsal to the anterior part of the alimentary canal but the pronounced downward dip of the oesophagus towards the ventrally-placed mouth brings the dorsal surface of the oesophagus into an anterior position, and the brain thus is topographically anterior to the oesophagus and mouth. The cerebral ganglia are connected to the sub-oesophageal ganglia by the circum-oesophageal connectives, which themselves are joined one with the other by a small transverse commissure running close to the posterior wall of the oesophagus. The sub-oesophageal ganglia forming the anterior and largest swelling of the ventral nerve chain are of compound segmental origin, and in them five pairs of ganglia are represented, and from them pass nerves to the mandibles, maxillules, maxillae, and first and second maxillipeds.

The six thoracic ganglia which follow are those of the third to eighth thoracic segments. These ganglia are joined by clearly separated longitudinal connectives which, between the fourth and fifth ganglia, are curved outwards to permit the passage of the sternal artery. In the abdomen there is a series of six abdominal ganglia, also linked by longitudinal connectives whose double nature, however, is not so obvious as in those of the thorax. From all the ganglia issue nerves which pass to the various structures of the body. There are always at least two pairs of nerves for each pair of ganglia and in them are found not only the processes from cells lying within the ganglia of the central nervous system, but also fibres which are direct prolongations of superficially situated afferent neurons or of receptor cells of the epidermis. As might be expected **statocyts, antennules**, and **antennae**, connect with the brain. Here, the fibres from the receptor cells of the sense organs synapse with the fibres of relay cells and with those leading into special association centres (**corpora pedunculata**). It is also interesting to note that some of the fibres entering from the eyes and from the antennules pass across to the opposite side of the brain; a point of resemblance with the decussation of fibres in the vertebrate brain (see p. 614).

The degree of nervous integration naturally shows an advance upon that of the annelids. The pattern of behaviour is at a higher level, indicating a greater degree of control by the higher centres of the nervous system, and although a comparable arrangement of superficial receptors, intermediary, and effector neurons exist, their interrelations are more complex. Reflex arcs involving many body segments instead of one or two only can be traced, the intermediary neurons extending their influence further up and down the ventral nerve chain. Great flexibility of behaviour and co-ordination between the organs of special sense is made possible by the association centres of the brain, injury to which produces a greater disturbance in behaviour than in annelids—a good indication of the greater measure of control of the body by the brain.

A general resemblance to certain annelids and to other animals which have a special escape movement in the face of violent, unpleasant stimuli, is the presence of a giant fibre system. Originating from two large cells in the brain are two giant fibres which traverse the length of the cord, rather near to its mid-dorsal line, to end in the sixth abdominal ganglion. Shortly before their exits from the brain they decussate and in each abdominal ganglion they enter into a synapse (megasynapse) with giant motor fibres which decussate and leave the cord to innervate the large abdominal muscles. In addition to the median pair of giant fibres there is another pair, one on each side of the ventral nerve cord, but in this pair the fibres are not continuous. They are built up of segmental contributions from cells in each ganglion from the sixth abdominal to the first thoracic, inclusive, and linkage with the brain is effected by fibres which run forward from the thorax into the circum-oesophageal connectives, and, finally, into the antennary nerve. These laterally placed giant fibres conduct nerve impulses forwards towards the brain, whilst the more median pair conduct in the reverse direction. Both pairs are concerned with relaying impulses to the giant motor cells of the abdominal ganglia and in this way evoking the rapid "flipping" movement of the tail fan and abdomen, by which danger may be avoided.

The innervation of the viscera is by way of the **stomatogastric** ("sympathetic") nervous system which, although connected to the central nervous system by two visceral nerves, yet lies mainly outside it, consisting of cell-bodies and fibres near to, or on, the viscera. The first visceral nerve is formed by fibres arising in the brain, as well as by some arising from two small ganglia in the circum-oesophageal connectives. The second visceral nerve consists of fibres from the last abdominal ganglion.

Interesting resemblances may be traced between the crustacean and vertebrate sympathetic nervous systems (p. 598). As in vertebrates the sympathetic nervous system consists of neurons lying in the central nervous system, connected by preganglionic fibres to peripheral neurons (sympathetic effector neurons) near the viscera, the nervous arcs being completed by afferent neurons whose cell-bodies also lie on the viscera, and whose axons run into the central nervous system. In the crayfish, however, fibres issuing from and running into the central nervous system are confined to two courses only, namely, the two visceral nerves.

The innervation of the muscles and control of muscular movement is interesting and provides several points of contrast with those of vertebrates. In the first place only a few (in contrast to many) motor neurons supply each muscle. In some instances only three motor axons supply a whole muscle. Moreover, each of the motor fibres is of a different type in the sense that stimulation of one type produces a different response in the muscle from stimulation of another. One type of fibre causes a fast contraction: another causes a slow contraction, whilst a third causes the muscle to relax by inhibiting contraction. Each type of axon branches repeatedly and sends several twigs to each muscle fibre, that is, a fibre is supplied by many nerve endings (polyneuronal). The different responses of the muscle to stimulation by the three types of motor axons are thought to be due to the release of different "transmitter substances" at the neuromuscular junction (see p. 477).

**Sense Organs.**—The eyes are borne on short movable stalks and lie between the bases of the antennules and the rostrum. Each eye is in reality a composite structure, being made up of a great number of individual optical units termed **ommatidia**, the whole collection thus forming a **compound eye.** This type of eye is found in many kinds of arthropods. The surface of the eye is roughly hemispherical and is divided into a large number of squarish facets, each facet corresponding to a single ommatidium. Each ommatidium is a complete apparatus for refracting light rays and initiating nerve impulses which are relayed into the optic nerve. Externally in each is the **cuticular lens**, a secretion of the lenticular (epidermal) cells which lie below it. Beneath the lens is a group of four cells called **vitrellae**, the inner borders of which are transparent and refractive, each thus contributing to form the so-called **crystalline cone** and **crystalline tract.** Beneath the vitrellae lies another group of cells, the **retinulae**, the inner borders of which also are refractive. These are called **rhabdomeres** and are referred to collectively as the **rhabdom.** They seem to consist of very fine tubules in which is contained the visual purple. The retinulae are not only partly refractive but are also the receptor cells of the ommatidium and the base of each retinula is prolonged into a nerve-fibre which runs a short way downwards to synapse with cells in the optic ganglion, a conspicuous mass of nervous tissue within the optic stalk. The large optic nerve passes from the optic ganglion into the brain. The outer parts of each retinula also contain black

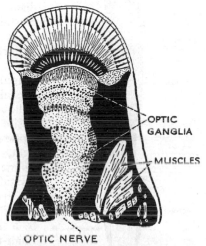
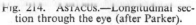

Fig. 214. ASTACUS.—Longitudinal section through the eye (after Parker).

pigment granules (the proximal pigment) whilst similar pigment (the distal pigment) is found also in a group of cells surrounding the crystalline cone and tract. Mainly under the stimulus of varying light intensity, these pigment granules can move and so alter the optical properties of the eye.

In bright light, the distal pigment migrates downwards to become concentrated around the crystalline tract whilst the proximal pigment migrates upwards to surround most of the rhabdom. In dim light, the pigment groups migrate in the reverse directions so that most of the proximal pigment lies beneath the basement membrane whilst the distal pigment moves up around the crystalline cone.

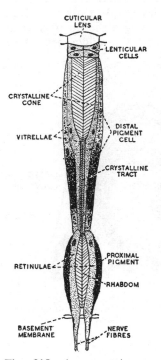

Fig. 215. ASTACUS.—An ommatidium in the "light-adapted condition".

It was formerly believed that in a dark-adapted eye (when the pigment is retracted away from the retinulae and crystalline tracts) the crystalline cones could collect up oblique rays so that a brighter composite image formed of many overlapping images was formed. In bright light, with the pigment tending to isolate, optically, each rhabdom and crystalline tract, the image formed by the eye as a whole consists of many "bits", each of which is contributed by a single ommatidium rather like a "half-tone" engraving such as is used for reproducing photographs in print. This was termed an **apposition** or **mosaic image** in contrast to the single, but blurred **superposition image**, thought to be formed by the dark-adapted eye. Recent work suggests that in types of crustacean eye which have a reflecting tapetal layer at the base of the ommatidia the **visual acuity** (that is, the ability to distinguish between objects close together in the visual field) is greater in bright light than in dim light and that this is due to the positioning of the proximal pigment in the light-adapted eye so that it prevents internal reflection by the tapetal layer from causing halation. In bright light the proximal pigment moves up above the basement membrane (Fig. 215) to mask the tapetal layer (not shown in the figure). The function of the distal pigment is obscure, but doubt must be cast upon the older view that, in conjunction with the proximal pigment, it is to prevent oblique rays from entering the ommatidium since each crystalline cone and tract, having a refractive index higher than that of the surrounding cells, would retain by internal reflection any light rays whose direction was nearly the same as that of the axis of the ommatidium, so that only that light which falls on a particular cuticular lens would reach its rhabdom. In the dark-adapted eye, some light passes from one retinula to the next so that the brightness is increased but the image is blurred.

Most larger crustaceans have organs which provide information about the position of the body in space, i.e. about the relation of the main axis of the body to gravity. These organs also detect accelerations of the body relative to the three planes of space, i.e. turning or angular accelerations, as does the vertebrate labyrinth. Thus, like the vertebrate ear, they function as gravity receptors and as accelerometers but clear evidence that they act also as organs of hearing, in the sense that they respond to vibrations in the water, has yet to be provided. These organs are the **statocysts**. That they are concerned with maintaining equilibrium is evident from the fact that if they are removed the animal loses all sense of balance and ability to orientate normally. Each consists of a sac-like cavity in the coxopodite of the antennule communicating with the exterior by an aperture. From the lining of the sac project hair-like processes, the bases of which are connected to nerve-fibres. These "hairs" are of two types each of which mediates one of the two different functions of the statocysts. Firstly, there are long, thread-like hairs which do not come into contact with statoliths and secondly there are shorter, hooked hairs (statolith hairs) on which statoliths impinge. Within the cavity are found numerous sand grains—put there by the animal after each ecdysis—which become cemented together to form a statolith. This is then fixed to the floor of the statocyst by a flexible secretion. With alterations in the orientation of the animal, the statolith comes into contact with varying groups of statolith hairs, the stimulus of contact evoking nerve impulses which pass along nerve-fibres of the antennulary nerve to the brain, there to be "interpreted" and to originate reflexes which enable balance to be maintained.

Conclusive evidence that the statoliths are gravity receptors comes from experiments on the prawn. Prawns which had just moulted (when the lining of the statocysts and the contained sand grains were cast off along with the rest of the cuticle) were kept in a tank and, in place of sand grains, were provided with iron filings. Lacking anything else the animals placed filings in the statocysts.

They were then subjected to the influence, in different positions, of a magnet. When the magnet was placed beneath the tank so that the pull exerted on the filings corresponded to that of gravity, then the animals behaved in a perfectly normal manner. When the magnet was placed in other positions, the prawns adjusted their positions accordingly and could even be made to swim upside down when the magnet was held above the tank.

The long sense hairs project freely into the cavity of the statocyst and are stimulated by movements of the fluid when the animal turns relative to the axes, *i.e.* to angular accelerations. Recording in the nerves connected to these hairs shows that they are spontaneously and continuously producing impulses. Movements in one direction increase the rate and in the other decrease the number of impulses which pass along nerve in unit time thus providing information about turning movements of the body.

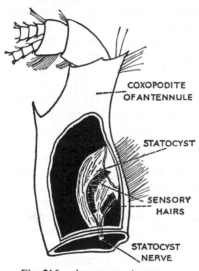

COXOPODITE OF ANTENNULE

STATOCYST

SENSORY HAIRS

STATOCYST NERVE

Fig. 216. ASTACUS.—A statocyst.

The sense hairs in the statoliths seem to be special examples of somewhat similar ones situated on many parts of the body, particularly on the limbs. These initiate nerve impulses when they are bent mechanically and so are collectively called **mechanoreceptors**. Many are tactile but some on the antennules detect water currents. Proprioceptors on the muscles and joints have been identified.

**Chemoreceptors**, organs which detect substances in solution in the water, are found (as shown by changes in behavioural responses following amputation of the parts mentioned) on the inner branch of the antennules, the chelae and the first two pairs of walking legs. The actual organs have not been identified with certainty but are probably special "hairs" and also innervated pores through the cuticle.

The crayfish, then, has far more complex receptor organs than other invertebrate animals so far mentioned. Although totally different in structure, yet they rival in complexity the sense organs of vertebrate animals. It may be inferred that the crayfish is able to respond to a greater range of stimuli than animals with simpler sense organs and less highly organised nervous systems.

**The Reproductive System.**—The sexes are separate. The gonads lie in the cephalothorax beneath the pericardium and above the gut. The ovaries and the testes are similar in shape and each consists of a hollow, three-lobed sac, two lobes lying anteriorly and one posteriorly. The gonoducts are paired, indicating that the single gonad is really a double structure, and are continuous with the gonad from which they arise at the junction of the paired lobes with the median one. The oviducts are short, thin-walled, and almost straight, passing vertically downwards to their openings on the coxopodites of the second walking legs. The vasa deferentia, on the other hand, are long and coiled, the straighter, terminal, muscular portions leading downwards to open on the coxopodites of the fourth walking legs.

The testes consist of numerous small vesicles or alveoli, where the spermatozoa are produced by proliferation and maturation of the lining cells. From each alveolus arises a short duct, the ducts from the various alveoli joining up with one another until eventually they open into the main cavity of the testis from which leads the vas deferens. The ovaries are likewise hollow and the eggs as they mature project from its walls and are finally released into the cavity from which they pass into the oviducts. The spermatozoa are remarkable objects, for, instead of the usual head and flagelliform tail,

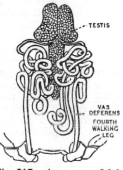

Fig. 217. ASTACUS.—Male reproductive organs (after Huxley).

each consists of a central nucleated disc-like portion from which radiate a number of curved stiff-pointed processes. The eggs are large—about one-eighth of an inch in diameter—owing to the presence of appreciable quantities of yolk and are yellowish or white in colour when laid. The yolk is centrally placed within a layer of cytoplasm and the eggs thus conform to a type termed centrolecithal.

**The Life-history.**—During the breeding season, which begins early in December and lasts into the early spring, the males seek out the females and during pairing, the male turns the female over and deposits **seminal fluid** in the form of strings of milky material on the sterna near the oviducal apertures. The spermatozoa are held together in masses by a secretion of the vasa deferentia and, as they issue from the apertures on the bases of the fourth walking legs, they pass into the tubular first pair of abdominal appendages which are used to direct the strings of seminal fluid on to the female.

The eggs, as they issue from the oviducal apertures, are not allowed to escape, but are attached to the hair-like processes of the abdominal appendages by a secretion of certain glands which open on the sterna. Fertilisation occurs

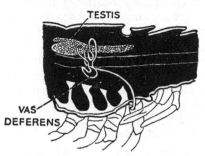

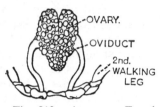

Fig. 218. ASTACUS.—Male reproductive organs from the side.

Fig. 219. ASTACUS.—Female reproductive organs (after Huxley).

immediately after egg deposition and after the spermatozoa have become attached to the eggs by the stiff projections, the protoplasmic portion of the disc is forced into the egg by expansion of the central cytoplasm.

As has already been mentioned, the egg is rich in yolk and this is enclosed in a layer of cytoplasm lying beneath the egg membrane. The egg nucleus, lying in the centre of the egg, is also enclosed in a thin capsule of cytoplasm. Shortly after fertilisation, the nucleus begins to undergo repeated mitotic divisions and the daughter-nuclei then migrate into the superficial cytoplasm, on the upper part of the egg, converting it into a sort of blastoderm (p. 307) above the still undivided yolk. Within this blastoderm it soon becomes possible to distinguish groups of cells which are the fore-runners of the endoderm, mesoderm, and genital rudiments, the rest of the blastoderm remaining to form ectoderm. Thus, on the upper side of the egg, roughly in the middle, is a plate of cells which will form endoderm; just in front of this endodermal plate is a crescentic group of mesoderm cells, whilst the cells surrounding these areas on all sides will give rise to ectoderm. The blastoderm gradually spreads in extent so that eventually it quite encloses the yolk.

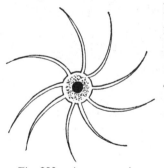

Fig. 220. ASTACUS.—A spermatozoon.

Gastrulation is begun by the endodermal plate sinking downwards, soon to form a sac which gradually extends and deepens allowing, as it grows, the yolk to pass through its walls into the cavity which thus fills with nutrient material. Later, the yolk contained by the endodermal sac (yolk sac) becomes arranged in a number of radiating pyramids. Mesoderm and genital rudiments are invaginated just in front of the opening (blastopore) of the endodermal sac, whilst other cells (mesenchyme) wander inwards from the superficial layers. The true mesoderm cells are not numerous but they soon begin to multiply and in numerical order from front to rear they form the mesodermal segments.

The ectoderm, meanwhile, begins to thicken in patches which then grow out to form the rudiments of the appendages into which mesoderm grows. The ventral nerve cord arises from paired, longitudinal thickenings of the ectoderm

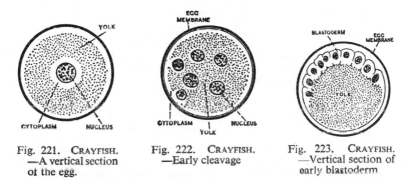

Fig. 221. CRAYFISH.—A vertical section of the egg.

Fig. 222. CRAYFISH.—Early cleavage

Fig. 223. CRAYFISH.—Vertical section of early blastoderm

and the "brain" develops from several ectodermal rudiments, one of which is median and pre-oral, and the others post-oral.

The endodermal sac, later to form the mesenteron and digestive diverticula, soon becomes shut off from the exterior by closure of the blastopore, but regains both an entrance and an exit when the stomodeal and proctodeal invaginations meet and fuse with it.

The vascular system and heart are formed from the mesenchyme, whilst the sense organs, like the nervous system, develop from the ectoderm.

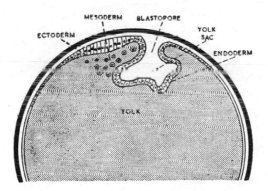

Fig. 224. CRAYFISH.—Longitudinal section of early gastrula.

Elaboration of the main parts of the embryo and growth in size, during which time the yolk becomes progressively used up, results in the formation of a small crayfish not greatly different, except in size, from the adult. These small creatures hatch from the eggs but, remaining clinging to the bristles of the abdominal appendages of the mother, they still retain a large measure of protection. After one moult they acquire full independence, and by successive moults gradually assume the adult form and size.

The life-history of the crayfish is far from being typical of Crustacea in general or, for that matter, of the order **Decapoda**, to which the crayfish belongs. From the eggs of most Crustacea hatch larvae totally unlike the adults and it is

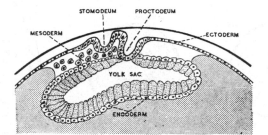

Fig. 225.   CRAYFISH.—Longitudinal section of early embryo.

only after a more or less lengthy life and a series of moults, resulting in a succession of different larval forms, that the adult stage is reached. The crayfish is also somewhat unusual in living in fresh water and in the provision of much yolk, making possible an abbreviated development and the omission of larval forms from the life-history. This has great survival value for not only would larvae in fresh water be hard put to it to find sufficient food in the scanty plankton, but they would be in continual danger of being swept downstream to unfavourable situations, perhaps even into the sea.

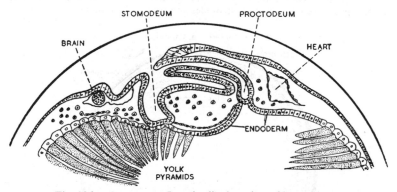

Fig. 226.   CRAYFISH.—Longitudinal section of late embryo.

**Other Crustacea.**—The account of the crayfish has provided a review of many of the important crustacean characteristics, but there are many other Crustacea which may be commonly found and which, because of the crayfish's many specialised features (it belongs to the highly differentiated subclass Malacostraca) differ in many important respects from it.

Anyone who collects animals from an ordinary pond will usually find in it water-fleas (*Daphnia*) quite abundantly. *Daphnia* has already been mentioned (p. 138) as the common food of Hydra and it will not be out of place to say something about it at this point. Its body, as Fig. 228 shows, is very different in shape from that of the crayfish and is entirely covered, apart from the head, by a carapace which, however, is incomplete ventrally so allowing the protrusion of the abbreviated abdomen. The head appendages are also very different from those of the crayfish for the antennae are proportionately much larger, and by their to-and-fro movements the animals row themselves along through the water in a series of rather ungainly jerks. The thoracic appendages are modified phyllopodia, bearing filtratory bristles which are rather comb-like in appearance, and have a food-collecting as well as a respiratory function. Prominent on the

Fig. 227. A young crayfish.

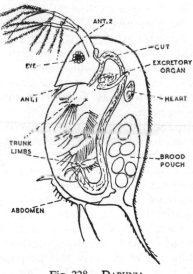

Fig. 228. DAPHNIA.

peculiarly shaped head are the compound eyes which have fused together in the median line. Also on the head are the small antennules. These carry sensory rods and bristles. The blood contains haemoglobin in solution, the amount depending on the oxygen concentration of the water: the less the oxygen, the more the haemoglobin. The eggs also contain haemoglobin.

One remarkable feature about this crustacean, but which is found also in certain other members of the subclass (Branchiopoda, "gill-footed") to which it belongs, is its ability to reproduce parthenogenetically. In fact, most of the specimens collected will prove to be parthenogenetic females. The unfertilised eggs, after issuing from the oviducts, pass into a chamber lying between the body and the roof of the shell and are prevented from leaving by a projection from the hinder end of the body. In this chamber, or brood pouch as it is called, the eggs develop where they may often be seen. At the appropriate time the closing mechanism is opened and the active young escape to begin life in the water of the pond.

In some species, generally towards the end of the summer, males appear in the *Daphnia* population and the eggs produced by the females are fertilised in the normal way but still pass into the brood pouch. After this, the female usually dies but may survive to the following spring. Before this has happened, however, an elliptical portion of the brood sac thickens and becomes darker in colour to form a structure known as the ephippium (because of its resemblance to a saddle) which receives the fertilised eggs from the brood pouch. At the next moult the ephippium becomes detached from the parent and surrounds the eggs until conditions once more become favourable to active life. Then the eggs hatch out parthenogenetic females which be-

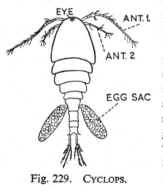

Fig. 229. CYCLOPS.

gin the cycle anew. It will be realised that this type of life-history is peculiarly well suited to maintain *Daphnia* in the ponds which it inhabits, the fertilised ("winter") eggs with their thicker coats and the protective ephippium serving as a dormant phase to withstand the severity of winter and the partheno-genetic phase of reproduction in the warmer months favouring a rapid increase as soon as favourable conditions return.

Among the hauls from the pond another small but interesting crustacean, the common *Cyclops*, may be found. It belongs to the subclass Copepoda ("oar-footed") and is a small animal with a body more closely resembling the crayfish in shape, in that there is a cephalothorax and an abdomen. The absence of paired eyes and the presence of a median eye persisting from the larva or nauplius, is a distinctive feature from which the animal derives its name. Like the water-flea, *Cyclops* propels itself through the water by means of one of the pairs of head appendages, but in this instance it is the antennules which serve this purpose, although the antennae may assist in swimming. The thoracic appendages are mostly of the biramous type and also assist in swimming but the abdomen is devoid of limbs and terminates in a caudal fork. Female copepods can be readily recognised, as a rule, by the large pair of egg sacs which protrude backwards from their attachment to the first abdominal segment.

Marine copepods are important and are often abundant members of the zooplankton and one genus, *Calanus*, forms the main food of the herring at certain times of the year.

Much more rarely, and only from particular ponds, yet another remarkable crustacean may be taken. This is the fairy shrimp (*Chirocephalus*) a member of the Branchiopoda but lacking any protective shell or carapace. It is quite large when compared

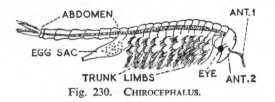

Fig. 230. CHIROCEPHALUS.

with *Daphnia* or *Cyclops* but is delicate and transparent. It habitually swims upside down. Swimming is the result of rhythmical movements of the foliaceous (phyllopod) limbs which, in addition, cause water currents bringing in small particles of food to the ventral side of the body and sieve them off through the bristles fringing the lobes of the limbs. The food particles entangled in mucus pass forward between the limbs to the mouth. The name *Chirocephalus* refers to the peculiar shape of the head which, with its stout antennae, somewhat resembles (particularly in the male) an outstretched hand. Both median (nauplius eye) and paired eyes are present on the head.

Skittering about on the mud at the bottom of the pond lives yet another small crustacean, *Cypris*, whose body and head is totally enclosed in a bivalved shell through whose ventral opening only

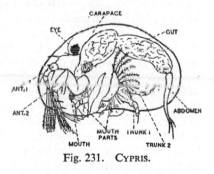

Fig. 231. CYPRIS.

the tips of the antennae and antennules and a small abdomen can be protruded. The two halves of the shell are hinged together dorsally by a ligament and kept in position by a strong adductor muscle running between them in much the same way as are the valves of the shell of a bivalve mollusc. The body, which is remarkably short and very indistinctly segmented, bears only seven pairs of appendages including those of the head region. Swimming is effected by vigorous strokes of the antennae, but *Cypris* can also crawl over the mud when its trunk limbs as well as the antennae come into play. As in so many other small Crustacea, the food consists of small particles and this is carried inwards towards the mouth parts by movements of the antennae. The cyprids and their allies belong to the sub-class Ostracoda, where also is included a host of marine as well as fresh-water forms.

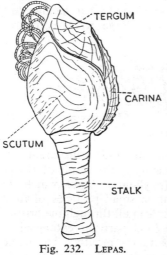

Fig. 232. LEPAS.

In fact, the majority of Crustacea are inhabitants of the sea where in many ways they occupy the niches filled by the insects on dry land. In so doing some of them acquire the most strange modifications of structure and life-history. Thus, in addition to more familiar inhabitants of the shore and shallow seas, like the lobsters, prawns, shrimps, sand-hoppers, crabs, and sea-slaters, there are others which are rarely seen and others, yet again, which although they may be common are so modified that at first sight they are difficult to recognise as crustaceans. Among these is the goose barnacle (*Lepas*) which is sometimes found growing on the bottom of ships or on driftwood. Unlike most other crustaceans the barnacles have adopted a sedentary mode of life with a success which can be judged by the abundance of the sessile or acorn barnacles (e.g. *Balanus*) on the piers and rocks around our coasts. The whole body of a barnacle has, when compared with the usual crustacean plan, become completely transformed to suit its way of life. The head has virtually become lost and the abdomen much reduced in size. What remains of the body is chiefly the thorax and this is enclosed within folds of the body wall, termed the mantle, strengthened by a number of calcareous plates so as to form a shell. In *Lepas* one of these plates is dorsal in position, whilst the remaining two pairs are laterally placed. This arrangement is modified in the acorn barnacles so that certain of the plates (commonly six) form a rampart around the body, whilst four others form the valves of a movable lid which can open to allow the appendages to emerge. Apart from the reduced mouth parts, all that remains of the

Fig. 233. BALANUS.

appendages are the six pairs of thoracic limbs. These are biramous and feathery and, whilst they can be completely retracted within the shell, can also be protruded between the valves of the carapace when they sweep rhythmically through the water and strain off the nutritious particles from it. These accumulate on the bristles of the limbs and are thence transferred to the mouth parts and mouth.

This remarkable form of the body is the climax of a lengthy developmental history during which several different types of swimming larvae occur. First in the series is the small nauplius, a minute larva with only three pairs of appendages, viz., antennules, antennae, and mandibles. Several ecdyses are required for the transformation into the larva termed, because of its resemblance to an Ostracod, the cypris larva. It is this larva which settles down on a suitable substratum to which its head becomes attached by a cement poured out through the anten-nules (traces of which may still be found at the base of the stalk of a goose barnacle). A barnacle is then permanently attached by its head region and it has been aptly said by T. H. Huxley that a barnacle is "a little animal that stands on its head and kicks food into its mouth".

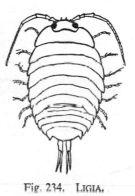

It has been emphasised that Crus-tacea are predominantly aquatic creatures, but there are a few which have abandoned this habitat to become terrestrial. For example, the sea-slater, *Ligia*, lives in crevices just above high-tide mark. Others are the familiar wood-lice which also belong to the order Isopoda.

Fig. 234.   LIGIA.

A common wood-louse or pill-bug is *Armadillidium* which is to be found under stones, in decaying vegetation and rotten wood, and like situations, often in such numbers as to constitute a pest. Quite commonly these animals are mistaken for insects but their many walking legs, the presence of antennules as well as antennae, and the absence of a clear division of the body into head, thorax, and abdomen belie this view. In fact, wood-lice are not fully adapted to life on dry land and bear unmistakable traces of their origin from aquatic ancestors. They still require dampness for their survival, although they approach the insects in their method of aerial respiration, breathing as they do by means of minute branching tubes invaginated from their integument into their abdominal limbs.

## THE COCKROACH (PERIPLANETA)

**Introduction.**—Of the roughly one million species of animals so far described, about three-quarters are insects. This fact alone bears witness both to the diversity of the group and to its "success" in the evolutionary sense. Many species, *e.g.* house-flies, are cosmopolitan; others occupy restricted ecological niches, whilst many more are intermediate in this respect. By and large, insects can be regarded as the supreme representative of arthropod organisation on dry land. A relatively few species have secondarily invaded fresh water and still fewer, the sea. Some terrestrial species have developmental stages which are obligatorily aquatic, in ponds, streams, or the sea. Most adult insects can fly and some undertake

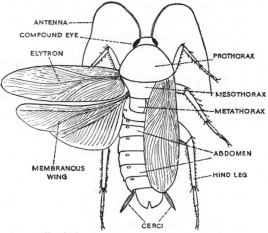

Fig. 235. PERIPLANETA.—Dorsal view.

migration over thousands of miles—which aids their dispersion. Part of their success is no doubt due to their ability to restrict the loss of water from their bodies in which they are aided by possessing a cuticle which, while being well water-proofed by an outer wax layer, is lighter than that of Crustacea because its strength comes mainly from tanned protein and not from the inclusion of calcium salts (p. 269).

It is impossible to select a single example for special study as a representative of the whole of this class. The choice of the cockroach is, perhaps, determined by the fact that formerly it was easily obtained and that in many respects it is not particularly specialised though, of course, it has its special features. It should be remembered that it is a true representative of only one insect order (the

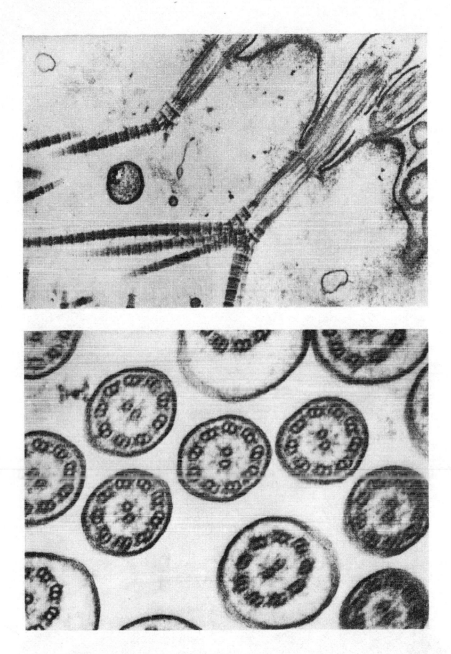

Electronmicrographs of flagella of *Deltatrichonympha*, a flagellate protozoan which is a symbiont in the gut of a termite. *Above:* Longitudinal section showing basal bodies and rootlets (× 12,000). *Below:* Cross-section of flagella showing the arrangement of the fibrils (× 12,000). (*E. H. Mercer.*)

PLATE III.

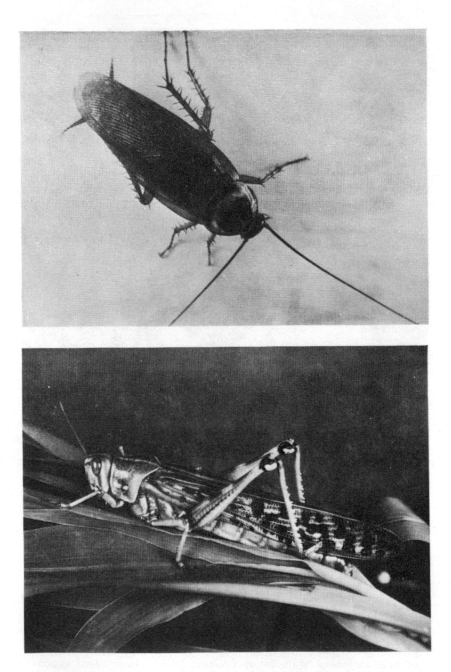

*Above:* COCKROACH. (*J. D. Carthy.*)   *Below:* LOCUST. (*Shell Photographic Unit.*)

PLATE IV.

Dictyoptera) just as the crayfish is of one order (the Decapoda) of the Crustacea.

**Habitat.**—Cockroaches are cursorial omnivorous insects with a wide range of tastes in food. They were until recent years very common insects in this country in places such as kitchens, boiler houses, etc., where there is warmth and food remains of various kinds. Both species found have been introduced into this country from elsewhere; *Periplaneta americana* is, because of its size, the one usually selected for practical examination. It differs from the other species, *Blatta orientalis*, not only in size but in being fully winged in both sexes, whereas in *Blatta* the wings of the female are vestigial.

Both species are now rare, thanks to the use of modern insecticides. Other species, found in the tropics and sub-tropics, are regarded as dangerous carriers of various diseases such as plague and leprosy.

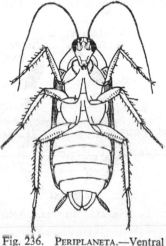

Fig. 236. PERIPLANETA.—Ventral view.

**The Form of the Body.**—The segmented body is elongated and divisible into three distinct regions— the **head, thorax,** and **abdomen.** Connecting the head to the thorax is a slender soft region, sometimes called the neck or **cervicum.** The head represents six segments, though all external indications of segmentation have been lost; the thorax, three segments; and the abdomen eleven, though only ten are recognisable and all these are not visible externally.

The head is held almost at right angles to the axis of the body so that its morphologically dorsal surface is anterior. It is flattened from front to back (dorso-ventrally) and rounded posteriorly. It bears, at its sides, the large **compound eyes** and four pairs of appendages: the **antennae,** the **mandibles,** the **maxillae,** and the **labium.**

The **thorax** consists of three segments: the **prothorax, meso-thorax,** and **metathorax** respectively. Each segment bears a pair of appendages, the **walking legs,** and both the mesothorax and metathorax carry a pair of wings.

The **abdomen** is flattened dorso-ventrally and has seven visible segments, the remainder being telescoped within the hinder end.

None of the anterior (visible) segments bears appendages, but those of the terminal ones are modified in connection with the reproductive apparatus. The last pair of appendages appear externally as the **cerci**. The terminal region of the abdomen differs in the two sexes. In the male, in addition to the cerci, two slender processes, the **anal styles**, are present. In the female the sternites of the seventh segment are prolonged backwards to form a keel-like structure, in the middle of which lies the entrance to the reproductive apparatus (see p. 330).

**The Cuticle.**—The whole of the body and limbs is encased in a cuticle which is formed of chitin with the addition of tanned proteins. The outermost layers are very thin but, nevertheless, complex; the most important of them are the wax layer (of hard wax, or

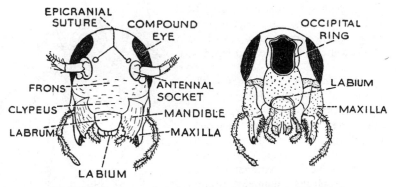

Fig. 237.  BLATTA.—Anterior and posterior views of the head.

a soft grease in the cockroach), responsible for water-proofing the cuticle and the outer cement layer, protective to the wax (p. 270). In each segment the cuticle is thickened and hardened in regions to form sclerites: **tergites** in the dorsal **tergum** region, **sternites** in the ventral **sternum** region and, in the thorax particularly, **pleurites** in the lateral **pleuron** region. In the thoracic region each tergite is termed a **notum**. Typically, between the individual sclerites of each segment and between the borders of contiguous segments are thinner flexible articular membranes. In some instances, however, the sclerites of a segment may fuse. Thus in the head, the outlines of the sclerites are indicated merely by **sutures**, the whole complex forming the **head capsule** made up of the **frons, clypeus**, and **labrum** in front, the **vertex** (composed of two epicranial plates) above, and the genae at the sides.

**The Internal Skeleton.**—As was seen in the crayfish, the exoskeleton is carried inwards in certain places to form apodemes for the attachment of muscles. The arrangement of apodemes (sometimes called the endoskeleton) is not so complex as that in the thoracic region of the crayfish, but in the head capsule there is found a series of struts abutting upon a transverse girder to form what is termed the **tentorium**, to which are attached many of the head muscles. The oesophagus passes over the body of the tentorium, while the cerebral and sub-oesophageal ganglia are arranged around it.

**The Appendages.**—The **antennae** are long, filiform, mobile structures, made up of a large number of podomeres and arise from the sides of the head near the inner margins of the eyes. They are set in depressions with flexible walls (the antennal sockets) and each consists of a large basal podomere, the **scape**, surmounted by a smaller one, the **pedicel**, and a long, many-jointed portion, the **flagellum**.

The remainder of the head appendages form the mouth parts and are grouped around the entrance to the alimentary tract. The **mandibles** are stout, heavily sclerotised structures articulating with the sides of the head capsule. They

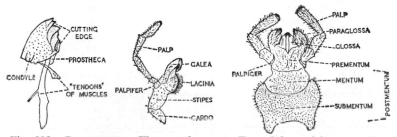

Fig. 238. PERIPLANETA.—The mouth parts. From left to right: mandible, maxilla, labium.

swing transversely across the head, bearing prominent teeth-like projections on their inner edges which bite against each other. On the inner side of each mandible there is a less heavily sclerotised portion, the **prostheca**. The mandibles are gnathobases and are used for crushing and cutting up food. The **maxillae** each consist of a basal portion of two podomeres, the **cardo** and **stipes**, bent at an angle to one another. From the stipes arises a jointed **palp** on the outer side, and on the inner a bipartite portion composed of an outer **galea** and an inner **lacinia**. At the base of the palp is a small sclerite, the **palpifer**. The galea is hooded in shape, while the lacinia terminates in a claw-like projection and has its inner margin fringed with stout bristles. The **labium** looks like a median structure but is really derived from the fusion of a pair of appendages. The broad basal portion—**submentum** and **mentum**—articulates by the proximal border of the submentum with the head capsule, and from it arises an obviously paired structure, each half of which is composed of an outer palp and an inner bipartite portion, exactly similar to the distal region of the maxilla. This palp has fewer podomeres than that of the maxilla, and the bipartite inner portion is made up of an inner **glossa** and an outer **paraglossa** which correspond respectively with the lacinia and galea of the maxilla and might be so named; but to keep in line with the names used in more specialised insect mouth parts the former terms are usually employed.

In addition to these true appendages, there are other structures, contributions from the head capsule, which may be included in the mouth parts. The **labrum**, which lies over the mandibles, has already been mentioned in connection with the sclerites of the head capsule, and on the inner side of it is the anterior wall of the entrance to the alimentary canal which forms the **epipharynx**. The posterior margin of the alimentary opening is prolonged into a flattened cylindrical structure lying between the maxillae and above the labium and termed the **hypopharynx**—sometimes called the lingua in the cockroach—at the front of which is the aperture of the salivary duct (see p. 322).

Each thoracic segment bears a pair of walking legs, made up of a basal portion, the **coxa**, articulating with the sclerites of the thoracic segment, followed by a small **trochanter**, which is freely movable on the coxa but is fixed to the femur which follows it. The **femur\*** is succeeded by the **tibia** and the distal

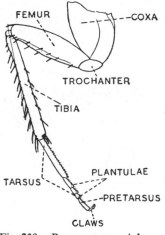

Fig. 239.   PERIPLANETA.—A leg.

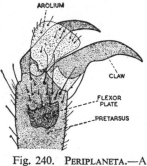

Fig. 240.   PERIPLANETA.—A pretarsus.

portion of the limb is termed the **tarsus**, made up of five movable podomeres and terminating in two claws. The claw-bearing portion is often called the **pretarsus**.

On the tibia numerous stout spines are present, the tibial spurs and the **tarsal podomeres** have many finer hairs, whilst at their lower edges are soft adhesive pads, the **plantulae**. Between the claws there is a delicate, hair-covered pad, the **arolium**.

Of the abdominal appendages only the terminal ones persist and these have mostly been incorporated as **gonapophyses** in the genital armature, which is too complicated to be dealt with in detail. The appendages of the eleventh segment form the **cerci**—two many-jointed projections from the end of the body.

**The Wings.**—Wings are found on the mesothorax and meta-thorax only. In structure, a wing consists of an upper and lower

\* The names of the distal portions of the limb have been borrowed from human anatomy and have no real significance here since the sclerites so named are exoskeleton, whereas the corresponding structures—bones—in man are endoskeleton.

membranous layer, really an extension of the cuticle, strengthened by a framework of branching veins. Developmentally a vein is a tubular prolongation of the body haemocoel, and in the early stages contains haemolymph. Running through the vein is also a trachea (see later, p. 325) and a fine nerve. Both upper and lower wing membranes are secreted by an underlying epidermis, but in the completed wing the epidermis degenerates and the two membranes are in close contact except at the veins. In the cockroach the anterior (mesothoracic) wings are more heavily sclerotised than the larger posterior (metathoracic) pair, which are membranous and much folded when at rest The anterior pair of wings are sometimes called the wing covers or **elytra**.

**Insect Flight.**—Cockroaches are not noted for their powers of flight. They depend for escape more on running fast. Nevertheless, like all pterygote insects they have wings and the muscles for moving them.

Most of what follows has been discovered in insects other than cockroaches and particularly in the true flies, the Diptera. Wings are pivoted so that the side wall of the meso- and metathoraces support the middle of the wing. In addition there are articulations at front and rear of each wing which are part of the dorsal part (the notum) of the segment. Between these main pivots there are a number of small plates, the axillary sclerites, which are an important part of wing articulation.

The muscles concerned with flying are of two kinds, direct and indirect. In most insects the direct muscles, attached to the wing base, are responsible solely for the fine control of wing angle and so forth as well as folding it when not in use. The indirect muscles supply the power for flight. Dragon-flies, however, use the direct muscles for power.

One set of indirect muscles pass on each side of the centre line from front to rear of each wing-bearing segment (the dorsal longitudinal muscles). The other set runs vertically from notum to sternum of the segments. Contraction of the longitudinal muscles shortens the segment pushing the notum upwards and lifting the anterior and posterior articulations while depressing the wings (Fig. 241). The vertical muscles contract in opposition to the longitudinal set depressing the notum and bring the posterior and anterior articulations below the central articulation, thus raising the wing (Fig. 241).

It was once believed that the wings moved smoothly through the cycle of elevation and depression under the influence of these two sets of muscles. But this does not take into account the fact that **the thorax is an elastic box**, the rigidity of whose sides can be altered

by the muscles running from the side to the base.   In fact the wings
of Dipterous flies and perhaps of all insects move quickly through
the down or up stroke but halt momentarily at the top and the
bottom of their strokes.   This is due to what is known as the
"click" mechanism.   When the vertical muscles contract the wing
does not move immediately.   The side of the thorax is pulled
inwards by the sternopleural muscles.   However, as the contraction
of the vertical muscles increases in strength there comes a point
when the axillary sclerites of the articulation of the wing click over

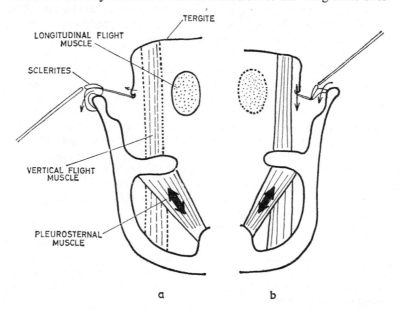

a                                   b

Fig. 241.   The click mechanism in an insect's wing movement:  (a) the end of the
downstroke when the dorsal longitudinal muscles have been contracting;  (b) the
end of the upstroke when the vertical muscles have been contracting.   The
pleurosternal muscles are in a continued state of tension.

the point of articulation on the side wall and the wing moves up.
The reverse happens on the down stroke.

The speed of wing movement may be very great (*e.g.* as high as
1,000 per second in some midges).   It is not possible for the muscles,
which are striped (p. 473), to contract and extend at this rate if it
depends upon a nerve impulse for each contraction.   The impulses
in the nerve would have to be so frequent that they would fall into
the refractory period of the previous ones.   Instead it appears that
the muscles react to stretching by the antagonistic set by themselves
contracting.   A nerve impulse is required to begin and end the chain.

Sensory control of flight is complex. Information from the eyes and from proprioceptors on the neck, as well as air current receptors on the head and on the wings themselves, aid in control of posture and direction. Among the true flies the hind wings are modified into halteres which function to detect rolling and yawing when in flight.

**The Alimentary Canal.**—As in most arthropods, the alimentary canal of the cockroach consists of a long stomodeum, a shortish mesenteron, and a long proctodeum. Both stomodeum and proctodeum have a chitinous lining continuous with the cuticle.

The ill-defined **mouth cavity** is bounded by the epipharynx and labrum anteriorly, the hypopharynx and labium posteriorly, and, at the sides, by the mandibles. At the base of this rather indeterminate cavity is the aperture—leading into the pharynx. The

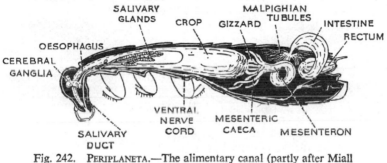

Fig. 242. PERIPLANETA.—The alimentary canal (partly after Miall and Denny).

tubular **pharynx**, which in the cockroach is not very clearly delimited from the succeeding region (the oesophagus), passes vertically upwards until the region of the tentorium is reached, where it makes a sharp bend and passes into the **oesophagus** which traverses the neck into the thorax. Soon the oesophagus dilates into the **crop**, a large pear-shaped sac which extends well into the abdomen and is by far the largest part of the stomodeum. At its rounded end, the crop opens into the **gizzard** or **proventriculus**, a small, thick-walled sac from the inner surface of which project thickenings of the cuticular lining. The gizzard marks the end of the stomodeum and the **mesenteron** consists of a narrow, soft tube from the anterior end of which arise eight blindly ending (hepatic or mesenteric) **caeca**. The posterior limit of the mesenteron is marked by a large number of extremely fine tubules, the **Malpighian tubules**, which, however, really arise from the beginning of the proctodeum. The proctodeum consists of a short proximal tubular portion, the **small intestine** (sometimes called the ileum), followed by an enlarged coiled **large intestine** (colon), which joins the **rectum**. The internal surface of the large intestine is corrugated and that of the rectum is thrown into six prominent longitudinal folds. The rectum terminates at the anus.

As accessories to the alimentary canal there should be mentioned the pair of **salivary glands** which lie in the thorax. Each gland consists of a bipartite glandular portion and a sac-like reservoir. From the glandular portion issues a duct which joins its fellow of the opposite side to form a common duct. Similarly, the duct from the reservoir joins its fellow to form a single duct and the two common ducts then join to give rise to the salivary duct which opens out into the "mouth cavity" on the hypopharynx.

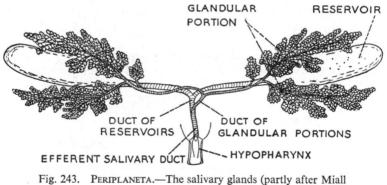

Fig. 243. PERIPLANETA.—The salivary glands (partly after Miall and Denny).

**Nutrition.**—For food, the cockroach will use any kind of organic material, animal or plant, devouring the dead bodies of its fellows, and even its own cast cuticle. The material is seized and cut up by the mandibles, assisted by the maxillae, and then pushed by them with the help of the labium into the "mouth cavity" and thence into the pharynx. During this masticatory process the saliva is poured on to the food from the aperture of the salivary duct on the hypopharynx. The moistened food is passed down the oesophagus into the crop, where it is retained; and since the salivary secretion of insects usually contains diastatic enzymes, digestion begins at once. Into the crop, proteolytic and other enzymes are regurgitated from the mid-gut. The gizzard, though it may by its contractions assist in breaking up the partially digested food material, most probably acts as a filter mechanism, allowing only food particles of the appropriate size to pass into the mesenteron.

Here, the food is subjected more completely to the enzymatic secretions of the mesenteric caeca and mid-gut epithelium, and thus the digestive processes are completed. Absorption of the digested food takes place in the mesenteron and also in the caeca; and during the passage of the remainder of the material through the intestine and rectum, particularly in the rectum, water is extracted from it. Absorbed food material, carbohydrates, proteins in the form of albuminoids, as well as fats, are stored in the diffuse fat body.

**The Blood Vascular System.—** The blood or haemolymph of the cockroach is a colourless fluid in which float numerous large white corpuscles. The oxygen necessary for respiratory activities is conveyed to the tissues and organs by other means (see p. 324). The blood vascular system mainly consists of a series of haemocoels, the circulation through which is maintained by the pulsations of the dorsally situated heart. The heart is a long tube, enclosed in a (haemocoelic) pericardial sinus or cavity, lying in the mid-line immediately beneath the terga of the thorax and abdomen. It is made up of thirteen segmentally arranged chambers, each of which communicates with the pericardial cavity by a pair of laterally placed ostia. Inserted in the ventral pericardial wall is a series of twelve paired alary muscles,

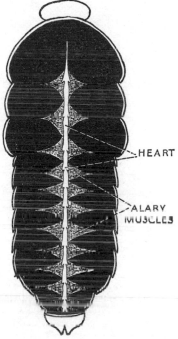

Fig 244 PERIPLANETA.—The heart (after Miall and Denny).

triangular in shape, the pointed ends of which take origin from the under sides of the terga. In the floor of the pericardium which forms a kind of diaphragm are small apertures by which the perivisceral haemocoels communicate with the pericardial cavity. The contractions of the alary muscles enlarge the pericardial cavity and thus cause the blood to pass from the perivisceral cavity into that of the pericardium; thence it enters the heart through the ostia, and, on the contractions of the heart, which take place in a wave from behind forwards, the ostia close and the blood is forced along the forward extension from the heart, which is called the dorsal aorta, to re-enter the haemocoelic system, and the circulation is completed. The heart is attached to the diaphragm whose movements (caused by the alary muscles) serve to dilate it after it has contracted.

The perivisceral cavity is divided into haemocoels by two horizontally disposed muscular septa: the dorsal diaphragm which forms the floor of the pericardial cavity, and the ventral diaphragm which lies immediately above the ventral nerve cord.

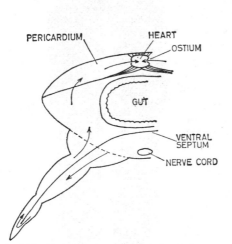

Fig. 245. PERIPLANETA.—Blood circulation in a body segment.

Fig. 246. PERIPLANETA.—The tracheal system (after Miall and Denny).

**The Respiratory System.**—The terrestrial insects, which use the oxygen of the atmosphere for their respiratory needs, have a system of air-tubes or **tracheae** which branch repeatedly and form a network carrying air to all parts of the body. The tracheae have a cuticular lining since they are in continuity with the exoskeleton and, like it, are shed at each ecdysis, and this lining has spiral thickenings which prevent them collapsing, although allowing them to bend when parts of the insect move relative to each other. The wall of each trachea is completed by the epithelium which secretes the lining. The main tracheal trunks communicate with the exterior by paired apertures termed **spiracles** or **stigmata,** which are segmentally arranged. There are ten pairs of spiracles, two in the thorax (pro- and metathoracic) and one pair in each of the first eight segments of the abdomen. In the cockroach the thoracic spiracles are larger than the abdominal ones and each lies in the pleuron, appearing as a small slit-like aperture in an oval sclerotised area. The abdominal spiracles are small and lie in the soft membrane between the terga and sterna. Each spiracular aperture is bounded by an annular sclerite and leads into a cavity, the **atrium** or tracheal chamber, from which the tracheal trunks arise. These main tracheae divide into dorsal and ventral trunks, the branches from which anastomose and penetrate to all parts of the body. The ultimate branches of the tracheae terminate in finer branching tubes called **tracheoles**, the cavities of which

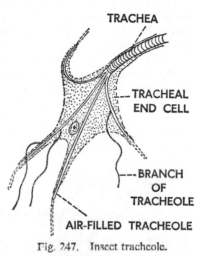

Fig. 247. Insect tracheole.

are *intra-cellular* (Fig. 247), have no cuticular lining, and, except when respiratory activities are high, are partly filled with fluid in which the oxygen dissolves. It is these tracheoles which closely

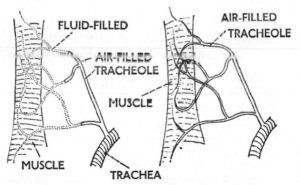

Fig. 248. To show the movement of the fluid in the ends of the tracheoles of an insect. Left, muscle at rest; right, after work.

surround the organs and tissues and even penetrate the living cells. Unlike tracheae, tracheoles are not broken down and renewed at ecdysis. They remain throughout the life of the insect but can grow to meet new physiological demands.

Since there is a continuous uptake of oxygen from the tracheoles by the living cells, an oxygen gradient is set up between them and the spiracular opening and oxygen will diffuse into the tracheae from the outside air. It has been shown that sufficient oxygen for normal respiratory needs can reach the tissues by this means. When, however, the metabolic rate is high, the fluid is withdrawn from the tracheoles due to a rise in the osmotic pressure of the cells counteracting capillarity, which usually keeps fluid in the tracheole, and the oxygen can dissolve in the tissue fluids themselves. The movement of air through the tracheal trunks is facilitated by respiratory movements, the dorso-ventral flattening and relaxation of the body segments, for example. Also the inward and outward passage of gases through each spiracular opening can be controlled by an occlusor mechanism. Although some carbon dioxide may diffuse out into the tracheae and leave the body by the spiracles, it is probable that the major part of it dissolves in the blood and then passes outwards through the cuticle. Also since the body cavity is haemocoelic, the blood is in close contact with the whole of the cuticle, which, particularly the thinner regions, though impervious to water, is permeable to gases and consequently will allow the carbon dioxide to pass through it.

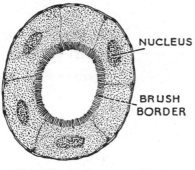

NUCLEUS

BRUSH BORDER

Fig. 249.   Transverse section of a Malpighian tubule.

**The Excretory System.**—It is now generally accepted that the main excretory processes of insects are carried out by the Malpighian tubules, though some other structures may participate. The excreted materials are urates and uric acid, which pass into the hind-gut and thence out of the body with the faeces. Each Malpighian tubule is an outgrowth from the beginning of the proctodeum and is lined by a glandular epithelium with a brush border or layer of microvilli. There are sometimes recognisable histological differences in the epithelium of the proximal and distal portions. The number of tubules varies in different insects, and the number in the adult male *Blatta orientalis* is 148, and 186 in the female, arranged in six groups.

Experiments on insects, other than the cockroach, have shown that the distal portion of each tubule is secretory, pouring into the lumen nitrogenous salts (chiefly urates) in solution. From this solution—probably through the agency of carbon dioxide—uric acid is precipitated in crystalline form. Though in some insects water and the inorganic base—in the form of bicarbonate—is restored to the blood by being resorbed in the proximal portion of the tubule, in the cockroach this—and the necessary acidification—probably takes place in the rectum. In this way a circulation of water and base occurs during the excretory processes, resulting in the elimination of the uric acid, etc., but conserving the water and base, which is an important necessity in a terrestrial animal (uricotelic excretion). This mechanism may be contrasted with that in the aquatic arthropods like the crayfish, where osmoregulation is a complementary function of the excretory organs, chiefly concerned with the elimination of surplus water from the body of the animal. The ability of the malpighian tubules to control the salt concentrations in the haemolymph is not great.

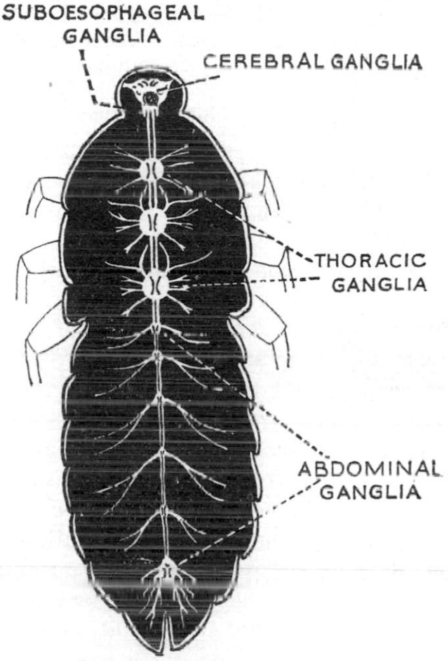

**The Neuro-sensory System.**—The central nervous system of the cockroach resembles in its main features that of the crayfish. The cerebral ganglia lie in the head above and around the

Fig. 250. PERIPLANETA —The nervous system (after Miall and Denny).

tentorium, and the circum-oesophageal connectives pass around the oesophagus to the sub-oesophageal ganglia, which also lie within the head capsule. From the sub-oesophageal ganglia pass the double longitudinal connectives, traversing the neck to the first (pro-) thoracic ganglia in the hinder end of the prothorax.

Then follow the **mesothoracic** and **metathoracic ganglia** and six **abdominal ganglia**, of which the first five lie in the first five abdominal segments and the last some short distance behind. This last paired ganglion is larger than the others and represents a number of fused ganglia. From each pair of ganglia issue nerves which pass to various parts of the segment in which they lie and form the peripheral nervous system. The nerves innervating the eyes and antennae join the cerebral ganglia; those of the mandibles, maxillae, and labium are connected to the sub-oesophageal ganglia; while in the thorax and abdomen the nerves are segmentally arranged, except that from the last abdominal ganglia arise nerves to the terminal abdominal segments.

The sympathetic nervous system consists of an **oesophageal sympathetic system** connected to the brain and supplying the heart and fore part of the intestine. To this system are connected behind the brain the important endocrine glands, the **corpora allata** (p. 331). The **ventral sympathetic system** is a longitudinal nerve between the longitudinal connectives giving rise in each segment to a pair of nerves supplying the spiracles. The **caudal sympathetic system** arises from the last abdominal ganglion and supplies the posterior part of the gut and the reproductive system.

Of the sense organs, the compound eyes are the most obvious. These are large and occupy a considerable portion of the sides of the head capsule. They are composed of ommatidia, similar in structure to those already described for the crayfish, except that the pigment between them does not appear to be retractable. Consequently, the cockroach can probably utilise only the mosaic (apposition) image. The sight of insects is undoubtedly, in some instances, very acute. In such cases the number of ommatidia is enormous, and the great convexity of the eye renders possible a wide "sphere" of vision. Although the extent to which insects "see" is doubtful, it is evident that some, at least, can distinguish differences in the colour and form of objects. The prime function of the eyes in undoubtedly to enable the insects to distinguish between differences in intensity of light and although slow movement in their immediate surroundings may be ignored, quick movements are at once appreciated. The function of the ocelli appears to be that of enhancing the sensitivity of the compound eyes; thus, when the ocelli are covered over, some insects fail to react to dim light which they walk towards when their ocelli are uncovered.

Insects react to many different stimuli and possess other sense organs, *e.g.* chemo-receptors and sound-receptors. Sensory hairs —of different structures and appearance—can be employed for both of these functions. Many of the hairs on the antennae are sensitive to air-borne chemicals in very small concentrations—they are **distant chemo-receptors**; while others on the maxillary palps and often on the tarsi detect the presence of chemicals in greater concentration in contact with them—they are **contact chemo-receptors**. The movement of hairs on the cerci by air-currents or sounds of low frequency is detected, but the better developed hearing organs occur on other insects, such as crickets. Here the movements of a diaphragm, sunk below the surface of the body, stretch special sensory structures, the **chordotonal** organs, and as a result the insect reacts to the sound. Other groups of sensory cells and epidermal cells, called **sensillae**, responding to a rise in temperature, occur mainly on the labial palps, antennae, cerci, and tarsi; but cold receptors are found chiefly on the tarsi.

**The Reproductive System.**—In the insects the sexes are separate. In the male cockroach the testes are paired and each is made up of a number of vesicles which are connected by short ducts with the vas deferens. This unites with its fellow of the opposite side

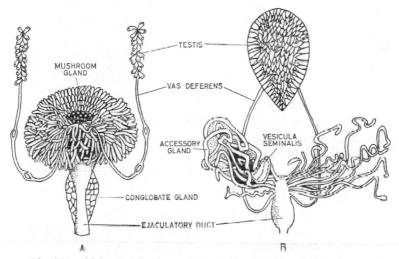

Fig. 251. Male reproductive systems of (A) cockroach and (B) locust.

to form a muscular ductus ejaculatorius, which opens out on a median rod-like structure which is part of the complicated genital armature. Opening into the vasa deferentia and the ejaculatory duct are found accessory glandular structures. Just before the vasa deferentia join, each dilates into a vesicle from which arise a large number of blindly ending diverticula, the whole arrangement forming what is termed the **mushroom-shaped gland**. On to another part of the genital armature opens the median **conglobate gland**, the function of which seems obscure. Although it has been stated that the testes are functional only in the young cockroach, they are still recognisable in the adult and spermatozoa may be found in them. Usually, however, the majority of the spermatozoa have accumulated in the diverticula of the mushroom-shaped gland.

The ovaries are also paired, and each consists of eight ovarian tubules in which the eggs are elaborated. A tubule shows a number of swellings, each of which contains a developing egg; and these become progressively larger and the tubule correspondingly wide from the narrow, pointed anterior end towards the broader posterior end. All the ovarian tubules of each side join to form an oviduct and the two oviducts unite to give rise to a median channel, sometimes called the vagina, which opens out by

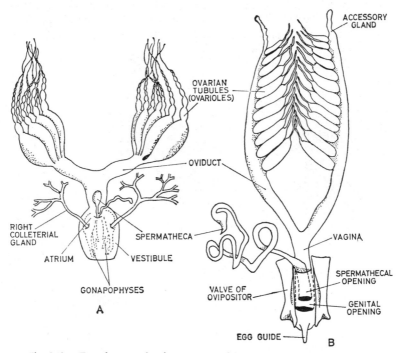

Fig. 252. Female reproductive systems of (A) cockroach and (B) locust.

a slit-like aperture into a **genital pouch** bounded by the terminal sclerites of the abdomen. Into this pouch open a pair of branched glands which are called the **colleterial glands**, which produce the substances used to form the ootheca. From this pouch also open inwards the spermothecae, of which only one is properly developed and functional.

**The Life-history.**—During pairing, the spermatozoa are introduced into the genital pouch and pass into the spermotheca, where they are stored until required.  Fertilisation evidently occurs after the eggs have been passed from the oviducts and in the genital pouch they are enclosed within an egg-case or **ootheca**.  This is formed solely of tanned protein without an admixture of chitin. The ootheca is purse-shaped, and the eggs are arranged in it in two rows, rather like cigarettes in a cigarette case.  For some time after the egg-case has been completed, it is carried about by the female, protruding from between the terminal sclerites of her abdomen. Eventually it is deposited and the young cockroaches complete their development within it.  They emerge from it as small **nymphs**.

**Metamorphosis.**—The nymph is very like the parent but devoid of wings.  As it feeds and consequently grows, it is necessary, as in all arthropods, for it to cast off its cuticle and form a new one.  In essentials, the process is much the same as that described for the crayfish (see p. 289), the new covering being formed by the reactivated

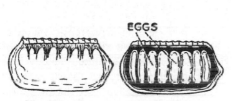

Fig. 253.  PERIPLANETA.—Oothecae.

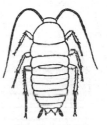

Fig. 254. PERIPLANETA.—
A young nymph.

epidermis (hypodermis) before the old is sloughed off and the increase in size taking place before the new covering has hardened. At the later ecdyses, on the dorsal surface of the mesothorax and metathorax, small projections from the posterior margins of the terga appear; these are the **wing pads**.  After each ecdysis the pads increase in size until, at the last moult—that which precedes the assumption of the adult characters, including the maturation of the gonads—the wings make their appearance fully formed.  When this has taken place, *no further increase in size* occurs.

Moulting is under the control of a hormone, the growth differentiation hormone, produced by the **prothoracic glands,** a small pair of glands in the first thoracic segment.  So long as a second hormone, the juvenile hormone, is secreted by the corpora allata (p. 328), the final moult into an adult is suppressed; only when this hormone is no longer being produced does the change from nymph to adult take place.  Neuro-secretory cells of the brain seem to determine the rhythm of activity shown by adult cockroaches, which are active by night and comparatively still by day.

The type of development found in cockroaches is called **incomplete** or **hemimetabolous metamorphosis** to contrast it with that termed **complete** or **holometabolous metamorphosis**, which occurs in many other types of insects where the egg hatches into a form very *unlike* that of the adult. In insects such as butterflies or beetles and many others, the stage which emerges from the egg bears no resemblance to the perfect insect and is termed a **larva**. The main role of the larva is to feed and grow and in it the gonads are quite undeveloped. Insect larvae present an almost infinite variety of forms from the familiar caterpillar of a butterfly to the active grub of the beetle or the headless maggot of the house-fly. But whatever

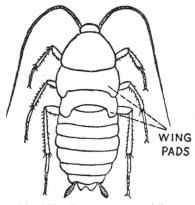

Fig. 255.    PERIPLANETA.— A later nymph showing wing pads.

the form, at the end of the larval period it passes into a quiescent stage known as a **pupa** or **chrysalis**, which is again different in form from the larva and yet does not have the adult features. Although this pupal stage remains immobile and apparently inactive, yet within it there is a scene of tremendous change and developmental activity. Most of the larval organs (with the exception of the central nervous system) are being broken down, largely by the action of phagocytic blood-corpuscles, into a kind of elemental cream for the nourishing of certain groups of cells which were early set aside and have retained their original embryonic capacity for development. From these cells the adult organs will be elaborated.

These groups of cells are called **imaginal discs** or **buds**, and are found distributed throughout the larval body in situations appropriate for the development of the adult organs. Those which will give rise to external structures such as appendages (which may not even be represented at all in the larva) lie in pockets in the epidermis, produced by an invagination of the epidermis at an early stage in the development of the larva. At the bottom of the invagination lies the imaginal bud. Other imaginal buds are lodged in relation with the internal organs. Each imaginal disc may produce a single adult structure, such as a limb, or a group of them may develop into some internal organ. In those instances where the imaginal disc gives rise to an external structure—such as an appendage—the cells multiply and become differentiated, so that the structure as it is elaborated projects outwards to the exterior through the channel—which is never closed—by which the original invagination was produced. In this way, eventually, the perfected structure appears in its appropriate place on the insect's body. Where an internal organ is to be produced, a similar process of multiplication and differentiation takes place to produce the new adult structure.

It is by this means that the adult, or **imago** as it is called, develops within the pupa, and, when the task is completed and the appropriate time arrives, the pupal covering splits and the perfect insect emerges.

The stages in the life-history of an insect with complete metamorphosis may therefore be summarised thus:—

Egg—larva—pupa—imago;

whilst those of an example with incomplete metamorphosis will show only:

Egg—nymph—imago.

A further point of contrast is that in incomplete metamorphosis the wings develop outside (Exopterygota) the body, whereas in complete metamorphosis they are developed from within (Endopterygota)—from imaginal buds. These distinctions were at one time used in classifying the insects but nowadays (Chapter XXIII) the nature of the adult wings and their venation form the basis of classification.

**The Migratory Locust** (*Locusta migratoria*) compared with the cockroach.

**External Features.**—The body of a locust is more cylindrical than that of a cockroach, though in its division into segments it is essentially the same. Locusts are herbivorous with strong jaws for biting and chewing foliage. Their mouth parts therefore have a general similarity to those of a cockroach. Their antennae are much shorter than the cockroach's. They are of course much the stronger flier of the two species of insect and swarms of locusts travel long distances on the wing.

The femurs of the hind legs of locusts are enlarged and contain the well-developed muscles responsible for quick extension of the whole leg causing the animal to leap but locusts can also walk like other insects. On the inner side of the femur is a longitudinal ridge (the stridulatory crest) which is moved back and forth across a row of pegs born on one of the veins of the forewing. This movement, which takes place with the wing folded along the body, sets the wing in vibration producing the stridulatory noises made by the male locust. A female has a stridulatory crest as well, but the wing pegs are reduced so that she cannot make a noise.

The auditory organs of locusts are two tympanal organs, one on each side of the first segment of the abdomen. Each has a membrane, the tympanum, stretched across a cavity. Sound waves cause the tympanum to vibrate and this stimulates sense organs attached to the inner face of the drum.

A female locust can easily be distinguished from a male by the valves of her ovipositor, four of which appear as tooth-like projections at the tip of the abdomen.

**Internal Features.**—The gut of a locust is in general like that of a cockroach; though at the beginning of the mid-gut, in place of the mesenteric caeca of a cockroach, are found six elongated gastric caeca. Each is in two parts, a longer forward-pointing blind ended tube and a shorter backward-pointing one. It is in these caeca that the major part of absorption from the gut takes place. The food in the mid-gut is contained in a peritrophic membrane, a thin chitinous sleeve produced by the walls of this part of the gut. Near the hind end of the mid-gut open the numerous Malpighian tubules. The walls of the rectum bear thickened pads, the rectal papillae.

As Figs. 251 and 252 show, the structure of the reproductive organs of locusts differ from those of cockroaches, but these differences are in detail only.

**Life-history.**—Sperms are conveyed in a spermatophore which is placed by the male in the genital opening of the female during mating. The eggs of a locust are laid in a frothy material which hardens into a cocoon or "pod". When egg-laying, the female extends her abdomen and pushes it down into the sand, so that the pod is deposited well below the surface. The young locusts emerge as nymphs, called hoppers, since they move by walking and hopping only for their wings are reduced, otherwise they resemble their parents. As they undergo subsequent moults the wings grow larger until the adult stage is reached.

Locusts reared in crowded conditions look darker and have slight differences, for example, of parts of the thorax, from those brought up in isolation. These are the gregarious and solitary phases of the locust respectively.

**Some Other Insects.**—Though the consideration of the anatomy and physiology of the cockroach can give a general picture of an insect, it can give little indication of the many variations on this basic plan. One of the factors of the success of insects is the modification of their mouth parts for various ways of feeding and thus for using various sorts of food. The cockroach and the locust will serve as examples of the biting type of mouth part, but in addition, insects pierce, suck, lap, and rasp their food. Some of these habits will become apparent when we consider the biology of a few insects in an endeavour to display the range of habit this class shows.

Fig. 256. A wingless female aphid.

**The Green-fly.**—Any rose tree will provide examples of "green-fly", a common garden pest. Green aphids and their relatives, "black-fly", "woolly aphis", and so forth, are also serious agricultural pests for they carry virus diseases from plant to plant. They are members of the order Homoptera, all members of which have mouth parts specially modified for piercing plant tissues and sucking up the fluids released. Fig. 257 shows that both mandibles and maxillae have been trans-

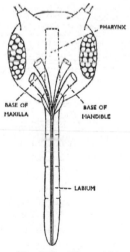

Fig. 257. The aphid mouth parts.

formed into needle-like stylets which can be thrust into the soft parts of the plant, while the labium has been modified to form a cylindrical "proboscis" or rostrum, on the anterior face of which is a groove which houses the stylets when they are not in use. The stylets lacerate the cells they pierce and they may even reach the vascular tissue. The fluid food is sucked up a canal formed by the apposition of two grooves, one on the face of each maxilla (see Fig. 258). Another similar canal provides a means whereby the salivary secretion is forced into the wound made by the stylets. This secretion probably prevents coagulation of the tissue fluids so that they may pass unimpeded up the suction canal.

The life-cycles of aphids are, in many instances, very complicated. Not only may any one species live on more than one host-plant but also there may be several forms of the insect during one complete life-history. The whole affair is still further complicated by the fact that some forms reproduce parthenogenetically—that is, the eggs develop without fertilisation—and, to make matters still more difficult, may produce their young alive. Again, these parthenogenetic viviparous females may be either winged or wingless. So, starting with a normal fertilised egg produced by sexual forms, male and female, the wingless female developing from the egg will produce viviparously and parthenogenetically a series of generations of exactly similar individuals. Then winged forms will appear which will fly away to infest other plants of the same species, or maybe, plants of a different species, and, still reproducing parthenogenetically, set in train a further series of wingless forms. Eventually normal sexual forms are produced, perhaps only on a return to the original food-plant, and the cycle

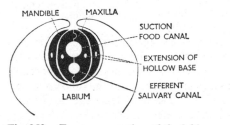

Fig. 258.    Transverse section of the labium of an aphid.

Fig. 259.    HOUSE-FLY.

completed. From this, some idea will be gathered of the complexity of life-cycle found among these insects whose economic importance is very great.

**The House-fly.**—An insect with very different habits is the house-fly, *Musca domestica*, or the blue-bottle fly, *Calliphora erythrocephala*. Unlike the green-flies, these insects have complete metamorphosis and only one pair of wings; they are placed in the order Diptera. The imago lives on fluid foods but its mouth parts are adapted for dealing with fluids exposed on a surface. The apparatus by which the fluid is imbibed is a very profound modification of the labium to form a proboscis shown in Fig. 260. Essentially the structure consists of a protrusible trunk at the end of which are two expanded lobes. The greater part is derived from the labium and on the under side of the expanded lobes are numerous incompletely cylindrical tubes—the pseudotracheae—opening downwards and all joining up to lead into a main canal in the trunk. These expanded lobes are placed on the exposed fluid which is collected in the pseudotracheae and passed thence to the food canal in the trunk, on its way to the alimentary canal. Although this method of feeding is the common one adopted by flies, they will also utilise solid material of an easily soluble kind as may be seen when a fly settles on a lump of sugar. The fly then exudes a drop of fluid from its proboscis on to the sugar, subsequently sucking it up again after some of the sugar has dissolved in it. This habit of regurgitating fluid from the alimentary canal is one of the ways in which flies are the disseminators of, particularly, enteric diseases. Flies are not in the least discriminating in the fluids which they will imbibe so long as they contain organic food material. It is quite likely that, before settling on exposed human food, they may have been feeding on fluids from decomposing material or even faecal matter. The regurgitated drop of fluid (as well as the sticky pads on the feet) is thus often contaminated with pathogenic organisms which are left behind on the food. They are often responsible for the spread of dysentery, diarrhoea, and other digestive troubles; sometimes the responsibility for the spread of typhoid can be laid at their door.

Although in this way flies are a menace to human welfare, they may bring compensation by their other activities. Female flies are attracted to all kinds of dead organic matter in order to lay their eggs therein. The larvae which emerge are headless, legless maggots with extraordinary mouth parts, unlike

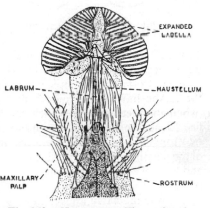

Fig. 260.  HOUSE-FLY.—The proboscis.

in every possible way the jaws of, say, a cockroach. They merely consist of a few chitinous hook-like structures with a complex musculature. Their food is largely liquid in form and solid organic matter is liquefied by the action of enzymes produced by the larvae. Fly larvae are, therefore, of importance in the disposal of dead and decomposing organic material. The obvious way of reducing the fly population in any district is the effective destruction of such organic material—mere burial is often not sufficient.

An interesting feature in the development of the fly is that when the larva undergoes its last ecdysis to give rise to the pupa, the last larval skin is not discarded but encloses the pupa to form the familiar dark-brown barrel-shaped puparium from which the adult fly emerges. To break open this investing case, the freshly emerged imago has a bladder-like structure—the ptilinum—protruding from a slit in the front of its head. This is really a flexible, distensible part of the head capsule and is filled with blood. By forcing more blood into it, considerable force can be exerted, and the fly not only uses this mechanism to free itself from the puparium but also to force its way through the earth to gain the surface, if the material upon which the larva fed and near which it subsequently pupated was buried after the eggs had been laid on it. The importance of the effective destruction of such material, and not mere burial, as mentioned above, in fly control will now be understood.

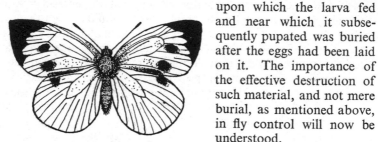

Fig. 261.   BUTTERFLY.—Large white.

**Butterflies and Moths.—** Any account of the insects must include some reference to the butterflies and moths (order Lepidoptera). Their life-history is an excellent example of complete metamorphosis (see p. 332) and, while the larvae have biting mouth parts, the adults feed on fluid food (when they feed at all). Their mouth parts are consequently modified for sucking up fluids but the suction tube—the familiar proboscis or "tongue"—is formed from the apposition of the extended galeae of the maxillae. Both mandibles and labium are either very much reduced or entirely absent.

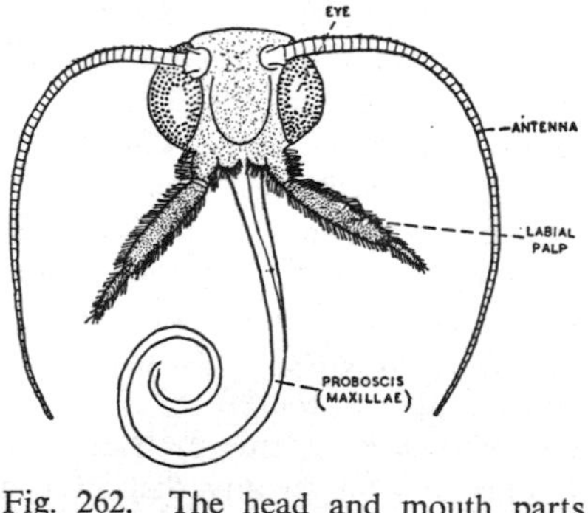

As was indicated above, the larval stage in insects with complete metamorphosis, is one which is almost exclusively devoted to feeding, thus laying up a store of reserve food for the elaboration of the adult organs during the pupal stage. If the larvae of, say, a cabbage-white butterfly and of a clothes-moth are considered, a good illustration of the very different kinds of food utilised by these

Fig. 262.   The head and mouth parts of a moth. (From a microscopical preparation.)

larvae will be afforded. The larva of the cabbage-white lives on the succulent leaves of the plant from which its name is derived, but the clothes-moth larva lives on wool, an animal derivative. Both larvae are provided with mouth parts of the biting type, similar to those of the cockroach, and by means of which the food is rendered sufficiently fine to be passed into the alimentary canal and digested. The succulent food of the cabbage-white will contain abundant water, but the clothes-moth larva will have no access to natural water and must conserve all the water it can. Further, whilst the ordinary complement of digestive enzymes will serve

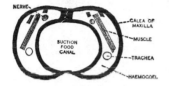

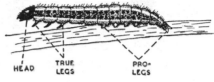

Fig. 263. Transverse section of the proboscis of a butterfly.

Fig. 264. Caterpillar of a cabbage-white butterfly.

adequately to deal with the organic compounds found in the cell contents of a green leaf, the clothes-moth larvae will require special enzymes to digest the very specialised proteins which constitute wool.

To most people the most striking feature about butterflies and moths is their often brilliant colouration. Instances do occur where the wing pattern and colouration may play an important part in the lives of the insects, in that by closely resembling their surroundings *when at rest,* the insects may escape detection by enemies. But as soon as they move their wings or fly away, the brilliant colours make them all the more conspicuous. To these conspicuous colours, in some instances, has been ascribed the function of acting as a warning to intending predators that the insect is distasteful. In other cases it is not easy to ascribe any particular purpose to the colour, though it should be borne in mind that the pigment may be a way in which the organism deals with unwanted excretory material.

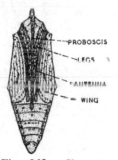

Fig. 265. Chrysalis of cabbage-white butterfly.

A useful, though not quite infallible, way of distinguishing between butterflies and moths is to look at the antennae. Most butterflies have knobbed antennae, whilst those of moths may be very varied in form, some almost whip-like, others plumose, and others again comb-like. As a general rule, too, butterflies are day fliers, while moths are more nocturnal in habit; also, many butterflies, when at rest, fold their wings over the back, but moths keep them outstretched. Although these differences are useful, it should be remembered that all Lepidoptera are distinguished by the presence

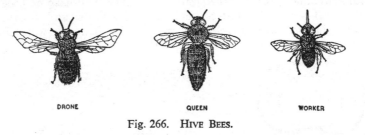

DRONE                    QUEEN                    WORKER

Fig. 266.   HIVE BEES.

of scales on the wings.   These scales are often beautifully shaped and have intricate markings on them, sometimes so fine as to break up reflected light into its constituent colours, so that some colours— particularly some deep blues and greens—are not due to pigment but are the result of this diffraction effect.

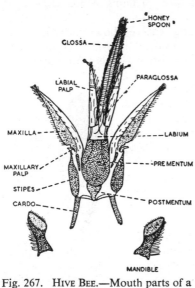

Fig. 267.   HIVE BEE.—Mouth parts of a worker.

**The Honey Bee.**—Much attention has been paid to the social communities formed by some insects.   These insects fall into two main orders, the Isoptera (white ants) and the Hymenoptera (ants, bees, and wasps).   Of these, perhaps the Honey Bee—*Apis mellifera* (= *A. mellifica*)—is the most familiar visitor to any garden because of its habit of collecting nectar and pollen from flowers.   This nectar is used by the insects to produce honey—an important food of the bee community—and, incidentally, when produced in excess, for the human owners and encouragers of the hives!

If a hive is examined during the late summer, three kinds of bees will be found in it.   By far the most numerous are the **workers**. These are often spoken of as neuters, but are really females in which the gonads are not developed and which consequently are sterile.

The centre of the community is a single **queen**. She is a mature female and her sole duty within the society is to produce eggs from which new members develop. At this time there will also be present a number of **drones**. These are mature males and are produced, for the continuation of the race, at the same time as virgin queens.

Each of these types of bee has its own structural characteristics which fits it to play its part in the life of the community. Since the queens and drones are concerned solely with reproductive activities, it is the workers who have to perform all the duties concerned with feeding the community, tending the young, producing wax, and building the

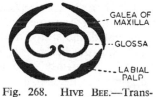

Fig. 268. HIVE BEE.—Transverse section of proboscis of worker bee.

combs, cleaning and ventilating the hive, and so forth. Thus the mouth parts are modified for collecting nectar and moulding wax, the limbs for collecting pollen, and the epidermis to produce special glandular areas which secrete wax. A glance at Fig. 267 will show that the mandibles, instead of being structures for crushing and cutting food, have almost completely lost their denticulate character, being spatulate, shaped much more like a

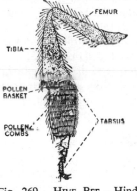

Fig. 269. HIVE BEE.—Hind leg of worker.

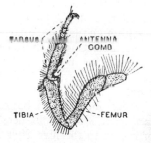

Fig. 270. HIVE BEE.—Fore leg of worker.

clay-modeller's moulding tool; also, that the maxillae and labium are even more profoundly modified. The maxillae have each lost the lacinia, and the galea has become elongated and blade-like; the maxillary palp has been reduced to a vestige. In the labium the paraglossae are very much reduced and the glossae are united, very hairy, and enormously elongated, with a mobile expanded termination—the "honey-spoon". The labial palps are also elongated. The galeae of the maxillae and the glossae and palps of the labium form, when applied to one another, a tubular structure which can be inserted deeply into the corolla of a flower to reach the nectaries and suck up their secretions. This sugary fluid is then stored temporarily in a special part of the alimentary canal to undergo some modification and, on return to the hive, it is regurgitated and stored in the cells of the comb as honey.

During the bee's visits to the flowers in its search for nectar, its body frequently becomes covered with pollen—an important factor in the cross-pollination of plants—and from time to time this pollen is removed from the hairs with which it is entangled and collected together for conveyance to the hive. Fig. 269 shows the hind leg of the bee, and it will be seen that the first tarsal podomere is elongated and furnished with rows of stiff bristles. By passing the tarsal region of these legs over the abdomen particularly, the bristles act as a comb and the pollen so removed is collected in a depression—the pollen-basket—between the tibia and the first tarsal podomere. Any pollen or other material adhering to the antennae is removed by a comb-like mechanism—the antennal comb—situated on the first leg at the junction of the tibia and the first tarsal podomere (Fig. 270). An examination of the under side of the abdomen of a worker bee will reveal four pocket-like depressions in the articular membrane between the segments. These are the wax-secreting glands and the wax thus produced, mixed with saliva, is moulded by the mandibles to form the characteristic bee's comb with its familiar hexagonal cells.

Although the mouth parts of the queens and drones are fashioned on the same plan as those of the workers, they are shorter and are not used for nectar collecting. Also the wax-secreting glands and the modifications of the hind legs for pollen collecting are absent.

Much more could be written about the economy of the hive than space will permit here. Suffice it to say that the queen on her return to the hive after her nuptial flight devotes the whole of her time to egg-laying. She also seems to be able to determine whether the egg, as she lays it, shall be fertilised or not. She thus lays two kinds of eggs. The unfertilised eggs *always* produce drones; the fertilised ones queens *or* workers. Whether the larva which emerges from the fertilised egg shall develop into a queen or worker is determined by the food with which it is supplied by the attendant workers. It is true that while the worker larvae are found in the ordinary sized cells of the comb and the drone larvae in rather larger sized ones, those destined to produce queens are housed in large specially shaped, rather conical cells built usually at the edge of the comb. But it is the food which is the determining factor. All larvae, on hatching, are fed for the first few days on a secretion—the so-called "royal jelly"—produced by special glands opening into the pharynx of the worker. After this period, larvae in the worker and drone cells are fed on a mixture of honey and predigested pollen, but those in the queen cells receive continuously the "royal jelly".

All workers are not employed upon the same tasks all the time. On first emerging they are concerned chiefly with the tasks of the hive—feeding larvae, cleaning, ventilation, etc.—but later on are promoted to pollen and nectar gathering and so fly away from and return to the hive. Though this is a general outline of tasks upon which a worker bee is engaged during her life, the order is not fixed rigidly. Bees can change their occupation to suit the immediate needs of a hive; thus, if all the forages are removed, young bees, not yet at foraging age, will nevertheless take the place of the missing older members of the hive.

Certain aspects of the behaviour of bees are of intense interest and serve to show that animals with a neuro-sensory system built on a totally different plan from that of our own are yet able to communicate information to one another in a way which recalls communication between human beings.

This they do, not of course, by speech but by a series of peculiar movements, usually described as "dancing". These dances by the bees returning from foraging for pollen or nectar are performed on the face of the honey combs and serve to inform other members of the hive not only of the distance of a supply of food but also of its direction. Near and distant feeding places are indicated in two quite distinct ways. Foragers working in the proximity of the hive on returning dance round in circles, sometimes to the right and sometimes to the left, whereas bees which have been gathering food at some considerable distance from the hive (from 100 metres to 5 kilometres) trace out a semicircle and then return along a straight line to the starting point. On the straight return run the abdomen is vigorously waggled. They then trace out a semicircle in the opposite direction and again return to the starting point, producing a figure-of-eight pattern. These manoeuvres form the so-called "waggle dance".

The distance travelled by the foraging bees bears a direct relationship to the proportion of the time spent waggling during one circuit of the dance. The slower the rate of turning: the longer the distance of the food from the hive. Or, to put it another way, the fewer the number of waggle runs per minute: the greater the distance. So the tempo of the dance indicates the distance. But the bees make allowance for head winds or following winds, a head wind being denoted in the same way as an increase in distance and *vice versa* if the wind is favourable. It is concluded, therefore, that it is the flight towards the feeding ground which determines the recording of the distance and this is related rather to the amount of energy expended than to the actual distance from the hive. It is now known that the dancing bee makes a low frequency sound by vibrating the wings during the straight part of the waggle dance. It may be through the duration of these sounds that information on distance is transferred rather than through the actual waggling.

Not only does the waggle run indicate the distance of the feeding place from the hive but its direction also for the direction of the straight waggle run on the vertical face of the comb bears the same relation to gravity as the foraging flight does to the direction of the sun. If, for example, the feeding ground lies in the same direction as the sun, the waggle run is vertically upwards on the comb. If the source of food is 60° to the left of the sun the waggle run is 60° to the left of the vertical—and so on.

It may well be asked—"How do the foraging bees returning to the hive indicate direction when the sky is overcast?" Provided some patch of blue sky is visible bees can still recognise the direction of the sun by utilising the pattern of polarised light in the sky, a pattern which bears a direct relationship to the position of the sun. It now seems quite certain that the eye of the honey bee is capable of analysing the direction of polarisation of light and thus of utilising the sky pattern directly. Bees are also sensitive to ultra-violet light and can detect the sun's position behind cloud by this means.

One of the painful associations with bees is their sting. It is only the workers and queens which can sting because this defensive weapon is in reality an ovipositor which has become modified to subserve this second function. The workers, since they are sterile and do not lay eggs, will use this weapon much more freely than the queens, for if the sting penetrates deeply into the tissues of the victim it cannot be withdrawn, and a queen deprived of her ovipositor could no longer perform her normal function of egg-laying.

There is not space here to discuss other habits of insects. But it should not be forgotten that a large and important group are the parasites, both of other insects and of other animals. The influence of lice as carriers of epidemic typhus on the history of Europe has been very great, though nowadays it is difficult to get specimens of human lice. Fleas also, as ectoparasites carrying bubonic plague, have played their part in human history. Other parasites are important to man for they prey on insects which are agricultural pests and thus can be used to control the pests.

# CHAPTER XIV

## MOLLUSCA

**INTRODUCTION**

In the account given of the Nematoda (p. 193) it was pointed out that a description of the anatomy of any one species would give a good general idea of that typical of all other members of the phylum. It is quite otherwise for the Mollusca since this huge phylum contains animals as diverse as an octopus and a slug. Indeed, it is by no means easy to recognise a common architectural plan for the phylum. Nevertheless, this has been attempted with considerable success by several workers and interest has centred very largely on how this plan has become altered so as to adapt molluscs to colonise so many major habitats and in each to occupy so many different ecological niches. To follow this story adequately would fill a large book, for molluscs are second only to the arthropods in their number of species and in the diversity of their ways of life. On the other hand, merely to describe the land-dwelling snail, *Helix aspersa*, one of the most specialised and highly-evolved members of the phylum, would omit mention of many important molluscan features possessed by this essentially marine group of animals and which have been lost during the long evolutionary history of the transition of one particular branch of molluscs to become adapted to live on dry land.

To begin with, then, it will be as well to mention the main characteristics of the phylum as a whole and later to examine the ways in which the main molluscan groups have each developed their own peculiarities.

All molluscs have soft, unsegmented bodies (the Latin word *mollis* means soft) broadly divisible into two main parts; a **visceral mass**, typically enclosed permanently within a **shell**, and a **head-foot** situated anteriorly and ventrally to the visceral mass. The head-foot is capable of being retracted within the shell but when the animal is active it is protruded. Then its main part, the foot, acts as a locomotory organ enabling the animal to creep over the surface. The anterior end of the head-foot, not clearly marked off as a rule from the foot, is the head itself and, as might be expected, it bears sensory structures such as tentacles and eyes. The mouth is usually an opening through a short snout at the front of the head and through it can be protruded by means of muscles a tongue or **odontophore** which lies in the floor of the buccal cavity and bears along its upper side a complex ribbon of teeth, formed of chitin reinforced by hardened protein, called the **radula**. This acts as a rasp by means of which the animal can scrape food into its mouth. The upper parts of the body wall enclosing the visceral mass are extended downwards so as to form a skirt-like flap, the **mantle**, reaching downwards and even over the head-foot to which its margin may fuse in places. It is this mantle which secretes the shell which, in its simplest form, is a low cone but which in some becomes spirally coiled, in others consists of right and left valves joined by a hinge, and in others again may become internal or even be lacking altogether. Between the mantle and the body is a space, the mantle

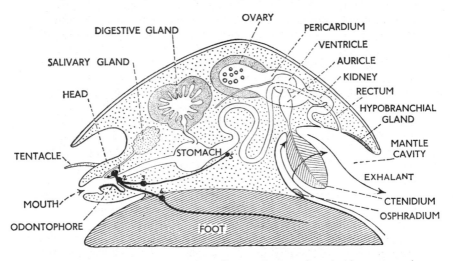

Fig. 271. Schematic drawing to show the organisation of a primitive gastropod: 1. cerebral ganglia; 2. pleural ganglia; 3. pallial ganglia; 4. pedal ganglia; 5. visceral ganglia.

cavity, which serves to house several important organs attached to the body wall. Thus on each side, hanging down into the cavity, are **ctenidia** or gills—up to five pairs in the recently discovered member of the class Monoplacophora, many more in the Polyplacophora, but restricted to a single pair in most living molluscs. Each ctenidium is composed of an axis carrying a row of ciliated leaflets or filaments so that it is not difficult to imagine how, in addition to acting as gills, the ctenidia can set up feeding currents, as indeed they do in many different kinds of molluscs. But primarily it must be imagined that the cilia were for clearing the gills of unwanted particles trapped in mucus and for maintaining a respiratory current of water. This water current entered the mantle cavity on each side below the gills and then, after passing between the filaments, converged to the mid-dorsal line before leaving the cavity again. The exhalant stream served to carry outwards the material leaving the anus and the "kidneys". But a scheme of this sort is not attained in any living mollusc. At best it helps us to visualise how more complex arrangements have arisen.

The internal anatomy of molluscs is extremely variable but they are unusual in having a very restricted coelom. It is, in fact, confined to the space, the pericardium, around the heart and to the cavities of the gonads. There is little reason for supposing that the coelom has become restricted during the evolution of the molluscs although they resemble the arthropods in having large haemocoelic spaces filled with blood. The **heart** usually consists of a ventricle and a pair of auricles opening into it. Oxygenated blood enters each auricle from the ctenidia and general body surface of that side, passes into the ventricle, and is thence pumped by arteries to the organs of the body which lie bathed in blood. From these haemocoels blood is collected eventually into veins which supply the ctenidia.

The form and extent of the gut varies greatly but usually it has a buccal cavity, housing a radula and odontophore, an oesophagus, a stomach containing in its walls complicated ciliary sorting areas, a large **digestive gland** (sometimes called the "liver"), and an intestine whose terminal part is the rectum.

The excretory organs are paired coelomoducts, usually called **kidneys**, which open from the pericardium into the mantle cavity. In many molluscs they serve also to convey gametes to the exterior since the gonads open into the pericardial cavity in many primitive molluscs.

The nervous system of the simplest molluscs consists of a nerve ring with ganglia around the oesophagus connected to two main pairs of longitudinal nerve cords running back in the body. The more ventral pair, the pedal cords, innervate the foot whilst the pair running laterally in the body wall supply the organs which project into the mantle cavity. But in most molluscs the bodies of the nerve cells become concentrated into well-defined ganglia on the plan shown in Fig. 271 and from these nerves run to the various organs of the body.

The time-honoured plan of constructing a diagram of a hypothetical animal which incorporates many of the main molluscan features has been followed and is shown in Fig. 271. It was for a long time commonly accepted that some such kind of animal must have been ancestral not only to Gastropoda but to the phylum as a whole, but the discovery of *Neopilina* (Fig. 279), a living representative of a very ancient molluscan group, has caused this view to be modified for in *Neopilina* some of the internal organs such as auricles, coelomic sacs, and kidneys are serially repeated along the body, an arrangement suggestive of metamerism. Indeed, to some it indicates a closer affinity than is usually supposed with annelids and arthropods. But attempts to link the molluscs closely with truly segmented animals remain unconvincing despite the evidence provided by *Neopilina*. Rather it seems reasonable to assume that the most primitive molluscs were bilateral animals resembling the Monoplacophora and Polyplacophora and which could have arisen by modification of the Platyhelminth plan of organisation.

An outline classification of the phylum Mollusca is given in Chapter XXIII and a brief reference to examples of the main classes follows an account of the snail, the molluscan "type" most usually chosen for study.

## THE GARDEN SNAIL, HELIX ASPERSA

**Introduction.**—Although in many respects highly specialised for life on dry land the members of the subclass Pulmonata, of which *Helix aspersa* is a common and well-known member, retain many of the essential features of the class Gastropoda as a whole—a huge assemblage of molluscs which includes the limpets, topshells, winkles, whelks, and many others familiar on our shores as well as the less well-known sea-butterflies and sea-slugs of various kinds, in addition to the land slugs and snails. Pulmonates, like the snails and slugs, are mainly distinguished from other gastropods by loss of the ctenidia whose respiratory functions are taken over by the thin vascularised roof of an anteriorly situated mantle cavity and by the concentration of all the main nerve ganglia in a ring around the oesophagus with a corresponding loss of the connectives between them. The evidence is very strong that the pulmonates arose from estuarine prosobranch gastropods and indeed, some prosobranchs, easily recognisable by the retention of an operculum (*e.g. Pomatias elegans*) are true land dwellers, whilst many more are amphibious in habit. These, as well as the true pulmonates, are often spoken of as snails with much justification. But in this country the word snail usually means creatures like the garden snails (species of *Helix*), the other land snails (e.g. *Cepaea nemoralis*), or the water snails (e.g. *Limnaea* species). All of these are pulmonates as are also the various slugs, distinguished from snails mainly by the great reduction or complete absence of a shell.

**External Features.**—When active, the part of the body protruded from the shell, the head-foot, can be seen to consist of a broad, muscular region, the sole, lying ventrally and on which the animal creeps along, and an anteriorly placed head bearing two pairs of tentacles. There is no sharp division between head and foot nor in fact between the dorsal part of the foot and the visceral mass or hump which always remains enclosed within the shell, although the lower margin of the visceral hump is thickened to form a collar, complete except for a small area near the anterior end on the right side where the mantle cavity (or lung as it is called in pulmonates) opens to the exterior by a hole, the pneumostome. The collar also marks the lower edge of the mantle which has, except for the pneumostome, joined with the body wall and it is this collar which secretes all except the innermost layers of the shell. The mouth is a small slit-like aperture just below the tip of the head. The reproductive aperture is a minute pore almost immediately behind and a little below the posterior tentacle on the right-hand side.

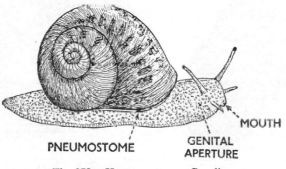

PNEUMOSTOME  GENITAL APERTURE  MOUTH

Fig. 272. HELIX ASPERSA.—Crawling.

**The Shell.**—As in most gastropods the shell is a spirally coiled structure which can be thought of as a hollow cone wound round a hollow central axis, the **columella**. This is itself formed of the thickened centrally placed edges of the cone (Fig. 274), and opens by a narrow slit, the umbilicus, so overlaid by the thickened margin of the shell that it is nearly concealed. The shell which is mainly secreted by the edges of the mantle is composed chiefly of layers of calcite forming the prismatic layer but on the outside is a thin layer, the periostracum, formed of a horny substance termed conchiolin. The innermost layer of the shell of most molluscs is formed of "mother-of-pearl", a translucent, shiny form of calcium carbonate, but this is secreted by the whole surface of the mantle. The shell whorls increase in size from the apex downwards and the fourth or lowest whorl is by far the largest and forms a chamber into which the body can be retracted by means of a muscle, the columellar muscle, which, as its name suggests, takes origin from the central pillar of the shell.

**The Mantle and Lung.**—The mantle is closely applied to the inner surface of the shell and joins the dorsal edges of the head-foot except for a short distance near the front on the right-hand side where there is the pneumostome. In front of the visceral hump it forms the roof of the mantle cavity or lung which, as befits a respiratory surface, is richly supplied with veins whose factors join up to form the main pulmonary vein which enters the auricle of the heart. The floor of the lung is formed of the body wall over the head-foot. It is arched up into the lung but can be flattened by the contraction of its intrinsic longitudinal muscles. Rhythmic arching and flattening coupled with the opening and closing of a valvular

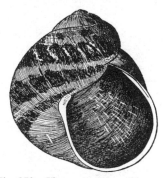

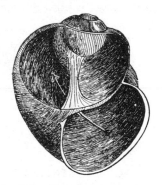

Fig. 273. HELIX ASPERSA.—The shell × 2.

Fig. 274. HELIX ASPERSA.—Shell with part of posterior side removed to show the columella.

arrangement on the pneumostome allows the air to be expelled and renewed.

Opening into the lung just behind the pneumostome is the anus, the rectum running around the right side of the lung before joining the intestine within the visceral hump. The ureter, the duct draining the single kidney, runs alongside the rectum and its opening lies just above the anus.

**The Alimentary Canal and Associated Structures.**—The terminal mouth opens into a **buccal cavity** in whose floor lies the buccal mass, a term given to the radula, odontophore, and the muscles which control its movements. In the roof of the buccal cavity is a transverse bar of tissue resembling cartilage called the **jaw** and it is against this that the radula works when the animal is feeding. The

**odontophore** is a pad of connective tissue, some of it of cartilage-like consistency, to which a complicated series of muscles are attached. Attached to its upper surface is the radula, a strong membranous strip bearing transverse rows of very numerous, recurved teeth formed of chitin and hardened protein. As it is used the radula is continually worn away and is as continually renewed by special cells lying in the wall of the radula sac into which the posterior end of the radula fits and to which it is attached (Fig. 275). The buccal cavity opens into a narrow oesophagus which widens out to form the crop, a large sac alongside which lie the salivary glands whose ducts open into the anterior end of the oesophagus. These produce

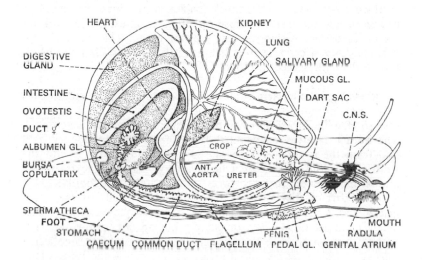

Fig. 275. HELIX ASPERSA.—Main organs seen from the right-hand side—highly diagrammatic.

a watery mucus which also contains diastatic enzymes. But the crop contains also fluid derived from the digestive gland and passed forward from its openings into the stomach. This fluid is brown in colour and contains a whole series of enzymes including a cytase or cellulose-splitting enzyme which digests plant cell walls and liberates their contents. The origin of this cytase is in dispute. For long it was believed that it was a product of the digestive gland but more recently it has been claimed that it is formed by bacteria which live in the intestine and crop. This view would certainly bring the snail more into line with most other animals (*e.g.* herbivorous mammals) but it must be remembered, nevertheless, that some molluscs, like the wood-boring shipworm (*Teredo*), do secrete cytases and in this are most unusual. The crop is followed by the stomach, a fairly capacious but simple sac closely surrounded by the large **digestive gland** which is brown in colour and forms a packing around most of the organs of the visceral hump. It consists of a mass of branching tubules ending blindly in clusters of cells which are of three main types; those which secrete the enzymes responsible for extra-cellular digestion; those which take in food particles conveyed there by the action of cilia on the walls of the tubules and digest them intra-cellularly as well as absorbing soluble products of digestion; and, thirdly, cells which secrete calcium carbonate and are perhaps analogous in function to the calciferous glands of the earthworms (p. 242). From the stomach an intestine runs a coiled course to join the rectum whose opening, the anus, lies just behind the pneumostome within the cavity of the lung. The buccal cavity, oesophagus and crop lie in a haemocoelic cavity within the head-foot and so are bathed in blood.

**Blood Vascular System.**—During the evolution of the pulmonates from more primitive gastropods one auricle of the heart has been lost so that the heart consists of a single auricle and a ventricle. The auricle receives oxygenated blood from the pulmonary vein and from the general surface of the body and pumps it into the ventricle. From this arises a large artery, the main aorta, which almost immediately forks to form an anterior aorta supplying the anterior part of the body and then passes ventrally as the pedal artery and a posterior aorta which runs to the visceral hump. These main arteries branch repeatedly but do not form capillaries. Instead they empty into haemocoels from which blood is collected into two circular veins, one above and one below the lung. It is from these veins that the vessels in the roof of the lung receive their blood supply and then return it to the auricle (Fig. 276). The heart lies in a pericardial cavity which is coelomic in origin. The blood contains the respiratory pigment haemocyanin and is distinctly blue in colour.

**Excretory System.**—There is a single excretory organ or **kidney**, which is a modified coelomoduct, opening from the pericardial cavity and discharging into a narrow ureter which runs a course parallel to and just above the rectum ending in a pore just above the anus. The main part of this kidney is a yellow coloured sac with folded glandular walls which secrete uric acid. This is, as in so many other animals adapted to a fully terrestrial life, discharged as solid particles with the minimum of water. Like reptiles, birds, and insects, land pulmonates are fully uricotelic and but little water is lost through the small reno-pericardial aperture and a good deal of this is resorbed by the kidney as may be judged by evidence provided by observations on the giant tropical snails, *Archachatina*.

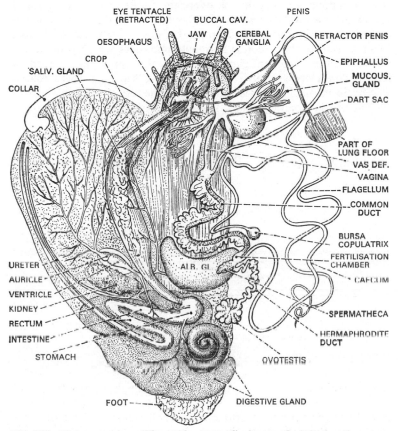

Fig. 276. HELIX ASPERSA.—Dissected so as to display, on the left, the alimentary canal, lung, etc., and, on the right, the reproductive system.

**Reproductive System.**—In the account given of the reproductive system of the earthworm (p. 252) it was pointed out that in addition to the gonads and their ducts there are accessory structures designed to make possible internal fertilisation of the eggs, to ensure cross-fertilisation between two partners, and to protect the eggs after they have been laid. Many of these features can be thought of as adaptations to life on dry land and similar considerations apply to the reproductive system of the snail, which is even more complicated than that of the earthworm.

As in all gastropods there is a single gonad and, like the opisthobranch molluscs (which include the sea slugs) the gonad of the snail is a hermaphrodite organ, the **ovotestis**, producing both eggs and sperm at one and the same time. It is a small, whitish gland embedded in the digestive gland at the apex of the visceral hump. From it runs a short, coiled **hermaphrodite duct** which runs downwards to the albumen gland where it then enlarges to form a **fertilisation chamber** into which opens a sac, the **spermatheca**. The fertilisation chamber is connected to the rest of the reproductive system by the **spermoviduct**. Except during copulation the sperm

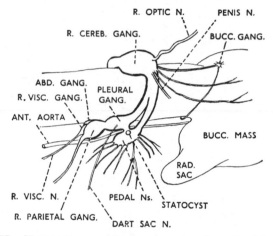

Fig. 277. HELIX ASPERSA.—The nervous system from the right side.

groove communicates with the oviduct throughout the length of the duct, the cells of which secrete fluids into the lumen. Eggs pass down the hermaphrodite duct from the ovotestis and they are fertilised in the fertilisation chamber by sperm stored in the spermatheca. Albumen is secreted around the fertilised eggs by the albumen gland and the whole is encased in a leathery calcareous shell secreted by the oviduct as they pass down towards the exterior. Spermatozoa are continually produced by the ovotestis and are stored prior to copulation in the hermaphrodite duct. During copulation sperms are conducted from the hermaphrodite duct down the sperm groove to the **spermatophore** which is formed in the **flagellum** and **epiphallus**. Spermatophores are exchanged during mating by the eversion and protrusion of the penis. After copulation the partners spermatophore is lodged in the bursa stalk and the most motile sperms migrate through its long stalk to be stored in the spermatheca. The function of the **bursa copulatrix** is to digest the spermatophore, excess sperm from the ovotestis and other products, which are conveyed to it by a strong peristalsis in the stalk.

**Nervous System.**—The ganglia joined by obvious commissures, so characteristic of the nervous system of lower gastropods, are in the snail so concentrated near the anterior end that they are difficult to identify. All the main ganglia and commissures contribute to form a ring of nervous tissue around the anterior end of the oesophagus but, nevertheless, it is possible to recognise dorsally the **cerebral ganglia**, ventrally the **pedal ganglia** and, lateral to these,

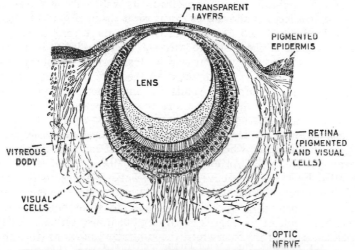

Fig. 278. GASTROPOD EYE.—Diagrammatic longitudinal section.

the **pleural ganglia.** Slightly behind these are the **parietal ganglia** and the median **abdominal ganglion** (Fig. 277). In front of the main nerve ring, on the dorsal side of the buccal cavity, lie the small buccal ganglia. They are joined to the cerebral ganglia by delicate connectives. From the main nerve ganglia nerves run to the organs of the body.

**Sense Organs.**—The main receptor organs are found at the anterior end. The first pair of tentacles are thought to be **chemoreceptors** whilst **eyes** of rather a complex nature are borne at the tips of the second pair of tentacles and may suffice for a limited degree of form vision. The tentacles are, however, fully extended only when the snail is active and they can be withdrawn by retractor muscles which turn them outside-in. The general surface of the body is sensitive to touch and probably to other stimuli. A pair of **otocysts** lies embedded in the pedal ganglia but are innervated from the cerebral ganglia by delicate nerves alongside the cerebro-pleural connectives.

**Locomotion and Behaviour.**—Snails are able to creep, rather slowly it is true, over a variety of surfaces and undertake quite long journeys in search of food. Usually, however, they return to the same spot in a protected situation. They are active only under moist conditions and crawl mainly at night. The organ of locomotion is the foot whose lower surface is thrown into a series of waves of minute amplitude by the muscles in its wall. These waves can be seen, if a snail is allowed to crawl on a sheet of glass, to begin at the hinder end and pass forwards. The mere throwing of the ventral surface of the foot into waves at right angles to its surface would not produce movement and a closer study has shown that the crest of the waves are directed to the rear, thus thrusting the animal forwards. In some snails the muscular contraction producing a wave does not reach more than half-way across the sole of the foot and alternating waves are produced on the two sides. Co-ordination of the locomotory waves is mainly by way of a nerve net, as in flatworms, whose gliding movement much resembles the creeping of a snail. As everyone knows, the path of a snail is signalled by a train of mucus left in the wake of the animal. This mucus is secreted by the pedal gland whose duct opens on the ventral surface just behind the mouth. Not only does this mucus protect the delicate sole of the foot from drying but it also serves to fix the expanded parts of the foot whilst the contracted portions (the crests of the waves) are moving forward. The shell tends to fall over to the right-hand side and is periodically jerked back in position by contraction of the columella muscle.

Snails feed on a variety of plants and can do great damage in a garden. The food is rasped off by the radula which also serves as a conveyor belt to take the food into the buccal cavity. In dry weather they become inactive (they are said to aestivate) and withdraw into the shell in a sheltered spot. Then the foot secretes a tough sheet of hardened mucus which seals the shell aperture and prevents undue desiccation. A similar period of inactivity termed hibernation occurs throughout most of the winter. The Roman or edible snail, *Helix pomatia*, secretes a tougher lid or epiphragm formed at first mainly of calcium phosphate, later to be replaced by carbonate and having a porous consistency rather like Plaster of Paris so that gaseous interchange is not completely prevented.

**Reproduction and Life-history.**—Copulation resulting in reciprocal fertilisation of two partners is preceded by a process which can be compared to the "courtship" activities of higher animals. Two snails in breeding condition approach each other and evert the genital atria so as to bring the male and female apertures to the exterior. When the partners are close together each shoots out a calcareous dart from the dart sac so violently that the darts pierce the body wall and become imbedded in various internal organs. Presumably this acts as a violent mutual stimulus to undertake the later stages of copulation for the two snails then come together and the penis of one is inserted into the vagina of the other. Spermatophores are then made in the glandular region above the penis, the epithallus, and the flagellum. Sperms are then transferred into the spermatophores and transfer starts about five minutes after insertion of the penis. After transfer of the spermatophore, which is placed in the bursal stalk, the penis is withdrawn and sperm starts migrating from the spermatophore to the spermatheca, a process which takes several hours. After four to five hours sperm migration is prevented by the long tail of the spermatophore (which is formed in the flagellum) becoming closed and the residual mass is digested by the bursa copulatrix. Eggs are produced in the ovotestis only during a short season in the late spring or early summer but are not laid until June, July, or August, some weeks after copulation. The eggs are laid in holes in the soil and miniature adults emerge from them after about three weeks.

## CLASSES OF MOLLUSCS

### CLASS AMPHINEURA

These animals are represented on British shores by several genera and species of chitons, all of which are small, sluggish creatures found in rock crevices or clinging to the undersides of stones. Their most obvious feature is the possession of a dorsal shell made up of a longitudinal row of eight plates. The head-foot has a broad creeping sole but the head is inconspicuous and devoid of tentacles or eyes. Attached to the body wall in the pallial groove between the mantle and the foot are numerous ctenidia—up to twenty on each side. The mouth is at the anterior end of the head and the anus at the extreme posterior end of the body. The paired kidneys, the general bilateral symmetry of the organs of the body, including the nervous system (in which localised ganglia are hardly evident), speak for the primitive nature of the Amphineura. *Lepidochitona cinerea* is the commonest British species.

## CLASS GASTROPODA

The most primitive group of Gastropoda are placed in the subclass Prosobranchia, a vast assemblage of mainly marine animals, including limpets, topshells, periwinkles, and carnivorous and scavenging "sea-snails". Despite a wide diversity of form and habit all are distinguished by having a mantle cavity, which houses a paired or single ctenidium, and into which anus and kidney apertures open. But this mantle cavity in the adult is at the anterior end and faces forwards to open above the head whereas in the larva (and it must be supposed in the ancestors of the prosobranchs), it occupied a posterior position. The change in position occurs in two stages, the first one taking place quite rapidly and being brought about by the contraction of a muscle taking origin from the right side of the larval shell and passing over the body to become inserted in the left side of the head-foot, and the second one slowly by differential growth. This torsion, as it is called, results in all organs above and behind the "neck" being reversed in position. Even parts of the nervous system are affected and the nerves joining the pleural to the pallial ganglia (the visceral loop) become twisted over one another in the form of a figure-of-eight. Moreover, organs of the left-hand side tend to be reduced or even lost, so that many prosobranchs have but a single auricle and kidney. No satisfactory explanation in terms of survival value has yet been advanced for this extraordinary process of torsion. Coiling of the gut and other viscera has nothing to do with torsion and takes place at a later stage.

Space does not permit of description of the anatomy and habits of prosobranchs but a few schematic figures are given which will serve as a basis of comparison with pulmonates and opisthobranchs.

The second subclass of Gastropoda is the Opisthobranchia of which nudibranch sea-slugs are the most familiar examples. All opisthobranchs are characterised by their bodies having undergone partial or complete de-torsion so that the organs of the body attain or approach bilateral symmetry again. But all bear witness to their evolution from prosobranch ancestors for none have paired ctenidia, kidneys, or auricles. At best there is only one of each of

Fig. 279. Diagrams to show the main features of representatives of the classes of molluscs.

    I. A, *Neopilina*—ventral view.  B, *Neopilina*—side view.
    II. A, *Lepidochitona*—side view.  B, *Lepidochitona*—dissection from side.
    III. A, *Littorina*—dorsal dissection.  B, *Littorina*—side view.
    IV. a Nudibranch—dorsal view.
    V. *Dentalium*—to show main organs and mode of life.
    VI. *Anodonta*—dissection from left side.
    VII. *Sepia.*    VIII. *Octopus.*

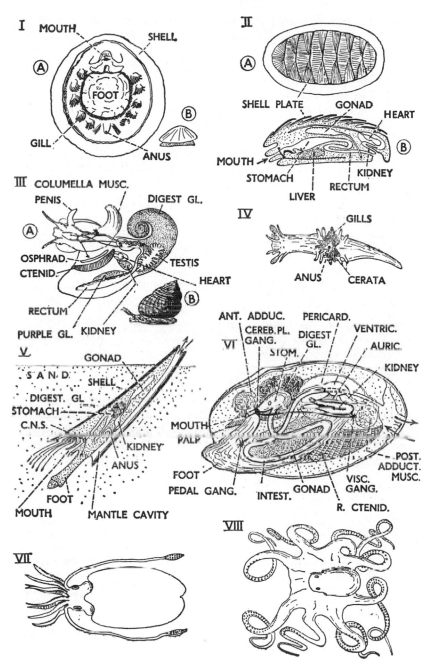

these originally paired organs and in some even the single ctenidium (of the right side) has disappeared to be functionally replaced by gills of other types.

The third subclass of Gastropoda is the Pulmonata—the land snails and land slugs as well as the water snails (e.g. *Limnaea*). All of these retain the condition of full torsion inherited from their prosobranch ancestors, but whilst evidence of this can still be seen in the single auricle and kidney, the nervous system has acquired a symmetry by a shortening of the connectives and the incorporation of the main ganglia in the nerve ring around the oesophagus. To a greater or lesser degree all excrete uric acid as well as ammonia and amino acids. This can be regarded as fitting them to life on dry land though some have become secondarily aquatic.

## CLASS SCAPHOPODA

In contrast to the gastropods, the scaphopod anatomy is very uniform. These are the tusk-shells, molluscs with a hollow shell open at both ends within which is an elongated body with a short extensible foot used mainly for burrowing into the sandy deposits on which the animals live. The head bears bunches of slender tentacles used for catching the foraminifera on which the animals feed. The mantle is long and tubular and acts as a respiratory surface for there are no ctenidia. There is one British genus, *Dentalium*, the elephant tusk-shell.

## CLASS LAMELLIBRANCHIATA

This class is constructed to include all the bivalved molluscs, oddities in some ways, although they are so numerous in genera and species, for they have during their evolution lost the head with its buccal mass, radula, and sense organs. They are characterised by the enormous size of the ctenidia which are no longer mere respiratory organs but are predominantly concerned with feeding, producing a feeding current, and numerous ciliary sorting tracts by the cilia which clothe them. Although mainly marine, some have colonised fresh water—as for example, the swan-mussels (*Anodonta*). In many the foot remains as a locomotory organ but in others it is minute. Many, as for example, *Mya*, the clam, are deep burrowers in sandy mud, retaining connection with the surface by means of inhalant and exhalant siphons which are posterior tubular extensions of the mantle serving to draw a respiratory and feeding current from the overlying water and to pass it again to the surface when oxygen and food have been removed by the ctenidia. Mussels (*Mytilus*), scallops (*Pecten*), oysters (*Ostrea*), cockles (*Cardium*), clams (*Mya*), and piddocks (*Pholas*) all belong here.

## CLASS CEPHALOPODA

Undoubtedly the cephalopods must be regarded as the apex of molluscan evolution and indeed in most ways rival in complexity and activity not only the larger arthropods, but even the lower vertebrates. All are active predaceous creatures able to swim and capture large crustacea such as crabs. At first sight they do not appear to conform to the molluscan plan. Yet closer inspection reveals that there is a large mantle cavity housing paired (or two pairs of) ctenidia and into which opens the anus and the kidneys; that there is a head-foot, the foot portion of which has grown out to form either eight or ten pairs of arms bearing rows of suckers. The mantle itself is very muscular and by its contractions a jet of water can be expelled giving, literally, jet propulsion but in a backward direction. The head bears large eyes whose parts—cornea, lens, and retina—resemble those of a vertebrate, although derived embryologically in quite a different manner—and are undoubtedly image-forming. Not only is the molluscan radula retained but the mouth is equipped with a pair of powerful horny jaws—rather like a parrot's beak. The shell of most modern cephalopods, like the squid (*Loligo*), has become internal within the mantle but in *Nautilus* and in many fossil cephalopods the shell is external and consists of many chambers, the animal occupying only the largest and terminal chamber. The octopus and its allies have lost the shell completely but one modern cephalopod, the pearly nautilus, *Argonauta*, has formed an external shell of a different nature secreted by expansions of two of the arms.

Most cephalopods have opening into the mantle cavity an ink-sac which can squirt out a dense cloud of black pigment under cover of which the animal can escape its enemies. Additional concealment is provided by the ability to change colour rapidly, the method of colour change being unusual amongst animals in that the coloured sacs can be expanded by muscles under nervous control.

The viscera conform in the main to the molluscan plan but are perhaps more complicated than most. A description of their anatomy is outside the scope of this account but some idea of their appearance may be obtained from the diagrams (Fig. 279, VII and VIII).

# CHAPTER XV

## THE CHORDATE ANIMALS

### INTRODUCTION

The examples of coelomate animals so far considered will have served to show that a high degree of specialisation can be attained within any one group. The arthropods have shown this in a definite way, especially the insects. Other equally successful forms of coelomate animals are found living at the present time and some have a long geological history. Among these may be mentioned the brachiopods (lamp shells), the echinoderms (the spiny-skinned animals), but most particularly the "vertebrates" or higher

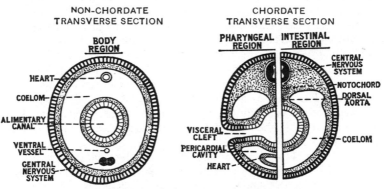

NON-CHORDATE TRANSVERSE SECTION

CHORDATE TRANSVERSE SECTION

ECTODERM · ENDODERM · MESODERM

Fig. 280. Diagram to show the main differences between non-chordate and chordate types of organisation.

chordates. In each main group, specialisation is entirely different as is at once clear if an octopus, a starfish, and a vertebrate are compared, and between them there is little in common apart from the fact that they are all triploblastic and possess a coelom.

Between the chordates and all the other highly organised coelomates, however, there are very fundamental differences and although there is some evidence that they may come nearer in their ancestral forms to those of the echinoderms, their actual origins are still matters of conjecture. Moreover, within the chordates themselves are included animals such as the sea-squirts and acorn worms in which many common chordate characters are lacking.

The phylum Chordata (or the chordate animals) includes all those animals which at some time or another during their life-history possess a notochord. A **notochord** is an axial stiffening rod extending along the length of the animal and lying immediately above the alimentary canal and below the central nervous system. It may persist throughout life (as in the lancelet, lamprey, and certain

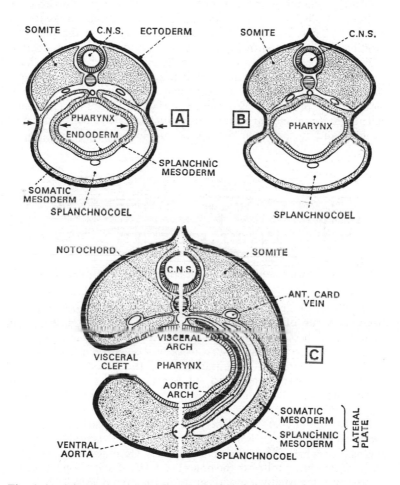

Fig. 281. Diagrams to show the development of visceral clefts with their associated structures.

A, transverse section through the pharyngeal region of a chordate embryo. B, a later stage. C, a still later stage; on the left through a cleft, on the right through an arch.

fishes) or be replaced, either wholly or partially, in the adult animal by a backbone or vertebral column. If this happens the animals are termed vertebrates, which include such forms as fishes, amphibians, reptiles, birds, and mammals. But all chordates have certain other characteristics by which they may be distinguished from all other animals.

Among the more important of these is that the pharynx, has, in its lateral walls, a series of paired openings communicating with the exterior. These apertures arise as outpushings from the pharynx, meeting and fusing with inpushings from the exterior; the intervening walls break down and free channels, termed **visceral clefts**, are formed. Such clefts appear during the development of every chordate animal, but in the aquatic forms all or some usually become modified to provide a respiratory surface and are consequently called **gill** or **branchial clefts**. In adult terrestrial forms all may close up and most of them disappear completely.

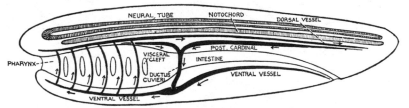

Fig. 282.   Diagram of the basic plan of the chordate vascular system.

The central nervous system is always *dorsal* in position and from its mode of development (see p. 559) *hollow* and *tubular* in structure. These are two points of contrast with the non-chordate animals in which the central nervous system is ventral and solid.

In most chordates the circulation of the blood in the blood vascular system is maintained by a central muscular propulsive organ, the heart. This heart is *ventral* in position and the flow of the blood in the dorsal vessel is from the anterior to the posterior end. In the higher chordates the heart is contained within a modified part of the coelom termed the pericardial cavity, the bounding wall of which forms the pericardium. Here again is a contrast with the condition found in the non-chordates where, when a "heart" is present, it is usually a modification of the dorsal vessel.

Associated perhaps with the aquatic habit of the primitive chordate animals, the posterior part of the body takes the form of a post-anal, metamerically segmented, tail—quite different from the unsegmented condition of the terminal portion of the body found in the non-chordates. This tail is extremely flexible and muscular and in the aquatic forms constitutes the main propulsive organ. In the terrestrial chordates it takes various forms, or may be very reduced in the adult.

Metamerism, though present, is frequently masked externally.

During development, each segment of the mesoderm becomes divided into the **somite**—from which the body muscles (myotomes) and other structures are produced—and the **lateral plate mesoderm**, the somatic and splanchnic layers of which bound the coelom, and from which the bulk of the unstriped muscles are developed. The body wall muscles, however, never consist of circular and longitudinal muscle layers as in many non-chordates.

## DIFFERENCES BETWEEN CHORDATE AND NON-CHORDATE ANIMALS

The differences between the chordate and non-chordate animals may be briefly summarised as follows:

| CHORDATE | NON-CHORDATE |
|---|---|
| Visceral clefts present in pharynx. | Visceral clefts absent. |
| Notochord present. | Notochord absent. |
| Central nervous system dorsal, hollow, tubular. | Central nervous system ventral, solid, double. |
| Heart ventral. | Heart dorsal. |
| Post-anal, metamerically segmented tail. | |

# CHAPTER XVI

## SIMPLE CHORDATE ANIMALS—ACRANIA

### INTRODUCTION

Although in by far the greater number of chordate animals the notochord has been partially or wholly replaced by a vertebral column (vertebrates or Craniata), there is a number of simpler forms (protochordates or Acrania) in which a vertebral column and skull are lacking. Most of the acraniate chordates are fairly obviously allied to the vertebrates, but none possesses the full complement of chordate characters, and in some even a notochord is lacking.

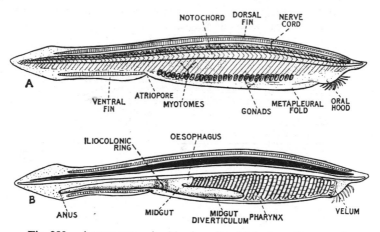

Fig. 283. AMPHIOXUS.—A, side view; B, with body wall removed.

The Acrania fall into three classes—**Hemichordata** (acorn worms), **Urochordata** (tunicates or sea-squirts), and **Cephalochordata** (lancelets); but as the members of the first two are peculiar in so many respects, a representative (amphioxus) of the Cephalochordata—which is extremely instructive in that it illustrates the simple chordate condition—alone will be described in any detail.

### CEPHALOCHORDATA—AMPHIOXUS

**External Features.**—Popularly known as amphioxus or the lancelet, *Branchiostoma lanceolatum* is a small (4 to 5 cm) marine inhabitant of shallow water. The creature bears a superficial resemblance to a small fish, for its elongated body is pointed at both ends and bears fins. One of these, the dorsal, extends as a fold of skin along the whole length of the dorsal surface and is continuous around the tip of the tail with the ventral fin which, however, extends forwards for only about a third of the length of the body. At the anterior end the dorsal and lateral parts of the body project forwards to form a hood, the **oral hood.** Above the mouth and from the edges of this hood arise some twenty rather stiff, tentacle-like processes, the **oral cirri.** Posteriorly the oral hood is continuous with the **metapleural folds.** These, in effect, are two downgrowths from the dorsolateral body wall which form the side walls of the **atrial cavity** or **atrium** as it is also called (see p. 373). The floor of the atrium is also derived from the tissues of the body wall and is termed the **epipleur.** The under side of the oral hood bears a complicated series of ciliated grooves and ridges, which collectively are referred to as the **wheel organ** because of the whirling currents of water that it sets up during life. Between the ciliated lobes of the wheel organ, towards the left-hand side of the notochord, is a glandular groove called **Hatschek's groove (Hatschek's pit)** which secretes mucus.

**Body Wall.**—The body is everywhere covered by an epidermis which, unlike that of higher chordates, is composed of a *single layer* of cubical epithelial cells which bear a thin but distinct cuticle. Below the epidermis is a layer of tough fibrous connective tissue

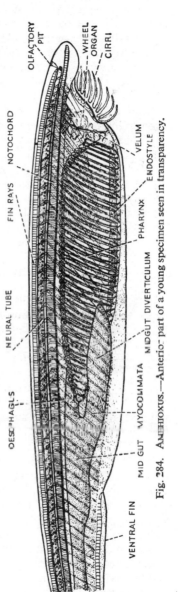

Fig. 284. AMPHIOXUS.—Anterior part of a young specimen seen in transparency.

continuous with that surrounding most of the important structures of the body, but the greater part of the body wall is formed by the segmentally arranged muscle blocks, the myotomes. There are about sixty of these and each is enclosed in a complete box or envelope of connective tissue. The anterior and posterior connective tissue walls are referred to as myocommata and are visible externally as a series of >-shaped lines along the sides of the body. These indicate that the myomeres also are >-shaped, having an upper and a lower limb. The inner connective tissue walls are formed by the layer beneath the parietal layer of peritoneum and the outer wall by the sub-epidermal connective tissue.

The muscle-fibres, which are of the striated type, lie parallel to the longitudinal axis of the body and at each end are inserted in the myocommata. It is by the contraction of these segmental muscles that the rhythmical side to side bending of the body, by which the animal swims, is effected. Each wing of a myotome is set at an acute angle to the sagittal plane of the body, the upper wing at an angle which falls forward, the lower wing at an angle which falls to the rear (Fig. 285). The body wall is completed internally by a layer of somatic peritoneum, but in the pharyngeal region this epithelium is restricted to certain small tracts (see below).

Fig. 285.—AMPHIOXUS.—Lateral view of a single myotome muscle.

**Skeletal System.**—The most conspicuous skeletal structure is the notochord, a large cylindrical rod extending the *whole length* of the body just ventral to the central nervous system and just above the gut. The notochord, which is surrounded by a thick layer of connective tissue (notochordal sheath), consists of alternate discs of fibrous and gelatinous cells. The turgor pressure of these cells acting against the resistance of the notochordal sheath gives to the notochord its stiff elastic nature and enables it to resist shortening of the body when the myotomes contract. Apart from the notochord, a distinct endoskeleton of stiff material is not present, but an endoskeletal function (see p. 417) is carried out by the system of sheets of dense fibrous connective tissue already mentioned. This is continuous throughout the body and forms a layer below the epidermis and somatic peritoneum, encloses the myotomes, and ensheathes the notochord and central nervous system. It thus serves as a packing tissue and for the attachment of muscles. In certain parts of the body the fibrous tissue is replaced by a gelatinous substance resembling soft cartilage. The oral hood is supported by a jointed hoop of this material, and articulating with the hoop are delicate rods which form the axes of the oral cirri. The gill bars and also the endostyle contain rods of a similar nature. The dorsal fin is stiffened by a single row of fin rays, whilst the ventral fin has a double (right and left) row of these fin rays. Each ray is, in effect, a connective tissue box containing a gelatinous substance.

**Alimentary Canal.**—Across the base of the oral hood lies a partition, the **velum**, which is perforated by an aperture, the "mouth" around whose margins are numerous slender ciliated tentacles **(velar tentacles)** which normally project backwards into the pharynx, forming a sort of strainer.

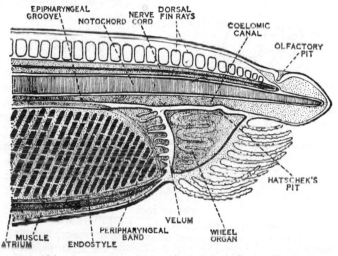

Fig. 286. AMPHIOXUS.—Parasagittal view of the anterior end.

The **pharynx** is a large sac-like region whose walls are pierced by very numerous **gill slits**. These, when first formed, are long narrow apertures, with their long axes roughly vertical, which open from the pharynx to the exterior. Between successive gill slits lie the **gill bars**. These are of two types, the primaries and the secondaries, which alternate regularly along the length of pharynx and differ not only in their structure but in their mode of development in the larva. A primary bar is formed from the tissue of the septum which remains between two successive slits after they have perforated to the exterior and is therefore composed partly of body wall and partly of the wall of the pharynx. It follows that each primary bar is covered on its inner (pharyngeal) face by endoderm and on its outer face by ectoderm whilst its core is of lateral plate mesoderm. The limits of the ectoderm and endoderm covering a gill bar are fairly easy to make out, for the endodermal epithelium is richly ciliated and is found not only on the pharyngeal surface but also at the sides, whilst the ectodermal (atrial) epithelium is practically non-ciliated and is restricted to the outer surface. The ciliation of the endoderm on the gill bars is, however, by no means uniform, the cilia on the anterior and posterior faces and in a narrow tract down the pharyngeal surface being especially prominent. The internal mesodermal core of a primary bar is represented by a skeletal rod (which forks at its ventral end and is perforated throughout its whole length by a narrow coelomic canal, see p. 374), by fibrous connective tissue, and by blood vessels.

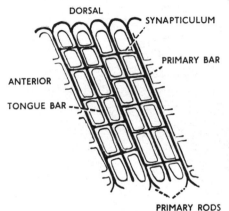

Fig. 287. AMPHIOXUS.—Portion of the pharyngeal wall to show the skeletal elements.

The secondary or **tongue bars** arise, after perforation of the gill slits, as down-growths from the dorsal wall of each slit. By their growth ventralwards the tongue bars divide each of the original slits ("primary" gill slits) into anterior and posterior halves ("secondary" gill slits). The secondary bars, like the primaries, are covered on their inner, anterior, and posterior surfaces by endodermal (pharyngeal) epithelium, and on their outer faces by the sparsely ciliated ectodermal (atrial) epithelium. Each bar contains a skeletal rod which, however, differs from that of a primary bar in that it does not fork at its ventral end and contains no coelomic space. Blood vessels and connective tissue complete the core of the bar. New gill slits, both primary and secondary, are added to the posterior end of the series until a very late stage in development is reached, and as a consequence the number of slits varies from specimen to specimen according to age. The primary and secondary bars are linked together by cross-connections (**synapticula**), which are traversed by blood vessels which connect with those in the bars. The synapticula, which, like the gill bars, are stiffened by skeletal rods, develop only after the gill slits have been completed, and their formation lends to the wall of the pharynx the appearance of a complicated basket-work, which in many respects resembles the branchial sac of tunicates.

In addition to the cilia on velar tentacles and gill bars, other important ciliated tracts, which receive special names, are found on the inner surface of the wall of the pharynx. Most prominent of these is the **endostyle**, a shallow gutter, lying in the mid-ventral line. It is lined by four longitudinal tracts of mucus-secreting gland-cells between which are tracts of ciliated cells, the median strip bearing especially long cilia. Below the cells of the endostyle

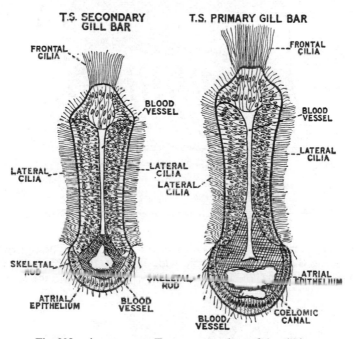

Fig. 288. AMPHIOXUS.—Transverse sections of the gill bars.

are two skeletal plates which lie just above a longitudinal coelomic canal. It is extremely interesting to find that an endostyle of almost identical nature is found in the tunicates and in the **ammocoete larva** of the lamprey (*Petromyzon*), an agnathous craniate. In the latter instance the endostyle is lost at metamorphosis but contributes to the formation of the **thyroid gland** of the adult. Indeed, like the thyroid of craniates, the endostyle concentrates iodine in itself whilst extracts of endostyle simulate the action of the thyroid hormone (p. 660).

The endostyle is linked at its anterior end, just behind the velum, by two ciliated tracts, the **peripharyngeal bands**, to a dorsal ciliated gutter, the **epipharyngeal groove**.    This leads to the posterior end of

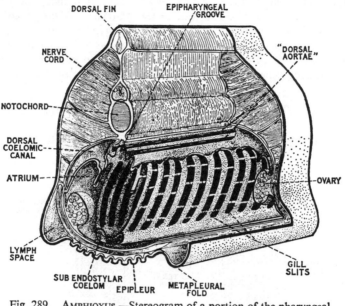

Fig. 289.   AMPHIOXUS.—Stereogram of a portion of the pharyngeal region.

the pharynx where lies the opening of the short, ciliated oesophagus. Following the oesophagus is the wide **mid-gut** and behind this the ilio-colonic region (**ilio-colonic ring**), in turn opening into the narrow

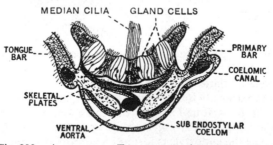

Fig. 290.   AMPHIOXUS.—Transverse section of the endostyle.

**hind-gut.**    This terminates at the anus, a small aperture on the left-hand side of the ventral fin a short way from the posterior end of the body.    At the junction of oesophagus and mid-gut there arises a large forwardly pointing blind pouch, the **mid-gut diverticulum**, which is, in fact, a digestive gland. The diverticulum pushes the coelom and a part of the body wall before it as it develops and so comes to lie on the right-hand side of the pharynx, protruding into the atrial cavity from which, however, it is separated by the thin layer of body wall enveloping it.

**The Atrium.**—The gill slits of the adult amphioxus, unlike most other chordates, do not open directly to the exterior but into a special cavity, the **atrium**. This, in turn, opens to the exterior near its posterior end by a small aperture, the **atriopore**, which lies on the ventral side of the body just in front of the ventral fin. The atrium is lined by ectoderm, for it is really a small portion of the

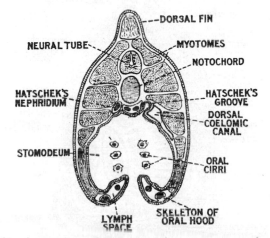

Fig. 291. AMPHIOXUS.—Transverse section through the region of the oral hood seen from the front.

outside world which has become enclosed by flap-like extensions of the body wall, the right and left **metapleural folds**, which are united ventrally by a transverse shelf, the epipleur. The atrium, which surrounds the pharynx and anterior part of the intestine laterally and ventrally, extends back on the right side behind the atriopore almost as far as the anus, whilst from a point level with the posterior end of the pharynx there arises on each side from the atrium a forwardly directed pocket which pushes into the dorsal coelomic canal. These pockets are termed the **brown funnels** or **atrio-coelomic canals**. Their function is unknown.

**The Body Cavity.**—The body cavity, like that of vertebrates, is a true coelom as is shown by its relations to the viscera and by

its mode of development in the embryo (pp. 700-3). Posterior
to the pharynx the coelom is a fairly spacious cavity in which the
mid- and hind-gut regions of the alimentary canal are suspended
by a mesentery. In the pharyngeal region the coelom has become
very much restricted in extent for, although in the young larva it
surrounds the pharynx (except in the mid-dorsal line where the
mesentery is found), it gradually becomes reduced in size as the gill
slits develop. In the adult all that remains of the coelom in the
pharyngeal region of the body is a pair of **dorsal longitudinal canals**,
a **mid-ventral sub-endostylar coelom**, and the vertically running

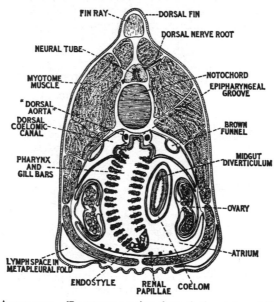

Fig. 292. AMPHIOXUS.—Transverse section through the region of the pharynx.

coelomic spaces in the primary gill bars which place the dorsal
and sub-endostylar coelomic spaces in communication with one
another. This condition, which at first sight appears peculiar, is
readily understandable when the mode of development of the gill
slits is recalled. It will be remembered that each slit arises as an
outpushing of the pharynx which meets and fuses with a corresponding ingrowth from the body wall, followed by perforation at the
point of fusion so that the pharynx is placed in communication with
the outside world. It follows that the coelom is obliterated except in
the dorsal and ventral regions and in the septa (primary gill bars)
between the gill slits (pp. 369, 370). In the higher chordates reduction

of the coelom goes still further, for it is obliterated altogether in the region of the pharynx.

**The Blood Vascular System.**—It would be natural to expect that the blood vascular system of a simple chordate of this type would follow the main lines that have been given as characteristic of the chordates as a whole. It so happens, however, that the vascular system of amphioxus is peculiar in that no heart is present and the blood contains no respiratory pigment.

There is a main **ventral vessel**, below the gut, which collects the blood and passes it forward below the pharynx, whence it is conveyed by vessels in the primary gill bars

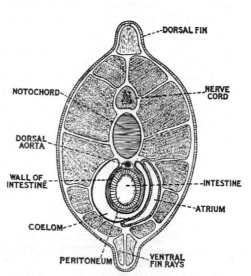

Fig. 293. AMPHIOXUS.—Transverse section through the region of the intestine behind the atriopore.

paired to vessels running longitudinally just above the pharynx. The vessels in the primary bars communicate with those in the secondaries by the vessels in the synapticula and the former also open into the dorsal vessels. The paired **dorsal vessels** join behind the pharyngeal region, forming a single main **dorsal vessel** which passes back in the body, distributing blood to the various organs. Circulation of the blood through this system of vessels is effected by peristaltic-like contractions of the ventral vessel and by small contractile bulbils at the bases of the "afferent branchial" arteries. Blood flows forwards in the ventral vessel and backwards in the body in the dorsal vessel.

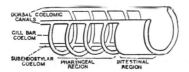

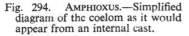

Fig. 294. AMPHIOXUS.—Simplified diagram of the coelom as it would appear from an internal cast.

It will be seen that, although there are differences, the blood vessels do in general conform to the basic plan common to all chordates (Fig. 295).   Nevertheless, some caution is required when comparing the vessels with those of the vertebrates, and in the above description the terms employed have purposely been non-committal.   Some further details of the vessels together with a nomenclature in common usage can now be given.

The ventral vessel, which in the region of the mid- and hind-gut is termed the "sub-intestinal vessel", is for the greater part of its length not a single vessel but rather a plexus of vessels receiving blood from the gut wall.   In the anterior mid-gut region the smaller vessels unite to form a short but wide vessel, the "hepatic portal vein", which passes along the ventral border of the mid-gut diverticulum and there again splits up into capillaries which ramify in the wall of the diverticulum, thus forming a system which although not strictly comparable with a hepatic portal system yet does indeed foreshadow a portal system such as is found in the vertebrates, for, just in front of the mid-gut diverticulum, the ventral vessel is

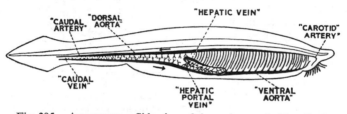

Fig. 295.   AMPHIOXUS.—Side view of the main vessels ("cardinal veins" omitted).

re-formed as the "hepatic vein".   This is joined almost immediately by a pair of vessels (right and left "ductus Cuvieri") which return blood from veins ("posterior and anterior cardinal veins") lying laterally in the body wall.   Blood is also returned to the ventral vessel by several smaller transverse vessels.   The ventral vessel then continues forward below the pharynx as the "ventral aorta" and gives rise to the paired vessels ("afferent branchial vessels") which pass dorsalwards in the primary gill bars.   At the base of each of the "afferent branchial vessels" is a small swelling which is contractile and assists in the circulation of the blood.   As has been stated above, the vessels in the primaries are linked to those in the secondary gill bars by vessels in the synapticula so that blood passes into the paired dorsal vessels ("dorsal aortae") by "efferent branchial vessels", both from the primary and from the secondary bars.   Behind the pharynx the "dorsal aortae" unite to form the median "dorsal aorta" which continues into the tail as the "caudal artery".   The "dorsal aortae" also continue forwards to the anterior end of the body as the "carotid arteries" which end blindly in the snout. The median "dorsal aorta" during its course gives off numerous small vessels to the gut and to the body wall, and from these situations the blood is collected into the "cardinal veins" and into the ventral vessel.

Many of the peculiarities in the vascular system of amphioxus are susceptible of a functional explanation. The absence of both renal and of true hepatic portal systems is of course due to the absence of kidneys and liver, but a more interesting correlation is between absence of heart and absence of kidneys. Kidneys, which are the typical excretory organs of vertebrates, depend for their functioning on a high arterial blood pressure which is developed and maintained by a muscular pumping organ, the heart. It seems that in amphioxus where kidneys are lacking, the low blood pressure developed by the peristaltic contractions of the ventral vessel suffices for an adequate circulation of the blood through the body,

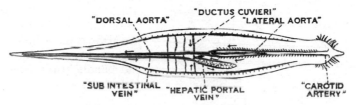

"DORSAL AORTA"    "DUCTUS CUVIERI"    "LATERAL AORTA"

"SUB INTESTINAL VEIN"    "HEPATIC PORTAL VEIN"    "CAROTID ARTERY"

Fig. 296. AMPHIOXUS.—Dorsal view of the main vessels (anterior part of the "ventral aorta" omitted).

whilst excretion is carried out by numerous nephridia of the closed type.

**The Excretory System.**—Although in practically all of its main features amphioxus is a typical chordate, the nature of its organs of nitrogenous excretion provides an interesting link with the non-chordate animals for, like so many of them, amphioxus is provided with **protonephridia** (closed nephridia) almost exactly like those found in certain polychaetes and like them developed from the ectoderm.

There are about ninety pairs of these nephridia. They lie in the pharyngeal region just above the gill slits and in fact one pair of nephridia is associated with each pair of slits. Each nephridium is a small bent tube opening by its upper limb into the atrium on the dorsal edge of the gill slit, whilst the other limb passes ventrally and terminates blindly. Numerous short branches arise from the sides of the nephridial tube and into each of these side branches open many fine tubules each terminating in a typical **solenocyte.** The tufts of solenocytes project into the dorsal coelomic canal and, owing to the breaking down of the coelomic epithelium immediately above them, they become directly bathed in coelomic fluid, a factor which no doubt facilitates the extraction of nitrogenous waste. In addition, the nephridia are well supplied with small blood vessels from which excretion also takes place in all probability, but the mechanism of excretion is not well understood. Apart from the paired series of nephridia already described, there is a single large nephridium, the so-called **nephridium of Hatschek,** situated above the roof of the pharynx laterally to the left dorsal blood vessel. Hatschek's nephridium in all essential respects resembles the paired nephridia. It is a narrow tube which opens at its posterior end into the pharynx just behind the velum, and then passes anteriorly to end blindly just in front of Hatschek's pit.

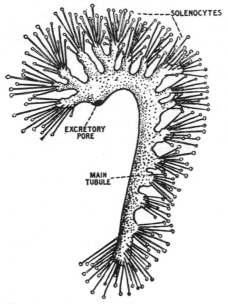

Fig. 297.  AMPHIOXUS.—Nephridium (based on Goodrich).

**The Nervous System.—** The **central nervous system (neural tube),** like that of higher chordates, is situated in the mid-dorsal line just above the notochord. Whilst simple in structure in that no anterior enlargement or brain is present, it resembles the vertebrate central nervous system in consisting of a mass of nervous material perforated throughout its length by a narrow **central canal.** The actual nerve-cells are grouped around the canal whilst the processes arising from the cells lie more superficially, an arrangement which differs from that in non-chordate animals where the nerve-cells occupy the superficial position, whilst most of the nerve-fibres are centrally placed. As in certain other elongated animals, giant cells and giant fibres are found in the central nervous system. The fibres appear to decussate before running longitudinally in the neural tube.

At the anterior end, the central canal in the neural tube enlarges to form the so-called **cerebral vesicle**. A single pair of sensory nerves from the snout and oral cirri enters the extreme anterior end, whilst a second pair, also from the same region, joins the neural tube just dorsal to the cerebral vesicle. The central nervous system, which runs practically the whole length of the body, terminates, after tapering considerably, a short way in front of the hinder end of the notochord. Regions of the body behind the cerebral vesicle are innervated by the nerves which arise from the central nervous system in a paired (right and left) segmental series. The nerves are functionally of two main types, afferent, or sensory, and efferent, or motor, and in every segment there is one afferent entering and several efferent nerve-roots leaving the neural tube on each side. The asymmetry, which has already been noticed in the arrangement of the myomeres, is reflected in a similar alternation of the members of the segmental nerves, the afferent roots of one side being opposite the efferent roots of the other. As in vertebrates the dorsal roots are chiefly composed of sensory fibres arising from receptor organs in the skin and the ventral roots, so far as is known, pass to the myotome muscles alone. Giant multi polar cells lying close to the central canal send axons which cross to the opposite side and contribute to giant fibres lying laterally in the neural tube. This system of cells and fibres probably provides a means for co-ordinating movements of the myotomes in swimming.

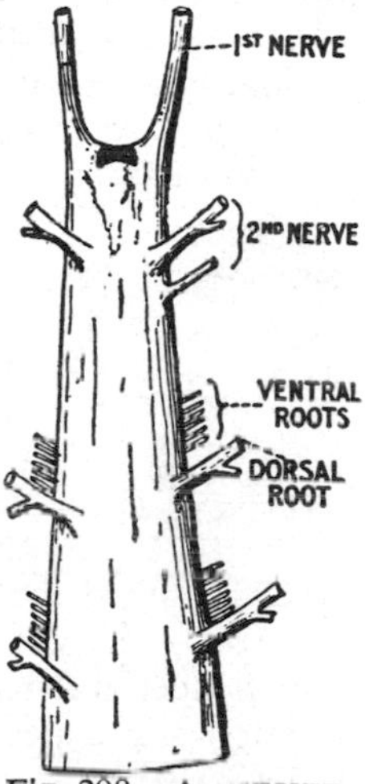

Fig. 298. AMPHIOXUS. —Anterior part of the central nervous system.

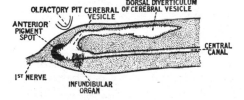

Fig. 299. AMPHIOXUS.—Vertical section of the the anterior part of the neural tube (modified after Boeke).

A system of motor and sensory neurons lies on each side of the atrial epithelium and is connected to the neural tube by way of the dorsal roots. This atrial nervous system is concerned with various reflexes, including the cough reflex and with sensory impulses from the wall of the gut.

**The Sense Organs.**—The most obvious receptor organs are the pigmented eyes which occur along the whole length of the neural tube. Each is composed of a photosensitive cell and a blackish-brown, cup-shaped pigment cell. The cytoplasm of the pigment cup is differentiated to form a striated layer which probably acts like a lens. A single nerve-fibre leaves the cell on the side opposite to that bearing the pigment. These photoreceptors are arranged along the walls of the central canal of the neural tube and in definite tracts, but their orientation is not the same on opposite sides of the body nor uniform along the length of the nerve cord. At the extreme anterior end of the central nervous system, that is, in the anterior wall of the cerebral vesicle, there is an exceptionally large pigment spot, but it lacks the structure of a typical simple eye and merely serves to shield the photoreceptor from frontal stimulation. It is found that when the animal is illuminated by a pencil of light from the front it does not respond by movement, but light from any other direction stimulates it to swim in a spiral fashion which may be related to the asymmetrical arrangement of the eyes. Sooner or later the animal buries in the sand with its "head" sticking out.

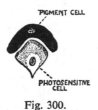

Fig. 300.
AMPHIOXUS.—Eye.

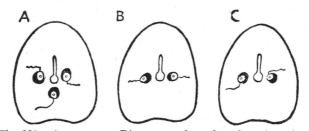

Fig. 301. AMPHIOXUS.—Diagram to show the orientation of the eyes in the central nervous system. A, anterior region; B, middle region; C, posterior region (after Franz).

In this position the large anterior pigment spot prevents further stimulation. On the floor of the cerebral vesicle is a patch of tall ciliated cells forming the **infundibular organ.** They secret a viscous material which solidifies to form a fibre extending along the central canal, whose function is unknown. Also of unknown function is the so-called "olfactory pit" (Kölliker's pit), a ciliated pocket of ectoderm which dips down towards the dorsal wall of the anterior end of the cerebral vesicle just dorsal to, and towards the left-hand side of, the median pigment spot. This pit marks the position of the **neuropore** (see Fig. 299). The true chemo-receptors are sensory cells grouped in papillae on the velar tentacles, which are also sensitive to touch. The oral cirri are sensitive to touch only.

As might be expected, the whole of the ectoderm covering the body bears scattered sensory cells which probably respond to tactile stimuli. They are particularly abundant on the dorsal surface. Each bears a short hair-like sensory process which protrudes through the cuticle, whilst from the other end of the cells a nerve-fibre passes inwards to contribute to one of the sensory nerves. Certain of these cells must be concerned with detecting

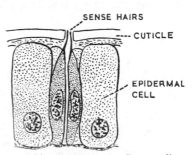
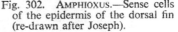

Fig. 302. AMPHIOXUS.—Sense cells of the epidermis of the dorsal fin (re-drawn after Joseph).

the nature of the deposits into which the animal will burrow, for it is known that it can and will avoid "sands" that are too fine.

**The Reproductive System.**—The sexes are separate, but males and females are not distinguishable apart from the nature of

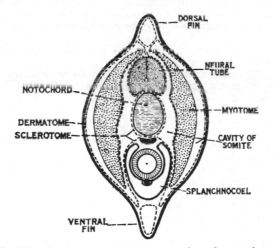

Fig. 303. AMPHIOXUS.—Transverse section of young lava.

the gonads. The gonads, either **testes** or **ovaries**, lie in the ventral part of the lateral body wall in the pharyngeal region where they form a series of bulges inwards towards the atrium. They follow a strict metameric arrangement, for there is one pair in each of segments twenty-five to fifty-one. There are no genital ducts and the ripe gametes are released, by rupture of the overlying tissue, into the atrium, whence they are carried to the outside by way of the atriopore.

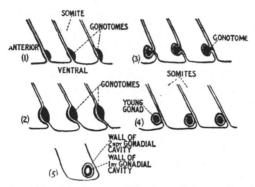

Fig. 304. AMPHIOXUS.—Diagram of the mode of development of the gonads.

Although, as would be expected, the germ-cells arise as proliferations of the coelomic epithelium, their relations to the coelom in the adult animal are by no means clear, and can only be understood fully when the various stages in the development of the gonad are known. In the young larva, after separation of the lateral plate, the dorsal metamerically segmented blocks of mesoderm (*i.e.* those parts of the mesoderm corresponding to the somites of vertebrates) undergo differentiation to form the **myotomes**, the **sclerotomes**, and the **dermatomes** (see pp. 703, 725). The myotomes form the segmental muscles of the body whilst the sclerotomes and dermatomes give rise to connective tissues. The somites retain for some time a narrow cavity, which was at an earlier stage of development continuous with the general perivisceral coelom and must also be considered as coelomic in nature. It is from the tissue forming the ventral wall of this cavity that the germ-cells develop. This tissue is really quite distinct from the tissue (embryonic muscular tissue) of the myotome and is often termed the **gonotome**. The gonotomes then grow out forwards and ventralwards so as to form small buds which push before them the posterior wall of the somite immediately in front. For a time each gonotome remains connected to its own somite by a short stalk, but eventually the stalk is severed and in this way a series of young gonads is formed. Each gonad then acquires a cavity—the **primary gonadial cavity**—and becomes almost completely surrounded by a **secondary** or **perigonadial cavity (gonocoel)** which is bounded by tissue derived from the wall of the somite (the one immediately in front of that giving origin to the gonad) into which the gonad has pushed. The germ-cells develop from the cells forming the outer wall of the primary gonadial cavity. They are separated from the

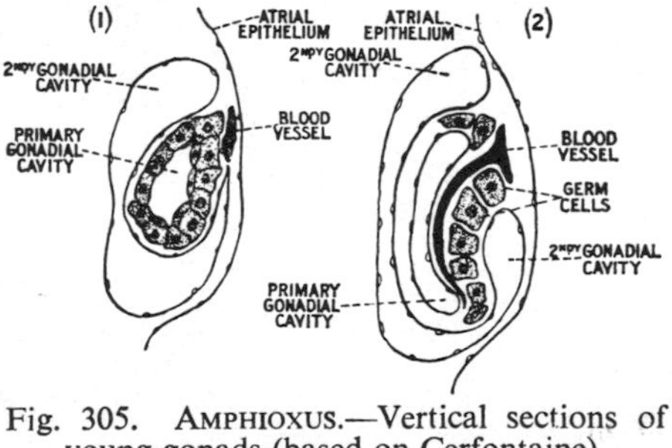

Fig. 305. AMPHIOXUS.—Vertical sections of young gonads (based on Cerfontaine).

atrial epithelium by the walls of the secondary gonadial cavity, whilst on their inner surface is a layer of **follicular epithelium** which also is derived from the cells of the wall of the primary gonadial cavity. Between the germ-cells and the follicular epithelium is a large blood space which communicates with the "posterior cardinal vein".

**Habits.**—Various species of amphioxus occur in many different parts of the world and may be so abundant locally that, as in Japan, they are used for food. But they are restricted to regions where the sea-floor is covered by deposits of a fairly coarse or "open" texture. For example, the main British source of amphioxus is the shell-gravel near the Eddystone Lighthouse. Experiments and studies on the ecology of the Nigerian species have shown that these animals will not burrow into very fine sand nor even into coarser deposits which contain silt. They can live successfully only in "open" deposits in which water can percolate freely between the sand grains.

Although able on occasions to swim rapidly, amphioxus does not move about actively in search of its food, for like many other marine animals it feeds on the minute organic particles which it strains off from the surrounding sea water by the filtering action of its pharynx.

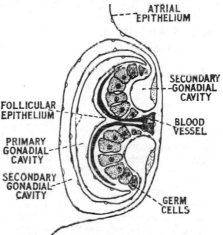

Fig. 306. AMPHIOXUS.—Vertical section of an older gonad (based on Cerfontaine).

**Feeding.**—When feeding, a continual stream of water bearing food enters the mouth and passes out into the atrium through the gill slits to be expelled at the atriopore, this water current being maintained by the action of cilia, for which reason amphioxus is said to be a **ciliary feeder.** The details of the feeding process are interesting, for they are almost identical with those found in the Urochordata and bear some resemblances to those of the ammocoete larva of the lamprey, where, however, the current of water is caused by muscular movements of the pharyngeal wall and velum. There is reason for supposing that the feeding method of primitive chordates may have been on similar lines to that of amphioxus.

It will be recalled that the anterior and posterior faces of each gill bar are carpeted with long powerful "lateral cilia", and it has been found that these cilia, by their lashing, cause a current of water to pass outwards from the pharynx through the gill slits into the atrium. It follows therefore that water is drawn in at the mouth and that particles in suspension are carried backwards and outwards from the mid-line towards the gill bars. The cilia of the atrial epithelium on the outer face of the gill bars and isolated patches of ciliated epithelium overlying the gonads assist in promoting this water current. The cilia ("frontal cilia") (so-called because they face the main water current), on the inner (pharyngeal) surface of the gill bars lash along the length of the bars in such a way that they cause a current to move upwards around the pharynx from the ventral towards the dorsal mid-line. In this way a stream of watery mucus (secreted by the endostyle and to a lesser extent by the pharyngeal epithelium) passes from the endostyle into the epipharyngeal groove, where it is carried back to the opening of the oesophagus.

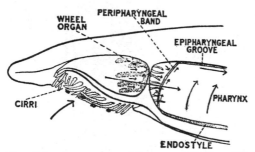

Fig. 307. AMPHIOXUS.—Diagram of feeding currents. Large arrows show main currents; small arrows, subsidiary currents (based on Orton).

The mucus secreted by the gland cells of the endostyle is transferred initially to the ventral wall of the pharynx by the lateral tracts of cilia, which run the whole length of the endostyle and lash outwards from the mid-line. The function of the median tract of long endostylar cilia is to supply the lateral endostylar cilia with a continual supply of mucus. They do not cause any appreciable stream of mucus to pass forwards to the peripharyngeal bands as is sometimes stated.

The ciliated structures in front of the velum and the peripharyngeal bands play only a subsidiary part, and the main features of the feeding process may be summarised in the following way. A water current is automatically maintained by ciliary action (chiefly by the "lateral" cilia on the gill bars), enters the pharynx, and passes out into the atrium via the gill slits. Particles of suspended matter are forced against the pharyngeal surfaces of the gill bars where they become entangled in mucus secreted mainly by the endostyle. The mucus layer lies between the bases of the cilia and suspended matter is forced into it and entrapped by the action of the cilia, which beat downwards on to, as well as causing a current in, a plane parallel to the epithelium. But for the thin sheet of mucus on the walls of the pharynx many of the food particles would be washed out through the gill slits but, as it is, the mucus with the entangled food passes, by the action of the "frontal" cilia dorsalwards to the epipharyngeal groove, the action of whose cilia carries it back to the oesophagus.

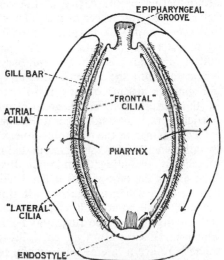

Fig. 308. AMPHIOXUS.—Diagrammatic transverse section of pharynx to show feeding currents (based on Orton).

Before entering the mouth larger particles in the water are sieved off by the buccal cirri which, when the animal is feeding, are folded over one another to form a mesh. These particles progressively accumulate on the cirri and so obstruct the current of water entering the oral hood and pharynx. They are removed by a violent current of water produced by the muscular movement of the floor of the pharynx. When this happens the artriopore is closed and the atrial floor is suddenly elevated by the contraction of its transverse musculature. This reduces the volume of the atrial cavity so that a powerful current of water is forced out through the pharynx and oral hood. At the same time the

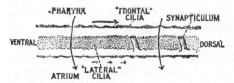

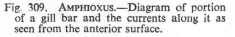

Fig. 309. AMPHIOXUS.—Diagram of portion of a gill bar and the currents along it as seen from the anterior surface.

cirri are flexed in and out of the oral hood and the particles they carry become dislodged by the outgoing water current.

Feeding is not, however, a continuous process and from time to time the lateral cilia cease beating or fail to beat in metachronal rhythm so that the inhalant current stops. The addition of anaesthetics to the water restores proper metachronal beating from which it is inferred that the lateral cilia (not the rest) can be inhibited by the nervous system, probably by way of multipolar neurons on the gill bars and beneath the endostyle which connect with the central nervous system.

Any particles which (possibly because of their higher specific gravity) drop out of the main water current are prevented from accumulating in the cavity of the oral hood by ciliary currents generated by the wheel organ. They become mixed with mucus and are passed back towards the pharynx when the bulk of them again enter the main stream. The peripharyngeal bands serve to collect up and pass to the epipharyngeal groove any particles which fall into the extreme anterior part of the pharynx (where gill slits are absent) whilst certain papillae (Fig. 292) on the floor of the atrium contain phagocytic cells which engulf small food particles that occasionally get wafted into the atrial cavity.

The passage of the food-plus-mucus down the gut is also effected by ciliary action, for there are ciliated tracts along the whole length of the gut, including the mid-gut diverticulum. By the time the food enters the oesophagus it has become rolled up with mucus to form a cord which is rotated as it slowly passes backwards along the gut. The rotation is caused solely by the action of ciliated cells in the ilio-colonic ring, and behind this region any portions which break away from the cord fail to rotate. The chief ciliated tracts here are a lateral tract on the left-hand side of the anterior part of the mid-gut; in the posterior part of the mid-gut a dorsal tract which links up with the lateral tract; and a dorsal tract in the hind-gut.

**Digestion and Absorption.**—In addition to ciliated cells, the lining epithelium of the gut contains cells which secrete digestive juices. These are mixed with the food as it passes along the gut. The secretions of the mid-gut diverticulum are driven out into the mid-gut by ciliary action, where further digestive juices are added to the gut contents by all the regions of the gut behind the oeso-phagus. It follows, therefore, that digestion is started in the mid-gut and continued until the gut contents are voided. It seems extremely likely that in addition to this extra-cellular digestion there is an intra-cellular process by which small particles are taken into certain of the epithelial cells and there digested. This process takes place mainly in the hind-gut region. Absorption of the digested food occurs chiefly in the hind-gut region and only to a lesser extent in the mid-gut.

**Gaseous Interchange.**—Unlike the pharynx of vertebrates, which is not as a rule concerned with the collection of food, and where the gill slits and arches are specialised to form respiratory organs (gills), the pharynx of amphioxus is mainly a feeding organ. It seems clear that some gaseous interchange between the blood and the sea water must take place in the walls of the pharynx, but it is open to doubt (in view of the respiration needed to release the energy necessary for the activity of the cilia) if the blood entering the "dorsal aortae" is more oxygenated than that in the "ventral aorta".

It has been suggested that oxygenation of the blood-stream and other body fluids and elimination of carbon dioxide takes place over the whole surface of the body, particularly in the walls of the atrium.

**Excretion.**—In the absence of experimental evidence it is reasonable to assume that the bulk of the nitrogenous excretion is carried out by the nephridia. These presumably function in much the same way as the protonephridia found in some non-chordate animals (p. 166) and, in addition, they probably act as the osmoregulators of the body.

**Reproduction.**—The eggs and spermatozoa are released into the atrial cavity and thence into the sea by way of the atriopore. Here fertilisation takes place and the zygotes develop into larvae. The details of these events are given later (p. 691).

## THE AFFINITIES OF THE CEPHALOCHORDATES

It will be realised from the description of amphioxus that it possesses practically the full complement of chordate characters, and whilst it differs from the craniates mainly in the relative simplicity of its bodily organisation, there are certain other important distinguishing features. Opinion is divided as to whether these are to be regarded as "secondary features" (that is, peculiar to amphioxus alone and in the nature of adaptations to its mode of life), or whether they are "primitive characters" (in other words, relics of characters originally common to all primitive chordates).

The large pharynx, for example, which is chiefly concerned with feeding, is frequently said to be a secondary or adaptive feature, but against this view it may be urged that all other protochordates as well as the ammocoete larva of the lamprey rely on their pharynx for the maintenance of a current of water from which the food is sifted. The collection of sufficient amounts of small particles of food by means of ciliary currents is not possible unless the volume of water passing over the collecting surface is very large, and the water must be free to escape before entering the digestive region of the gut. The action of the pharynx in the ammocoete is of great interest in this connection, because it seems

to represent a condition intermediate between that in the protochordates and adult craniates (see p. 482). The inhalant current,
which is both respiratory and nutritive, is due to the pumping action
of the pharynx, but the food particles are entrapped in mucus and
conveyed to the oesophagus by ciliary currents.

In most ciliary feeders the separation of food particles from
the water is effected by means of ciliated tracts on processes
("tentacles") *outside* the body, so that the mucus-plus-food and
a minimum quantity of water alone enter the mouth. This method
is, then, quite different from that adopted by the protochordates,
where (with few exceptions) sorting of the food does not take place
until after the food-containing stream has entered the pharynx.
It may well be that gill slits arose in connection with this type of
feeding to provide a suitable outlet for the excess water. If this
is true, then the respiratory gill slits of the vertebrates represent
a secondary condition. The number of slits in the protochordate
pharynx is variable, but very much greater than in the vertebrate
pharynx. This is a feature related to the feeding process, for
numerous small gill slits and gill bars allow of more ciliated tracts
than would a few larger ones and also prevent the escape of food
particles from the pharynx.

The endostyle, which plays an important part in feeding, is
also a primitive feature. The great extension of the notochord,
on the other hand, is probably a secondary feature, for in no other
chordate does the notochord reach in front of the anterior end of
the central nervous system.

The absence in amphioxus of many important structures found
in craniates is indicative of a simpler and probably more archaic
type of chordate organisation, for it is doubtful if the earliest
chordates possessed a skull, a brain, a vertebral column, or a true
heart. Certain features may well be taken as relics from a *pre-
chordate* ancestor. Among these may be mentioned the nephridia
and perhaps also the curious asymmetries of the body; for similar
asymmetries are found in the echinoderms, which of all the non-
chordate groups are thought to be the most nearly related to the
chordates. The almost complete lack of cephalisation and the
retention of obvious metamerism are also very noteworthy features.

On the whole then, amphioxus may best be regarded as a
primitive type of chordate.* Its structure throws much light on
the probable structure of the ancestral chordates, but unfortunately
gives practically no hint of the way in which they might have
evolved from any known type of non-chordate animal. We are,

* This thought has been very neatly expressed when it was said that "If
amphioxus had not been discovered it would have to have been invented".

at the outset, so to speak, confronted by a simple but completely chordate animal, and it may be fairly said that, up to the present, all attempts to derive the chordates from one or other of the non-chordate phyla, although ingenious, are far from convincing.

# CHAPTER XVII

## CRANIATA

### INTRODUCTION

The **vertebrates**, as their name implies, are those chordates which possess a **vertebral column** or backbone. This consists of a series of bones or cartilages embedded in the great muscles of the back and replaces the notochord of less complex chordates and of the vertebrate embryo. Animals which have a backbone always have other skeletal structures, for example, those developed around the brain and sense organs to form a **cranium**. Such chordates are therefore termed craniates, an alternative name for vertebrates.

In addition to the cranium, all craniates have present in the wall of the pharynx, lying between the visceral clefts, a series of **visceral arches**, to which the muscles causing the movements of the pharynx are attached. In some craniates certain of these visceral arches are modified to form jaws bordering the mouth. Such vertebrates are grouped together and are said to form the Gnathostomata to distinguish them from those, the Agnatha, which do not possess jaws, and in the modern representatives of which (the Cyclostomata), the mouth is round and suctorial.

The gnathostomes invariably possess **limbs**, and these have their particular skeletal structures forming the **limb skeleton** which articulates with a **limb girdle**.

Most of the other chordate features are clearly recognisable in the craniates. The hollow **central nervous system** is always well developed and the anterior part enlarged to form a **brain**, the remainder constituting the **spinal cord**. The ventral **heart** is contained in a **pericardial cavity** and leads into a **ventral aorta** from which arises a series of **aortic arches** passing along the visceral arches to **dorsal aortae** above the pharynx. In many craniates both the liver and the kidneys have a double source of blood-supply, an artery and also a "portal" vein bringing blood which has been collected up from other organs.

The respiratory system always includes a respiratory surface developed from the pharynx—either **gills** (in aquatic forms) or **lungs** (in terrestrial animals).

The excretory organs of craniates are true **kidneys**, collections of renal tubules of mesodermal origin.

The **gonads** are compact bodies and their ducts are closely associated with those of the kidneys.

Such, then, are the generalised features of a craniate and are shown in Fig. 309. But the diversity of form is so great that in an elementary treatment of the group it is impossible to do more than consider the main trends which can be traced from their history in time. Undoubtedly, the earliest craniates were aquatic animals, and many examples of these forms are known from their fossil remains. It is equally certain that, eventually, some migrated from the water to the land with consequent modifications of structure and physiology to adapt themselves to the terrestrial environment.

Consequently, three examples are usually selected for consideration: a relatively simple aquatic form such as a fish (*dogfish*); an intermediate form presenting some features correlated with the transition from an aquatic to a terrestrial environment such as are found in an amphibian (*frog*); and lastly, a thorough-going terrestrial form such as a mammal (*rabbit*). Although from these three examples an intelligible picture of the progressive complexity of form and function can be painted, yet it must not be imagined that they are directly related one to another.

## HISTOLOGY

### INTRODUCTION

In order to understand fully the activities of a complex animal such as a rabbit or frog it is necessary not only to deal with the gross anatomy, such as has been done in the subsequent pages, but also to study the minute structure of the organs and of the tissues of which they are composed. Such a study is called **histology**, or might equally well be called microscopical anatomy. The term "tissue" is often used somewhat loosely and no hard-and-last meaning can be given to it. In general, it is applied to any aggregate of more or less similar cells together with any intercellular secretion produced by them. For convenience in description, the various types of tissue are often grouped together into four main classes: **epithelia, connective tissues, muscular tissues**, and **nervous tissues**. These groups are in turn subdivided, as will be seen later. All the types of tissue mentioned are of regular occurrence in the body of a vertebrate, and many of them can be recognised in the invertebrates. As might be expected, however, there is a certain amount of variation in the tissues of different types of animals and for this reason the descriptions given will apply to mammalian tissues, except in those instances where it is otherwise stated.

Commonly histology is dealt with as a separate study but there is much to commend an arrangement in which the finer structure is discussed immediately after or alongside a presentation of the gross anatomy and functions of the organs of the body. In the main this is the plan which has been followed in this book but, since they enter into the composition of so many different structures and organs, epithelia and simple connective tissues are described in the section which follows.

**Epithelia.**—The word **epithelium** means a covering, and epithelia are tissues bounding or covering the surface of the body as a whole, the surface of various organs, forming the lining of the gut and glands, and in fact, wherever a free surface is to be found. Those epithelia which line internal spaces, such as the body cavity, or the blood vessels, are frequently referred to as **endothelia.**

The various types of epithelia can be classified and named according to two very different criteria: firstly according to the structure and arrangement of the cells, and secondly according to the function of the tissue. Thus, for example, columnar epithelium is a term applied to an epithelium which consists of tall pillar-like cells: that is, the term is indicative of the form of the cells. The term glandular epithelium refers to the secretory properties of the cells: that is, to their function. Now it frequently happens that columnar epithelium has a glandular function, and therefore the terms columnar and glandular are not mutually exclusive. Confusion will be avoided if this dual nomenclature is realised at the outset.

It may be said in general that epithelia consist of cells arranged in sheets, the cells being joined together by a small amount of intercellular cement substance (a product of the epithelial cells themselves) and almost invariably the cells rest on a delicate, structureless, or fibrillar **basement membrane.** Epithelia and other close-packed tissues are thus to be contrasted with the more open connective tissues in which the intercellular spaces filled with fibrous and gelatinous secretions of the cells predominate. In sections of epithelia examined in the electron microscope, the plasma membranes of the component cells (see Frontispiece) are seen as dense parallel lines separated by an intercellular gap of rather constant width—about 200 Å—which probably represents the intercellular cement. Most epithelial cells are specialised in some respect or other but all contain in variable proportions the intracellular organelles described earlier. Many epithelia also contain bundles of fine filaments which contribute towards the mechanical strength of the tissue and are particularly dense in the cells of the external layers of vertebrates where they may keratinise to form a protective covering. The cell surfaces are often studded with dense deposits from which tufts of intra-cellular fibrils may extend. The adhesion between the cells at such points is particularly strong and greatly reinforces the strength of the whole cellular formation. Such deposits are referred to generally as **desmosomes** but may have other names, such as terminal bars—where they run completely around the edges of cells facing an open cavity—or intercalated discs in striated muscle (see Fig. 402)—where they hold the muscles cells together. As can be seen from the diagram, the cell membranes are continued through a desmosomal deposit and the two cells remain separate although at low magnifications the tuft of filaments may seem to run continuously from one cell to another, forming an "intercellular bridge". Desmosomes are extremely frequent in keratinised tissues; they also form on the internal surface of cells facing a basement membrane in keratinised tissues.

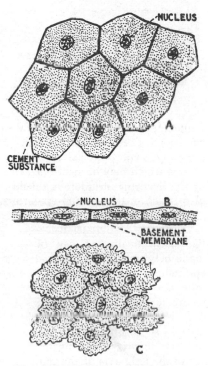

Fig. 310. Squamous epithelium.— A, surface view; B, vertical section; C, tessellated variety.

Under the electron microscope, basement membranes are seen to consist of an amorphous layer about 400-600 Å thick running parallel to the plasma membranes of the cells. Beyond this layer one finds deposits of collagen fibrils laid down by the fibrocytes of the underlying connective tissue (see below).

Epithelia may be composed of a single layer or of many layers of cells, and this fact allows at once for the division of the epithelia into two main groups: **simple epithelia** and **compound epithelia**.

The four main types of simple epithelia, into which can be fitted most varieties, and the two types of compound epithelia are described in the two following sections.

**Simple Epithelia.** SQUAMOUS EPITHELIUM.—This type consists of a single layer of flattened plate-like cells whose edges fit closely together to form a sort of mosaic, and in fact an alternative name, "pavement epithelium", is sometimes given to it. With special staining the cell outlines may be seen in surface view. These frequently have a roughly hexagonal outline (*e.g.* Bowman's capsule), but in some types (*e.g.* lining epithelium of the blood vessels, peritoneum) the cell outlines are wavy and the epithelium is said to be **tessellated.** In a vertical section, it is seen that the thickness or depth of the cell is appreciably increased in one place to accommodate the somewhat flattened nucleus.

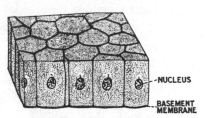

Fig. 311. Cubical epithelium.

CUBICAL EPITHELIUM.—In many respects this resembles the previous type but, as the name implies, the cells are cubical in shape. The nucleus is spherical and lies centrally in the cell. In surface view the cell outlines are polygonal. Cubical epithelium is found lining many glands, among which may be mentioned the **thyroid, sweat glands**, and parts of the **liver**, the **uriniferous tubules**, and certain other ducts. It is also found in the gut and epidermis of many invertebrates.

COLUMNAR EPITHELIUM.— No sharp distinction can be made between this and cubical epithelium, but the cells are taller and, in extreme instances, so much so that they resemble tall pillars or columns. The nucleus is variable in position but is invariably elongated along the long axis of the cell. It is common to find some differentiation at the free surface, *i.e.* the surface of the cell away from the basement membrane. Thus, many of the

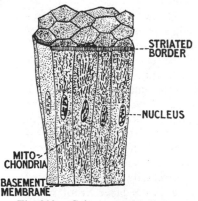

Fig. 312. Columnar epithelium.

cells lining the small intestine have a **striated border**. These striated, or the slightly different, brush borders found on many cells are in reality a series of thimble-like protrusions (microvilli) of the cell membrane which greatly multiply the surface area of the cell available for absorption or secretion. The alimentary canal from the stomach to the rectum is lined by columnar epithelium and in this instance the cells are glandular (or secretory) or absorptive. The **gastric glands** of the stomach are lined by an epithelium of this type, as also are the various **intestinal glands**. Mitochondria and granules representing the precursors of the secretion can usually be seen particularly well in the secretory cells.

CILIATED EPITHELIUM.—The free surface of these cells which are usually columnar in form, has numerous cilia arising from it. The base of each cilium has one or more darkly staining granules. The behaviour of the cilia gives an obvious clue to their func- tion. In small animals, which have a ciliated outside surface, it is to move the animal through the surrounding me- dium: when present as part of a ciliated epithelium in the cavities of larger animals, it is

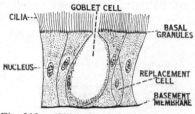

Fig. 313. Ciliated epithelium in vertical section.

to cause a current in a fluid. It is to be noted that ciliated epithelium is found only in moist surroundings and, almost invariably, in between the ciliated cells are found cells which secrete mucus. The main func- tion of a ciliated epithelium is often to shift a stream of small solid particles, but this effect is always carried out indirectly, the particles being first of all entangled in mucus. Good examples of epithelia carrying out this function are those found lining the **nasal cavities**, **trachea**, and **bronchi**. Par- ticles of dust from the air are entangled in the mucus, a continual stream of which passes upwards to the buccal cavity and nos- trils so that the delicate tissues of the lungs are kept free from damage.

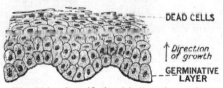

Fig. 314. Stratified epithelium in vertical section.

Cilia may occur on types of epithelia other than columnar.

**Compound Epithelia.**—The epithelia so far described cover sur- faces which are moist and on which there is little wear and tear. The arrangement of the cells in a single sheet does not afford as ready a means of rapid replacement as is required in situations where there is considerable friction, such as the outside surface of a land- living animal. To meet these conditions compound epithelia have been developed.

STRATIFIED EPITHELIUM.—This type of epithelium is built up of several layers of cells. The lowermost layer, *i.e.* the layer next to the basement membrane, consists of roughly spherical cells whose nuclei stain darkly with haematoxylin. To this layer the name **germinative layer** may be given, whilst the name **stratum Malpighii**, after its first describer, is applied to it in descriptions of the epidermis. The germinative layer is always in active cell division, so that cells are always being produced to renew the overlying layers. The outermost layers of the epithelium consist of dead flattened cells whose nuclei no longer stain deeply. Friction at the surface causes these dead cells to be sloughed off so that the deeper layers, which themselves are being continually replaced

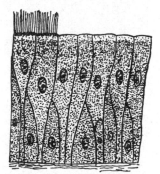

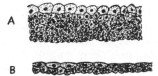

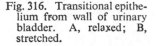

Fig. 315. Vertical section of pseudostratified epithelium. On the left are shown two ciliated cells such as often occur in epithelia of this type.

Fig. 316. Transitional epithelium from wall of urinary bladder. A, relaxed; B, stretched.

from below, in time come to occupy a superficial position. A gradual alteration in the shape of the cells between the cornified cells and those of the germinative layer is found in the intermediate layers. (To the outer, cornified layers, the misleading term "squamous" is sometimes applied, but the term is better kept for the true squamous epithelium.)

Stratified epithelium is found in the epidermis of the skin of most vertebrates, and also lines the oesophageal and buccal cavities.

PSEUDOSTRATIFIED EPITHELIUM.—In this type the cells are not uniform in size and their nuclei appear at different levels in the epithelium. Indeed, many cells do not reach as far as the free surface but there are no definite layers of cells such as are found in stratified epithelia. Pseudostratified epithelium occurs only rarely in the vertebrate body (*e.g.* lining the larger ducts of certain glands) but is found in many ciliated regions of invertebrates.

TRANSITIONAL EPITHELIUM.—This type is found, for example, in the male urethra and bladder (Fig. 316) and allows stretching,

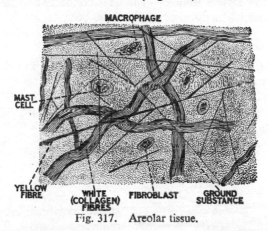

Fig. 317.   Areolar tissue.

when the epithelium becomes transformed to a layer not more than one or two cells thick.

**Connective Tissues.**—A connective tissue may be contrasted with epithelia in that it is the interstitial (intercellular) matrix which forms the major part of the tissue, whereas in epithelia the amount of intercellular substance is insignificant. It is, however, difficult to give a precise definition of the function of a connective tissue, for whilst certain types do indeed connect up struc-tures one with another, others have as their chief role the for-mation of a packing around organs, and others again are skeletal in function.   Moreover it is common to find that several different kinds of tissue enter into the formation of what is at first sight a homogeneous connective tissue.

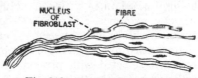

Fig. 318.   White fibrous tissue.

The various types of connective tissue that are distinguished differ widely in structure, but they have in common one important feature, namely, they are composed of cells together with a **matrix** secreted by those cells.

Connective tissues fall into three main groups: **connective tissues proper, skeletal tissues**, and **blood**. Skeletal tissues and blood will be discussed in the sections dealing with the skeleton and blood vascular system, respectively. Obviously, connective tissues of invertebrate animals may differ radically from these.

**Connective Tissues Proper.**—The term connective tissue naturally implies tissues which connect parts of the body together. Of these, the areolar and fibrous tissues most nearly fulfil this function and may therefore be called connective tissues proper.

AREOLAR TISSUE.—This is the most widely distributed and commonest connective tissue found in the body. It forms a continuous layer beneath the skin, fills out the spaces between many organs, acts as a packing between muscles, is found in the peritoneum and mesenteries, and even surrounds blood vessels when they penetrate into the organs and body cavity. When isolated, it is a white sticky mass which under the microscope is seen to consist of interlacing bundles of very fine fibres, between whose meshes is a thin jelly-like **ground substance** in which are scattered numerous cells. The fibres are of two types: **white fibres** which are wavy and are arranged in bundles and **yellow fibres**, fewer in number and more slender than the white, not arranged in bundles but branching and joining up with one another to form a delicate network. The white fibres, which are resistant to stretching, are formed of a scleroprotein called **collagen**, which is soluble in dilute acetic acid and which on boiling with water yields a solution of gelatin. With suitable staining, extremely elongated cells arranged in a single row running along the fibres, can be seen. These fibrocytes are secretory cells characterised by an elaborate development of the internal cytomembranes associated with protein synthesis (Fig. 23). The soluble collagen precursor (tropocollagen), a long (2,400 Å), thin (15 Å) molecule, accumulates in large vacuoles in the Golgi region of the cells, which open to the cell surface to release their contents. The long molecules of tropocollagen then aggregate laterally to produce fibrous collagen which is characterised by cross-markings 640 Å apart. The regular, repeating pattern is a consequence of the peculiar overlapping manner in which the component tropocollagen molecules aggregate in bundles (Fig. 24). The yellow fibres, on the other hand, are made of **elastin**, a protein which is easily stretched

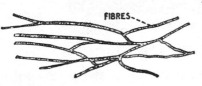

FIBRES

Fig. 319.   Yellow elastic fibres.

(*i.e.* elastic in the popular sense) but is resistant to boiling and insoluble in acetic acid. The fibroblasts do not lie adjacent to the yellow fibres which they secrete. It will be realised that the matrix is represented by the two types of fibres as well as by the amorphous ground substance.

Many types of cells are found in the amorphous ground substance. Among the commonest are the fibroblasts, which are large flat stellate cells. Certain kinds of **white corpuscles** are also found, *e.g.* **macrophages**—large irregularly-shaped cells with an ovoid nucleus. **Mast cells,** large amoeboid cells with a granular cytoplasm and a large nucleus, are also present. They have three main functions, (1) the production of the ground substance, (?) the secretion of heparin (p. 537), and (3) the secretion of histamine, a substance which causes the blood vessels to dilate and which is released when tissues are injured. In many regions of the body, notably just below the skin and in the mesenteries, there is a tendency for fat to be deposited in the loose connective tissue. The fat is laid down in special large **fat cells** which usually occur in groups and swell to an enormous size as the fat accumulates. When the quantity of fat is appreciable, the tissue is called **adipose tissue.**

In addition to its role of packing material, areolar tissue also plays an important part in combating foreign toxins, for it can react locally to give inflammation. The number of cells and blood-supply is then increased and the macrophages in particular become **phagocytes,** *i.e.* engulf the foreign bodies and destroy them.

WHITE FIBROUS TISSUE.—White fibres similar to those in areolar tissue are found in an almost pure state in **tendons.** These inelastic cords are found wherever a strong pull has to be transmitted from one structure to another. Thus muscles are connected to bones by tendons which are inserted into the connective tissue sheath (**periosteum**) surrounding the bone.

YELLOW ELASTIC TISSUE.—Yellow fibres, as a rule a good deal thicker, but otherwise similar, to those found in areolar tissue, contribute to a variable degree to form the **ligaments** of the body. Because of their elastic nature ligaments are to be found in situations where sudden strong stresses are not met with and where some "give" is required. Thus ligaments form the great cords which run down the dorsal side of the neck (**ligamentum nuchae**), and also connect bones together. The term ligament, however, as used in anatomy is a very wide one and in certain ligaments the fibres are mainly collagenous.

## EXTERNAL FEATURES OF THE SELECTED VERTEBRATES

### *THE DOGFISH*

Most free-swimming fishes, including the dogfish, have a stream-lined shape. That is, the anterior end is smoothly rounded and reaches its maximum girth about one-third of the way from the front end, tapering gradually to the end of the tail. A body of this shape moves through a fluid with the minimum of turbulence or eddies, so wasting very little energy. In a dense medium, like water, this is important even at relatively low speeds, whereas in air, stream-lining is unimportant except at high speeds. The head, it will be noticed, merges insensibly into the trunk, there being no neck.

**Head.**—In shape, the head is blunt and widens in all directions backwards from the snout, and since the head is the most anterior

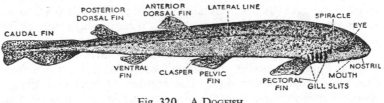

Fig. 320. A DOGFISH.

part of the body and is directed forwards during movement, the organs of special sense are situated in it. Of these, only the eyes and nostrils are visible externally, the eyes at the sides and the nostrils on the ventral surface just anterior to the mouth. The nostrils are connected with the mouth, which is ventral, by grooves, the **oro-nasal grooves**. Ramifying over the surface of the head are a number of **canals** which open on to the surface by pores. This sensory system is part of the acoustico-lateralis system and is concerned with the perception of vibrations in the water. One branch of this system is a long canal which, although originating in the head, passes along the whole length of the body almost to the end of the tail, forming the **lateral line canal**. In addition to the sense organs in these canals, there are others which lie at the bottom of flask-shaped depressions (ampullae) scattered over the surface of the snout and have other functions also.

Just behind each eye is a small aperture, the **spiracle**, which is a modified gill cleft, and at the sides of the body are the five elongated functional **gill clefts**.

**Trunk and Tail.**—The trunk extends from the head to the region of the **cloacal aperture** which is on the ventral side of the body, a little less than half-way along its length. In the trunk the body is firm on the back and dorsolateral surface because of the presence of the thick body muscles around the vertebral column. But on the belly or ventral and ventrolateral surfaces the body wall is soft, since here the muscles are thin, and within is the body cavity in which lie the viscera.

The tail forming the posterior half of the body contains the caudal portion of the vertebral column and spinal cord. Its muscles are very powerful since the tail is the main propulsive organ of the animal.

**Appendages.**—The appendages of the dogfish are fins which are extensions of the body wall in which skeletal structures have been laid down. In the trunk region are to be found the paired fins: the **pectoral fins** anteriorly and the **pelvic fins** posteriorly. The pectoral fins project outwards from the sides of the body just behind the gill region. The pelvic fins lie one on each side of the cloacal aperture closely against the under side of the body. They afford a means whereby the sexes may readily be distinguished. In the male dogfish the posterior angle of the fin is continued into a grooved intromittent organ known as a **clasper.**

Projecting from the tail are the median fins: dorsally an **anterior dorsal fin** at the root of the tail and a **posterior dorsal fin** about half-way along its length. On the ventral side in line with the space between the anterior and posterior dorsal fins is the **ventral (anal) fin.** Around the terminal region of the tail, which is usually upturned, are the two lobes of the **caudal fin**, the upper lobe being smaller than the lower.

## THE FROG

The body of the frog, in common with all the members of its order (Anura), has been shortened by the loss of the tail. But here again, as is to be expected in an animal which swims in water, there is no sharp distinction between head and trunk.

**Head.**—The head has no projecting snout and the wide mouth is terminal. Above the mouth are the **external nostrils**, and the **eyes** are situated almost on the top of the head. The prominence and mobility of the eyes are noticeable features and enable the animal to "cover" a considerable portion of the area around it. At each side of the head behind the eye is to be found a circular, deeply pigmented patch of skin. This is the **tympanic membrane.**

The under side of the throat is soft and in the living animal its movements associated with breathing can be seen.

**Trunk.**—The trunk is short and compact, tapering off in its distal half from the "hump" or sacral prominence. The cloacal aperture is terminal.

**Appendages.**—Like most terrestrial vertebrates, the frog is provided with two pairs of limbs. The fore limbs are short but are divisible into regions by the presence of movable joints. Thus the part by which the limb joins the trunk is the upper arm to which is joined the fore arm, and distal to this again the wrist and hand. When compared with many four-footed animals, the wrist is reduced as is also the number of digits in the hand, there being only four (the first having been lost) in place of the usual five.

Fig. 321.    A FROG.

The hind limbs are similarly divisible into regions; the thigh, shank, ankle, and foot. Both thigh and shank are long, but the ankle is exceptionally long.

## THE RABBIT

The rabbit has the typical mammalian form of body, consisting of head, neck, trunk (which is further subdivided into thorax and abdomen), and tail.

**Head and Neck.**—The mouth is terminal and the upper lip is divided, thus exposing the front (incisor) teeth. The nostrils are immediately above the mouth and the eyes are at the sides of the head. From the sides of the upper lips project the sensitive tactile **vibrissae** (whiskers). The tympanic membrane is no longer superficial as in the frog because in the mammal an outer ear, consisting of a movable **pinna** and an **external auditory meatus,** is present.

The neck is an extension of the body interposed between the head and the trunk. It allows of great mobility of the head in all directions.

**Trunk and Tail.**—The trunk is divisible into **thorax** and **abdomen**. In the thoracic region a bony cage is formed by the ribs and sternum within which lie the important organs, the heart and lungs. Internally the cavity of the thorax is separated from that of the abdomen by a muscular septum, the **diaphragm**. The ventral wall of the abdomen is soft and distensible. At the posterior end of the trunk lie the external apertures of the alimentary canal and renal and reproductive organs, for in this type of mammal (Eutherian) the anus is separated from the urino-genital aperture.

The tail of the rabbit is short and in the wild forms has on the under side a prominent tuft of white hairs to which has been ascribed the function

Fig. 322. A RABBIT.

of acting as a warning signal to others of the community to flee from enemies. Fig. 412 shows some of the more important internal organs and their relations to the rest of the body.

**Appendages.**—The fore limbs are short and can take the shock of alighting at the end of a leap. The upper arm is directed backwards and the fore arm forwards in the resting position, and the wrist and five-clawed hand are permanently in the pronate position. The hind limbs are by far the longer and divided into thigh, shank, ankle, and foot (which has only four claws). When at rest, the thigh is directed forwards, the shank backwards, and the ankle and foot forwards.

## LOCOMOTION

### THE DOGFISH

The dogfish, like other fishes, swims by causing lateral undulations to pass down its flexible body (including the tail) in a tailward direction (Fig. 323). These waves are due to the serial and alternating contraction of the right and left sets of myotome muscles, the vertebral column resisting the tendency of the body to shorten whilst at the same time providing flexibility. Mechanically the

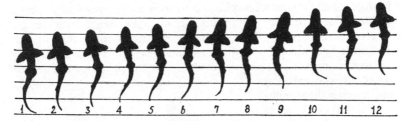

Fig. 323. DOGFISH.—Diagram to show changes in the flexures of the body (at intervals of $\frac{1}{10}$ second) during forward swimming (after Gray).

undulations can be regarded as a series of inclined planes moving laterally in the water so causing it to flow to the rear and at the same time producing a forward thrust (Fig. 324). The tail with its caudal fin is particularly effective in that it has a large surface and oscillates through the widest arc. The other median fins and the

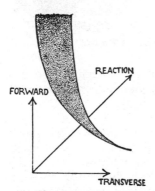

Fig. 324. DOGFISH.—Diagram to show the forces acting on the tail of a fish to cause forward swimming. Horizontal view, tail moving to left.

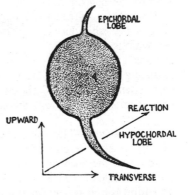

Fig. 325. DOGFISH.—Diagram to show forces acting on a heterocercal tail, causing it to lift.

paired fins also play an important part in swimming. So does the shape of the body whose streamlined shape not only reduces turbulence, but by being so much thicker at its anterior end offers a large resistance to lateral movement and so reduces the tendency for the fish to swing from side to side. The median fins give stability during swimming for they tend to prevent "yawing" (unstable movements about a vertical axis) and also reduce "rolling".

The role of the paired fins is not only to act as planes whose inclination can be varied, so altering the swimming level, but when set at their normal angle to produce a steady upthrust at the front end of the body. This upthrust is offset by another upthrust automatically produced by the tail when it sweeps from side to side. This force tends to raise the tail and to cause a nose-dive. Thus the upward force by the tail and the upthrust on the pectoral fins result in horizontal swimming, an arrangement which is necessary because the dog-fish is heavier than sea water. In this way a heavy fish automatically swims clear of the bottom as soon as it begins to move forward. The tail is enabled to pro-

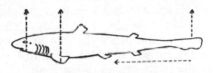

Fig. 326. DOGFISH.—Diagram to show upward and forward thrusts when the fish is swimming.

duce a lift as well as a forward thrust because its lower lobe is larger than its upper one, and because, being flexible it becomes bent as it sweeps sideways (Fig. 325). The degree of flexibility can be varied by special muscles in the lower lobes. If this were not so, a dogfish could not swim horizontally, except at low speeds but would tend to rise as its velocity increased. Such a tail is called **heterocercal** and contrasts with that found in most bony fishes, which, because they have an air-bladder, can adjust their specific gravity until it equals that of the water. In consequence they have symmetrical (or **homocercal**) tails which produce a forward thrust only. But some primitive actinopterygian fishes, of which the sturgeon (*Acipenser*) is one, have a heterocercal tail which is very like that of their fossil ancestors.

## THE FROG

As is well known, a frog has three methods of progression. It can swim, it can walk, and it can leap or hop. In all these activities the limbs act as levers which transmit propulsive forces arising from reactions against the water or the ground. As will be better understood when the skeleton of the limbs has been considered (p. 448), each limb is a series of levers which joint or articulate one with another so that the various parts can be flexed and straightened by the muscles which clothe them. These muscles (p. 465) are arranged in opposing sets, the contraction of one set (extensor muscles) causing the limb to straighten whilst the other set (flexor muscles) causes it to bend. There are, in addition, muscles running from the limb to the girdle (p. 450), which move the limb forwards (protractor muscles) and others which swing it backwards (retractor muscles).

The fore limbs play little part in swimming where the propulsive force is obtained from the long and powerful hind limbs. The large foot with its webbed digits sweeps through the water so providing a moving inclined plane whose action is not greatly different from that of the tail of a fish.

When walking, the propulsive thrust is due to the reaction of each foot against the ground, each limb being lifted, flexed, swung forward, and then planted on the ground again. As a foot reaches the ground the limb begins to straighten so that it carries a proportion of the weight of the body. Then the retractor muscles come into play and since the foot grips the ground firmly, the body is swung forwards, additional thrust being given by a continuation of the straightening of the limb. Now it is clear that if the body is not to topple over, only one leg at a time can be lifted from the ground as a frog walks along. Moreover, the centre of gravity of the body must lie above a triangle made by the three feet resting on the ground. It is to satisfy these conditions that the limbs are lifted in a definite sequence which is the same for all four-footed animals when they are walking. This stepping sequence is said to follow a diagonal pattern by which is meant that the limb to be lifted after the right fore foot is the left hind foot; next the left fore foot is lifted and this is followed by the lifting of the right hind foot, after which the cycle is repeated. The timing of the movement is such that there are always three feet on the ground and one clear of it. A frog can cover ground only rather slowly when it is walking. For fast progression it hops or leaps and if its leaps are compared with its size it must be rated as a very good jumper indeed. This is because it has powerful hind limbs whose system of long levers makes possible the application of an upward force over some considerable time so giving the body a good acceleration. This upward acceleration is produced by the straightening of the fully flexed legs and, of course, takes place before the feet have left the ground. Then the feet, so to speak, follow the body in its upward flight and the legs are again flexed, whilst the frog is still in the air. At the same time the fore limbs are straightened so as to take the shock of landing.

## THE RABBIT

A rabbit can hop and leap, and can cover the ground very fast when the occasion arises. It never runs by using its limbs for "striding" like a dog, for example, and only very rarely seems to walk. It is much more fully suited to move on land than is the frog, and its limbs have become adapted mainly for this purpose, although its fore limbs are used also for digging its burrow. As in other mammals its limbs are, when compared with the primitive position (p. 451), brought more under the body so that they form nearly vertical levers so making unnecessary many of the muscles required to keep the belly clear of the ground, the thrust due to the weight of the animal being taken more nearly along the axis of each limb.

The principles underlying walking do not differ materially from those outlined for the frog, but when a mammal runs, different considerations apply. Then, stability of the body is sacrificed for speed. As the speed of walking increases each foot is lifted clear of the ground before the previous one is planted down. This can be seen when a cat or a dog runs, but rabbits do not run like this. Indeed, when a rabbit is "flat out" only two feet are on the ground at a given instant, so that the body is in unstable equilibrium. An increase in speed is also obtained by the use of the back muscles which cause the back to arch and so lift all four feet off the ground at the same time. The animal thus bounds along, both fore limbs thrusting backwards at the same time, whilst both hind limbs are pointing forwards. This enables the hind feet to hit the ground at a point in front of that which the forelimbs have just left.

## THE SKIN

### INTRODUCTION

The bodies of all vertebrates are invested by an outer covering called the **skin**. This is attached to the underlying body muscles by connective tissue (see p. 397) but is frequently partially movable over the muscles.

The skin is composed of two distinct layers of different embryological origin; externally is found the **epidermis** (ectodermal in origin) and beneath this the **dermis** (mesodermal), each layer having a composite structure.

The epidermis is many cells thick and is made up of a stratified epithelium, the cells becoming progressively flattened the nearer the surface is approached. The basal layer of the epidermis is composed of cuboid cells lying on a basement membrane and constitutes the Malpighian layer from which all the epidermal cells are derived. Passing outwards from the Malpighian layer, the cells are found to become more and more flattened until when the surface is reached the cells are flat, easily detached, and practically dead with only faintly staining nuclei (p. 396).

The dermis is a composite layer including muscle-fibres, connective tissue, capillary blood vessels, nerves, and frequently, between it and the underlying muscles, fat.

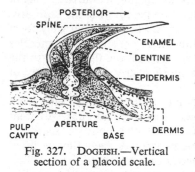

POSTERIOR→
SPINE
ENAMEL
DENTINE
EPIDERMIS
PULP CAVITY
APERTURE
BASE
DERMIS

Fig. 327. DOGFISH.—Vertical section of a placoid scale.

Although this general structure is common to the skin of all vertebrates, important differences in detail occur in different types. These are most obviously seen in the various types of skin derivatives —scales, glands, feathers, hairs, and so on. Receptor organs of various kinds are always found in the skin. Paradoxically the skin serves not only to isolate the body from its surroundings but also to keep it in touch with the environment. It also plays an important part in retaining the shape of the animal or, indeed, by its "subcutaneous" layers, in modifying the shape.

## THE DOGFISH

It is a characteristic feature of most fishes that **scales** of one sort or another are present in the skin. In some instances, particularly in primitive bony fishes, the body is literally encased in an armour of thick, bony scales (e.g. *Lepisosteus*), but in most modern bony fishes (teleosts) the scales are thin, overlapping, nearly circular plates of bone which are easily rubbed off so that at best their protective function is slight. Some teleosts, *e.g.* the eel, have only microscopic scales.

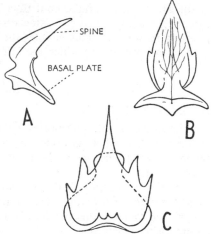

SPINE
BASAL PLATE
A
B
C

Fig. 328. DOGFISH.—Isolated dermal denticles. A, from the dorsal surface, side view; B, the same in face view; C, a tooth, face view.

The skin of the dogfish is quite firm to the touch and if stroked with the finger from head to tail, appears to be smooth. If, however, the finger is moved in the opposite direction, it meets with considerable obstruction and the skin seems rough. This roughness is due to the scales. Each of these scales (which are called **placoid**

**scales** or **dermal denticles**) consists of a four-lobed or rectangular basal plate from which arises a pointed, backwardly directed spine. The base is embedded in the dermis and the spine projects through the epidermis beyond the surface. In vertical section, each spine is seen to be covered on the outside with a hard, transparent, shiny layer, the **enamel** within which is the **dentine** through which ramify numerous fine canals (**canaliculi**). In contrast to enamel, dentine has a considerable basis of organic material and, therefore, is not so fully soluble in mineral acids. The centre of the spine is occupied by a cavity—the **pulp cavity**—in which lie blood vessels, nerves, and the dentine-forming cells (**odontoblasts**), the cytoplasmic prolongations of which are continued into the fine canals of the dentine. The cavity in the spine is continued into and through the base which

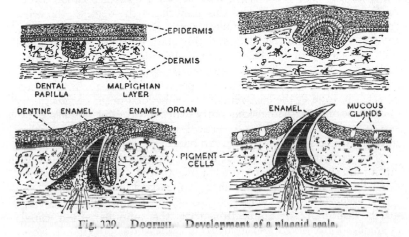

Fig. 329.  Diagram.  Development of a placoid scale.

is thus perforated by a small aperture to admit the blood vessels and nerves. The base itself is claimed by some authorities to be composed of bone, but whilst this may or may not be so, it does not show the usual Haversian systems (see p. 461) but consists of a kind of loose trabecular calcified material.

The form of the spine varies in different parts of the body. Scales from the ventral surface have spines with one point and from the dorsal surface, three points. The scales in the skin which covers the lips bounding the mouth and which is continued into the buccal cavity over the jaws are larger and stronger. They have a trilobed base and a five-pointed spine and in this situation function as **teeth**. In this connection it is interesting to note the similarity in structure and development between the dermal denticles and the teeth of higher vertebrates (see p. 442).

The first sign of a developing scale is a condensation of mesenchyme cells in the dermis to form a **dental** or **dermal papilla**. This becomes capped by a cone-like down-growth of the epidermis, the layer next to the papilla forming a single layer of columnar cells termed the **enamel organ**. The outermost cells of the papilla form collagen fibres which are the organic basis of the scale. Details of the way in which this organic basis becomes calcified are not clear but it seems that calcification begins at the top and gradually extends down the sides, between the enamel organ and the outermost layers of the dental papilla. It is probable that this material, which is the enamel layer, is formed by the cells of the papilla but that the enamel organ in some way is essential for its deposition. In mammals, on the other hand, the enamel organ actually secretes the enamel. The scale is then thickened by further calcification on the inside of the cone and this material is the dentine, the cells secreting it being called **odontoblasts**. But a cavity, the pulp cavity, is left within the scale and this communicates by a channel with the cavity in the basal plate which also gradually becomes calcified. As the scale grows in size its spine pushes up through the epidermis and as the latter's superficial layers are worn away emerges beyond it, the base remaining embedded in the dermis.

The scales are being constantly worn away and equally constantly replaced, so that in a vertical section of the skin denticles in various stages of development can be seen.

The dermal denticles are set closely together in the skin and so strong are the "enamel"-covered spines that the dried skin of some dogfish and sharks, known as shagreen, is used for polishing.

At intervals in the epidermis are to be found **mucus-secreting glands**, especially in connection with the ampullary and canal systems.

In the dermis, just beneath the Malpighian layer of the epidermis, numerous **pigment cells—chromatophores**—are found which contain a black pigment, and it is partly from the varied distribution of these chromatophores that the variation in the colour of the skin arises. They are much more numerous on the dorsal surface than on the ventral, so that the back of the animal is darker than the belly. The remainder of the dermis is largely muscle and connective tissue.

Also among the derivatives of the skin may be considered the **horny fibres—dermotrichia**—which form the flexible margins of the fins.

## THE FROG

The frog's mode of life is very different from that of the dogfish. The animal does not live in deep water and spends much of its time on land. The skin, besides acting as a protective outer covering, also serves (for reasons which will be apparent later) as a respiratory surface, and in order to fulfil this function efficiently must be continually moist even when the animal is out of the water. To maintain this condition the skin is provided with numerous mucus-secreting glands. Each gland is flask-shaped and is formed by an inpushing of the epidermis deeply into the dermis, the narrow "neck" of the flask opening on to the surface. The globular portion of the gland is lined by cubical or columnar glandular cells continuous with the Malpighian layer. The watery mucous secretion

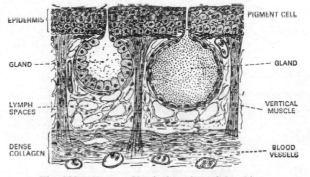

Fig. 330. FROG.—Vertical section of the skin.

is continually poured out on to the surface of the skin which is consequently moist and slimy. An additional advantage arises from this condition, for the slippery nature of the skin makes the animal difficult to hold. Some of the glands may produce a distasteful and possibly poisonous substance which serves as a further protection against capture by enemies. This is certainly true of the toad.

The outer layer of the epidermis is made up of flattened cells and this may be sloughed off from time to time as a "cuticle". In the dermis, **pigment cells** are present by the expansion and contraction of which alterations in colour result. The remainder of the layer is composed of muscle and connective tissue.

The skin is attached to the underlying muscles only at intervals so that it is extremely loose, the subcutaneous spaces being **lymph spaces**.

In the breeding season the male frogs develop a deeply pigmented cutaneous thickening of the "ball" of each thumb, forming what are termed **nuptial pads** which are used for grasping the female during the deposition of the eggs and their fertilisation. Also, the skin becomes thickened over the tips of the digits to form a kind of claw.

## THE RABBIT

The environmental conditions under which the mammals live are so very different from those of the fish and frog that it is not

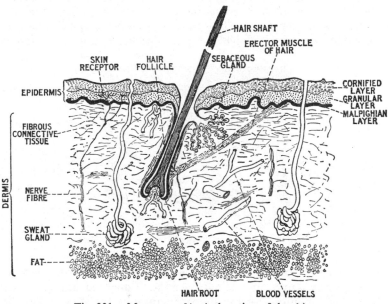

Fig. 331. MAMMAL.—Vertical section of the skin.

surprising to find that the structure of the skin is entirely different. Except for those parts of the body which are in contact with the substratum, there is very little friction between the external surface and the air, and therefore the development of protective structures such as scales has given way to other more important needs. Also the skin is not used as a respiratory surface, so that the necessity for mucous glands no longer arises. But the mammals are "warm-blooded" animals; that is, their metabolic activities are such that their body temperature is usually higher than that of their surroundings—a condition which is known as **homoiothermic** in contrast to the **poikilothermic** or "cold-blooded" animals—and moreover, the body temperature is, within narrow limits, maintained constant. In the mammals, therefore, the skin plays an important part in the regulation of body temperature. All mammals have **hairs** present in their skin, which, by entangling air among them, form an insulating layer and thus limit the loss of heat from the body and consequent lowering of temperature. Imbedded in the dermis are coiled, tubular glands, the **sweat** or **sudorific glands**. They produce a watery secretion which is poured on the surface of the skin through narrow ducts and there evaporates, affording a means by which the temperature of the body can be lowered in a controlled way, for although the secretion of sweat and its evaporation from the skin is a continuous process, the amount produced varies according to the

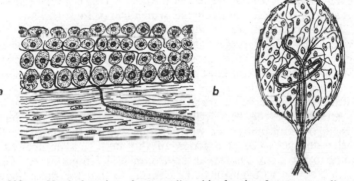

Fig. 332. *a*, Vertical section of mammalian skin showing free nerve endings in the epidermis. *b*, An encapsulated nerve ending.

rate of heat production by the body. Normally, droplets do not appear on the skin and only do so when excessive amounts of heat—as in exercise—are produced. The distribution of sweat glands differs in different mammals. In man and the rabbit, for example, they are found everywhere in the skin. In the dog they are absent from the general surface, and the areas where the main heat loss takes place are the nose, tongue, and buccal cavity.

The mammalian skin, therefore, is a relatively complex structure; the two main layers, epidermis and dermis, are still evident, but the epidermis is more complex. The outermost layer, or **stratum corneum**, is composed of hardened flattened cells, and in places where there is considerable friction this layer may become very thick. Beneath this is the **stratum granulosum**, the deepest layers of which form the familiar Malpighian layer. The **dermis** is an intricate arrangement of muscles, connective tissue, blood vessels, nerves, and in the deepest portions, fat cells.

The hairs are mainly epidermal structures but each has at its base a conical concentration of vascular dermal tissue which forms the hair papilla which projects into the base of the hair. The shaft of the hair is a slender cylindrical structure firmly embedded in a pit-like invagination of the epidermis called a hair follicle, the base of the shaft being continuous with the wall of the follicle. The shaft of the hair, which of course projects beyond the surface, is made up of an outer cornified layer, the cortex, the outer surface of which may be modified to form a thin cuticle. The central part of the shaft constitutes the medulla or pith in which air is frequently present. Into the follicle open the sebaceous glands—really glandular outpushings of the follicular wall—and these secrete an oily substance on to the hairs which keeps them supple and prevents them becoming wetted with water. This is a very necessary precaution, especially in mammals which live or hunt in water.

The hairs are movable, and while they normally rest with the greater part of the projecting shaft almost parallel with the skin surface, contraction of the muscles in the dermis will bring the shaft at an angle to the skin, thereby increasing the effective thickness of the air-layer entangled by the hairs. This is a natural reaction to a low atmospheric temperature so as to reduce the amount of heat loss.

Attached to the hair follicle is a special muscle, the erector muscle, the contraction of which brings the shaft almost vertical to the surface. The contraction of this muscle is controlled by stimuli other than changes in environmental temperature. Often it is a reaction to fright or acute danger, so that the hairs literally "stand on end", so making the animal appear larger.

Receptor structures in the skin (Fig. 332) consist of (a) free nerve endings, often forming complex network between the cells of the epidermis and also in the dermis, some of which run along the walls of the small blood vessels, whilst others supply the hairs and sweat glands, and (b) encapsulated nerve endings in the dermis. The encapsulated nerve endings form a plexus among the surrounding cells and the general form of these little sense organs is rather variable. The modern view is that they cannot be classified into morphologically distinct types nor can the varieties be related to particular sensations such as heat, cold, pain, or touch.

Although the above description covers any typical portion of the skin, in certain regions modifications may occur. In the region around the mouth the hairs may become very strong and long, forming the **vibrissae,** and variations in the colour and character of the hairs give to the animal its characteristic markings. In regions where the skin is subjected to considerable friction, such as the under surface of the feet, the stratum corneum becomes extremely thickened, forming the familiary pads of animals and in human beings, callosities. At the tips of the digits strong **claws** or **nails** may be present and in some mammals the digits terminate in **hooves.** All of these are formed of especially hard kinds of keratin.

Another characteristic of the mammals, that from which they derive their name, are the **mammary glands.** These glands have been derived from sebaceous glands which have been highly concentrated and enlarged. They secrete the fluid milk upon which the young mammal lives for some time after it has entered the world.

From the above it will be seen that although the skin has the same basic structure in all vertebrates, it reflects the conditions

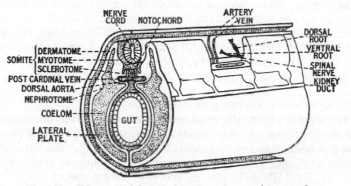

Fig. 333. Diagram of the trunk region of a vertebrate to show metamerism.

under which the animal lives and also that cutaneous derivatives are characteristic of the main group to which the animal belongs. In general terms, the skin of a fish is scaly, that of an amphibian soft and moist, while that of a mammal is hairy.

## METAMERISM

In many adult vertebrates there is very little external indication of metameric segmentation, yet in dealing with many of its important systems it frequently happens that a full understanding of the anatomy involves some knowledge of the extent to which metamerism is present.

A definite segmentation is recognisable in the embryo extending throughout the greater part of the body. It first becomes evident during the differentiation of the mesoderm because of the development of a linear series of paired **mesodermal somites** (see Fig. 333) extending backwards from the region of the hind-brain and throughout the trunk.

At first the metamerism is confined to the mesoderm, but soon it becomes imposed upon the peripheral nervous system and the blood vessels.

Each mesodermal somite in the trunk region has three main parts. The **myotome** gives rise to main body muscles; another, the **sclerotome**, to the vertebrae; a third, the **dermatome**, contributes to the dermis. Lying just at the base of each somite is the **nephrotome** which gives rise to kidney tubules. The mesoderm lying laterally to the nephrotomes is not divided into segments and forms the **lateral plate mesoderm**. It gives rise to the **somatic** and **splanchnic** layers (see p. 746) bounding the splanchnocoel. In the trunk region, therefore, the mesodermal contributions to a body segment are the somite, the nephrotome, and a portion of the lateral plate mesoderm. As these structures are differentiated, nervous and blood supplies are developed and it follows naturally, therefore, that these, too, must have a metameric arrangement imposed upon them. For example, every segmental muscle must have two blood vessels and a motor nerve in connection with it, and in relation with each motor nerve a sensory nerve is present.

In this way, then, the arrangement of the muscles of the back and the blood vessels, the vertebral column, and the spinal nerves have a definite metameric segmentation, but since all these structures are internal, the adult animal shows little external evidence of segmentation. Even the internal metamerism may become obscured by the condensation and telescoping of parts. For example, the main nerve to a limb arises from a collection of nerves called a plexus, an indication that the limb bud from which the limb developed arose from several trunk segments. Again, the adult kidney appears as a compact structure, but developmentally, it arises as a series of metamerically arranged renal tubules.

In the head region, owing to the profound changes which have occurred in connection with the elaboration of the brain, jaws, sense organs, etc., the metamerism is largely obscured in the adult.

The visceral clefts give an appearance of a metameric series but, as will be remembered, they develop in the gut and lateral plate region, well below the somites and are not primarily related to them. Nevertheless, each segmental dorsal root nerve in the head forks round each of the visceral clefts so that secondarily they become part of the segmental plan of the head.

## THE SKELETON

### INTRODUCTION

Most of the skeleton lies within the principal body muscles and is called an **endoskeleton**; it provides a system of rigid levers to which muscles can be attached. This must be regarded as its chief function. During the phylogenetic history of the vertebrates there has been a strong tendency for the exoskeleton of the primitive forms to become reduced, though some of the dermal bones in the anterior region have become incorporated with the endoskeleton to form a protective cranium around the brain and sense organs. Secondarily, too, the endoskeleton may take on a protective function by being developed around such organs as the heart, lungs, etc. (girdles and ribs).

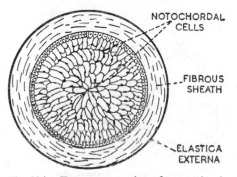

Fig. 334.  Transverse section of a notochord.

Some of the lower vertebrates have a skeleton composed entirely of cartilage, though parts of it may be made more rigid by the deposition in it of calcareous salts. In the higher vertebrates, although a great part of the skeleton appears in the embryo as cartilage, this in the adult is largely replaced and added to by bone.

For convenience the skeleton may be considered in two parts. First, that which lies in the long axis of the body—the **axial skeleton** —and second, that which is associated with the appendages (fins or limbs) and forms the **appendicular skeleton.**

### AXIAL SKELETON

This in the Craniata consists of the **skull** and **vertebral column** of which it is perhaps best to consider the vertebral column first.

**Vertebral Column and Notochord.**—In all vertebrates the vertebral column is preceded in the embryo by the **notochord,** and in the adult consists of a series of metamerically arranged components termed vertebrae. Now, although the vertebral column is always preceded by the notochord, it is never formed from it but merely around it, and in the process of its formation the notochord is either wholly or in part obliterated.

The notochord is composed of vacuolated cells around which is secreted the **notochordal sheath**. This is composed of two layers, the outer **elastica externa**, internal to which is the thicker **fibrous sheath** formed of collagen. The turgor pressure of the vacuolated cells acts against the sheath, and the notochord thus forms a stiff but flexible axial rod along the length of the animal from the region of the infundibulum to the tail.

The vertebrae arise from mesenchymal cells of the sclerotomes which migrate up and around the notochord and spinal cord. Each

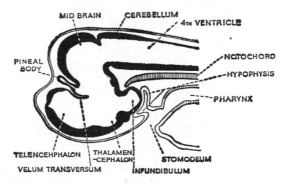

Fig. 335. Longitudinal section of the head of a dogfish embryo.

vertebra occupies an intersegmental position with respect to the muscles derived from the sclerotomes.

In the adult vertebral column in all but a few rare instances each component vertebra is made up of a **centrum** (which develops in the position occupied by the notochord), a **neural arch** around the spinal cord, and various **processes** (e.g. transverse processes) arising from the vertebra. The centrum may be formed by cells from the skeletogenous layer penetrating through the notochordal sheath and invading the notochord; or it may be formed by downgrowths from the neural arch region surrounding the notochord and eventually obliterating it. When each centrum is formed by an invasion of the notochord by cartilage-forming cells (chondroblasts), so that the notochord is replaced by the centrum, it is called **a chordal centrum,** but when the notochord is merely obliterated by being squeezed out and surrounded by down-growths from the neural arch, it is termed an **arch** or **perichordal centrum**.

The processes from the vertebrae are always developed in the connective tissue layers which separate and enclose the great body muscles, and serve for their attachment. Thus, their arrangement varies with the changes in the form of the body muscles. The **transverse processes**, however, usually the most constant and obvious of these processes, lie in the horizontal septum which separates the dorsal from the ventral portion of the myotomal muscles. Frequently the distal portion of the transverse process may become separate from, and articulate with, the process. This separate portion is then called a **rib**.

## *THE DOGFISH*

The vertebral column of the adult dogfish is composed of a series of separate vertebrae, but presents the appearance of an intricate series of cartilaginous plates bound together by fibrous connective tissue. To understand this arrangement satisfactorily it is necessary to refer to its development.

In the practically continuous skeletogenous layer around the notochord and spinal cord, in each body segment (as indicated by the original somite), condensations of cartilage-producing cells, **chondroblasts**, appear, which eventually lay down four pairs of cartilaginous plates arranged dorsal and ventral to the notochord. In each segment the condensations in the posterior part are larger than the anterior ones, and are termed the **basi-dorsals** and **basi-ventrals** respectively. The smaller anterior ones which usually do not reach the elastica externa of the notochord are the **inter-dorsals** and **inter-ventrals**. In the mid-dorsal line, in the spaces between the basi-dorsals and inter-dorsals, two condensations per segment are formed, producing the **supra-dorsals**, which are median and not paired as are the other vertebral elements.

Then the chondroblasts penetrate the notochordal sheaths and invade the notochord itself, later forming a cartilaginous centrum in continuity with the basi-dorsals and basi-ventrals, mainly in an intersegmental position. As the basi-dorsals increase in size they grow up on each side of the spinal cord, forming the sides of the neural arch, the keystone of which is formed by median blocks of cartilage developed from the supra-dorsals.

The basi-ventrals also grow out into the transverse septum between the upper and lower parts of the myotome, but the inter-ventrals become vestigial.

The depth of penetration of the notochord by the chondroblasts during the formation of the centrum is not uniform, the greatest amount of chondrification occurring in line with the middle of the basi-dorsals and basi-ventrals with the result that the centrum is hour-glass shaped with a conical depression on both its anterior and posterior faces and is consequently termed **amphicoelous.**

If the adult vertebral column is now considered, it will be seen that each biconcave centrum is surmounted by a neural arch made up of a **vertebral neural plate** (basi-dorsal) on each side and that the space between consecutive vertebral neural plates is occupied by the **intervertebral neural plate** (inter-dorsal), the arch being completed by a series (two per vertebra) of "neural spines" (supra-dorsals). Projecting from each side of the lower part of the centrum in the

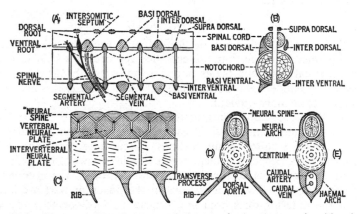

Fig. 336. Dogfish.—Development of the vertebral column. A, side view of an early stage; B, sections, on the left through basi-dorsal and basi-ventral; on the right through inter-dorsal and inter-ventral; C, side view, adult, trunk region; D, end view, trunk region; E, end view, tail region.

trunk region is a **transverse process** (basi-ventral), the distal portion of which forms the slender **rib** lying in the transverse septum. Each vertebra is thus clearly a compound structure. Between the intervertebral and vertebral neural plates are the separate apertures for the dorsal and ventral roots of the spinal nerves.

Externally, very little of the notochord is seen, the only remnants being the fibrous rings or pads (developed from the notochordal sheath) between consecutive centra. Internally, however, as will be seen if a sagittal (median vertical) section of the column is examined, a considerable portion of the notochord persists as a jelly-like material filling the lens-shaped inter-central spaces. These spaces even communicate with one another by a small canal (through the centre of the centrum) left by the failure of the chondrification to penetrate completely to the centre of the notochord. Thus the column as a whole forms a flexible structure embedded in the myotomal muscles, which allows of the side-to-side bending causing locomotion, but which resists shortening of the body when the muscles contract. The vertebrae are not properly jointed one with another and the spinal cord is enclosed in a continuous canal formed from the neural arches because of the presence of the intervertebral neural plates.

Two departures from the condition in the trunk region are found in the tail. Here, the transverse processes, instead of projecting laterally, are bent inwards beneath the centrum and meet and fuse in the mid-ventral line to form an arch—the **haemal arch**—and ribs, as such, are therefore absent. Within this arch are found the caudal artery and vein. Also, in this region there are two vertebrae per body segment, a condition known as diplospondyly.

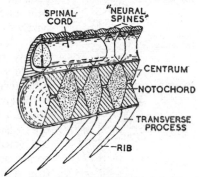

Fig. 337. DOGFISH.—Sagittal half of vertebral column, trunk region.

It is a characteristic feature of the group of fishes (Sclachii) to which the dogfish belongs, that the skeleton remains cartilaginous in the adult. Other fishes have a bony skeleton in the adult, but even when the vertebrae are thus ossified, their centra are bound together by fibrous material representing the remains of the notochord although articulating facets (zygapophyses) of a rather simple kind are present on the neural arches.

### THE FROG

The frog's vertebral column presents a very different picture from that of the dogfish. The vertebrae are solid bony structures and only limited movement between them is possible, either in a lateral or a vertical plane. Nor is much flexibility required, for locomotion depends solely on the movements of the limbs and not on flexures of the body as it does in fishes, tailed amphibia, and most reptiles. But the vertebral column has to carry most of the weight of the body and also to withstand thrusts from the limbs. The muscles to which the vertebrae are attached (mainly by their transverse processes) have as their chief function the keeping of the bones in position; instead of being *bound* together the vertebrae *articulate* with one another by means of articular surfaces and processes which are characteristic features of all the vertebrae and which are so arranged that "sagging", due to the weight of the body, is limited. Further, the relationships of the vertebral column

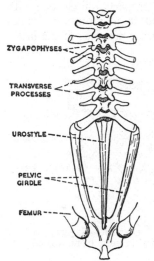

ZYGAPOPHYSES

TRANSVERSE PROCESSES

UROSTYLE

PELVIC GIRDLE

FEMUR

Fig. 338. FROG.—Vertebral column, dorsal view.

to other parts of the skeleton involve modifications in some of the vertebrae for particular purposes. A feature peculiar to the frog and its allies is the shortening of the vertebral column, partly due to the complete absence of a tail. The vertebral column consists merely of nine vertebrae at the hinder end of which is found a slender shaft of bone termed a **urostyle**.

There is no point in going into the details of the development of the vertebrae. Suffice it to say that each vertebra is developed from the same fundamental condensations as described previously (pp. 418-21), but that the centra are perichordal centra and the inter-dorsals and inter-ventrals become incorporated into the centrum.

If a vertebra from the middle of the column is examined, it is found to be composed entirely of bone and to consist of a cylindrical **centrum** surmounted by a **neural arch** from *the base of which* (in contrast to the condition in the dogfish) two laterally directed **transverse processes** project. The centrum is concave in front and convex behind (a condition which is known as **procoelous**). This arrangement provides an efficient ball and socket joint between the centra so that the column could be flexed in any direction. Were it not for articular processes arising from the neural arch, which, although allowing a little freedom of movement in the vertical direction and more laterally, prevent the vertebrae rotating

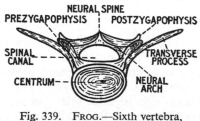

NEURAL SPINE

PREZYGAPOPHYSIS | POSTZYGAPOPHYSIS

SPINAL CANAL

TRANSVERSE PROCESS

CENTRUM

NEURAL ARCH

Fig. 339. FROG.—Sixth vertebra, anterior view.

completely upon one another, the vertebrae might be displaced. These processes are the **zygapophyses**, of which there is a pair both anteriorly and posteriorly. The anterior pair fit beneath the post-zygapophyses of the preceding vertebra and the posterior pair above the pre-zygapophyses of the succeeding one. The articular surfaces of the pre-zygapophyses consequently face upwards and those of the post-zygapophyses downwards. The neural arches show no individual components since complete fusion between the parts has taken place and each terminates in a pointed neural spine—a point of contrast with that of the dogfish. Between consecutive neural arches are spaces, intervertebral notches through which the spinal nerves issue. The transverse processes project out laterally from the base of the arch but have no ribs in connection with them. The above description may be taken as that of a typical bony vertebra.

Certain of the vertebrae show departures from the condition just described. The most modified is the first, which, because it carries the skull, is called the **atlas vertebra**. It consists of a ring of bone comprised mainly of neural arch, the centrum being very much reduced. On its anterior face there are no zygapophyses,

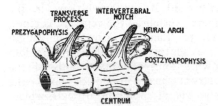

Fig. 340. FROG.—Fifth and sixth vertebrae from the left side.

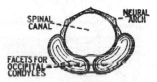

Fig. 341. FROG.—Atlas vertebra, anterior view.

but, instead, paired concave depressions into which the knob-like condyles of the occipital region of the skull fit. Posteriorly there are the normal post-zygapophyses. This arrangement permits the head to be moved (within limits determined by the muscle and other attachments) on the vertebral column.

The other region of modification is the posterior end. The centrum of the eighth vertebra is concave on both its surfaces, *i.e.* it is amphicoelous. Naturally the anterior surface of the centrum of the ninth vertebra is convex and on its posterior face are two rounded knobs or prominences for the articulation of the urostyle. It may therefore be termed acoelous. On this ninth or **sacral vertebra**, the transverse processes are stouter than in all the others and to their extremities the ends of the long ilia of the pelvic girdle are firmly held by ligaments. This provides a flexible girder system to which the thrusts brought about by the movements of the hind limbs during leaping and swimming are transmitted.

The terminal part of the vertebral column consists of a single shaft of bone, the **urostyle**, which probably represents a number of fused vertebrae and articulates with the sacral vertebra by two concave depressions on its anterior face. The shaft of the urostyle lies in the space between the two ilia of the pelvic girdle.

## THE RABBIT

In the mammals in general the number of vertebrae in each region of the body is fairly constant. Thus the column consists of **cervical** (seven vertebrae), **thoracic** (twelve or thirteen), **lumbar** (seven), **sacral** (three or four), and **caudal** (variable) regions. In each region the vertebrae are all more or less alike and all types can be looked on as modifications of a generalised type of vertebra. Vertebrae are less modified than in others in the **lumbar** region.

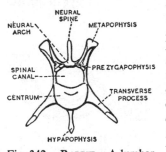
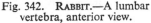

Fig. 342. RABBIT.—A lumbar vertebra, anterior view.

Here the vertebrae have attached to them the powerful muscles of the back. As would be expected then, in the lumbar vertebrae the centrum is stout, and to facilitate growth in length three centres of ossification are laid down. Thus the main body of the centrum has a plate of bone, the **epiphysis**, at each end, beneath which, in the young animal, cartilage still persists, but which, in the fully grown animal, fuses completely, so that even a suture between the epiphysis and the body of the centrum is difficult to make out. A feature which is characteristic of all mammalian vertebrae is that the articular surfaces of the epiphyses are flat, so that there are no ball and socket joints between them as exist in the frog. So, to bind the vertebrae together and to act as cushions there is between one centrum and the next a pad of fibro-cartilage—the intervertebral disc—in the centre of which is a soft area—the nucleus pulposus—which represents the sole remnant of the embryonic notochord. This arrangement permits of a considerable degree of flexibility in the column.

The neural arch arises from the centrum, is surmounted by a prominent neural spine, and encloses the spinal canal in which lies the spinal cord. Arising from the arch are the pre- and post-zygapophyses, the articular surfaces of the former facing upwards and inwards and those of the latter outwards and downwards. Although the intervertebral discs allow of movement of one vertebra relative to the next the presence of the zygapophyses, while still permitting a certain amount of movement between them, limits flexibility. From the sides of the vertebra project the **transverse**

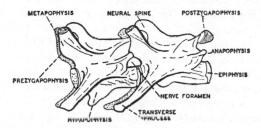

Fig. 343. RABBIT.—Two lumbar vetebrae from the side.

**processes**, which are stout, long, flattened, and directed obliquely downwards and forwards.

In addition to transverse processes and neural spines other bony projections are present for the attachment of muscles. At the anterior end, rising well above the pre-zygapophyses, are a pair of stout **metapophyses**, and below the post-zygapophyses, the less prominent **anapophyses**. On the first two or three lumbar vertebrae a ventrally directed projection, the **hypapophysis**, on the under side of the centrum, is present. In the other lumbar vertebrae this is merely represented by a ridge.

In the **thoracic** region, each vertebra itself has few points of special interest apart from the modifications to provide for the

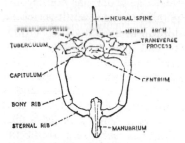

Fig. 344. RABBIT.—First thoracic vertebra and ribs from the front.

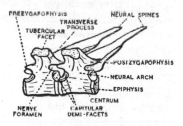

Fig. 345. RABBIT.—Third and fourth thoracic vertebrae from the side.

articulation of the ribs whose movements are essential for breathing. Each of the anterior ribs consists of a slender curved shaft of bone terminating dorsally in two articular projections so that it has the appearance of a **Y**, one limb of the fork being shorter than the other.

The shorter projection—the **tuberculum**—articulates with the short, stout transverse process of the vertebra which has on its ventral side an articular facet. The longer projection—the **capitulum**—articulates with the centrum, but the place of articulation is where the centrum almost touches the one in front of it. At this point in both centra there is a concave depression termed a **demi-facet**, so that, when the vertebrae are fitted together as in life, the two demi-facets form a hemispherical depression into which the rounded head of the capitulum fits. The last three ribs have reduced tubercula and their articulation is solely with the centrum. Consequently the transverse processes in these vertebrae are reduced. Also the articular facets lie completely on the body of the centrum and not at the juxtaposition of two centra. Thus, in the thoracic region the

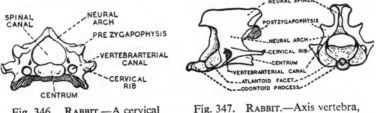

Fig. 346. RABBIT.—A cervical vertebra, anterior view.

Fig. 347. RABBIT.—Axis vertebra, side and front views.

vertebrae have been modified in order to provide articulations for the ribs.

The vertebrae of the **cervical** region present many peculiarities, to understand which it is best to consider one which has not become very specially modified. If any of the third to the seventh vertebrae are examined it will be seen that in addition to the flattened **centrum**, and large **neural arch** with its **pre-** and **post-zygapophyses**, the vertebra has a canal—the **vertebrarterial canal**—on each side of the centrum, apparently running through the length of the vertebra. This canal is a characteristic feature of the cervical vertebrae and is formed by the fusion of a **cervical rib** with the vertebra, so that what is commonly called a cervical *vertebra* is really a vertebra plus a pair of very short ribs. From what has been said about the relationships of the vertebrae and the ribs in the thoracic region, and a glance at Fig. 344, it will be clear that even in that region a space exists between the rib and the vertebra bounded by the forked proximal end of the rib, the transverse process, and the lateral surface of the centrum. If it is now imagined that the shaft of the rib has disappeared and that the tuberculum and capitulum have completely fused on to the vertebra then the condition in a cervical vertebra is arrived at. Through the vertebrarterial canal thus formed pass the vertebral blood vessels of the neck. The size and shape of the cervical ribs vary in the different vertebrae. To them are attached the dorsal muscles of the neck.

The second or **axis vertebra** has special features which provide a mechanism by which the head can be rotated on the vertebral column. When viewed from the posterior aspect the axis vertebra has all the features already enumerated except that the neural spine is ridge-like, being flattened laterally, and extends much further forwards than is usual in the other neck vertebrae; also the cervical ribs are not prominent. Anteriorly, however, much alteration is apparent. There are no pre-zygapophyses and projecting from the front of the centrum is a pointed, conical, peg-like projection called the **odontoid process**, which developmentally is really the greater part of the centrum of the first or atlas vertebra and which has become fused on to the centrum of the second. This peg, in life, lies in its correct positional rela-

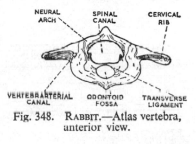

Fig. 348. RABBIT.—Atlas vertebra, anterior view.

tionship to the neural arch of the first vertebra but is in no way fastened to it so that the atlas vertebra can rotate freely on it and since the atlas vertebra carries the skull, the head, too, rotates. To facilitate this movement there are present on the centrum, on each side of the odontoid peg, rounded articular surfaces on which appropriately placed facets of the atlas vertebra can slide.

The first or **atlas vertebra**, as is that in the frog, is specially modified to provide for articulation with the skull. The vertebra seems to consist of neural arch practically alone with a reduced neural spine since the main part of the centrum has been appropriated by the axis vertebra; so that it is composed merely of a ring of bone. Projecting from the sides, however, are the much enlarged flattened cervical ribs with the usual vertebrarterial canals. On the anterior face of the atlas there are no pre-zygapophyses, but the bone is hollowed out to form two shallow concave depressions to receive the rounded occipital condyles of the skull. This articulation permits an up and down (nodding) movement of the head. On the posterior face, too, there are no zygapophyses. Instead there are two stout facets which move over the articular surfaces on the rounded centrum of the axis vertebra. The space occupied by the spinal cord is separated from that into which the odontoid peg projects by a **transverse ligament** which divides the canal into an upper spinal canal and a lower **odontoid fossa**. The transverse ligament prevents the odontoid peg pressing into the spinal cord during the nodding of the head.

Proceeding now to the other end of the column, the lumbar region is succeeded by a very specially modified region made up of the **sacral vertebrae**. Opinions differ as to the number of sacral vertebrae found in the rabbit for, since only one is actually fused with the pelvic girdle, it is considered that this alone is the true sacral vertebra and that the succeeding ones belong to the caudal region. Suffice it to say that many of the features found in the first of the series are also found in the succeeding ones, and that at

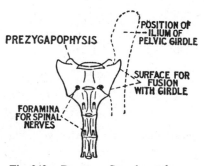

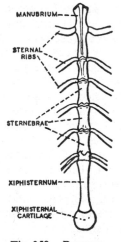

Fig. 349. RABBIT.—Sacral vertebrae.      Fig. 350. RABBIT.— The sternum.

least three and sometimes four vertebrae are joined to form the composite structure known as the **sacrum**. The first of these is very stoutly built with a broad, flattened centrum, from the sides of which arise two massive bony projections to which the name "transverse processes" is usually given. It is to these processes that the ilia of the pelvic girdle are jointed, and there is some evidence that the actual junction is with sacral ribs and not the transverse processes, so that if the usual name is used the above *proviso* should be borne in mind. Following from the junction of this vertebra with the pelvic girdle and with the succeeding vertebrae, the usual intervertebral position of the foramina through which the spinal nerves issue is interfered with and the foramina are found on the ventral side of the sacrum. This condition is found in the three succeeding vertebrae, which gives some justification for regarding them as sacral and not caudal vertebrae. All the vertebrae in the sacrum have prominent neural spines, on both sides of which are tubercle-like projections which indicate the positions of the zygapophyses.

The **caudal** vertebrae call for little comment. In the rabbit they are not very numerous, and as the tip of the tail is approached, the neural arch, zygapophyses, etc., become progressively reduced until the terminal ones consist solely of the rounded cylindrical centrum.

**The Ribs and Sternum.**—The ribs of a mammal form with the sternum a bony cage completely surrounding the thorax. Not only do they articulate with the vertebral column dorsally, but their ventral extremities, which are cartilaginous, are attached to a median bony skeletal structure, the **sternum**. Each rib therefore has a **vertebral portion** and a **sternal** or **costal portion**, and it is the latter which is cartilaginous. The first seven pairs of ribs are attached directly to the sternum, but the sternal parts of the eighth and ninth are joined to those of the seventh pair, so that the connection is indirect. These are commonly called the **false ribs**. The last three pairs of ribs have no connection with the sternum at all and are termed the **floating ribs**.

The sternum itself consists of a series of **sternebrae**, articulated together, and the sternal ribs are joined to it at the junctions of the sternebrae. The first sternebra is called the **manubrium**, and the first pair of sternal ribs are attached about half-way along its length. The manubrium is followed by a series of sternebrae, making seven in all, of which the sixth is very small and the seventh or **xiphisternum** is long and slender and terminates in an expanded plate of cartilage, the **xiphisternal cartilage**.

This bony cage has an appropriate musculature by which the ribs can be moved in an antero-posterior direction, thus varying the internal capacity of the thorax and playing an important part in the breathing mechanism.

**The Skull.**—In ordinary language, the term skull denotes the skeleton contained within the head. In considering the vertebrates as a whole, however, it has already been seen that, particularly in the aquatic forms, it is sometimes difficult to define the limits of the head. It seems better, therefore, to approach the question from another aspect and examine the structures which can be considered to contribute to the skull, and the delimiting of the head region will follow.

Morphologically, the skull consists of two parts, the **cranium**—which encloses the brain and the organs of special sense—and the **visceral skeleton**—which comprises the jaws and allied structures. Developmentally, the two skeletal contributions are quite distinct,

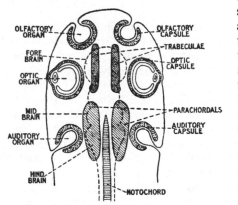

Fig. 351. Diagram of the main cartilaginous elements entering into the composition of the cranium.

and in a lower vertebrate such as a dogfish, the actual connection between them is not very close, so that they can be easily separated. The best method of approach, therefore, is to consider each separately, and later, the connection between them.

The first indication of the development of the **cranium** is the appearance in the embryo of two pairs of cartilaginous plates in the mesoderm on the under side of the brain. Two of these plates—the **para-chordals**—lie one on each side of the anterior end of the notochord, and the other two—the **trabeculae** (the equivalents of a premandibular arch)—further forwards. At about the same time around the developing sense organs—olfactory, optic, and auditory—enclosing capsules called **sense capsules** are laid down. Without going into elaborate detail it may be said that the parachordals and trabeculae increase in size and coalesce to form a

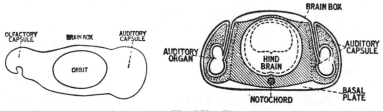

Fig. 352. Side view of a young        Fig. 353. Transverse section through the
           cranium.                              hinder region of a cranium.

**basal plate** beneath the brain, whilst other cartilage is laid down at the sides and above the brain until it becomes enclosed in a cartilaginous box, the **brain box** or **neurocranium**. Dorsally the roof of the brain box may be incomplete, spaces termed **fontanelles** being left, and ventrally a transitory foramen around the pituitary body (see p. 657) is formed, but may become closed up later. Posteriorly, the cartilage grows around the spinal cord, leaving a large foramen, the **foramen magnum**, in the cranium. Of the sense capsules, the olfactory and auditory capsules are cartilaginous and as they develop they fuse on to the neurocranium, the olfactory anteriorly and the auditory capsules posteriorly. The optic capsules, however, *never fuse* with the cranium but form the **sclerotic** coats of the eyeballs, which later come to lie in laterally situated depressions termed the **orbits**.

The **visceral arch skeleton** is laid down in the pharyngeal wall between successive visceral clefts, being developed, curiously enough, from those neural crest cells which do not contribute to the formation of the dorsal root ganglia (see p. 579). In all embryos it consists of a series of visceral arches, one behind the mouth and one behind each of the visceral clefts. But, as has already been explained, the remnants of a premandibular arch are represented by the trabeculae. Each arch encircles the pharynx except on the dorsal side where the two halves of the arch do not meet, and each is made up of a series of paired pieces bound together by fibrous material.

The relationships of the visceral skeleton to the cranium vary in different vertebrates and one possible arrangement can best be realised by considering a relatively simple example such as the dogfish.

## THE DOGFISH

The cranium in the dogfish consists of a brain box with an incomplete roof, a pair of thin olfactory capsules anteriorly, a pair of auditory capsules posteriorly, and at the sides are the shallow **orbits**. Because the cartilaginous condition is retained into adult life, it is called a **chondrocranium**.

The visceral arch skeleton is made up of a series of visceral arches which develop in the pharyngeal wall (see pp. 364 and 482) one behind each visceral cleft and of the jaws. Typically, each visceral arch is an incomplete ring of cartilage consisting of a basal piece (probably of paired origin) in the mid-ventral line with which articulates a series of pieces on each side extending upwards almost to the mid-dorsal line. It is clear, therefore, that this system is independent of the chondrocranium and, strictly speaking, is not even part of the axial skeleton. The number of arches is determined by the number of clefts. In the dogfish, seven obvious visceral arches are developed, and it is the first two which warrant special attention. The first, which is called the mandibular arch, borders the mouth and has been modified to form the jaws. It lies in front of the first pair of existing visceral clefts, the spiracles. The second, the hyoid arch, lies between the spiracles and the first pair of functional branchial clefts. The succeeding five arches are usually termed branchial arches because they lie one behind each of the successive branchial clefts.

The **mandibular arch** has lost its original arch form and consists of four pieces of cartilage, two on each side of the mouth and all joined together by ligaments. An upper jaw (**palato-pterygo-quadrate bar**) above and a lower jaw (**Meckel's cartilage**) below, meet behind the mouth. The two rami of the upper and lower jaws articulate anteriorly in the mid-line. Thus the jaws have no connection with the chondrocranium except through the hyoid arch.

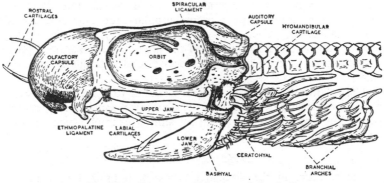

Fig. 354. DOGFISH.—Side view of the skull.

The **hyoid arch** is a hoop of cartilage made up of five pieces. A median **basihyal** lies between the rami of the lower jaw, and from this arise two stout **ceratohyal** cartilages which lie in the anterior walls of the first gill clefts. Dorsal to the ceratohyals are two **hyomandibular** cartilages which fit, at their upper ends, into depressions in the auditory capsules of the chondrocranium to which they are attached by ligaments. On each side of the body, the hinder ends of the upper and lower jaws are bound to the hyomandibular cartilages by ligaments. Thus the hyomandibular cartilage forms, on each side, the suspensorium of the upper and lower jaws, and a skull with this type of jaw suspension is termed **hyostylic**. Although the only skeletal connection between the jaws and the cranium is by means of the hyomandibular cartilages, the upper jaw is connected with the under side of the cranium on each side by ligaments. Towards the anterior end is the short **ethmo-palatine** ligament and posteriorly is the **spiracular** ligament.*

The remainder of the visceral arches lie in relation with the branchial clefts and are termed **branchial arches.** Typically, each consists of nine pieces of cartilage, a median **basibranchial** cartilage to which are attached on each side a **hypo-, cerato-, epi-,** and **pharyngobranchial** cartilage. The cartilages do not lie in a straight line with one another but are inclined as is shown in Fig. 355. Although each branchial arch consists typically of the cartilages enumerated above, in the dogfish a considerable amount of fusion and reduction has taken place. The basibranchials are represented by a single piece of cartilage. The hypobranchials of the first branchial arch are attached to the basihyal and in the fifth arch are absent. The pharyngobranchials of the fourth and fifth arches are

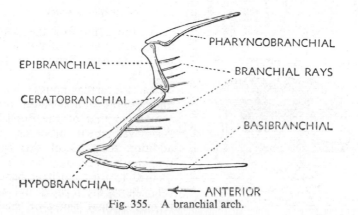

Fig. 355.  A branchial arch.

joined together. All these modifications may be seen by reference to Fig. 354.

In addition to the essential cartilages which have been dealt with above, in the skeleton of the dogfish certain *extra* cartilages are found. These are the **labial cartilages** at the corners of the mouth and the **extrabranchial cartilages** in connection with the gills. Also three **rostral cartilages** project from the front of the cranium into the snout.

From all these considerations it will be clear that the skull is a combination of the cranial and visceral arch skeletons. In an animal like the dogfish they are entirely cartilaginous, and there

* In most cartilaginous fishes this ligament is a pre-spiracular ligament, but it has long been established that in the dogfish it lies behind the spiracle, and the pre-spiracular ligament is absent.

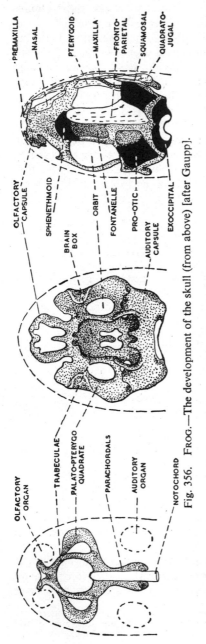

PREMAXILLA

NASAL

PTERYGOID

MAXILLA

FRONTO-PARIETAL

SQUAMOSAL

QUADRATO-JUGAL

OLFACTORY CAPSULE

SPHENETHMOID

BRAIN BOX

ORBIT

FONTANELLE

PRO-OTIC

AUDITORY CAPSULE

EXOCCIPITAL

OLFACTORY ORGAN

TRABECULAE

PALATO-PTERYGO QUADRATE

PARACHORDALS

AUDITORY ORGAN

NOTOCHORD

Fig. 356. FROG.—The development of the skull (from above) [after Gaupp].

is no fusion between the two. It might be argued that the skull by comparison with that of a mammal consists merely of the cranium and the mandibular and hyoid arches. But all the muscles of the mandibular and hyoid arches are innervated by cranial nerves arising from the brain. On these grounds, therefore, it is clear that the skull region extends as far back as the last gill arch.

### THE FROG

In the adult frog the skull, along with the rest of the skeleton, is in part, bony. In the tadpole, however, it is laid down as a cartilaginous cranium and visceral arch skeleton very similar to that of the dogfish. As development proceeds, in addition to parts of this cartilaginous skeleton becoming ossified to form **cartilage bones,** other bones are formed and are termed **dermal bones** or **membrane bones** since they arise in ensheathing membranes of connective tissue.

Fig. 356 shows the form of the cartilaginous skull of the tadpole and from it some points of difference from the condition in the dogfish are at once apparent. Each half of the upper jaw or **palato-pterygo-quadrate bar,** instead of being separate from the cranium, has become fused to it at its two ends, forming a lower margin to the **orbit.** With the quadrate portion of this bar articulates the lower jaw. Skulls with this type of jaw suspension are termed **autostylic** as distinct from the hyostylic condition seen in the dogfish and represent a stage in the progressive fusion of the upper jaw with the cranium, the highest development

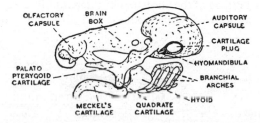

Fig. 357. FROG.—The skull of a tadpole from the side.

of which will be seen later in the mammal. In the autostylic skull the hyomandibular cartilage is no longer concerned with jaw suspension, and becomes part of the auditory apparatus (see p. 627) as an ear ossicle, the **columella** (or **stapes**). Apart from this difference, the remainder of the visceral skeleton is somewhat similar.

Before the adult condition is attained, certain parts of the cartilaginous skull ossify to form cartilage bones. The cranium

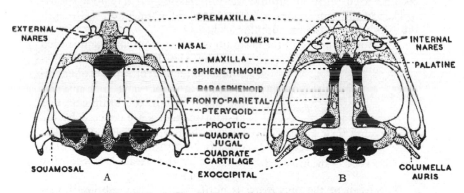

Fig. 358. FROG.—The skull. A, dorsal view; B, ventral view.

around the front part of the brain ossifies to form the ring bone or **sphenethmoid**. The anterior faces of the auditory capsules produce the **pro-otic** bones, whilst in their posterior sides, on each side of the foramen magnum, the **exoccipital** bones are laid down. These five bones are the only large cartilage bones visible in the adult skull, but certain other parts ossify, though they become hidden later. The **palatine** and **pterygoid** regions of the palato-pterygo-quadrate bar become bony, but are overlaid later in development by dermal bones. In the lower jaw two small portions of Meckel's cartilages on each side of the median symphysis, ossify to form the **mentomeckelian bones**.

The head region of the ancestors of the Amphibia was ensheathed in a heavy armour of dermal bones, and although with the evolu-

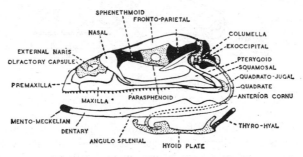

Fig. 359. FROG.—The skull from the side.

tionary progress of the group many of these have been lost, certain of them have been retained and "plastered" on to the chondro-cranium. These membrane bones of the skull, have therefore become fixed on to or around the pre-existing cartilaginous frame-work. Membrane bones are developed in connection with both the cranium and the mandibular arch.

The dorsal surface of the cranium is largely unossified and roofed over by the **frontoparietal\*** bones which extend from the spheneth-moid to the pro-otics. Ventrally, the under side of the brain case and auditory capsules are covered by the dagger-shaped **parasphenoid**. The olfactory capsules are reinforced dorsally by the **nasals** and ventrally by the **vomers**.

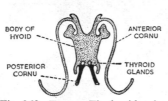

Fig. 360. FROG.—The hyoid arch.

The form of the upper jaw is entirely due to membrane bones which extend forwards in front of the olfactory capsules and back-wards to the angle of the jaw. On each side, are the **premaxilla** and **maxilla** (both of which bear teeth) followed by the **quadratojugal**, the maxilla and the quadratojugal curving out from the ends of the

\* There is good evidence that these bones are **frontals** only, the parietals having been lost owing to the shortening of the skull in the evolution of the modern Amphibia. The name has, however, been in use for so long that it is retained here.

pterygoid and quadrate portions of the palato-pterygo-quadrate bar. The palatine and pterygoid portions of this cartilage become ossified, but are overlaid by membrane bones, the slender **palatine** extending along the ventral side of the front of the wide orbit from the sphenethmoid to the pterygoid, whilst on the ventral side of the skull the **pterygoid** takes the form of a Y-shaped bone, the stem of the Y lying along the inside of the maxilla, one prong of the fork being attached to the auditory capsule and the other ensheathing the quadrate cartilage. Dorsally, the connection between the quadratojugal and the auditory capsule is formed by the hammer-shaped **squamosal** bone, which together with the prongs of the pterygoid forms an efficient rigid girder system for the articulation of the lower jaw. Projecting from the sides of the auditory capsules will be seen the partially ossified hyomandibular cartilage which now forms the **columella auris**.

It will be recollected that the only ossifications in the cartilages of the lower jaws are the mentomeckelian bones, the remainder of Meckel's cartilages being entirely ensheathed by membrane bones. These are, on each side, the **dentary** (which in spite of its name bears no teeth) anteriorly and the **angulosplenial** posteriorly, the angular portion of which serves as the articular surface with the quadrate region of the upper jaw.

Of the rest of the visceral skeleton, since it remains largely cartilaginous, little is seen in the usual dried specimens used in the laboratory. The remainder of the hyoid arch is represented by a cartilaginous plate in the floor of the pharynx termed the body of the hyoid, and this has a secondary connection with the auditory capsules through the long anterior cornua. Projecting from the posterior margin of the cartilage are the partially ossified posterior cornua, which lie one on each side of the larynx. The branchial arches of the tadpole are modified during metamorphosis to form the **laryngeal cartilages**.

## THE MAMMAL

Apart from the facts that the skull of a mammal is bony, is autostylic and, of course, has those very general features common to the skull of all bony vertebrates, there is very little resemblance to that of a modern amphibian, such as a frog. Nevertheless, it is possible by a study of fossil forms to trace a fairly complete series which shows how the mammalian skull condition has arisen from that of the earliest known reptiles, which themselves can be seen to have arisen from a group of fossil Amphibia. It will not be possible to discuss this here. Nor, to begin with, will the skull of any particular mammal be emphasised. Instead, the general arrangement

of the bones will be described and reference should be made
first to Fig. 361 and then to the figures of the dog's skull, which is
usually chosen for study.

To accommodate the large brain, the brain box has been
enormously enlarged, with the result that the auditory capsule on
each side has sunk into the cranium and is embedded in the cranial
wall in which is also included the squamosal. The next feature is
the separation of the respiratory channel from the food canal by
the development of a hard, bony (secondary) palate, which necessitates many alterations in the arrangement of the bones in the upper
jaws and the nasal capsules. Lastly, the change in the mode of
articulation of the lower jaw, viz. with the squamosal direct, is
accompanied by the inclusion of the **quadrate** of the upper jaw and

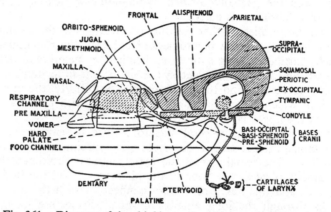

Fig. 361. Diagram of the chief bones found in a mammalian skull.

the **articular** of the lower jaw in the auditory apparatus as ear
ossicles additional to the modified hyomandibular (p. 627).

The relation between the cartilage bones (indicated by C) and
membrane bones is very much closer than in the skull of the frog.
In the embryo a chondrocranium and a visceral skeleton of the
pattern already outlined are laid down, but soon both cartilage and
membrane bones are formed and it is not easy to distinguish between
the two types of bone. This is because the membrane bones no
longer overlie the cartilage bones, but their edges adjoin, forming
sutures between them.

Considering the cranium first, from the basal plate of the
embryonic chondrocranium three bones are developed which form
what are termed the **bases cranii**, namely the **basioccipital** (C), the
**basisphenoid** (C), and the **presphenoid** (C). These three bones form

the bases of three rings of bone forming the greater part of the cranium, the rings being sometimes spoken of as the occipital, parietal, and frontal "segments" of the skull (the word segment is not used in a metameric sense here), between which gaps occur. These gaps are filled either by other bones or provide exits or entrances for structures such as nerves and blood vessels.

The **occipital ring** is made up of the **basioccipital** (C), paired **exoccipitals** (C), and the **supraoccipital** (C). Between the exoccipitals lies the **foramen magnum**, at the sides of which are the **occipital condyles**. The **parietal ring** includes the **basisphenoid** (C), paired **alisphenoids** (C), and **parietals**. (In the dog there is a small interparietal partially separating the parietals.) In the **frontal ring** are

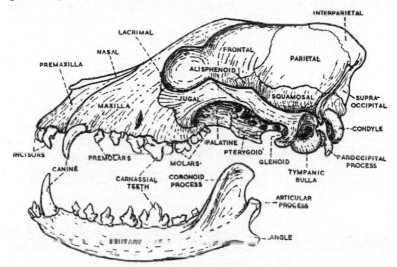

Fig. 362. DOG.—The skull from the side.

the **presphenoid** (C), paired **orbitosphenoids** (C), and **frontals**. These three rings of bone comprise the greater part of the cranial wall, the space between the exoccipital and alisphenoid on each side being filled by the cranial portion of the **squamosal**. The front of the cranial cavity is closed by the **ethmoid** (C) in which is the **cribriform plate**, through the perforations of which pass the fibres of the olfactory nerves. Between the alisphenoid and orbitosphenoid, foramina develop for the exit of the third to sixth cranial nerves.

Almost completely embedded in the wall of the cranium, as will be seen from Fig. 361, is the rather small **auditory capsule** in which is enclosed the membranous labyrinth of the inner ear. It consists of a single bone, the **periotic** (C), which is perforated by various apertures. Lying outside the cranium against the periotic between the basisphenoid and the squamosal, is the **tympanic** bone, the outer part of which forms the **external auditory meatus** whilst its inner part encloses the **middle ear** or tympanic chamber. Within the tympanic chamber are the **auditory ossicles** (see p. 627).

The development of a separate respiratory channel with which the olfactory capsules have become incorporated and in part enclosed, has involved the inclusion of some of the bones of the upper jaw, not only in the floor of the channel in the form of the **palate**, but also in the sides giving rise to the muzzle of the head. In the bony palate ingrowths from the **premaxilla** and **maxilla** extend anteriorly in front of the **palatines** whilst the lateral walls of the channel where it enters the pharynx are formed by the **pterygoids**. Externally, the sides of the channel are formed by upgrowths of the premaxilla and maxilla and the roof by the **nasals**. In this way the

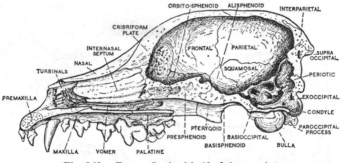

Fig. 363. DOG.—Sagittal half of the cranium.

buccal cavity is completely separated from the respiratory channel, a condition which confers upon the mammals the privilege of being able to retain food in the buccal cavity and breathe at the same time. (A similar condition occurs in the crocodiles.) This is important for the mammals are the only vertebrate animals which can chew their food.

The respiratory channel is divided into right and left halves by the partly cartilaginous **ethmoid** (C), the anterior end of which helps form the **internasal septum** or mesethmoid which rests in the V-shaped gutter on the upper margin of the **vomer**, which, in section, is Y-shaped, the stem of the Y being attached to the upper surface of the palate. The greater part of the cavity is occupied by the very much folded **turbinal** or **scroll bones** which have arisen as ingrowths from the maxillae, nasals, and ethmoid, in passing between the folds of which the air is warmed and filtered.

Extending between the maxillary and cranial regions of the skull, on each side, there is an arch of bone, the **zygomatic arch**, the anterior part of which forms the lower margin of the orbit, while

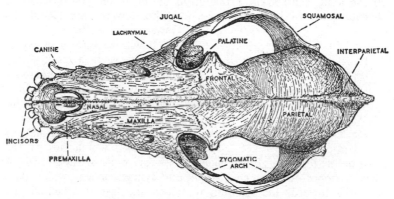

Fig. 364. Dog.—The skull, dorsal view.

between its posterior part and the cranium lie some of the powerful jaw muscles. The anterior part of the arch is formed by the **jugal bone** and the posterior by the **zygomatic portion or process of the squamosal**, the two bones meeting in an oblique suture.

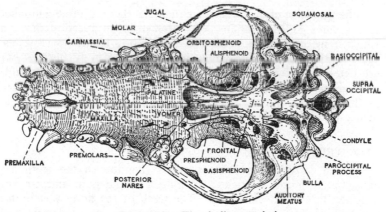

Fig. 365. Dog.—The skull, ventral view.

The bones which comprise the functional upper jaw, the premaxillae and maxillae, have already been referred to, and lodged in sockets in these bones, teeth are found.

The lower jaw consists of a single membrane bone, the **dentary**, on each side, and the two together are sometimes spoken of as forming the mandible. These bones were formed around the distal portion of the original Meckel's cartilages of which no trace is seen in the adult skull. The articulation of the lower jaw is with the **glenoid fossa** of the squamosal. In the dentary are found the lower set of teeth.

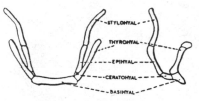

Fig. 366. DOG.—The hyoid arch.

It will be natural to expect that in such a profoundly modified skull the remainder of the visceral arches will likewise be modified. The greater part of the hyoid arch is represented by a plate, the **body of the hyoid**, which lies at the root of the tongue. From each end of this plate arises a series of slender bones—**cerato-, epi-, stylo-**, and **tympanohyals**—which pass upwards towards the cranium to which the last is attached just behind the tympanic bone. These represent the anterior cornua. The posterior cornua consist of a single bone, the thyrohyal, on each side; other parts of the hyoid arch go to form the **thyroid** cartilages of the larynx. The rest of the **laryngeal** cartilages—**cricoid** and **arytenoid**—are derived from the posterior visceral arches of the embryo.

**The Foramina of the Skull.**—In all skulls there is found a number of apertures called foramina. A **foramen** is an aperture through which some other structure such as a nerve or a blood vessel passes. The positions of these foramina are by no means fortuitous, but are determined by the way in which the vessel or nerve develops in relation to the skull. This of course happens at the same time as the cartilage or bone is being laid down, and the skeletal tissue forms *around* the nerves or vessels leaving the foramen.

## THE TEETH

In the cartilaginous fishes such as the dogfish the teeth are exaggerated dermal denticles, are lodged in the skin, and are not attached to the jaw cartilages. The teeth of the bony vertebrates are also skin structures derived partly from the epidermis and partly from the dermis. Although their structure is much the same in the various classes, that of a mammal is usually taken as an example, to understand which, a knowledge of the structure of a fully-formed tooth is necessary.

If one of the front teeth of a dog is considered, part of it is seen to project up above the skin covering the jaw and the remainder is lodged in the bone or covered by the gum. From the vertical section shown in Fig. 367 it will be seen that the tooth can be divided into three regions: the **crown**, which projects above the gum; the **neck**, between the crown and the jaw bone; and the **root**, embedded in the bone. Externally the crown is covered with hard, white, shiny **enamel**, but the outer wall of the neck and root is composed of **cement**, a tissue approaching in its histological structure that of bone. The bulk of the body of the tooth is made up of **dentine** (ivory), a tissue which is permeated by numerous fine canals in which lie the protoplasmic processes from the **odontoblasts** found along the periphery of the **pulp cavity** which occupies the centre of the tooth. Here are found blood vessels, connective tissue, and nerve-fibres, some of which enter the dentine. Since the jaw-bone and the tooth de-velop at the same time, the root of the tooth be-comes enclosed in a bony socket, the **alveolus**, and between the tooth and the alveolar wall lies a con-tinuation of the perio-steum of the bone forming the vascular **alveolar** (or **periodontal**) membrane.

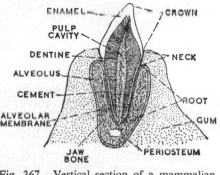

Fig. 367. Vertical section of a mammalian tooth in its socket.

The early development of the tooth takes place below the surface of the gum. In the embryo the skin covering the jaw bears a longitudinal groove or furrow—the **dental groove**—which marks a vertical intucking of the epidermis to form a **dental lamina**. At intervals, outgrowths from the lamina grow towards condensations of dermal tissue, the **dental papillae**, and these constitute the tooth rudiments. Without going into great detail it will be seen from Fig. 368 that the relations between the papilla and the epidermal layer are precisely those found during the development of a dermal denticle (Fig. 329), and it will suffice to say that the epidermal layer forms an **enamel organ** which merely secretes the enamel covering to the crown, and that the dentine and cement layers are produced from the odontoblasts of the dental papilla which persist in the tooth. Eventually all the essential parts of the tooth are laid down whilst it is still within the gum, and as the tooth grows the bone of the jaw is formed around its root. Finally, the fully-formed tooth becomes pushed up in a short space of time beyond the surface of the gum, and the alveolus—which when first formed was much too large—closes around its root so that the tooth is fixed in position in the jaw.

The teeth of the frog—which occur only in the premaxilla and maxilla of the upper jaws and in the vomers—are all alike, that is, **homodont.** Each is conical in shape, and is cemented to the side of the jaw and not lodged in a socket. In a mammal such as the dog, however, the teeth are not all of the same form and are said to be **heterodont.** Moreover, in many mammals there are two successions of teeth, that of the young animal or the milk dentition,

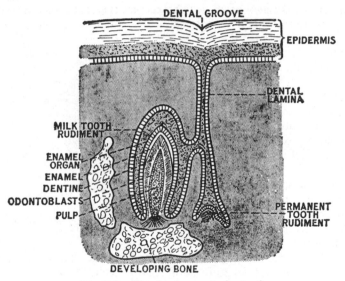

Fig. 368. The development of a tooth.

and that of the adult or the permanent dentition. This is called a **diphyodont** type of dentition as opposed to the **monophyodont,** in which there is only one set of teeth. The number of teeth in the two successions is not the same, there being usually fewer in the milk dentition.

Where the teeth are heterodont the different types of teeth are given distinctive names. Those in the front of the jaws are called **incisors,** followed on each side by the **canines,** and then by the **cheek teeth,** made up of the **premolars** and the **molars.** The distinction between premolars and molars is not always one of form, but the premolars are the only cheek teeth represented in the milk dentition.

In many mammals the **incisor** teeth (the upper ones of which are always borne on the premaxillae) have a sharp chisel edge and are used for biting off pieces of food. This is true, for example, of the rabbit, but in the dog and most carnivores, the shearing of the flesh is performed by specially modified cheek teeth and the incisors are used for prehension. The **canines** are usually large, often conical and pointed, and get their name from their condition in the dog-like mammals, where they are used for fighting and seizing the prey. In the rabbit they are absent. The **cheek teeth** usually have broader crowns and the root has more than one fang. In some mammals they are modified for grinding—as in the rabbit and ruminants— but, as has been mentioned, the last pair of premolars in the upper and the first molars in the lower jaws in the dog are pointedly ridged and sharp edged, and are termed the **carnassial teeth**. They are used for shearing through the food, both flesh and bone. The number of teeth of each type varies in different mammals, and to indicate this, a kind of equation called a **dental formula** is used. This shows the number of teeth on each side in upper and lower jaws, the type of tooth being indicated by the initial letter. A full dentition is shown thus:

$$i. \tfrac{3}{3}, c. \tfrac{1}{1}, pm. \tfrac{4}{4}, m. \tfrac{3}{3} = 44.$$

The pig is an example of a mammal which possesses this number of teeth but most have fewer. For comparison the dental formulae of the dog, rabbit, and man are given.

*Dog:*    $i. \tfrac{3}{3}, c. \tfrac{1}{1}, pm. \tfrac{4}{4}, m. \tfrac{2}{3} = 42.$

*Rabbit:*   $i. \tfrac{2}{1}, c. \tfrac{0}{0}, pm. \tfrac{3}{2}, m. \tfrac{3}{3} = 28.$

*Man:*    $i. \tfrac{2}{2}, c. \tfrac{1}{1}, pm. \tfrac{2}{2}, m. \tfrac{3}{3} = 32.$

In many mammals, ourselves included, when once the teeth have been completely erupted no further growth takes place and they gradually wear away. This cessation of growth is due to the constriction of the aperture at the base of the root leading into the pulp cavity, so that the amount of food conveyed by the blood vessels is restricted, and while sufficient to maintain life, does not allow for additional growth. In other cases where from the nature of the food the wear on the teeth is intense, as for instance in the incisors of the rabbit, the pulp cavity remains wide open and the teeth continue to grow throughout life. Such teeth are said to have persistent pulps.

The skull of vertebrates can be thought of, then, as a composite structure made up of a cranium surrounding the brain and the sense organs, and a visceral arch skeleton. Of the visceral arches, the second (mandibular) gives rise to the jaws. The third (hyoid), after being partly pressed into service for jaw suspension (hyostylic skull), loses its connection with the jaws and contributes partly to the auditory mechanism and partly to structures (hyoid apparatus) for the attachment of muscles for movement of the floor of the buccopharynx. The first, or premandibular arch, remains as such in the adults of the Agnatha but, in all gnathostomes, becomes reduced to form the trabeculae and is incorporated in the floor of the skull. In the line leading to the tetrapods and in the tetrapods themselves, the upper jaw fuses with the cranium and the skull is then said to be autostylic. Moreover, the number of cartilage and membrane bones decreases as the vertebrate series is ascended, whilst the membrane bones become more closely incorporated in the skull. At the same time the cranial capacity increases to accommodate the larger brain. In the reptilian line leading to the mammals, the bones of the lower jaw other than the dentary become fewer, or relatively much smaller, so that in mammals the dentary articulates with the squamosal. Then the former articular quadrate joint is preserved but the two bones form additional ear ossicles which receive the names of malleus and incus (p. 627).

## THE APPENDICULAR SKELETON

The appendicular skeleton includes that of the paired limbs and their girdles. In the fishes, the paired limbs consist of the **pectoral** (anterior) and **pelvic** (posterior) **fins**, and in connection with these, embedded in the muscles of the body wall, are the **pectoral** and **pelvic girdles** which act as more or less rigid girders on which the fin skeleton can move.

### THE DOGFISH—LIMBS AND GIRDLES

Both the median and the paired fins of the dogfish develop from folds of the embryonic epidermis into which mesenchyme from the dermis soon penetrates. The folds extend over many more segments of the embryo than enter into the composition of the adult fin and, indeed, there is for a short time a nearly continuous fin fold along the back, around the tail, and along the ventral surface nearly as far as the cloaca. Similarly, the paired fins develop from lateral folds which are relatively extensive in the embryo. The mesenchyme of the fin lays down the cartilaginous fin rays which are typically three-jointed rods of cartilage, whilst into the fin grow small buds from each of the myotomes. These buds form the **radial muscles**, one on each side of each **radial cartilage**. But this segmental arrangement is soon lost and the fins get narrower at their bases relative to the rest of the body which grows at a faster rate. For this reason there is a good deal of fusion between the radials and also between the muscle buds, fusion taking place particularly at the anterior and posterior ends of the fins and leading to the formation of larger cartilages termed **basals**. The girdles also chondrify in mesenchyme which for some time is continuous with the sheet which lays down the radials in the paired fins. Distally in each of the fins, but overlapping the radials, a double series of horny fin rays are secreted to complete the skeleton of the fin.

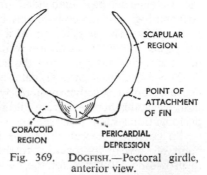

Fig. 369.   DOGFISH.—Pectoral girdle, anterior view.

In the pectoral region of the dogfish the halves of the **pectoral girdle** have met and fused in the mid-ventral line — a feature which is peculiar to the dogfish and its allies—to form a hoop of cartilage embedded in the ventrolateral body muscles. Each half of the girdle may be called a scapulo-coracoid cartilage. Dorsally is found the tapering **scapular region** and ventrally the broader coracoid region. In the mid-ventral portion of the girdle, where the two coracoid portions have fused together, is a well-defined depression—the pericardial depression—in which, in life, lies the ventral part of the pericardium containing the heart. The fin articulates with the girdle by three basal cartilages the **pro-, meso-,** and **meta-pterygium**. Distally to these are the radial cartilages or cartilaginous fin rays. These each consist typically of three pieces and those of the proximal row, although varying somewhat in size,

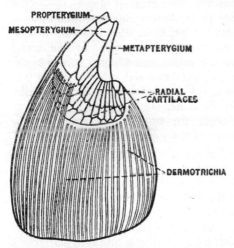

Fig. 370.  DOGFISH.—Skeleton of left pectoral fin from above. The proximal parts of some of the dermotrichia have been removed.

are uniform in arrangement but the second and third pieces take the form of interfitting polygonal plates. The periphery of the fin is formed by an upper and lower series of horny fin rays or

**dermotrichia**, the proximal ends of which overlap the cartilaginous fin skeleton.

The **pelvic girdle** merely consists of a simple bar of cartilage—the **ischiopubic bar**—embedded in the ventral body muscles. Articulating with the ends of this bar is the cartilaginous skeleton of the fins. This is made up of a single basal cartilage, the **basipterygium**, and a fairly uniform series of **radials**. As in the pectoral fins, the peripheral portion of the fins is formed of dermotrichia.

The pelvic fins show a sexual dimorphism. In the male the two fins are joined together posteriorly and in connection with each basipterygium there is a grooved, backwardly directed cartilaginous prolongation which forms the axis of the **clasper**.

So far, the pectoral girdle has been considered only from the point of view of its relations to the pectoral fins, but in addition to providing a fulcrum for the movements of the fins and for the attachment of muscles connected therewith, it also serves as a place of attachment (origin) of other muscles. These run forwards to become inserted in the ventral parts of the visceral arch skeleton. In fishes, these muscles play an important part in swallowing and breathing movements. Thus in the dogfish, arising from the coracoid region of the girdle are the **coraco-mandibular**, **coraco-hyoid**, and **coraco-branchial** muscles which pull the lower parts of the mandibular, hyoid, and branchial arches down and back, so opening the mouth and enlarging the cavity of the pharynx.

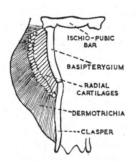

Fig. 371. DOGFISH.—Pelvic girdle and fin skeleton.

ISCHIO-PUBIC BAR
BASIPTERYGIUM
RADIAL CARTILAGES
DERMOTRICHIA
CLASPER

**The Pentadactyl Limb.**—The transition from the water to the land during the phylogenetic history of the vertebrates has been accompanied by great modifications of the bony skeleton, possessed by fish-like ancestors, and in the appearance of an entirely different type of limb and limb skeleton. The terrestrial animal, then, is provided with two pairs of limbs (from which the name **Tetrapoda** is derived), which have a skeleton based upon a common plan which is called the **pentadactyl** limb skeleton, since the limb terminates in *five* digits or fingers. No existing animals have exactly this type of limb skeleton, and the special interest in the limbs of the two animal types under consideration is to see how far the limb skeleton complies with, or departs from, this condition.

Both fore and hind limbs are divisible into regions movable on one another; the fore limb into **upper arm**, **fore arm**, **wrist**, and hand, and the hind limb into **thigh, shank, ankle,** and **foot.** The skeleton of the upper arm consists of a single long bone, the **humerus,** and that of the fore arm of two, the **radius** and **ulna.** In the wrist or **carpus** nine small bones, the **carpal bones,** are present, arranged in three rows. In the proximal row there are three bones, the **radiale** (next to the radius), the **intermedium** in the middle, and the **ulnare** (next to the ulna). The middle row consists of a single **centrale.** The distal row is made up of five bones, **carpale** one to five. In the hand (or **manus**), two series of bones are found; first a series of five **metacarpals** (the palm of the hand), followed by a

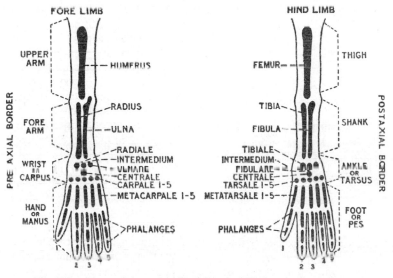

Fig. 372. Plans of the generalised fore and hind pentadactyl limbs.

series of **phalanges** arranged in rows and forming the skeleton of the separate digits or fingers. The number of phalanges varies in the digits, the first or **pollex** having two and the others three each.

In the hind limb the bones have precisely the same arrangement but are differently named. The thigh bone is the **femur,** the tibia and **fibula** lie in the shank. In the ankle, the **tibiale, intermedium,** and **fibulare** form the proximal row, **centrale** the middle, and **tarsale** one to five the distal row of tarsals. The bones of the foot or **pes** are termed **metatarsals** and **phalanges** respectively, and the first digit is called the **hallux.**

In both instances, the side of the limb which bears the first digit is spoken of as the **pre-axial** side or border and the other as the

**post-axial**. These terms refer to the original or primitive relationships of the limb to the body in which the axis of the limbs was at
right angles to that of the body, so that one margin of the limb was
in front of, and the other behind, its axis so far as the orientation of
the animal as a whole was concerned. These relationships become
important when considering the ways in which limbs are modified
(see Fig. 374).

**The Girdles of Tetrapods.**—The girdles can also be represented
in a generalised form. In the cartilaginous phase of development,
the pectoral girdle arises as a single plate on each side of the body,
the plates either meeting or overlapping ventrally. In each side of
these cartilages, ossifications appear which may be single, producing
a scapulo-coracoid bone in
which lies the glenoid fossa;
or a dorsal scapula and a ventral coracoid may ossify
separately, each contributing
to the glenoid cavity. In
those reptilian groups leading
to the mammals the cartilage
may become tripartite by the
development of a fenestra between an anterior **precoracoid**
and a posterior **coracoid**, all of
which ossify. In the primitive
Tetrapoda a number of mem-
brane bones—legacies from their fish-like ancestors—overlie the
scapula and coracoid, the pectoral girdle, of which the **clavicle**
covering the anterior (precoracoid) region seems usually to persist.

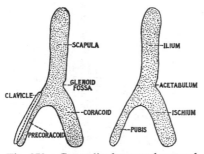

Fig. 373. Generalised tetrapod pectoral
and pelvic girdles from the side.

The pelvic girdle is also tripartite in the cartilaginous stage and
in it three centres of ossification are laid down, producing the dorsal
**ilium**, and ventrally the anterior **pubis** and posterior **ischium**, all
meeting at the **acetabulum**. No membrane bones are added to
this girdle.

Between the girdle and the limb and between the different
regions of the limb, joints occur (see p. 458).

In any adult animal the long bones of the limbs show characteristics by which those of the fore and hind limbs can be distinguished from one another. The **humerus** usually has a broad,
rather flattened **head** at its proximal end where it articulates with
the girdle, and the distal end is shaped like a pulley, called the
**trochlea**, to allow extensive movement of the ulna upon it. The
ulna is distinguished by the presence of a projection, the **olecranon**

process, extending beyond the depression by which it articulates with the humerus. In the hind limb the **femur** has usually a smaller, more spherical **head**, which is joined to the main shaft of the bone by a narrower portion called the **neck**. The main articulation between the shank and thigh is by the tibia and, in the mammal, in the tendon which runs from a ridge on the tibia—the **cnemial crest**—to a muscle attached to the femur, a small sesamoid bone (formed by an ossification within the tendon), the knee-cap or **patella**, is present. The distal end of the femur bears a groove— the patellar groove—in which this bone moves. Again, on the various bones, projections are present in the form of, for example, the **tubero-sities** of the humerus and the **trochanters** of the femur for the attachment of muscles. These and other features of the bones peculiar to the particular animal are best noticed when the actual bones are handled and examined in detail.

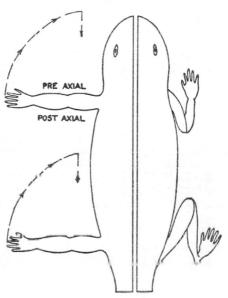

**Modifications of the Pentadactyl Limbs and Girdles.**—It is usually assumed that the earliest tetrapods had limbs more-or-less at right angles to the body axis and moved by wriggling along much as does a newt, the limbs serving mainly to raise the

Fig. 374. Diagrams to show the forward adduction and flexure of the tetrapod limbs.

body from the ground. From the curvature of the body during such movement, the axes of the limb-pairs will be inclined to one another at the end of each "wriggle", so that the fore limb of one side and the hind limb of the *opposite* side would be simultaneously directed forwards and the other partner of each pair, backwards (see p. 406). The nervous co-ordination of these movements has become so established that the tactile stimulation of, say, the hind foot of the *left* side, inducing a forward movement of the limb, is accompanied by a corresponding movement of the fore limb of the *right* side, and *vice versa*. Thus a definite walking pattern has been established, which, interestingly enough, has been retained throughout the tetrapods.

The presence of joints in the limbs and between the limbs and girdles allows of three possible lines of modification. Firstly, by bending the limbs the body can more efficiently be raised above the ground, giving greater freedom of movement. Secondly, the limbs can be reorientated in relation to the body axis, so that instead of being at right angles to it they can be held roughly parallel to and beneath the body. Thirdly, the limb can be modified so far as its skeleton and musculature is concerned according to the mode of progression assumed, such as walking, running, leaping, swimming, or flying.

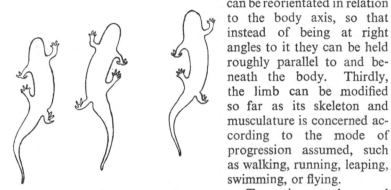

Fig. 375. The diagonal walking pattern of tetrapods.

To arrive at the usual positions the fore limbs are moved as a whole forwards through roughly ninety degrees, so that the hands are now pointing in the same direction as the head, and the pre-axial border of the limb lies near and parallel to the side of the body. The upper arm is then bent backwards, the fore arm forwards, while the hand rests palm downwards on the ground. Similarly, the hind limbs are moved forwards so that the pre-axial border is near the body, but the thigh remains directed forwards, the shank is bent backwards, and the plantar surface of the foot is in contact with the ground.

In these positions various modifications of the limb can be visualised to meet the various modes of progression, as is shown

Fig. 376. Diagram to show the raising of the body from the ground by the flexure of the limbs.

particularly well in the mammals. The palm of the hand and the sole of the foot can remain in contact with the ground, when the limb is said to be **plantigrade**, as in bears, rabbits, or ourselves. Increased speed is attained by the raising of the proximal portion of the hand and foot off the ground so that the animal walks or runs on its fingers and toes—the **digitigrade** type of limb as in the dog.

An extreme example of this condition is found in the **unguligrade** type of limb where the animal progresses on the tips of its fingers and toes (*e.g.* the horse). But these are examples of modifications for a purely terrestrial habit. Tetrapods, however, also live in water, some fly in the air (birds and bats), etc., and the appropriate modifications can be traced in their limbs, both as regards the limb skeleton and the associated structures.

Similarly, modifications in the girdles will be apparent in accordance with the type of progression assumed and, in the terrestrial tetrapods where frequently the greater part of the impetus is provided by the hind limbs, greater rigidity in the fulcrum upon which they work is brought about by the pelvic girdle forming a firm joint with the sacral region of the vertebral column. The

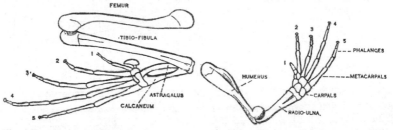

Fig. 377. FROG.—The skeleton of the hind limb.    Fig. 378. FROG.—The skeleton of the fore limb.

pectoral girdle, however, never joins with the column, the rigidity imparted to it by being embedded in muscle being sufficient.

## THE FROG—LIMBS

The swimming movements of the frog are chiefly those of the hind limbs, and the increased relative length of its regions is a distinct advantage in this connection. Consequently the **femur** is long, as also are the tibia and fibula, but these two bones are fused together to form a **tibio-fibula** bone. In the ankle and foot great modifications of the skeleton are apparent, which, with the web between the toes, increase the effective surface of this region in the thrust against the water. The proximal row of tarsals is represented by two longish bones, the **tibiale** (astragalus) and the **fibulare** (calcaneum), which is rather stouter than the tibiale. The centrale is absent and there has been reduction in the distal row of tarsals, which is represented by two bones only, in which are included tarsale one and two-plus-three. The metatarsals are elongated, as also are the phalanges. On the pre-axial side of the foot skeleton is an additional "digit", the **prehallux**. On land these same modifications in the hind limb skeleton are an advantage in leaping movements.  At the end of the leap, however, a great part of the shock of alighting has to be taken by the fore limbs.  These are in fact short and stout, and there has been fusion of bones to provide additional strength.  The **humerus** is short and stout, as also are the radius and ulna which have fused together to form a **radio-ulna**.  The **carpus** is reduced to six small bones; in the proximal row are the radiale, intermedium, and ulnare.  The centrale is absent and the distal row is represented by three bones, carpale one, two, while three, four, and five have fused together.  The five metacarpals are short and the phalanges of the first digit are absent.

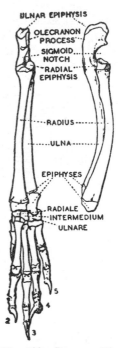

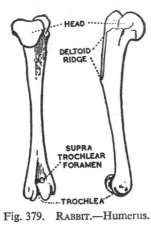

Fig. 379.  RABBIT.—Humerus.

Fig. 380.  RABBIT.—The bones of the fore arm, wrist, and hand.

## THE RABBIT—LIMBS

The fore limb has a stout **humerus**, and the **radius** and **ulna**, although still separate bones, are so tightly bound together that they cannot move over one another as can those of many mammals, and the limb is fixed with the palm downwards, that is, in the prone position.  This is an advantage in burrowing.  The carpus is made up of eight bones.  In the proximal row are the usual radiale, intermedium, and ulnare; the centrale is small, and in the distal row are four bones, the fourth and fifth carpals having fused together to form a single bone. In the hand there are the five meta-carpals and five digits with the typical number of phalanges.

In the hind limb the forwardly directed **femur** is stout and the **tibia** and **fibula** of the shank are fused together, the fibula being reduced to a slender slip of bone (brooch bone), and takes no part in the articulation with the femur. In the **tarsus** there has been a certain amount of fusion, the proximal row consisting of two bones, the astragalus—probably representing tibiale-plus-intermedium—

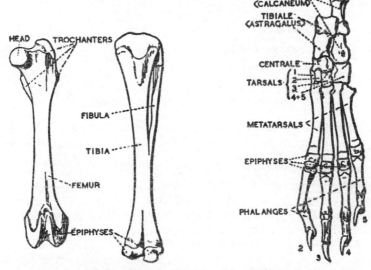

FIG. 301. RABBIT. The bones of the hind limb.

and the calcaneum (fibulare). The latter has a backwardly pro-jecting spur and forms the heel bone. The centrale has been displaced so that it lies in line with the astragalus. In the distal row there are only three distinct bones representing the second, third, and fourth-plus-fifth tarsals, the first having fused with the second metatarsal. These modifications are associated with the complete suppression of the first digit, so that there are only four metatarsals and each of the second to fifth digits has three phalanges.

## THE FROG—GIRDLES

The pectoral girdle of the frog is a strongly built structure and in development each half appears as a tripartite cartilaginous plate, a precoracoid portion being separated from the **coracoid** by a fenestra which is closed ventrally by connective tissue. In this cartilaginous plate two centres of ossification are laid down, but the whole of the cartilage does not ossify, with the result that in the dorsal part a **scapula** is produced surmounted by a cartilaginous **supra-scapula**, whilst ventrally there is a **coracoid** bone. The precoracoid cartilage, however, does not ossify, but is overlaid by the membrane bone,

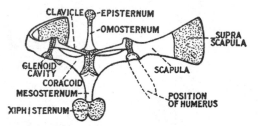

Fig. 382. FROG.—The pectoral girdle from below. The left scapula and supra-scapula have been flattened out.

the **clavicle**. The lower margin of the coracoid fenestra is formed by a cartilaginous **epicoracoid**. In the frog and its allies, the two halves of the girdle meet and fuse in the mid-ventral line, where, as development proceeds, other cartilages are added in front and behind to form the so-called **sternum**. Of these cartilages two portions ossify to form the **omosternum** in front and the **mesosternum** behind. The cartilaginous portions form the **episternum** anteriorly and **xiphisternal** or **xiphoid cartilage** posteriorly.

The whole girdle forms a system for the articulation of the fore limbs and for the attachment of the strong muscles upon which most of the strain falls at the moment of alighting at the end of a leap.

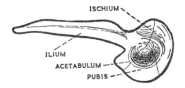

Fig. 383. FROG.—The pelvic girdle.

The **pelvic girdle** is peculiar in many ways. The whole of the ischio-pubic portion and the proximal part of the ilium of the two halves have coalesced, with the obliteration of the space between the ischium and pubis, to form a disc of bone set vertically in the body. In the centre of each side of the disc is an acetabulum. The dorsal parts of the two ilia extend upwards and their ends meet the tips of the transverse processes of the sacral (ninth) vertebra. The ilia are inclined at an angle to the axis of the vertebral column. This arrangement provides a very rigid fulcrum on which the two femurs can move. On each side the disc-like portion of the girdle is made up of the lower part of the **ilium**, the **ischium**, and the incompletely ossified **pubis**, all of which contribute to the formation of the **acetabulum**.

## THE MAMMAL—GIRDLES

In the higher mammals (which include the rabbit), the pectoral girdle has become profoundly altered from the generalised condition.

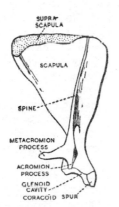

Fig. 384. RABBIT.—Right scapulo-coracoid of a young animal.

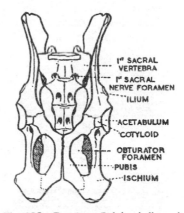

Fig. 385. RABBIT.—Pelvic girdle and sacrum from the ventral aspect.

Almost the whole of the coracoid region of each side has disappeared, and the girdle consists of two cartilage bones, the shoulder blades or scapulae, and of two membrane bones, the clavicles. If the scapula of an adult rabbit is examined, then it is seen to consist of a triangular-shaped bone at one end of which is a depression forming the **glenoid cavity**, and on the outer surface of the blade a ridge—the spine—which has two projections, the **acromion** and **metacromion** processes, all for the attachment of muscles. Overhanging the glenoid fossa, however, is a spur of bone, the **coracoid spur**

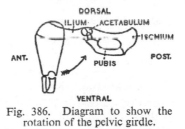

Fig. 386. Diagram to show the rotation of the pelvic girdle.

or process, and in the young animal there still exists a suture between the spur and the scapula, while the spur can be seen to be made up of two pieces representing the precoracoid and coracoid bones. The so-called scapula is therefore really a scapula *plus a very much reduced coracoid region* and more correctly should be called a

**scapulo-coracoid.** In the rabbit the **clavicle** is very feebly developed and is a slender curved bone attached by ligaments to the acromion process and the manubrium of the sternum.

The **pelvic girdle** is a well-formed structure in which in each half—known as the **innominate** bone—the three bones **ilium, ischium, and pubis** are easily recognisable in a young animal. In the adult these appear to form a single bony structure on each side, the two halves of the girdle meeting in the mid-ventral line of the **pubic symphysis.** The ischium and pubis are separated by a wide aperture, the **obturator foramen,**\* to the formation of the lower margin of which the ischium also contributes. The ilia are stout and flattened and at their inner surfaces connect with the processes of the sacral vertebra. There is, however, no fusion of bones at this point for the ilium always remains separated from the sacrum by a thin layer of fibro-cartilage. The **acetabulum** lies at the common meeting point of the three bones, but the pubis does not contribute to it, for its ventral wall is formed by the small accessory bone known as the **cotyloid.**

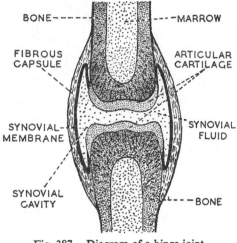

Fig. 387. Diagram of a hinge joint.

## JOINTS

Movements of the parts of the skeleton relative to one another are made possible by **joints.** Since the movements are varied, so are the joints. Some permit the bones to slide over one another (as in the ankle); others allow of movement in (almost) all directions when they are fashioned on the ball-and-socket principle (hip joint); or the parts may move on one another like the two halves of a hinge (elbow joint); or one part may rotate on the other (atlas vertebra on the axis). But whatever may be the type of joint, the jointing surfaces are always covered with a layer of cartilage. Any joint

---

\* This wide so-called foramen is really a fenestra, but there is included in it the obturator foramen proper through which the obturator nerve and vessels pass.

permitting free movement is enclosed in a fibrous capsule which is stronger and thicker in certain places and keeps the bones together. The capsule is lined by a serous synovial membrane which also covers the end of each bone, except where it is replaced by articular cartilage, and secretes the synovial fluid which fills the synovial cavity. This thick sticky fluid acts as a lubricant so that the whole arrangement permits of free movement of the parts upon one another. Joints which do not allow of much movement of the bones, as between successive vertebrae, or between the pelvic girdle and sacrum, are composed of pads of fibro-cartilage.

## HISTOLOGY OF CARTILAGE AND BONE

CARTILAGE. **Hyaline Cartilage.** Macroscopically this is a clear, bluish-coloured, glassy substance. It is familiarly known as

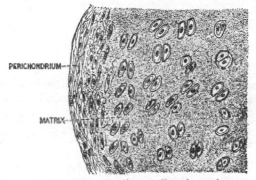

PERICHONDRIUM

MATRIX

Fig. 388.   Hyaline cartilage in section.

"gristle". It consists of a matrix of **chondrin**, in which a delicate network of collagen fibres can be demonstrated by means of special staining technique. The cells (chondroblasts) which secrete the chondrin lie in small groups, each cell in a fluid-filled space called a lacuna. Cartilage is always bounded by a fibrous membrane, the **perichondrium**, which bears numerous blood vessels from which nutritive substances diffuse through into the cartilage. Growth of cartilage is carried out by new layers being added to the outside, and, in the layers nearest to the perichondrium, the lacunae lie much closer together and the cells which they contain resemble the undifferentiated connective tissue cells from which they were derived.

Hyaline cartilage is found as the costal cartilages joining the ventral ends of the ribs to the sternum, thus forming an attachment sufficiently elastic to allow the ribs to move slightly; on the ends of the bones, where they form joints so that friction between the two bone surfaces is reduced; and in the cartilages of the larynx and trachea. The bulk of the embryonic skeleton is also cartilaginous, but is later to a great extent replaced by bone. Cartilage, although not so strong as bone, compares favourably with it in resisting tension or compression, properties which reside in its collagen component (including the perichondrium) and contained fluid. The matrix is slightly elastic and cartilage provides a means of lessening friction and "cushioning" the bones where they meet at joints. But cartilage can, as in the vertebrae of sharks, become highly calcified when it loses much of its elasticity and resembles bone in its mechanical properties. **Elastic cartilage** is fundamentally similar to hyaline and owes its great flexibility to the presence of numerous yellow fibres which run through the matrix in all directions. Elastic cartilage is found in the pinna of the ear, the end of the nose, in the epiglottis, and certain other regions. **Fibro-cartilage,**

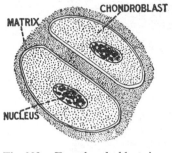

Fig. 389.  Two chondroblasts in lacunae.

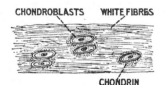

Fig. 390.  Fibro-cartilage.

typically found in the intervertebral discs, differs from the previous types in that the abundant fibres in the matrix are of the white variety. This tissue may be regarded as a transitional type between hyaline cartilage and areolar tissue.

BONE.—a typical limb bone such as the femur consists of a long shaft which is hollow for part of its length, and which is swollen out at the ends which bear articulating surfaces and various knobs for the indirect attachment of muscles. The cavity in the shaft is filled by a soft fatty tissue, the **marrow,** which in certain regions contains highly vascular sinuses and is called **red bone marrow** because of the blood it contains. (It is in these sinuses that the red blood-corpuscles and the granular white corpuscles are produced.) The bone is surrounded on the outside by a fibrous membrane, the **periosteum** (cf. perichondrium), whilst the cavity of the bone is lined by a more delicate membrane, the **endosteum.**

The texture of the bone itself is by no means constant, for while the shaft is formed of dense bone, towards the end of the shaft below the epiphysis the bone is spongy, being formed of an interconnecting system of bony bars with the spaces between them in continuity with the marrow. The extreme ends of the bone are again composed of dense bone. A longitudinal section through the entire bone reveals that the layers of bone are so arranged that they

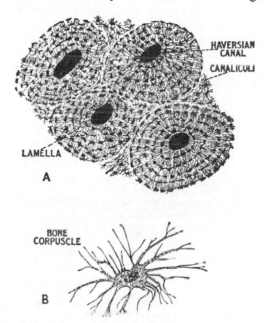

Fig. 391. BONE.—A, Haversian systems in a transverse section of dried material; B, a bone corpuscle.

resist any alteration in shape and give maximum rigidity for the least bulk of bone substance.

The microscopic structure of dense bone is fairly constant. The organic matrix contains small amounts of white fibres, but the matrix is so heavily impregnated with calcium salts, chiefly phosphates, that the fibres are somewhat difficult to detect. The importance of the organic matrix must not, however, be overlooked, for it forms about 30 per cent. to 40 per cent. of the weight of living bone, and its presence is clearly seen when bone is decalcified, when the histological appearance of the bone does not seem to be materially altered. The matrix is everywhere perforated by a series of fine canals (**Haversian canals**) which, although they join up with one another and with the blood vessels of the periosteum and bone marrow, yet run for the most part in a direction parallel with the long axis of the bone. Each of these Haversian canals transmits a small artery and vein for, unlike cartilage, the matrix is impermeable, and so the bone corpuscles or osteocytes receive their nourishment from near at hand. The osteocytes are arranged around the Haversian canals in series of concentric circles, each series together with the canal being called a **Haversian system**. It follows from this arrangement of the cells that the matrix is secreted in such a way that it forms a series of thin concentric cylinders around the Haversian canals. These cylinders are the bone lamellae. The osteoblasts themselves lie in small spaces, the lacunae, and these link up with neighbouring lacunae by a series of fine canaliculi

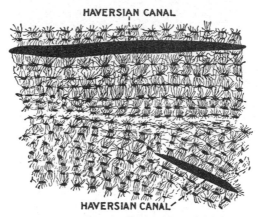

Fig. 392. BONE.—Longitudinal section of dried material.

which contains fine processes from the osteocytes, so that an organic protoplasmic network is formed, penetrating the matrix in all directions.

**Ossification.**—It has already been mentioned that certain craniates (*e.g.* Selachii) retain their cartilaginous skeleton throughout life; but in most instances the cartilaginous skeleton of the embryo is almost completely replaced in later life by bone. In addition, many bones are found in the adult which are never represented in the embryo by cartilages. It is thus usual to distinguish two kinds of bones—**cartilage bones,** which replace the embryonic cartilages, and **membrane bones,** which are not preformed in a cartilaginous model, but arise more superficially from specialised clusters of connective tissue cells.

The first stage in the ossification of a cartilage is that the cartilage cells multiply rapidly and, moving through the matrix, become arranged in columns roughly parallel to the longitudinal axis of the future bone. Then, beginning in the centre and gradually working towards the outside, the cartilage becomes calcified and many of its cells degenerate and die. From this stage onwards the perichondrium becomes known as the periosteum. The calcified cartilage is then invaded by embryonic connective tissue which passes through the periosteum. The vanguard of the invading tissues consists of large multinucleate amoeboid cells termed **osteoclasts** and, in company with certain of the surviving cartilage cells, they actively erode the calcified cartilage, so making channels for blood vessels and mesenchyme cells which follow in their wake. The erosion of the cartilage then proceeds from the centre outwards towards the periosteum. Certain of the invading mesenchyme cells, termed **osteoblasts,** then begin to lay down a series of bony bars (trabeculae), which for the most part are parallel to the long axis of the future bone, but are also interconnected by other bony strands. The

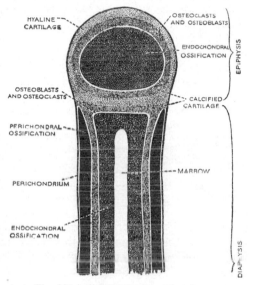

Fig. 393. Ossification in a long bone.

spaces in between the trabeculae are filled with blood vessels and connective tissue—the embryonic bone marrow. Ossification of this type—**endochondral ossification**—then gradually spreads outwards towards the periosteum. Whilst this process has been continuing, osteoblasts just below the periosteum have begun to lay down a layer of dense bone around the spongy endochondral ossification. Thus an outer layer of bone—**perichondral bone**—is formed. In its formation certain of the osteoblasts become enclosed in lacunae after which they are called osteocytes in the bone substance but, as the bone grows in thickness, they remain connected by way of canaliculi with the others which remain in a layer on the outside of the bone. For some time, however, Haversian systems are not formed.

In this way the cartilage gradually becomes replaced by bone which, approximately at any rate, is the same shape as the cartilaginous model. Bone formation does not, however, end here and frequently the adult bone, as the result of rearrangement of the bony tissue, departs quite materially from the form of its cartilaginous precursor. In some instances the final remodelling seems to require the stimulus provided by stresses set up by muscular action. Any gross transformation of the shape of the bone is accompanied by histological changes involving alternating activities of osteoclasts and osteoblasts, the bone being removed and then rebuilt in a different form. Quite apart from any remodelling it is usual to find that the layer of dense perichondral bone becomes invaded at numerous points by osteoclasts which eat channels into it, so paving the way for the incursion of blood vessels. The intercommunicating channels formed in this way are the Haversian canals. Moreover, the original endochondral ossification is destroyed by osteoclasts and is replaced by spongy (cancellous) bone by a new generation of osteocytes. In the limb bones a central cavity remains unossified to house the marrow.

Between the epiphyses and the main ossification of the shaft (diaphysis) a narrow band of cartilage persists and makes possible the growth of the whole bone *in length*, for whilst growth in thickness is simply effected by the addition of new bone beneath the periosteum, growth at the ends is not possible once muscular movement has begun. The ends of the bone always have a very definite form in relation with the joint, and this would be interfered with by the simple addition of bone at the ends. Growth in length is therefore carried out by the intercalation of new bone tissue on both sides of the cartilage separating epiphysis from diaphysis. Osteoclasts and osteoblasts behave in the same manner as in other parts of the bone. Eventually, when growth in length has reached its maximum, the whole of the cartilage is broken down and replaced by a layer of bone which firmly cements the epiphyses and diaphysis together.*

Other bones are never preformed in this way but are laid down direct in the embryonic connective tissue as follows: Connective

* The above account applies to the commonest type of ossification of a cartilage such as occurs in a limb bone. In smaller bones the process is simpler in that the layer of perichondral bone is not formed, and growth in thickness is due to the progressive expansion of the endochondral ossification during which process new cartilage is continually laid down below the periosteum. Strictly speaking, only this type of cartilage replacement gives rise to a true cartilage bone, for in the limb bones the perichondral ossification is really a layer of membrane bone, for it has no cartilaginous precursor.

tissue cells aggregate together and form a delicate fibrous matrix and then secrete small interconnected bars of calcified matrix. Some of the cells become enclosed in lacunae, whilst others remain on the surface of the newly secreted bone which thus grows in size by addition of new layers to its surface. The osteoblasts form for a long time a continuous sheet of cells over the surface of the growing bone, but they retain their connection with the cells which have been enclosed in the matrix by means of canaliculi. Bone which has been formed in this way remains of the spongy variety for a considerable time, although, as a rule, typical Haversian systems develop later on as a result of a rearrangement of the bone tissues.

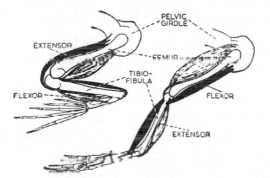

Fig. 394. Diagram of leg of frog to show flexor and extensor action. Contracted muscles grey; relaxed muscles—black.

In some instances, however (some of the bones of Teleost fishes), they are never present.

Bones thus laid down in condensations of embryonic mesodermal tissues (embryonic membrane) are termed membrane bones or dermal bones since many are formed in the deeper layers of the dermis.

## MUSCLES

Muscles are of three kinds: the skeletal muscles (striped muscle); visceral muscle (unstriped muscle) as, for example, in the wall of the gut; and heart muscle (cardiac muscle) in the wall of the heart.

It is important to remember that a muscle works in one way only—by its contraction. The movements of, say, a jointed limb, are not effected by one set of muscles only, but often by several, depending upon the kind of movement of which the parts are capable. Thus, the simple bending movement of parts articulated together by a hinge joint will be brought about by two opposing sets of muscles. One will flex the parts and the other will extend them. A simple example of opposed **flexors** and **extensors** is shown in Fig. 394. But a joint may permit of more complicated movements, combining both flexion and extension with rotation. In such instances the muscle complex will be more intricate and consist of groups of muscles, by the co-ordinated contractions of which the complex movements are brought about.

Each region of the hind leg of a frog, for example—thigh, shank, etc.—has its bones enclosed in a compact sheath of muscles. If the muscles are separated from one another, it will be seen that they are arranged in groups or disposed singly and that each group of muscles is attached at each end to the skeleton. But the arrangement of the muscles in relation to the skeleton will vary, as between the different groups, so that each group (or single muscle) will exert a pull in one

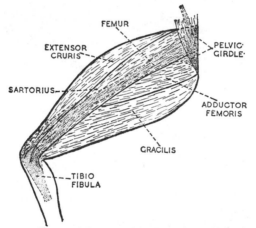

Fig. 395. Muscles of the ventral surface of the thigh of a frog.

direction only and therefore bring about only one kind of movement. Usually, in such a movement, the muscle responsible is attached at one end to a portion of the skeleton, which remains or is maintained (by other muscles) rigid, and the other end to the part which moves as a result of the contraction. The end which is attached to the rigid structure is called the **origin** of the muscle and that to the movable part, the **insertion**. Muscle origins are often broad and attached directly to the surfaces of bones, whilst insertions are frequently brought about by the white fibrous sheath of the muscle being extended to form a cord-like **tendon** which itself is attached—often to a projection on the bone.

Returning now to the hind leg of the frog, if, after the removal of the skin, it is viewed from the ventral surface, it will be seen that there are several bands or masses of muscle partially overlying one another and disposed in different directions. Some of these are evidently attached to the pelvic girdle at one end and, at the other, to either the femur or the tibio-fibula. The uppermost is a narrow band of muscle—the **sartorius**—which takes origin from the symphysised ventral parts of the iliac bones and is inserted into the inner (tibial) side of the tibio-fibula. In the middle of the thigh, and lying along the inner border of the sartorius to pass beneath its distal end, is a larger muscle, the **adductor femoris**. This muscle has its origin on the ischiopubic symphysis and is inserted into the distal portion of the femur. Still nearer the inner border, and partly overlying the adductor femoris, but distally passing beneath the sartorius, is the **gracilis**. This is made up of two slips of muscle, a major and a minor portion, the origin and insertion of which

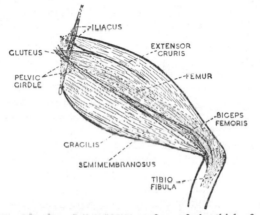

Fig. 396. Muscles of the dorsal surface of the thigh of a frog

is much the same. The outer part of the thigh is formed by a compound muscle, the **extensor cruris** (triceps femoris), which also extends to the dorsal surface. It is made up of three muscles combined together, as is evidenced by the three origins. The ventral portion arises from the ventral and anterior margins of the acetabulum; the middle or outer portion from the inner border of the proximal part of the ilium; and the dorsal from the hinder edge of the crest of the ilium. All these portions join together to form a single muscle mass which is inserted by a broad tendon into the head of the tibio-fibula.

On the dorsal surface, the two dorsal components of the extensor cruris occupy the anterior half of the thigh. Along the inner border of the extensor is seen the **biceps femoris**, posterior to which lies the broad **semimembranosus**. Along the inner border is found the edge of the gracilis. The biceps femoris takes origin from the crest of the ilium and is inserted into the head of the tibio-fibula. The semimembranosus arises from the ischial symphysis and is inserted into the head of the tibio-fibula. In this view, part of the iliacus can also be seen.

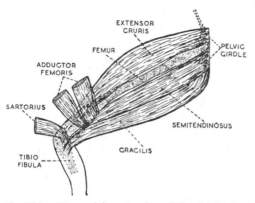

Fig. 397. Deeper-lying muscles of the thigh of a frog—from the ventral surface.

All the above named muscles can be seen on the surface of the thigh, but others lie beneath them. If, on the ventral surface, the sartorius is removed to reveal the insertion of the adductor femoris and then this, too, is cut away from its origin, the **semitendinosus** will be seen. This is a long, muscle made up of two bands with different origins; one from the ischium near the acetabulum; the other from the ischial symphysis. The two parts pass into a tendon which is attached to the tibio-fibula. The removal of the semitendinosus brings into view a group of small muscles related to the proximal portion of the femur. These are: the **psoas** muscle, arising from the hinder ventral part of the ilium and inserted into the anterior face of the femur about half-way along its length; the **pectineus,** arising from the pubis and attached to the ventral side of the femur a little nearer to the body than the psoas; and the **retractor femoris**, a tripartite muscle taking origin on the ischium and inserted into the posterior face of the femur. The most posterior part of this muscle is also visible from the dorsal surface.

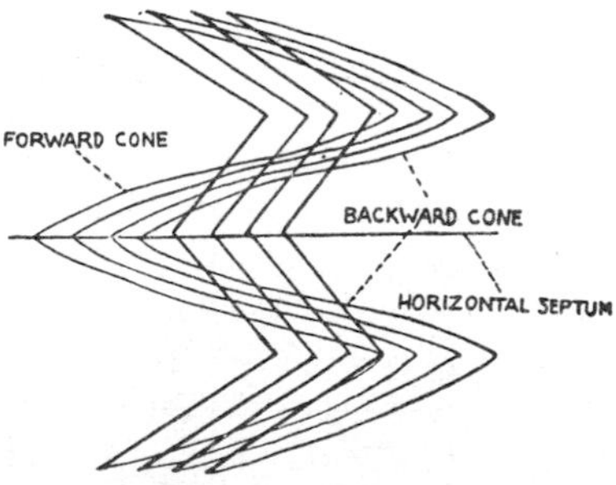

Fig. 398. DOGFISH.—Diagram to show the arrangement of the trunk myotome muscles—from the side.

The removal of the extensor cruris, biceps femoris, and semimembranosus from the dorsal side of the thigh brings into view the **gluteus**, the **caudofemoralis**, and the posterior component of the **retractor femoris** already seen from the ventral side. The **iliacus** can now fully be seen, originating from the dorsal border of the ilium and inserted into the posterior face of the femur.

One muscle remains, but it can only be seen by cutting away all the others. This is the **obturator internus**, which has a wide origin around the acetabulum and is inserted into the dorsal side of the femur.

Although the arrangement and disposition of these muscles gives some indication of the direction of the pull which they exert, it must be understood that, in life, the limb is moved in all sorts of ways and that these movements are brought about by the co-ordinated, combined interaction of several muscles or groups of muscles. Thus the sartorius, by its contraction, pulls the femur forwards (protracts it) and at the same time depresses its distal end. Similarly, the extensor cruris can also protract the femur but at the same time extends the shank. Again the biceps femoris protracts but raises the end of the femur and at the same time bends the knee. Opposed to these movements, the femur is depressed by the adductor femoris, gracilis, semimembranosus, and semitendinosus, the last three forming the flexor cruris as opposed to the extensor cruris since they also flex the shank. From the form of the hip joint, the femur can also be rotated—inwards by the iliacus and outwards by the obturator internus. Of the group of internally lying small muscles, the gluteus and caudofemoralis rather feebly pull the femur inwards and the retractor femoris in the opposite direction. Similarly, the psoas protracts and the pectineus depresses the femur. Thus by all sorts of co-ordinated combinations of the contractions, the limb can be moved in any direction.

The muscles of the limb of the frog have been dealt with in some detail in order to show the principles of interaction between them and the lever system formed by the limb bones. All of these muscles are derived from modifications of the embryonic myotomes but, as has been mentioned, in the fish these, to a large extent, preserve a fairly obvious segmental arrangement and the skeleton with which they interact during swimming is the vertebral column. The myotomes, it is true, give outgrowths to form the fin muscles but the bulk remain relatively unaltered to form the propulsive muscles, the chief modification which they undergo being a complex bending which is marked externally by a series of ⋛-shaped lines along the sides of the body. These lines indicate the insertion of the myocommata into the skin but do not give a true picture of the complexity of their folding in the deeper layers of the muscles. In reality each myotome is bent so as to form a series of cones which fit snugly into those of adjacent myotomes (Fig. 398). In the dogfish there are two forwardly pointing cones and two back-wardly projecting ones.  Each of these is obliquely attached to the median sheet of connective tissue and to the vertebral column in such a way that the contraction of the muscle-fibres in the anterior cone tends to bend the body backwards whilst the posterior cones tend to bend it forwards, but in all the cones the muscle-fibres are nearly parallel with the longitudinal axis.  The contractions of the successive but closely applied muscular cones therefore cause the body to become concave since on the opposite side of the body the corresponding cones are relaxed.  Thus when the successive myotomes contract the body is thrown into a series of undulations which result in swimming (p. 404).  At first sight this arrangement would seem to be unnecessarily complicated but it has the effect of producing smooth curves in the body, since one myotome over-laps several vertebral segments, and of reducing internal friction in the muscles.

## THE FUNCTIONS OF THE SKELETON

So far only one aspect of the interaction between the skeleton and the muscles has been emphasised, namely, that the various cartilages and bones form a system of levers which can be moved relative to one another by the contraction of opposing sets of muscles.  But, particularly in terrestrial animals, which, so-to-speak, cannot "make use of Archimedes' principle", much of the skeleton together with the muscles, tendons, and ligaments can be looked on as a system of girders whose function is, in part at any rate, to resist deformations due to the weight of the body.  Muscles and skeleton thus co-operate not only to cause movements but to maintain the shape of the body and the appropriate relations of its parts one to another when the animal is moving or at rest.

Most of the living girders of the body are designed to resist tension and shearing stresses as well as compression.  Usually compression and shearing are resisted by bone and cartilage: tension stress by muscles and ligaments.  For example, the vertebral column of the rabbit, together with the muscles and ligaments attached to it, can be regarded as a beam or girder carrying the weight of the body and supported near each end on the limbs.  Of course it has other more active roles to play and its bones and muscles actively contribute to hopping and leaping, during which activities the stresses on the girder continually change.

The compression members of this system are the centra and intervertebral discs.  These resist shortening of the beam.  The tension members are the sets of muscles and ligaments connecting the vertebrae and, at the hinder end, the pelvic girdle.  This system is much too intricate to describe in detail but a hint of how it works will be given by considering a few of the more obvious muscles associated with the lumbar and thoracic regions.

Appearing externally as a median mass of muscle above the vertebral column is a series of fibres running from the transverse processes and metapophyses of one vertebra forwards to become inserted on the neural spines of more anteriorly placed vertebrae. On each side of this series is a large muscle originating on the ilium and metapophyses of the posterior lumbar vertebrae and inserted on the transverse processes of those in front and, in front of these again, inserted on to the ribs. The long neural spines of the

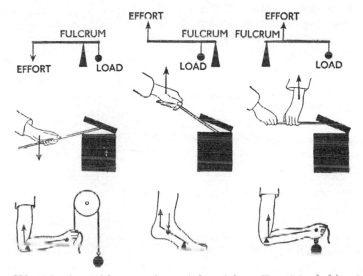

Fig. 399. Muscles and bones as levers (adapted from *How Animals Move* by Sir James Gray).

thoracic region are kept in place by a strong ligament and by fibres of a median dorsal set of muscles reaching back to the lumbar region.

The seven cervical vertebrae, although provided with muscles and ligaments, act mainly as a beam which resists compression whilst at the same time allowing of considerable movement. The head itself is held in position and its movements mainly controlled by muscles arising from the neural spines of the thoracic vertebrae and inserted into the postero-dorsal part of the skull.

The ways in which the limb bones function as levers of different kinds is fairly easy to understand and a few examples are shown in Fig. 399.

## HISTOLOGY AND PHYSIOLOGY OF MUSCLES

Leaving aside flagella, cilia, and pseudopodia, which have already been considered, the mechanical work of an organism is carried out by contractile tissues which form the muscles of the body. Here reference will be made primarily to those of vertebrates in which on histological grounds three main kinds of muscle can be recognised, namely, unstriped (smooth), cardiac (found only in the heart), and striped (striated) muscle which bulks by far the largest and of which most, but not all, is derived from the myotomes of the embryo.

**Unstriped Muscle.**—Of these kinds of muscle, unstriped is by far the simplest for it consists of elongated cells bound together by delicate connective tissue. It originates for the most part

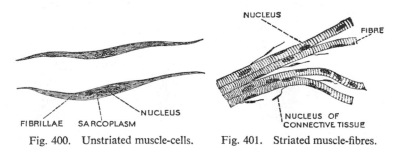

Fig. 400.  Unstriated muscle-cells.     Fig. 401.  Striated muscle-fibres.

from the lateral plate mesoderm and is usually arranged in sheets around hollow organs such as the gut or other viscera but it occurs also around the blood vessels and in the skin. It will be remembered that unstriped muscle (not histologically identical with but similar to that of vertebrates) is the only type of muscle found in many invertebrate animals, such as worms and many molluscs. Usually, vertebrate unstriped muscle has a double innervation, one set of nerves coming from the sympathetic and the other from the parasympathetic nervous system. Presumably this is bound up with the physiological properties of unstriped muscle for, even when all its nerves are cut, it will, under suitable conditions undergo "spontaneous" contractions and relaxations, sometimes rhythmically. Thus, both contractions and relaxations need to be controlled. When isolated from the body, smooth muscle is very extensible. A small applied force will cause it to stretch to several times its original length. Adrenalin causes some unstriped muscle to contract; acetylcholine causes it to relax and the effects are dependent on the source of the muscle. Thus these two drugs parallel the action of the sympathetic and parasympathetic nerves, respectively. The distribution of unstriped muscle gives some hint of its functions for it has the power of causing slow but sustained contractions which are carried out independently of the "will".

**Striped Muscle.**—The histological picture of striped muscle is much more complex than that of smooth muscle and is fundamentally similar in vertebrates and in arthropods. Striped muscle is so-called because under the microscope its fibres present a banded appearance and the reasons for this will be considered later. A muscle in the anatomical sense— say, our biceps—consists of a vast number of fibres united in bundles by connective tissue (perimysium) whilst these bundles or fasciculi are in turn bound together by the thick muscle sheath (epimysium) continuous with the tendons uniting the muscle to the skeleton.

The physiological units of striped muscle are the muscle fibres, ranging in diameter from 0·01 to 0·1 mm in diameter and, despite the fact that they are very long, only rarely running the whole length of a muscle. Each fibre is a mass of special cytoplasm in which many nuclei are embedded. Moreover, each fibre can be split into a number of fibrils and it is these which seem to be the essential contractile elements. The banding seen in a fibre is a property of the fibrils for they are built up of alternating light

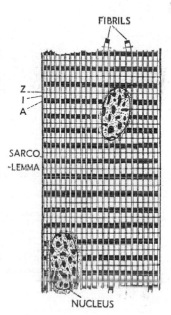

Fig. 402. Diagrammatic longitudinal section of part of a single muscle-fibre.

and dark discs with a periodicity of 2·3 μm in the resting condition and since these discs or bands are "in register" for all the fibrils in a fibre, the fibre as a whole has a regularly striped appearance. The dense bands are birefringent, or anisotropic (and are hence termed *A* discs). The less dense bands are practically non-birefringent, or isotropic (and are therefore designated the *I* bands or discs). Each *I* band is cut across its middle by a dense membrane, appearing as a line in microscopic preparations and termed the *Z* line or membrane, and this membrane continues right across the whole fibre to join at its edges the delicate membrane (sarcolemma) surrounding the fibre. The units between successive $Z$ lines are termed sarcomeres. The $Z$ membranes serve to hold the sarcomeres and fibrils together and each probably corresponds to two adjacent cell membranes reinforced by desmosomes so that a fibre can be regarded as a number of cells placed end to end. The $Z$ membrane may also have the important function of carrying the signal for contraction of the fibrils inwards from the surface of the fibre. Each $A$ disc is crossed by a somewhat lighter zone, the $H$ zone (Fig. 403).

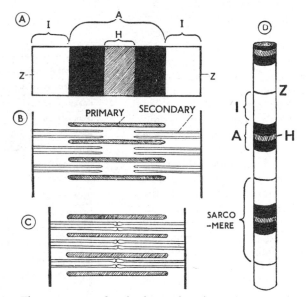

Fig. 403. Fine structure of striped muscle. A, surface view of a single sarcomere; B, L.S. single sarcomere to show position of filaments—relaxed state; C, L.S. single sarcomere with filaments partially contracted; D, part of a fibril (after H. E. Huxley).

So much can be seen with an ordinary microscope but much more has been revealed by electron microscopy and these details provide evidence for a working hypothesis for the mechanism of contraction. Now it is known that each fibril is built up of two quite distinct kinds of filaments. In cross-section it is seen that in the denser parts of the $A$ disc (*i.e.* the regions on each side of the $H$ zone) there is a hexagonal arrangement of primary filaments about 100 Å in diameter and spaced at intervals of about 200–300 Å. Between these are thinner, secondary units, about 40–50 Å in diameter so arranged that each primary filament is surrounded by 6 secondary ones. But sections through the central part of each *A* disc (*i.e.* in the *H* zone) show only the hexagonal sets of primary filaments. On the other hand, sections through the *I* discs show only the thinner, secondary filaments. From this it is concluded that each fibril is composed of two overlapping series of filaments. The primary filaments are confined to the *A* disc or band and confer on it its high optical density and birefringence. The secondary filaments extend in each direction from the *Z* line through the *I* band and for some distance into the *A* band where they overlap with the primary filaments. Thus the striated appearance is in reality due to the contractile substance being composed of two different kinds of overlapping filaments.

Fig. 404. Diagrammatic transverse sections through part of a muscle-fibre. A, in the *I* band (secondary filaments only); B, in region of overlap in the *A* band (primary and secondary filaments); C, in the *H* zone (primary filaments only).

By the use of phase contrast microscopes it has been possible to measure the changes in length undergone by the various bands when a muscle-fibre contracts. These are a fraction of a $\mu$ but are additive throughout the length of a fibre. On contraction the distance between successive *Z* lines shortens but the length of each of the *A* bands remains the same. The *I* zones and the *H* bands also shorten and finally disappear. These effects can be explained

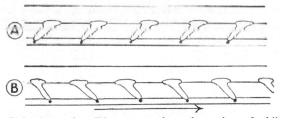

Fig. 405. Striped muscle. Diagram to show the action of oblique bridges which pull secondary filaments inwards. A, relaxed state; B, contracted state. (Based on a figure by H. E. Huxley.)

if it is assumed that the two sets of filaments remain of constant length but that the secondary filaments slide inwards between the primary ones.

Muscles contain about 20 per cent. protein and about 80 per cent. dilute solutions of salts. About 65 per cent. of the protein is of a fibrous nature whilst the remainder is largely composed of enzymes concerned with converting by a series of steps (p. 547) glycogen into $CO_2$ and water. But adenosine triphosphate, a compound which breaks down into adenosine diphosphate, supplies the energy for contraction. About half the weight of the fibrous proteins is composed of myosin whilst the rest is mainly a protein called actin. Remarkably, it was found that myosin also acts as the enzyme which catalyses the breakdown of ATP to ADP. Actin and myosin can be extracted separately from muscle, and microscopic studies reveal that removal of myosin causes the almost complete disappearance of the dense material of the $A$ bands, that is, of the primary filaments. If afterwards the actin is removed the secondary filaments can no longer be seen. This shows how the two different proteins are localised. When in solution together actin and myosin combine to form actomyosin from which fibres can be precipitated. These will contract if supplied with ATP immediately, suggesting a parallel with what may happen in the living muscle, especially as electron-micrographs of longitudinal sections of fibrils reveal that the primary (myosin) filaments and the secondary (actin) filaments are linked by a complicated system of oblique cross bridges. It is believed that these cross bridges pull the secondary filaments inwards, return to their original position, and so on, many times a second, their combined action being analogous to a sort of ratchet energised by the breakdown of ATP to ADP. When the ATP is no longer being broken down the bridges no longer combine and the muscle can relax and can be elongated if stretched.

**Electrical Properties of Muscles.**—If an insulated electrode of minute size is inserted within a fibre and another one placed on the surface of the sarcolemma, a potential difference of the order of 50-100 millivolts can be detected. This is called the resting potential. The outside of the fibre is positive with respect to the inside and in this respect muscle-fibres resemble nerve-fibres and many cells. This resting potential results from the differential permeability of the surface membrane to anions and cations so that the inside of the fibre has an excess of anions which are too large to pass out through the membrane. Moreover the membrane is more permeable to potassium than to sodium and more potassium than sodium occurs inside the muscles and *vice versa* outside. In some way not understood, sodium ions are actively pumped out of the fibre. When a muscle is stimulated to contract, either artificially or by the arrival of nerve impulses, there is a sudden change in the electrical potential. The resting potential drops to zero and "overshoots" so that momentarily the inside of the fibre becomes positive with respect to the outside. This change can be recorded and is called the action potential. The changes are due to a sudden increase in the permeability of the membrane so that sodium ions rush in along the concentration gradient and the electrical gradient. In the recovery phase, when the membrane becomes repolarised, sodium ions are "pumped" out again. The sudden depolarisation of the membrane is rapidly propagated along the fibre by local electrical currents and inwards to the fibrils *via* the Z membranes. Presumably it is these currents which in some way cause the breakdown of ATP and lead to contraction.

**Neuromuscular Junctions.—** When a motor nerve-fibre reaches a muscle-fibre the myelin sheath is lost so that the nerve-fibre itself comes into contact with the muscle-fibre membrane. The nerve-fibre may branch there to form fine terminal twigs or, as in mammals, terminate in compact motor end plates. Electrodes placed within motor end plates show that when a nerve is stimulated an end plate potential arises due to a local alteration in the permeability of its membrane under the influence of acetylcholine released in minute amounts by the nerve endings. The end plate potential then stimulates the spread of the action potential along the muscle

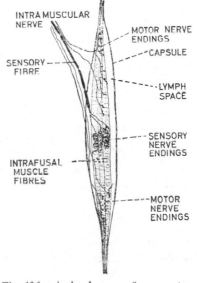

Fig. 406. A simple type of mammalian muscle spindle. Surrounding muscle-fibres not shown.

membrane. Acetylcholine is rapidly destroyed by the enzyme, choline esterase, so that a muscle is not continuously excited. Certain fibres in striped muscle are modified to form proprioceptors which are supplied by fibres of afferent neurons and so provide information about the degree of contraction of the muscle. The tension of the proprioceptive fibres can be altered by efferent neurons so altering the sensitivity of the sense organ. These proprioceptors are called **muscle spindles** in mammalian muscle and are shown in Fig. 406, but proprioceptors similar in function occur in other vertebrates and in arthropods.

**Physiology of Whole Muscles.**—Striped muscles contract only when stimulated—either directly as by mechanical, chemical, or electrical means, or indirectly by a nerve. Many of the properties of muscles can most conveniently be studied by electrical stimulation and a suitable apparatus for recording the contractions. It is found that a current of a particular strength must flow for a certain time to cause a contraction. The smallest current, no matter how long it flows, which is necessary to cause contraction is called the threshold stimulus or rheobase. This will vary according to the type of muscle chosen and the excitability of a particular muscle can be measured by the shortest time for which a current of twice the rheobasic strength must flow to produce a response. This measure is known as the chronaxie. For the muscles of a frog's leg it is of the order of 0·7 milli-seconds. When two subliminal stimuli are given close together they may together give a response. This is called the summation of stimuli. A muscle does not contract immediately a stimulus is applied. There is a short delay, the latent period, which is of the order of a few milli-seconds, before the onset of the active state which, itself, is not fully attained until after about 20 ms (frog's sartorius) after which it declines for a period of about 40 ms. But the latent period varies with temperature, condition of the muscle, and load applied. The greater the tension the longer the muscle takes to develop that tension. If a muscle is stimulated by a single short electric shock it gives a brief contraction termed a twitch, that is, a short contraction followed by a relaxation. If the load is very slight and the muscle can contract freely the response is called an isotonic contraction but under large loads which prevent free contraction but allow tension to develop, the contraction is said to be isometric for the length of the muscle does not alter.

As far as individual fibres are concerned they either contract maximally or not at all. This is the "all or none rule", but as is very obvious, a gradation of response can and does occur in whole muscles. In striped muscles this is brought about by a variation in number of fibres brought into play at any one time. The "all or none rule" applies only under the conditions of the experiment. For example, if a fibre is stretched beyond its normal length it contracts more strongly when stimulated. If a second stimulus is applied during the beginning of the excitatory state there is an additional contraction but the sum of the two contractions is less than twice that of the first. A series of stimuli gradually increases the size or strength of contraction until the muscle no longer momentarily relaxes but remains contracted, a state known as tetanus. Of course, if the stimuli are too widely spaced in time relaxations do occur and no tetanus results.

In fact there is a short period following a contraction when a muscle is inexcitable, the refractory period, which is about 5 ms for frog's sartorius muscle. This feature together with that of the threshold value serves to explain why muscle-fibres obey the "all or none rule", for as soon as the threshold value stimulus is applied contraction begins and no further increase in the stimulus can have any effect since the fibre becomes refractory.

When a muscular movement is made the motor centres in the central nervous system send impulses via the motor nerves to the muscles which contract in the appropriate sequence and in such a manner that the appropriate degree or strength of contraction is attained. In a very general sort of a way the principles by which this is attained can be compared to machines employing a "feedback" principle—like the governors on a steam engine. Proprioceptors (muscle spindles) excite sensory nerves which "inform" the central nervous system of the state of contraction and the response is modified or, as we say, "monitored" *via* the nervous arcs of the somatic motor system. The stronger the contraction required, the greater is the number of fibres brought into play.

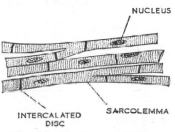

Fig. 407. Cardiac muscle-fibres.

A single somatic motor fibre supplies many muscle-fibres scattered throughout the muscle, the total number of fibres supplied constituting a motor unit. For muscles giving gross movements only, one nerve-fibre supplies many—up to several hundreds of muscle-fibres. But for fine discrimination as few as ten fibres per motor unit are found, as for example, in the eye muscles.

**Cardiac Muscle.**—As its name implies, this is found in the walls of the heart and there alone. Its structure is in some respects intermediate between that of unstriped and striped muscle for, although composed of separate cells, its fibres are united by bridges to form a contractile network. Its fibres are cross-striped but not surrounded by a definite sarcolemma. Where the cells meet end-to-end, they form intercalated discs, rather like Z membranes in striped muscles and which are cell membranes plus desmosomes (Fig. 407). Cardiac muscle has the property of contracting rhythmically, even when isolated from the body and has a very long refractory period. It is immune to fatigue or to tetanus by repeated stimulation at short intervals. During embryonic development the muscular wall of the heart arises from the splanchnic mesoderm cells just below the pharynx and almost from the first

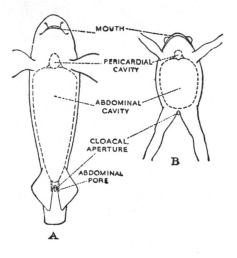

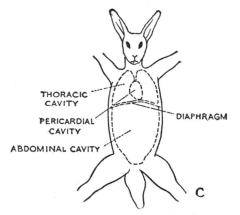

Fig. 408. Diagrams to show the body cavities in A, a dogfish; B, a frog; C, a rabbit.

begins its rhythmic contractions. Only later does it receive its nerve supply from the sympathetic and parasympathetic nervous system which modifies its strength and speed of contraction. Its beat is therefore said to be intrinsic or myogenic.

## THE BODY CAVITY

In the craniates, the internal organs or viscera are suspended in a body cavity—the **splanchnocoel**—which is derived from the embryonic coelom. The viscera protrude freely into this cavity but are completely covered by the coelomic epithelium. This, in some instances, may become pulled out into double sheets between which connective tissue is laid down to form mesenteries, by which various organs may be slung from the dorsal wall of the body cavity or connected with one another.

As will be seen when the blood vascular system is considered (p. 510), during the development of the heart in all craniates, a portion of the coelom surrounding the heart becomes separated from the main cavity to form the pericardial cavity enclosed by the pericardium (pericardial membrane). In the lower vertebrates the remainder of the coelom remains undivided, but in the mammals the abdominal coelom (splanchnocoel) becomes separated from the thoracic by the development of a transversely-running muscular septum, the dia-

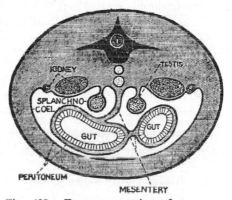

Fig. 409. Transverse section of a young generalised vertebrate to show the relations of the coelomic epithelium to the viscera.

phragm. Further, the thoracic cavity itself becomes subdivided by the folding of the peritoneum around the organs to form the two laterally situated **pleural cavities** surrounding the lungs (the coelomic epithelium—or peritoneum—forming the pleura) and a median cavity, the **mediastinal cavity,** surrounding the median structures such as the blood vessels, oesophagus, etc.

Although in the embryo the coelom forms a spacious cavity surrounding the developing gut, in the adult its dimensions become considerably reduced by the enlargement of the various organs as they grow rapidly in size and push the peritoneum in front of them. Yet a cavity still persists and the coelomic fluid secreted by the epithelium acts as a lubricant, enabling the viscera to slip easily over one another.

# THE ALIMENTARY SYSTEM

## INTRODUCTION

The alimentary canal is a tube beginning at the mouth and passing through the body to the anus. The greater part of the tube is lined by endoderm, forming a glandular epithelial lining of a variable nature termed the **mucosa**, but at the anterior and posterior ends ectoderm has been invaginated to form a stomodeum and proctodeum respectively. The wall of the alimentary canal is completed by tissues of mesodermal (splanchnic) origin and consists mainly of muscle and connective tissue together with the blood vessels and nerves. Outwardly the tube is invested by the peritoneum (coelomic epithelium), which also covers the mesenteries by which it is suspended in the body cavity.

Primarily, the alimentary canal is for the reception of food, but in all vertebrates part of the anterior region plays a part in respiration. It originates in the embryo as a simple straight tube, but in the adult it is differentiated into regions which differ according to the particular work carried out therein. The tube is usually much longer than the body, so that since the two ends are fixed, it is thrown into coils.

The main regions of the alimentary canal are the **buccal cavity, pharynx, oesophagus, stomach**, and **intestine**.

**The Buccal Cavity.**—This is derived from the stomodeum and consequently is lined by ectoderm and the mouth aperture opens directly into it. As derivatives of the skin covering the jaws, teeth are present, the form and functions of which differ in different vertebrates. In the higher craniates a variable shaped, movable, muscular tongue is found on the floor of the buccal cavity. In many instances, even in the lower vertebrates, there is no clear line of demarcation between the buccal cavity and the next region, the pharynx, but developmentally the former is lined with ectoderm and the latter by endoderm. The two regions can be called the bucco-pharyngeal region.

**The Pharynx.**—It is in this region that, in the embryo, the visceral clefts make their appearance. In the fishes they persist in the adult as branchial clefts, but in the higher vertebrates for the most part they close up and disappear. It is from the hinder end of this region that the lungs, the respiratory organs found in air-breathing vertebrates, originate. The lungs may arise directly from a laryngeal chamber or when a neck is present, as in the mammals, they are situated at the end of a tubular trachea. In those vertebrates where the majority of the visceral clefts have been lost, the pharyngeal part of the first cleft is retained as part of the auditory apparatus to form the Eustachian tube and tympanic chamber.

**The Oesophagus.**—Following the pharynx is the oesophagus in which, either by the flattening of the tube or the folding of its lining, the lumen is considerably reduced. The oesophagus forms a connecting channel between the pharynx and the next region, the stomach.

**The Stomach.**—In this region the tube is dilated to form a receptacle in which the food can accumulate during feeding and also where some digestive processes can take place. The form of the stomach varies in different animals, according to their type of food and mode of life, but usually regurgitation is prevented by a valvular arrangement, the **cardia**, where the oesophagus enters the stomach. The premature escape of the food from the stomach into the intestine is guarded against by the development in the wall of the distal (pyloric) end of the stomach of a ring of muscle termed a sphincter, the **pyloric sphincter**, which closes the aperture and only opens under certain conditions. The position of this sphincter is usually marked externally by a constriction, the pyloric constriction. In the wall of the stomach numerous simple glands, the peptic or gastric glands, are present which secrete the gastric juice.

**The Intestine.**—Following the stomach comes the longest part of the alimentary canal, the intestine. In most vertebrates the intestine can be divided into regions distinguishable by morphological and histological features. Into the first part, which immediately follows the stomach, open the ducts of the two important accessory glands, the liver and pancreas, and this region is called the **duodenum**. The next part of the intestine is the region where a great part of digestion and most of the absorption of the digested food takes place, and is appropriately modified in different vertebrates (see pp. 484-91). The lining epithelium provides numerous glands, the products of which are poured out on to the food. The form of this region varies considerably in different vertebrates as also does its terminology, and details must be left to the description of individual examples. The terminal portion of the intestinal region is the **rectum**, which includes the proctodeum and terminates at the anus.

The passage of the food through the alimentary tube is caused by waves of muscular contraction of a rhythmical character which pass along the length of the tube, pushing the food in front of the constricted region. This rhythmical contraction is termed **peristalsis**.

**The Accessory Glands.**—In the higher vertebrates **salivary glands** are present. The salivary secretion or saliva is poured on to the food during mastication. In other instances, where salivary glands are absent, the lining epithelium of the buccal cavity and pharynx contains many mucus-secreting glands which help to lubricate the food during its passage through these regions. The **liver** and **pancreas**, already mentioned as opening by their ducts into the duodenum, arise in some animals (*e.g.* chick) as outgrowths of the endoderm of the alimentary canal, but soon mesodermal derivatives become incorporated with these outgrowths to form the adult glands, the connection between the gland and the intestine being retained as the duct. The secretions of these glands, bile and pancreatic juices, play an important part in the digestive processes.

In addition to these special glands, the lining epithelium of all regions of the alimentary canal contains numerous mucous glands.

## THE DOGFISH

The mouth of the dogfish is wide and ventral in position. In the skin covering the jaws are several rows of enlarged dermal denticles which serve as teeth, and are used for the prevention of escape of living prey. The buccal cavity, into which the mouth opens, is wide from side to side but compressed dorso-ventrally, and the lining membrane is rich in mucous glands. In the floor of the bucco-pharyngeal cavity is the so-called tongue, really a pad of tissue over the basihyal cartilage. This is of use in swallowing by pressing the food backwards against the hard floor of the cranium which forms the roof of the buccal cavity.

The pharynx is wide laterally and compressed dorso-ventrally, and there is no clear indication where the buccal cavity ends and the pharynx begins. Opening into the pharynx are the internal apertures of the branchial clefts, which take the form of elongated slits and the spiracle. In the wall of the pharynx in between the clefts is the visceral arch skeleton with its musculature, by means of which, by the lowering of the floor, the cavity can be enlarged and by the subsequent raising, reduced.

Following the pharynx is the short, wide oesophagus, the lining of which is much folded, so that normally the cavity is practically closed, preventing the entry of the water which is continuously passing through the pharynx. When, however, food is pushed during swallowing into the beginning of the oesophagus, the foldings allow of considerable dilatation and the food is passed quickly into the stomach.

The stomach is bent into the form of a **U**, the **cardiac limb**, into which the oesophagus opens, being wider than the **pyloric limb**.

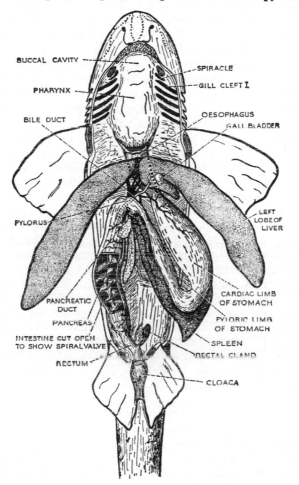

BUCCAL CAVITY

PHARYNX

BILE DUCT

PYLORUS

PANCREATIC DUCT

PANCREAS

INTESTINE CUT OPEN TO SHOW SPIRAL VALVE

RECTUM

SPIRACLE

GILL CLEFT I

OESOPHAGUS

GALL BLADDER

LEFT LOBE OF LIVER

CARDIAC LIMB OF STOMACH

PYLORIC LIMB OF STOMACH

SPLEEN

RECTAL GLAND

CLOACA

Fig. 410. DOGFISH.—The alimentary canal.

At the distal end of the pyloric limb is the pyloric sphincter, and beyond this the alimentary tube bends backwards to pass in an almost straight line to the cloaca.

Externally, the intestine increases considerably in width as it passes backwards, though there is very little demarcation into regions until the narrower rectum is reached. However, the narrower beginning is called the duodenum, since it receives dorsally the **bile duct** and ventrally the **pancreatic duct**. It is quite short and passes insensibly into the next region, which is usually called the **ileum**. Within the ileum is found a remarkable structure for increasing the absorptive surface of the intestine without adding to its length. Here the lining of the tube—together with the underlying tissues—becoming partially separated from the wall and twisted into a spiral to form the so-called **spiral valve** through the turns of which the food slowly passes. The ileum continues into the rectum, which passes back directly to the cloaca. Near the beginning of the rectum is a small brown body, the **rectal gland**, which excretes excess sodium chloride taken in with the food.

The **liver** is a large bilobed organ occupying much of the body cavity. Anteriorly it is suspended from the pericardio-peritoneal septum by a short, strong double sheet of peritoneum called the falciform ligament. The left lobe is frequently shorter than the right and is subdivided anteriorly to form a smaller median lobe. The gall bladder is usually almost completely embedded in the small median lobe. From the gall bladder arises the **bile duct**, which also receives a tributary duct from each main lobe of the liver. The bile duct passes backwards in the mesentery and enters the posterior end of the duodenum on its dorsal side.

The **pancreas** has two lobes, a larger elongated dorsal lobe—which lies between the pyloric limb of the stomach and the intestine —and a smaller ventral lobe lying close against the duodenum. From the ventral lobe arises the pancreatic duct which runs along in the wall of the intestine before entering it.

Although it has no physiological relation with the alimentary canal the spleen may be noted here, attached as a fringe to the pyloric limb of the stomach.

## THE FROG

In the frog the mouth is terminal and wide. Teeth are present only in the upper jaw and on the vomers in the roof of the buccal cavity. In the floor of the bucco-pharyngeal cavity a protrusible tongue is present, the arrangement of which is rather unusual. Instead of being attached at the back of the cavity and projecting forwards, its point of attachment is anterior, immediately within the margin of the lower jaws, so that it lies with its bifid end directed backwards towards the pharynx. This is of advantage in the method of feeding (see p. 502) adopted by the animal. The buccal cavity itself is wide from side to side and in its floor, beneath the tongue, lies the body of the hyoid, by the raising and lowering of which the internal capacity of the bucco-pharyngeal cavity can be increased and decreased—an arrangement which is used in breathing. At the anterior end of the buccal cavity are the two internal nostrils or nares, which communicate with the exterior through the olfactory organs and the external nostrils. The lining epithelium of the bucco-pharyngeal cavity, besides containing numerous mucous glands, is richly ciliated.

The pharynx is imperfectly delimited from the buccal cavity, but opening into it are the two wide, laterally placed apertures of

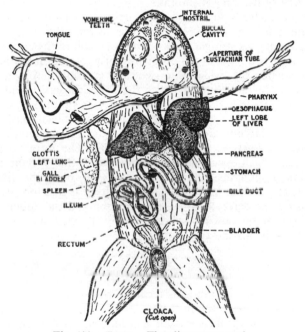

Fig. 411. FROG.—The alimentary canal.

the **Eustachian tubes**—leading to the tympanic cavity of the middle ear—and in the floor the slit-like glottis opening into the laryngeal chamber. Dorsal to the glottis is the opening into the oesophagus.

The oesophagus is short and its lining is folded longitudinally to allow for expansion during the passage of food along it to the stomach into which it merges almost insensibly.

The stomach is a thick-walled, slightly dilated portion of the alimentary tube and in the lining epithelium, which is folded longitudinally to provide for expansion as the food accumulates, glands are present. The termination of the stomach is marked by the pyloric constriction but there is no bending of the tube or division into regions.

Following the stomach is the intestine, the first part of which, though not clearly demarcated from the rest, is the duodenum into which the common hepato-pancreatic duct opens. The next region, variously called the small intestine or the ileum, is the longest part of the alimentary canal and is thrown into several coils to accommodate it within the body cavity. This increase in length adds considerably to the absorptive area which is further enhanced by the folding of its lining membrane.

The ileum leads to the wide, short rectum (sometimes termed large intestine), the transition from one to the other being abrupt. The rectum passes straight back to and opens into the cloaca.

The liver is a large organ consisting of two main lobes of which the left is again subdivided into two, and is suspended by a mesentery. Between the main lobes lies the spherical gall bladder from which arises the bile duct, which also receives supplementary ducts from the main lobes of the liver. On its way backwards to the duodenum, the bile duct passes through the substance of the pancreas, the ducts of which open into it. The terminal portion of the duct which opens into the duodenum is therefore an hepato-pancreatic duct conveying the products of both the liver and pancreas to the intestine. The pancreas itself is an elongated body lying in the mesentery between the stomach and the duodenum.

The spleen is a dark-red spherical body attached to the mesentery between the ileum and the rectum.

## THE RABBIT

The mouth in the rabbit is terminal and relatively small. It is bounded by the movable lips, the upper of which is divided in the middle to expose the incisor teeth, a condition which is an advantage in gnawing. As has been explained in dealing with the skull, the bucco-pharyngeal cavity is separated from the respiratory channel by the **palate** (made up of hard and soft portions), and in both upper and lower jaws teeth are present. Along the floor of the buccal cavity lies the muscular protrusible tongue which is attached just in front of the body of the hyoid and is directed forwards. Into the buccal cavity open the ducts of the salivary glands of which, in the rabbit, there are four pairs; the **parotid glands** in the cheek immediately beneath the external auditory meatus, the **zygomatic glands** deep in the cheek, the **submaxillary** (or better **submandibular**) **glands** on the inner side of the angle of the lower jaws, and the **sublingual glands** beneath the tongue. Dorsally, the roof of the food channel is continued backwards as the membranous soft palate, the free border of which projects into the pharyngeal cavity.

The pharynx is relatively short and wide, and there open into it the posterior nares of the respiratory channel. Posteriorly, the pharynx leads into the oesophagus dorsally and the larynx ventrally. The entrance to the larynx, the glottis, is guarded by the epiglottis, a cartilage also covered by the pharyngeal lining. Into the pharynx open also the internal apertures of the Eustachian tubes from the middle ear.

The oesophagus is a long narrow tube leading from the pharynx through the neck and thorax to the abdomen. Its lining membrane of stratified epithelium is very much folded, and at the opening leading from the pharynx its lumen is almost closed to prevent the swallowing of air during breathing. Immediately after its entry into the abdomen it opens into the stomach.

The stomach is the most dilated portion of the alimentary

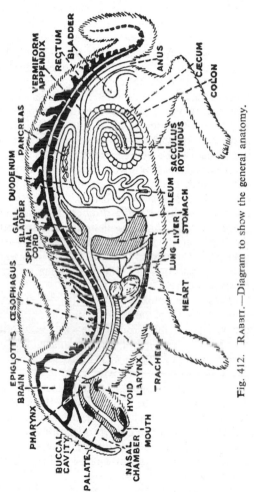

Fig. 412. RABBIT.—Diagram to show the general anatomy.

canal and takes the form of an ovoid sac or bag which lies transversely in the abdomen. Where the oesophagus enters the stomach there is a valvular arrangement called the cardia to prevent the regurgitation of food, and this strongly rounded region of the stomach is consequently called the cardiac region. The somewhat narrower pyloric region has at its termination the pyloric constriction and sphincter. In the lining of the stomach are the gastric glands and the wall is strongly muscular, the fibres passing in different directions around the organ.

Passing from the stomach is the proximal part of the intestine, the small intestine whose first part is the duodenum, which takes the form of a U-shaped loop, in the mesentery of which is the pancreas, a somewhat diffuse gland in the rabbit but solid and compact in many mammals. The bile duct opens into the somewhat expanded beginning of the duodenum and the pancreatic duct into the distal limb of the U. Following the duodenum is the rest of the small intestine, which is by far the longest part of the alimentary canal and forms a series of loops and coils. This great length provides an extensive area for absorption, but this is further considerably increased by the development of folds and, from its inner surface, of innumerable tiny finger-like projections called villi. Each villus has its own extensive blood and lymphatic supply. In the rabbit the small intestine shows little differentiation, but in some mammals the region following the duodenum is termed the jejunum and the more distal portion the ileum. In the wall of the duodenum, at the bases of the villi, are special glands—Brünner's glands—which open into the crypts of Lieberkühn. In the ileum, interspersed at intervals among the villi, are areas of lymphoid tissue termed Peyer's patches.

From the small intestine the direct course of the alimentary canal is into the large intestine or colon, but in mammals there occurs at the junction of the two, a blindly ending diverticulum called the caecum which terminates in a narrower vermiform appendix. This three-way junction is marked by a swelling, the sacculus rotundus, into which the ileum opens and within which is a valvular arrangement, the ileo-caecal valve, by which the contents of the ileum can be directed into the caecum before passing into the colon. The form of the caecum varies in different mammals and in the rabbit, whose food is almost exclusively vegetable, the caecum is large, consisting of a series of pouch-like dilatations. Terminally, the caecum narrows somewhat abruptly to form the smooth vermiform appendix. The colon is narrower than the caecum but resembles it externally in that its proximal part is sacculated. Distally it passes into the rectum.

The **rectum** in the rabbit is a long simple coiled tube passing backwards from the colon to the anus.

The **liver** is very large, and is made up of five lobes, the right and left central, the left lateral, the caudate, and the Spigelian lobes.

Its detailed structure is given on p. 495. The gall bladder is large and lies between the right and left central lobes. Arising from it is the main bile duct, or the cystic duct, which receives subsidiary ducts from the various liver lobes and opens into the rather swollen beginning of the duodenum.

The spleen is a dark-red body lying close to the posterior cardiac region of the stomach.

## THE HISTOLOGY OF THE ALIMENTARY CANAL AND ASSOCIATED GLANDS

**Tongue.**—The tongue is largely composed of interlacing bundles of striped muscle. It is covered by a layer of stratified epithelium,

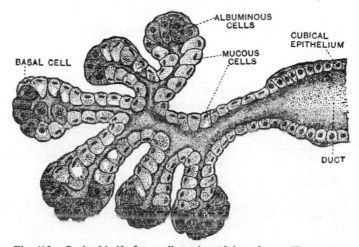

Fig. 413. Sagittal half of a small portion of the sub-mandibular gland.

which is smooth on the lower surface but raised up on the upper surface of the tongue to form different kinds of ridges and papillae. Below the epithelium is a layer of loose connective tissue, and embedded in the epithelium of the papillae are **taste buds**. These are **chemo-receptors**, *i.e.* they respond to stimuli of chemical substances in solution. The structure of the tongue is shown in Fig. 417.

**Sub-mandibular Gland.**—This is chosen as an example of a compound racemose gland and, apart from finer points of difference, other salivary glands have the same essential structure. It is a gland of the "mixed" type, producing both a watery, albuminous secretion and a mucous secretion (see Fig. 413).

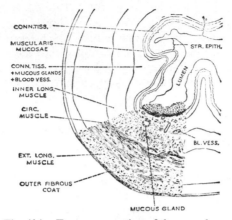

Fig. 414.   Transverse section of the oesophagus.

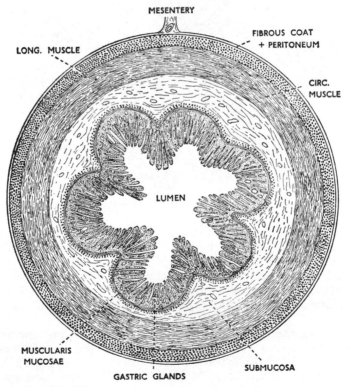

Fig. 415.   Diagrammatic transverse section of a frog's stomach.

**Oesophagus.**—The structure of the oesophagus is shown in transverse section (Fig. 414). An interesting point of variance to be noted is that, whereas the muscles in the anterior third are of the

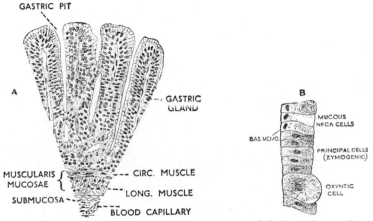

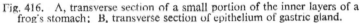

Fig. 416. A, transverse section of a small portion of the inner layers of a frog's stomach; B, transverse section of epithelium of gastric gland.

striped variety, in the middle third they are partly striped and partly unstriped, whilst, finally, in the posterior third they are entirely unstriped.

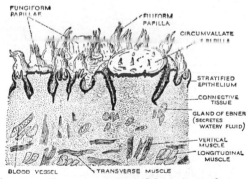

Fig. 417. TONGUE.—A part of the upper surface.

**Stomach.**—The wall of the stomach is composed of the layers shown in the diagram (Fig. 415), which is a transverse section of the stomach of a frog chosen for illustration since it is conveniently small. The epithelium which lines the stomach is of the columnar type and glandular in function. The gastric glands (Fig. 416) are to be regarded as devices for increasing the surface area of the glandular epithelium and so producing a large volume of gastric juice. The nature of the juice varies in different regions of the

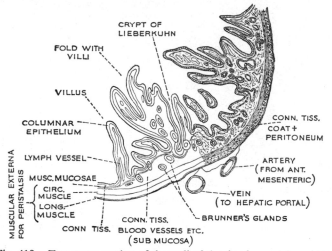

Fig. 418. Transverse section of the wall of the duodenum of a rabbit.

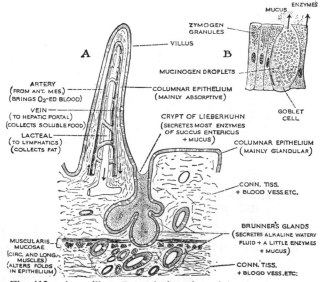

Fig. 419. A, a villus; B, vertical section of duodenal epithelium.

stomach and the various constituents of the juice are secreted by different types of cells. Mucus is secreted by the surface cells of the epithelium and by those lining the ducts of the gastric glands, hydrochloric acid by oxyntic cells, and pepsin by the peptic cells. Oxyntic cells are absent from the glands of the pyloric region, and juice secreted in this region contains no free hydrochloric acid.

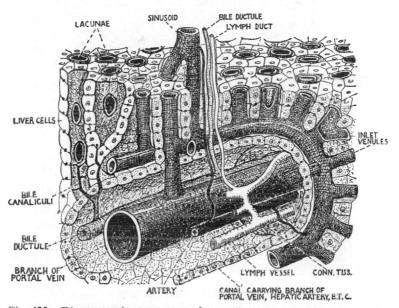

Fig. 420. Diagrammatic stereogram of a small portion of liver to show one of the portal canals carrying branches of the hepatic portal vein, hepatic artery, bile duct, and lymph vessels. From the portal vein inlet, venules run through the liver and branch to form capillaries or sinusoids in the liver lacunae. The smallest branches of the hepatic artery also empty into the network of sinusoids. (Based on and redrawn from a figure by Child from Elias.)

**Duodenum.**—The structure of the duodenum in most respects resembles that of other parts of the intestine, but is different from the large intestine in the possession of villi and Brünner's glands (see Figs. 418, 419).

**Liver.**—The liver is the largest gland in the body and it is of interest that such a large and important organ has recently received a description which shows its structure in an entirely new light. According to this account the liver is chiefly built of a continuous mass of cells tunnelled by a labyrinth of spaces or lacunae through which run blood-capillaries, the liver sinusoids (Fig. 420). Or it

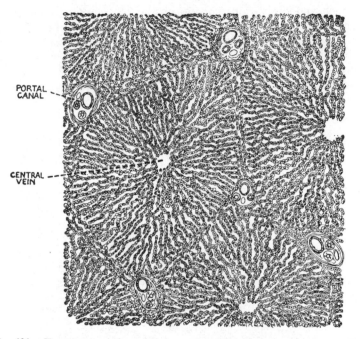

PORTAL
CANAL

CENTRAL
VEIN

Fig. 421. Transverse section of liver to show histological appearance. Each portal canal transmits a branch of the hepatic portal vein, hepatic artery, bile duct, and lymphatic vessels.

might be said that the polygonal liver cells join to form a series of walls, one cell thick, around the lacunae so that the whole liver might be compared to a building with a multitude of interconnecting rooms (the lacunae). These rooms have walls consisting of the single layer of liver cells and they are lined (or wall-papered, so to speak) by the endothelial cells of the capillaries or sinusoids (Fig. 423). The sinusoids are derived from branches of the hepatic portal veins and empty into factors of the hepatic vein (the central veins, Fig. 421). Thus the liver is a spongy organ interposed between the large vessels carrying blood from

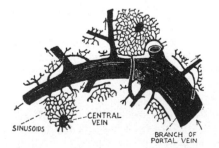

SINUSOIDS

CENTRAL
VEIN

BRANCH OF
PORTAL VEIN

Fig. 422. Diagram to show supply of blood to the liver sinusoids from the branches of the hepatic vein. (Based on Elias.)

the gut and those carrying it into the heart. The sinusoids receive blood also from the finer branches of the **hepatic artery**. All the blood vessels in the liver, including the sinusoids, have powers of constricting their lumens and thus varying the supply of blood to any part of the liver.

Sections of the liver of many mammals (Fig. 421) give the appearance of liver cells being grouped in polygonal lobules bounded by portal canals and having as their axis one of the central veins, that is, a factor of the hepatic vein. Each lobule is thus delimited by branches of the portal vein which connect with the central vein by tor-

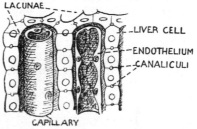

Fig. 423. Diagram to show capillaries (sinusoids) lying in liver lacunae. (Based on Elias.)

tuous, but roughly radially-arranged, sinusoids. In a few animals, e.g. the pig, the liver lobules are further defined by a layer of connective tissue but in practically all other animals, the arrangement of the lobules can be quickly altered by changes in the blood pressure of the vessels. For example, "normal lobulation" is abolished if the pressure in the hepatic veins is raised. Then lobules appear which have as their axes branches of the *hepatic portal* veins instead of hepatic veins which then appear to run around the periphery of the lobule. Liver lobules are thus strictly temporary effects and depend on the pressure gradients in the vessels traversing the liver.

Surrounding each liver cell are minute tubules, the **bile canaliculi**, into which bile is secreted. They join to form bile ductules which in turn unite to them for main bile ducts. Each canaliculus is formed from the cytoplasm of the liver cell it surrounds (Fig. 423).

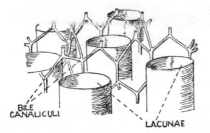

Fig. 424. Diagram to show the relation of the bile canaliculi to the lacunae. Liver cells are not shown in this drawing. (After Elias.)

At intervals, the liver is transversed by larger canals (Fig. 424) lined by connective tissue continuous with that on the outside of the organ. These carry branches of the hepatic portal vein, the hepatic artery, lymph vessels, and branches of the bile ducts. They are termed **portal**

**canals.** Similar canals serve to convey factors of the hepatic vein to the outside of the liver as well as housing bile ducts and lymph vessels.

**Pancreas.**—As will have been gathered from what has gone before, the appearance of the pancreas in different vertebrates varies a good deal. In many, as for example the dogfish, frog, and some mammals, it is a compact gland but in the rabbit it has a fluffy, diffuse appearance. In all instances, however, the essential structure is the same (Fig. 425), the pancreas consisting of a series of lobules

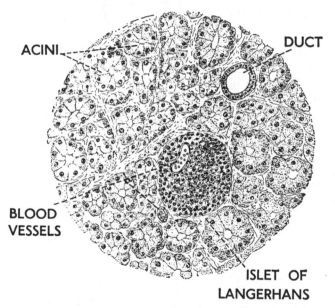

Fig. 425. Transverse section of mammalian pancreas.

each made up of small clusters or acini of cells which secrete the pancreatic juice into small ductules which join larger ducts and eventually the main pancreatic duct. Here and there, between the acini are more compact masses of cells, the islets of Langerhans, which secrete the hormones, insulin, and glucagon (p. 666), directly into the small blood vessels which permeate the gland. Thus, the pancreas is a compound structure having an exocrine portion concerned with producing the pancreatic juice and an endocrine portion, the islet cells. Delicate strands of connective tissue bind the acini and lobules together.

# NUTRITION

## FOOD

It is common knowledge that vertebrate animals select different kinds of material on which to feed. Some feed exclusively upon vegetable material, the leaves and roots of plants, grass, etc., and are said to be **herbivorous**; others feed only on the bodies or flesh of other animals and are **carnivorous**; to yet others the products of both animals and plants are equally acceptable and these are **omnivorous**. But whatever form the food-stuff may take, it must contain the essential materials for the building up of new tissues and for the supply of energy for vital processes.

To supply the needs of most animals the food must contain the following types of chemical substances:

| | | |
|---|---|---|
| **Carbohydrates.** | **Fats.** | **Water.** |
| **Proteins.** | **Mineral salts** (inorganic). | **Vitamins.** |

Among the familiar carbohydrates are sugars, starches, and cellulose; proteins comprise a large part of the solids of the contents of plant cells, the flesh (muscle) of vertebrates, milk, cheese, eggs; fats occur in animal and vegetable oils, eggs, milk, and most animal tissues.

The chemistry of the three main classes of food-stuffs can receive only scant treatment here but the following points are of interest as showing their essential qualities.

The simplest **carbohydrates** are the hexose sugars or **monosaccharides**, such as glucose and fructose, with the empirical formula of $C_6H_{12}O_6$. They may be combined with the loss of a molecule of water to form disaccharides, such as cane sugar ($C_{12}H_{22}O_{11}$) or further polymerised to form polysaccharides, such as starch and cellulose. Before absorption into the body all carbohydrates are split into monosaccharides but few animals have the enzymes in their gut to split cellulose and so it is usual to find that animals having a diet containing large amounts of this substance have symbiotic bacteria or protozoa which break it down into simpler compounds by means of enzymes called **cellulases** or **cytases**. Such symbionts occur, for example, in the rumen and reticulum of artiodactyls, the caecum of rabbits and other mammals, in the gut of wood-eating termites and so on.

**Fats**, which often serve as reserves of energy, are esters of the trihydric alcohol, glycerol, which has the formula $CH_2OH.CHOH.CH_2OH$. Each of the three hydroxyl groups can be replaced by fatty acids, such as palmitic acid $CH_3(CH_2)_{14}.COOH$, stearic acid $CH_3(CH_2)_{16}.COOH$, and oleic acid $CH_3(CH_2)_7.CH.CH.(CH_2)_7.COOH$. During digestion fats are split into fatty acids and glycerol which are recombined after absorption.

**Proteins** are highly complex compounds of **amino acids**, often combined with inorganic radicals. Some thirty amino acids are known. All of them contain a carboxyl group —COOH and an amino group —NH₂, the latter being usually, although not always, attached to the carbon atom next to the —COOH group. Any amino acid can also be thought of as a derivative from a saturated fatty acid in which an H atom of the hydrocarbon residue has been replaced by an $NH_2$ group. Thus, for example, a simple fatty acid, acetic acid, with the formula $CH_3.COOH$, leads to the amino acid, glycine, with the formula $CH_2.NH_2.COOH$. It is the hydrocarbon residue which by its rather variable nature introduces the main variety into the amino acid series. Combinations and permutations of the known amino acids can lead to vast numbers of different proteins and these may again be multiplied by the addition of inorganic radicals such as iron (as in the chromoproteins, of which haemoglobin is an example), or with nucleic acid (as in nucleoproteins), or with carbohydrates (as in various mucoid proteins). Some amino acids (*e.g.* phenylalanine) cannot be synthesised in the animal body, and ten of them are therefore considered as essential. Amino acids are present in proteins of animal origin (except gelatin and collagen) in proportions similar to those of tissue proteins and thus contain enough essential amino acids; proteins from plants often lack these essential constituents of animal diet and to obtain a sufficient range and quantity of amino acids a large quantity and wide variety of plants must be taken in any solely vegetarian diet. This is usually impractical or inconvenient and some animal protein (as milk, for example) should be included in any balanced diet. The various combinations of amino acids form huge molecules linked in various ways which in many respects often parallel their macroscopic properties. Thus inextensible fibres, such as silk, consist of straight chains of long protein molecules, those of rubber are in zigzag chains, allowing of stretching, and so on.

**Mineral salts,** particularly the chlorides and sulphates of sodium and potassium, carbonates and phosphates of calcium, potassium, magnesium, and sodium, and various salts of iron are used in many metabolic processes and in the constitution of organs and tissues. **Water** is also an important metabolic requirement, for, in addition to forming a large proportion of the composition of protoplasm, it is used in the body as a solvent for many materials to be conveyed about the body and plays an important part in many vital processes. Of **vitamins** it may be stated here that these complex organic substances or their precursors, which are normally taken in with the food, are essential to the maintenance of health and for growth, and without them the animal does not appear to be able to utilise to the full the food it takes in. Thus, animals fed on a complete synthetic diet containing the requisite amounts of carbohydrates, fats, and proteins, will not thrive and often develop obvious signs of ill-health. This is all the more surprising since vitamins are often present in only the minutest quantities in natural foods. Although some are present in animal tissues, the ultimate source of most of them seems to be the green plants. Many vitamins are now known and most have been isolated and analysed chemically. The convention adopted by researchers in this field has been to use a letter of the alphabet to designate the different vitamins as they were discovered, but this has led to difficulties because what was once thought to be a single vitamin has been found to be a complex, and again, a new discovery indicated by a new letter, has been found later to be part of the complex of an already known vitamin. So far as man is concerned, the most important are Vitamins A, B, C, D, and E. Vitamins K and P, as well as some others, specific to only one kind of animal, are also known.

Vitamin A (axerophtol) appears to be necessary in order that resistance to infection may be maintained through the healthy activity of various epithelia, particularly those of the eyes, respiratory passages, and alimentary canal. Its absence also results in impaired growth and vision (night-blindness). It is a fat-soluble substance and is present in milk, butter, cream, and various animal fats, particularly the oil from the liver of fishes such as cod and halibut. Its original source, often as carotin, is usually photosynthetic plants.

Vitamin B is now regarded as a complex and has been subdivided into $B_1$ $B_6$, $B_W$, $B_M$, $B_V$, of which perhaps, $B_1$ and $B_2$ are the most important. $B_1$ (aneurin or thiamin) used to be called the anti-beri-beri vitamin because its absence induces the development of a disease in man affecting the nerves and known by that name. It is a water-soluble substance and is present in most vegetables and fruits as well as more abundantly in malt-extract, oatmeal, leguminous seeds, etc. $B_2$ is also water-soluble and contains two active principles: riboflavin (lactoflavin) and nicotinic acid. Riboflavin appears to be essential for efficient tissue respiration and is present in yeast, eggs, salads, rose-hips, and a variety of other foods. The absence of nicotinic acid is responsible for a skin infection known as pellagra. Nicotinic acid occurs in yeast, malt-extract, and liver, but does not occur in dairy produce. Of the other components of the B complex, $B_3$–$B_5$ are related to deficiencies in the growth of animals such as birds and rats; $B_6$ has affinities with nicotinic acid; $B_W$, $B_M$, and $B_V$ are concerned with either growth or dermatitis in animals and sometimes man. Recently it has been shown that $B_{12}$ is a factor necessary for the formation of red blood-corpuscles in man. Its absence causes pernicious anaemia.

Vitamin C has now been isolated as the water-soluble substance, ascorbic acid, and is present in many fruits and fresh vegetables. Animal substances and cereals are deficient in it, as sailors on the long voyages of the old sailing ships discovered, its deficiency resulting in scurvy.

There are many forms of Vitamin D since many of the sterols become anti-rachitic in their effect after being irradiated with ultra-violet light. That which occurs naturally ($D_3$) is produced by the irradiation of dehydro-cholesterol. This vitamin seems to be essential to efficient calcium and phosphorus metabolism, and its absence induces rickets and dental caries. It is produced in the human skin by exposure to sunlight or ultra-violet light. Vitamin D occurs (in association with A) in animal fats such as egg-yolk, butter, and particularly fish-oils. It is absent from cereals, meat, and many vegetables and fruits.

Of Vitamin E (anti-sterility), all that may be said is that rats fed on food deficient in this vitamin failed to reproduce. It is present in salads and seeds.

As will have been seen, most of these vitamins have been recognised in relation to the human or animal diseases which ensue if they are *absent*, but their importance is that they prevent in the healthy organism those lesions or metabolic disturbances which are manifested as disease.

## INGESTION

The way in which the food is taken into the alimentary tract differs in different vertebrates and the three examples under consideration illustrate three distinct methods. The food of the dogfish consists mainly of smaller fishes and crustaceans, molluscs, and other animals, so that it may be said to be carnivorous. It is a voracious feeder, swimming about in search of food which it does not bite or chew up but swallows whole. The teeth covering the jaws are merely used to seize and hold the food during the swallowing process and not for dividing it into pieces. In effect then, the dogfish "bolts" its food, the passage through the pharynx being as rapid as possible so as not to interfere with the respiratory movements.

The frog feeds chiefly on insects which it ingests whole and may therefore also be considered to be carnivorous (insectivorous). Its method of catching its prey is rather remarkable since it can catch insects on the wing. When feeding, it squats in a suitable place frequented by insects and when one flies near to it, it opens its mouth and flicks out its tongue and strikes the insect with it. The tongue is covered with a sticky secretion so that the insect adheres

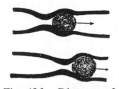

Fig. 426. Diagram of the passage of a bolus down the oesophagus.

to it and, by a return flick, is brought to the back of the pharynx and passes, by the contraction of the pharyngeal wall into the oesophagus and thence by peristalsis to the stomach. The movement of the tongue is amazingly rapid and its position of attachment (at the front of the buccal cavity) and muscular character enable it to be projected to some distance from the mouth and returned in about $\frac{1}{10}$ second. Frogs will also, on occasion, take small animals from the ground, but it is essential for them to be moving to attract the frog's attention. If any small fragments are left in the bucco-pharyngeal cavity they become entangled in mucus secreted by cells in the lining epithelium and are carried by ciliary currents to the oesophagus.

The rabbit gnaws its food. By means of the chisel-shaped incisor teeth, small portions of vegetable matter are sliced off, and after being seized by the mobile lips are taken into the buccal cavity. Here, the food is further subjected to the action of the grinding cheek teeth and at the same time is thoroughly mixed with saliva from the salivary glands. During this process which is termed mastication, the food is turned round and round in the buccal cavity by movements of the tongue until the whole forms a pulpy mass or bolus. The saliva is largely water but contains mucus and digestive enzymes. Mastication completed, the bolus is then swallowed by being pushed into the pharynx by the raising of the tongue, prevented from entering the posterior nares by the soft palate, and, by the contraction of the pharyngeal wall, forced into the beginning of the oesophagus. The muscles of the pharyngeal end of the oesophagus are voluntarily dilated to receive the bolus, but once it has entered its further progress is involuntary. During the contraction of the pharynx and the entry of the bolus into the oesophagus, food is prevented from entering the larynx through the glottis by the flap-like epiglottis.

## DIGESTION

With the exception of the mineral salts and some of the sugars most of the solid food of vertebrates is insoluble or, even if capable of mixing with water, forms a colloidal solution, and is consequently incapable of passing through the lining membrane of the alimentary canal. The food must therefore be rendered soluble and diffusible before it can be of use to the organism, and this process is called **digestion**. Most of the work which has been done on this subject has been carried out on mammals but much of it can be applied to other vertebrates.

In all animals, digestion consists of a process of hydrolysis, during which the molecular size of the substances is progressively reduced until a true solution is possible. This has the equally important consequence that those simple molecules which are required for later synthesis into complex ones, specific for any given animal, are produced. These changes are brought about by organic catalysts known as **enzymes**. These complex protein substances produced by the living cells induce with ease and rapidity chemical reactions which would be difficult and lengthy under ordinary chemical conditions in, say, a chemical laboratory. Each particular enzyme acts only upon a particular substance or group of substances—called the **substrate**—and, at the end of the reaction, it itself remains unaltered. Enzymes can be rendered inactive or "killed" by heat and other means. They act best at a particular temperature—called the optimum temperature—and require that the medium in which they work shall have a certain alkalinity or acidity ($pH$ value). The chemical reaction with which they are concerned can, in their presence, be reversed so that the substrate will be reformed, the direction in which the reaction proceeds being determined by the relative concentrations of the substrate and the end products. The rate at which the reaction takes place is affected by the amount of the enzyme present, but however small the quantity, the reaction always takes place.

Enzymes are classified in accordance with the type of substance upon which they act. Thus, taking the chief classes of food materials, carbohydrates are digested by **diastatic** or **amylolytic** enzymes (carbohydrases), proteins by **proteolytic** (proteases), and fats by **lipolytic** enzymes (lipases). In each group (as we shall see) are numerous enzymes named according to their source and the particular substances upon which they act.

In the mammal, because of its capacity to retain and chew its food in the buccal cavity, digestion usually begins there. During the process of mastication already referred to, the food is mixed with the saliva which usually contains a diastatic enzyme, **ptyalin** (its presence has been determined in the saliva of pigs, some rodents, and all primates, but not carnivores). This acts on certain of the polysaccharides or starches, turning them into compound sugars such as dextrin and maltose. Generally speaking, the saliva on standing is slightly alkaline (though its $pH$ value may vary from time to time and in accordance with the food taken), but ptyalin acts best in a neutral or very slightly acid medium. The extent to which digestion occurs in the buccal cavity depends upon the length of time food is retained there and the thoroughness of the mastication. However, after the bolus has been swallowed and the food has entered the stomach, the ptyalin digestion does not necessarily cease at once, though it does eventually.

In the stomach, the food is subjected to the action of the gastric juice, the flow of which is partly induced by the hormone gastrin and which is secreted by the gastric glands. The secretion of gastric juice is also promoted by appetite. These responses are mediated by the nervous system. The chemical nature of the food also acts directly on the gastric mucosa as also does the presence of any food in the stomach by stimulating it mechanically. In this juice—which contains over 90 per cent. water—two enzymes are usually present, pepsin and rennin. **Pepsin** is a proteolytic enzyme acting upon proteins and reducing them to proteoses and peptones by attacking the protein chains in the middle parts; **rennin** coagulates milk. For the efficient action of the pepsin it is necessary for the medium to be acid, and this is produced by the hydrochloric acid secreted by the oxyntic cells (see p. 495) of the peptic glands and included in the gastric juice. This acid fluid, as it penetrates each bolus of food received from the buccal cavity, gradually inhibits any further action of the ptyalin and also helps to sterilise the food, thus giving protection against some pathogenic bacteria (*e.g.* cholera). In the stomach, then, there is a copious addition of fluid to the food, and the tissue surrounding fat is dissolved away, liberating it in the form of free globules. Diastatic and lipolytic enzymes are sometimes found in the stomach but are probably regurgitated from the duodenum, though in some mammals they may be parts of the gastric juice. The presence of the pyloric sphincter results in the food being retained in the stomach for some time (in man as long as from two to three hours), during which time the rhythmical contractions and relaxations of the stomach muscles churn the food, ensuring a thorough mixing with the gastric juice, until it takes the form of a thick, creamy, acid fluid called chyme.

When the food has been reduced to the condition of chyme, the duodenum has been stimulated to produce a hormone secretion —secretin—which induces the pancreas to become more active and the liver to pour out bile. Then the pyloric sphincter relaxes and by the contraction of the stomach the chyme spurts in small jets into the duodenum. Here, the food meets further additions in the form of bile from the liver, the pancreatic juice from the pancreas, and the succus entericus from the wall of the intestine.

Bile is a greenish alkaline watery fluid which, however, contains no digestive enzymes, but it adds water to the food and the salts present in it are essential to complete intestinal digestion, while it is also stated to prevent putrefaction. The salts are sodium bicarbonate, glycocholate, and taurocholate. Sodium bicarbonate reduces the acidity of the intestinal contents and the two other bile salts activate the pancreatic lipase and lower the surface tension of the fats so that they become emulsified. The bile pigments— bilirubin and biliverdin—are, in a measure, excretory products as they are produced from the breakdown of the haemoglobin of red blood-corpuscles which have been replaced; the iron from the haemoglobin is mainly retained in the body and used for the formation of more blood pigment

The pancreatic juice is a watery, alkaline fluid, very rich in enzymes. It contains trypsin and chymotrypsin (proteolytic), amylase (diastatic), and lipase (lipolytic). In the form in which they are produced, the trypsin and chymotrypsin are relatively inactive, but become much more active when they are mixed with the succus entericus which contains an activator.*

The succus entericus is produced by the glands of the duodenum and small intestine already mentioned (p. 490). It is an alkaline watery fluid containing as its most important constituents many enzymes, but it also contains mucus. These enzymes complete the various digestive processes of all kinds of food substances;

---

* It is a common occurrence, especially with proteolytic enzymes, for them to be produced in a form which requires to be activated by another enzyme called a -kinase before becoming fully active. Another device to prevent auto-digestion is the secretion by the cells of the organ (*e.g.* stomach) of an anti-enzyme which prevents the action of the enzyme. Thus the fragility of the wall of the stomach of a dead rabbit during dissection is due to the auto-digestion of the wall by the pepsin, and is more evident if the animal was killed some short time after a meal.

the more important ones are aminopeptidases and dipeptidases, parts of a mixture once called erepsin (proteolytic), enterokinase (activator), lipase (lipolytic), invertase, maltase, and lactase.

During its passage through the duodenum, then, the food has received still more water from the various secretions so that it is now a quite fluid emulsion called **chyle**, which is slowly forced by peristalsis through the long jejunum and ileum, the enzymatic action continuing the whole time. All the enzymes of the pancreatic juice and succus entericus require an alkaline medium in which to work at a maximum efficiency, and the sodium bicarbonate of the bile and pancreatic juices to a large extent neutralises the hydrochloric acid of the stomach contents and enables the enzymes in the intestine to maintain a fairly high efficiency.

A single protein molecule has a molecular weight varying from tens of thousands to millions and has a "backbone" consisting of a long chain (not a straight chain in two dimensions as the diagram suggests, but folded in space of three dimensions) in which the grouping —CO—NH— recurs. This grouping is called the "peptide link". $R_1$, $R_2$, etc., represent various alkyl or aryl groups. At one end of the molecule is a free —COOH group and at the other a free —NH$_2$ group.

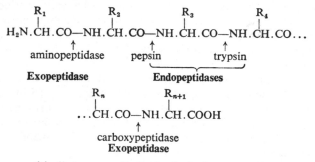

The peptide link can split by hydrolysis, an —OH group going on to the carbon atom and a hydrogen on to the nitrogen atom.

$$—CO—NH— + HOH \rightarrow —COOH + H_2N—$$

The result of such a hydrolysis is to split the molecule into two parts, each of which still has a —COOH group near one end and an —NH$_2$ group near the other. When all the peptide links of a protein have been hydrolysed the final products are relatively simple compounds, each still containing a —COOH and an —NH$_2$ group, but substances that are crystallisable and soluble in water. These are amino acids, the simplest of which is glycine, aminoacetic acid CH$_2$ (NH$_2$) COOH. Only about twenty different amino acids are found in proteins, but variations in the order in which they are arranged along the molecule render possible an almost infinite number of proteins.

Hydrolysis of peptide links is brought about by the catalytic action of various enzymes. Those acting only on terminal peptide links are known collectively as exopeptidases, and are called amino-peptidases if they act on the terminal peptide link nearest the free —$NH_2$ group, carboxypeptidases if they act on the peptide link nearest the free carboxyl, and dipeptidases if they act on single peptide links in the molecules of dipeptides, compounds which hydrolyse to two amino acid molecules only. Enzymes, such as trypsin and pepsin which act on internal peptide links are known as endopeptidases.

Taking the pancreatic enzymes first, the trypsin, after it has been activated by the enterokinase of the succus entericus, acts on the undigested proteins. The amylase converts polysaccharide starches (also glycogen) into maltose. The lipase splits certain fats into glycerol and fatty acids.

In the succus entericus, the enterokinase activates the pancreatic enzymes, trypsin and chymotrypsin. The lipase, as before, hydro-lyses fats, and of the remaining enzymes, invertase turns sucrose (cane sugar) into glucose and fructose; maltase turns maltose into glucose; and lactase, lactose into glucose and galactose.

Thus, as a result of this enzymatic action, the proteins are all hydrolysed to amino acids, the carbohydrates to glucose or similar monosaccharide sugars, and the fats to glycerol and fatty acids. As the brush border of the cells lining the intestine contains much of the invertase, maltase and aminopeptidase within its microvilli, it is very possible that the final stages of digestion take place there, that is, they are intra-cellular (p. 142).

Although this account of digestion applies chiefly to mammals and to man in particular, the essential processes are similar in the dogfish and frog, though, of course, in these animals, since the food is not retained in the buccal cavity and in the absence of salivary glands, digestion cannot occur there, but begins in the stomach.

## ABSORPTION AND ASSIMILATION

Very little absorption of digested food takes place either in the buccal cavity or the stomach. Certain drugs and alcohol may be absorbed in the stomach, but these may scarcely be classed as ordinary foods. By far the greater part of absorption takes place, in the mammal, in the jejunum and ileum, where, digestion completed, the food is in a form in which it will pass through the walls into the capillary blood vessels.

As has already been stated, the absorptive area of the small intestine is enormously increased by the development of villi in this region.   Each villus is well supplied with capillary blood vessels and as each villus is bathed in the fluid food, absorption readily takes place.   In the centre of each villus is a vessel called a **lacteal** —also part of the vascular system, but instead of containing blood it contains lymph and is part of the extensive lymphatic system.

The digested carbohydrates (as monosaccharide sugars, chiefly glucose) and proteins (as amino acids) are absorbed into the villous capillaries and these lead to factors of the (hepatic) **portal vein.** As will be seen later (p. 515) the portal vein does not pass directly to the heart, but enters the liver where it breaks up into capillaries in the liver substance.   Thus all digested carbohydrates and proteins must pass through the liver before entering the general circulation. In this respect the liver acts as a regulating mechanism, for only the required amount of blood-sugar is allowed to enter the blood, leaving the liver by the **hepatic vein.**   Any excess is turned into glycogen and stored.   If the systemic circulation becomes depleted of its sugar content, the glycogen is turned once more into glucose and released.   In this way the liver, besides acting as a regulator, also acts as a storehouse for carbohydrates; but animals cannot similarly store proteins.   The amino acids, which form the basis for elaborating the body proteins, are liberated in suitable quantities into the blood-stream, but any excess is subjected in the liver to the process of deamination, to form a *relatively* harmless substance, urea, with the formula $CO(NH_2)_2$, and into other products which can enter the glycolytic cycle (p. 547).   The liver has many other important functions.   For example, it synthesises vitamin A and blood proteins.

The digested fat as glycerol and fatty acids is reformed into fats in the intestinal cells and passes directly into the lacteals where it is transformed at once into minute globules of fat, so that the vessel contains a whitish fat emulsion from the milky appearance of which it gets its name.   Since the lymphatic vessels join the systemic circulation by way of the thoracic duct and the right lymphatic duct (see p. 539) the fat enters the blood-stream by these routes.   It may be that some of the digested fat enters the villous blood-capillaries, and again that fat as fat is taken up by the cells of the epithelium covering the villi and by that route enters the lacteals *direct*.   This is, however, only subsidiary to the main absorption by the lacteals.

Conveyed by the blood, the various absorbed food products reach the tissues and are assimilated, either to build up new protoplasm or to provide a source of energy for vital activities.

The removal from the intestinal contents of the digested carbohydrates, proteins, and fats leaves a fluid mixture containing the undigested and indigestible portions of the food material, the remains of the various secretions and organic detritus. In herbivorous mammals the solid content consists largely of cellulose which is not digested by any of the ordinary diastatic enzymes. In these animals, cellulose digestion occurs in the especially large caecum into which the fluid intestinal contents are directed. In the caecum large numbers of bacteria and protozoa are present—really symbionts—and these and the cytases in the plants themselves facilitate the digestion of cellulose.

Whether a large caecum is present or not, the fluid contents of the alimentary canal now pass into the colon. In terrestrial animals one of the biggest problems which has to be solved is the conservation of water. In all the digestive processes, water, as the main part of the various secretions, has been poured out upon the food, and much of this water has come from the body organs and tissues. If the intestinal contents merely pass unaltered along the terminal portion of the alimentary canal and outwards to the exterior through the anus, the water will be lost to the animal and the tissues become seriously depleted of their water content. To obviate this loss, a large proportion of the water is extracted during the passage of the fluid through the colon, so that by the time the rectum is reached, the material is semi-solid and forms the faeces. These consist to a very large extent of dead bacteria, dead cells, and mucus. They are egested at intervals through the anus.

The process of absorption in other mammals and in the lower vertebrates differs from the account given above in accordance with the type of food and the form of the alimentary canal. In all herbivorous vertebrates the alimentary canal is relatively longer than in the carnivorous ones; for example, the alimentary canal of a tadpole is much longer in relation to the length of the body than that of the adult frog; that of a cow or sheep longer than that of a dog or cat; and in addition, the herbivorous mammals have a larger caecum and colon than have the carnivorous ones. This is because the large amount of cellulose present in a vegetable diet forming the walls of the cells containing the nutrient material has to be broken down before complete digestion and absorption can take place. Also, as has already been pointed out, a large amount of cellulose digestion and absorption takes place in the caecum.*

* It has been shown that the rabbit, in order to obtain the maximum amount of nutriment from its food, has adopted the habit of coprophagy. The faeces produced during the night are soft and moist, differing from the hard dry pellets produced during the day. The nocturnal faeces are eaten and consequently are subjected once more to all the digestive and absorptive processes. This coprophagous habit has a distinct analogy with that of "chewing the cud" found in ruminant ungulates.

Animals utilise carbohydrates and fats as sources of energy and proteins are used for building up the tissues, in addition, after deamination, to supplying energy-containing substances. Thus, during the process of protein digestion these complex materials are reduced to the simpler form of amino acids in which they were originally synthesised by the plants, and from these relatively simple basic components the specific complex proteins of each individual animal are built up.

## THE BLOOD VASCULAR SYSTEM

### INTRODUCTION

In vertebrates, as in all coelomates, the blood vascular system is the transport system of the body. By its means various materials are conveyed from one part of the body where they are formed or are taken in, to another where they are required or are got rid of. The conveying medium is a liquid tissue, the blood, and this is transported throughout the tissues and organs by tubular channels, the blood vessels, the circulation being maintained by the contractions of a central pumping station, the heart.

The heart is a muscular organ consisting of chambers separated by valves so that, by the contraction of its walls, the contained blood is forced through it in one direction and thence into the arteries. The heart lies in a special region of the coelom, the pericardial cavity.

The blood is really a tissue—closely allied to the connective tissues (see p. 397)—being made up of cells—the red and white corpuscles—and a fluid matrix—the plasma. It differs from other connective tissues in that the matrix (plasma) is not wholly produced by the cells (corpuscles). The details of the structure of blood are given later (p. 533), but some of the important features of its mode of functioning may be considered here.

The vessels conveying true blood are arranged in two systems, those carrying blood from the heart to the tissues, termed arteries, and those which bring the blood back from the tissues to, or towards, the heart, termed veins. Thus the blood in an artery can be termed arterial blood and that in a vein, venous; but this is no indication of its oxygen content or other physiological condition nor of its "purity" or "impurity". Arteries have thicker walls than veins and there are also slight differences in the histology of the layers forming the walls (Fig. 427), the muscular and elastic coats being less well-developed in the veins. As the arteries branch in the various organs of the body they get smaller and smaller, their walls get thinner, and below a certain size they are termed arterioles.

These in turn branch to form capillaries (Fig. 429), minute vessels whose walls consist solely of a delicate layer of tessellated epithelial cells through which exchanges between the blood and the tissue take place. It is important to notice, however, that the exchanges are not direct, since between the tissues and the walls of the capillaries, there is the interstitial fluid which itself is collected into definite drainage channels, the lymph vessels, and these in turn empty back into the venous system at definite places (p. 539). But not all the blood traverses the system of capillaries into the minute veins (venules) for there are certain through channels (Fig. 429) between the arterioles and venules and these are particularly plentiful in structures, such as the skin, which have a fairly constant blood-supply

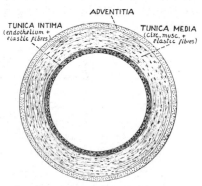

Fig. 427. Diagram to show the layers in the wall of a medium-sized artery. Veins have a similar structure but the muscular layer (tunica media) is thinner.

whereas in tissues like muscles, which have a more variable blood-supply, the capillaries dominate. By contraction or relaxation of small sphincter muscles, at the origin of the capillaries from the arterioles, the amount of blood passing along the direct route from arterioles to venules can be varied. These through channels probably account for the appreciable hydrostatic pressure of the blood in the veins.

The general transport function of a blood vascular system has already been stressed, but the particular functions of the blood in vertebrates might usefully be summarised here. It participates in nutrition, conveying nutrient material (dissolved in the plasma) from the alimentary canal to the tissues; in respiration by carrying oxygen (chiefly through the agency of the

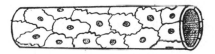

Fig. 428. Diagram of a blood-capillary to show the tessellated epithelial cells forming its walls.

haemoglobin of the red corpuscles) from the respiratory surfaces to the tissues and carbon dioxide (mainly as sodium bicarbonate dissolved in the plasma) in the reverse direction; in excretion by conveying waste excretory products (mainly urea dissolved in the plasma) to the kidneys; in regulating body temperature in warm-blooded animals by distributing heat from its place of origin throughout the body; in the control of metabolic activities by transmitting the hormones (products of the endocrine organs, see p. 655) through the body and in defence against disease by (*a*) carrying antibodies and (*b*) by circulating white corpuscles which can ingest bacteria.

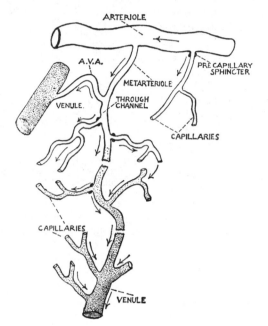

Fig. 429.   Diagram to show possible connections between arterioles and venules *via* through-channels, capillaries, and metarteriole-venular anastomosis (A.V.A.).   (Based on Chambers and Zweifach.)

## THE BASIC PLAN OF THE VERTEBRATE VASCULAR SYSTEM

In dealing with the arrangement of the blood vascular system it is convenient to consider the heart first.   The heart, in all vertebrates, arises as a simple tube of splanchnic mesoderm and during its formation becomes enclosed in a special portion of the coelom, the pericardial cavity, bounded by coelomic epithelium forming the pericardium or pericardial membrane.   Thus, in the frog, the heart, as will be seen from Fig. 430, appears beneath the pharynx (see also pp. 754-5).   Here the upper part of the coelom has been, for the most part, obliterated by the development of the visceral clefts, but in the ventral region the coelom, formed by the fusion of the coelomic spaces, persists. Between the pharyngeal endoderm and the splanchnic mesoderm a group of cells which are destined to form the lining epithelium or endothelium of the heart appears in the mid-ventral line. These cells eventually become arranged in the form of a tube, the endocardiac tube. This gradually enlarges and as it does so the splanchnic mesoderm folds around it (Fig. 430) to form the cells from which the muscular wall of the heart (myocardium) and outer epithelial covering (epicardium) are derived. Soon the right and left folds of splanchnic mesoderm meet above

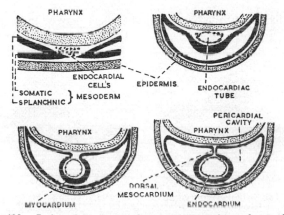

Fig. 430. Series of diagrammatic transverse sections of an embryo to show the development of the heart.

the endocardiac tube which thus becomes suspended in the anterior part of the coelom by a kind of mesentery, the dorsal mesocardium. A transverse septum then grows across to shut off the pericardial cavity from the splanchnocoel.

This straight tube, contained within the circumscribed space of the pericardial cavity, increases in length and, since its ends are fixed, it becomes twisted into a characteristic S-shaped curve. Also the lumen becomes divided into chambers by the development of valves which are formed by foldings of the lining membrane with some inclusions from the wall. In this way a simple heart is formed, consisting of four chambers: the contractile, but only feebly muscular **sinus venosus**, into which the main veins of the body open; the **atrium**, slightly more muscular, which receives blood from the sinus venosus; the **ventricle**, which is very muscular and receives the blood from the atrium; and the **bulbus cordis** which is also muscular, though not to the same extent as the ventricle, the blood passing from the ventricle into it to be distributed to the arteries.

Thus, from the posterior to the anterior end of the heart, the chambers are progressively more muscular, the contraction of one chamber forcing blood into the next one and causing its relaxed musculature to stretch, thus raising the blood pressure in a stepwise fashion from the low values it has progressively reached during circulation round the body.  This is necessary since cardiac muscle requires to be stretched before it can contract.  The sinus venosus is dilated merely by blood entering it at low pressure from the main veins of the body and this is why its walls are only slightly muscular.

The division of the heart into chambers may be looked upon as

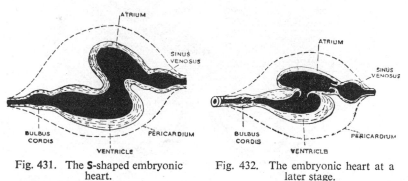

Fig. 431.  The S-shaped embryonic        Fig. 432.  The embryonic heart at a
          heart.                                   later stage.

a mechanism for increasing the pressure of the blood by the dilatation of chambers of progressively increasing muscularity until, finally, a sufficiently high pressure is reached which will ensure the circulation of the blood around the body.

As the simple heart develops, arteries and veins are laid down in the mesoderm in connection with the various organs and tissues. Taking the arteries first, from the bulbus cordis a ventral aorta passes forwards beneath the pharynx and from it arises a series of **aortic arches** which run in between the visceral clefts, on each side, passing dorsalwards to open into paired **dorsal aortae**.  These aortae pass backwards above the alimentary canal and beneath the notochord to the hinder part of the body, and from them arise arteries to the various organs and tissues.  As development proceeds, there is a tendency, particularly in the post-pharyngeal region, for the paired aortae to fuse together to form a single dorsal **aorta such as is found in the adult animal.**

Similarly, a system of veins is elaborated, collecting blood from the various structures and organs. The main factors, apart from those from the gut, ultimately open into paired longitudinally running veins, the **anterior cardinal veins** in the front part of the body and the **posterior cardinal veins** posteriorly. In the region of the pericardium, where the pericardial cavity is separated from the general coelom by the **pericardio-peritoneal septum (transverse septum)**, the anterior and posterior cardinal veins open on each side of the body into a transversely running vessel, the **Cuvierian vein (ductus Cuvieri)**, which, running in the transverse septum (the only

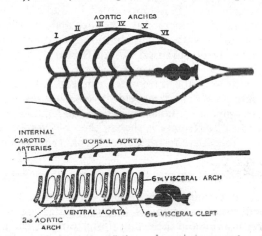

Fig. 433. Plan and side view of the embryonic heart and aortic arches.

possible situation in the body for such a vessel), enters the pericardial cavity and joins the sinus venosus.

In this simple blood vascular system the circulation of the blood is therefore from the heart, through the ventral aorta and aortic arches to the dorsal aorta (or aortae), and thence to the tissues by the various arteries; from the tissues, by the various veins, to the anterior and posterior cardinal veins, into the Cuvierian veins, and back to the heart, the hepatic veins opening directly into the sinus venosus.

Two departures from the direct course of the blood to the heart, however, make their appearance with the formation of what are known as the **portal systems**. The blood brought from the tail by the caudal vein does not pass at once to the posterior cardinal vein but is conducted to the kidneys by the **renal portal veins**. After passing through the capillaries of the kidneys the blood enters the posterior cardinal veins by the **renal veins**. Similarly, blood collected from the alimentary canal is first conveyed to the liver by the **hepatic portal vein,** and after passing through the capillaries of the liver, passes—usually directly to the sinus venosus—to the heart by the **hepatic veins.** A **portal system,** then, is one in which blood collected from one set of organs or tissues is conveyed to another organ through whose capillaries it passes before entering the heart.

This simple form of blood vascular system appears during the development of all gnathostomatous craniates and forms the basis of the particular types of vascular system met with in the various groups of vertebrates. It is represented in Plate V along with the other characters of a generalised craniate, and affords an obvious contrast to the blood vascular system in the coelomate invertebrates. In those animals the dorsal vessel with its forwardly directed flow acts as a collecting vessel and the ventral vessel as a distributing vessel. The reverse is true of the vertebrates where the dorsal aorta is the main distributing vessel and the ventral vessel has become broken up with the development of the heart and of the hepatic portal system. A further point of difference is that in the vertebrates, commissural vessel (aortic arches) are restricted to the pharyngeal region.

## THE VASCULAR SYSTEM IN DIFFERENT CRANIATES

The most important factors which have induced modifications in the blood vascular system of the vertebrates are their requirements for respiration and their migration from the water to the land with the introduction of a second type of respiratory organ. In an aquatic vertebrate such as a dogfish where oxygen is obtained from the water, the respiratory organs are gills (p. 551). To provide for the blood-supply of these gills all that is necessary is the interposition of a net work of thin-walled vessels in those aortic arches supplying the modified visceral clefts. Now, the deoxygenated blood passing from the heart by way of the ventral aorta reaches the gills by **afferent branchial vessels,** takes up oxygen during its course through the small vessels, and is collected up again by **efferent** and **epibranchial vessels** and conveyed to the dorsal aorta. Thence it is distributed to the tissues and organs. With this modification, the respiratory and ǫther needs of the animal are fulfilled by the simple **single** circulation as it is called, in which the course of the blood is from the heart to the gills, thence to the tissues, and back to the heart (Fig. 434).

Even before the vertebrates began their migration from the water to the land another type of respiratory structure—the lung—which could utilise the oxygen of the atmosphere, was developed.

Although the lungs are derivatives of the hinder end of the pharynx, their blood-supply is quite different from that of the gills. Also a capillary system, more extensive than that in the gills of fishes, is required in the lung, and this appreciably lowers the pressure in the blood circulating through it. Although a moderately low aortic pressure suffices for the fishes, the tetrapods, when compared

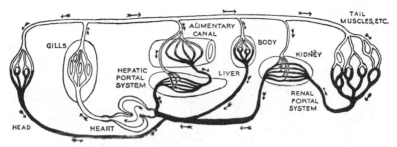

Fig. 434. The single circulation as in a fish. Arteries, white; veins, black.

with them, are much more active animals, especially the mammals, and therefore require a more rapid circulation of the blood. The blood, oxygenated during its passage through the pulmonary circulation, is therefore returned to the heart in order that its pressure can again be raised, whence it can be pumped at high pressure into those arteries supplying the body. Provision for the

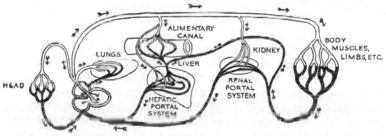

Fig. 435. The double circulation as in an amphibian. Arteries, white; veins, black.

reception and separation of the oxygenated and deoxygenated blood involves the division of the atrium into two halves by the development of an interauricular septum. Thus, in the amphibians such as the frog, right and left auricles are present, the systemic veins discharging by way of the sinus venosus into the right auricle, and the pulmonary veins into the left. The ventricle, however, is still undivided and some mixing of the blood must take place.

Passing through some of the reptiles to the mammals (and the birds) it is possible to trace a progressive division of the ventricle by an interventricular septum until in the mammals there exists a complete **double circulation.** In these animals, the two sides of the heart are completely separated from one another, one side containing deoxygenated blood and the other oxygenated, so that the blood in a complete circulation through the body passes twice through the heart, thus: heart to tissues, thence back to the heart, thence to the lungs, and once more back to the heart.

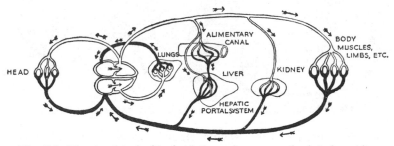

Fig. 436. The complete double circulation as in a mammal. Arteries, white; veins, black.

In general, then, this is the line of modification in the blood vascular system in the vertebrates, and the three examples under consideration illustrate three phases in it and may now be described in detail.

## THE DOGFISH

The heart of the dogfish consists of four chambers, the **sinus venosus,** the single **auricle** (derived from the embryonic atrium), the **ventricle,** and the **conus arteriosus** (derived from the embryonic bulbus cordis), and is contained in a pericardium, the ventral wall of which rests on the inside of the coracoid region of the pectoral girdle. The thin-walled sinus venosus is triangular in shape, the apex marking the opening into the auricle. Between the sinus venosus and the auricle are the **sinu-auricular valves.** The auricle is moderately muscular and lies above the very thick-walled muscular ventricle. Between these two chambers are the **auriculo-ventricular valves.** The ventricle tapers anteriorly to the thick-walled conus arteriosus which, perforating the anterior wall of the pericardium, leads to the **ventral aorta.** Within the conus arteriosus are two sets of **semilunar valves** which prevent the regurgitation of the blood into the ventricle. In each set there are one dorsal and two ventrolateral valves.

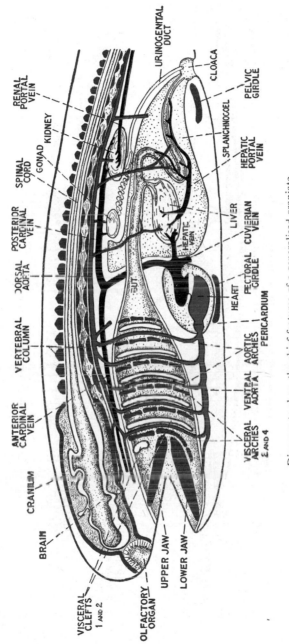

BRAIN

CRANIUM

ANTERIOR CARDINAL VEIN

VERTEBRAL COLUMN

DORSAL AORTA

POSTERIOR CARDINAL VEIN

SPINAL CORD

GONAD   KIDNEY

RENAL PORTAL VEIN

URINOGENTIAL DUCT

CLOACA

PELVIC GIRDLE

SPLANCHNOCOEL

HEPATIC PORTAL VEIN

LIVER

CUVERIAN VEIN

HEART

PECTORAL GIRDLE

PERICARDIUM

AORTIC ARCHES

VENTRAL AORTA

VISCERAL ARCHES 3 AND 4

GUT

HEPATIC VEIN

VISCERAL CLEFTS 1 AND 2

OLFACTORY ORGAN

UPPER JAW

LOWER JAW

Diagram to show the chief features of a generalised craniate.

FLATE V.

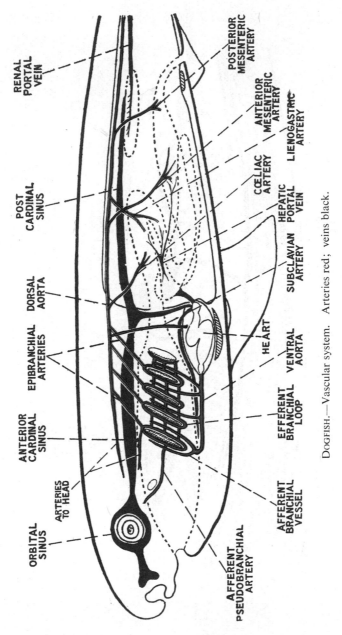

RENAL PORTAL VEIN

POSTERIOR MESENTERIC ARTERY

ANTERIOR MESENTERIC ARTERY

LIENOGASTRIC ARTERY

POST CARDINAL SINUS

CŒLIAC ARTERY

HEPATIC PORTAL VEIN

DORSAL AORTA

SUBCLAVIAN ARTERY

EPIBRANCHIAL ARTERIES

ANTERIOR CARDINAL SINUS

HEART

VENTRAL AORTA

EFFERENT BRANCHIAL LOOP

ARTERIES TO HEAD

ORBITAL SINUS

AFFERENT BRANCHIAL VESSEL

AFFERENT PSEUDOBRANCHIAL ARTERY

DOGFISH.—Vascular system.   Arteries red; veins black.

PLATE VI.

The ventral aorta passes forwards in the median line beneath the pharynx between the hypobranchial musculature to about half-way between the anterior end of the pericardium and the mouth. From it are given off five pairs of **afferent branchial arteries** passing to the gills. The two anterior pairs arise as bifurcations of a pair of single vessels—sometimes called the **innominate arteries**—which come off at right angles to the ventral aorta at its front end. The third pair is situated at about the middle of the length of the ventral aorta and the posterior two pairs arise close together just anterior to the conus arteriosus. An afferent vessel runs along the ventro-lateral aspect of the hyoid arch and one along each of the four succeeding branchial arches, giving off small vessels to the filaments of the gills. Blood from the gills is collected up by **efferent branchial**

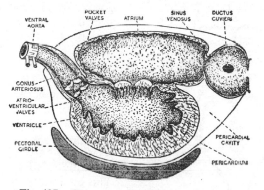

Fig. 437. DOGFISH.—Sagittal half of the heart.

arteries which take the form of four pairs of complete loops encircling the first four pairs of gill clefts and one pair of half-loops along the anterior face of the fifth pair of gill clefts. The loops are connected to one another by longitudinally running connecting vessels. Leading from each complete loop, on each side, is a single vessel—the **epibranchial artery**—which opens into the single median **dorsal aorta**, which passes backwards along the whole length of the body beneath the vertebral column.

From this dorsal aorta arises a series of arteries, some paired and some single and median, which pass to the various organs and tissues of the body. The chief of these are, the paired **segmental arteries** to the myotomes; the **subclavian arteries** to the pectoral fins; median **coeliac, anterior mesenteric, lienogastric**, and **posterior mesenteric arteries** to various parts of the alimentary canal; paired **renal** and **gonadial** arteries to the kidneys and gonads. Each main artery divides into branches to supply the various organs and structures.

Anteriorly, arising from the dorsal aorta, are arteries which supply the brain and associated structures. The terminology of these arteries varies in different accounts and many of the names do not fall in line with that adopted in the higher vertebrates. Suffice it to say that the brain receives blood from the so-called **internal carotid arteries**\* which arise on each side from the anterior limb of the first efferent loop (and not as a forward continuation of the dorsal aorta as in the higher vertebrates) entering the brain case through an aperture in the floor of the cranium and also from continuations of the so-called **hyoidean arteries (efferent pseudobranchial)**. The forward continuations of the dorsal aorta itself consist merely of two insignificant vessels on the under side of the cranium which join the so-called internal carotid arteries.

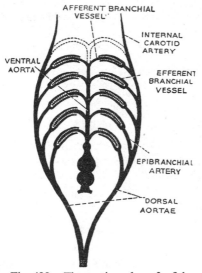

Fig. 438.   The aortic arches of a fish.

The **venous system**, by means of which the blood is returned to the heart, consists for the most part, not of narrow tubular veins, but of large blood spaces or sinuses. This condition, which is found in all fishes, may be correlated with necessity for offering as little resistance as possible to the flow of blood back to the heart, for, as explained earlier (p. 517), the blood pressures in both arteries and veins in a fish are rather low when compared with those in animals, such as mammals, with a double circulation.   When the ventricle and conus contract blood is driven into the ventral aorta

---

\* This peculiar arrangement in the fishes, in which they differ from the terrestrial vertebrates, is due to their utilisation of the gills for respiration. Since all the blood entering the heart is deoxygenated and does not become oxygenated until it has passed through the gills, the supply of blood rich in oxygen to the brain can only be by vessels arising from the efferent branchial system.   In terrestrial vertebrates, this supply is provided by the paired carotid arteries, the external carotids being formed as forward continuations of the ventral aorta, which forks, and the internal carotids are derived from the forward extensions of the (originally paired) dorsal aortae.   Thus there are no true external carotids in fishes.

and if an equivalent volume did not enter the sinus venosus and auricle the heart as a whole would occupy a smaller volume. But the walls of the pericardial cavity are fairly rigid and tend to keep the volume of it constant. So blood must be "sucked in" to the sinus venosus from the main sinuses entering it, a mechanism necessary in view of the low hydrostatic pressure in the blood returning to the heart. The disposition of these sinuses follows the general basic plan already outlined. The blood from the anterior part of the body is collected into the **anterior cardinal** and **inferior jugular sinuses** on each side; that from the posterior part in the **posterior cardinal** and **hepatic sinuses**. On each side of the body, the anterior cardinal, inferior jugular, and posterior cardinal sinuses open into the transversely running **Cuvierian sinus**, which, like its fellow of the opposite side, passes through the pericardium and opens into the basal angle of the sinus venosus. The hepatic sinuses open directly into the middle of the base of the sinus venosus.

The **portal systems** are well represented. In the **hepatic portal system** the factors of the veins from the various parts of the alimentary canal join to form a common **hepatic portal vein** which, before entering the liver, divides into three main branches to the liver lobes. The **renal portal system** consists of two **renal portal veins** formed by the bifurcation of the caudal vein which, running on the inside of the kidneys, break up into branches in the kidney substance. The blood is then collected up in the **renal veins** which open into the hinder portions of the posterior cardinal sinuses.

The **lymphatic system** has not been well worked out in fishes such as the dogfish, but appears to consist of paired longitudinal trunks into which most of the lymphatic vessels from the tissues and organs open, and which themselves eventually open into the cardinal sinuses.

In considering this arrangement of the blood vascular system in the dogfish in relation to the basic or embryonic system, it will be seen that the major difference is that apparently only five aortic arches are represented. If it is postulated that an aortic arch is developed in front of each visceral cleft, then the number should be six, and actually in most vertebrate embryos parts at least of six arches do appear. That in front of the hyoidean cleft, however, is liable to considerable reduction or modification, and in the dogfish is probably represented merely by that part of the "hyoidean" artery in relation with the spiracle. The peculiar condition of the efferent branchial system, taking the form of loops around the clefts, can be explained by the bifurcation and anastomosis of the efferent part of the aortic arch.

## THE FROG

To understand the way in which the adult vascular system in this animal, and particularly the aortic arrangement, is derived from the basic system, it is necessary to refer to the changes which take place during the life-history of the organism. The early stages are passed in the water as a free-living larva or tadpole, and adaptive modifications in the vascular system can be traced as the animal passes from a purely aquatic to a partially terrestrial habit of life. The most obvious ones occur in the form of the aortic arches, though changes occur also in the venous system.

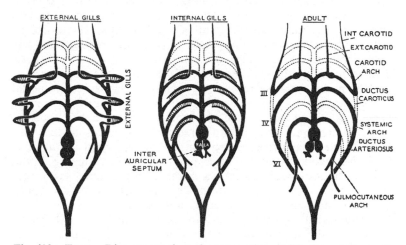

Fig. 439. FROG.—Diagrams to show the progressive modification of the aortic arches during development.

The first type of respiratory organs which appear in the young tadpole are external gills, mere vascular extensions or outgrowths of the skin at the dorsolateral region of the first three branchial (third to fifth visceral) arches. These, however, are soon replaced by internal gills of the fish-like type, developed within the four gill clefts (third to sixth visceral clefts). The blood-supply of the gills is from the third to the fifth aortic arches for the external gills and from the third to the sixth for the internal gills. Just as in the dogfish, the embryonic aortic arches become divisible into a ventral (afferent) part, a capillary system, and a dorsal (efferent) part. Although at first it is only in four visceral arches that the aortic arches are complete, yet there are traces of the first and second aortic arches. The dorsal aorta in the region of the pharynx is double and by the development of transverse-connecting vessels

Game 1
Diagram

≥ form of a circle in this region (circulus
body region the dorsal aorta is single. At
≥ made up of a sinus venosus, a single atrium,
uncus arteriosus (shortened ventral aorta plus

breathing by internal gills, the aortic system is
ame as that of a fish, and the animal is dependent
eds upon the water. At the same time, however,
e formed the lungs have grown and a pulmonary
not concerned with respiration—is established.
ry arises from the efferent region of the last
nd the **pulmonary vein** opens into the left side
s now becoming divided into two auricles by an
m. As soon as the lungs have become
they begin to take over from the gills, and this
he short-circuiting of the branchial capillary
oment of connecting vessels between the afferent
of the aortic arches. In this way, some of the
ly from the heart to the dorsal aorta without
of the gills. Thus, for a time, the tadpole uses
d pulmonary types of respiratory mechanism,
short-circuiting of the branchial capillary system
lungs play an increasing part in respiration.
forced to come to the surface of the water to
n its lungs from the atmosphere.
osis is nearly complete the branchial respira-
perseded. Then what amounts to four aortic
t interruption from the truncus to the dorsal
ed and from these the adult arterial arches
rches are the third to the sixth aortic arches
first of these, together with the anterior part
licus, forms the **carotid arch** system; the
onic) with the posterior part of the circulus
**≥ arch**; the third (fifth embryonic) disappears;
connection with the circulus cephalicus, forms
**rch**. This separation of the various aortic
he lumen of the vessel connecting the parts
that it becomes solid, and examples of such
seen in the **ductus caroticus** (ductus Botalli),
and systemic arches, and the **ductus arte-**
es called the ductus Botalli), between the
nary arches (which, however, is not dis-
lt). Thus, when metamorphosis is completed
ers assumed, the arterial arches are three in

number, the carotid supplying the head, the systemic the rest of the body, and the pulmo-cutaneous the lungs and the skin, for in the frog the skin is also a respiratory surface, equal in importance to the lungs.

The chief change which occurs in the venous system is the suppression of the posterior cardinal veins and their replacement by a single median vessel, the posterior vena cava. This arises as a separate outgrowth from the sinus venosus and grows back receiving the hepatic veins, and between the kidneys, the renal veins. The suppression of the posterior cardinal veins takes place by their lumina becoming blocked and in the adult they are difficult to distinguish except in abnormal specimens where, for some reason or other, they have failed to become solid and are present along with the posterior vena cava. Anteriorly, the Cuvierian veins become the anterior venae cavae, the anterior cardinal veins the internal jugular veins, and the inferior jugulars the external jugular veins. The portal systems remain unchanged and become those of the adult.

In the adult then, the vascular system consists of a five-chambered heart, from the truncus arteriosus of which arise three arterial arches and of systems of arteries going to, and veins coming from, all parts of the body.

The heart is roughly pear-shaped, and it and the roots of the great vessels are enclosed completely within a membranous pericardium. The S-shaped curvature of the heart is now so pronounced that the auricles lie anterior to the ventricle, the sinus venosus being dorsal in position, and the truncus arteriosus ventral. The sinus venosus is ▽-shaped, each point of the triangle receiving a main vein, the opening into the right auricle being in the middle of the triangle. The larger right and smaller left auricles are completely separated from one another by the interauricular septum, and the pulmonary veins open into the left auricle by a common aperture. The entrance of the sinus venosus into the right auricle is guarded by the **sinuauricular valves** but that of the common pulmonary vein into the left auricle has no valves. The vessel takes an oblique course through the wall of the auricle and is closed when the wall contracts upon it. Both auricles open into the single ventricle by the auriculo-ventricular aperture which is bounded by four flap-like **auriculo-ventricular valves**. The movements of two of these valves, the dorsal and ventral ones, are controlled by cords of tissue—the **chordae tendineae**—connecting them to the thick muscular wall of the ventricle. The wall of the ventricle is spongy with numerous projections into the cavity, once thought to prevent undue mixing of oxygenated and deoxygenated blood. From the ventral side of the ventricle, set rather to the right, arises the short,

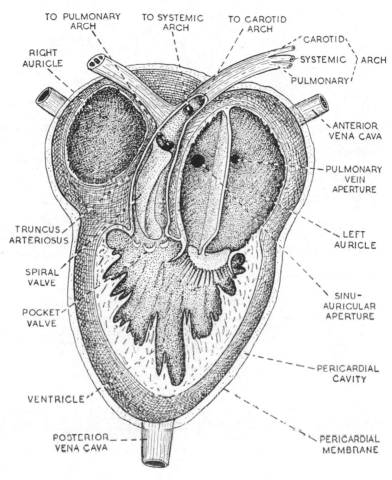

TO PULMONARY
ARCH

TO SYSTEMIC
ARCH

TO CAROTID
ARCH

RIGHT
AURICLE

CAROTID

SYSTEMIC ) ARCH

PULMONARY

ANTERIOR
VENA CAVA

PULMONARY
VEIN
APERTURE

TRUNCUS
ARTERIOSUS

LEFT
AURICLE

SPIRAL
VALVE

SINU-
AURICULAR
APERTURE

POCKET
VALVE

VENTRICLE

PERICARDIAL
CAVITY

POSTERIOR
VENA CAVA

PERICARDIAL
MEMBRANE

Frog.—Ventral dissection of the heart.

## PLATE VII.

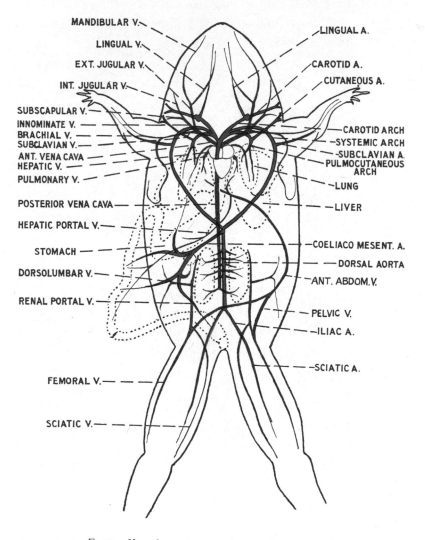

MANDIBULAR V.

LINGUAL V.

EXT. JUGULAR V.

INT. JUGULAR V.

SUBSCAPULAR V.
INNOMINATE V.
BRACHIAL V.
SUBCLAVIAN V.
ANT. VENA CAVA
HEPATIC V.
PULMONARY V.

POSTERIOR VENA CAVA

HEPATIC PORTAL V.

STOMACH

DORSOLUMBAR V.

RENAL PORTAL V.

FEMORAL V.

SCIATIC V.

LINGUAL A.

CAROTID A.

CUTANEOUS A.

CAROTID ARCH
SYSTEMIC ARCH
SUBCLAVIAN A.
PULMOCUTANEOUS
ARCH

LUNG

LIVER

COELIACO MESENT. A.

DORSAL AORTA

ANT. ABDOM. V.

PELVIC V.

ILIAC A.

SCIATIC A.

Frog.—Vascular system.   Arteries red;  veins black.

PLATE VIII.

spirally twisted truncus arteriosus which, as has already been mentioned, represents a condensed or telescoped ventral aorta plus bulbus cordis. At the entrance to the truncus are three semilunar valves and within the cavity of the lower part is a spirally twisted septum or flap, commonly called the spiral valve. Immediately anterior to this valve is an aperture leading to the pulmo-cutaneous arteries and the truncus then divides into right and left halves. Soon after the bifurcation are the apertures leading to the systemic arteries and more anterior still those of the carotid arteries. These distal portions, therefore, although appearing externally to be single vessels, are made up of three channels which ultimately separate into the three aortic arches on each side. There is at present no satisfactory functional explanation for the complicated arrangement in the truncus, but this is discussed on p. 527.

From this description of the constitution of the truncus arteriosus it will be gathered that the main arteries of the body arise as branches from three aortic arches: the carotid arch supplying the anterior part of the body, the systemic which supplies the trunk and hinder part, and the pulmo-cutaneous which gives off arteries to the lungs and skin.

The carotid arch curves outwards and forwards from the truncus arteriosus and gives off, on each side, the external carotid artery (also very commonly called the lingual artery) and then swells immediately into the carotid labyrinth. In it the lumen of the vessel is divided up by projections from the walls into small channels which, however, join up again so that the single channel is re-established. Beyond the carotid labyrinth the single artery, the internal carotid, continues around the side of the head and enters the orbit. Here it gives off branches to the under side of the orbit—in effect, part of the "palate" or roof of the buccal cavity—and to the brain.

The **systemic arch** curves round to the dorsal surface of the body cavity, where the two halves unite to form the single **dorsal aorta**. From each side of the arch arise the **oesophageal artery** to the oesophagus and the **occipito-vertebral artery** to the occiput and vertebral column; laterally, the large paired **subclavian arteries** are given off to the fore limbs. Where the two halves of the arch join beneath the backbone to form the **dorsal aorta** there arises a large vessel, the **coeliaco-mesenteric artery**, which soon divides into coeliac and mesenteric branches, the subsidiaries of which supply the alimentary canal and spleen. This condensation of the blood-supply to the alimentary canal, so that it is derived from one main artery instead of several arising at intervals (as in the dogfish), is associated with the shortness of the trunk in the frog. From the median dorsal aorta **renal arteries** pass to the kidneys, **gonadial arteries** to the gonads, and, at the hinder end of the body cavity, the dorsal aorta bifurcates to form the **iliac arteries** which pass to the hind limbs. After entering the limbs these arteries become known as the **sciatic arteries.**

The pulmo-cutaneous arch divides on each side into two main branches, the pulmonary artery to the lungs and the cutaneous artery to the skin.

In the **venous system** the blood from the anterior part of the body is returned to the heart by the two anterior venae cavae and from the posterior part by the posterior vena cava. On each side the lingual vein from the throat and the mandibular vein from the lower jaw join to form the external jugular vein; the **subscapular vein** from the dorsolateral pectoral region and the **internal jugular vein** from the region of the angle of the jaw unite to form the **innominate vein**; the **brachial vein** from the fore limb and the **musculo-cutaneous vein** from the skin of the side and back join to form the **subclavian vein.** These three main trunks, external jugular, innominate, and subclavian veins enter the **anterior vena cava** simultaneously, and this opens into the sinus venosus. The posterior vena cava receives the **renal** and **gonadial** veins from the kidneys and gonads and the **hepatic veins** from the liver.

The two portal systems are curiously linked up with one another. In the renal portal system the femoral and sciatic veins from each hind limb join within the body cavity to form a common trunk. From the femoral vein arises the pelvic vein—which passes ventrally to the mid-line of the abdominal wall, where it joins its fellow of the opposite side to form the anterior or median abdominal vein. Each renal portal vein passes from the junction of the sciatic and femoral veins to the outer edge of the kidney of that side receiving the dorso-lumbar vein from the lumbar region of the body wall. Reverting now to the anterior abdominal vein, this runs forwards along the mid-ventral line of the abdomen to the region of the xiphisternum, where it passes upwards in between the lobes of the liver to join the hepatic portal vein. The hepatic portal vein receives factors, the gastric, splenic, and intestinal veins from the alimentary canal. After the junction of the anterior abdominal and hepatic portal veins, the common trunk divides into branches which pass to the liver lobes.

The blood from the lungs is returned direct to the left auricle by the common pulmonary vein, formed by the union of the vein from each lung.

The lymphatic system in the frog is very extensive, for, in addition to the lymph spaces of the body between the organs and also of the organs themselves, large subcutaneous spaces are present which account for the apparent looseness of the frog's skin; but at intervals the skin is attached to the body muscles by sheets of tissue or septa so that the space is divided up into chambers on the dorsal, lateral, and ventral surfaces. Where the lymphatics join the venous system are pulsating vessels called **lymph hearts** of which there is an anterior pair beneath the transverse processes of the third vertebra, opening into the vertebral vein, and a posterior pair at the sides of the urostyle which open into a transverse connection between the proximal ends of the femoral and sciatic veins.

The action of the heart in distributing the blood about the body is far from being fully understood. It was once believed that by the action of the spiral septum in the truncus arteriosus deoxygenated blood was caused to enter the pulmo-cutaneous arch, mixed blood to enter the systemic arch, and oxygenated blood to enter the carotid arch, the carotid labyrinth so raising the pressure in it that only highly oxygenated blood leaving when the ventricle was near the end of systole could enter it.

As a result of X-ray photography, it is now known that all the arches fill with blood practically simultaneously and it is extremely doubtful if there is any significant separation of oxygenated from deoxygenated blood in the heart. This view is rendered all the more likely when it is remembered that the lungs are not the sole, nor even, perhaps, the most important, of the respiratory surfaces in the frog. Particularly when the animal is in water, the skin functions as the main site of gaseous interchange whilst the mucous membrane lining the buccal cavity is always of great importance.

It follows, therefore, that the musculo-cutaneous vein, whose blood is emptied into the right auricle by way of the subclavian vein and sinus venosus, may contain blood with a higher oxygen content than that of the pulmonary vein. On the other hand, work on the heart of the clawed-toad, *Xenopus laevis*, suggests that blood from the right auricle does not get to the left side of the ventricle and during systole goes almost entirely into the pulmo cutaneous arch. But the pressure in the pulmonary circulation is lower than in the carotid and systemic arches and its circulation is much more rapid. Therefore this arch receives blood also from the left auricle *via* the ventricle. The pressures in the carotid and systemic arches are similar and when the lungs are operating both receive oxygenated blood from the left side of the heart. Perhaps the heart of *Xenopus* functions rather differently from that of the frog.

In some ways the frog's heart, which appears to be an "inefficient" organ (in the sense that it does not separate the oxygenated from the deoxygenated blood), can be seen to be well adapted to the amphibious mode of life. When, for example, the frog is breathing mainly through its skin and buccal cavity, the oxygenated blood from the cutaneous vein is directed into the right auricle and, if the heart were completely divided into right and left halves as it is in birds and mammals, this oxygenated blood would have to pass through the pulmonary circulation before becoming available to the tissues.

## THE RABBIT

It is impossible to deal here in detail with the progressive evolutionary changes in the vascular system which have resulted in the mammalian arrangement. However, the vascular system is first laid down on the now

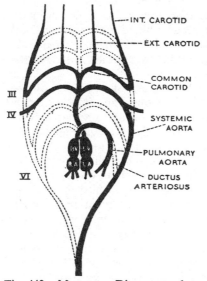

Fig. 440. MAMMAL.—Diagram to show the origin of the adult aortic arches.

familiar basic plan and it is possible to see the way in which this is modified to give rise to the adult arrangement. Although at no time during the development of a mammal do the visceral clefts ever take on a respiratory function, in fact none ever perforate the pharynx, yet they are present in a rudimentary way. Between these vestigial visceral clefts and encircling the pharynx, aortic arches make their appearance, and although some may never be completely developed, there is evidence that the six are present, of which only the third to the sixth participate in the formation of the adult system.

Again, although the heart is formed from a simple cardiac tube, and when first differentiated consists of the usual four chambers, sinus venosus, atrium, ventricle, and bulbus cordis, the atrium soon becomes divided by the interauricular septum into two auricles and this septum is continued as the interventricular septum, dividing the ventricle also into right and left halves. Associated with this is the longitudinal division of both the now condensed and twisted bulbus cordis and ventral aorta, to produce two aortae—the carotico-systemic aorta and the pulmonary aorta—arising directly and separately from the ventricles and *not* from a common trunk as in the lower forms. Later on most of the right systemic (4th aortic) arch disappears. The sinus venosus becomes incorporated in the wall of the right auricle, forming the sinuauricular node ("pacemaker") which starts the heart beat. Thus the adult heart has only four chambers, two auricles, and two ventricles.

The first and second aortic arches disappear, but the portions of the ventral aorta and the dorsal aortae in connection with them are appropriated by the third arch, which then becomes the carotid arch with its internal carotid (forward continuation of the dorsal aorta) and external carotid (forward continuation of the ventral aorta) on each side. Dorsally the carotid and systemic arches remain joined by the solid **ductus caroticus** on each side, but ventrally the portion of the ventral aorta between them persists as the **innominate artery** by which the carotid arteries arise from the carotico-systemic aorta. The systemic arch is derived from the left half of the fourth aortic arch which also appropriates the posterior part of the foetal left dorsal aorta. The right side of the arch is represented by the proximal part of the right subclavian artery, the left subclavian arising in the normal manner from the left portion of the dorsal aorta now taken over by the systemic arch. The fifth arch disappears and the sixth forms the pulmonary arch or aorta already separated from the systemic by the division of the bulbus cordis and ventral aorta, and the flow of blood between the two is interrupted by the development of the ductus arteriosus, which is the only one of these occluded vessels to persist in the adult. But in the foetus it functions as an important channel short-circuiting the lungs since it allows blood to pass freely from the right ventricle *via* the pulmonary aorta into the systemic aorta and thence to the body.

In the venous system the main lines of change in the basic plan follow those seen in the frog. The Cuvierian vein on each side becomes the anterior vena cava, which receives the main veins of the anterior body region. The blood from the hinder part is returned to the heart by the posterior vena cava which takes over the function of the posterior cardinal veins. The anterior portion of the right posterior cardinal vein, however, is retained as the **azygos vein** which drains blood from both sides of the chest wall. The left posterior cardinal loses its connection with the Cuvierian vein and acquires an opening into the azygos vein.

In the adult mammal there is only one portal system, the hepatic, the renal portal system having been lost.

Having got a general idea of the main lines of the mammalian vascular system, that of the rabbit may be examined in detail.

The ovoid or pear-shaped heart lies in the centre of the anterior part of the thorax, completely enclosed by the membranous pericardium. As was noticed in the frog, the S-shaped curve of the cardiac tube is exaggerated so that the two auricles lie anterior to the ventricles. The left ventricle is larger than the right and is also more muscular since it has to drive the blood completely round the body whereas the right ventricle is only concerned with the pulmonary circulation. If compared with the ventricles, the auricles are relatively thin-walled but yet are muscular. Into the right auricle the three venae cavae open by separate apertures. Between the apertures of the two anterior venae cavae, and to some

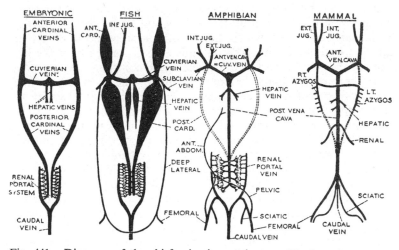

Fig. 441. Diagrams of the chief veins in vertebrates. The hepatic portal system has been omitted.

extent guarding that of the posterior vena cava, is the Eustachian valve, a remnant of the right sinuauricular valve, which also contributes to the Thebesian valve protecting the left anterior vena cava. Laterally, the margins of the auricles project to form the auricular appendices. The pulmonary veins open by separate apertures into the left auricle, the common trunk having been incorporated into the auricular wall. The cavities of the auricles are completely separated from one another by the interauricular septum, in the middle of which is an ill-defined depression—the **fossa ovalis**—which marks the position of a connection in the foetus between the posterior vena cava and the left auricle by the **foramen ovale** which, usually, becomes closed shortly after birth.

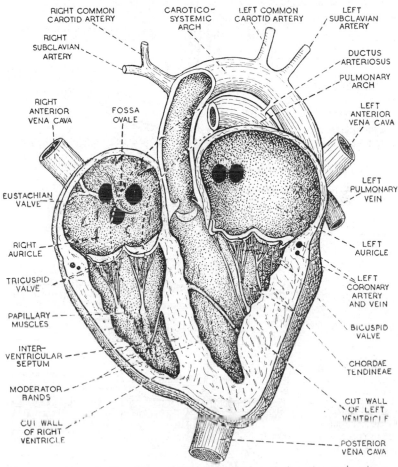

RIGHT COMMON CAROTID ARTERY

CAROTICO-SYSTEMIC ARCH

LEFT COMMON CAROTID ARTERY

LEFT SUBCLAVIAN ARTERY

RIGHT SUBCLAVIAN ARTERY

DUCTUS ARTERIOSUS

PULMONARY ARCH

RIGHT ANTERIOR VENA CAVA

FOSSA OVALE

LEFT ANTERIOR VENA CAVA

EUSTACHIAN VALVE

LEFT PULMONARY VEIN

RIGHT AURICLE

LEFT AURICLE

TRICUSPID VALVE

LEFT CORONARY ARTERY AND VEIN

PAPILLARY MUSCLES

BICUSPID VALVE

INTER-VENTRICULAR SEPTUM

CHORDAE TENDINEAE

MODERATOR BANDS

CUT WALL OF LEFT VENTRICLE

CUT WALL OF RIGHT VENTRICLE

POSTERIOR VENA CAVA

MAMMAL.—Ventral dissection of the heart. (For clarity, the pericardium has been omitted.)

PLATE IX.

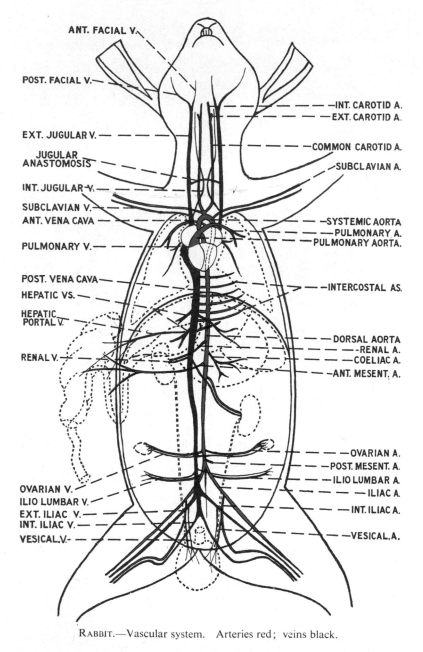

ANT. FACIAL V.

POST. FACIAL V.

EXT. JUGULAR V.

JUGULAR
ANASTOMOSIS

INT. JUGULAR V.

SUBCLAVIAN V.

ANT. VENA CAVA

PULMONARY V.

POST. VENA CAVA

HEPATIC VS.

HEPATIC
PORTAL V.

RENAL V.

OVARIAN V.

ILIO LUMBAR V.

EXT. ILIAC V.

INT. ILIAC V.

VESICAL. V.

INT. CAROTID A.
EXT. CAROTID A.

COMMON CAROTID A.

SUBCLAVIAN A.

SYSTEMIC AORTA
PULMONARY A.
PULMONARY AORTA.

INTERCOSTAL AS.

DORSAL AORTA
RENAL A.
COELIAC A.
ANT. MESENT. A.

OVARIAN A.
POST. MESENT. A.
ILIO LUMBAR A.
ILIAC A.
INT. ILIAC A.

VESICAL. A.

RABBIT.—Vascular system.   Arteries red;   veins black.

PLATE X.

The ventricles are very thick-walled and muscular, the bundles of fibres being so arranged as to form longitudinal ridges—**columnae carneae**—projecting from the wall into the ventricular cavity. The interventricular septum is not entirely median but is inclined to the right, where it joins the right wall of the heart, leaving the apex to the left ventricle. The auriculo-ventricular apertures are guarded by flap-like valves, the edges of which are anchored to muscular projections of the ventricular wall—**musculi papillares**—by cords—**chordae tendineae**. The valve guarding the right auriculo-ventricular aperture has three flaps and is called the **tricuspid valve**, while that between the left auricle and ventricle has only two flaps and is termed the **bicuspid** or **mitral valve**. The entrances from the ventricles into the aortae—both systemic and pulmonary—are provided with semilunar valves.

As has already been explained, the two aortae *appear* to arise directly from the ventricles, the systemic from the left and the pulmonary from the right, and because of the way in which the bulbus cordis became twisted before its division into two, the bases of the aortae are also twisted around one another. Consequently the systemic aorta arising from the anterior end of the left ventricle passes dorsal to the pulmonary aorta and then curves round to the dorsal side of the heart to lie beneath the backbone, continuing thence backwards as the dorsal aorta. The pulmonary aorta arising from the right ventricle curves over to the left side of the heart beneath (ventral to) the systemic and passes through the arch of the systemic aorta to the dorsal side of the heart and thence direct to the lungs. Where the pulmonary aorta passes beneath the arch of the systemic is found the ductus arteriosus, a solid cord connecting the two aortae.

From the systemic arch and dorsal aorta arise the main arteries of the body, the general distribution of which is shown in Plate X. The reason for the apparent origin of the carotid arteries from the systemic aorta has already been given, but a further complication in the course of the carotid arteries is introduced by the presence of a neck between the thorax and head. In order to reach the head these arteries have to traverse the neck, so that long right and left common carotid trunks are present in this region before the division into internal and external carotid arteries is reached. Usually the right and left common carotid arteries arise from the **innominate artery**, but this is not invariable and the left may arise directly from the aortic arch. The right subclavian artery arises either from the innominate or from the base of the right carotid artery; the left subclavian takes its origin from the left side of the arch; the reasons for this asymmetry having already been made clear (p. 528).

The principal arteries given off from the dorsal aorta are: the **intercostal arteries** to the wall of the chest; the coeliac artery to the liver and stomach; the anterior mesenteric artery to the intestine; the **renal arteries** to the kidneys; the genital arteries to the gonads; and the posterior mesenteric artery to the rectum. In the pelvic region, the dorsal aorta divides to form two iliac arteries which pass into the legs to become the femoral arteries. A small caudal artery continues in line with the dorsal aorta into the tail.

From the pulmonary aorta, after it has curved over on to the dorsal side of the heart arise the two pulmonary arteries which subdivide into branches to the lobes of the lungs.

The blood from the head region is returned to the heart by factors opening into the external and internal jugular veins of which the former are the larger; that from the fore limbs by the sub-clavian veins. The external and internal jugular and the subclavian veins on each side meet almost at a point to open into the anterior vena cava of that side. The right anterior vena cava also receives two more veins: the anterior intercostal vein and the asymmetrical azygos vein, both of these draining the wall of the thorax. In the rabbit there are two anterior venae cavae, but in some mammals, including man, only the right is present, the two external jugular veins being connected by a transverse anastomosis.

Into the posterior vena cava, bringing blood from that part of the body posterior to the heart, open the external and internal iliac veins from the outside and inside of the hind limbs; the genital veins from the gonads; the iliolumbar veins from the posterior abdominal wall; the renal veins from the kidneys; and the hepatic veins from the liver.

The portal vein receives factors (duodenal, anterior, and posterior mesenteric veins) from the intestine and rectum, from the stomach and spleen (lienogastric vein), and then passes to the liver where it breaks up into several branches which enter the liver lobes.

**The lymphatic system** in a mammal is very extensive. In addition to that part of it already mentioned in connection with the villi of the intestine, every organ and limb has its complex network of lymphatic vessels. All these vessels eventually open into two main collecting ducts or vessels of which the one on the left is the **thoracic duct**, which opens into the left subclavian vein at the point where it joins the left external jugular vein. That of the right side opens into the right subclavian at approximately the same position and is called the **right lymphatic duct**. The flow of the lymphatic fluid is always toward the main collecting vessels, its direction of flow being maintained by valves in the larger vessels.

**Changes in the Vascular System after Birth.**—As already has been mentioned, important changes in the vascular system of mammals takes place shortly after birth. These are in the main necessary because the lungs supersede the placenta as respiratory organs. In the foetus oxygenated blood from the placenta is brought to the foetus by the umbilical vein which empties into the posterior (or inferior) vena cava. This then pours blood into the right auricle but in the foetus about 50 per cent. of the blood in the posterior vena cava passes through an aperture directly into the **left** auricle. This aperture is the foramen ovale. By this means some oxygenated blood from the placenta is directed *via* the **left** auricle into the left ventricle and thence to the body of the foetus. Of course, at this time the lungs are not functioning and are collapsed. So their blood vessels offer a high resistance to the passage of blood. Being unused as a respiratory surface the lungs can be by-passed; blood passes from the pulmonary aorta into the systemic aorta through the ductus arteriosus. At birth the lungs become inflated with air, resistance to blood flow drops and they now function as respiratory surfaces whilst, with severance of the placental layers, the placenta becomes useless. Then by constriction of its walls the ductus arteriosus becomes closed (and soon occluded by fibrous tissue) whilst by increased hydrostatic pressure in the left auricle the flow of blood from the posterior vena cava *via* the foramen ovale is reversed. Later on still the foramen ovale becomes occluded and the adult circulation is confirmed. Then oxygenated blood from the lungs *via* pulmonary veins returns to the left auricle and deoxygenated blood from the three venae cavae into the right auricle.

It will be seen, therefore, that the most striking development in the blood vascular system associated with the migration of the vertebrates from an aquatic to a terrestrial environment is the progressive elaboration of a double from a single circulation. All the evidence suggests that the main factor leading to the development of the separate pulmonary circulation is the lowering of the blood pressure by that circulation below the level required for "efficiency" in the systemic circulation. Therefore the blood is returned to the heart where its pressure is once again raised before it enters that circulation. The partial separation seen in the amphibians and reptiles is, therefore, carried a step further and an anatomical separation with a less extensive and muscular part of the embryonic ventricle is utilised to drive the blood through the lungs.

**Histology and Physiology of Blood.**—Blood is a liquid tissue which, although usually described under the heading of connective tissue, differs from it in two important respects. As has been seen, these tissues consist of cells together with a matrix which always has a fibrous basis, although the fibres may be masked by the deposition around them of chondrin as in cartilage or of calcium salts as in bone. Blood, in the living condition, on the other

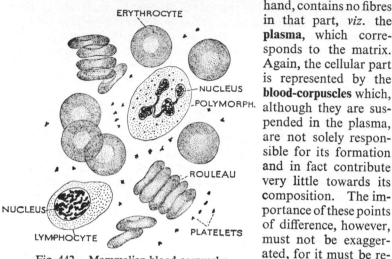

hand, contains no fibres in that part, *viz.* the **plasma**, which corresponds to the matrix. Again, the cellular part is represented by the **blood-corpuscles** which, although they are suspended in the plasma, are not solely responsible for its formation and in fact contribute very little towards its composition. The importance of these points of difference, however, must not be exaggerated, for it must be remembered that there is

Fig. 442.   Mammalian blood-corpuscles.

a constant interchange of plasma and of cells between the surrounding connective and other tissues and the blood. Moreover, the blood vessels as well as the blood they contain are, in the embryo, derived from the same system of mesodermal (**mesenchyme**) cells which are responsible for the laying down of the embryonic connective tissues proper.

Blood, then, consists of a fluid plasma in which are suspended cells of several different types. Other more minute bodies, termed blood-platelets (thrombocytes), which are not cells but are cytoplasmic fragments of large, multinucleate cells, megakaryocytes, found in red bone marrow, are also present.

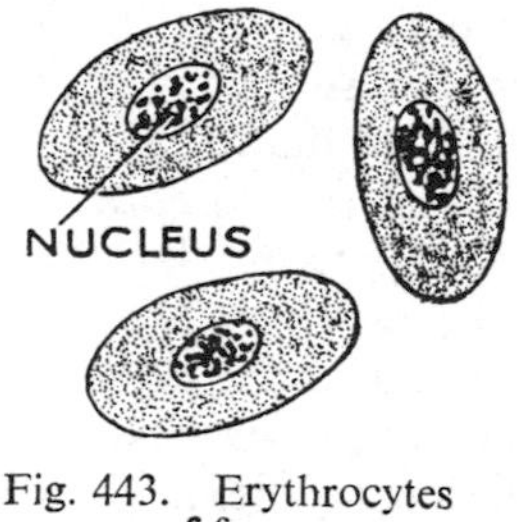

The plasma consists of a complex mixture of inorganic salts in true solution and blood proteins in colloidal sol. The

Fig. 443.   Erythrocytes of frog.

total salts present amount to about 1 per cent. by weight of the plasma, and, of this, chloride and bicarbonate of sodium form by far the greater part, whilst the remaining salts are chiefly chlorides and bicarbonates of potassium and sulphates and phosphates of sodium and potassium. (It is interesting to note, however, that the salts present inside the corpuscles are chiefly salts of potassium and not sodium.) It is the presence of these salts which gives the blood its slightly alkaline reaction. The blood proteins which are present in appreciable amounts are of three types—albumins, globulins, and fibrinogen; others are present as mere traces. Other substances normally present in the plasma include such "useful" substances as glucose, fats (especially evident soon after a meal), amino acids, and lecithin, as well as excretory products like urea. Further substances whose presence cannot be detected by analysis, but only by physiological tests, always circulate in the blood. Of these substances the hormones (secretions of the endocrine organs) and those concerned with the neutralisation of foreign toxins are worthy of special mention.

The corpuscles normally present in the blood fall into two well-defined groups: the red corpuscles (**erythrocytes**) and the white corpuscles (**leucocytes**). None of these corpuscles divide or multiply in the blood, certain of them being formed in the red bone marrow and others in the nodules of the lymphatic system (see below).

As has been mentioned above, erythrocytes in mammals are formed in the adult in the red bone marrow. They arise here from cells, termed **erythroblasts**, which lie in thin-walled sinuses, and several generations of erythroblasts arise by mitotic division; but eventually a generation of cells is formed which lose their ability to divide and after extruding the nucleus, they become fully formed erythrocytes. Their life is, however, short, and after a few weeks (100-120 days in man) of circulation in the blood-stream they are destroyed by phagocytes in the blood itself and in the spleen in particular. This short life is possibly connected with the absence of a nucleus; but the reason for the enucleation, which results in a continual renewal of the red corpuscles being necessary, is difficult to see, for, although the iron liberated from the haemoglobin is retained by the spleen, other break-down products contribute to the formation of bile pigments and are passed out with the bile, so that most of the substance of the corpuscles is lost.

The development of the red blood-corpuscles in vertebrates other than mammals has not been so well worked out. That they are formed in a different situation is quite certain, for only the bones of mammals contain red marrow. It is probable that the liver and the spleen in addition to destroying old erythrocytes also form new ones as they are required, but it is a fair inference that, because the red corpuscles of the lower vertebrates have a nucleus, they retain their functional activity and are in circulation for a considerably longer time, so that their renewal takes place much less frequently than in mammals and perhaps specialised renewal regions are therefore unnecessary.

The erythrocytes of a mammal are biconcave circular discs. They are far more numerous than the white corpuscles and, in man, have a diameter of about 7·5 μm and a thickness of about 2 μm. They are bounded by an elastic envelope, about 1 μm thick, which is composed of a very thin outer lipoid layer and a thicker inner layer of radially orientated protein molecules, so that they are easily distorted by pressure, but when the pressure is removed they return at once to their normal shape, and this elasticity is of importance in allowing the blood to pass through the smaller capillaries, whose diameter is frequently less than that of the corpuscles. The cytoplasm of the erythrocytes is apparently of a homogeneous nature and in it is dissolved the respiratory pigment, haemoglobin, which, although of a yellowish colour when seen in the corpuscles under the microscope, yet appears red when seen in bulk. Although absent from the erythrocytes of the adult mammal, nuclei are present in those of all vertebrates lower than the mammals, so that almost at a glance one is enabled to say of any sample of adult blood whose red cells contain nuclei, "This is non-mammalian blood". (Consideration of other factors such as the size of the erythrocytes and the nature of the white cells present also make it possible to identify blood as human or non-human, a fact that has considerable medicolegal importance.) A detailed description of the white corpuscles (such as is required by the medical profession) is out of place here, but the following more important points may be mentioned. The white cells are of two main types: the non-granular leucocytes, or lymphocytes as they are sometimes called, and the granular leucocytes or granulocytes. The former, as their name implies, have a non-granular cytoplasm and take origin from special cells in the lymphatic nodules. They are extremely variable in size, and form about a quarter of the total number of leucocytes normally present in human blood. The granulocytes, in contrast to the lymphocytes, have in their cytoplasm granules which react to certain stains, and are formed in the closed sinuses of the red bone marrow from special cells called myelocytes, and only later enter the bloodstream when the sinuses acquire an opening into the nearby vessels.

The classification of the different types of granulocytes depends upon the staining reactions of the cytoplasmic granules. Thus cells whose granules stain with "acid" stains are acidophils, those which stain with "basic" dyes are the basophils, whilst a third type whose granules stain differently according to the species of animal being studied are called heterophils. In man the heterophils stain well with most microscopic dyes and are often termed neutrophils. The nucleus of the different types of white corpuscles is also variable, but by far the commonest condition is that of an irregularly lobed form when it is said to be polymorphous but large monocytes are numerous when tissues are invaded by foreign micro-organisms.

Blood-platelets, which break up into still more minute granules when a blood smear is prepared, are not seen by this method of examination.

CLOTTING OF BLOOD.—Perhaps one of the best-known properties of blood is its ability to clot after it escapes from a vessel and the importance of this to the animal is at once obvious, for by means of the clot the wound is sealed and further loss of blood is prevented. The blood of certain human beings does not possess the ability to clot, with the result that even a small puncture of the skin may cause a dangerous haemorrhage. The disease is called **haemophilia**, and not the least interesting thing about it is that it is transmitted from parent to offspring in a relatively simple Mendelian fashion (see Chapter XX).

The mechanism by which clotting is brought about is somewhat complicated, but the net result is that the fibrinogen, present in the undamaged vessel as a colloidal sol, is precipitated in the form of a dense meshwork of fine threads of fibrin to which are added blood-platelets which early on join up to form a plug to the damaged vessels. In the meshes of the threads the corpuscles of various types become entangled to form a clot. The precipitation of the fibrinogen, like so many other processes taking place in the body, is effected by the action of an enzyme. It appears that the enzyme causing clotting is not present in an active form in the circulating blood, but only as an inactive precursor (pro-thrombin) which requires the presence of calcium ions to form the active enzyme, thrombin. Apparently the reason why clotting does not take place in a normal healthy blood vessel is that any thrombin which may be formed is neutralised by an anti-thrombin (heparin) so that clotting cannot take place. When, however, a blood vessel is injured, a substance, thrombokinase, is released from the disintegrating blood-platelets and from the walls of the vessels. This substance renders the anti-thrombin ineffective. The pro-thrombin becomes activated and causes the platelets to form the plug and the fibrinogen to be precipitated. Thus clotting depends on the presence of platelets, pro-thrombin, and thrombokinase together with calcium ions, and the absence of any of these factors will prevent clotting. It is frequently necessary to use samples of blood which will not clot, and the simplest way of obtaining these is to add to the blood a substance which will remove or render inactive the calcium ions. The addition of citrates or oxalates precipitates the soluble calcium salts of the blood, and therefore after such treatment the blood will no longer clot. A similar effect is obtained if the enzymes are precipitated by the addition of magnesium sulphate, or if some antagonistic enzyme, such as is present in leech extract, is added to the blood.

AGGLUTINATION.—It has been found that when the blood from two different types of vertebrate animals is mixed, the red blood-corpuscles run together and form clumps. This clumping is usually spoken of as agglutination and is more marked when the blood-samples come from animals not closely related in the animal kingdom; but, even when the blood of animals of the same species is mixed, some agglutination may take place if the two samples do not come from animals of the same "blood-group".

Agglutination has been shown to be due to the interaction of two substances, one of which (the antigen or agglutinogen) is present in the red corpuscles, whilst the other (the antibody or agglutinin) is found in the plasma. These agglutinins seem to belong to the same general class of substances produced in the blood and termed antibodies, examples of which are those which react with certain pathogenic bacteria. Now it is also known that there are two types of corpuscle substances (agglutinogens), which are termed A and B, and two kinds of plasma substances (agglutinins), $\alpha$ and $\beta$. The corpuscles of any one individual may contain A or B, or both A and B, or neither A nor B, whilst the plasma may contain either $\alpha$ or $\beta$, both $\alpha$ and $\beta$, or neither. From this it follows that the blood of human beings falls into four groups which are designated A, B, AB, or O respectively. Group A possesses A corpuscles and $\beta$ plasma; group B possesses B corpuscles but $\alpha$ plasma; group AB possesses both A and B corpuscles but no plasma factor at all; whilst group O has no corpuscle factors but both $\alpha$ and $\beta$ plasma factors. Hence group AB is agglutinated by A, B, and O groups; group A is agglutinated by groups B and O; group B by groups A and O; and group O by neither groups A, B, nor AB.

The importance of these phenomena lies in the now common practice of blood transfusion during which agglutination must at all costs be avoided, and the choice of the correct blood-group is all important. Since the corpuscles of the O group are not agglutinated by the plasma of any of the other groups, persons of the O group are said to be "universal donors", whereas persons with blood of the AB group are termed "universal recipients" since their plasma will not agglutinate the transfused corpuscles from any of the other groups. It is important to notice that, from the point of view of blood transfusion, what matters is that the antigens (or corpuscles) of the transfused blood should be compatible with the serum of the patient receiving the transfusion. If the corpuscles are incompatible they become agglutinated and then rapidly destroyed. But the kidneys cannot excrete the products of destruction and it is these that give rise to unpleasant and sometimes fatal symptoms. On the other hand, but little harm results if the transfused blood has antibodies in its serum which agglutinate the host corpuscles. This is due to the antibodies in the transfused blood becoming so diluted by the serum of the recipient that they affect a few red corpuscles only. It is also thought that the incompatible antibodies combine with the cells of the blood vessels and so become neutralised without affecting the recipient's corpuscles. When incompatible corpuscles are transfused then the recipient continues to produce antibodies in sufficient quantity to agglutinate all the corpuscles which have been received.

LYMPH.—Lymph is a colourless fluid which resembles the plasma of the blood in composition, in the fact that lymphocytes and other white corpuscles are suspended in it, and that it has the power to form a clot. It arises by a process of ultrafiltration: diffusion under pressure of fluid outwards through the walls of the blood-capillaries, and comes to surround and bathe all the tissues of the body and so form a sort of "middle-man" which hands on substances (e.g. oxygen, food) from the blood to the tissues. The outward passage of water and dissolved crystalloids takes place mainly at the arterial end of a capillary since there the hydrostatic

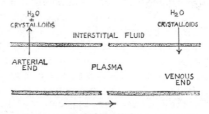

Fig. 444. Diagram to show the process of lymph formation and the exchange of substances between the blood—the capillaries and the interstitial fluid.

pressure exceeds the colloid osmotic pressure due to the blood proteins. Further along to the venous end of the capillary the hydrostatic pressure is lower and the colloid osmotic pressure begins to exceed the hydrostatic pressure so that water and solutes pass in from the surrounding tissue fluid.

Excess fluid is drained away from the fluid spaces into the capillaries of the lymphatic system of vessels and eventually, after passing through the larger lymphatic vessels, it is returned to the blood system by the thoracic duct, which joins the base of the left subclavian vein, and the right lymphatic duct, which joins the base of the right subclavian vein. The larger lymph vessels are provided with valves to prevent the backward flow of the fluid, and a muscular coat may also be present. At intervals even the larger lymph vessels are interrupted by a system of lymph capillaries which form, on the vessel, a swelling or nodule.

### Defence against Infection

Many of the diseases from which Man and other animals suffer arise from the invasion of their bodies by particular kinds of micro-organisms termed pathogens—viruses, bacteria, protozoa—but by no means all micro-organisms found in or on the body are harmful. Most are harmless commensals and some may be obviously beneficial, *e.g.* the bacteria and ciliates in the rumen of ungulates which digest cellulose; or the trichonymphid flagellates in the gut of wood-eating termites. The microbial flora of the gut of Man, although usually rated as harmless commensals may be beneficial in less obvious ways by competing with or actually destroying pathogens or by synthesising vitamins.

The tissues and cells of the body, however, are normally kept free from micro-organisms by a variety of mechanisms ranging from very general ones, like the provision of a body surface through which micro-organisms cannot pass to the production of proteins which kill only particular species or strains of pathogens. Of paramount importance in this connection are the white corpuscles of the blood and lymph. The large monocytes and polymorphs act as phagocytes. That is, by amoeboid action they engulf and digest foreign micro-organisms. These white corpuscles can pass through the walls of the vessels and accummulate around sites of infection in the tissues to form pus. Thus they are a mechanism of defence against many pathogens. Many pathogens call forth a more specific response for they act as antigens, causing the lymphocytes to produce and release into the blood specific antibodies—proteins which kill the invaders and sometimes neutralise the toxins they produce. Some viruses are neutralised by a substance, interferon, in a way which is more immediate but is not well understood.

# THE RESPIRATORY SYSTEM

## INTRODUCTION

In the earlier chapters (*e.g.* pp. 214-22) it has been pointed out that the tissues and individual cells of the body of a complex animal have no immediate access to oxygen from the outside and that the transport of oxygen from the source of supply to the tissues and the conveyance of carbon dioxide in the reverse direction, introduces structural and physiological problems which are somewhat alike, and consequently have been solved in a basically similar way in a great variety of animals. Vertebrates have been much studied from this point of view and it is now convenient to review the general principles of respiration showing how they apply in particular to vertebrates.

Large animals, like vertebrates, require a large respiratory surface for gaseous interchange and without considering such respiratory organs and associated structures in detail, it is instructive to suggest the probable ground plan they will follow and the factors of importance in its formulation.

The rate at which oxygen can enter the body and at which carbon dioxide can be got rid of depends upon the gradient of concentration of these gases between the exterior and the interior of the animal and on the surface area available for gaseous interchange. The need, or demand, for oxygen and the rate of production of carbon dioxide depends on the mass of the respiring tissue and the intensity of metabolism (metabolic rate) which in turn depends on the activity and so varies greatly even in one animal. In a general way it seems that small animals have a higher metabolic rate than large ones provided that comparisons are made between animals of a similar grade of organisation. Except for warm-blooded animals the reasons for this are not known.

For warm-blooded animals, however, it is obvious that a proportion of their respiration is concerned with supplying energy in the form of heat and that because of the surface/volume ratio law (the smaller the animal the bigger its surface) the heat loss from a small animal will be more rapid than from a big one. Consequently, small animals have a higher metabolic rate than large ones. To sustain this they have to eat relatively larger amounts of food. A small mammal, like a shrew for instance, eats more than half its own body weight of food every twenty-four hours. As has been mentioned above the mass of an object increases as the cube of its linear dimensions whereas the surface increases only as the square of these. Therefore when considering a series of animal forms of increasing size a stage is reached, way down in the invertebrate scale, when gaseous exchange by simple diffusion over the general surface of the body does not suffice because the distance between the deeper tissues and the surface becomes too great, so to make up for this inadequacy, structural and physiological alterations and additions take place. These developments can be listed as follows:

(1) *Movement*.—An animal which is continually on the move and so never stays in a place, where the oxygen supply can be materially depleted, can maintain a gradient of diffusion. This applies to many small animals. Equally the development of a mechanism which causes water or air to flow over the respiratory surface can have the same effect. Often both these mechanisms operate together, as for example, in pelagic fishes such as the mackerel.

(2) *Alteration of Body Shape.*—With increase in the mass of the body the area of the surface set aside for gaseous interchange is enlarged by foldings or the formation of filaments and so on to form gills (pp. 434-7).

(3) *Transport Mechanisms.*—The cells and tissues of complex animals such as vertebrates have, as has been pointed out, no direct contact with an outside source of oxygen and so, as in most coelomate animals, the carriage of gases takes place by a blood vascular system whose "efficiency" in this respect when compared with the water vascular system of such animals as coelenterates and echinoderms is increased by a respiratory pigment (always haemoglobin in vertebrates) which carries oxygen in reversible chemical combination. Transport of carbon dioxide as carbonates and bicarbonates ensures the carriage in greater amounts than is possible by mere physical solution.

In sum, large animals, such as vertebrates, require for their respiratory needs a source of oxygen, a respiratory surface of large area, a current of air or water over the respiratory surface, a respiratory medium (blood), a respiratory pigment (haemoglobin) and a means of carrying more carbon dioxide than is possible by mere physical solution. All the features and processes so far mentioned are concerned with facilitating gaseous exchanges between the cells and tissues of the body and the external environment, that is, with external respiration or breathing, but the release of energy takes place within the cells and is termed internal or tissue respiration (p. 546).

It has already been pointed out that one of the chordate characteristics is the visceral clefts. In the aquatic vertebrates certain of these clefts become modified for respiratory purposes and are then called branchial or gill clefts. From them is developed, by a folding of the lining, a type of respiratory surface which is called a gill. Gills are highly vascular and are bathed in oxygen-containing water passing from the pharynx to the exterior, so that all the requirements set out above are fulfilled. This type of respiratory mechanism is found in the fishes and in the early (tadpole) stages of amphibians. In the partially or wholly terrestrial vertebrates, however, where the source of oxygen is the atmosphere, a second type of respiratory organ, the lung, is developed. Lungs arise as outgrowths from the modified posterior region of the pharynx which becomes partially separated off to form a laryngeal chamber communicating with the pharynx by the slit-like glottis. The lungs are air-tight sacs, the lining of which is much folded and highly vascular, and air is alternately taken into and passed out from them.

All vertebrates, apart from one group of deep sea fishes have the respiratory pigment **haemoglobin**, a complex, iron-containing organic compound in the red corpuscles. Haemoglobin, in conditions of high oxygen concentration (*e.g.* at a respiratory surface), combines with this gas to form bright-red oxyhaemoglobin; but in situations of low oxygen concentration (*e.g.* in the deoxygenated tissues where the medium is slightly acid, due to the accumulation of carbon dioxide) it gives up most of its oxygen, returning to its normal purply-red colour (the Bohr effect). It thus acts as an efficient oxygen carrier, enabling the blood to carry more oxygen than it could in physical solution, but a small proportion of oxygen will also dissolve in the watery blood-plasma.

That this is by no means an insignificant amount is shown by the fact that fishes will stay alive even if all their haemoglobin is rendered inert by carbon monoxide provided that they are not stimulated to swim vigorously.

As has been mentioned haemoglobin crops up sporadically, as and when required, in many groups of the animal kingdom, not only in the blood vascular system but also in various organs and tissues such as muscles, nerve ganglia, and even in eggs. In all instances the haemoglobin molecule has the same general form and properties but there are important differences in detail between the haemoglobins of different groups of animals. All haemoglobins consist of a prosthetic group, **haem**, a porphyrin containing an atom of divalent iron, and a protein, the prosthetic group being the same in all instances but the protein being different. Vertebrate haemoglobins have smaller molecules than those of invertebrates and it is probably for this reason that they are contained within the red corpuscles whose walls are impermeable to them. If this low molecular weight haemoglobin was free in the blood plasma it might pass out from the body through the walls of the capillaries and kidney tubules, which are permeable to it. The differing protein components of haemoglobin molecules in different animals control the way in which they combine with or release oxygen at different partial pressures and also the way in which those reactions are affected by carbon dioxide (the Bohr effect). From Fig. 445 it will be seen that the partial pressures of oxygen at which various haemoglobins are 95 per cent. saturated ("loading tension") and those at which they are only 50 per cent. saturated with oxygen ("unloading tension") are variable. These variations can often be related to the conditions under which the animals live. Thus, animals which live in water which is not fully oxygenated require a haemoglobin which has a low loading tension. Although the blood circulates through the tissues it always remains within the blood-capillaries and normally never penetrates into the tissues. The cells of the tissues are, however, bathed in tissue fluid which seems to be practically identical in composition with the lymph and which, owing to the blood pressure and the permeable nature of their walls, passes out of the blood-capillaries by a process of ultra-filtration.

The carbon dioxide liberated as the result of the respiratory activities within the cells is conveyed back to the respiratory surface chiefly in the blood-stream. Carbon dioxide is more soluble than oxygen, and in the quantities in which it is normally liberated, dissolves freely in the tissue fluid surrounding the cells and passes thence to the plasma within the blood-capillaries. By far the major

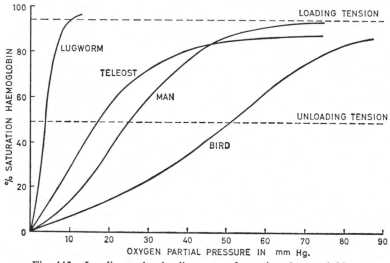

Fig. 445. Loading and unloading curves for various haemoglobins.

portion of the carbon dioxide is not carried in simple solution in the blood or as carbonic acid. In fact, the amount so carried is estimated at only about 10 per cent. of the total. A further 10 per cent. travels in combination with the amino-groups of the haemoglobin and other blood proteins and the remainder (about 80 per cent.) as bicarbonate ion which is combined with sodium in the plasma and with potassium within the corpuscles. All the reactions in which carbon dioxide combines with substances in the blood are, of course, reversible. But, the reaction, carbonic acid ⇌ carbon dioxide and water is accelerated by an enzyme contained in the red corpuscles, carbonic anhydrase. Also, the liberation of oxygen from haemoglobin in the tissues makes it possible for more carbon dioxide to be carried by the blood. This is because reduced haemoglobin is less strongly acid than oxyhaemoglobin and dissociates less, producing fewer hydrogen ions and thus permitting the formation of more bicarbonate. That is, it shifts to the right the equilibrium expressed by the equation $H_2CO_3 \rightleftarrows HCO_3' + H'$. At the respiratory surface the haemoglobin when it becomes oxygenated becomes more strongly acid, thus favouring the decomposition of bicarbonates to release carbon dioxide. Other changes which occur in the blood in connection with the transport of oxygen and carbon dioxide involve an exchange of basic and acidic radicals between the red corpuscles and the plasma. During oxygenation of the blood, bicarbonate ions diffuse into the corpuscles (there to be decomposed by the action of carbonic anhydrase) and, in accordance with the Donnan effect, other ions must move out to compensate for this movement. But the envelope of the corpuscles is relatively impermeable to cations such as sodium and potassium and so the ions which pass out into the plasma are chlorine ions. Thus oxygenated plasma contains more chlorine ions than does deoxygenated blood. A movement of ions in the reverse direction takes place at the tissues, chlorine ions diffusing into the corpuscles. At the same time, carbon dioxide diffuses into the corpuscles where it forms carbonic acid which then dissociates.

$$CO_2 + H_2O \rightarrow H_2CO_3$$
$$H_2CO_3 \rightleftarrows HCO_3' + H'.$$

Some of the bicarbonate ions then diffuse out into the plasma.

The above changes are sometimes known as the "chloride shift".

This is by no means the complete story of what happens, but it will suffice to give a general idea of the main lines of the reactions.

During the transference of the oxygen from the blood to the tissue cells an oxygen gradient is established between the blood-plasma and the cells *via* the lymph. The oxygen tension in the tissues is considerably lower than that in the plasma so that a continual diffusion of oxygen from the plasma to the cells is set up. Similarly, a carbon dioxide gradient in the reverse direction arises between the cells and the blood-plasma. The concentration of carbon dioxide in the cells is higher than that in the plasma, where it is being continually reduced by passing out of solution during the formation of bicarbonates as explained above. It follows that there is a continual flow of carbon dioxide from the tissues to the blood. At the respiratory surface, the gradients are reversed so that there is a continuous movement of oxygen inwards and a similar shift of carbon dioxide outwards.

Although terrestrial animals are commonly stated to obtain their oxygen direct from the atmosphere, yet, before it can enter the body of the organism it must pass into solution. In the lungs, the inner surface of the air chambers is lined with mucus and the oxygen from the air within the lung dissolves in this layer before it can pass into the blood-stream. The amount of oxygen dissolved in the mucus layer is governed by the solubility of oxygen in it and by the rate at which oxygen can diffuse into the blood-stream of the pulmonary capillaries. The conditions are very different from those which exist in the gill where there is a continual stream of oxygen-laden water passing over the surface, for the mucus layer, although thin, has an appreciable thickness and consequently the respiratory surface is not itself in contact with the pulmonary air. Moreover, the lungs are internal structures and air has to be passed into and out from them. Thus the air in the air-sacs and alveoli always contains a lower oxygen content than atmospheric air, and a certain amount of residual air is always present in the lungs.

Activities resulting in the release of energy take place in the tissues themselves and are designated tissue respiration. This release of energy is the result of a complex chain of chemical reactions, each of which is catalysed by a special respiratory enzyme within the cells. In many instances the chemical compound, or **substrate** as it is termed, whose stepwise oxidation provides the energy is glucose itself; but sometimes the substrate is only in part derived from the glucose and, as will be explained later, is provided by fatty acids and amino acids made available by the digestion of fats and proteins respectively. Whatever the substrate may be it is brought to the tissues by the blood.

Now in the laboratory, glucose can be burnt to give carbon dioxide and water with the release of energy according to the equation

$$6O_2 + C_6H_{12}O_6 \longrightarrow 6CO_2 + 6H_2O + 690,000 \text{ cals. of Energy}$$

per g. mol.

but in the cells the glucose is in dilute aqueous solution, and such a simple straightforward oxidation by atmospheric oxygen is impossible. Furthermore, the oxidation must take place at the relatively low temperature of the body. It is clear, therefore, that, whilst in most instances the end products of the respiratory oxidative processes are indeed carbon dioxide and water, tissue respiration must not be thought of as the direct oxidation of the substrate by oxygen.

Nevertheless, the respiratory quotient (R.Q)* when carbohydrates are being utilised as a source of energy is approximately one and is in agreement with the equation given.

$$* \; RQ = \frac{\text{Vol. of } CO_2 \text{ given out}}{\text{Vol. of } O_2 \text{ taken in.}}$$

The very complicated series of chemical reactions, each catalysed by a particular enzyme, which are involved in the release of energy, can be thought of as a device whereby molecules of a compound, adenosine triphosphate (ATP, for short) containing energy-rich bonds are synthesised stepwise and then provide a source of energy which is released when they break down to form adenosine diphosphate (ADP) and free phosphate, both of which are then available for re-synthesis into ATP. The energy for the build-up of ATP is ultimately derived from the oxidation of glucose. All the reactions involved take place at temperatures, pressures, and pH values at which tissues normally operate. Apart from working at low temperatures, etc., such a system has other advantages. By having ATP distributed in the cells energy is always immediately available for protein synthesis, movement, etc., whereas if energy had to be obtained by the oxidation of glucose some delay would be inevitable. As will be seen, the oxidative reactions in cells are mainly brought about by the enzymatic removal from the substrate of pairs of hydrogen atoms, by enzymes called dehydrogenases, followed by a combination of those with an atom of oxygen to form water. The oxidation takes place a step at a time and in this way energy is liberated in the small amounts, and at a rate which cells can deal with. This contrasts with, say, the sudden liberation of energy when hydrogen is directly exploded with oxygen from a gram-molecule of which 690,000 calories are released as heat, a form which could not be utilised by cells and would, indeed, be harmful to organisms.

Nevertheless, in the end stepwise oxidation in the cells yields the same amount of energy as direct oxidation of an equivalent amount of substrate.

Full details of the ways in which foodstuffs are dealt with are matters for a biochemist but a simplified outline of them is given below.

Firstly the point should be made that the main source of energy in most animals is carbohydrate of one sort or another. In general, these are, by the process of digestion, hydrolysed to form molecules of glucose but all organic constituents of cells are potential fuel. Indeed there is a small continual breakdown of cell constituents alongside the main breakdown of food substances supplied to the cell. Part of the energy produced is required for the replacement of the cell material destroyed. The bonds in the molecules of all these initial respiratory substrates are low energy ones and it is the pooling of the energy in them into a high energy bond attaching a phosphate residue which enables energy to be transferred with the phosphate and to be stored in adenosine triphosphate.

Glucose, then, enters the cells and a phosphate group is attached to the molecule (phosphorylation). This has at first a low energy bond with the sugar molecule, but by dehydrogenation of the sugar portion, internal rearrangement, and further addition of phosphate, the energy content of the phosphate bond is raised to a high level. This energy-rich phosphate is transferred to adenosine diphosphate (ADP) to form adenosine triphosphate (ATP).

The end product of the glycolysis of a molecule of glucose is two molecules of pyruvic acid and since the reactions occur in the absence of oxygen this phase is called **anaerobic glycolysis**. The energy so released is used to synthesise two molecules of ATP. But under some conditions the breakdown of fats and proteins occurs and those also yield pyruvic acid or enter the cycle at other points. Under anaerobic conditions (such as may temporarily occur when an animal is utilising energy at a greater rate than can be sustained by the maximum rate at which oxygen can reach the cells) pyruvic acid is broken down to lactic acid so releasing more energy.

Tissues can for a time build up an oxygen debt and can withstand a certain amount of lactic acid within them. However, there is a limit to this and muscle fatigue is largely the consequence of accumulation of lactic acid. As soon as oxygen becomes available in the cells, the debt is repaid by the remainder of the aerobic cycle being completed removing the lactic acid.

But as oxygen reaches the cells each molecule of pyruvic acid is, in the **tri-carboxylic cycle**, oxidised stepwise providing energy for the synthesis of six molecules of ATP. Further oxidative reactions in the presence of oxygen eventually yield thirty molecules of ATP, this last series being called **oxidative phosphorylation (Kreb's or citric acid cycle)** because in the final steps, oxygen carried to the cells reacts with the respiratory enzyme, **cytochrome**, in its reduced form to yield water.

The reduced cytochrome is the end link in a chain of reactions in which hydrogen, derived from the oxidative dehydrogenations in the glycolytic and citric acid cycles, is passed from substance to substance. This chain supplies a great deal more energy for the formation of high energy bonded phosphate since the free electron forming part of the hydrogen atom is passed from substance to substance releasing an amount of its bond energy at each step. Thus though we represent the process as one of handing on hydrogen atoms it is in reality electron transfer which is going on.

This series giving electron energy supplies the greater part of the energy produced in aerobic respiration. The hydrogen is first accepted by diphosphopyridine nucleotide (DPN) which is reduced as a result. Flavin adenine dinucleotide (FAD)* then receives the hydrogen restoring DPN to its unreduced form. FAD hands the hydrogen to the cytochrome.

When the aerobic part of the cycle is able to be completed, thirty-eight molecules of ATP are produced for each molecule of glucose in place of the two which are the sole result of anaerobic respiration.

To sum up, the complex reactions of tissue respiration consist of the stepwise oxidation of glucose, *the energy released being utilised to synthesise molecules of ATP in the cells where it acts as an immediate source of energy.*

The synthesis of one gram-molecule of ATP from inorganic phosphate and ADP takes 10,000 calories, and it is this energy which is available to the cells and is released enzymatically.

The whole process is set out in simplified form overleaf.

In short, the energy-yielding reactions of the body are those which involve the hydrolysis of organic phosphate compounds which have a "high energy" phosphate bond. This energy is required not only for movement but also for the synthesis of complex compounds from the simpler products of digestion, these syntheses being endergonic reactions (requiring energy). They contrast with the endergonic reactions carried out by the organic chemist in that they proceed at much lower temperature and less extreme pH conditions. A further point of difference is that the intermediate compounds are usually quite different in the two instances but there is a similarity for, both in the organism and in the laboratory, energy is required, although in the latter instance it is provided by heat, "strong" reagents, etc.

## THE DOGFISH

Since the dogfish is an aquatic animal, it is dependent upon the oxygen dissolved in the sea water for its respiratory needs. Water, as opposed to air as a source of oxygen, has two main features. It is 800 times as dense and contains much less uncombined oxygen. Air contains $\frac{1}{5}$ oxygen by volume and $\frac{4}{5}$ nitrogen, approximately. That is, it has about 250 mg/litre at 20° C. Depending on whether it is salt or fresh, water at 20° C contains 7·6 mg/litre or 9·4 mg/litre, when fully saturated with oxygen. Water is more viscous than air and more muscular effort is required to ventilate the respiratory surfaces. Only rarely is sea water not fully saturated with oxygen. Carbon dioxide is much more soluble than oxygen and is relatively easily got rid of. A good deal of it is immediately combined with bases in the water to form carbonates and bicarbonates.

There are five pairs of functional branchial clefts lying between the hyoid and successive branchial arches. The first pair of visceral

*The fact that these two hydrogen acceptors are nucleotides stresses the general biological importance of this class of organic compounds (see also p. 17).

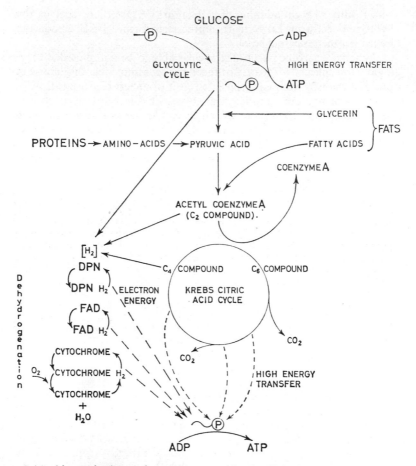

Aerobic respiration cycle, many steps omitted. Dehydrogenation, removal of hydrogen, indicated by →; energy transfer by - - - →; phosphorus attached by energy-rich bond by ~Ⓟ.

clefts—those between the mandibular and hyoid arches—have given rise to the **spiracles**. Each branchial cleft consists of an internal pharyngeal opening, a spacious branchial or gill pouch containing the gills, and an external opening, the gill slit or cleft on the surface of the head. Between successive clefts there is present on the inner (pharyngeal) side, the branchial arch, and extending obliquely backwards from the arch to the surface of the body, the inter-branchial septum. This septum is covered by an epithelium and contains blood vessels, nerves, etc., and originally, a portion of the lateral plate mesoderm and coelom which, however, is occluded in the adult. Arising from the arch and extending through the septum into the gill are the cartilaginous gill rays. The gill proper is formed of leaf-like foldings of the septal tissue, each fold constituting a gill lamella or filament, the distal ends of which are free. From each branchial arch two groups of lamellae arise, one on the anterior and the other on the posterior face of the septum, and each is termed a **hemibranch**, the two arising from the same arch, constituting a **holobranch**. The lamellae in turn bear small leaflets, the secondary lamellae. Each of the first four pairs of branchial pouches contains a hemibranch on both the anterior and posterior wall, but the last pair has a hemibranch on the anterior wall only. In each spiracle a vestigial gill termed a **pseudobranch** is present. It has no respiratory function.

Within each gill lamella is an extensive system of sinusoids which receives blood from the **afferent branchial vessels**—which run along the outer side of the arch—and passes it to the **efferent branchial loop** at the sides of the arch (Fig. 448). During the passage of the blood through the network it becomes oxygenated.

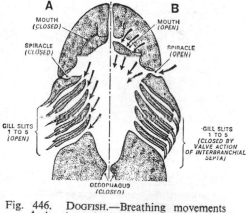

Fig. 446. DOGFISH.—Breathing movements during A, expiration; B, inspiration.

To maintain a continuous supply of oxygen to the gills, a current of water through the gill pouches is set up by movements of the pharyngeal wall. Water is taken into the bucco-pharyngeal cavity through the mouth (and also through the spiracle) by the lowering of the floor of the pharynx. This is brought about mainly by the contraction of the coraco-hyoid and coraco-branchial muscles. The bucco-pharyngeal cavity is made smaller mainly by the contraction of a system of superficial constrictor muscles but these are aided by small, internal constrictors running between the epi- and cerato-branchial cartilages. With the closing of the mouth, the raising of the floor then forces the contained water (the entrance to the oesophagus being closed by the contraction of its muscles) through the internal openings of the branchial clefts into the gill pouches, bathing the gill lamellae, and thence to the exterior through the gill slits. During this process, the pharynx is acting as a sort of force pump which maintains a current of water in one direction only—in through the mouth and out through the gills. It is therefore obvious that some system of valves is required. The lips serve to seal the mouth when it is closed and prevent water rushing out when the pharynx contracts. The flaps of skin along the anterior border of each gill slit prevent the entry of water through the slits when the mouth is open and the pharynx is expanding. The gill pouches themselves also actively assist in maintaining the respiratory current by alternate contractions and relaxations of their intrinsic musculature attached to the gill arches in rhythm with the movements of the pharynx. But an essential part of the mechanism of ventilation of the gills is a suction pump. This is largely due to expansion of the space just outside each gill owing to the elastic nature of their walls.

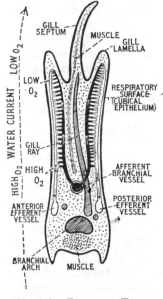

Fig. 448. DOGFISH.—Transverse section of a gill.

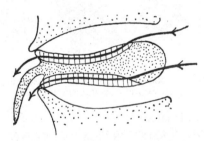

Fig. 447. DOGFISH.—Diagram to show water circulation through a gill.

By these means a higher hydrostatic pressure is maintained on the pharyngeal side than on the outside of the gills and although varying in speed, there is a continuous unidirectional flow of water over the gills. The lamellae of adjacent hemibranchs overlap with the edges of each slit (Fig. 447) and water has to pass through a fine meshwork made by the lamellar folds. As a consequence water is brought into close contact with the respiratory epithelium and probably about 50 per cent. or more of all the oxygen in the water passing through is extracted, an efficiency which is aided by the direction of blood flow which ensures a continuous oxygen gradient across the respiratory epithelium (Fig. 448), the most fully oxygenated blood being opposite fully oxygenated water. This is an example of a "counter current exchange system" such as appears also in the vertebrate kidney and in many other situations. The principle of these systems may be explained as follows.

If a chemical substance or heat is to be transferred from one fluid to another, the transference will be most efficient if the fluids can be made to flow alongside each other, but in opposite directions, separated by tissue through which the ions or heat may pass. Under these conditions the diffusion gradient or the difference in temperature will be almost equal along the length of the two tubes, thus

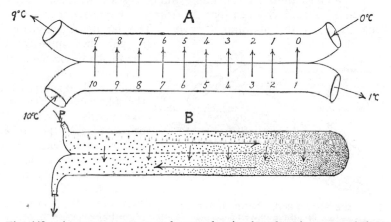

Fig. 449. A, counter current exchanger showing how heat is conserved since at any one point there is very little difference in temperature between the two limbs. The principle would apply to differences in concentration of dissolved substances in ingoing and outgoing streams. B, counter current multiplier showing how, with very little difference in concentration of dissolved substances at any one point in the two limbs of the hair-pin, a high concentration is built up and maintained at the blind end. The broken line separating the two limbs represents a semipermeable membrane. The ingoing liquid is under a hydrostatic pressure, P. Other forces such as active transport of ions could produce a similar effect.

ensuring maximum transfer all the way. As the concentration (or temperature) in one current drops, the concentration in the other rises. The efficiency of this arrangement is well shown in the fish's gill for a bony fish can take up 80 per cent. of the oxygen in the water passing over its gill but if this flow is artificially reversed so that the water flows parallel to the blood, oxygen uptake drops to 10 per cent.

Periodically the direction of the respiratory water current is violently reversed in order to clear the gills of foreign particles— the so-called "coughing reflex".

Breathing rate is controlled as in other vertebrates by a respiratory centre in the medulla oblongata (p. 562) which responds, not to decreased oxygen, but to increased concentration of carbon dioxide. This is why when oxygen is given by doctors it is always mixed with some carbon dioxide. Otherwise the patient would stop breathing.

## THE FROG

The frog has three respiratory surfaces, the skin, the lungs, and, of considerably less importance, the lining of the buccal cavity.

The lungs are relatively simple and consist of two elastic, highly vascular sacs which arise directly from the **laryngeal chamber** and lie freely in the abdominal cavity. The laryngeal chamber is roughly hemispherical in shape, lying between the posterior cornua of the hyoid and opening into the pharynx by the slit-like **glottis**. In its walls are cartilages—the **cricoid** and paired **arytenoid** cartilages—which are derived from the branchial arches of the tadpole. At each side of the laryngeal chamber arises a lung, the walls of which are thin,

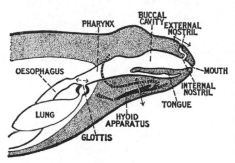

Fig. 450. FROG.—Diagram to show breathing movements.

only slightly muscular, and well supplied with capillaries. Internally the walls are hollowed out into numerous simple chambers so that when distended, the lungs have a bubbly appearance.

The passage of air into and out of the lungs—sometimes called the ventilation of the lungs—is effected by the floor of the bucco-pharyngeal cavity acting as a buccal force pump. Unlike the dog-fish, the olfactory organs have acquired openings—the internal nares—into the buccal cavity and these channels admit the respiratory current. The external nares are situated just above the premaxillae, which are movable, so that the external apertures can be closed. With the premaxillae lowered and the external nares open, the body of the hyoid in the floor of the bucco-pharyngeal cavity is lowered, thereby drawing air through the nostrils into the cavity. The premaxillae are then raised and the external nares closed, when, by a raising of the hyoid apparatus, the air in the bucco-pharyngeal cavity is forced through the glottis—the entrance to the oesophagus being closed by muscular contraction—into the lungs. During all these operations the mouth remains closed. To empty the lungs the procedure is reversed. With the external nares closed the floor of the bucco-pharyngeal cavity is lowered, the air from the lungs is drawn into the cavity through the glottis, aided by the contraction of the elastic walls of the lungs. This set of movements is usually repeated several times so that the same volume of air is passed into and out of the lungs more than once, a procedure which, although lowering the efficiency of gaseous interchange, prevents undue loss of water from the respiratory surface. Finally, the glottis is closed, the external nostrils are opened, and the floor raised so that the air is forced out.

The animal appears to use its lungs only when the need for oxygen is great. Ordinary requirements seem to be met by the use of the skin and the buccal mucous membrane. In buccal respiration the air in the bucco-pharyngeal cavity is changed by the

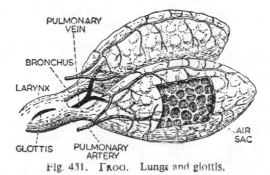

Fig. 431. Frog. Lungs and glottis.

rhythmical raising and lowering of the floor of the cavity, air being drawn into and expelled therefrom through the narial canal, the glottis remaining closed. More convulsive, gulping movements are seen when air is being forced into the lungs. These movements of air to and from the lungs cause two flaps of epithelium in the laryngeal chamber to vibrate and to give the sound of the "croak." The male croak is amplified by the vocal sacs. The rhythmical movements of the "throat", so easily seen in a living frog, are probably mainly for olfaction ("sniffing") rather than respiration.

## THE MAMMAL

In the mammal the respiratory mechanism is more complicated, and the lungs—which provide the sole respiratory surface—are more complex in structure. In these animals, too, as has been previously emphasised, the respiratory channel in the head is separated from the food canal by the development of a secondary palate. This has resulted in the internal nares opening directly

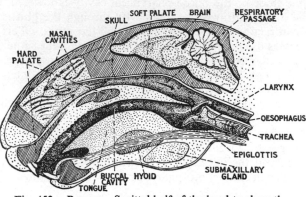

Fig. 452. RABBIT.—Sagittal half of the head to show the respiratory and food passages.

into the pharynx in the immediate vicinity of the glottis. The development of a neck, too, brings with it the need for a respiratory tube, the trachea, extending between the larynx and the lungs which lie in the thorax. A characteristic of the mammals is the complete separation of the thoracic cavity from that of the abdomen by the presence of a muscular septum, the diaphragm.

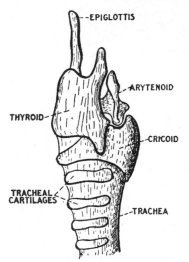

Fig. 453. MAMMAL.—The larynx.

The **larynx** is a cylindrical, box-like structure in the walls of which are four cartilages: the **thyroid cartilage**, which forms an incomplete ring around the larynx, and the **cricoid cartilage**, which is a complete ring around its base, whilst projecting inwards, on its dorsal aspect are the paired **arytenoid cartilages**, stretching from which to the thyroid cartilage are two pairs of membranes, the false vocal cords anteriorly and the true vocal cords posteriorly which can be set in vibration by the passage of air over them to produce the "voice"—in rabbits mainly a squeak. These membranes close the glottis except for a slit between their free edges, and it is through this gap that air enters during the respiratory movements. Projecting over the glottis is the cartilaginous flap-like **epiglottis** which serves to prevent food entering the glottis during swallowing.

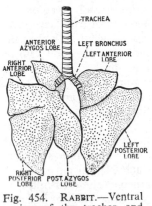

Fig. 454. RABBIT.—Ventral view of the trachea and lungs.

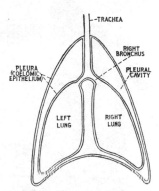

Fig. 455. MAMMAL.—Diagram of the pleural cavities and diaphragm in a young animal.

From the larynx extends the **trachea**, which is a tube lined by ciliated epithelium. In its wall are rings of cartilage which, however, are incomplete dorsally where the trachea lies against the wall of the oesophagus. These cartilages keep the tube from being constricted. After traversing the neck, the trachea enters the thorax and there divides into right and left **bronchi**, which have the same structure as the trachea. Each bronchus passes to a lung, which is a soft spongy mass of tissue lying in a special cavity—the **pleural cavity**—within the thorax. Each lung is invested by a fold of coelomic epithelium which is in contact with the organ, and the cavity in which the lung lies is

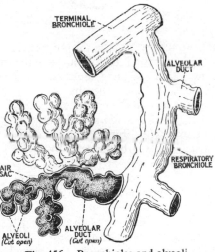

Fig. 456. Bronchioles and alveoli.

similarly lined. These membranes are the pleura and fluid is present between them to enable them to slip easily over one another. Each bronchus, as it enters the lung, divides and subdivides into finer and finer branches—the bronchioles—the ultimate branches terminating in small dilated air-sacs bearing alveoli. Around each alveolus is a network of capillary blood vessels which link the pulmonary artery and vein of that lung. The alveoli communicate with one another by apertures in their walls.

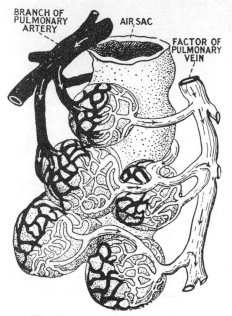

Fig. 457. Blood-supply of alveoli.

The ventilation of the lungs is due to movement of the diaphragm and ribs. Each pleural cavity is an air-tight chamber and it is only the interior of the lung (alveoli, bronchioles, etc.) which is in communication with the exterior. The diaphragm, against which the pleural cavities abut, is, in the position of rest, strongly domed, the curvature projecting in towards the thoracic cavity.

To take air into the lungs—inspiration—the diaphragm, by the contraction of its muscles, becomes flattened, thereby increasing the internal capacity of the thorax. This is enhanced by the forward* movement of the ribs which, by lowering the wall of the chest, has the same effect (see Fig. 458). Consequently, as the diaphragm flattens and the ribs are moved, air is drawn in along the air passage described above and enters the bronchioles and alveoli.

Fig. 458. Diagrams of the movements of the chest wall and diaphragm during breathing. A, from below; B, from the side.

---

* In ourselves, because of our erect posture, these movements result in the ribs and the wall of the chest being *raised*, but in a four-footed animal the ribs move forward and the chest wall is lowered.

With the return of the ribs and diaphragm to their normal positions, air is expelled outwards along the same route—expiration. However, these respiratory movements never completely empty the lungs, there always being some residual air left in them.

## THE NEURO-SENSORY SYSTEM

For convenience the nervous system of the vertebrates is considered in two parts: the **central nervous system**—comprising the brain and the spinal cord—and the **peripheral nervous system**—the cranial and spinal nerves and autonomic nervous system. Both are derived from a special part of the ectoderm of the embryo. The central nervous system is the part which co-ordinates and controls the activities of the animal; while the peripheral system forms the connecting links between the organs and tissues and the central nervous system.

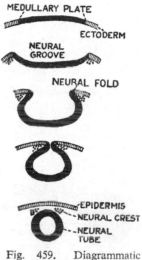

### The Central Nervous System

The **brain** and **spinal cord** are derived from a dorsally situated tract of ectoderm of the embryo called the neural plate (medullary plate). By a process of differential growth and cell migrations, the edges of the plate become raised into the neural folds which approach one another in the middle line and eventually fuse. This fusion results in the formation of a tube—the **neural tube**—

Fig. 459. Diagrammatic transverse sections to show formation of neural tube.

in the mid-dorsal line immediately beneath the now reformed epidermal surface of the embryo and lying above the notochord. In some instances the fusion is not at first complete along the whole length of the tube and an anterior neuropore, by which the cavity of the tube communicates with the exterior, may persist for some time, but always closes eventually. Posteriorly, the cavity of the neural tube may communicate temporarily with the archenteron by the neurenteric canal. At first, the neural tube is a simple tube, though usually it is not of the same diameter throughout. The anterior part—owing to the shape of the medullary plate—is usually wider than the more posterior part and it is from this wider portion that the brain is developed, the remainder of the tube giving rise to the spinal cord (see Fig. 460).

**The Brain.**—The first step in the differentiation of the brain from the neural tube is the moulding—by differential growth—of the anterior wider part into three dilated regions termed the three primary cerebral vesicles. These may be conveniently called the **fore-, mid-,** and **hind-brain,** and the various parts of the brain are

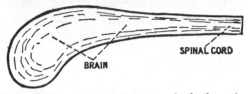

Fig. 460. The anterior end of the neural tube from the side.

formed by modification—thickenings and foldings—of these three primary cerebral vesicles.

THE FORE-BRAIN.—This region becomes divided into two parts by the appearance of a further constriction forming the **telencephalon**

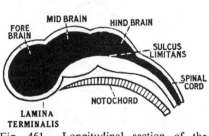

Fig. 461. Longitudinal section of the embryonic brain to show the three primary cerebral vesicles.

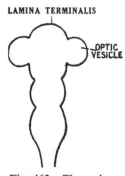

Fig. 462. The embryonic brain from above.

anteriorly and the **thalamencephalon** posteriorly. From the telencephalon on each side a lobe grows forward towards each olfactory organ. These lobes are the **olfactory lobes** and they receive the backwardly growing nerve-fibres from the olfactory cells. The original anterior wall of the telencephalon remains as a useful landmark and forms the lamina terminalis; the roof becomes the **pallium,** whilst the ventrolateral walls thicken to form **corpora striata.** In all vertebrates, but to a very variable extent, both the roof and side walls of the telencephalon later bulge outwards and forwards past the lamina terminalis, forming paired **cerebral hemispheres.** Thus the roof of each hemisphere is formed by the pallium and its ventrolateral wall by a corpus striatum and each contains a cavity, the lateral ventricle, which communicates with the median ventricle by an aperture, the foramen of Monro. In the higher vertebrates, and particularly in the mammals, the cerebral hemispheres become very much enlarged.

In the thalamencephalon, outgrowths appear from both the dorsal and ventral surfaces. That from the dorsal surface gives

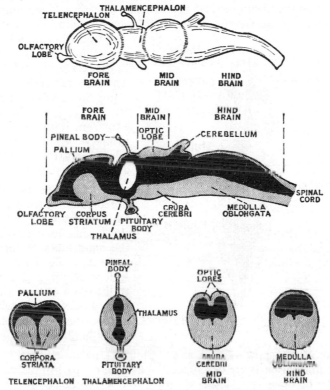

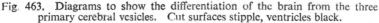

Fig. 463. Diagrams to show the differentiation of the brain from the three primary cerebral vesicles. Cut surfaces stipple, ventricles black.

rise to the **pineal organ**, whilst that from the ventral—the **infundibulum**—becomes attached to and fuses with an upgrowth from the stomodeum—the **hypophysis**—and these together form the **pituitary body**. Both pineal and pituitary bodies, though arising from the developing brain, and therefore potentially nervous, eventually lose their nervous character and become modified for other purposes. In some vertebrates, notably the reptiles and lampreys, the pineal organ—which may even be double—assumes the form of an eye, but these cases are exceptional, and generally the pineal organ consists of a knob-like structure lying beneath the roof of the cranium, connected to the thalamencephalon by a slender stalk. Its function is unknown. The pituitary body is an endocrine organ and secretes hormones (see p. 657). The roof of the thalamencephalon remains thin and non-nervous; it becomes much folded and pushed inwards into the cavity—the third ventricle—of this portion of the brain, carrying with it the vascular investing membranes to form the **anterior choroid plexus**, which plays an important part in the nourishment of the brain and the secretion of the cerebro-spinal fluid. The sides of the thalamencephalon become thickened to form the thalami, and from it also arise the primary optic vesicles which contribute to the developing eye (see p. 612).

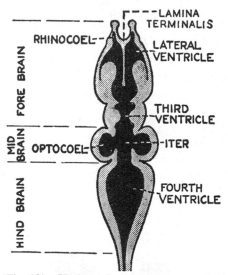

Fig. 464.   Horizontal section of a generalised brain to show the ventricles.

THE MID-BRAIN.—The dorsolateral surfaces of the mid-brain become thickened to constitute the **optic lobes** or **tectum opticum**, while the ventral wall thickens to give rise to the **crura cerebri** which form a nervous tract linking the thalamencephalon with the hind-brain.

THE HIND-BRAIN.—From this region of the developing brain two important parts are developed. The anterior part of the dorsal surface becomes thickened to form the **cerebellum**, while the posterior part of the roof remains thin, is non-nervous, folded, and in-pushed to give rise to the **posterior choroid plexus**. The remaining portion is very much thickened and becomes the **medulla oblongata**.

THE VENTRICLES.—As has already been indicated, the cavity of the neural tube, although, as a result of the various thickenings of the wall, becoming occluded to some extent, remains to form the cavities or **ventricles** of the brain. These are, the **lateral ventricles** within the cerebral hemispheres, the **third ventricle** in the thalamencephalon, and the **fourth ventricle** within the medulla oblongata.

Connecting the third and fourth ventricles and passing through the mid-brain is a canal, the **aqueduct of Sylvius** or the **iter** (in full, the *iter a tertio ad quartum ventriculum*). Sometimes the optic lobes are hollow, each containing an optocoel, and these open out from the iter. Where there are pronounced olfactory lobes, these are frequently hollow, enclosing a rhinocoel, each of which communicates with a lateral ventricle within the cerebral hemisphere. Each lateral ventricle connects with the third ventricle by a foramen of Monro.

COMMISSURES.—In all craniates, the two sides of the brain are linked up by bands of fibres, the commissures, the number and prominence of which varies in different examples. The anterior commissure (which originates in the lamina terminalis) lies at the front of the third ventricle. The dorsal (superior) and posterior commissures lie in the roof of the thalamencephalon; the dorsal one in front of the pineal body and the posterior one at the junction of the thalamencephalon and the mid-brain. These three commissures are present in most craniates, but in the higher vertebrates, in the floor of the third ventricle, lies the so-called middle or third commissure (massa intermedia) which is really a transverse linking of the thalami. Also in mammals, the corpus callosum, a dorsal extension of the anterior commissure, and other commissures are present.

THE CEREBRAL MEMBRANES.—During development, the brain becomes enclosed by investing membranes (the meninges); the tough **dura mater** lines the cranial cavity, and within this lies the more delicate, vascular **pia mater**. It is this layer which forms the vascular part of the choroid plexuses. In the higher vertebrates, a third layer, the delicate **arachnoid layer**, may be present between the dura and pia mater. In the spaces between these layers and within the cavity of the spinal cord and ventricles of the brain a serous fluid is present, the cerebro-spinal fluid, which is sometimes sampled by a "lumbar puncture" when diagnosing certain diseases.

BRAIN FLEXURE.—In the development of the central nervous system, owing to the shape of the embryo at the time of the laying down of the medullary plate, the neural tube is much bent in the mid-brain region, giving to it the curvature known as the cranial flexure. As development proceeds, however, with the thickening of the various parts, this flexure becomes obscured so that it is not obvious in the adult.

In considering the types of brain presented by the three representative vertebrates, it will be well to bear in mind the correlation between the form of the brain and the habits and behaviour of the animal as far as this can at present be assessed.

## THE DOGFISH

The dogfish, living as it does in relatively deep water where the light intensity is low, depends not so much upon its eyes in seeking its prey as upon its sense of "smell". The olfactory organs are relatively enormous and to accommodate all the fibres from them the olfactory lobes of the brain are correspondingly large. But the optic lobes and pallium are of only moderate size. Nevertheless, they are the main co-ordinating centres of the brain. The cerebellum is large, as it is in most active animals, for it acts as a co-ordinating centre for information from extero-ceptors and proprioceptors with motor impulses.

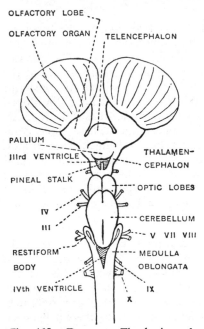

Fig. 465. DOGFISH.—The brain and olfactory organs from above.

THE FORE-BRAIN. —The telencephalon is largely repre-sented by the relatively enor-mous olfactory lobes which are closely applied to the olfactory organs. The pallium merely consists of small paired pro-tuberances on the dorsal surface of this region.

The thalamencephalon narrows from the wide telen-cephalon and from its dorsal surface arises the slender pineal stalk which terminates in the pineal body lying closely against the roof of the cranium. The greater part of the dorsal surface of the thalamencephalon is occupied by the anterior choroid plexus which dips down into the third ventricle. On the ventral side of the thalamencephalon is the optic chiasma (formed from the junction of the nerves from the eye-balls), and also the pituitary body. This is relatively large, for in addition to the infundibulum and hypophysis there are the accessory lobi inferiores of the infundibulum. The sacci vasculosi, which are extensions from the floor of the brain and are present only in fishes, lie at the sides of the pituitary body.

THE MID-BRAIN.—This region is only moderately developed. Dorsally are the two optic lobes and ventrally the crura cerebri which are hidden by the posterior part of the pituitary body.

THE HIND-BRAIN.—The cerebellum is the most prominent feature of this region. It projects forwards—partially covering the optic lobes—and backwards over the roof of the fourth ventricle. The posterior part of the roof of the hind-brain is ▽-shaped, the triangle marking the position of the posterior choroid plexus. The medulla oblongata is well developed.

THE VENTRICLES.—The cavities of the brain are fairly spacious. Each olfactory lobe contains a rhinocoel which communicates with its lateral ventricle within the cerebrum. The third ventricle is large, and each lateral ventricle opens into it by a foramen of Monro. Extensions of the third ventricle penetrate into the

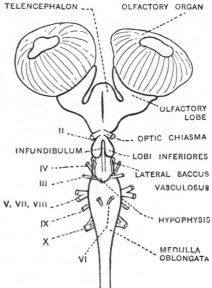

Fig. 466. DOGFISH.—The brain and olfactory organs from below.

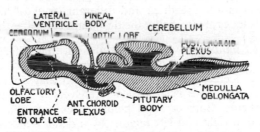

Fig. 467. DOGFISH.—Vertical section of the brain.

infundibulum, into the base of the pineal stalk, and also into the sacci vasculosi. Since the iter has opening into it the large optocoels within the optic lobes, it does not exist as a narrow canal, but, nevertheless, forms the connection between the third and fourth ventricles. The fourth ventricle lies within the medulla and opening into it is the cavity of the cerebellum.

The anterior, dorsal, and posterior commissures are present but are not easy to distinguish.

### THE FROG

Compared with that of the dogfish, the brain of the frog shows many points of contrast.

The smaller size of the olfactory lobes and the prominence of the optic lobes indicates a greater reliance on sight rather than smell. The smallness of the cerebellum is associated with a relative decrease in muscular activity, but the development of the cerebral

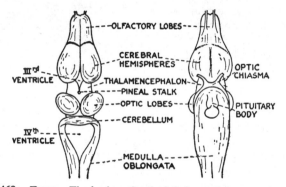

Fig. 468.    FROG.—The brain.    On the left from above; on the right from below.

hemispheres suggests greater possible complexity of activities such as those of hibernation, breeding, and different kinds of locomotion. But the optic lobes are probably the dominant co-ordinating centres.

THE FORE-BRAIN.—In the telencephalon, the olfactory lobes are relatively smaller and project forwards in front of the now clearly defined cerebral hemispheres of which they appear to be a continuation, but there is a shallow transversely running groove separating them. Both cerebral hemispheres and olfactory lobes are divided into right and left regions by a definite median furrow. The thalamencephalon is narrower than the telencephalon and bears the pineal body on its dorsal side. Ventrally lie the optic chiasma and pituitary body which, however, is relatively smaller. Neither lobi inferiores nor sacci vasculosi are present. The roof of the thalamencephalon is occupied by the anterior choroid plexus.

THE MID-BRAIN.—The optic lobes are well developed and take the form of two rounded swellings on the dorsal side which project laterally so that they are also visible from the ventral aspect. The crura cerebri are underlain by the posterior part of the pituitary body.

THE HIND-BRAIN.—The cerebellum is small and is reduced to a narrow transversely running ridge on the anterior dorsal region of the hind-brain. In the roof of the posterior part is found the posterior choroid plexus. The medulla oblongata merges into the spinal cord posteriorly.

THE VENTRICLES.—Both olfactory lobes and cerebral hemispheres are hollow, the cavities within the latter being the lateral ventricles. The cavities within the olfactory lobes, or rhinocoels, are not clearly demarcated from the lateral ventricles and appear as forward prolongations of them into the olfactory lobes. The third ventricle is somewhat compressed laterally because of the development of the optic thalami, and the lateral ventricles open into it, where they approach the median line, by the foramen of Monro. From the third ventricle the aqueduct of Sylvius passes backwards through the mid-brain and into it open the cavities— optocoels—of the optic lobes. The fourth ventricle lies in the medulla oblongata and into it projects the posterior choroid plexus. All three (anterior, dorsal, and posterior) commissures are present, of which the first and last are, perhaps, the more prominent.

## THE RABBIT

In contrast to the brains of the dogfish and frog, in the mammal, the cerebral hemispheres are much enlarged, as is also the cerebellum; there are four lobes on the roof of the mid-brain; and the olfactory lobes are relatively small, though many mammals, perhaps most, have an acute sense of smell. One has only to think of the behaviour of a dog to realise this.

Although the form of the brain of the rabbit shows the main characteristics of that of most mammals, taking the mammals as a whole, many differences will be found.

THE FORE-BRAIN.—The most striking feature of this region, and in fact of the whole brain, is the relatively enormous development of the cerebral hemispheres. So large are they that they completely overshadow the olfactory region and also project backwards to such an extent that they overhang and hide from view, in the dorsal aspect, the thalamencephalon and part of the mid-brain. They are the dominant co-ordinating centres of the brain and if damaged the behaviour of the animal is much more adversely affected than it is in the dogfish or frog.

The telencephalon, then, consists of these large cerebral hemispheres—forming the cerebrum—and the relatively small olfactory lobes lying beneath them. The two cerebral hemispheres are separated from one another by a deep groove, the median fissure, and their surface is further divided into regions by other fissures. Thus the shallow oblique Sylvian fissure at the side divides each hemisphere into an anterior frontal lobe and a lateral temporal lobe. Ventrally, a longitudinally running fissure demarcates the

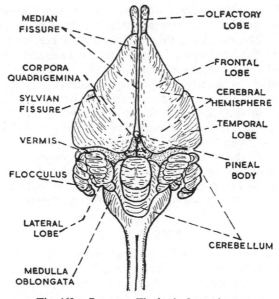

Fig. 469. RABBIT.—The brain from above.

hippocampal lobe. Posteriorly, the olfactory lobes are continuous with the olfactory tracts, the outer margins of which are defined by the rhinal fissures.

The small thalamencephalon is narrow and has the stalk of the pineal body arising from its dorsal surface, while on the ventral side are found the optic chiasma and the pituitary body. The stalk of the pituitary is prolonged posteriorly into a rounded swelling, the corpus albicans or corpus mammillare. The anterior choroid plexus arises in the roof of the thalamencephalon and extends through the third ventricle into the lateral ventricles. The sides of the thalamencephalon are thickened to form the optic thalami.

THE MID-BRAIN.—The optic lobes are represented by four rounded eminences, the corpora quadrigemina, on its dorsal surface. Ventrally are found the crura cerebri.

THE HIND-BRAIN.—The cerebellum, like the cerebral hemispheres, is very large, projecting forwards until it meets the posterior margin of the cerebrum and also extending laterally. It is clearly divided into regions, being made up of a median lobe, the vermis, and two lateral lobes, each of which bears laterally the flocculus. Ventrally is the much thickened medulla oblongata. Extending

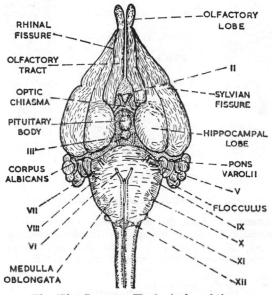

RHINAL FISSURE

OLFACTORY LOBE

OLFACTORY TRACT

II

OPTIC CHIASMA

SYLVIAN FISSURE

PITUITARY BODY

HIPPOCAMPAL LOBE

III

CORPUS ALBICANS

PONS VAROLII

V

VII

FLOCCULUS

VIII

IX

VI

X

XI

MEDULLA OBLONGATA

XII

Fig. 470. RABBIT.—The brain from below.

transversely across the under side of the medulla and connecting the two sides of the cerebellum is a tract of fibres forming the pons Varolii. In the roof of the hind-brain is the posterior choroid plexus.

THE VENTRICLES.—The brain of the rabbit appears to be a much more solid structure than that of the frog and dogfish, yet the cavities or ventricles still persist. The rhinocoel in the olfactory lobes is not very obvious and the lateral ventricles, too, are no longer the large spaces which were seen in the other brains considered. The thickening and modification of the walls of the cerebral hemispheres has reduced their size but has not obliterated them.

The third ventricle is narrow and joins the equally narrow iter into which open small cavities in the corpora quadrigemina representing reduced optocoels. The iter opens into the wider fourth ventricle in the medulla oblongata.

**Special Features of the Mammalian Brain.**—Although the surface of the cerebral hemispheres of the rabbit is relatively smooth except for a few main fissures, in the higher mammals and particularly the primates to which man

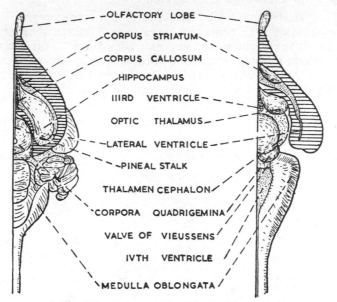

—OLFACTORY LOBE—

CORPUS STRIATUM

CORPUS CALLOSUM

HIPPOCAMPUS

IIIRD VENTRICLE

OPTIC THALAMUS

LATERAL VENTRICLE

PINEAL STALK

THALAMEN CEPHALON

CORPORA QUADRIGEMINA

VALVE OF VIEUSSENS

IVTH VENTRICLE

MEDULLA OBLONGATA

Fig. 471. RABBIT.—Dissections of the brain.

belongs, the surface is much folded, producing many curved grooves—sulci —giving the area a convoluted appearance. This folding increases the outer or cortical region of this part of the brain, thus allowing an increase in the number of nerve-cells without adding to the size. Also in the primates,

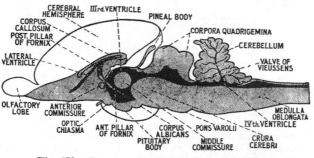

CEREBRAL HEMISPHERE   IIIrd. VENTRICLE   PINEAL BODY   CORPORA QUADRIGEMINA

CORPUS CALLOSUM   CEREBELLUM

POST. PILLAR OF FORNIX   VALVE OF VIEUSSENS

LATERAL VENTRICLE

OLFACTORY LOBE   ANTERIOR COMMISSURE   MEDULLA OBLONGATA

OPTIC CHIASMA   ANT. PILLAR OF FORNIX   CORPUS ALBICANS   PONS VAROLII   IVth. VENTRICLE

PITUITARY BODY   MIDDLE COMMISSURE   CRURA CEREBRI

Fig. 472. RABBIT.—Vertical section of the brain.

and particularly man, with the progressive increase in the size of the frontal lobes of the cerebral hemispheres the main axis of the cerebrum tends to become at right angles to that of the remainder of the brain.

The relatively small size of the olfactory lobes has already been noticed, but they are clearly defined and demarcated by the rhinal fissure. Within the telencephalon, however, the greatest developments are in connection with the cerebral hemispheres. The corpora striata occupy the great part of the anterior regions of the floor of the lateral ventricles, but in each hemisphere another thickening of the wall lying dorsal and medianly lateral to the corpus striatum forms the hippocampus within the hippocampal lobe. The extensive dorsal thickening of the telencephalon which forms the greater part of the cerebral

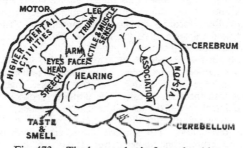

Fig. 473. The human brain from the side.

hemispheres arises as an upgrowth of the lateral portion of the pallium, on each side, between the hippocampus and the basal structures and extending dorsally and medianly. This is termed the **neopallium.**\* It follows, therefore from their mode of development that the cerebral hemispheres are separated by a deep groove or median fissure. In the higher mammals, too, definite regions are recognisable in the hemispheres, the significance of which will be apparent later. In addition, the two hemispheres are joined by a sheet of transversely running fibres, the corpus callosum, which is not found in the lower

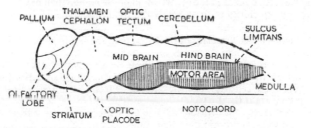

Fig. 474. Diagram to show main zones of an embryonic brain.

mammals such as the monotremes and marsupials. In addition the hippocampi are linked by a median commissure, the psalterium or lyra, while the corpora striata are connected by the fibres of the anterior commissure. The two sides of the brain are thus interconnected at a number of points.

Between the corpus callosum and the fornix is the **septum pellucidum**, separating the two cerebral hemispheres in this region. In the septum is a cavity

---

\* In this respect the mammalian brain offers a marked contrast with that of the bird where the greatest development is in the corpora striata.

the **pseudocoel,** sometimes called the fifth ventricle, although it is not in any way a true ventricle.

In the thalamencephalon lie the prominent **optic thalami** which fuse together in the third ventricle forming a large mass, the so-called **massa intermedia.** In front of the pineal stalk, the roof of the thalamencephalon is thin and vascular, forming the **velum interpositum (transversum),** which is continuous with the choroid plexuses which penetrate into the lateral ventricles. A depression in the floor of the thalamencephalon marks the position of the infundibulum.

The division of the optic lobes into the corpora quadrigemina has already been mentioned, and between the two anterior lobes is the transversely running delicate band of fibres forming the **posterior commissure.** The **crura cerebri** are very pronounced thickenings of the floor of the mid-brain, which in this situation is of considerable thickness.

In the hind-brain the cerebellum is connected with various other regions by tracts of fibres termed the **peduncles.** The **anterior peduncles** pass between the cerebellum and the posterior lobes of the corpora quadrigemina; the **middle peduncles** into the pons Varolii; the **posterior peduncles** establish connection with the dorsal part of the medulla oblongata. The fourth ventricle is a spacious cavity into which projects the posterior choroid plexus, but it does not extend into the cerebellum. The surface of the cerebellum is much folded like that of the cerebrum in the higher mammals, so that, in section, the thickness of the cerebellum is traversed by branching lines termed the **arbor vitae.** From medulla to thalamus, extends the reticular system, connecting posteriorly to the grey matter of the cord and anteriorly with the hypothalamus. This important feature of the brain has many functions including those of maintaining blood temperature and hence controlling body temperature, and of control of the neuro-secretory system. But its activity is also essential for arousal of the brain to activity and for wakefulness, while it influences phasic and tonic muscular control in the body and determines modification of the reception of incoming sensory signals so that some are accepted and others ignored by the brain. Though present in all vertebrates, the evolution of the reticular system into the commanding place it holds in the organisation of the mammalian brain is another reason for the superiority of these animals' mental functions.

**The Evolution of the Vertebrate Brain.**—At first, the brain region of the neural tube resembles, in its finer structure, the remainder of the tube from which the spinal cord is derived. The cell-bodies of its neurons lie grouped around the central canal, their fibres for the most part lying superficially. Also, just as in the spinal cord, the cell-bodies of the efferent neurons lie in the ventral part of the grey matter, whilst those which synapse with the axons of the afferent neurons lie in the dorsal parts—the division between the sensory and motor regions being marked by an internal groove, the sulcus limitans, on each side. This sulcus— also a feature of the adult brain—does not, however, extend into the fore-brain, but curves downwards across the mid-line just in front of the top of the noto-chord, thus indicating that in the main, the fore-brain receives fibres from sense (olfactory and optic) organs only.

As will be seen later, both afferent and efferent fibres occur in the cranial nerves, the former terminating in sensory centres in the dorsal parts of the brain, the latter arising from motor centres in the ventral parts. The internuncial or relay cells grouped around the terminations of the afferent nerve-fibres form the primary correlation centres of the brain and connect the sensory and motor centres together into reflex systems. In addition to these, there are other cells also grouped in certain regions of the brain (optic tectum, cerebellum, thalamus, corpora striata, and pallium) where they form the association centres so-called since they are not dominated completely by any one type of sense

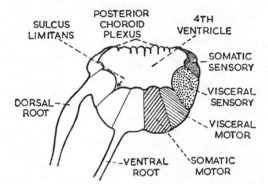

Fig. 475. Transverse section of the medulla of a selachian fish to show the relative positions of main types of cell-bodies.

organ but are concerned with the linking together of various primary correlation centres, so providing the anatomical basis for reflexes of a more complicated kind. It is these higher centres which are greatly elaborated in the higher vertebrates. The pallium and cerebellum reach their highest development in

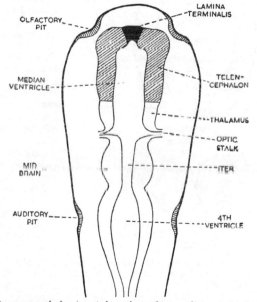

Fig. 476. Diagrammatic horizontal section of an embryonic brain (with cranial flexure straightened out) to show the relations of its main parts before formation of cerebral hemispheres.

the mammals, for they grow outwards from the primary brain to such an extent that they hide from view, when seen from above, practically all else and they develop, in addition, an extra layer of neurons which lies superficially to the original fibre layer and forms the cerebral and cerebellar cortex in their respective regions. This great enlargement of the association centres provides for further linkages, making possible a more complicated pattern of behaviour. Further, it is extremely likely that possible associations between cells in the pallium may be progressively increased during and until late in life, so that a greater degree of modifiable, *learned* behaviour is possible.

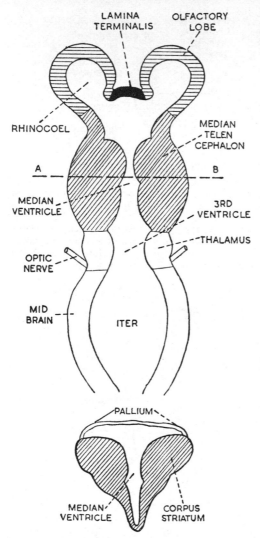

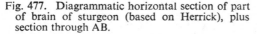

Fig. 477. Diagrammatic horizontal section of part of brain of sturgeon (based on Herrick), plus section through AB.

The role of the cerebellum cannot be summarised in a few words. It receives fibres from the proprioceptive organs of all parts of the body, from the acoustico-lateralis system as well as from the higher centres of the brain, whilst from it, fibres run to motor neurons controlling the main effectors of the body; also, the cerebellum apparently co-ordinates muscular movement as well as controlling the orientation of the body in space; again, any responses initiated in the higher centres are carried out through the cerebellum. The cerebellum is best developed in those animals which carry out a wide range of active movement.

In evolution, the most striking changes to be noticed in the brains of different types of vertebrates concern the cerebral hemispheres. In a general sort of way it is thought that the brains of some of the more primitive vertebrates still living at the present day may resemble stages comparable to those in the evolutionary history of the phylum, and this belief is to some extent substantiated by the examination of casts of the cranial cavities of fossil vertebrates.

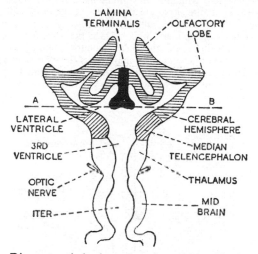

Fig. 478. Diagrammatic horizontal section of part of brain of dogfish.

In the lamprey and sturgeon the olfactory lobes are practically the only regions formed by a "ballooning out" of the front end of the telencephalon. In the dogfish and its allies the olfactory lobes have been developed still further and do not join the median telencephalon directly, but by way of stalks which arise from smallish lateral out-pushings from the telencephalon. These outpushings are the cerebral hemispheres. Their roof is the pallium, while their lateral walls may be termed corpora striata. For the most part these primitive cerebral hemispheres are concerned with the correlation of impulses relayed from the olfactory organs, linking them with other parts of the brain and providing for responses to the sense of "smell". A small

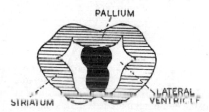
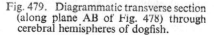

Fig. 479. Diagrammatic transverse section (along plane AB of Fig. 478) through cerebral hemispheres of dogfish.

lateral area in the striatal region of each hemisphere receives impulses relayed from other sense organs.

In mammals and to a lesser extent in amphibians and reptiles two distinctive areas are found in the pallium. One, which is solely concerned with the correlation of olfactory impulses, corresponds with the entire pallium of an animal like the dogfish and may be termed the olfactory pallium. The other area is an association centre in which fibres from all the main receptors are represented and is the non-olfactory pallium or neopallium. It is the neopallium which forms the greater part of the mammalian cerebral hemispheres, and a study of the amphibians and reptiles shows that in these animals it is not extensively developed. The neopallium of the amphibian brain is represented by a small area in the centre of the dorsal convexity of each hemisphere and is thus surrounded by the olfactory pallium. In reptiles the centrally placed neopallium increases in extent in all directions and separates the olfactory pallium into two parts, one lying along the median wall of the hemisphere next to the striatum, and the other along its dorsolateral wall. These two olfactory areas are often known as the archipallium and the palaeopallium respectively. The archipallium is the forerunner of the hippocampal lobe of the mammalian hemisphere, whilst the palaeopallium gives rise to a ventrolaterally situated region in close relations with the olfactory lobe and called the pyriform area or pyriform lobe.

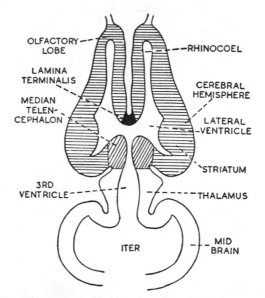

Fig. 480. Diagrammatic horizontal section of anterior part of brain of an amphibian.

In the eutherian mammals the neopallium increases vastly in extent and displaces both the archipallium and the palaeopallium so that they come to lie partially internally in each hemisphere. Also, although the corpora striata are well-developed structures within the cerebral hemispheres, they are functionally subordinate to the pallium. In the lower mammals (monotremes and marsupials) the neopallium is not so extensively developed and the cerebral region is not so pronounced.

In some reptiles, and particularly in birds, it is the corpora striata which are exaggerated and form the bulk of the hemispheres.

This difference in make-up between these two types of brain may be correlated with important differences in the behaviour patterns of their possessors. The corpora striata are believed to control innate or instinctive behaviour, that is, responses common to all members of the species and which are not susceptible to modification during the lifetime of the individual. The neopallium, however, is believed to govern modifiable or learned behaviour, resulting from the

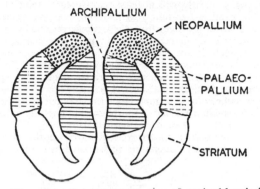

Fig. 481.  Diagrammatic transverse section of cerebral hemispheres of an amphibian.

experiences of the individual animal.  Thus, whilst birds and reptiles have the capacity for carrying out diverse and often very intricate actions, yet they have only limited powers of learning and in this sense are not particularly intelligent

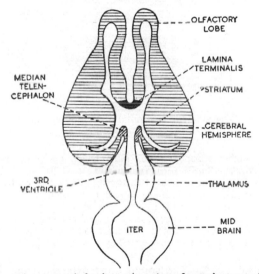

Fig. 482.  Diagrammatic horizontal section of anterior part of brain of reptile (based on Herrick).

animals.  Mammals, on the contrary, are the most intelligent of all animals, and their ability to perform a large proportion of their actions is acquired through experience and learning.

Attention has so far been centred mainly on the evolution of the cerebral
hemispheres, but it would be misleading not to mention comparable changes
which have been taking place in other parts of the brain, modifications which
are not, perhaps, so clearly evident from the structural point of view. In the
mid-brain it is of interest to note that in fishes it is predominantly visual in
function, only a few of the fibres from the optic nerves ending in the thalamus,
to be relayed thence to the telencephalon. In higher vertebrates, more and more
optic fibres terminate in the thalamus and fewer in the optic tectum or optic
lobes. Finally, in the mammals, only the anterior part of the optic tectum
(anterior corpora quadrigemina) receive visual fibres, the posterior parts
(posterior corpora quadrigemina) receiving many fibres from the cochlea,
which has now assumed greater importance as the organ of hearing. The bulk
of the optic fibres end in the thalamus for the relay of impulses to the neopallium,
and mammalian thalamus is called the optic thalamus.

So far as the hind-brain is concerned, it is perhaps in the medulla oblongata
that the main evolutionary changes are to be noted in the transition from the

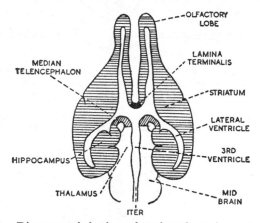

Fig. 483.    Diagrammatic horizontal section of anterior part of brain
of mammal.

fishes to the higher tetrapods. The main trend, first seen perhaps in the
amphibians, is the progressive shortening of the medulla relative to the rest of
the brain. This is to be attributed to the loss of certain primary centres, in
particular those associated with the lateral line system and with the nerves
supplying the branchial arches, by the tetrapods on their assumption of a
terrestrial existence. Consequently, the branchial portions of the ninth and the
lateralis components of the seventh and tenth cranial nerves are absent as are
also the cell-bodies with which they would synapse in the medulla.

**The Spinal Cord.**—The spinal cord extends from the medulla
oblongata of the brain almost to the end of the vertebral column.
In the four-footed animals the cord is slightly thicker in the brachial
and lumbar regions, forming the brachial and lumbar enlargements,
but otherwise there is no external distinction between the various

parts. At its end the cord narrows, forming, in the frog, the **filum terminale**, and in mammals the **conus terminalis**, from which the filum arises.

### The Peripheral Nervous System

**The Spinal Nerves.**—The spinal nerves are found in connection with the spinal cord and are segmentally arranged. Each nerve has two roots, dorsal and ventral, and on each dorsal root is a swelling, the dorsal root ganglion.

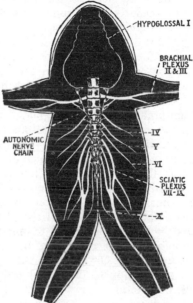

To understand the reason for the presence of these ganglia in this position, it is again necessary to refer to the development of the nervous system. In all craniates, with the closure of the neural tube, there appears on each side between the newly formed tube and the covering epidermis, a slip of cells extending along its whole length which forms the **neural crest**. It is from cells of these neural crests that the cell-bodies of the sensory neurons in the dorsal root ganglia are formed, and from each cell-body two fibres arise— one passing medially into the dorsal horn of the spinal cord and the other outwards to form a sensory fibre connected with one of the superficial receptors. The aggregate of these sensory fibres forms the dorsal root of the nerve. These developments do not take place along the whole length of the crests, but only at intervals corresponding with the trunk somites (see p. 725), the intervening portions contributing to the formation of quite different structures (see p. 723). The ventral root is made up of fibres which grow out from the cells in the ventral horn of the cord to pass to some effector structure (usually muscle). Beyond the developing vertebral column the two sets of fibres become bound together by connective tissue to form a spinal nerve. In this way, typically, there is a pair of spinal nerves for each trunk segment.

Fig. 184. FROG.—Diagram of the spinal nerves.

In the dogfish, where the number of trunk segments is large, the number of spinal nerves is correspondingly so. In the frog, however, where there has been considerable reduction and telescoping of terminal segments associated with the suppression of the tail, the number is reduced to ten pairs. In the mammal, although there may be some suppression of some terminal segments, there is usually a pair of spinal nerves for each vertebra.

Although, typically, the nerves issue separately from the cord and each appertains to its particular segment, in various parts of the body the nerves are grouped together to form **plexuses**. This is particularly so in connection with the paired limbs (indicating their origin from several somites) with the formation of brachial and sacral (sciatic) plexuses. But other plexuses may arise. In fishes, for example, the post-occipital spinal nerves innervate the hypobranchial muscles on the under side of the pharynx, foreshadowing the distinctive distribution of the hypoglossal nerve (first spinal) as in the frog. In the mammal this nerve has become incorporated with the cranial nerves, as will be explained later. In the mammals, there is also a cervical plexus, branches from the fourth, fifth, and sixth cervical nerves on each side combining to form the phrenic nerves which innervate the diaphragm.

In the fishes, where the notochord extends to the hinder end of the body, the spinal cord does so also and the terminal spinal nerves are normally arranged; but in the four-footed animals, the cord ceases at some distance from the end of the vertebral column, and in the mammals the terminal nerves, together with the filum terminale, form a bunch of nerves termed the cauda equina from its resemblance to a horse's tail.

In addition to the dorsal root ganglia, other ganglia occur in the peripheral nervous system. In the trunk region, lying usually on each side of the dorsal aorta, is a double chain of ganglia connected to one another by longitudinally running connectives, and each ganglion is connected to a spinal nerve by a short connection called a ramus communicans. Other isolated ganglia are also found in certain of the viscera. These ganglia with their interconnecting strands were formerly considered to form an anatomically separate system, called by the older anatomists the sympathetic system. It is now known that this system, now called the autonomic system, is intimately connected with the rest of the peripheral system but, nevertheless, is usually considered separately (see p. 597).

**The Cranial Nerves and Metamerism in the Head.**—The cranial nerves are much more constant in number than the spinal. In the dogfish and frog there are ten pairs and in the mammal, twelve.

Each nerve arises from a specific region of the brain and has generally the same distribution. Unlike the spinal nerves, however, the cranial nerves are not all composed of both sensory and motor fibres. Some are purely sensory, some purely motor, and others mixed. Some of them have ganglia near their roots, recalling the dorsal root ganglia of the spinal nerves. In general, it may be said that the arrangement of the cranial nerves—with certain important exceptions—resembles that of the spinal nerves, and like them is correlated with the disposition of the mesodermal structures and of the receptor organs.

To these cranial nerves, both numbers and names have been given, neither of which have much value in comparative morphology, apart from the mammals, since they come from human anatomy. The names indicate the distribution in a mammal and the numbers the order in which they arise from the brain, but conditions in the lower vertebrates are very different. Thus the VIIth nerve is called the facial; but it cannot, by any stretch of the imagination, be said to innervate the "face" of the fish.

It has already been explained that the vertebrate body is designed on a metameric plan, which, although obscured in the adult, particularly in the head region, is nevertheless of great interest here, since it renders intelligible the distribution of the cranial nerves.

It is supposed that primitively, in each embryonic head segment there were paired somites, a portion of the lateral-plate mesoderm, and, arising from the brain, a pair of ganglionated dorsal and a pair of ventral nerve-roots. But in the craniates, particularly in the gnathostomes, the primitive arrangement has been interfered with by the great development of the organs of special sense and by modifications concerned with feeding and breathing.

In general, however, the segments which enter into the composition of the head have much the same nature as those of the trunk, but, as might be expected, there are important differences. Thus, whilst in the embryo the development of the nerves follows the same course as in the trunk, namely, that the dorsal root ganglia are derived from the neural crests and the ventral roots grow out from the neural tube, the two roots, however, do not fuse to form a common trunk, but remain separate.* Also, whereas, in the trunk region, somatic sensory and visceral sensory fibres enter the cord by the dorsal roots, and somatic motor and

---

* The junction of the dorsal and ventral roots to form a common nerve in the trunk region is merely an incidental feature of development in the gnathostomes. In some cyclostomes (Lamprey) and in amphioxus the two roots remain separate, and it is of interest that this condition is retained in the head of all vertebrates.

visceral motor fibres issue by the ventral roots, in the head region the visceral motor fibres of the hind-brain nerves issue by way of the dorsal roots which, however, are still mainly sensory. The cell-bodies of the different types of fibres do, however, occupy the same relative positions as in the spinal cord, and a further point of similarity between the spinal and cranial nerves is that, in both, the ventral roots are made up of motor fibres.

The visceral clefts, which it will be remembered, perforate the lateral-plate mesoderm of the embryo (and are therefore probably not primarily metameric) do become segmentally arranged, whilst in between them blood vessels, visceral arches, and nerves also come to follow a segmental arrangement.

Evidence for the existence of a head segment consists of the presence of a somite or its derivatives, and of the development of segmental nerves associated with it. Always, however, there is found an anterior region in the head in which no evidence of segmentation can be detected, so that the head region consists of a pre-segmental and a segmented region. It is found that the number of segments in the head varies in different groups, and without going into elaborate explanations it may be said that in the lower gnathostomes (such as fishes like the dogfish) eight segments can be considered as contributing to the head, three of which will always lie in front of the otic (auditory) region. Thus, the fibres linked with particular kinds of end organ (component fibres) are differently distributed in cranial and spinal nerves. Moreover, certain nerve components are found only in cranial nerves. (See table, p. 584.)

The 1st head segment lies just in front of the mouth and, in the embryo, its somites lie right and left of the anterior tip of the notochord. The myotome of each of these somites (first pro-otic or premandibular somites) gives rise, by its subdivision, to four out of the six extrinsic eye muscles (see p. 606), namely, the anterior, superior, and inferior recti and the inferior oblique muscles. These muscles, which are voluntary muscles, are innervated by branches of a single nerve, the IIIrd or oculomotor, derived from the ventral root of this segment, since they have arisen from a single somite. The dorsal root of the 1st segment is the ophthalmicus profundus nerve which, although a separate nerve in the cyclostomes, becomes joined up with the Vth or trigeminal nerve in higher craniates and is much reduced in some (e.g. the dogfish). It innervates general cutaneous sense organs on the snout.

The myotome of the somite of the 2nd (second pro-otic or mandibular) segment gives rise to only one of the eye muscles, the superior oblique. This is innervated by the (ventral root) nerve called the IVth or patheticus. The corresponding dorsal root forms the Vth or trigeminal nerve, so-called because it usually has three main branches or rami. The main nerve divides into a superficial ophthalmic (ramus ophthalmicus) running to the snout over the orbit; a ramus maxillaris passing along the upper jaw and a ramus mandibularis passing to the lower jaw. Thus its more important branches fork around the mouth, but there are other branches also.

The somite of the 3rd (third pro-otic or hyoidean) segment gives origin to the last of the eye muscles, the posterior rectus. This muscle is innervated by the ventral root of the segment which becomes the VIth or abducens nerve, whilst the dorsal root forms the VIIth or facial nerve which, like the trigeminal, has several branches. Certain of these are distributed in a manner recalling

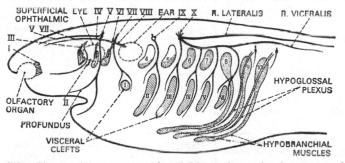

Fig. 485. The cranial nerves, numbered I-X, and associated structures in a generalised craniate. Dorsal roots black; ventral roots cross-hatched; somites of segments numbered 1-10.

those of the Vth, and the resemblance is made all the more apparent if it is realised that a premandibular visceral cleft, primitively related to the mandibular arch, has been incorporated in the mouth (see p 430). Thus, two important branches of the facial nerve, the pre- and post-spiracular nerves traverse the regions in front of and behind the spiracle (hyomandibular cleft). These nerves recall the maxillary and mandibular branches of the trigeminal in their distribution. Other branches of the facial supply structures in the bucco-pharyngeal region.

The development of the ear almost immediately behind the 3rd segment causes a complete or almost complete suppression of the myotomes of the somites of the 4th and 5th (first and second met-otic) head segments. As a consequence, the ventral roots of these segments are lacking in the adult. The dorsal roots are, however, fully developed. That from the 4th segment forms the

IXth or glossopharyngeal nerve, which forks just above the first gill cleft, one branch—the pre-trematic—passing in front of, and the other—the post-trematic—behind, the cleft. The dorsal root of the 5th segment contributes to the formation of the Xth or vagus nerve.

The 6th to 8th (third to fifth met-otic) segments retain their myotomes, and from them are partly derived the epi- and hypobranchial muscles concerned with the movements of the visceral arch skeleton of the pharynx. The ventral roots of these segments form the hypoglossal nerve which, together with contributions from some of the spinal nerves, makes the hypobranchial plexus. Thus, although these segments contribute to what may be termed the "head" region, in lower craniates the motor nerves supplying

## SEGMENTAL NATURE OF VERTEBRATE HEAD

| SEGMENT | DERIVATIVE OF SOMITE | DORSAL ROOT | VENTRAL ROOT | VISCERAL CLEFT AND ARCH |
|---|---|---|---|---|
| Pre-seg. | None. | None. | None. | None. |
| 1. Premandibular. | Ant., sup., inf. rectus and inf. obl. | Profundus. | Oculomotor. | Premandibular, trabeculae. |
| 2. Mandibular. | Sup. oblique. | Trigeminal. | Patheticus. | Mouth, mandibular. |
| 3. Hyoid. | Post. rectus. | Facial plus auditory. | Abducens. | Hyoidean, hyoid. |
| 4. — | Cyclostomes only, hypobranch. musc. | Glossopharyngeal. | None except in cyclostome. | 1st branchial. |
| 5. — | Cyclostomes only, hypobranch musc. | Vagus. | None except in cyclostomes. | 2nd branchial. |
| 6. — | Hypobranchial muscles. | Vagus. | Hypoglossal. | 3rd branchial. |
| 7. — | Hypobranchial muscles. | Vagus. | Hypoglossal. | 4th branchial. |
| 8. — | Hypobranchial muscles. | Vagus. | Hypoglossal. | 5th branchial. |

Somites 1-3 are termed pro-otic somites whilst the remainder are met-otic or post-auditory somites. The number of met-otic somites entering into the composition of the head is variable, being 2 or 3 in Amphibia, probably in most amniotes there are 3 or 4, whilst in selachians there may be 5-6.

muscles derived from their somites arise, in the adult animal, behind the cranium, and so are not included in the cranial nerves. In higher craniates (reptiles, birds, and mammals) certain primitively post-occipital nerves have been incorporated very completely in the head region and these motor nerves have thus become cranial nerves.

The dorsal roots of the 5th to 8th segments form a composite nerve, the Xth or vagus. Four of its branches behave like those of the more anterior dorsal roots, for each, in turn, forks into a pre- and post-trematic branch around the last four gill clefts. Another branch, the ramus lateralis (or lateral line nerve), innervates the lateral line sense organs, whilst yet another, the ramus visceralis, passes into the body cavity to innervate the viscera.

On all the cranial nerves derived from dorsal roots, ganglia are present as in the spinal nerves. These cranial ganglia are given special names: that in connection with the trigeminal nerve is termed the Gasserian ganglion (to which is fused the profundus ganglion); that with the facial, the geniculate ganglion; those with the remaining dorsal roots receive the same names as their nerves, namely, the glossopharyngeal and the vagus ganglia.

So far, no mention has been made of the nerves connected with the organs of special sense, the olfactory, optic, and auditory organs. The nerves which pass to the brain from the olfactory and optic organs belong to the pre-segmental portion of the head and are purely sensory. They arise as special developments in connection with the differentiating fore-brain. The Ist is the olfactory nerve and the IInd, the optic (developmentally an evaginated part of the brain). The VIIIth or auditory nerve is really a part of VII, having been appropriated by the ear, which may be regarded as a modified part of the lateral line system in the region supplied by the facial nerve. Thus none of these nerves are "segmental".

In brief, then, from this consideration of the segmental make-up of the head region a basic series of ten cranial nerves can be derived; I and II are pre-segmental; III, IV, and VI are formed from the ventral roots of the 1st, 2nd, and 3rd segments; and V, VII, VIII, and IX from the dorsal roots of segments 1-4, whilst the Xth nerve is made up of the dorsal roots of the 5th-8th segments.

Keeping this basic plan of the cranial nerves in mind, it will now be possible better to understand the particular features shown by the three types under consideration.

### THE DOGFISH

I. *Olfactory Nerve.*—The olfactory nerve is not a sensory nerve of the usual type but consists of a group of separate fibres which arise from the olfactory cells in the olfactory organ and pass directly into the olfactory lobes of the brain.

II. *Optic.*—This nerve lies in the stalk of the eyeball and is made up of fibres which leave the retina and enter the brain at the optic chiasma, where there is a crossing over of fibres. The reason for this arrangement will become apparent when the development of the eye is dealt with (see pp. 613, 614).

III. *Oculomotor.*—Arising from the floor of the mid-brain, this nerve passes directly into the orbit to supply its four eye muscles.

IV. *Pathetic or Trochlear.*—Although its origin is in the floor of the mid-brain, this nerve issues dorsally and passes at once into the orbit to the superior oblique muscle.

V. *Trigeminal.*—There are three main branches to the fifth nerve, the **ramus ophthalmicus, ramus maxillaris,** and **ramus mandibularis.** The profundus nerve, which is usually incorporated with the trigeminal, is minute in the dogfish, though large in its near relative, the skate. The fifth nerve arises from the side of the medulla oblongata and at once gives off the ophthalmic branch which

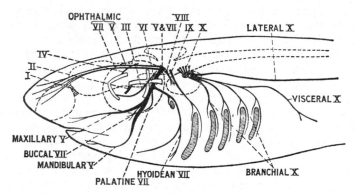

Fig. 486. DOGFISH.—Diagram of the distribution of the cranial nerves.

passes through the orbit (alongside a similar branch of VII) to cutaneous sense organs on the snout. The maxillary and mandibular branches arise from a common stem which enters the orbit by a separate foramen from that of the ophthalmic branch, and both pass to similar sense organs on the upper and lower jaws.

VI. *Abducens.*—This nerve arises from the floor of the medulla near to the mid-line and supplies the remaining eye muscle, the posterior (external) rectus.

VII. *Facial.*—At the side of the medulla oblongata, immediately behind V, is found the root of the seventh nerve. Although primarily it is related to the spiracle it has several branches. The **ophthalmic branch** penetrates the orbit close to the similar branch of V and runs alongside it to cutaneous sense organs on the snout. From the main root arise a **buccal branch,** which runs on the under side of the main ramus of V; a **palatine branch,** which crosses the floor of the orbit; and a **hyomandibular branch,** which lies in the posterior wall of the orbit and passes in the direction of the spiracle. The spiracle being reached, the hyomandibular nerve gives off a **pre-spiracular branch** in front of the spiracle and a **post-spiracular branch** behind it. The main ramus of the post-spiracular branch continues closely beneath the skin to the lower jaw as the **external mandibular branch** and down the hyoid arch as the **hyoidean branch.**

## COMPONENTS IN CRANIAL NERVES

| COMPONENT | NERVE | END ORGAN | CENTRE IN BRAIN |
|---|---|---|---|
| General cutaneous + Proprioceptive. | Profundus, max., mandib. and superfic. V, hyomandib. VII, glossopharyngeal vagus, N. termin., oculomotor. | Various types of skin receptors as in spinal nerves. Proprioceptors in muscles, etc. | Dorsolateral region ("skin brain") in hinder part of medulla and to cerebellum. |
| Special cutaneous. | Superfic. ophth. VII, hyomandib. VII, buccal VII, sub. temp. IX, vagus auditory. | Lateral line or neuromast system and ear. | Acoustico-lateralis area ("ear brain") in anterior part of medulla. |
| General visceral sensory. | Hyomandibular and palatine VII, IX, and X. | Receptors in mucous memb. of gut. Other viscera. | Visceral centre ("taste brain") of medulla. |
| Special visceral sensory. | Branches of VII, IX, and X. | Taste buds in mouth or in skin. | Visceral centre of medulla. |
| Olfactory. | Olfactory "nerve". | Olfactory cells. | Olfactory lobes. |
| Visual. | Optic "nerve". | Eyes. | Optic tectum and thalamus. |
| Somatic motor. | Oculomotor, patheticus, abducens, hypoglossal. | Extrinsic eye muscles and hypobranchial musc. | Ventral parts of mid-brain and medulla. |
| Autonomic. | Nerves III, VII, IX, and X. | "Viscera" of head and by way of X to trunk. | Cell bodies of preganglionic fibres lie near sulcus limitans. |
| Special visc. | Max-mandib. V, palatine and hyomandib. VII, IX, and X. | Striped lateral plate musc. of jaws and visceral arches. | Near sulcus limitans in medulla. |

N.B.—Not all nerve branches are mentioned. The table refers mainly to the condition found in selachians.

VIII. *Auditory.*—This is a very short nerve which leaves the auditory capsule and passes to the medulla oblongata alongside VII.

IX. *Glossopharyngeal.*—This name is quite inappropriate in the dogfish and the course of the nerve, which arises from the medulla oblongata closely behind VIII, is to the first gill cleft over which it divides into a **pre-trematic branch** and a **post-trematic**, the latter passing down the first branchial arch. There is also a small branch to the pharynx.

X. *Vagus.*—The alternative name for this nerve (pneumogastric) has little significance in the dogfish. The nerve has three main branches: a **lateralis branch** to the lateral line sensory system; a **branchialis branch** to the last four gill clefts; and a **visceralis branch** to the viscera. The main stem arises from the side of the medulla immediately behind IX by a series of rootlets and the lateralis branch leaves it quite early to pass upwards beneath the skin below the lateral line to its termination. The branchialis branch gives off subsidiary nerves to the second to the fifth gill slits, each nerve dividing over its respective cleft into a **pre-** and a **post-trematic branch**, the latter running right down the gill arch.

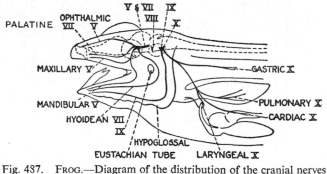

Fig. 437. Frog.—Diagram of the distribution of the cranial nerves (after Marshall).

The visceralis branch continues backwards into the body cavity where it gives off branches to the heart and viscera.

From this description it will be realised that the cranial nerves of the dogfish follow very closely the basic plan and represent the arrangement usually found in an aquatic gnathostome. Many of its features are related to the innervation of the lateral line sensory system and the gill clefts, and are the concomitants of an aquatic mode of life. It will be expected, therefore, that the chief lines of modification found in other types will be associated with the evolutionary migration from the water to the land and the adaptations incidental thereto.

## THE FROG

In the tadpole stage of the frog, much the same conditions occur as in the fish and nerves are present in connection with the gills and the lateral line sensory system. On metamorphosis, however, the gill clefts close up and disappear as such and the lateral line system is lost. In the adult frog, therefore, it will be found that ten pairs of cranial nerves are present and the distribution of I-VI and of VIII is much the same as that in the dogfish, but the course of VII, IX, and X is modified. The hyomandibular cleft of the tadpole has, in the adult, closed up and an outgrowth from part of it has been modified to form the tympanic chamber and Eustachian tube of the middle ear. The lateralis sensory components of VII have disappeared and only the visceral sensory and motor elements are represented. The complete suppression of the remaining visceral clefts has also altered the distribution of IX and X. Here, again, the branchial and lateral line contributions have been lost, but the visceral components remain. The main branch of IX passes forwards on the under side of the "throat" to the tongue. The branchial and lateral line branches of X have disappeared and only the visceral nerve persists with its branches to the heart, lungs, stomach, and other viscera.

## THE MAMMAL

The general trend of modification seen in the frog is carried further in the mammal, but here certain post-occipital nerves become intracranial, so that the number of cranial nerves is increased to twelve pairs. The additional nerves are XI, the **spinal accessory**, and XII, the **hypoglossal nerve** Both are motor

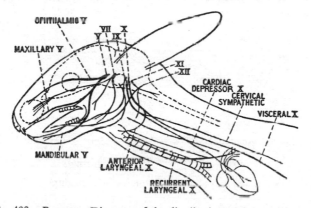

Fig. 488. RABBIT. Diagram of the distribution of the cranial nerves.

nerves, but XI includes both somatic and visceral motor fibres, while XII is purely somatic motor. The spinal accessory nerve innervates certain pharyngeal, laryngeal, and cervical muscles, and (along with X) thoracic and abdominal viscera. XII supplies the tongue muscles.

A further complication has been introduced, which particularly affects the course of certain branches of X. In a "neckless" animal like the frog, all aortic arches are close to the larynx, but with the development of a neck, arches 4-6 are moved posteriorly and the common carotids elongate. Thus the posterior nerve to the larynx in a mammal loops on the right side around the subclavian artery and on the left, around the ductus arteriosus. This is determined in the early embryo with the result that while the nerve arises at its normal position on the main trunk of the vagus and recurves around the subclavian artery or the ductus arteriosus, it then has to elongate and runs forwards up the neck alongside the trachea to the larynx. Because of this remarkable course it is called the **recurrent laryngeal nerve**. Thus it is sometimes said that the longest nerve known to the anatomist is the recurrent laryngeal of the giraffe!

## HISTOLOGY AND PHYSIOLOGY OF NERVOUS TISSUE

**Nervous Tissue.**—In the general description (pp. 559-85) of the nervous system of vertebrate animals, the distinction was drawn between the central nervous system—the co-ordinating centre—and the peripheral system—the connecting links and lines of transmission between the central nervous system and the organs and structures of the body. The true relationships of the two parts of this system to one another are revealed when their micro-anatomy is examined, and, complicated as the gross anatomy may have appeared, its microscopic anatomy is infinitely more complex. A detailed account of it would fill several textbooks, so that only the bare outlines are included here. Despite the complications hinted at, it is possible to make one important generalisation, namely, that the whole of the effective part of the nervous system is composed

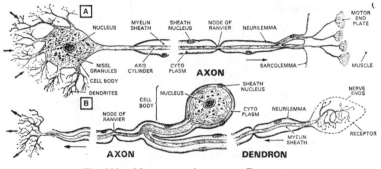

Fig. 489. NEURONS.—A, motor; B, sensory.

of nerve-cells and the processes arising from them. There are two main types of cells in nervous material, the more or less undifferentiated supporting cells which are collectively termed neuroglia and which form a packing around the true nerve-cells or neurons as they are called. It is these latter cells, the neurons, that are the essential elements of the nervous system, though the glial cells may be more important than has hitherto been suspected. Typically each neuron consists of the cell-body from which arises a system of branching processes or fibres.* The cell-bodies, which resemble any other cells in consisting of cytoplasm and a nucleus, are to be found only in the central nervous system itself, or else aggregated

* The number of processes arising from the cell-body is variable and forms the basis for a morphological classification of neurons. Thus, where there are two fibres joining the cell-body very close together, the neuron is called (misleadingly) **unipolar**; where there are but two fibres, one of which arises at the opposite pole of the cell to the other, the neuron is of the **bipolar** type; whilst neurons having many processes are called **multipolar** neurons.

in certain swellings termed ganglia. Their cytoplasm contains granules (**Nissl granules**—small masses of RNA) which readily stain with methylene blue, and also to be seen in the cytoplasm is a network of neurofibrillae, some of which appear to pass into the processes arising from the cell-body. Of these processes, one is as a rule especially long, whilst the other processes are relatively short. This distinction cannot always be clearly drawn, but a functional discrimination may be made, for it is held that the processes through which the nerve-cell receives its impulses are called dendrites and it transmits these impulses through its axon. Nerve impulses are conveyed or relayed from cell to cell across small gaps between the interwining branches of the processes by which the linking up of the cells is effected, such a close juxtaposition without actual organic connection being termed a synapse. The whole of the nervous system consists of chains of neurons linked together by synapses in a most complicated fashion.

In most instances the long nerve process, or axis cylinder as it is frequently called, is surrounded by a fatty myelin sheath, outside which is a thin membrane, the neurilemma, whilst just below the neurilemma lies a very thin layer of cytoplasm in which nuclei are embedded at intervals (tubes of Schwann or Schwann cells). These nuclei are termed sheath nuclei, for they, and the cytoplasm associated with them, are concerned with the laying down of the neurilemma, and also with the secretion of the myelin sheath. In addition, these cells remain after degeneration of a nerve and guide the regenerating nerve along the path followed by the original one. They are specific in some way, for, during regeneration a pre-ganglionic fibre will not follow the line of Schwann cells left by a post-ganglionic fibre and efferent fibres will not grow along the tracks of afferent fibres. The myelin sheath is not continuous along its whole length, but is interrupted at intervals, so that in these places the nerve-fibre appears constricted and the constrictions are termed the nodes of Ranvier. At these nodes, therefore, the layer of cytoplasm just below the neurilemma comes into direct contact with the axis cylinder, which may be either an axon or a dendron. Nerve-fibres of this type are said to be medullated, whilst those which lack the fatty sheath are non-medullated.

A nerve, in the anatomical sense, that is to say the nerves that are met with during dissection, consists of numerous fibres united in bundles by coats of white fibrous connective tissue (**perineurium**), the bundles themselves being bound together by yet another fibrous membrane (the **epineurium**).

It is important to notice that nerve-fibres are in some respects comparable to a "one-way-traffic" system, since any given nerve-fibre conducts impulses in one direction only, for though an axon can be shown under experimental conditions to be able to conduct in either direction, the impulses usually originate at the cell-body end of it and then pass from cell-body to axonic ending. It follows from this that the peripheral nervous system consists of at least two different kinds of fibres which, as regards their direction of conduction, are opposite. These two main types of fibres are **afferent fibres**, which carry sensory impulses from the receptor organs to the central nervous system, and **efferent fibres**, which carry impulses from the central nervous system to the various effector organs.

Afferent fibres have their peripheral origins in **receptor organs**— that is, in organs which are affected by external influences such as light, change of temperature, sound-waves, contact of external bodies, pressure, etc.—and the afferent fibres convey impulses set up by these external agents to the central nervous system. Efferent fibres, on the other hand, end in connection with muscle-fibres, gland-cells, and other cells which do *work* of any kind, and convey the impulses which cause them to contract, secrete, etc., or, in the case of unstriped muscle-fibres, impulses that tend to increase *or to decrease* their rhythmical contraction. Efferent fibres and efferent nerves are often called "motor", but this is a misleading term. Efferent fibres are motor when distributed to striped muscles, accelerator or inhibitor when distributed to unstriped muscles, secretory when distributed to glands. Even this does not quite exhaust the varieties of efferent nerve-fibres, but it covers most of them. For it is becoming increasingly evident, for example, that many sense organs are under efferent control by which their sensitivity can be altered. It is known that this functional division of the fibres into components has a definite anatomical basis, as may be seen from a study of the spinal cord.

**Spinal Cord** (see also p. 578).—From the examination of a transverse section of the spinal cord the general circular shape is at once apparent, and the central canal, which runs the whole length of the cord, finally linking up with the fourth ventricle of the brain, can usually be seen. Extending from just above the dorsal part of the canal, upwards to the dorsal surface of the cord, is a septum of fibrous tissue, continuous with the pia mater, which ends at the dorsal fissure and marks the point of fusion of the embryonic medullary folds (see p. 559). A ventral fissure, in the form of a groove along the ventral surface of the cord, is also present, but is formed in quite a different way. It results from unequal growth of certain parts of the embryonic spinal cord, for soon after the neural tube is separated from the overlying epidermis (see p. 559), the ventrolateral parts grow much more rapidly than the rest and finally grow towards each other in the mid-ventral line of the cord, to produce the fissure. This may be distinguished from the dorsal fissure when seen in section by the fact that it contains two relatively large blood vessels.

The spinal cord is surrounded by two fibrous membranes, a thick outer one, the dura mater, and a thin inner one, the pia mater. The membranes (meninges) are continued into the cranium where they invest the brain, the pia mater bearing in certain regions a more plentiful supply of blood vessels to form the so-called choroid plexuses.

The central canal is lined by a ciliated epithelium, whose function is presumably to circulate the cerebro-spinal fluid lying in the canal and cavities of the brain.

In each part of the spinal cord corresponding to a body segment there arises a pair of spinal nerves and each member of the pair has two roots, a dorsal and a ventral one, the two fusing a short distance from the cord to form the spinal nerve of that side. In a transverse section it is seen that the cell-bodies from which nerve-fibres arise are situated in the centre of the cord in an II-shaped area, which constitutes

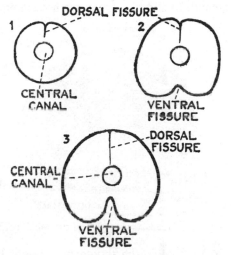

Fig. 490. Diagrams to show the development of the spinal cord from the neural tube.

the so-called "grey matter". It is surrounded by the "white matter", which owes its name to the fact that it consists of axons with their white myelin sheaths. Each dorsal root bears a ganglion which lies just outside the cord itself, and these segmental dorsal ganglia also contain cell-bodies of neurons. It has been discovered that all the afferent fibres enter the cord by way of the dorsal root and that the cell-bodies of the afferent fibres lie not within the cord, but in the dorsal ganglia. The efferent fibres, on the other hand, leave the cord by the ventral roots, and the cell-bodies of the fibres *do* lie in the grey matter of the cord. Afferent impulses, therefore, are conveyed into the central nervous system by the afferent neurons of the dorsal roots, whose axons pass into the cord and there enter into synaptic connection with the dendrites of a relay or internuncial neuron (association neuron) whose cell-body lies, of course, in the grey matter. One of the processes of the internuncial neuron then passes ventralwards and synapses with the dendrites of the efferent or motor neuron, so that a system is established by which an impulse entering the cord can be relayed to the motor neurons lying in the ventral part of the grey matter. Other processes from the internuncial neuron pass both anteriorly and posteriorly in the cord, so that they make synapses with cells in different "segments" of the cord. Such an arrangement of linked neurons is termed a reflex arc. These reflex arcs form the physiological basis of many of the apparently automatic responses of animals to stimuli. They are not confined to the spinal cord.

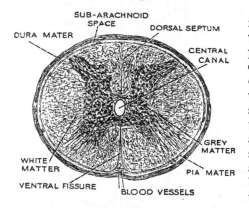

Fig. 491. Transverse section of the spinal cord.

A familiar example in man is the knee jerk where, in the sitting position with the legs crossed, the lower part of the uppermost limb will kick when the under side of the knee-cap is tapped smartly. This is a reflex action, but not simple. A simpler example is the involuntary blinking of the eyelids when an object suddenly approaches the eye, or the contraction of the pupil on the approach of bright light. Here the connection between afferent and efferent pathways is made in the brain.

Although, as was indicated above, a reflex arc involving only three neurons is theoretically possible, in vertebrates usually more than one intermediate neuron is utilised. The introduction of these intermediate neurons between the receptor and effector neurons gives to the whole mechanism a great elasticity and allows its utilisation not only for unconscious reflex activities but also for purposive actions. Such actions will usually involve not only the spinal cord but the brain also. It has already been seen that the arrangement of the grey matter in the cord divides the white matter into three columns on each side. In these columns there is a segregation of the fibres recalling that found in the different regions of the spinal nerves. In the dorsal columns the tracts of fibres convey impulses along the length of the cord in the direction of the brain; those in the ventral columns convey impulses away from the brain; while in the lateral columns there is a two-way traffic. Within the brain itself there are also definite tracts connected with those in the cord. The fibres conveying impulses to the brain are linked up in **spino-thalamic tracts**, and many impulses travelling to the brain pass to the thalamus. From there they may pass to the cortex of the cerebral hemispheres by the **thalamo-cortical tracts** or by other tracts to other parts of the brain. In the mammals only, from the cerebral cortex return impulses are conveyed by the **cortico-spinal tracts** to the appropriate part of the cord. In the anterior part of the cord there is a considerable crossing over of fibres from one side to the other so that in the more complex nervous activities, both sides of the brain may be involved. Into all these main lines of communication there are many subsidiary

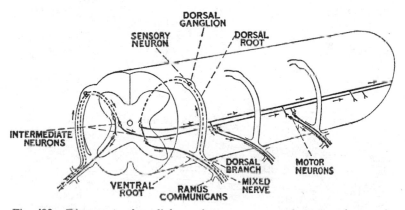

Fig. 492. Diagram to show linkages between neurons along the spinal cord.

tracts within the brain which are beyond the scope of the present treatment.

This mechanism renders possible the closest co-operation between the various parts of the central nervous system and also permits the carrying out in the higher vertebrates, and particularly in man, of the many extremely complicated and complex nervous activities. Even the simplest of our everyday actions employs large numbers of neurons in various parts of the brain and spinal cord.

As mentioned above, the anatomical differences of the nerve-roots of the spinal cord are correlated with functional differences, the dorsal roots being sensory or afferent, the ventral ones motor or efferent. The mixed nerve formed by the fusion of the two roots contains, therefore, fibres of at least two different types, but by means of special techniques, such as the study of nerve degeneration, it has been found possible to classify the various fibres in a mixed nerve into further subdivisions, based on functional considerations, for it has been found that a particular type of fibre is always associated with a particular type of end-organ. Four main systems of fibres are thus recognised, two sensory and two motor, and they are respectively known as the **somatic sensory**, the **somatic motor**, the **visceral sensory**, and the **visceral motor** systems. The cell-bodies of the four kinds of fibres occupy definite relative positions in the spinal cord, as is shown in the diagram; so that

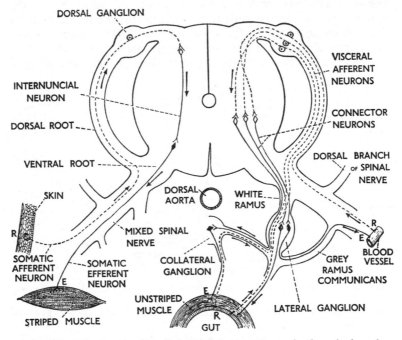

Fig. 493. Diagram to show linkages between neurons in the spinal cord.
E = effector; R = receptor.

any nerve is composed of several different kinds of fibres which are called nerve components—as in cranial nerves.

The somatic sensory fibres are those which carry impulses from the superficial receptors and those in the muscles and joints to the central nervous system, whilst the visceral sensory fibres carry impulses from receptors (taste buds, tactile organs, etc.) lying in the wall of the alimentary canal. The somatic motor fibres innervate the skeletal muscles of the body, including the extrinsic eye muscles. Such muscles are developed from the myotomes (see

p. 725) of the embryo and are under the control of the "will". Visceral motor fibres, on the other hand, supply muscles and glands developed from lateral-plate mesoderm. These muscles, which are of an unstriped variety, form the involuntary muscles of the body, good examples of which are the muscles of the gut and of the blood vascular system. (A few of the muscles of the lateral-plate mesoderm are of the striped variety and are under the control of the "will", although supplied by visceral motor fibres. Examples of these muscles are certain of the jaw muscles and the muscles of the hyoid apparatus.) The visceral efferent fibres and their cell-bodies constitute the "involuntary" nervous system sometimes called the sympathetic system or, better still, the **autonomic** nervous system. They also innervate other effector organs such as glands.

Morphologically, the autonomic nervous system is built up of peripheral neurons lying in ganglia which receive impulses from the central nervous system by special branches (rami communicantes) from the spinal and certain cranial nerves. The peripheral cell is called the sympathetic (autonomic) neuron, whilst the neuron whose cell-body lies in the central nervous system and whose axon runs out to the neuron in the ganglion is spoken of as the connector neuron, whilst its axon is the pre-ganglionic fibre. These fibres alone of all those in the autonomic system are medullated. The connector neuron may be regarded as a special internuncial neuron whose axon has had to elongate considerably in order to link up with a motor neuron (sympathetic neuron), which instead of occupying its usual position in the ventral part of the grey matter, has migrated out and come to lie in a ganglion some distance away from the central nervous system. From the sympathetic neurons situated in the autonomic ganglia, efferent fibres are in turn given off which run to the various effector organs under the control of the autonomic system, and, in addition, other fibres run out to connect up the various ganglia of the system with one another, and so relay impulses up and down the body.

From the above description of the autonomic system it will be realised that it is essentially an efferent system carrying efferent impulses to effector organs not under the control of the will. These impulses are, as a rule, evoked in a reflex way by visceral afferent impulses entering the spinal cord by way of the dorsal roots. The receptor organs of these visceral afferent nerves lie, as a rule, in or near to the effector organs, and the afferent fibres are generally bound up with efferent autonomic fibres as they leave the spinal nerve by way of the ramus communicans. It will be seen, therefore, that the anatomical basis for an autonomic reflex arc is essentially the same as that of the somatic reflex arc (Fig. 493).

The sympathetic autonomic ganglia are grouped in a similar way in all the higher vertebrates, but have been studied most fully in mammals. The chief ganglia are paired ganglia in the head connected with certain cranial nerves, paired segmental (lateral) ganglia in the neck and trunk connected to the spinal nerves, and median (collateral) ganglia in the mesenteries and smaller ones in the viscera. It may be assumed that, primitively, the lateral ganglia were segmental, that is each spinal nerve was associated with one sympathetic ganglion, but in mammals there are but two ganglia on each side in the cervical region. These are known respectively as the **anterior** and **posterior cervical sympathetic ganglia**, and they are joined by the longitudinal **cervical sympathetic nerve**. The reduction of the ganglia to two pairs has almost certainly arisen by a running together of ganglia which formerly had a segmental arrangement. A similar process of condensation gives rise in the anterior thoracic region to the **stellate ganglion**, which represents three segmental lateral ganglia. Behind this region the ganglia preserve their segmental arrangement, and form a chain right and left of the dorsal aorta. The most conspicuous of the collateral ganglia are grouped together immediately ventral to the dorsal aorta near the point of origin of the anterior mesenteric artery and form there the **solar plexus**.

The autonomic nervous system is now regarded as comprising two sub-systems: the **sympathetic** system and the **parasympathetic** system, and most of the organs not under the control of the "will" usually receive a double autonomic innervation, one set of nerves from the parasympathetic and the other from the sympathetic. The reason for this is that normally the two sets of fibres have antagonistic effects, one being excitatory, the other inhibitory. A few of the more obvious examples of organs controlled in this way will make this clearer. The iris is dilated by stimulation of the sympathetic, but contracts when the parasympathetic is stimulated; stimulation of the sympathetic also reduces the tone of the intestinal muscles, inhibits the secretion of the lachrymal gland, quickens the heart-beat, reduces the secretions of the genital glands and the glands of the mucous membrane and tongue (drying of the mouth), but increases the secretion (adrenalin) of the adrenal bodies. In the above-mentioned instances stimulation of the parasympathetic and sympathetic has opposite effects, inhibitory and excitatory, but this antagonism between the two systems, parasympathetic and sympathetic, does not always hold good—certain organs receiving only one set of fibres (*e.g.* erector muscles of the hairs receive sympathetic fibres only), whilst the salivary glands, although innervated by both systems, are not inhibited from secreting by one or the other set of fibres. Instead, the nature of the secretion is altered; stimulation of the sympathetic causes a slight but viscous secretion, whilst the parasympathetic evokes a large amount of watery fluid.

Everyone is familiar with the symptoms which accompany fear, the "drying of the mouth", the "prickly feeling at the back of the neck", and that "uneasy feeling" inside, and it is doubtful if these effects will be rendered any less unpleasant if, when they next occur, the thought enters our heads that our sympathetic system is being stimulated. Like many other objectionable sensations, however, the responses of our bodies to fear have their uses. The increased production of adrenalin results in an increase in tone of the skeletal muscles and in increased blood-sugars so that, with the speeding up of the heart, we are enabled to carry out muscular work at a greater rate. Our available horse-power is increased and we are better able to fight or to flee as discretion demands.

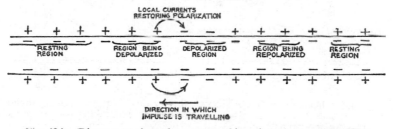

Fig. 494.   Diagram to show the passage of impulses along a nerve-fibre.

The athlete or the artiste who does not suffer from "stage fright" usually puts up an indifferent performance.

PHYSIOLOGY OF NERVE IMPULSES.—So far, nervous activity has been explained in terms of the transmission of impulses, but no attempt has been made to explain what an impulse is. It has been shown that when an impulse is passing along a particular section of a nerve-fibre, that part becomes electro-negative. These changes in the fibre are accounted for on the supposition that the surface of the axis cylinder in the resting condition is polarised, that is, the outside of the surface is +ly charged, whilst the inner surface bears an equal number of − charges. An electrical potential (**resting** or **injury potential**) of the order of fifty millivolts can then be detected by a galvanometer connected to two electrodes, one of which is placed on the outside and the other on the cut end or, in sufficiently large fibres, on the inside of the membrane (Fig. 496). Stimulation of the fibre at one point induces a local depolarisation which causes a difference of potential between the polarised and depolarised regions. Thus, if two electrodes are placed on the outside of the fibre it can be noticed that as a nerve impulse passes along the fibre the resting potential falls to zero and reverses in sign for a brief moment (a few thousandths of a second) so that *A* becomes negative with respect to *B*. When the impulse reaches *B*, then *B* becomes negative with respect to *A* for *A* has by this time become repolarised (Fig. 495). This electrical gradient causes + ions to move from the polarised to the depolarised regions and thus to a spreading of the depolarisation.

A nerve impulse may thus be visualised as a wave of depolarisation which may be detected by the electrical disturbances it sets up as it passes along the fibre. After the wave has passed, the surface of the fibre again becomes polarised (which involves the expenditure

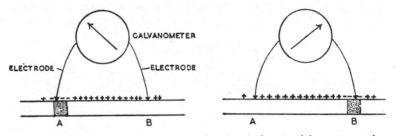

Fig. 495. Diagram to show changes in electrical potential accompanying the passage of an impulse along a nerve-fibre. Potential differences and time scale are approximate only.

of energy), and until this is accomplished the fibre is insensitive (refractory period).

It is believed that the resting potential of nerves, or other excitable tissues such as muscle, is due to polarisation of the surface membrane and this, in turn, is due to the unequal distribution of ions across it. Thus, positive ions are in excess on the outside and negative ions in excess on the inside of the fibre. This unequal distribution of ions results from a differential permeability of the membrane to positive and negative ions and can be checked by various ways which show that, as in many cells, there is an excess of potassium on the inside and an excess of chloride and sodium ions on the outside. The potassium ions inside the nerve are probably balanced by organic anions which, being large, cannot so readily diffuse outwards. It is also believed that the membrane is moderately permeable to potassium and chloride ions but much less so to sodium ions. When a fibre is stimulated electrically through electrodes or by the arrival of a nerve impulse the differential permeability of the membrane is altered so that sodium can then diffuse inwards (as it tends to do in order to equalise the concentrations inside and outside the nerve fibre) and so makes the surface relatively negative and the inside more positive. This change in potential is recorded as the **spike** or **action potential**. It is soon followed by an outward diffusion of potassium ions whose positive charge restores the polarisation. During the refractory period, in which the nerve is recovering, sodium ions are again pumped out of the nerve and this process requires the expenditure of energy and is accompanied by a small but detectable rise in temperature (heat of recovery). Presumably potassium is also actively resorbed by the fibre.

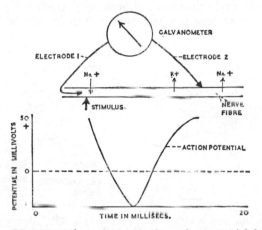

Fig. 496. Diagram to show changes in the surface potential during the passage of a nerve impulse along a fibre.

Whilst it can be shown experimentally that nerve-fibres *can* conduct equally well in either direction, normally, whilst in the body, a fibre is activated (by receiving impulses handed on from a neighbouring neuron or else from a receptor cell) at one particular end. Impulses are thus conducted away from the end which is activated — that is, a fibre normally conducts in one direction only.

THEORY OF CHEMICAL EXCITATION OF TISSUES BY NERVES.—
It has been found that when the parasympathetic fibres of the heart are stimulated, a substance (**acetylcholine**) is released at their peripheral terminations in the cardiac muscle, and it is thought that it is this substance which is the direct cause of the slowing of the heart-beat. Acetylcholine is also released at the endings of the other parasympathetic fibres, and probably by the terminal twigs

of the somatic motor nerves. It is also released at the terminations of the pre-ganglionic fibres, where its function is to stimulate the post-ganglionic neurons. The terminations of the sympathetic fibres, on the other hand, release a substance (**sympathin**) whose effects are practically identical with those of **adrenalin** and are antagonistic to those of acetylcholine, so that this theory, in addition to providing an intelligible picture of the way in which the effects of the autonomic system are produced, also suggests that a nerve impulse may be carried across a synapse by a rapidly diffusing chemical substance of an identical or similar nature to acetylcholine. It also brings certain other effects into line, for, as will be seen (see p. 655), those important organs, the endocrine organs, also elaborate chemical substances (hormones) which have important effects on remote organs when carried there by the blood-stream.

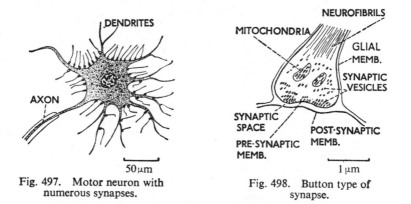

Fig. 497.  Motor neuron with numerous synapses.

Fig. 498.  Button type of synapse.

These effects are sometimes, but not always, similar to those evoked by stimulation of the autonomic system, but, because they are carried by the blood-stream, of course take place much more slowly. To take but one example, adrenalin causes acceleration of the heart, increase in blood-sugars, and in fact has (with a few minor exceptions) the same effects as stimulation of the sympathetic, so that the hormone system might be compared to sending a message by letter-post and the sympathetic system to sending the same message by telegraph.

NERVE ENDINGS.—The axons of efferent nerve-cells terminate in different ways according to the type of effector organ they supply. The medullated fibres of the somatic efferent system which carry efferent impulses to the striped muscles of the body lose their myclin sheath soon after entering the muscle, but the neurilemma remains for some distance after this point and finally becomes continuous with the sarcolemma. Below the sarcolemma the now-naked nerve-fibre branches repeatedly and the twigs in mammals end in a special mass of sarcoplasm, the **motor end plate.** The non-medullated fibres of the autonomic system, which supply cardiac muscle, unstriated muscle, and glands, branch repeatedly on the surface of the organs, the finer branches terminating in small, flat swellings.

SYNAPSES.—The detailed structure of synapses is variable. Indeed, by many the term synapse would include also the neuro-muscular junctions and other contacts between motor axons and effectors. Usually, however, it means junctions between nerves—between axons and axons, dendrites and dendrites, or between

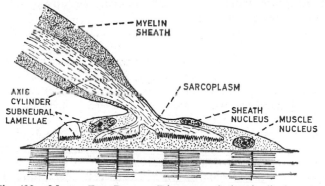

Fig. 499. MOTOR END PLATE.—Diagrammatic longitudinal section.

dendrites and cell-bodies. In all instances there are two elements to consider, one pre-synaptic and one post-synaptic for an essential property of a synapse is that it is "polarised". That is, it acts as a sort of valve, allowing excitation to pass in one direction only and acting in an "all or nothing" manner. Commonly a synapse consists of small button-like swellings separated from the surface of the post-synaptic membrane by a narrow space 0·01-0·05 μm in width (Fig. 498). The pre- and post-synaptic membranes bounding this cleft are even thinner. Small vesicles which occur in the cytoplasm of the pre-synaptic nerve and discharge into the space between the synaptic membranes are probably the source of the neuro-hormone, acetylcholine (p. 601).

## THE SENSORY SYSTEM

**Introduction.**—The senses of vertebrates, as judged by our own experiences, are those of touch, taste, smell, sight, and hearing.

Some of these—smell, sight, and hearing—are commonly considered to be associated with *special* organs—the nose, the eyes, and the ears—which are therefore termed organs of special sense. In contrast to these there are, lying in the skin, more closely together in some regions than in others, various minute sense organs which respond to stimuli such as changes in temperature or pressure (giving corresponding sensations) and, if the stimuli are excessive, others which give rise to the sensation of pain. These cutaneous sense organs are referred to collectively as the general sensory system, for the parts played by the various types of receptors in the skin are not at all clear, and it is doubtful if any given type can be related to a single sensation. The organs of special sense, on the other hand, are clearly designed to respond to particular types of stimuli to which indeed they are maximally sensitive, and their functions are better understood.

In the last analysis all sense organs whether in vertebrates or invertebrates have a basic plan and each contains a specially modified cell (or cells), termed the receptor cell, which receives the stimulus and then initiates and transmits a nerve impulse by way of a nerve-fibre, either directly, or indirectly through a chain of neurons, to the brain or spinal cord where the impulses are "interpreted" and appropriate action is made possible.

In doing this the cell changes the energy of the stimulus into the electrical energy of the impulse. This process is called transduction but the details of how it occurs are not by any means clear. In some way the depolarisation of the cell membrane which initiates an impulse or series of impulses must take place. Thus, every type of sense organ when stimulated gives rise to a similar chain of events, namely, to nerve impulses which are relayed to the central nervous system. As each nerve impulse in any one fibre is the same size as any other, intensity of stimulation cannot be signalled by an increase in the amplitude of the impulse but can only be rendered as an increase in the *frequency* of the impulses. This is the basic pattern of the code which reaches the central nervous system from the sense organs. The quality of the sensation depends, therefore, not on the nature of the nerve impulses, which are thought to be identical in all cases, but on their frequency, the position and structure of the sense organ and its connection to a particular part of the central nervous system.

Sensation is a process which takes place in the central nervous system. The place to which the impulses go determine how they will be interpreted; for example, impulses in the optic nerve are accepted as due to light whether they in fact come about by stimulation of the retina by light or by the pressure of a blow (which causes the receiver to see "stars"—disorganised visual sensation!). Thus, an animal sees, hears, tastes, and so on "with its brain" and not with its sense organs, which are concerned with the reception of the stimulus and not with its "interpretation". This is clearly shown by certain lesions of the brain when, although the sense organs are left unimpaired, the animal is deprived of sensation.

Receptor cells may be found in practically any situation in the body and structurally they are very variable. They give rise to diverse sensations but are described as exteroceptors when they receive stimuli from the outside world, as enteroceptors when they lie in the various parts of the alimentary canal, and as proprioceptors when they are concerned with the reception of stimuli from the muscles, joints, and deeper-lying parts of the body. Enteroceptors and proprioceptors are not sense organs in the usual meaning of the word, for they are not identifiable with any sensation and are merely concerned with the proper functioning of the organs in which they are situated. The exteroceptors, or sense organs in the commonly accepted sense, provide the means by which an animal is made aware of changes in its external environment and, like the central nervous system itself, they are developed from the ectoderm of the embryo. They vary in structure from the small groups of simple receptor cells, which remain in the skin and make up the organs of the general sensory system, to the larger organs of special sense in which the sensory epithelium has sunk away from the surface of the body to occupy a deeper position and to become enclosed in coats derived from the mesoderm. In addition, sense organs may be classified according to the type of stimulation they receive, into **radio-receptors** (those receiving radiant energy, *e.g.* light and heat), **chemo-receptors** (responding to chemical stimuli) or **mechano-receptors** (sensitive to mechanical stresses and strains). However, classification by stimulus is not always possible for the transduction process may be unknown. The hygro-receptors, for example, of land invertebrates may work by the action of water on some chemical system (chemo-receptor) or by the mechanical effect of hydration and dehydration of a structure (mechano-receptor). Nevertheless these are convenient classifications.

These special sense organs are, for the most part, concentrated on the head and their afferent nerve-fibres run into the brain by way of certain of the cranial nerves. In many respects their development is capable of reduction to a common plan, for each, in the first instance, arises from a special patch of ectoderm (placode) which sinks inwards so as to form a sac or vesicle in whose walls the receptor cells are later differentiated. The embryonic sense vesicle later undergoes considerable complication and becomes partially or wholly enclosed in one or more protective layers laid down by the mesoderm of the head. At first sight the eye does not appear to fit in with this scheme, for whilst the sensory layer of the eye does indeed, in the early stages of its development, form the lining to a vesicle, yet this vesicle arises, not as an invagination of the superficial ectoderm, but as an outpushing from the fore-brain. This peculiarity is due to the fact that the receptor cells of the eye occupy a special part of the embryonic ectoderm, namely the neural plate, and when this rolls up to form the neural tube, they therefore come to lie in the wall of the fore-brain, next to its cavity, the third ventricle. Thus, the relations of the optic vesicles to the third ventricle are much the same as those of the olfactory sacs and auditory vesicles to the outside world. The olfactory sac in all cases retains its embryonic opening to the surface as the external nostril, the auditory vesicle in most types loses its connection to the outside but retains it in some instances (*e.g.* dogfish), whilst the optic vesicle remains open for some time into the third ventricle (which itself may be regarded as a small bit of the outside world which has been enclosed by the walls of the neural tube). Thus in a very general sort of way, the early developmental stages of the special sense organs are comparable one to another.

**Visual Organs—the Eyes.**—STRUCTURE.—Apart from minor variations, the eye of vertebrates is remarkably constant, not only in the structure of the eyeball itself but in its innervation and extrinsic musculature, though the nature of the eyelids, glands, and mechanism of accommodation is variable.

EYEBALL.—The eyeball in, for example, a mammal is roughly spherical and is lodged in the eye socket or orbit. It has a complicated structure, but the various parts are intelligible as components of a system analogous to a camera. Thus, there is a light-proof box, a diaphragm, a lens system, and a surface sensitive to light. The outer wall of the eyeball, which corresponds to the capsule surrounding the olfactory and auditory organs, is formed of a tough, fibrous material. For the most part it is opaque, but in the front of the eye it is quite transparent. These two parts of the outer coat are respectively termed the **sclerotic** and the **cornea**. It is to the sclerotic that the **extrinsic muscles**, responsible for moving the eyeball in the orbit, are attached. The sclerotic (but not the cornea) is lined by a vascular, pigmented layer, the **choroid**, which in the front of the eye is modified to form the **iris**. The iris is perforated by an aperture of variable size, the **pupil**. Lying in the choroid coat around the outer margin of the iris is the **ciliary body**. This is a thickened zone of the choroid which contains larger blood vessels, glands, and muscle-fibres. The muscles run both in a circular and a meridional direction with respect to the eyeball, and the meridional muscles are attached at their inner ends to the choroid and at their outer ends to the point where the cornea joins the sclerotic. The iris also contains muscle-fibres. These control the size of the pupil and they are arranged in two groups, a set of circular fibres and a set of radial fibres. Immediately behind the iris lies the lens, a transparent biconvex body. It is attached by the margins of its transparent **capsule** to the ciliary body by the fibrous **suspensory ligament**. The part of the eye in front of the lens is filled with a watery fluid, the **aqueous humour**, whilst the chamber behind the lens contains a jelly-like substance, the **vitreous humour (vitreous body)**. This is perforated

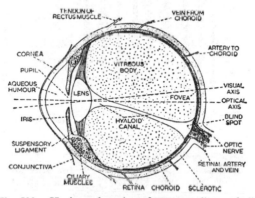

Fig. 500. Horizontal section of a mammalian eyeball.

from a point near the origin of the optic nerve to the lens by a narrow canal, the **hyaloid canal**, the site of an embryonic blood vessel which later disappears. The innermost coat of the eyeball is the **retina**, but it does not form a complete lining to the eyeball for it is continued only as far forward as the iris, whose hindermost layer it thus forms. The retina is chiefly concerned with the reception of visual stimuli and with relaying the nerve impulses so generated to the optic nerve. Beginning from the outside (*i.e.* next to the choroid) and passing inwards, the retina is composed of several layers as follows:

A layer of cubical epithelial cells, bearing on its inner surface cytoplasmic processes which form a sort of fringe. The cells contain numerous granules of a dark-brown pigment (melanin). It is this layer alone which is continued forwards to cover the posterior surface of the iris. Beneath these pigmented cells lies the receptor layer of the retina which is composed of visual cells (photoreceptors) of two types, the **rods** and the **cones**. The rods are single cells which bear at their outer ends a long slender cylinder with its long axis perpendicular to the surface of the retina. The cylinder contains a purple pigment (rhodopsin or visual purple), a substance derived from Vitamin A. The inner part of each rod is swollen out to contain the nucleus, and from this end of the cell a nerve-fibre passes inwards to link up with the dendrites of the cells of the next layer. The cones, like the rods, are also single cells, but instead of a cylinder each bears a short tapering process, whilst

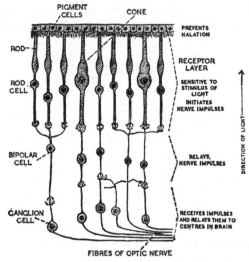

Fig. 501.　Diagram of the layers in the retina.

from the inner end of the cell arises a nerve-fibre as in the case of the rods. Rods and cones are not evenly distributed over the retina. Cones alone are found in a small depressed area, the **fovea centralis**, which lies at the hinder end of the eyeball on its main visual axis. Away from this area a mixture of rods and cones are found, and towards the periphery of the retina the rods very much outnumber the cones.

The visual elements link up by synapses with the dendrites of the cells of the next layer, a layer of bipolar nerve-cells which in turn, by their very short axons, relay the nerve impulses received from the visual cells to a layer of large ganglion cells whose axons are the component fibres of the optic nerve. These axons pass over the inner surface of the retina in a meridional fashion and converge to a point at the back of the eyeball. Here they pierce the retina, acquire myelin sheaths, and become bundled together to form the optic nerve. At the place where the fibres of the optic nerve issue from the retina, visual cells are absent, and it follows therefore that reception of stimuli at this point is impossible. It therefore receives the name of blind spot. Such, then, are the main retinal components, but there are also certain cells which link up the bipolar cells. Others have a supporting or packing function. There are the neuroglial cells and also the important cells termed Müller's fibres, which fill all the spaces between the visual and nervous elements of the retina.

From this description of the retina it will be realised that any light passing into the eyeball must, before reaching the receptor layer of the retina, *pass through the nervous layers*. In most regions it also has to penetrate a thin layer of retinal arteries. A retina of this type is characteristic of the vertebrates and is an "inverted retina". The meaning of this extraordinary arrangement is explicable only when the development of the eye in the embryo is understood (pp. 612-16).

CONJUNCTIVA.—The front surface of the eyes is separated from the outside world by a thin transparent layer, the conjunctiva. This is really an area of modified epidermis and is continuous at its periphery with the epidermis of the eyelids.

EYELIDS.—These are somewhat variable structures, having as their function the protection of the delicate front surface of the eye. Two lids, an upper and a lower, are invariably present. Each in essentials is a fold of skin and is provided with muscles so that it is capable of movement. Despite this, movement may be extremely limited, and whereas in the dogfish it is the lower lid alone which moves to close the eyes to any degree, in some mammals it is the upper one. In the rabbit and in many other vertebrates, yet a third lid is present. This is referred to as the nictitating membrane, and this membrane can move so as to close completely over the front surface of the eyeball so protecting and cleaning it.

GLANDS.—In addition to minute glands on the lids themselves, the eyes of land-living animals (notably the mammals) are provided with special glands (lachrymal glands) which pour their secretions (tears) by means of ducts on to the front surface of the eye. Tears are a watery secretion whose purpose is to keep the surface of the eye moist and to wash it free from dust. Excess tears are drained from the eye into the nasal sinuses by means of the lachrymal duct.

Tear glands are absent from the eyes of aquatic vertebrates, for the surrounding medium bathes the eyes and such glands would therefore be superfluous.

**Vision.—The Physiology of the Eye.**—The sense of sight is generally understood to mean something more than the mere ability of an animal to react to changes in the intensity of the light falling on its photoreceptors. Vision or sight implies the formation of more or less distinct images of the external objects on the photosensitive layer. These images are then "interpreted" by the central nervous system and the animal is made aware of certain aspects of its external environment. Sight in man comprises at least three sensations, colour, light intensity, and the form of objects. The nature of the interpretation of the visual stimuli, that is, the subjective aspect of vision, is far from being understood, and great caution must be adopted when crediting animals with the same powers of sight as in man. The physiological and physical properties of the eyes may in both instances be nearly identical, but the subjective processes which accompany the reception of visual stimuli are almost certainly in no way comparable.

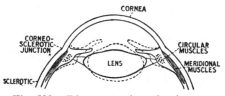

Fig. 502.    Diagram to show the changes during accommodation.

This word or two of caution applies with equal force to other receptor organs and the sensations associated with them in man.

FORMATION OF AN IMAGE ON THE RETINA.—The first essential requirement for an image to be formed on the retina is that light reflected from external objects shall be brought to a focus on the retina. This is effected by the refractive (optical) system of the eye, whose radius of curvature is capable of adjustment or accommodation so that images of both distant and near objects can be formed.

Light entering the eye is first of all refracted towards the normal by the cornea, this refraction being constant for a particular eye. The light is then focused by the lens so that a sharp inverted image is formed on the retina. The lens, it will be recalled, is a biconvex one of elastic nature enclosed in a tough but transparent lens capsule to which the suspensory ligament is attached. When the eye is at rest, or trained on distant objects, the tension in the suspensory ligament is high and keeps the lens flattened against its own natural elasticity, as is shown by the fact that a lens removed from the eyeball is more nearly spherical than before removal. The tension in the suspensory ligament is maintained by the pressure of the fluid in the eye against the coats of the eyeball. When near objects are viewed the radius of curvature of the lens decreases, *i.e.* the lens gets fatter, and the pupil contracts. This accommodation of the eye for near vision is brought about in the following way. The tension in the suspensory ligament is reduced by the contraction of the two sets of muscles in the ciliary body. The circular muscles cause a reduction in the diameter of the zone to which the ligament is attached, whilst the meridional muscles pull the part of the choroid to which their posterior ends are attached towards the corneo-sclerotic junction.

As a result of the relaxing of the ligament, the lens, by its own elasticity bulges, and the bulging is most pronounced in the central region of the front surface, for in this region the lens capsule is appreciably thinner and has a smaller restraining influence on the lens. The peripheral regions of the lens remain relatively flat, but the distortion of the retinal image which must result from the uneven curvature of the front surface of the lens is minimised by the contraction of the pupil, which results in the flattened peripheral regions being covered by the iris from the incident light. It should be emphasised that in terrestrial animals most of the refraction (convergence) of the rays entering the eye takes place at the cornea, the lens acting as a fine adjustment to ensure accurate focusing of the image on the retina.

The stimulus of light falling on the retina produces as its end result the sensation of sight and is accompanied by several well-defined responses in the retina itself. When an eye, after being kept for some time in the dark, is exposed to light, the brown pigment granules of the pigment layer of the retina move down the fringe in between the rods and the rods themselves become shorter. Both the brown pigment and rhodopsin (which is present only in the rods) become bleached, and the reaction of the retina becomes acid. One of the effects of the light stimulus is, then, the dissociation of the photosensitive rhodopsin by a series of steps into retinene and visual yellow which in the dark recombine to form rhodopsin. The immediate stimulus, as in many other receptor organs, is a chemical one. Rhodopsin, which is continually being regenerated from its intermediate compounds, is absent from the cones and the means by which they are stimulated is not known, but is probably by other visual pigments, of which there may be a series.

It is generally accepted that the rods and the cones have different parts to play in the retinal processes associated with sight. The cones are concerned with the reception of stimuli of high intensity ("daylight vision") and are also responsible for the detection of differences in the wave-length of the incident light; that is, they are associated with "colour vision". The rods, on the other hand, come into operation chiefly when the light is less intense and detect differences in the intensity of the light. They are concerned with "twilight vision", and it is of extreme interest in this connection that the eyes of nocturnal creatures, or of animals living in the depths of the sea or in other situations of low light intensity, have rods but no cones in the retina.

The mechanism by which different colours are discriminated is not definitely known, but one of the most popular theories postulates that three different types of cone are present in the retina, each of which responds to one only of the three fundamental colours, red, green, and violet. These three colours, if mixed in appropriate intensities, can give rise to any other shade of colour or to white light, and the sensation of colour is due to an integration by the brain of the impulses arising from the three different kinds of cone. Recent work has implied that there are more than three types of cones present in the retina and that each responds to a restricted wave-length only. The interpretation of the impulses relayed to the brain, so that the subjective effect of colour is produced, may depend on different fibres and different brain centres.

**Development of the Eye.**—The eye of the adult is developed from three main sources in the embryo—the brain, the mesenchyme of the head, and the embryonic epidermis. Soon after the closure of the neural folds to form the neural tube an outpushing appears on each side of the ventrolateral parts of the fore-brain. These outpushings or evaginations soon come to form prominent bulges, the primary optic vesicles, which continue their growth outwards towards the epidermis. The primary optic vesicles then become constricted on their inner (mesial) sides so that they remain connected to the thalamencephalon only by a narrow neck, the optic stalk. Meanwhile, a patch of epidermis on the side of the embryo immediately opposite each optic vesicle has been thickening to form the rudiment of the lens. The outer and lower (i.e. the ventrolateral) portion of the wall of each primary optic vesicle invaginates, thus transforming the primary into a secondary optic vesicle or optic cup, as it is usually called. The outer layer of the optic cup, that is, the part of the optic vesicle which is not invaginated, retains its epithelial character, and eventually forms the pigment layer of the retina. The layer of cells lining the optic cup forms the rest of the retinal layers, its inner (i.e. deeper) cells later differentiating into the rods and the cones whilst its more superficial cells form the nervous layers. Owing to the invagination, the cavity of the primary optic vesicle disappears and the two main layers of the retina become closely applied to one another and are, of course, continuous at the rim of the optic cup. The lens rudiment, meanwhile, invaginates, separates from the epidermis, and sinking inwards, forms a thick-walled vesicle lying within the cavity of the optic cup. The lens rudiment later becomes a solid body and its cells form a mass of transparent refractive fibres, which are arranged

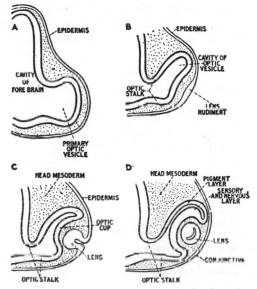

Fig. 503. Vertical sections through the embryonic head to show the development of the eye.

chiefly in a meridional fashion. The rim of the optic cup now constricts, but remains as the retinal pigment layer of the iris, whilst the opening of the optic cup persists as the pupil. The epidermis covering the eye after separation of the lens then becomes modified to form the transparent conjunctiva.

By this time mesenchyme has invaded the head around the eyes and from it are formed the choroid (including the ciliary muscles), the sclerotic, and the cornea. The six extrinsic eye muscles are developed from mesodermal condensations which correspond to the first three head somites. The optic nerve is formed from fibres which grow back from the ganglion cells of the retina. These fibres converge over the surface of the retina towards the end of the optic stalk, along which they grow to link up with the brain. Not all of the fibres are concerned with sight. Some, for example, are associated with reflexes of the iris. In the lower vertebrates all the visual fibres of each optic nerve cross over (decussate) at the optic chiasma and pass to the side of the brain opposite to the eye from which they originated. In mammals, however, only half of these fibres decussate; thus in man, where the eyes face forward, all the fibres from the right side of *both* retinae run in the right optic tract to the right side of the floor of the fore-brain, whence they are relayed to the cerebral cortex. Conversely, the left optic tract contains fibres from the left side of both retinae. The non-visual fibres do not decussate, and many of them run to the anterior part of the optic lobes of the same side. This arrangement of the eyes

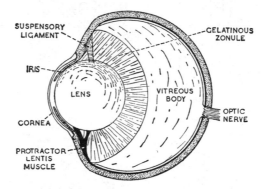

Fig. 504. DOGFISH.—Vertical section of eyeball to show mechanism of accommodation. The lens is pulled forward closer to the cornea so accommodating for near vision. (Based on Franz.)

and their brain linkages is associated with what is known as stereoscopic vision, by which we human beings are able to appreciate the three-dimensional aspect (depth of vision) of objects.

The eye of the dogfish, although consisting of much the same parts as that of the eye of a mammal, has certain important differences (Fig. 504). The lens, like that of other fishes, is nearly spherical and incompressible, and so accommodation for near vision cannot be brought about by altering its shape. At first sight it might be thought that a spherical lens would cause considerable spherical and chromatic aberration of the retinal image but direct viewing through freshly excised fish lenses show that nearly perfect images are formed. This is a result of the peculiar way in which the lens is constructed for its refractive index is not uniform but increases steadily from the surface to the centre. This feature also results in it having a much shorter focal length (F = about 2·5 R) than if it had a uniform refractive index throughout. This is a feature required by the eyes of all aquatic animals since practically no convergence takes place at the cornea which has a refractive index only slightly above that of water. It is suspended by a **gelatinous zonule** which is strengthened dorsally to form a supporting suspensory ligament. On the ventral side of the lens is the **protractor lentis muscle** which moves the lens when accommodation is required. The retina lacks cones and the eye is adapted for near vision in dim light. The sensitivity of some fishes which normally live in the dimly lit regions below the surface layers of the sea is increased by the presence of a reflecting layer or tapetum in the choroid.

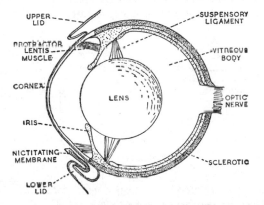

Fig. 505. FROG.—Vertical section of eyeball. (Based on Franz.)

This is composed of plates of guanin so orientated that they lie at right angles to the longitudinal axis of the rods in all parts of the retina. Thus a light ray reaching the retina passes through a rod and back again through the *same* rod after reflection by the tapetum. In this way more visual pigment molecules can be dissociated without loss of visual acuity. (Similar reflecting tapeta are found in the eyes of some nocturnal mammals.) Sensitivity is also increased by having a visual pigment, **chrysopsin**, which preferentially absorbs light in the blue-green part of the spectrum since light of this wavelength (about 470 nm) alone penetrates far below the surface.

In the eye of the frog (Fig. 508) there is a complete fibrous zonule or suspensory ligament and the protractor lentis muscle is also complete so that the supporting structure of the lens could be moved.

**Explanation of the "Inverted Retina".**—In many animals, *e.g.* Amphibia, the areas of the neural plate which will later give rise to the optic vesicles are recognisable before the formation of the neural tube. In each of these patches the outermost layer is already pigmented and, although not yet differentiated, the sensitive and nervous elements are disposed in a way which is found in all other adult sense organs. That is, the receptor cells are on the outside, the conducting (nervous) cells on the inside away from the source of stimulation. Clearly, this arrangement is reversed when the neural plate rolls up to form the neural tube, for then the receptor layer lies towards the *cavity* of the fore-brain, that is, away from the source of light. The curious "inverted" arrangement of the components of the adult retina is now explicable in terms of embryonic development, for it will be recalled that, because of the invagination of the primary optic vesicle to form the optic cup, it was the layer lining the cup which gave rise to the sensory and nervous elements of the retina, the nerve-cells lying nearer the skin whilst the rods and the cones were in the deeper layers next to the pigment layer.

**Organs of Chemical Sense—Chemo-receptors.**—These comprise a system of **taste buds** and the **olfactory organs**. Both kinds of organs

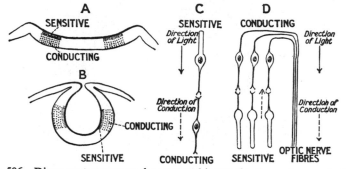

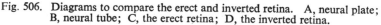

Fig. 506. Diagrams to compare the erect and inverted retina. A, neural plate; B, neural tube; C, the erect retina; D, the inverted retina.

are stimulated by substances in solution in an aqueous medium which is provided by the surrounding water in the case of aquatic animals, but in land-living animals by a solution of watery mucus over the sensory epithelium in the mouth cavity or olfactory organs. The two senses, **taste** and **smell**, are always closely associated but a distinction that can usually be drawn between them is that the olfactory organs respond to a much lower concentration of chemical substances in the surrounding medium. Also, stimulation of the olfactory organs usually leads to a somatic response, *i.e.* to a movement of the whole body towards or away from the stimulus. During this movement the animal appreciates a gradient of concentration in the stimulating substance and is thus enabled to detect the direction from which it is diffusing. Taste buds, on the other hand, respond only to higher concentrations of chemical substances and their stimulation leads to visceral responses, secretion by the buccal glands, swallowing, and the like. Thus the olfactory sense gives a sense of awareness of substances at a distance; taste is chiefly concerned with testing substances entering the alimentary canal and may be thought of as a later stage in chemical discrimination.

THE TASTE BUD SYSTEM.—The structure of the taste buds is more or less uniform in all vertebrates. They consist of groups of sensory cells ensheathed in supporting cells and embedded in the stratified epithelium of the epidermis or that lining the buccal cavity. The sensory cells are long and narrow and their free surfaces bear short, hair-like processes, whilst around their basal portions there are nerve-fibres. These fibres contribute to the

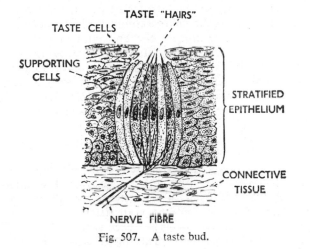

Fig. 507. A taste bud.

seventh, ninth, or tenth cranial nerves according to the position of the taste buds in question. In most animals the taste buds lie in the epithelium lining the mouth cavity, but in many fishes they occur also on the skin, particularly on the lips and barbels.

In man, where most of the taste buds are found grouped on the tongue, it seems that there are only four primary sensations of taste, viz., sweet, bitter, sour, and salty, and that the diverse flavours which may be experienced are due to a combination of the four primary sensations and to a mental integration of these with olfactory sensations (odours). Many substances also provoke sensations in the mouth (e.g. "burning" sensation of spirits) by stimulating the numerous tactile organs present, thus contributing to the sensation commonly termed "taste". Receptors responsive to water have been located on the tongues of cats and frogs.

Not all the taste buds are concerned with the sensation of taste. Those on the epiglottis of man, for instance, or around the opening into the swim bladder of certain fishes, have as their function the testing of substances in these regions and of promoting reflexes which guard against solid particles entering the air passages, thus preventing food from "going the wrong way".

THE OLFACTORY ORGANS.—The olfactory organs of all vertebrates are similar in essential structure but are frequently associated with accessory structures of variable nature. They consist of paired cavities (olfactory or nasal sacs) situated at the anterior end of the head and lined by an epithelium which is continuous with the ectoderm from which it is derived in the embryo. The olfactory sacs are partially enclosed and protected by the olfactory capsules which are attached to the neurocranium. Their openings are the nares or nostrils. The area of the epithelium lining the sacs is increased by the presence of the folds and ridges on the wall of the sac. This epithelium is of the columnar type and is rich in mucous gland cells, and in certain regions patches of receptor (olfactory) and supporting cells are found. Each olfactory cell bears a single prolongation from its outer surface bearing at its tip a bunch of short "sensory hairs", whilst from the basal end of the cell there arises a nerve-fibre which runs back to the brain as one of the component fibres of the olfactory nerve.

In all air-breathing animals each olfactory organ has two openings: an external and an internal nostril. The latter primitively opens into the front of the buccal cavity, but in certain creatures (e.g. crocodile and the mammals) the olfactory organ opens into an air passage which has been shut off from the buccal cavity by the secondary palate. This air passage opens at its hinder end into the back of the mouth cavity (pharyngeal region), its posterior openings being then termed the internal nares or choanae.

In general, the olfactory organs are better developed in the lower than in the higher vertebrates, but the sense of smell is still of great importance in many mammals. On the other hand, with the exception of the kiwi, birds seem largely devoid of an olfactory sense.

Smell.—The olfactory organs are stimulated by chemical substances, which in truly aquatic animals are dissolved in the surrounding water, but which in land animals are carried into the organs by the respiratory current of air and are then dissolved in the mucus covering the olfactory epithelium. By no means all substances are odoriferous, that is, capable of stimulating the olfactory organs, and many of the substances with the strongest odours are practically insoluble. Moreover, it is impossible to relate odorous properties with any type of molecular structure, and many substances, *e.g.* vanilla or prussic acid, which have a strong and distinctive smell for many people, are odourless to others. Many pungent substances, *e.g.* ammonia or the vapour of mineral acids, produce their effects, in part at any rate, by stimulating other nerve endings which are found in the nasal epithelium in addition to the olfactory cells.

In man the sense of smell is easily fatigued, but whilst the nose may become insensitive to one particular odour, it usually remains sensitive to others of a different nature. Smell and taste are closely allied senses and certain animals (some reptiles and mammals) possess an organ of chemical sense (**Jacobson's Organ**) lodged in a diverticulum from the dorsal boundary of the buccal cavity, which may be described as having the function of "smelling the food in the mouth".

**Organs of Hearing and Balance—the Acoustico-Lateralis System.**—The acoustico-lateralis system of receptors is found developed to a greater or lesser degree in all vertebrates. It comprises a system of **neuromast organs**, disposed in definite tracts at or near

Fig. 508. Diagram to show the main branches of the lateral line canals in a fish.

the surface of the body, and the **membranous labyrinth** (inner ear) of the ear. The actual receptor cells of the neuromast organs and of the membranous labyrinth are fundamentally similar in structure, and both are adapted to detect stimuli from the movements of the watery medium which surrounds them. Both have a similar origin in the embryo and both are innervated by afferent nerves terminating in neighbouring regions in the medulla oblongata. For all these reasons, therefore, the neuromast system and the inner ear are regarded as complementary parts of a single acoustico lateralis system, and it is also thought that the inner ear has evolved from an enlarged and specialised neuromast organ.

THE NEUROMAST SYSTEM (LATERAL LINE SYSTEM).—A system of neuromast organs is found only in the cyclostomes, fishes, and the aquatic stages of the Amphibia. The neuromasts are disposed on or near the epidermis according to a fairly regular plan which is indicated externally as a series of lines. A generalised diagram of these sensory lines is shown in Fig. 508, but their pattern is subject to important variations in different types of vertebrates.

In the dogfish and its allies the neuromasts lie at intervals in the floor of an ectodermal tube, which has sunk into the dermis but opens on to the surface of the body by a series of vertically running tubes. The ectodermal epithelium of the sensory tube or canal is rich in mucous gland cells. Each neuromast organ consists of a group of receptor cells, which bear a distinct resemblance to those

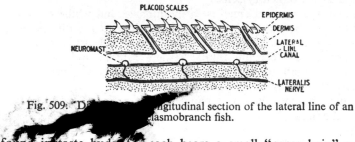

Fig. 509. Diagrammatic longitudinal section of the lateral line of an elasmobranch fish.

found in taste buds; for each bears a small "sense hair" and is surrounded by nerve-fibres and a number of supporting cells. The canals and their sense organs develop in the embryo from patches of specialised ectoderm—the **dorsolateral placodes**—which extend along the side of the dorsal surface of the head, the lateral lines proper later growing back from this region along the sides of the trunk.

The way in which the lateral line system functions, and its full

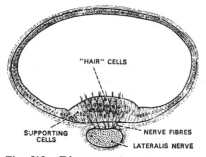

Fig. 510. Diagrammatic transverse section of the lateral line of an elasmobranch fish.

significance to the animals in which it is found, is not fully understood, but it has been shown that the sensory cells are stimulated by movements of the fluid which surrounds them in the canals. Presumably such currents bend the hairs and as a result nerve impulses are initiated. It is known that slow vibrations (low frequency) are especially effective in evoking a response.

It will be plain that if the intensity of the vibrations reaching different parts of the body can be compared, the position of the object setting up the vibrations can be located (Fig. 511). It is interesting that though the system stretches the length of the body, it is not innervated from spinal dorsal roots, as might be expected, but from a single cranial nerve; thus all impulses will pass to one centre making simultaneous comparison the easier. Deep-sea fish which seem to depend upon the lateral line system to find their prey, tend to have long bodies; the longer the lateral line system is the more accurate will be location of objects by this means. But in addition to serving to locate prey the lateral line system may serve to pick up vibrations of other fish so that shoals of fish can stay together even without visual stimuli. It may also serve as a means by which obstacles in the water may be avoided, for any solid object will tend to reflect the vibrations set up during swimming and will alter the nature of the vibrations falling on the neuromasts.

Fig. 511. A source of vibrations is detected near the head end of the lateral line and near the tail end. The greater the angle α the more accurate is the localisation of the source.

The scattered **ampullary sense organs (Lorenzini's ampullae)** found on the snout of elasmobranch fishes are not really part of the acoustico-lateralis system. They have been shown to be temperature receptors, but this is hardly likely to be their main function since the receptor cells lie at the base of a tube containing jelly—a poor conductor. Most probably they are designed as receptors of changes in the electrical field set up by "weak" electrical organs common in many groups of fishes and which play a part in navigation and in the detection of prey.

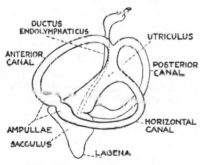

Fig. 512. The membranous labyrinth of a shark.

THE EAR.—The essential part of the ear, that is to say the part actually concerned with the reception of stimuli, is the **membranous labyrinth** (inner ear). It is the sole type of ear found in fishes but, in the Amphibia and in the Amniota, accessory structures (**middle** and sometimes **external ears**), which increase the sensitivity of the ear to sound, are present in addition to the inner ear. The membranous labyrinth is built on a plan common to all vertebrates, but is of a simpler nature in the lower than in the higher forms, which have a more highly developed sense of hearing.

## FISHES

In the dogfish and its allies the membranous labyrinth consists of a vesicle partially divided by a constriction into an upper chamber, the **utriculus**, and a lower chamber, the **sacculus**, together with outgrowths from both chambers. From the sacculus there arises a narrow tube, the **ductus endolymphaticus**, which runs to the dorsal surface of the skull where it dilates to form the **saccus endolymphaticus**. This may open by a small pore on to the surface of the head. The upper chamber, the utriculus, opens into three tubular loops, the three **semi-circular canals**, which are mutually at right angles, one being horizontal, the other two vertical. At one end, where it joins the utriculus, each semi-circular canal is swollen out to form an **ampulla**. The walls of the membranous labyrinth have an elastic

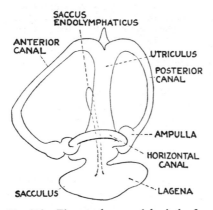

nature, for they are formed of a dense fibrous tissue lined by an epithelium of ectodermal origin, which in certain places is modified to form the receptor organs of the labyrinth. There are six such patches: one in each ampulla, one in the utriculus, and two in the sacculus. The ampullary receptors are termed **cristae**; those of the utriculus and sacculus, **maculae**. The cristae and the maculae have, however, a similar structure and they also resemble the receptors of the lateral line system,

Fig. 513. The membranous labyrinth of a bony fish.

for they consist of groups of elongated receptor cells—each provided with a tuft of "sensory hairs". In the case of the cristae the hairs, similar to cilia in structure, are embedded in a gelatinous pad, the **cupula**, which fits tightly across the cavity of the ampulla. Each macula, however, is overlain by an otolith covering all or part of the sensory cells. This otolith is composed of several small particles in the dogfish but a single larger one in bony fishes. The nerve-fibres are, of course, the component fibres of the eighth (auditory) nerve. The membranous labyrinth is completely filled with a fluid, endolymph.

The membranous labyrinth is enclosed in the auditory capsule, but, except at a few points, it is separated from it by a space filled with a watery fluid, the **perilymph**. The perilymphatic space is traversed by irregular strands of connective tissue which carry small blood vessels and also suspend the membranous labyrinth from the lining (perichondrium or periosteum) of the auditory capsule.

In most fishes one of the maculae is enlarged and accommodated in a bud-like outgrowth of the sacculus called the **lagena**, whilst in the carps and their allies a thin patch in the wall of the auditory capsule is linked by a chain of small bones—the **Weberian ossicles**—to the anterior end of the air-bladder. Both these features are associated with a more acute sense of hearing.

## *AMPHIBIANS*

The inner ear of the frog, and of other amphibians, is much like that of a bony fish, for it has semi-circular canals, a utriculus, and a

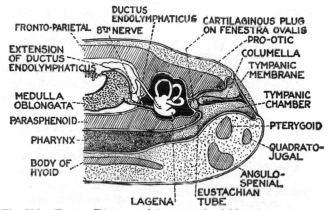

Fig. 514. FROG.—Diagram of the ear and neighbouring structures.

sacculus bearing a lagena. The ductus endolymphaticus does not, however, run only to the dorsal surface of the skull but also passes down the spinal canal and connects up the calcareous nodes lying alongside the vertebrae. These serve as stores for calcium. In addition to the inner ear, a middle ear or tympanic chamber is present, derived from the first visceral cleft. This is a pouch, filled with air, which separates the outer wall of the auditory capsule from the skin. Its lining membrane is of endodermal origin. From the lower end of the tympanic chamber a tube, the **Eustachian tube**, runs downwards and opens into the pharynx, but its lower opening is normally kept closed by a small valve. The outer (lateral) wall of the chamber is fused to the skin to form the **tympanic membrane**, which is visible externally as a more or less circular patch on the side of the head, just behind the eye. Attached to the inner surface of the tympanic membrane is a slender bone, the **columella auris (stapes)**, which bridges the middle ear and is connected at its inner end to a membrane closing a small aperture (**fenestra ovalis**) in the wall of the auditory capsule.

The auditory apparatus of reptiles and of birds is similar to that of the amphibians, but the lagena is much larger and is a coiled structure which might well be dignified by the name of cochlea.

## MAMMALS

It is in the mammals that the ear reaches its highest degree of development for not only are the membranous labyrinth and middle ear more complex than in lower types, but an additional part, the **external ear**, is present.

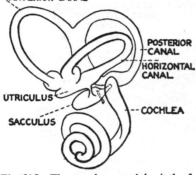

ANTERIOR CANAL

POSTERIOR CANAL

HORIZONTAL CANAL

UTRICULUS

SACCULUS

COCHLEA

Fig. 515. The membranous labyrinth of a mammal.

The membranous labyrinth differs from that of previously described examples in many minor details and in one important respect also, namely, that instead of a bud-like lagena, the sacculus connects with a spirally coiled tube which, together with part of the auditory capsule which encloses it, is called the **cochlea**. A transverse section of the cochlea gives the impression of a bony tube—the **cochlear canal**—lined by connective tissue and crossed by two membranous partitions so that it is divided into three chambers, an upper, a middle, and a lower chamber. The middle chamber is termed the **cochlear duct** or **scala media** and is the extension from the sacculus. It is filled with endolymph. Its upper and lower walls, which are the two membranous partitions referred to above, are, of course, merely parts of the wall of the membranous labyrinth, but they receive special names. The upper one is called **Reissner's membrane**; the lower, the **basilar membrane**. The upper chamber of the cochlea, which lies between Reissner's membrane and the wall of the bony canal, is the **scala vestibuli**. It is a perilymphatic space which connects at the base of the cochlea with a space below the **fenestra ovalis**—a small area in the wall of the auditory capsule which is membranous (not bony) and which abuts against the cavity of the middle ear. At the tip of the cochlea the scala vestibuli connects by a narrow duct, the helicotrema, with the lower chamber of the cochlea, the **scala tympani**. This is also filled with perilymph and ends blindly at its lower end against the fenestra rotunda.

The somewhat complicated arrangement of the canals in the cochlea may be visualised if the scala media be imagined as growing out as a spiral tube from the sacculus and, as it does so, pushing in front of it a portion of the perilymphatic space and the wall of the auditory capsule so that the coiled tubular extension of the sacculus becomes enclosed in a bony tube, from which it is separated by

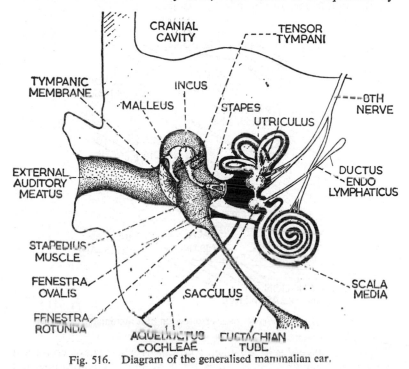

Fig. 516. Diagram of the generalised mammalian ear.

perilymph. Then, except at the tip, the scala media fuses along its upper and lower surface with the periosteum lining the bony canal, and the perilymph thus becomes confined to two coiled chambers (the scala vestibuli and scala tympani) which communicate with each other only at the tip of the cochlea.

Like the rest of the membranous labyrinth, the scala media is lined by an epithelium of ectodermal origin which, where it overlies the basilar membrane, is modified to form the supporting and receptor cells of the cochlear sense organ (**organ of Corti**). This consists of a collection of tall pillar-like supporting cells and four longitudinal rows of receptor cells, which are very similar to those of the maculae found in other parts of the membranous labyrinth. The "hair cells" are overlain by a gelatinous ribbon-like membrane, the **tectorial membrane**, attached along both its edges and in which, during life, the "sensory hairs" of the receptor cells are actually embedded. The basilar membrane, upon which the organ of Corti rests, lies across the cochlear canal. Though not under tension, it varies in rigidity along its length. It is narrowest at the base, by the middle ear, and broadest at its apex.

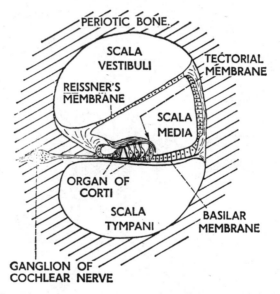

Fig. 517. Diagrammatic transverse section of a mammalian cochlea.

The middle ear is an air-filled space of irregular shape contained within the tympanic bone. It is lined by mucous membrane and connects with the back of the bucco-pharyngeal cavity by the Eustachian tube. Its outer wall is fused to the skin to form the tympanic membrane which is usually stretched across a spout-like projection from the tympanic bone, the external auditory meatus. The inner wall of the middle ear is largely formed by the wall of the auditory capsule, in which there are two apertures closed only by membrane. The upper one is the **fenestra ovalis** (oval window) and the lower one the **fenestra rotunda** (round window). The tympanic membrane is linked to the fenestra ovalis by a chain of three small bones, the auditory ossicles, which are, named in order, the **malleus**, the **incus**, and the **stapes**. They are connected one with another by definite articulations (like other bony joints), whilst the malleus is attached to the tympanic membrane and the stapes to the fenestra ovalis. Tension on the tympanic membrane is maintained by a small muscle (**tensor tympani**) which runs from the wall of the tympanic chamber to become inserted in the periosteum of the malleus. A small muscle is also attached to the stapes.

The outer ear or **pinna** is a trumpet-like structure of variable shape and size which projects outwards from the auditory meatus. It is formed very largely from skin but is supported by elastic cartilage, and in many mammals is provided with muscles so that it can be moved to point in different directions ("pricking" or "cocking" of the ears).

**Hearing and Balance—Physiology of the Ear.** —The ear, although usually described structurally as a single or unit sense organ, is concerned with the perception of totally different kinds of stimuli and fulfils two distinct purposes in the life of the animal. It is an organ by means of which the animal is made aware of its movements and of gravity, and is so enabled to orientate itself and maintain its balance. The ear is also a sound receptor, but the extent to which the sense of hearing is developed is very variable. The two distinct functions, orientation and hearing, are associated with different parts of the membranous labyrinth. The upper part of the labyrinth, including the utriculus and semi-circular canals, is concerned with the detection of changes in velocity (and so can be termed an accelerometer) and with reactions to gravity, whilst the sacculus and its outgrowth, the lagena or cochlea, if they are present, contain the sound receptors. Some hint of the mode of action of the semi-circular canals and their ampullae may be gained by recalling that the canals are so arranged that each lies in one of the planes of space, that is, there are two vertical canals at right angles to each other and a horizontal canal. This, in itself, suggests that any change in the direction of the head will produce a movement in the fluid of the canal system and might stimulate the receptors in the ampullae. Observations on fishes (pike and cod) confirm this view for the cupulas overlying the ampullary cristae have been seen to move with movements of the endolymph. Moreover, impulses in single fibres in the nerve leading from an ampulla have been recorded in the skate and it has been shown that impulses continually pass along them and that they increase in number when the head is rotated in one direction and decrease when it is rotated in the opposite direction.

Thus, the three canals and the cristae provide a mechanism whereby changes in acceleration in any of the three planes of space can be detected by an animal since the canal which most nearly coincides with the plane of rotation is most strongly affected. The maculae in the utriculus (and possibly also of the sacculus) are thought to be chiefly receptive of gravitational stimuli, for they initiate impulses most rapidly when the body is in its normal position with respect to gravity, and bring about reflexes by which muscular tone and body posture are maintained. They also respond to linear accelerations. The cristae in the ampullae, as has been mentioned, react to changes in angular velocities, that is, to rotations. In

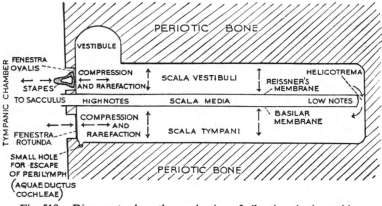

Fig. 518. Diagram to show the conduction of vibrations in the cochlea.

both instances, the immediate stimulus is probably due to movements in the endolymph which produce a shearing force in a plane parallel to the surface of the sensory cells.

The mechanism by which sound is detected in the lower vertebrates is far from being fully understood but it has been shown, quite definitely, by the establishment of conditioned reflexes, that certain fishes, e.g. minnows, can react to a very wide range of sound and can distinguish between notes separated by less than an octave in pitch. In the minnow and other carp-like fishes the sensitivity of the ear is increased by the Weberian ossicles, which act in a way analogous to the auditory ossicles of higher animals, for they convey vibrations, set up by sound-waves from a vibrating structure, the air-bladder, to the perilymph. In other types of fishes, where the ossicles are absent, vibrations presumably reach the perilymph through the wall of the auditory capsule and the tissue overlying it. The middle ear apparatus provides a mechanism by means of which waves falling on the superficially placed tympanic membrane can be conveyed to the more deeply situated inner ear. The tympanic membrane is set in vibration and the vibrations are transmitted by the auditory ossicle or ossicles to the fenestra ovalis, which in turn sets up vibrations in the perilymph. These then cause vibrations in the elastic wall of the sacculus which are handed on to the maculae by the endolymph.

The physiology of hearing has been most fully investigated in mammals, but our knowledge is incomplete in many respects and no attempt has been made in this account to discriminate between several theories of hearing, each of which finds many supporters. The mode of functioning of the outer and middle ear apparatus is a less controversial topic than is the physiology of the cochlea, but even so, there is a good deal of disagreement about the details. The outer ear, a vestigial organ in man, is probably of use in most mammals for detecting the *direction* of the source of sound. It is almost certainly of little use as a sound collector, for it does not fulfil the condition that, for the efficient reflection of waves, a surface must be large in comparison with the wave-length of the vibrations falling on it. But the dimensions and shape of the external meatus in man cause it to have a resonant frequency which is the same as the frequency to which the ear is most sensitive (about 3,000 cycles per second). Thus at this frequency, the pressure on the tympanic membrane is higher than the pressure at the external ear. So the meatus acts as an amplifier for this frequency. It also serves to protect the eardrum from changes of temperature and humidity, in order that its elastic properties may remain unchanged.

The tympanic membrane is so constructed that it has no inherent natural frequency and responds to forced vibrations only, immediately coming to rest when they cease. That is, it is an aperiodic membrane. Its peculiar properties are partly due to the arrangement of its constituent fibres and partly to the way in which the handle of the malleus is attached to its inner surface. The fibres run both in a circular and in a radial direction in the tympanic membrane and the tension in them varies in different parts of the membrane. The periodicity of the fibres varies, therefore, from place to place and the membrane as a whole is aperiodic. The difference in tension in the fibres is due to the fact that the malleus, which is pulled inwards by the tensor tympani muscle, and so in turn pulls on the membrane which it maintains in the form of a cone, is attached, not centrally, but along nearly a complete radius of the membrane.

The Eustachian tube has as its function the equalisation of the air pressure on both sides of the membrane. It acts as a sort of safety valve, for its lower aperture is made to open into the pharynx by the act of swallowing. The ear ossicles form a system of levers which transmit the vibrations induced by sound falling on the tympanic membrane to the membrane of the fenestra ovalis. They gear down the movements of the tympanic membrane by two-thirds, thus increasing their force by one-and-a-half. These movements are very small indeed. Ordinary conversation causes displacement of the eardrum of man of the same order as the dimensions of a hydrogen molecule ($10^{-8}$ cm ). Sensitivity is further increased by the larger size of the tympanum. When compared with that of the fenestra ovalis in man, the pressure is twenty-two times greater on the fenestra ovalis than on the tympanum. When the fenestra ovalis vibrates it causes alternate increases and decreases in pressure (of the same frequency) in the perilymph of the scala vestibuli and, since the wall of the cochlea is made of a rigid substance (bone) and fluid is incompressible, Reissner's membrane is caused to vibrate up and down. The result of this is that the pressure changes are communicated to the endolymph of the scala media and the basilar membrane vibrates in unison. This, in turn, hands them on to the perilymph of the scala tympani and the yielding fenestra rotunda is alternately bulged out into, and withdrawn from, the tympanic chamber. Thus the action of the fenestra rotunda may be likened to a pressure relief valve and it is clear that any inward movement of the fenestra ovalis is accompanied by a corresponding outward movement of the fenestra rotunda to which the pressure change is transmitted in turn through the perilymph of the scala vestibuli, the endolymph of the scala media, and the perilymph of the scala tympani. It seems that the helicotrema is so minute that, although forming a through channel from the scala vestibuli to the scala tympani, it is unable to transmit the rapid changes in pressure set up by vibrations of audible frequency.

When vibrating, the basilar membrane probably executes a sort of rocking motion, its attachment nearest to the bony axis of the cochlea remaining relatively stationary. This type of movement causes a rhythmical distortion of the "sensory hairs" of the organ of the Corti, whose tips are imbedded in the tectorial membrane which always remains practically motionless. These distortions of the "sensory hairs" are thought to be responsible for the initiation of nerve impulses in the component fibres of the branch of the auditory nerve which supplies the cochlea. The brain interprets the impulses arriving from the auditory nerve as sound to which it attributes pitch (an aspect of frequency), loudness (to amplitude), and quality or timbre (to the nature of the wave form).

The discrimination of pitch is thought to take place in the cochlea itself and to be made possible by the peculiar nature of the basilar membrane. A note of any given frequency, although setting a considerable length of the basilar membrane in motion, calls forth a *maximal* vibration only from a particular region.

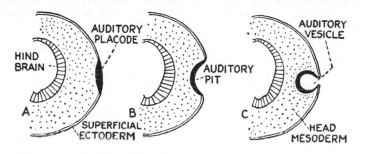

Similarly, the "hair cells" of this region alone are adequately stimulated and nerve impulses are initiated only in the corresponding nerve-fibres. This localisation of maximal response is thought to provide the physical basis by means of which the pitch of a note is "appreciated" by the brain but this so-called "place theory" cannot explain a number of the phenomena of hearing, for example, man's ability to discriminate changes of less than 1 per cent. in the frequency of sounds around 1,000 cycles per second; localisation on the basilar membrane is not sharp enough for this.

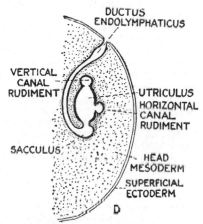

Fig. 519. Diagrammatic transverse sections through the embryonic hindbrain region to show the development of the membranous labyrinth.

**The Development of the Ear.**—In most vertebrate animals the development of the ear follows the scheme outlined previously, and the first sign of the rudiment of the membranous labyrinth is that a patch of ectoderm on each side of the head, in the region of the hind-brain, thickens to form the auditory placode. The placode becomes depressed below the level of the rest of the ectoderm, forming at first a saucer-like area and later a pit. A continuation of this process of invagination and a growing together of the rim of the pit leads to the formation of a vesicle, the auditory vesicle, embedded in the mesoderm of the head and connected to the exterior by a narrow tube. In the dogfish and certain other types, this tube, which marks the primary connection of the auditory vesicle with the outside world, remains as the ductus endolymphaticus. But in other animals the ductus endolymphaticus arises as an outgrowth from the auditory vesicle after it has become completely separated from the superficial ectoderm. This is the state of affairs in the frog where, from the first, development follows a slightly different course from that described above, the ear rudiment never occupying a position on the surface of the ectoderm, but forming entirely from the deeper-lying ectodermal cells. Because of this the auditory vesicle never at any stage opens to the exterior, but from the beginning is a closed sac.

The auditory vesicle then becomes partially divided by a constriction into an upper chamber, the utriculus, and a lower one, the sacculus, the latter receiving the ductus endolymphaticus. From the utriculus, one by one, the three semi-circular canals are separated off. Each canal rudiment is at first delimited by two parallel grooves which deepen and then grow towards each other. Eventually they meet and fuse, so forming a septum which separates off a tubular portion from the rest of the utricular cavity. This tube, which remains in connection with the cavity of the utriculus at each end, then grows and, as it does so, bends outwards and becomes a semi-circular canal.

The lagena of the fishes, amphibia, and reptiles, or its homologue the **cochlea** of birds and mammals, arises as a bud or diverticulum from the ventral wall of the sacculus.

The middle ear and the Eustachian tube of the tetrapods (four-footed creatures) is a derivative of the hyoidean visceral pouch, which corresponds to the spiracular pouch of the dogfish, whilst the columella auris or stapes is a direct modification of the upper element (hyomandibular) of the embryonic hyoid arch. Thus, with an evolutionary change from an aquatic to a terrestrial life, the course of development of the hyoidean pouch has been switched over so that, instead of forming part of the respiratory apparatus, it takes on the new role of a sound-conducting chamber. The modification of the hyomandibula to form an auditory ossicle is made possible because, unlike that of most fishes, it plays no part in the suspension of the upper jaw which, in all tetrapods, is firmly fused to the cranium.

The way in which the middle ear and Eustachian tube are derived from the hyoidean pouch is subject to variations in detail in different types of animals, but the general outline of this process is that the external opening of the pouch (if present) becomes closed and a diverticulum grows out dorsally and fuses with the skin on the side of the head to form the tympanic membrane. The diverticulum is, of course, the tympanic chamber, and the remainder of the visceral pouch, retaining its connection with the pharynx, undergoes slight modification to form the Eustachian tube.

In the mammals the two additional auditory ossicles, the malleus and the incus, are formed from bones which are homologous with the articular and quadrate bones of reptiles, whilst the ring-like bone which supports the tympanic membrane—the tympanic bone—represents the modified angular bone found in most lower vertebrates. The gradual modification of these bones, originally part of the jaw apparatus, receives full confirmation from the fossil history of the mammal-like reptiles.

Of the outer ear or pinna it may be said that it is formed chiefly of skin tissues supported by elastic cartilage, whilst the auditory capsule, at first laid down as cartilage and later replaced by bone, is formed by the skeleto-genous tissue derived from the head mesoderm. The mesoderm is also responsible for the formation of the connective tissue which invests and strengthens the membranous labyrinth.

## THE RENAL AND REPRODUCTIVE SYSTEMS

### INTRODUCTION

It is usual to consider these two systems together, for, although each is concerned with a different activity of the animal and during development arises separately, in the adult animal their ducts become intimately related.

The organs of nitrogenous excretion in the vertebrates are true kidneys, usually fairly compact bodies made up of a large number of tubules called renal or uriniferous tubules and lying above the dorsal part of the coelom. Primitively, each tubule consists of a ciliated peritoneal funnel opening from the splanchnocoel by a coelo-mostome, a Malpighian body, and a convoluted ciliated tube which ultimately opens into a longitudinal collecting duct communicating with the exterior. Each Malpighian body is composed of a chamber—Bowman's capsule—and a knot of capillary blood vessels—the glomerulus. The tubules arise in pairs segmentally, one pair for each segment, though the number per segment may be increased as development proceeds.

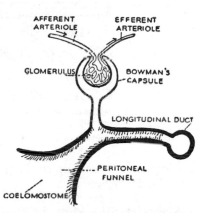

Fig. 520. An embryonic renal tubule.

In the adult, the renal tubules become elongated and coiled, and their segmental arrangement is lost. Eventually they become aggregated and surrounded by a connective tissue capsule to form the kidney.

**Origin of the Kidney Tubules.**—Each tubule is derived from a special part of the mesoderm lying between the somite and the lateral-plate mesoderm. This is called the intermediate cell mass or nephrotome, the cavity within which is the nephrocoel. During development the tubules grow out away from the median line and their ends unite to form a longitudinal duct which extends, on each side of the body, backwards to the embryonic cloaca.

Although, in a wide view, the complete series of tubules may be regarded as constituting a single kidney, in modern vertebrates it does not develop at once, but in three stages or phases which are termed the **pronephros, mesonephros**, and **metanephros** respectively.

The pronephros is the first to appear and lies at the front end of the splanchnocoel, the mesonephros and metanephros following in sequence behind it. Between the pronephros and mesonephros there is little difference developmentally, and it is doubtful if a pronephros can be distinguished except in larval anamniotes and a few adult teleosts, but in the metanephros the tubules never develop peritoneal funnels and its duct arises differently. Also, the segmental origin often obvious in the pronephros and mesonephros is not evident in the metanephros, its tubules arising from

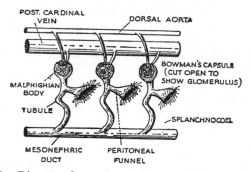

Fig. 521. Diagram of a portion of the mesonephros from the side.

nephrogenous tissue formed from the amalgamated nephrotomes of the hinder region.

**Kidney Ducts.**—When the pronephric tubules arise, the longitudinal duct into which they drain continues to grow backwards on each side of the body outside the somatic mesoderm to the cloaca. These pronephric tubules rarely develop independent Malpighian bodies, but the nephrocoels amalgamate to form a nephrocoelar chamber into which projects a knot of capillary blood vessels, the glomus. Consequently, separate peritoneal funnels are not developed, and the tubules open out from the splanchnocoel by the common nephrocoelar chamber.

As development proceeds and the mesonephric tubules arise, they acquire openings into the original longitudinal duct and the mesonephros retains this hinder part of the duct right back to the cloaca as its functional duct. This portion of the longitudinal duct thus becomes the mesonephric duct, or, as it is frequently called, the Wolffian duct, since the mesonephros is sometimes called the Wolffian body, whilst that part of the duct into which the pronephros opened is distinguished as the pronephric duct. Although the mesonephric tubules arise segmentally, the number of tubules in the mesonephros is usually increased by the development of supplementary ones.

Where a metanephros is developed (as in the Amniota) the tubules arise from the nephrogenous tissue posterior to the mesonephros and their ends grow outwards from the median line. At the same time the metanephric duct (ureter) appears as a forwardly projecting outgrowth from the dorsal side of the posterior end of the mesonephric duct and eventually the metanephric tubules acquire openings into it.

Although the kidney of vertebrates arises in these three phases, they are never all functional at the same time. In the Anamnia the pronephros is the functional kidney of the embryo and early larva and is superseded in the adult by the mesonephros. In the Amniota, the pronephros is rarely functional and usually only a few imperfectly developed tubules indicate its presence. The longitudinal duct, however, is formed in the usual way. The mesonephros functions during embryonic existence, but, as has been indicated above, the adult kidney is a metanephros. It is in the fate of these kidney ducts that the intimate relations between the renal and reproductive systems become evident.

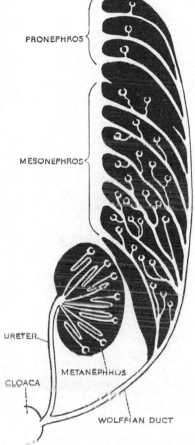

Fig. 522. A plan of the pronephros, mesonephros, and metanephros.

**Gonads.**—The essential reproductive organ, the gonad—**ovary** or **testis** as the case may be—is developed from a specialised tract of coelomic epithelium situated, on each side, nearer to the median line than the nephrotome and forming the genital ridge. This ridge may extend along almost the whole length of the splanchnocoel in lower vertebrates, but further development may be restricted to a small region in higher members of the group. It is in accordance with the type of gonad produced—whether ovary or testis—that the form of the relationship to the kidney ducts is determined so that the relationship differs in the two sexes.

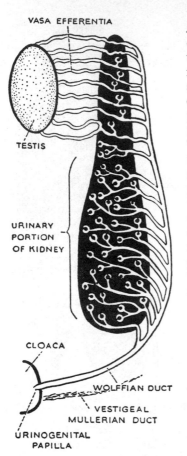

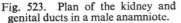

Fig. 523. Plan of the kidney and genital ducts in a male anamniote.

**Male Genital Ducts.**—In the Anamnia the adult kidney is frequently differentiated into an anterior genital portion and a posterior renal portion. In the genital portion some of the tubules lose their excretory function and become continuous with the seminiferous tubules of the adjacent testis. There are thus formed slender tubules, the vasa efferentia, connecting the testis with the genital portion of the kidney. Along these vasa efferentia the spermatozoa pass on their way to the exterior by way of the anterior renal tubules and the mesonephric duct, which thus becomes a vas deferens as well as a urinary duct. In the male anamniote, therefore, the mesonephric or Wolffian duct is a *urinogenital* duct conveying both urine and spermatozoa to the exterior.

In the Amniota, however, where the excretory function in the adult is taken over by the metanephros, the mesonephric duct is completely appropriated by the testis and becomes solely a genital duct, the vas deferens, a coiled portion of the duct forming the epididymis of the testis, while the renal portion of the mesonephros degenerates.

**Female Genital Ducts.**—The sequence of developmental events in the female is rather different. Since the eggs are discharged directly into the splanchnocoel it is essential that the oviduct also should open out from it to the exterior. Although the origin of the oviducts differs in different vertebrates, the end result is the same. In the region of the pronephros a coelomic funnel is developed by the folding over of the sides of a groove formed in the epithelium alongside the nephrotomal region. This groove—except at its anterior end—becomes closed to form a tube which extends backwards on the outside of the mesonephric duct to the cloaca, thus attaining, in addition to its internal coelomic opening, an exit to the exterior. This tube is termed the Müllerian duct and becomes the functional oviduct of the female. A similar duct appears in the male, but only vestiges of it remain in the adult. In the selachian fishes the Müllerian duct is formed by a splitting of the longitudinal duct.

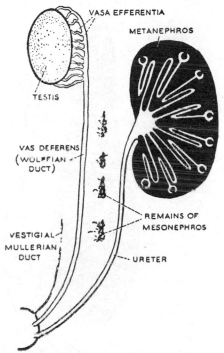

In the female anamniote, therefore, both Müllerian and Wolffian ducts are present, the former acting as the genital (ovi-) duct and the latter as a urinary duct.

In the female amniote similar developments occur, but when the metanephros assumes its adult excretory activities the mesonephros and its ducts are no longer required to discharge their urinary function and they degenerate, being represented in the adult merely by a vestige known as the parovarium. In viviparous

Fig. 524. Plan of the kidney and genital ducts in a male amniote.

amniotes, i.e. the mammals, a portion of the oviduct becomes modified to form a uterus in which the embryos develop.

## THE DOGFISH

**Male.**—It is perhaps rather unfortunate that, in the male dogfish (Fig. 527), the basic plan outlined above is somewhat obscured by specialisation so that it does not afford as simple an illustration as could be desired for a first example. The mesonephric kidney is very fully differentiated into genital and renal portions. The functional kidneys lie, one on each side, close to one another in the hinder end of the coelom and covered by toughish peritoneum. In the genital region, which extends right forwards to the front end of the body cavity beneath the peritoneum, excretory tubules are absent. The mesonephric duct, however, is well developed and is thrown into an intricate series of coils overlying the genital portion of the kidney and forming the vas deferens on each side. Opening into the anterior end of the vas deferens are found the delicate vasa efferentia which run in the mesorchium (folds of peritoneum suspending the testis in the body cavity) to the front end of the adjacent testis. The testes are elongated bodies extending more than halfway backwards from the front end of the coelom and covered by peritoneum.

It is at the posterior that the end ducts become most specialised and complicated. Each vas deferens, as it approaches the hinder end of the body, becomes swollen to form a vesicula seminalis and the two vesiculae seminales open into a common urinogenital sinus, which in turn opens into the cloaca by an aperture situated

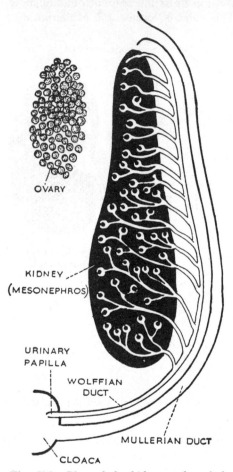

OVARY

KIDNEY
(MESONEPHROS)

URINARY
PAPILLA

WOLFFIAN
DUCT

MULLERIAN DUCT

CLOACA

Fig. 525. Plan of the kidney and genital ducts in a female anamniote.

at the end of a short prominence, the urinogenital papilla. From the urinogenital sinus open out two elongated blindly ending sacs which are termed the sperm sacs, but whose function is unknown. Practically the whole of the Wolffian duct has been appropriated for genital purposes and it will therefore not be surprising to find that special collecting ducts—usually termed "ureters"—have been developed in connection with the functional kidneys. From each kidney arise five collecting ducts into which the renal tubules open and the ducts join to form a large "baggy" single channel, the "ureter", which opens into the urinogenital sinus, the aperture of the "ureter" of each side lying closely alongside that of the vesicula seminalis. As would be expected, the Müllerian ducts are not functional, but vestiges of them are found in the anterior part of the coelom on the ventral side of the oesophagus.

The arrangement of the ducts is very specialised, yet, basically, the renal and reproductive systems comply with the plan outlined for the Anamnia since the kidney is a mesonephros and the mesonephric duct, where it forms the urinogenital sinus, is ultimately a urinogenital duct.

In the dogfish, fertilisation of the eggs is internal and the spermatozoa are brought into the cloaca of the female by grooved elongations of the pelvic fins, termed claspers, and the so-called siphon. The siphon is a muscular sac lying beneath the skin of the ventral surface in the pelvic region and has two

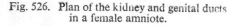

Fig. 526. Plan of the kidney and genital ducts in a female amniote.

channels emerging from it, each of which opens into the groove of the clasper of its side. The use of this siphon is obscure, though to it has been ascribed the function, by squirting out the contained sea water, of flushing spermatozoa accumulated in the claspers into the female cloaca. The spermatozoa ascend the oviduct and fertilise the egg in its upper part.

**Female.**—In the female of the dogfish (Fig. 529) the arrangement is not quite so specialised. The kidney shows the same differentiation into an anterior and posterior portion, and only the latter is excretory. The mesonephric (Wolffian) ducts commence in the anterior part and continue backwards where, in the region of the functional kidney, each duct dilates and joins its fellow of the opposite side to form a urinary sinus. This sinus opens into the cloaca at the tip of the urinary papilla, and into it open *by separate*

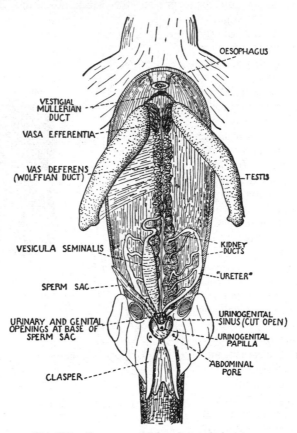

Fig. 527. DOGFISH.—Male urinogenital system.

*apertures* a series of collecting urinary ducts from the functional part of the kidney.

Only a single ovary is present in the mature fish and, although it is apparently median in position, suspended in the splanchnocoel by the mesovarium, it is actually the right ovary, the left not having been developed. The Müllerian ducts are large and constitute the oviducts. Each oviduct is a stout tube and the internal coelomic openings of the two have coalesced to form a single wide aperture situated on the under side of the oesophagus at the anterior end of the body cavity. From this common aperture each oviduct narrows slightly and, lying beneath the peritoneum, passes backwards at the sides of the coelom to the cloaca. At about one-third of the way along from its anterior end each oviduct swells out, owing to a thickening of its wall, to form an oviducal gland, which is responsible for the secretion of the egg-case or shell within which each egg is enclosed. The walls of the oviducts also contain glands which secrete albumen with which the egg is coated before being enclosed within the egg-case. The oviducts open into the cloaca by separate apertures, situated one on each side of the urinary papilla. In the immature female these openings are closed by sheets of tissue continuous with the roof of the cloaca and these form the hymen.

. If the ovary of a mature female is examined, in it will be seen eggs in various stages of development and varying in size from that of small peas to large marbles, the increase in size being due to the accumulation of yolk within the egg. When fully developed they are released from the ovary and fall into the cavity between the viscera. They are then transported, by the action of the cilia borne on definite tracts of peritoneal cells, to the anterior end of the

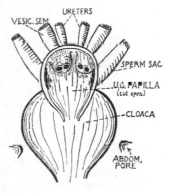
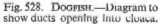

Fig. 528. DOGFISH.—Diagram to show ducts opening into cloaca.

body cavity where they enter the large oviducal aperture. During its passage down the oviduct each egg is coated with albumen and also enclosed within an elongated egg-case of a resistant material resembling hardened chitin or horn. The egg-case is roughly rectangular in shape and extending from the four corners are long, coiled, filament-like structures which are very elastic. When the female is depositing these egg-cases—which are familiarly known as mermaid's purses—she swims in and out among seaweed or other objects, and as the egg-case emerges from the cloacal aperture the elastic coiled threads at the corners become entangled in the seaweed, etc., so that the case is securely anchored and remains so until the young fish emerges. Thus, although the dogfish does not produce eggs in such large numbers as other fishes, the provision of much reserve food material in the form of yolk in the egg and the protection afforded by the egg-case, ensure that the young animal enters the world in an advanced stage of development with a much greater chance of survival.

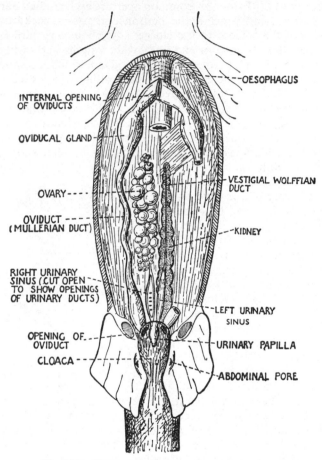

Fig. 529. DOGFISH.—Female urinogenital system.

## THE FROG

**Male.**—The renal and reproductive systems in the frog (Fig. 530) present fewer specialisations than were found in the male dogfish. The mesonephric kidneys are elongated, ovoid, compact bodies, and from the outer side of each arises a (Wolffian) duct which passes directly back to the cloaca, into which it opens at the tip of a slender urinogenital papilla. The testes lie adjacent to the kidneys and from each testis a number of delicate vasa efferentia pass to the tubules of the anterior part of the corresponding kidney. About two-thirds along its length, in the mature male, the duct dilates into a vesicula seminalis. Thus the mesonephric (Wolffian) duct, although it is commonly called a ureter, is a urinogenital duct conveying both urine and spermatozoa to the exterior. In the frog, however, the kidney, although physiologically composed of genital

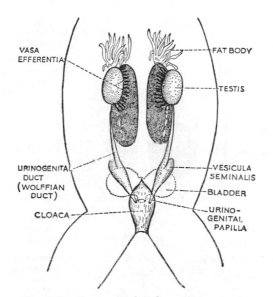

VASA EFFERENTIA

FAT BODY

TESTIS

URINOGENITAL DUCT (WOLFFIAN DUCT)

VESICULA SEMINALIS

BLADDER

CLOACA

URINO-GENITAL PAPILLA

Fig. 530.   FROG.—Male urinogenital system.

and renal portions, does not from the outside show any morphological differentiation.

Arising from the floor of the cloaca and extending forwards into the body cavity is a large thin-walled sac, the bladder. From its mode of origin it is sometimes called an allantoic bladder (for the development of an allantois, see p. 774) and to it, in the frog, is ascribed the function of urine storage. In other more definitely terrestrial amphibians, such as the toad, there is good evidence that the watery fluid contained in it serves as a reserve to prevent desiccation, and whatever urinary function it may serve in the frog there is no reason why it should not also serve on occasion, as a reservoir for water. Its walls are provided with bands of unstriped muscle and the entrance into the cloaca is closed by a sphincter. At times, as when the animal is roughly handled, the contents of the bladder may be suddenly discharged. The Müllerian ducts become solid and vestigial.

**Female.**—The arrangement of the kidneys and their (Wolffian) ducts in the female frog (Fig. 531) is much the same as in the male except that no vesiculae seminales are present and there is, of course, no connection between the kidneys and gonads. The ovaries are

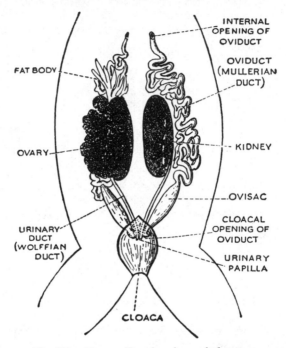

Fig. 531. FROG.—Female urinogenital system.

large and in the mature female occupy a considerable portion of the coelom. The thin-walled internal openings of the oviducts are situated at the front end of the splanchnocoel near the base of the lung, but the greater part of the long, coiled oviducts is thick-walled and glandular. At the hinder end of the body, each oviduct dilates into a thin-walled region, the ovisac, then narrows again and opens into the cloaca on the expanded base of the urinary papilla of its side. The oviducts are formed from the Müllerian ducts.

The eggs of the frog are small, pigmented spheres about 1·6 mm in diameter and contain a small amount of yolk. In the mature female the large size of the ovaries is an indication of the large number of eggs produced. As the eggs ripen, they are released into the splanchnocoel, pass forwards by ciliary action to the front end of the body cavity, and enter the oviducts by their coelomic funnels. During the passage of the eggs down the oviducts they are coated with albumen and accumulate in the dilatable egg-sacs until the time comes for deposition. Fertilisation takes place outside the body. During the breeding season, when the frogs always return to the water, the male mounts the female and clasps her behind her fore limbs, its grasp being facilitated by the nuptial pads developed on the metacarpal region of the first digits of his fore limbs. The pair remain in this position for some time, and may swim about together in the water. As the eggs are discharged from the cloacal aperture of the female, the male pours out his seminal fluid on them and fertilisation ensues at once. When the eggs come

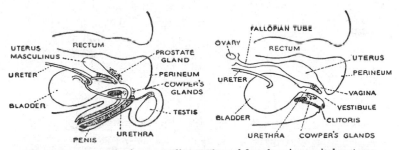

Fig. 532.   Generalised mammalian male and female urinogenital systems.

into contact with the water, the coating of albumen swells up and the eggs adhere together in masses, forming the familiar frog spawn.

## THE MAMMAL

**Male.**  In the rabbit the metanephric kidneys are asymmetrically placed in the body cavity, the right kidney being anterior to the left. Each kidney is a bean-shaped body, and from the indentation on its inner side issues the metanephric duct or true ureter. The two ureters do not open, in the adult, into a cloaca* as they do in both the frog and dogfish, but into a urinary bladder, although such a cloaca does exist in the embryo. The urinary bladder is derived

---

* The rabbit belongs to the eutherian mammals in which, together with the marsupials (forming the **Ditremata**), the anal aperture is separated from the renal and reproductive aperture by the **perineum**, and the common cloaca receiving both rectum and urinogenital ducts is lost. In the lowest group of mammals, the **Monotremata**, a cloaca with (as the name implies) a single aperture to the exterior is present.

from the stalk of the allantois (see p. 774) and it communicates with the exterior by a channel, the urethra, which, for reasons which will be apparent later, is long. In the adult, each testis lies in a sac-like outpushing of the body wall termed the scrotum. Each testis develops in the normal way in the coelom of the abdomen, but, as the animal matures, migrates through a channel—the inguinal canal—connecting the body cavity on each side with the scrotal sac. Attaching each testis to the scrotal wall is a band of tissue, the

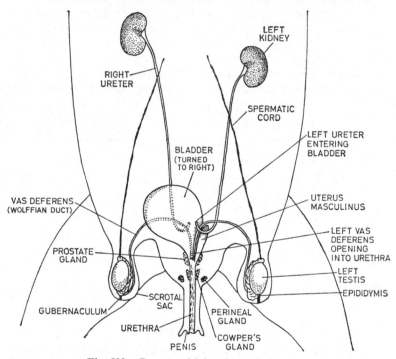

Fig. 533. RABBIT.—Male urinogenital system.

gubernaculum, by the contraction of which the withdrawal* of the testis from the abdomen into the scrotum is effected. The epididymis shows two enlargements, the caput epididymis at the anterior (abdominal) end of the testis and the cauda epididymis at the opposite end. From the cauda epididymis emerges the vas deferens (mesonephric duct) which passes forwards into the inguinal canal—

* In some mammals, particularly some rodents, this descent of the testes takes place periodically at each breeding season, the testes returning to their abdominal position on the cessation of sexual activity.

where it lies alongside the spermatic cord formed by the spermatic artery, vein, and nerve—and thence into the abdominal cavity. Here each vas deferens curls around the end of the ureter of its side and passes on to the dorsal side of the bladder to open ultimately into the urethra. At the place where the vas deferentia and urethra join is a blindly ending sac, the uterus masculinus, which is the sole vestige in the male of the Müllerian ducts.

In the male, because it has to subserve a urinogenital function, the urethra is long and traverses an erectile intromittent organ, the penis. At ordinary times, the urethra acts as a channel for conveying the urine from the bladder to the exterior. During sexual

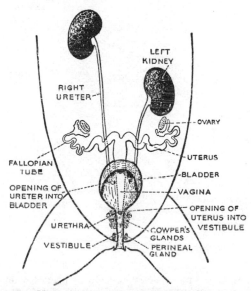

Fig. 534. RABBIT.—Female urinogenital system.

activity, however, the erect penis is introduced into the female reproductive tract to deposit therein the spermatozoa. Surrounding the urethra, therefore, in the body of the penis, are three tracts of erectile tissue, which, previous to copulation, become gorged with blood and consequently turgid. These tracts are the paired corpora cavernosa, anteriorly, and the corpus spongiosum, through which the urethra passes, posteriorly. Normally the tip of the penis— the glans penis—is retracted within a sheath of skin, the prepuce.

**Female.**—The arrangement of the kidneys and their ducts is the same as in the male with the exception that in this case the urethra is short and is solely urinary in function.    The ovaries are small ovoid bodies and lie one on each side of the dorsal side of the abdomen.    The oviducts have their internal openings, which are funnel-like, closely applied to the surface of the ovaries so that the eggs—which are very minute—will pass directly into them when they are discharged.    Each oviduct, the upper part of which is termed a Fallopian tube, passes backwards and inwards towards the median line, where it joins its fellow of the opposite side to form a median channel, the vagina.    Before this junction occurs, however, each Fallopian tube dilates to form a uterus within which

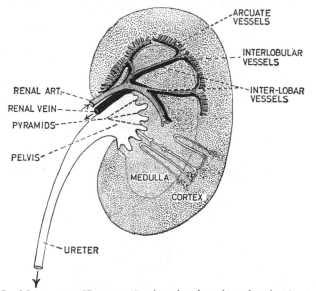

Fig. 535.    MAMMALIAN KIDNEY.—Section showing plan of main blood vessels (upper half) and nephrons (lower half)—parts not to scale. Note the different lengths of the loops of Henle.

the embryos develop.    The vagina is joined by the urethra and is continued as a common urinogenital chamber, the vestibule, which opens out to the exterior by the slit-like vulva.    At the entrance to the vestibule is a small rod-like body, the clitoris, projecting into it. The clitoris is erectile and is the homologue of the penis, but the urethra does not pass through it.

## HISTOLOGY AND PHYSIOLOGY OF KIDNEYS

**Kidney.**—The glandular substance of the kidney (which is enclosed in a connective-tissue capsule) is composed chiefly of a complex system of uriniferous tubules. These small tubules, which receive the urine as it is formed, join collecting tubules, which in turn unite and open into the pelvis of the kidney. The main

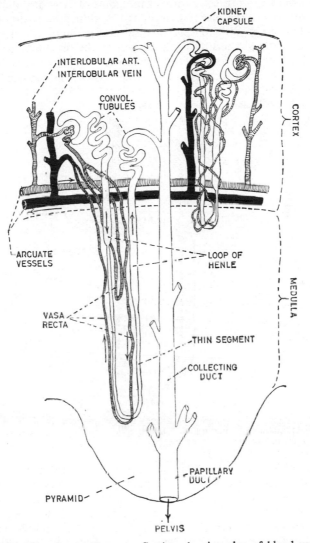

Fig. 536. MAMMALIAN KIDNEY.—Section showing plan of blood vessels and nephrons. Arrows show direction of flow in vessels and tubules. Note the two kinds of nephrons and the differing pattern of blood vessels around their Henle loops. In reality the structure is much more complicated than is suggested by the diagram.

collecting tubules or ducts are bound together to form one or more pyramids. The blood-supply is by the renal artery, and blood leaves the kidney by the renal vein. The anatomical relations of these parts are shown in Fig. 536. From the point of view of the formation of the urine, the essential unit of the kidney is a uriniferous tubule (Fig. 537). Blood is conveyed under high pressure (see Fig. 536) to the glomerulus by a branch of the renal artery. The blood in the glomerular vessels is separated from the cavity of the capsule only by the endothelial cells and those of the capsular wall.

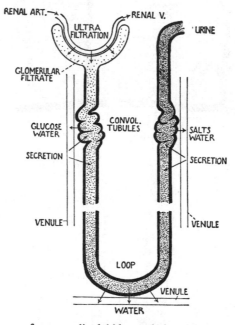

Fig. 537. Diagram of a generalised kidney tubule to show its mode of functioning. Differences in shading are an indication of changing concentration of the contained fluid.

These layers together form a membrane which is permeable to small molecules. The hydrostatic pressure of the blood exceeds the colloid osmotic pressure due to the blood proteins, and so water and all small molecules in true solution in the blood-plasma pass through the capsular membrane into the lumen of the tubule. This fluid is called the glomerular filtrate and is thus identical with blood-plasma minus its proteins, as has been shown by analysis of the glomerular filtrates of amphibian kidneys, some of which have sufficiently large capsules for the insertion of micro-pipettes. From this fluid, substances "useful" to the body, *e.g.* glucose and salts, are selectively resorbed, chiefly through the convoluted tubules, whilst water is resorbed also through the loop of the tubule. In this way the composition of the fluid becomes altered as it passes along and it also becomes more concentrated. More constituents are added to it by active secretion into the tubule, and by the time it reaches the collecting tubules it can be termed urine. This

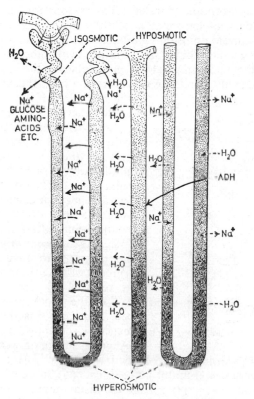

Fig. 538. Counter-current system in kidney.

account applies chiefly to the mammalian kidney. In other types, *e.g.* certain fishes, active secretion is the chief mechanism of urine formation.

In most types of vertebrates a uriniferous tubule consists of two functionally distinct parts—a glomerulus-plus-Bowman's-capsule, energised by the heart, since it functions by filtration under pressure; and convoluted tubules and loops which when relatively long as in birds and mammals, are called loops of Henle which actively resorb and secrete, both of which require the local expenditure of energy.

The close approximation of the two limbs of the loop of Henle and of the collecting tubules and the distribution of the blood vessels are strongly suggestive of a counter-current system (such as operates in the gills, p. 553). Modern work suggests that such a mechanism is involved in modifying the urine as it passes along the tube by the resorption of water. The reasons for this assumption reside not only in interpretation of the hair pin-like loops of Henle as counter-current diffusion multipliers but in the discovery that the electrolytes in the tubules are isosmotic with those in the fluid of the medullary interstitium at any given place in the kidney but that there is a steadily increasing concentration from the cortex towards the medulla. How such a system could function is shown in Fig. 538.

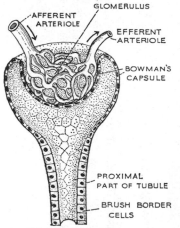

Fig. 539. Sagittal half of a Malpighian body.

Just before they enter the glomerulus the afferent arterioles come close to the distal tubules; at this point the cells of both arteriole and tube are thickened to form the **juxta-glomerular complex**. This can in some way detect a change in sodium concentration in the blood. When the concentration falls (due, for example, to an increase in the amount of water in the blood) **renin** is released by this complex. It causes the production of **angiotensin** from the inactive **angiotensinogen** circulating in the blood. Angiotensin not only controls blood pressure (by causing vaso-constriction) but also affects the production of **aldosterone** from the adrenals (p. 665), more of which causes more sodium reabsorption from the distal tubule. This continues until the level of sodium rises to its normal level when renin ceases to be produced. Thus, these inter-connecting control mechanisms are a good example of negative feed-back control. They work just as a governor on an engine does, as the product of the process increases (in this case, increased sodium reabsorption) so the process is reduced in vigour.

This mechanism interplays with control of water excreted by the **anti-diuretic hormone** (ADH) from the neurohypophysis of the pituitary. An increase in the total osmotic pressure of the plasma means that too much water is being excreted and causes ADH to be produced. In this way the amount of water excreted is cut down (p. 660). The water content of the plasma is, of course, also reflected in the concentration of sodium.

## HISTOLOGY OF GONADS

Testis.—The male gonad consists essentially of a mass of coiled seminiferous tubules between which are small amounts of connective tissues (fibres, blood vessels, etc.) and certain peculiar cells,

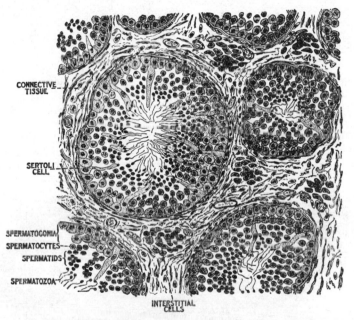

Fig. 540.   Transverse section of the testis of a mammal.

the interstitial cells, which secrete the hormones influencing the secondary sexual characters. The spermatozoa are derived by maturation divisions from the germinal epithelium lining the seminiferous tubules, and a section of testis shows developing sperms in all stages of maturation (see p. 675). In between the germinal cells certain larger cells (Sertoli cells), which have the role of supplying nourishment to the developing sperms, may usually be seen. The testis is enclosed in a fibrous capsule, the tunica albuginea.

**Ovary.**—The ovary consists of a mass of connective tissue (fibrous) and of spindle-shaped cells, the two together constituting the stroma. Lying in the stroma are egg-cells in various stages of

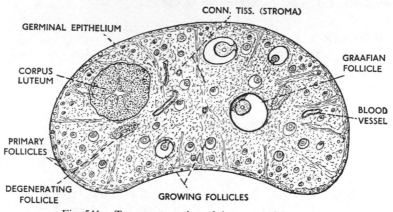

Fig. 541.    Transverse section of the ovary of a mammal.

maturation, each surrounded by a nourishing epithelial layer, the follicle. The size and appearance of the follicles depends on the stage of maturation they have reached after their detachment from

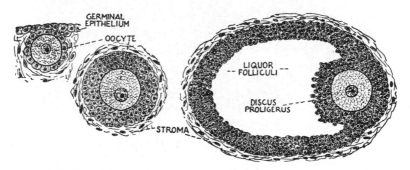

Fig. 542.    Three stages in the development of Graafian follicles.

the layer of germinal epithelium which forms the outer coat of the ovary.

The primary follicles arise from ingrowths into the stroma of the germinal epithelium. One of the group of cells enlarges to form the developing oocyte, whilst the others form a single layer of cells, the follicle, around it. The follicle and the oocyte slowly move deeper into the stroma and become larger. Later, the oocyte and cells around it become separated by fluid-filled spaces from the rest of the follicle cells (which become several layers thick) except at one point.

Further enlargement of the oocyte and follicle as a whole, results in the production of a mature follicle (**Graafian follicle**). The Graafian follicle then migrates to the surface of the ovary and ruptures, releasing the oocyte (which may have undergone one of the maturation divisions and so become a secondary oocyte) into the Fallopian tube, but some follicles fail to complete their maturation and degenerate. After discharge of the oocyte, the follicle cells undergo proliferation and change in structure to form a yellow body (**corpus luteum**) which secretes hormones, influencing certain stages in the oestrous cycle. In the human being, if the ovum is not fertilised and implanted in the uterus, the corpus luteum undergoes gradual absorption with the onset of the next menstruation, but in the event of pregnancy taking place, the corpus luteum enlarges still further and the hormones that it secretes cause the mammary glands to enlarge and produce milk. Its disappearance takes place only after birth of the child. Other hormones, one of whose functions is the control of the female secondary sexual characters, are secreted by the ovary as a whole (p. 668).

# THE ENDOCRINE SYSTEM—THE DUCTLESS GLANDS

## INTRODUCTION

In various situations in the vertebrate body there are found groups of specialised secretory cells which, although often aggregated to form quite large obvious glands, yet have no duct. They are therefore termed endocrine organs or ductless glands, for, unlike other types of glands (exocrine glands) they pour their secretions, not into a duct, but directly by diffusion, into the blood vessels which supply them. For the same reason they are also known as internally secreting glands. The secretions are variously called internal secretions or hormones. Chemically they fall into two main groups, steroids and amino acid derivatives. Some hormones have now been synthesised in the laboratory.

The secretions of the endocrine organs play an important part in the regulation of the functions of the body, and some of them act directly on the effector organs such as muscles and glands in a way that is analogous to the action of nerve impulses. Other hormones are concerned, not with calling forth a response from a particular effector organ or system, but with the regulation of metabolic activities and with the control of the growth of the animal. Some endocrine organs are themselves stimulated to activity by hormones from other organs, but others are under the control of the nervous system and are thus enabled to respond very rapidly in an appropriate way to changes in the external and internal environment.

Hormones for some time were thought of as chemical messengers which produced their effects on "target organs" through the bloodstream and thus contrasted with the rapid excitation through the nerves. We now know that the distinction between nervous and hormonal control is by no means a sharp one and that they are perhaps aspects of a single integrative system, the similarities being greater than the differences between the two, both acting in essentially the same way.  Both eventually produce their results by delivery on the target organ of a chemical substance which elicits the appropriate response.  Nervous conduction is, so to speak, a device which allows of a rapid spread of excitation resulting in the release of small amounts of an unstable, rapidly destroyed substance at the terminal twigs of the nerve.  It is this substance which stimulates the tissue to respond as long as a volley of nerve impulses along the nerve promotes its release.  An endocrine organ produces a concentration of hormone in the blood which causes more lasting effects on a target organ.  In a few instances, *e.g.* noradrenalin, the excitatory chemicals released at the nerve endings are practically identical with that formed in an endocrine organ (adrenal medulla).

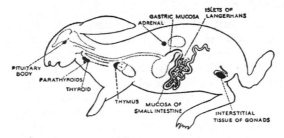

Fig. 543.  Diagram to show the position of the main endocrine organs in a male rabbit.

Moreover, as will be mentioned again, some hormones are produced by neuro-secretory cells lying in the central nervous system—cells which are modified neurons and which provide a means whereby the hormonal balance of the body can be altered *via* the neuro-sensory system by changes in the environment.  For example, increasing day length influences the gonads of birds to enlarge and their germ cells to mature because it causes a release of hormones from the pituitary body; copulation in some mammals (*e.g.* rabbit) stimulates the release of pituitary hormones which cause ovulation—and so on.

The procedure for investigating the action of endocrine organs has been:

(1) to remove the organ and note the effects on the animal;

(2) to graft the organs back into an animal from which they have been removed to see if the previous symptoms are reversed;

(3) to inject extracts of an endocrine organ into experimental animals and to compare the results with normal animals;

(4) the chemical analysis and synthesis of hormones and the study of their effects when injected into the body.

For descriptive purposes it is convenient to deal with the endocrine organs separately. The account which follows relates mainly to those of mammals but those of other vertebrates, although differing in detail, are basically similar in structure, embryonic origin, and in the hormones they yield. Firstly the pituitary body may be considered.

## The Pituitary Body

**Anatomy and Embryonic Development.**—The pituitary body is a fairly large gland lodged in a pit in the base of the skull just below the thalamencephalon immediately behind the optic chiasma, to whose floor (hypothalamus) it is attached by a short stalk. It is a composite organ and arises from two distinct embryonic contributions. Firstly its larger part develops from cells forming the walls of an upward evagination of the embryonic stomodeum towards the embryonic hypothalamus. This evagination is the **hypophysis** and gives rise in the adult to several parts of the **adenohypophysis**.

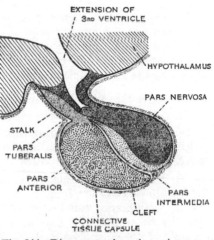

Fig. 544 Diagram to show the main parts of the pituitary body of a mammal.

The second contribution is a downgrowth from the embryonic hypothalamus. This is the infundibulum which forms the **neurohypophysis** of the adult. Thus although the pituitary body appears in most adult vertebrates as a compact gland, yet it arises from two separate sources in the embryo. The hypophysis soon loses its contact with the stomodeum but the infundibulum retains its connection with the hypothalamus and when it differentiates to form the neurohypophysis it remains penetrated by an extension of the third ventricle.

Modern endocrinologists recognise that the adenohypophysis generally has three main parts—the *pars tuberalis* (not present in all vertebrates), a thin layer of cells around the pituitary stalk; the *pars intermedia* and the *pars distalis*, the main bulk of the adeno-hypophysis. The neurohypophysis has two parts only—*the median eminence* (closely linked with the adenohypophysis by way of a common blood-supply) and the *pars nervosa*. The arrangement of the parts in a mammal is shown in Fig. 544.

The two main parts of the pituitary body are quite distinct histologically. The adenohypophysis consists of branching cords of secretory cells between which run blood vessels. The neuro-hypophysis, on the other hand, is composed of neuroglial cells (pituicytes), blood vessels, and the axons of neuro-secretory cells situated in the hypothalamus. It is these neuro-secretory cells which form the various hormones which pass droplet-wise along their axons to accumulate in the neurohypophysis.

**Physiology.**—The adenohypophysis is known to produce seven hormones and the neurohypophysis, two only. These are listed below along with the various parts with which they are related. All of them are composed of chains of amino acids forming molecules of varying size and complexity, most of which are polypeptides.

### Hormones of the Adenohypophysis

*A. Pars distalis.*—

(1) The thyroid-stimulating hormone (TSH), a glycoprotein which influences the thyroid gland to develop in the embryo and to secrete its own hormone, thyroxin.

(2) The follicle stimulating hormone (FSH) which stimulates the growth of Graafian follicles (p. 655) of the mammalian ovary and of seminiferous tubules (p. 653) of the testis. Its secretion tends to be stopped by hormones produced by the gonads, an example of "negative feed-back".

(3) The luteinising hormone (LH) has several effects and is probably identical with the interstitial cells-stimulating hormone (ICSH). It induces the formation of *corpora lutea* (p. 655) and also maintains the interstitial tissue of the gonads in an active condition. It also stimulates the development of some secondary sexual characters. Because of their effects on the gonads, FSH and LH are often spoken of as gonadotropic.

(4) Adrenocorticotropic hormone (ACTH) which stimulates the activity of parts of the cortex of the adrenal body but when its output reaches a certain level a negative feed-back *via* the median eminence causes a decrease in its production.

(5) Lactogenic or mammotropic hormones (LTH or MH). This has several functions, the best known of which is to start milk production. In birds it stimulates the secretion of the crop, also used for feeding the young.

(6) Growth or somatropic hormone (STH), a protein which, as its name implies, stimulates growth and which in excess leads to the formation of giants. A second effect is to stimulate the release of glucagon from the pancreas (p. 666).

*B. Pars intermedia.—*

This produces a single hormone, the melanophore-stimulating hormone (MSH) or intermedin. It has the effect of causing the

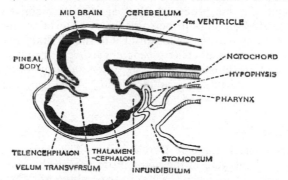

Fig. 545. Longitudinal section through the head region of an embryo dogfish to show the development of the pituitary body.

melanophores of fishes and amphibia to expand and to darken the skin. Over long periods it causes the melanophores to increase in number.

*C. Pars tuberalis.—*

This does not, so far as is known, produce any hormones but has a rich blood-supply from which run vessels to the median eminence of the neurohypophysis thence down the pituitary stalk to the pars distalis, so linking the two main parts of the hypophysis each of which also receives a direct blood-supply from branches of the internal carotids.

### Hormones of the Neurohypophysis

Unlike the adenohypophysis, the neurohypophysis does not secrete, but stores hormones produced by neuro-secretory cells situated in the hypothalamus, the secretions being carried droplet-wise along their axons which end in the neurohypophysis thence to be released into the blood, for which reason the neurohypophysis is said to be a neurohaemal organ, such as occurs in many invertebrates (pp. 671-2). It is generally agreed that the droplets which can be seen to pass along the axons are not the hormones themselves but are proteins to which the active substances are attached.

Hormones released from the neurohypophysis are:

(1) Oxytocin, which is released when the mammary glands are stimulated by suckling and causes ejection of milk. It also causes contraction of the uterine muscles and hence plays an essential part in the birth of the young.

(2) Antidiuretic hormone (ADH) which increases the resorption of water and sodium from the uriniferous tubules and hence decreases the amount of urine produced, thus aiding water conservation in mammals. In amphibia it increases water absorption through the skin. If for any reason, such as injury to the pituitary body, ADH is deficient a disease called *diabetes insipidus* results. This is characterised by intense thirst and excessive formation of urine which can be relieved by administering ADH (in man by inhaling it).

### The Thyroid

**Anatomy and Embryonic Development.**—The thyroid is a relatively large gland lying in the mid-ventral part of the hinder region of the pharynx from which it arises in the embryo as a diverticulum. In many animals it has a bilobed structure, the two lobes being connected across the mid-line by a narrow isthmus. In the mammals, owing to the development of a neck, it occupies a position some way behind the head, its two lobes lying right and left of the trachea. The thyroid is homologous with the endostyle of the lower chordates, as is well shown by its development in the lamprey (*Petromyzon*). In this animal the larva (*Ammocoete* larva) is a ciliary feeder and has a well-formed endostyle, but at metamorphosis the endostyle for the most part atrophies, but certain of its cells proliferate to form the vesicles of the adult thyroid gland. This change is associated with an alteration in the feeding habits, for the adult is a predator and feeds on the flesh of living fishes.

The histology of the thyroid is fairly simple, for it consists of a collection of small vesicles or follicles enclosed in connective tissue. The vesicles are lined by a single layer of cubical epithelial cells and are filled with a homogeneous jelly-like substance—the so-called colloid. The thyroid is very richly supplied with blood vessels which penetrate all through its substance and lend a dark-red coloration to the gland.

**Physiology.**—The thyroid has been spoken of as a gland whose chief characteristic is its ability to accumulate iodine and combine it with an amino acid (tyrosine) to form the main active principle of the thyroid, thyroxin (tetraiodothyronine). Another hormone, triiodothyronine, is formed but in lesser amounts. Many other organs have the property of concentrating iodine. Indeed, this ability is widespread in the animal kingdom and has certainly been demonstrated in such lowly phyla as coelenterata where, however, it is not associated with functional hormones.

The thyroid hormones are attached to a protein (thyroglobulin) which forms the "colloid" inside the follicles of the thyroid. From here they are released into the blood by the action of a special proteolytic enzyme.

The general effects of thyroid hormones on the body may be summarised by stating that they speed up most of its activities. For example, they increase the rate of those metabolic activities which are mainly concerned with the maintenance of the life of the animals at rest (**basal metabolism**, in contrast to that which has to do with external work such as movement). It produces these effects by speeding up the oxidative processes in the energy release cycle (p. 547). The thyroid secretion in the young animal stimulates growth and, if deficient, development is defective, giving rise, in man, to the condition known as **cretinism**. Extirpation of the thyroid from the larvae of Amphibia prevents metamorphosis, but it can be induced by dosing with thyroxine or thyroid extract. Metamorphosis of the intact tadpoles can be hastened by the addition of thyroid or even traces of inorganic iodine to the water. The disease of man termed **myxoedema**, which is characterised by a general sluggishness of metabolism, and by a swelling of the subcutaneous connective tissue, is associated with atrophy of the thyroid gland in later life. The condition is improved by the administration of thyroid extract. Over-activity of the thyroid, usually accompanied by enlargement of the gland, causes symptoms which are the reverse of myxoedema. The basal metabolic rate is excessively high, the heart-beat is rapid, there is wasting of the body, and a general condition of restlessness. The disease is known as **exophthalmic goitre**. Deficiency of iodine in the diet results in an enlargement of the thyroid but not in any very marked disturbances of metabolism. The condition receives the name of **simple** or **hyperplastic goitre** or, since it was at one time common in Derbyshire, of **Derbyshire neck**.

### The Parathyroids

**Anatomy and Embryonic Development.**—The **parathyroid bodies** are two pairs of glandular masses which lie at the sides of the thyroid from which, in some animals, they are difficult to separate. They consist of cords of epithelial cells between which are found numerous blood sinuses. Parathyroids are absent, as such, in fishes, but are possibly represented by patches of cells placed just behind the last branchial cleft. In the Amphibia and in the amniotes, the parathyroids are usually developed from cells of the tissue of the third and fourth visceral clefts.

**Physiology.**—Extirpation of the parathyroids causes tetanus followed shortly by death. The parathyroid hormone (para-thormone) is a protein and its function is, in conjunction with vitamin D, to maintain an adequate concentration of calcium in the blood. Injection of parathormone increases calcium in the blood by promoting absorption of calcium through the wall of the intestine and also by abstracting it from the bones and other sources. When bone is drawn on to restore the level of blood calcium, phosphate is released and the excess is passed out in the urine. It seems likely that the parathyroid secretes a second hormone which acts in the opposite sense by increasing calcium excretion by the kidney, thus lowering the blood calcium level if it gets too high. The "milk fever" experienced by many dairy cattle shortly after calving appears to be due to a sudden drop in the calcium content of the blood. It is as if the parathyroids cannot cope quickly enough with the increased calcium demands of the mammary glands during the onset of milk production.

### The Thymus

**Anatomy and Embryonic Development.**—The thymus is most easily seen in mammals, especially in young animals, where it takes the form of large masses of glandular tissue wrapped around the bases of the great arteries and the anterior end of the heart. In the Amphibia it is represented by a small mass of tissue near the angles of the jaws, whilst in fishes it is recognised as rather diffuse patches of tissue along the dorsal border of each gill slit. In mammals the thymus consists of a network of epithelial cells between which are masses of cells resembling lymphocytes. The organ is richly vascularised and is therefore reddish in colour. The thymus arises in the embryo from cells which proliferate from the epithelium of the visceral clefts (except the hyoidean cleft) and, as mentioned above, it retains its embryonic relations in the adult stage of the fishes.

**Physiology.**—The main function of the thymus is to provide a source of lymphocytes for the young animal. The cells so formed are then distributed to organs such as the spleen, lymphatic nodes, and Peyer's patches (p. 490) where they proliferate. Having, so to speak, seeded such organs, the thymus degenerates as the animal gets older. Thus the thymus is essential for the development of the immunological reactions which play such an important part in the defences of the body against bacterial and viral diseases as is shown by the fact that if the thymus is removed from a new born mouse it cannot produce antibodies against foreign substances. But the thymus does more than this, it produces a hormone which stimulates lymphoid tissues to form lymphocytes.

## The Suprarenal and Adrenal Bodies

**Anatomy and Embryonic Development.**—These two endocrine organs are often considered together because in amniotes, particularly in mammals, they are united to form compact organs, the adrenal bodies (although retaining their separate and distinct functions). The suprarenal bodies of fishes are small masses of cells surrounding the sympathetic ganglia in the trunk. Nowadays they are often referred to as the chromaffin tissue. In selachians (like the dogfish) the chromaffin tissue occur as patches of cells just above the dorsal wall of each cardinal sinus, the hinder ones being embedded in the kidney. Each consists of cells which have wandered out from the neural crests and so have the same embryonic origin as the cells in the sympathetic ganglia.

Interrenal bodies, often nowadays called the adrenocortical tissue, are composed of glandular cells forming masses in the midline between the kidneys. They arise from embryonic coelomic epithelial cells in the region of the mesonephros (p. 633). The adrenal bodies of mammals lie near the kidneys and may actually be situated on their surfaces. In the mammals the equivalent of the chromaffin tissue surrounds the adrenocortical tissue and the two regions of the composite adrenal body so formed are known respectively as the cortex and the medulla. An intermediate condition is seen in creatures such as the frog, where the adrenal bodies, appearing as yellowish patches on the ventral surface of the kidney, consist of anastomosing strands of both kinds of tissue.

The adrenal medulla is supplied by medullated nerve-fibres which are preganglionic fibres of the sympathetic nervous system which innervate medullary cells directly—at first sight a most surprising arrangement, but, as has been pointed out, adrenalin is released at the terminal twigs of the post-ganglionic fibres of the sympathetic system so there is a physiological equivalence between the medullary cells and the ganglion cells in the sympathetic ganglia. The resemblance is enhanced when it is remembered that both have a similar embryonic origin from neural crest cells. Perhaps the medullary cells should be thought of as neuro-secretory cells.

**Physiology.**—The composite nature of the adrenal bodies of the higher vertebrates, as shown by the embryonic development and by the condition in lower vertebrates, is fully borne out by what is known of the physiology of these organs, for they play two quite distinct parts in the life of the animal. The **chromaffin tissue** produces two hormones, adrenalin and noradrenalin, which resemble each other chemically and in their effects on their target organs. Both are catecholamines but noradrenalin has no terminal methyl group. The relative amounts in which the two hormones are produced varies in different species of vertebrates. Both can be formed in tissue other than chromaffin tissue and even occur in invertebrates. Both hormones have the effect of increasing the blood pressure but they do it in different ways. Adrenalin increases the rate at which the heart beats and also the output of the heart, so bringing more blood to the muscles, etc. Noradrenalin constricts the arteries and arterioles by stimulating the smooth muscle in their walls. Both of them together increase the amount of glucose in the blood by causing muscle and liver glycogen to be broken down. This is accompanied by increased oxygen consumption and increased heat production as well as by increased muscular tone (mainly due to noradrenalin), dilatation of the pupils of the eye, relaxation of the smooth muscle of the bronchi and alimentary canal. Adrenalin causes symptoms of anxiety and dilatation of the arterioles and its effects in general are to reinforce the action of the sympathetic nervous system (p. 597). It is now known that the **adrenocortical tissue** produces several hormones, all of which are steroids called "corticosteroids". Many biologically active steroids can be extracted from adrenocortical tissue by means of fat solvents, but it is by no means certain that all of them are normally released into the blood. Some may be merely compounds necessary for the synthesis of the operative hormones. At present it is generally accepted that the more important adrenocortical hormones are:—

(1) Corticosterone—which accounts for 75 per cent. of these hormones in mammals. Its chief action is to increase the level of glucose in the blood but at the same time to increase the storage of glycogen by the liver. This it does by increasing the rate at which protein and fat are converted to compounds which enter the glycolytic cycle (p. 548). In the jargon of endocrinology it is said to be a glucocorticoid hormone or to have a glucocorticoid action. As will be seen, it also (in common with other corticosteroids) plays a part in regulating the balance of electrolytes in the blood by

influencing kidney action—mineralocorticoid action. Other hormones with similar effects on the body are—

(2) Cortisol and

(3) Cortisone.

Others like—

(4) Aldosterone are mainly mineralocorticoid, facilitating the retention of sodium chloride and bicarbonate by the kidney but—

(5) A hormone favouring sodium excretion but the retention of potassium is also present.

A balance between these two opposing hormones (4 and 5) is essential for maintaining the correct ionic balance in the blood.

But, in addition to their glucocorticoid and mineralocorticoid effects, corticosteroids have an androgenic effect. That is, they can induce male secondary sexual characters. Glucocorticoids increase the break-down of collagen and hinder its replacement. Mineralocorticoids, on the other hand, increase the replacement of collagen. Here again, it is the balance which is important. Corticosteroids also influence immunological responses, partly by releasing globulins from the breakdown of lymphoid tissue and partly by stimulating the production of phagocytic cells. On the other hand, they decrease the severity of allergic reactions. Many other reactions are known. Although all the effects of adrenocortical secretions are imperfectly understood, it is clear that they are, in sum, essential for life; if the adrenocortical tissue is removed, the animal dies. If it is deficient, "Addison's disease" results. This is characterised by decreased ability to withstand changes in the external environment such as heat and cold and by adverse reactions to infections, wounds, or prolonged exercise. In short, it seems that adrenocorticoids are essential for the animal to withstand stress. If an animal is subjected to severe injury or to stress it displays symptoms similar to those of adrenocortical deficiency. This reaction is followed by increased production of corticosteroids and, if the stress is prolonged, by hypertrophy of adrenocortical tissue.

## The Pancreas

**Anatomy and Embryonic Development.**—In addition to secreting digestive juices (p. 505), the pancreas produces two protein hormones, insulin and glucagon, which are essential for the maintenance of the correct carbohydrate balance of the body. Both have been isolated and their chemical structure is known. The tissue—**islets of Langerhans** or interstitial tissue—which secretes them consists of small masses of irregular cords of cells richly supplied by capillaries, lying in between the acini or follicles of cells which secrete juices into the finer branches of the pancreatic duct. In many animals the pancreas arises as an outgrowth from the embryonic mid-gut and so the pancreatic hormones may be compared with those secreted by the gastric and intestinal mucosa (p. 671). Indeed, in cyclostomes the islet cells lie in the wall of the intestine. Interstitial tissue is formed mainly of two types of histologically distinct cells one of which forms insulin and the other glucogon.

**Physiology.**—The two hormones, insulin and glucogon, are absorbed in the hepatic portal system of veins and hence are conveyed to the liver which is their main target organ in which they are, for the most part, destroyed by the liver acting as a partial barrier between the pancreatic hormones and the rest of the body just as it does for many substances absorbed from the alimentary canal (p. 508).

The glucose concentration of the blood depends on the rate at which it is being used as a source of energy by the tissues and on the rate at which glucose is released by the breakdown of glycogen mainly in the liver but to a lesser extent in other organs (*e.g.* muscles). These reactions are controlled (in a way which is not understood in detail) by the balance between insulin and glucogon.

Insulin lowers the level of glucose in the blood by increasing the rate at which it is converted to glycogen in the liver. Glucogon acts in the opposite direction by increasing the rate at which glycogen in the liver is broken down to form glucose.

If the islet tissue is deficient, or if the pancreas is removed, from an animal the disease *diabetes mellitus* ensues. The blood sugar rises to such a high level that the kidney tubules cannot resorb all that appears in the glomerular filtrate and so glucose remains in the urine. The liver becomes depleted of glycogen and tissue proteins are broken down to form glucose so that the body loses weight. The fat reserves are also drawn on excessively and fat is synthesised from carbohydrate at a decreased rate. As a result of excessive fat utilisation, ketones accumulate in the liver, blood, and urine. There is also an increased flow of urine and despite increased water and food intake the animal dies in an emaciated state.

The diabetes can be alleviated by injections of carefully controlled amounts of insulin or if some islet tissue remains functional by certain urea derivatives which when swallowed produce a rapid and sustained lowering of the level of the blood-sugar. These drugs probably act by inhibiting the insulin-destroying enzyme normally present in tissues.

Many hormones besides glucogon tend to raise the level of blood-sugar. Only insulin tends to lower it. This is why the balance is upset in a diabetic the way when islet tissue is deficient. The release of each of the two hormones is directly governed by the level of glucose in the blood.

## The Gonads

Whilst the production of germ-cells is the primary (and in most animals the only) role of the gonads, these organs are, in vertebrates at any rate, the site of formation of several important steroid hormones and so must be reckoned as endocrine organs. Broadly speaking, the functions of the gonadial hormones are—

(1) to ensure that germ-cells are produced in an active condition at that particular season of the year most suitable for reproduction;

(2) to influence those structural and behavioural features of the two sexes whose end result is fertilisation of the eggs and,

(3) in viviparous species to control events concerned with growth of and leading up to birth of the young.

Few only of the metazoan animals breed throughout the year, even in the tropics where for all practical purposes seasons are not detectable. Very commonly a particular species has a short breeding period, perhaps—as in many marine species—restricted to one or two tides per year. Others may have a longer, though restricted, seasonal breeding period whilst some mammals, for example, may have several breeding periods per year which alternate with periods of inactivity. Whether seasonal or not, reproductive activities are usually periodic or cyclical in nature and in vertebrates are controlled by gonadial hormones. This is not to say that other hormones do not play a part. They do, and of prime importance in linking sexual cycles to events in the external environment is the adenohypophysis.

This is influenced via the hypothalamus by seasonal changes in day length, temperature, rainfall and so on, to produce varying amounts of gonadotropic hormones as well as affecting other endocrine organs, such as the adrenal cortex. Thus through the sense organs and brain the adenohypophysis hormones are varied and, in this way help determining the breeding seasons of vertebrates.

Man is one of the few species of mammals which in all climates breeds throughout the year and thus is an example of where the reproductive cycle is virtually independent of the external environment. The following account refers almost entirely to the gonadial hormones of mammals and to those of Man in particular. The structure and development of male and female gonads have been described in previous sections.

### The Testis

Special connective tissue cells, those called interstitial tissue, located between the seminiferous tubules, produce a series of steroid hormones (androgens). Possibly other cells including the Sertoli cells also secrete hormones whilst, as has been mentioned (p. 665) the adrenocortical tissue forms androgens similar to those of the testis. Androgens can also be extracted from the placenta and ovary. Whilst differing in chemical nature all androgens (the chief one of these is called testosterone) seem to play fundamentally similar roles which include—

(1) The maintenance of the male ducts and associated glands in functional condition.

(2) The development and maintenance of male secondary sexual characteristics, e.g. horns and antlers in many ungulates, plumage features, and wattles in birds, voice characteristics in many birds and mammals as well as a host of behavioural features in various groups of vertebrates.

(3) The stimulus to spermatogenesis.

(4) Various metabolic effects including an increased rate of protein build-up.

(5) In the embryo to influence the differentiation of maleness, both primary and secondary.

### The Ovary

Like the testis, the primary function of the ovary is to produce the gametes, in this instance the female ones or oocytes. But at the same time the vertebrate ovary also secretes hormones which influence the cyclical activity of the female ducts and also play an essential part in the determination of the female secondary sexual characters. Also, like the testis, the various activities of the ovary are controlled by the adenohypophysis. The cyclical activities are also influenced by the luteotropic hormones (LTH) whilst feedback mechanisms by ovarian hormones in turn influence the production of hormones by the adenohypophysis.

Ovarian hormones are secreted by the follicle cells and by the corpora lutea and are called oestrogens, progestins, and relaxins, the first two being steroids and the last a protein hormone.

Oestrogens are secreted by the follicle cells around the oocytes and around the corpora lutea. Progestins are formed by the deeper lying or granulosa cells of the corpora lutea when they have become "luteinised" to form the corpora lutea after the oocytes have been discharged. The source of relaxins (and androgens) is uncertain.

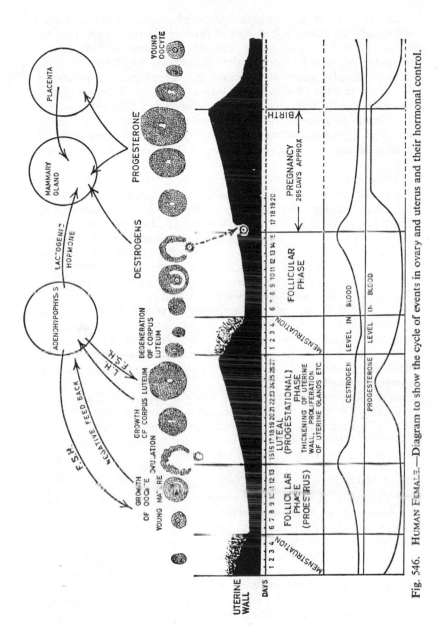

Fig. 546.—HUMAN FEMALE.—Diagram to show the cycle of events in ovary and uterus and their hormonal control.

Oestrogens not only have important effects on the female ducts, particularly the uterus, but on many other organs also. They increase the rate of cell division of the vaginal epithelium, and its supply of glucose and blood; increase the thickness of the uterine mucosa, its blood-supply, water and electrolyte content, rate of anaerobic glycolysis, aerobic glycolysis, and so on. In short they activate the mucosa of the reproductive tract so preparing it for pregnancy.

It is the oestrogens which cause enlargement of the mammary glands and other secondary sexual characters, like the deposition of adipose tissue, voice characteristics, skeletal characteristics, and so on.

Progestins, such as progesterone, act as a second stage in the preparation of the reproductive tract for pregnancy by causing the mucosa to become more secretory. The glands of the vagina and uterus proliferate and become more active; the mammary glands proliferate and are activated to secrete milk under its influence whilst the uterine muscles become relatively insensitive, favouring retention of the foetus. Relaxin causes dilation of the uterine cervix and inhibits uterine contractions. It also causes relaxation of the pubic symphysis. All these effects favour parturition.

As is well known, all the activities described have a cyclical occurrence and are controlled by the interactions of ovaries, pituitary body, and central nervous system. Maturation of the oocytes is followed by their discharge, preparation of the reproductive tract, fertilisation, implantation of the ovum, and activation of the mammary glands.

If fertilisation or implantation are not successful the cycle is repeated. The cycle in human beings is set out in diagram form on p 669. Important differences occur in the oestrous cycle of other mammals. Often their cycles are seasonal and are controlled by temperature, day length, and so on.

**Pineal Body**

The pineal body of many lower vertebrates has an eye-like structure and has been shown experimentally to be a photoreceptor mediating responses to light and dark conditions. Recently it has been shown that in mammals it is an endocrine organ innervated by fibres from the cervical sympathetic system. It produces a substance called melatonin, so called because when injected into amphibia it causes the skin to become lighter in colour. In mammals melatonin seems to inhibit growth and maturation of the gonads but its effects are antagonised by exposing the animals to light. It is now thought that this provides a means whereby seasonal sexual activity can be controlled—variations in daylight affecting pineal secretion *via* the sympathetic system. Various diurnal rhythms and the oestrous cycle may also be partly controlled by corresponding rhythms in pineal secretions.

### The Mucous Membrane of the Stomach

The structure of the stomach and the histology of its lining membrane have been dealt with elsewhere and it remains at this point to state that the first products of gastric digestion stimulate the cells of the lining epithelium to produce a hormone—**gastrin**—which after circulation in the blood-stream again reaches the stomach and evokes a further flow of digestive juice.

### The Mucous Membrane of the Small Intestine

The mucous membrane of the small intestine (pp. 495, 505), in addition to secreting the juices of the succus entericus (exocrine secretion), also produces a hormone—**secretin**—which, when it reaches the pancreas *via* the blood-stream, evokes a copious secretion of pancreatic juice. The stimulus for the release of secretin appears to be the presence of the acid contents of the stomach coming into contact with the intestinal mucosa.

### The Endocrine System in Invertebrates

At one time it was thought that endocrine organs were to be found only in vertebrate animals but it is now known that this is not so, and that most invertebrates that have been studied from this point of view possess hormones which play an important part in the regulation of the activities of the body. Indeed, it is probable that the phenomenon of internal secretion is widely spread throughout the animal kingdom, but investigation is made difficult in many invertebrates by their small size and, as a rule, by the fact that the cells which produce hormones are not localised to form obvious glandular structures. It has been found, for example, that pigment cells to which prawns owe their coloration, contract under the influence of a hormone located in the "sinus gland" in the stalk of the compound eye. When this happens, the colour of the animal as a whole gets lighter. Conversely, expansion of the pigment cells, with a corresponding darkening of the body colour, is caused by another hormone in the sinus gland. But the hormones are not secreted *in* the sinus gland but in the "brain" and other parts of the nervous system, and are conveyed to the gland as droplets along the axons of nerves which supply it. The cell-bodies of these nerves are therefore called neuro-secretory cells and many different kinds have been identified in Crustacea and insects. The stimulus for the release of the substance causing contraction of the pigment cells is that of light falling on the eyes and, if these are covered, there is therefore no response in the pigment cells to an increase in the intensity of light. The mechanism of colour change, whereby a prawn or shrimp becomes lighter in colour in lighter surroundings, is therefore due, in part, to the action of hormones. By rearrangement of pigment cells of different colours the animals can change the pattern of the body coloration so that they become practically invisible against any given background.

The larvae of certain insects, among them the common blow-fly, depend for their further development—pupation—on the stimulus provided by a hormone secreted by cells in the head. Moulting in insects is brought about by hormones produced by the corpora alata, small bodies related to the stomogastric nervous system (see pp. 331-2 for details).

The moulting of crustacea is also controlled by an endocrine system which is influenced by the environment. In the anterolateral part of the thorax of crabs, for example, is a pair of small bodies called Y organs. These produce hormones which stimulate moulting. A contrary effect is produced by the X organ, a group of neurosecretory cells in the eye-stalk ganglion, which pass their secretion along their axons to the sinus gland (pp. 289-90).

Recently it has been shown that some invertebrates show clearly defined responses to vertebrate hormones when they are injected into the body. Conversely, adrenaline has been extracted from gastropod molluscs and from *Paramecium*. Other physiologically active substances seemingly allied to vertebrate hormones have also been detected in a variety of non-chordate animals. It seems likely that the endocrine system of organs in vertebrates is not something peculiar to that group of animals but that it represents the climax in the differentiation of a mechanism whereby certain parts of the body are activated by substances carried to them (usually by the blood-stream) from others.

# CHAPTER XVIII

## THE GERM-CELLS—GAMETOGENESIS

### INTRODUCTION

The germ-cell or gametes of metazoa are always of two kinds. The female gamete or **ovum** is invariably larger than the spermatozoon or male gamete, is immobile where the other is active, and varies a good deal in size and constitution. The spermatozoon is, with few exceptions, flagellate, swimming actively in a fluid medium by undulations of its flagellum or "tail" in order to come into contact and fuse with the ovum. Both types of gametes are produced in gonads, from cells which form the germinal epithelium, the ova in ovaries, and the spermatozoa in testes.

**The Ovum.**—A generalised ovum is a single cell, consisting of a nucleus within a spherical mass of cytoplasm, and separated from it by a nuclear membrane. Frequently a nucleolus is present, and, up to the last phases of its development, a centrosome. The condition of the cytoplasm varies very considerably, but in all ova, present among the various cell inclusions, is a nutritive material termed yolk or deutoplasm, destined for the nourishment of the future developing organism. Eggs are always enclosed in a membrane or membranes (p. 689).

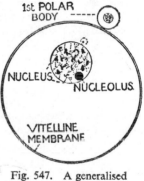

Fig. 547. A generalised chordate egg.

**The Spermatozoon.**—Although very varied in form in the Metazoa, the vertebrate spermatozoon is made up of a head, a middle piece, and a tail or flagellum. The head contains the nucleus, which forms the greater part of this region since it is merely invested with a thin layer of cytoplasm which is continued forwards into a pointed projection, the acrosome. In the slender middle piece are found the centrosome and mitochondria, and from it extends the vibratile tail or flagellum which has the same structure as a cilium of a protozoan.

Although this description gives the main characteristics of a spermatozoon, many variations are found in the various vertebrates. This applies particularly to the length of the head and middle piece and the form of the tail. In all instances, however, the spermatozoa are always discharged from the body suspended in a fluid secreted partly from the seminiferous tubules and partly by the accessory glands of the reproductive apparatus, to form the seminal fluid or semen. The spermatozoa are always produced in very large numbers, and are always much more numerous than the ova.

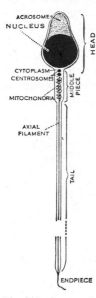

Fig. 548. A generalised chordate spermatozoon.

## GAMETOGENESIS

The series of changes through which cells of the germinal epithelium pass in order to give rise to the gametes is termed **gametogenesis**, and in spite of the fact that the two types of gametes are so different from one another, the phases of development through which they both pass are essentially similar. These phases are three in number: first a phase of multiplication, second a phase of growth, and third the phase of maturation.

This sequence of events can be best illustrated by considering first **spermatogenesis**, the development of the spermatozoa. The cells of the germinal epithelium can all, potentially, develop into spermatozoa, but all do not do so. Some, which may be called the primordial germ-cells,* do so, but others, such as the cells of Sertoli, found in mammalian testes, are concerned with the nourishment of the developing spermatozoa. When a cell in the germinal epithelium is destined to become a functional germ-cell, its fate is indicated by its entering at once into the phase of multiplication and undergoing repeated divisions, all of the mitotic type. Thus from the primordial germ-cell are produced numerous cells, termed **spermatogonia**. The spermatogonia also continue to divide for some time (many more divisions taking place than those indicated in Fig. 549), but this is eventually succeeded by the growth phase during which each spermatogonium slightly increases in size and comes to occupy a more superficial position in the epithelium lining

---

* It has been shown in many animals that the primordial germ-cells are set aside very early in development. Only later do they migrate to the gonads from the situations in which they arise.

the seminiferous tubule. It is now called a primary spermatocyte.
Following the growth phase is the important one of maturation,
during which two successive cell divisions take place, the first of
which is of a type different from mitosis, for during it the number of
chromosomes in the nuclei is reduced to half that possessed by the
somatic cells and also by spermatogonia, that is, to the **haploid**
number. If the normal complement (**diploid** number) of the somatic

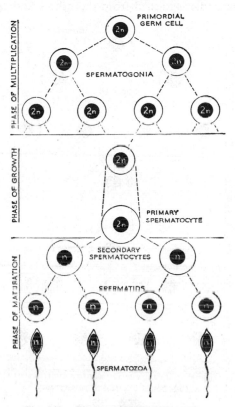

Fig. 549.  Phases of spermatogenesis.

cells is represented by **2n**, then as a result of these reduction or
meiotic divisions, each spermatozoon will possess only **n** chromosomes. (For details of the process of meiosis the reader should
see p. 678.) Each primary spermatocyte then divides by a meiotic
division, producing two secondary spermatocytes, each of which
had **n** chromosomes. Each secondary spermatocyte next divides
mitotically. Each of these cells is called a **spermatid** and undergoes

no further division, but grows into a **spermatozoon**. From each primary spermatocyte there are, therefore, produced four *functional* cells, a point of difference between spermatogenesis and oogenesis.

During oogenesis, after the follicle has sunk into the stroma of the ovary, the cells, which have been undergoing a process of multiplication during the formation of the follicle, continue to increase in number. Each is potentially an oogonium, but, soon, one cell becomes distinguished from the others and is destined to

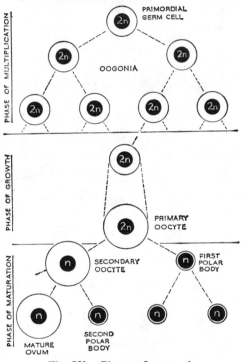

Fig. 550. Phases of oogenesis.

become the ovum, the others remaining as follicular cells. The selected oogonium grows and becomes a **primary oocyte**. The maturation phase in the vertebrates commonly takes place after the oocyte has been released from the ovary, by the bursting of the follicle, and has entered the oviduct. Indeed, it is the rule that the completion of maturation requires the stimulus of the entry of the spermatozoon. In essentials, however, the maturation of the ovum, like that of the spermatozoon, consists of two cell divisions involving a reduction of the chromosome number to half. During these divisions, however, unlike those in spermatogenesis, the division of the cytoplasm is unequal. Thus the primary oocyte divides into a large secondary oocyte and a small (first) polar body, each cell having **n** chromosomes. The secondary oocyte then divides, again with an unequal division of the cytoplasm, into the mature ovum and the second polar body. In some instances the first polar body may also divide so that at the end of oogenesis, there are produced from each primary oocyte, a single mature ovum and two much smaller polar bodies, the first of which may have divided again, but all have only **n** chromosomes. Normally, the polar bodies take no further part in the reproductive processes and eventually disintegrate and disappear.

The production of a mature ovum and three polar bodies from one oocyte has often been regarded as a theoretical or idealised state of affairs for it was difficult to demonstrate that the first polar body divides after separation from the oocyte. It has been convincingly shown that three polar bodies are formed during the maturation of the tapeworm egg (Fig. 551).

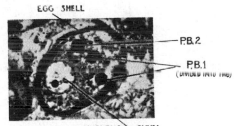
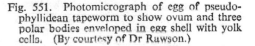

Fig. 551. Photomicrograph of egg of pseudophyllidean tapeworm to show ovum and three polar bodies enveloped in egg shell with yolk cells. (By courtesy of Dr Rawson.)

## MEIOSIS

An important difference between mitosis (see pp. 51-4) and meiosis lies in the behaviour of the chromosomes during the prophase. At the onset of meiosis, the full (diploid) number of chromosomes appear, that is become "fixable", in the same way as in mitosis, but they do not show any indication of division into chromatids as in mitosis. A further difference is that each chromosome had a definite granular appearance due to the formation of chromomeres. This phase is said to constitute the leptotene stage. The chromosomes now come together in homologous pairs (that is, chromosomes usually of the same shape and size, the one of paternal and the other of maternal origin), corresponding chromosomes lying side by side. The first contact between a pair is usually at the centromeres, but may occur almost anywhere along the length of the chromosomes. When this (the zygotene) stage is completed, so intimate is the connection between the paired chromosomes, that only the haploid number is visible, but it must be realised

that each apparent chromosome is really double or bivalent.   The
relation between the chromosomes now becomes still more intimate,
for, in addition to an apparent increase in thickness due to condensation and consequent shortening in length, they become coiled
around one another.   At the end of this stage, called the pachytene
stage, the spindle attachments, or centromeres, of the chromosomes
become more distinct and the formation in each individual chromosome of two chromatids becomes evident.   The two partners in
each bivalent chromosome now tend to separate, and each can be
seen to consist of two chromatids as in mitosis.   But the separation
is by no means complete, for along the length of the chromosomes
are to be seen chiasmata, where the individual chromatids cross one

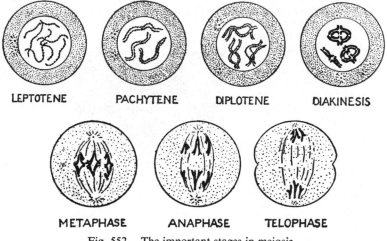

LEPTOTENE          PACHYTENE          DIPLOTENE          DIAKINESIS

METAPHASE          ANAPHASE          TELOPHASE

Fig. 552.   The important stages in meiosis.

another, often in rather a complicated manner.   This is called the
diplotene stage, and explains the phenomenon of "crossing over"
in genetics (see p. 790).   A further shortening and thickening of the
chromosomes now becomes evident and by this time the nuclear
membrane has disappeared, the centrosomes have separated, and a
nuclear spindle has formed exactly as in mitosis.   The chromosomes then travel to the periphery of the nucleus and arrange
themselves on the spindle, this stage being known as diakinesis.
All these changes correspond to the prophase of mitosis and the
nucleus now enters the metaphase.   At this stage each bivalent
chromosome is joined to the nuclear spindle by the two spindle
attachments (one for each partner chromosome), one attachment
above and the other below the equator.   The chromatids of each

individual chromosome still remain attached to one another. The two chromosomes, each consisting of its two chromatids, are now drawn apart, travelling along the spindle, each towards its corresponding centrosome. This represents the anaphase. In some cases a true telophase ensues in which each daughter-nucleus (with, however, only n chromosomes) may pass into a resting condition, the chromosomes becoming invisible owing to the loss of their fixability. Even where this condition, called an interphase, occurs, it may be very short, and frequently it is omitted altogether and the second division follows at once. The termination of the nuclear

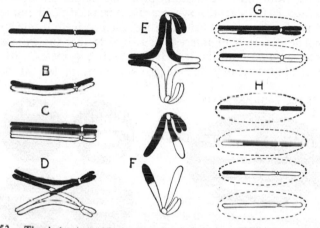

Fig. 553. The behaviour of chromosomes during meiosis. A, homologous chromosomes pairing (leptotene); B, homologous chromosomes paired (pachytene); C, chromatid formation; D, chiasmata and crossing over; E, separation of homologous chromosomes (diakinesis); F, later stage; G, end of first meiotic division; H, end of second division.

changes is, however, always followed by the division of the cytoplasm as in ordinary mitosis.

In the second division, if there has been an interphase, there may be a short prophase, during which the chromosomes, each consisting of two chromatids joined together at the spindle attachment, again appear, the centrosome divides, and a nuclear spindle is formed. If, however, there is no interphase, then the prophase is scarcely represented beyond the formation of the new spindle. The chromosomes then become arranged on the spindle and the metaphase follows at once. This is exactly as in mitosis, the division of the spindle attachments preceding the separation of the chromatids and the anaphase. This, and the succeeding telophase, show no differences from ordinary mitosis.

Meiosis might, therefore, be regarded as a special type of mitosis in which two cell (and nuclear) divisions take place. In the first, half of the chromosomes pass into each daughter-nucleus, and in the second, each chromosome divides into two chromatids such as occurs in an ordinary mitosis. The end result, however, is that four cells are produced from the original gametocyte, each cell having **n** chromosomes. The predestined fate of the gametes is that they shall meet in pairs and fuse to form zygotes, the fusion being primarily between the nuclei. This process is called fertilisation and is a characteristic feature of sexual reproduction in the Metazoa.

Apart from other considerations, one important feature of fertilisation is that, in the zygote, the normal diploid number of chromosomes is re-established, since each gamete contributes **n** chromosomes. It is of interest also to note that the zygote does not receive equal contributions from the two gametes. Its cytoplasm is almost wholly derived from the ovum, but its nucleus

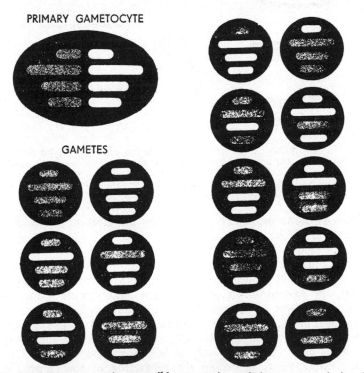

Fig. 554. Diagram to show possible segregations of chromosomes during the formation of gametes where the somatic number is eight. (Paternal chromosomes white; maternal grey.)

receives **n** chromosomes from each pronucleus and its centrosome from the spermatozoon.

## The Significance of Maturation and Fertilisation

The most obvious reason for the halving of the chromosome number during maturation is to maintain the somatic number of chromosomes constant from one generation to the next. If this process, or something comparable to it, did not occur, then with each successive generation the chromosome number would be doubled. The introduction into the zygote of nuclear material from different parents is a mechanism favouring variations. The segregation of the chromosomes during the meiotic divisions of maturation provides a means for a regrouping of the chromosomes upon fertilisation, for during gametogenesis it need not, and usually does not, happen that all the chromosomes of paternal origin pass into one gamete and those of maternal origin into the other. Even where the somatic number is small, say eight, it will be seen from Fig. 554 that sixteen different combinations of the chromosomes are possible in the gametes. When it is considered that each chromosome may carry the genes for many characters (see pp. 780-93) the immense possibilities of variation may be envisaged, particularly when the chromosome number is large.

# CHAPTER XIX

## THE DEVELOPMENT OF THE VERTEBRATES— CHORDATE EMBRYOLOGY

### INTRODUCTION

Embryology is that branch of zoology which deals with the development of animals. It is a study of those changes repeated in every generation, through which an organism must pass before it assumes its adult characteristics. In the chordate animals, with which this chapter is particularly concerned, and in which (with the exception of certain protochordates) reproduction is entirely sexual, each new organism originates from a fertilised egg and its development involves a gradual increase in complexity of structure from the single-celled zygote to the attainment of the adult condition. These changes are, for purposes of description and generalisation, usually considered to fall into a series of phases or steps which are common not only to the chordates, but are recognisable in a general sort of way in the life-histories of many other animals. They may be said to consist of **fertilisation, cleavage, gastrulation,** and **organogeny.** Each of these processes is in itself a complicated business, but in general terms it may be said that fertilisation is the fusion of male and female gametes to give a zygote; cleavage the division of the zygote into daughter-cells; gastrulation the rearrangement of these cells into definite layers (germ-layers), certain of which surround or enclose the others; whilst organogeny is the laying down of the organ systems of the body.

Chordate embryology has now become such a vast and specialised subject that it has acquired a nomenclature of its own, so that it is essential at the outset to explain some of the more important terms that will be encountered. Some of these and some general considerations relating to development and differentiation have been discussed in Chapter II, to which reference should be made.

**Fertilisation.**—The embryonic period is usually held to begin with the fertilisation of the egg by a spermatozoon. The details of this process are variable, but the chief features are usually as follows. By random swimming movements sperms reach the egg and one (in certain instances more than one), by means of its acrosome, penetrates tbe **vitelline membrane.** The tail of the spermatozoon is cast off and the head and middle piece are drawn deeply into the egg cytoplasm. The first obvious effect of the entry of the sperm is that the cytoplasm of the egg appears to shrink and a fluid-filled space appears between it and the vitelline membrane, which therefore becomes more clearly defined and is often spoken of as the fertilisation membrane. It seems clear that in all eggs the sperm entry causes profound changes in the egg cytoplasm, although they are not always so obvious as the rearrangements produced in the egg of amphioxus (pp. 691-4). Even in vertebrate eggs, however, visible changes may be produced, as for example, the formation of the grey crescent in the egg of amphibians (pp. 705-9). In nearly all instances an extremely rapid chemical change is produced in the cytoplasm which renders the egg incapable

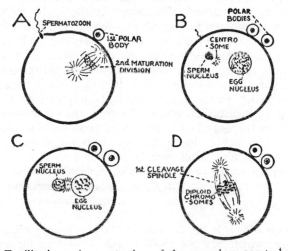

Fig. 555. Fertilisation. A, penetration of the secondary oocyte by a spermatozoon; B, passage of the head and middle-piece through the cytoplasm; C, approximation of the male and female nuclei; D, the first cleavage spindle.

of penetration by supernumerary sperms, but in occasional types polyspermy is of normal occurrence. These sperms, however, play no essential part in fertilisation, and they soon degenerate.

Soon after the entry of the sperm into the egg cytoplasm the egg nucleus completes its maturation divisions and the second polar body is extruded. The centrosomes associated with this division then disappear. The head of the spermatozoon, which contains the male nucleus, and the middle piece containing the centrosome (which may have already divided) then advance towards the egg nucleus. After entry the sperm rotates so that the middle piece precedes the head during the advance. As the male nucleus approaches close to the egg nucleus the haploid number of chromosomes appears in each of the nuclei and the nuclear membranes break down. An aster forms around the centrosome, which is released from the middle piece of the sperm, and a typical spindle appears. The paternal and maternal chromosomes become distinct and then arrange themselves on the spindle. In this way the diploid number of chromosomes is restored, the zygote (as the egg may now be called) having received both a paternal and a maternal set. A nuclear membrane is not, however, formed around the chromosomes, for they immediately appear as double structures consisting of two chromatids preparatory to the next stage in development—cleavage.

Without entering into details, the main results of fertilisation may now be summarised. The entry of the sperm into the egg activates it to undergo further development, the first obvious effect being the production of the fertilisation membrane. It also seems to provide the stimulus for the egg to complete its maturation. Fertilisation restores the diploid number of chromosomes and introduces the centrosome which is lacking from the mature egg. The significance of the nuclear changes has already been dealt with. In some instances, e.g. in the frog, the point at which the sperm enters the egg plays a part in determining the plane of the first cleavage, which is also the plane of bilateral symmetry. There is, however, much evidence in favour of the view that the plane of bilateral symmetry is, at least partially, determined before fertilisation and that the sperm entry path usually happens to coincide with it.

**Cleavage.**—This is the division of the zygote into many cells or **blastomeres** which become smaller and smaller as cleavage progresses. The size of the nuclei is not, however, diminished at the same rate as the cytoplasm, with the important result that the ratio nucleus-to-cytoplasm gradually increases until it reaches a constant which is, presumably, the optimum. Cell division continues, of course, throughout life, but the process of cleavage is arbitrarily held to end with the completion of the embryonic stage termed the **blastula**. This, in its simplest form, is a hollow sphere built up of a layer of blastomeres enclosing a central cavity, the **blastocoel,** but in many specialised types it is difficult to recognise a distinct blastula stage.

Cleavage does not seem to be primarily concerned with the differentiation of the different parts of the future embryo, for, in extreme instances of so-called **mosaic eggs** (*e.g.* amphioxus), definite cytoplasmic areas are evident even before the egg is fertilised. Each of these areas (**organ-specific areas, organ-forming** substances) gives rise to quite definite structures in the embryo and, because cleavage follows a definite pattern, the different types of cytoplasm become distributed to their correct positions in the later embryo. Cleavage of mosaic eggs is therefore said to be **determinate**. In contrast to eggs of this kind there are others **(regulation eggs)** which show but little or no cytoplasmic differentiation after fertilisation, and different types of cytoplasm are not separated out until cleavage is well advanced. Thus the differentiation of the cytoplasm does not take place until *after* it has been divided up and enclosed within the cell membranes of the blastomeres. Cleavage of eggs of the regulation type is said to be **indeterminate**, for it is not concerned with the distribution of organ forming substances pre-localised in the zygote. Eggs, and therefore cleavage, intermediate between the two types mentioned are well known but cannot be described here, and mosaic and regulation eggs are best regarded, not as illustrating fundamentally different phenomena, but as the extreme ends of a graded series

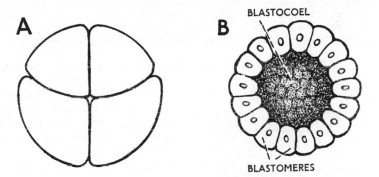

Fig. 556. Cleavage. A, 8-celled stage; B, sagittal half of a blastula.

of different types of eggs which differ chiefly in the *time* at which differentiation sets in.

**Gastrulation.**—In gastrulation a rearrangement of the cells of the blastula takes place with a result that certain cells, which are by this time potentially, and sometimes visibly, differentiated, come to occupy a position inside the embryo and are enclosed by the others. The deeper-lying cells form the **endoderm** and **mesoderm**, whilst the superficial layers form the **ectoderm**. Gastrulation is therefore a process which results in the **germ-layers** occupying their definitive positions in the embryo. It frequently happens that during gastrulation the blastocoel is obliterated and the deeper-lying cells form the wall of a new cavity, the **archenteron**, which later contributes to the formation of the gut cavity. The archenteron usually opens to the exterior by an aperture, the **blastopore**, whose margins are referred to as the **lips of the blastopore**.

The process of gastrulation, then, converts the blastula into a gastrula, a stage which was formerly of considerable theoretical importance, for, by an application of the Recapitulation Theory (p. 815) it was thought to represent in a modified form a "two-layered ancestor"; that is, an ancestor at the coelenterate level of organisation consisting of two cell-layers, the ectoderm and endoderm, surrounding a central gut cavity which opens to the exterior by way of the mouth. It is now known that any resemblance between a gastrula and a coelenterate is merely superficial, for although a "two-layered stage" is found, for example, in amphioxus, it must be emphasised that the embryo is not composed of the two primary germ-layers, ectoderm and endoderm, alone, as was formerly supposed, for the mesoderm is from the beginning of gastrulation a histologically distinct layer of cells, although temporarily accommodated in the wall of the archenteron. In fact, it is only in the much-modified gastrula stages of reptiles and birds that a "two-layered embryo" in this sense is found. In all other types the delimiting of the mesoderm goes on hand in hand with the formation of the endoderm. Gastrulation may therefore be best thought of as a kinetic process concerned with rearrangements, migrations, and foldings (formative movements) of the cell-layers. The blastopore cannot therefore be thought of as a "primitive mouth", nor the archenteron as a "primitive gut". Both are merely the result of cell migrations from the surface inwards, and probably have no evolutionary significance whatsoever.

The end of gastrulation is marked by the complete enclosure of the endoderm and mesoderm by the ectoderm, and, in those more primitive forms where a distinct archenteron and blastopore are recognisable, by the closure of the blastopore, part of which may, however, persist as the anus.

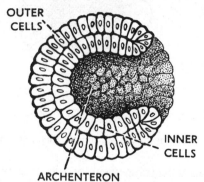

Fig. 557.   Sagittal half of a gastrula.

But gastrulation is not a separate process in itself, for, during the later stages at any rate, the rudiments of some of the main organs of the body are already distinguishable. This is particularly true of the central nervous system and notochord, which are delimited early on and share in the cell movements of gastrulation. For this reason also it is difficult to find in the chordates a counterpart of a two-layered, coelenterate-like ancestor. All chordates except in the earliest gastrula stages are provided with distinct notochordal and neural rudiments which are typical of chordates alone. Gastrulation overlaps, then, with the next stage in development, organogeny.

**Organogeny.**—This implies the laying down of the organs and systems of the body. It takes place step by step until the adult stage is reached.

It will be realised that the above description is, in general, a comparative one and pays attention to only some aspects of chordate embryology. It serves to emphasise the importance of the general similarity of the developmental processes of all chordates and is of particular value for an understanding of the early stages. On the other hand, when dealing with the way in which the young animal lives and the manner in which it adapts itself to its surroundings, it is customary to refer to its developmental stages in a different way.

Thus, all the early stages from fertilisation onwards which take place within the egg membranes constitute the embryonic period, and the young organism is said to be an embryo. It frequently happens that the young animal hatches from the egg and begins to fend for itself in a form which differs materially from that of the adult. It is then said to be a larva. This, after a period which varies in different examples, changes into the adult by a metamorphosis. Usually, however, an adolescent period may be distinguished in which the chief difference between the young and the adult organism is that the former is not sexually mature.

Thus treated, it will be realised that embryology leaves out of account one important chapter in the life-history of the individual, for it omits to deal with the origin of the germ-cells. These cells are also the end result of a remarkable series of processes which are essential preliminaries to sexual reproduction and have been discussed previously.

**Types of Eggs and Polarity.**—Although subject to a great deal of variation in size, the ovum is always large when compared with the spermatozoon, and indeed, the largest animal cells known are the ova of birds. The large size is due, not to a general exaggeration of all the parts normally found in an animal cell, but to the large quantity of cytoplasm that is present. This is *always*, although to a varying degree, charged with yolk (deutoplasm), which serves as a source of nourishment for the young organism until it can fend for itself or, alternatively, exploit the food provided by the parent. The yolk present in the eggs of chordates is, as a rule, aggregated chiefly towards one pole of the cell, the **vegetative pole**, whilst the egg nucleus lies near the other pole, the **animal pole**, in a region relatively, or completely, free from yolky substance. Eggs of this type are said to be **telolecithal**. Telolecithal eggs themselves differ, however, in the extent to which the cytoplasm is charged with yolk. In the moderately telolecithal egg of a frog, for example, the yolk is practically confined to the vegetative hemisphere of the egg. In the eggs of fishes, reptiles, birds, and some mammals

(Monotremata), the amount of yolk present is so great that the non-yolky portion is restricted to a small cap at the animal pole. The eggs of amphioxus and of the eutherian mammals are exceptional in that only minute amounts of yolk are present—for which reason, although showing polarity, they are termed **microlecithal**.

**Origin of Polarity in the Egg.**—It will be understood from the above description that the eggs of vertebrates, even whilst still in the ovary, have acquired a distinct polarity, which is evident from the fact that the yolk is deposited in the cytoplasm chiefly towards one (the vegetative) end of the cell whilst the nucleus lies nearer to the other end (animal pole). In contrast to the eggs of invertebrates, where the end of the egg nearest to the coelomic fluid becomes the animal pole, it is the attached end, that is the end attached to the follicle cells of the ovary, which forms the animal pole in the vertebrate eggs. This is probably because the animal pole develops in the region of higher oxygen tension and the polarity of the egg is therefore not, apparently, due to factors inherent in the egg but is induced by factors outside it.

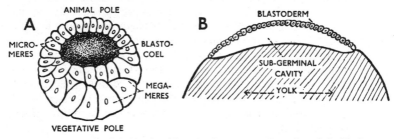

Fig. 558.  A, sagittal half of a blastula from a moderately telolecithal egg; B, vertical section of a blastoderm.

**The Influence of Yolk on Early Development.**—The quantity of yolk present and the nature of the yolk gradient—that is, the sharpness of separation of the yolk from the cytoplasm—play important parts in determining the type of cleavage undergone by the fertilised egg, for yolk is an inert material and when present in quantity has a noticeable effect in damping down the rate at which the cleavage planes pass through the egg. If yolk was evenly distributed, then its presence would merely slow down division, but its uneven distribution in telolecithal eggs causes the cleavage planes to be impeded chiefly in the vegetative hemisphere. This leads to the production of blastomeres of unequal size, those nearer the animal pole being smaller (because they divide more rapidly) than those nearer the vegetative pole. A further result is that the blastocoel becomes excentric in position and lies nearer the animal pole. In many of the larger eggs (those of fishes, reptiles, and birds) where yolk is present in great quantities and is separated sharply from the cytoplasm (which is confined to a small cap, the **blastodisc**, at the animal pole), cleavage is inhibited altogether in the yolky parts of the egg, the cytoplasmic blastodisc alone being capable of cell division. This type of cleavage, in which the furrows do not pass completely through the egg, is spoken of as **partial** or **meroblastic** in contrast to the **total** or **holoblastic** cleavage experienced by eggs having a smaller concentration of yolk (those of amphioxus, amphibians, and most mammals). The cleavage of a blastodisc produces a cap of cells, the **blastoderm**, which soon comes to be separated from immediate contact with the yolk by a sub-germinal cavity. This cavity, at any rate during early development, before distinct germ-layers are apparent, is usually held to be the equivalent of a blastocoel. Its physiological significance is probably that it contains a fluid, rich in products from digested yolk, which assists in the nourishment of the overlying blastoderm, and also, when the blastoderm becomes "two-layered", it allows of free movement of the layers relative to one other and makes possible certain of the later formative movements.

The provision of reserve food material in the egg is an important factor in determining the age at which the young animal emerges into the world by the hatching of the egg. The greater the quantity of yolk, the more advanced the stage of development reached prior to hatching and the greater the chance of survival. A similar end is also achieved by the embryo developing within the body of the parent so that it is born alive. This is termed **viviparity** and may be of two kinds: **ovoviviparity**, where the egg merely develops within the oviduct and, apart from the protection and possibly respiratory and excretory facilities afforded by this situation, derives little further assistance from the parent; and **true viviparity**, where organic connection is established between the embryo and the tissues of the oviduct in a specially modified portion forming a uterus. Here the provision of a potentially unlimited supply of food material enables the foetal life to be prolonged. Many gradations between ovoviviparity and true viviparity may be traced, particularly interesting series being found in the selachian fishes.

**Egg Membranes.**—Eggs are always surrounded by at least one, and sometimes more, membranes, which serve to protect and sometimes to act as supplementary food for the embryo. They persist until they are functionally replaced by other structures. With very few exceptions the ovum is immediately surrounded by a thin **vitelline membrane** which is a secretion of the oocyte itself and is therefore often termed a **primary egg membrane**. Outside the vitelline membrane a **secondary membrane** may be found. It is secreted by the follicle cells which surround the egg when it is in the ovary. One or more **tertiary membranes** are frequently deposited around the egg during its passage down the oviduct. Albumen, shell membrane, and shell, are membranes of this kind. Tertiary

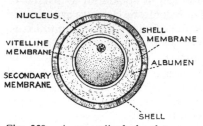

NUCLEUS

VITELLINE MEMBRANE

SECONDARY MEMBRANE

SHELL MEMBRANE

ALBUMEN

SHELL

Fig. 559. A generalised chordate egg.

membranes are seen at their maximum development in eggs of those animals (certain reptiles and birds) which lay their eggs on dry land, but the horny case of the dogfish egg (Mermaid's purse) provides an interesting example of a tough shell in an aquatic vertebrate. It seems, in general, that tough tertiary membranes are best developed around large-yolked eggs; and the probable reason is that such eggs can be produced only in relatively small numbers and a larger measure of protection is essential if the race is to survive. Another interesting process to be correlated with the presence of tough shell membranes is internal fertilisation, for it is clear that the sperm must penetrate the egg whilst it is in the oviduct before the membranes are secreted (selachians, reptiles, and birds).

**Orientation of the Egg and Embryo.**—It is important to realise that both the egg and the embryo may be orientated in two different ways. The usual convention is to orientate the egg with its animal pole uppermost so that the main axis of the egg, which passes through both animal and vegetative poles, is vertical. This is merely a convention and in no way implies that the upper surface of the egg will give rise to the upper (dorsal) side of the embryo and, at any given stage, the main axis of the embryo is usually inclined to the main

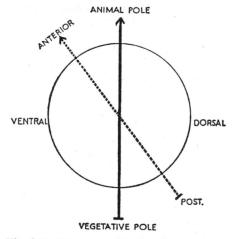

ANIMAL POLE

ANTERIOR

VENTRAL

DORSAL

POST.

VEGETATIVE POLE

Fig. 560. Diagram to show the main axes of egg and embryo.

axis of the egg to which, in any given type, it bears a constant relationship. When once this is known it is then possible to orientate the egg and early stages in development with respect to the axis of the later embryo; that is, to define the future anterior, posterior, dorsal, and ventral regions of the egg and early embryo before the organ rudiments make their appearance. During gastrulation of most types, the cell-layers (endoderm, chorda, and mesoderm) which become enclosed by the outer layer (ectoderm) for the most part migrate forwards from the place where they leave the surface of the embryo; that is, from the margins of the blastopore (or its equivalent) which lie at the hinder end of the embryo. It follows, therefore, that the first material to pass inwards comes to lie at the anterior end of the embryo, whilst the last material enclosed lies at the posterior end. In this sense, therefore, the mesoderm, chorda, and endoderm undergo a reorientation as a result of the formative movements of gastrulation.

In the following accounts of the development of selected chordate types the terms "vertical", "horizontal", "meridional", and "latitudinal" always refer

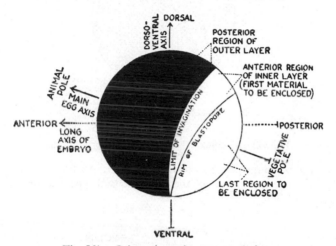

Fig. 561. Orientation prior to gastrulation.

to the egg or embryo so orientated that the animal pole is uppermost. The terms "anterior", "posterior", "dorsal", and "ventral" refer to the egg or embryo orientated with respect to the later embryo or the adult axis.

## THE DEVELOPMENT OF AMPHIOXUS

**The Egg.**—The oocytes are released by rupture of follicular membranes and overlying tissues so that they then find their way first into the atrium and thence, *via* the atriopore, into the sea. By this time the first polar body has been separated off and comes to lie just outside the thin vitelline membrane. The structure of the secondary oocyte, which has a diameter of about 0·12 mm, is particularly interesting, for it is almost identical with that of many ascidians (another group of protochordates). Just beneath the vitelline membrane is a peripheral layer of cytoplasm free from yolk granules, but containing numerous granular mitochondria. Inside this layer is a mass of yolky cytoplasm which, however, lies chiefly towards the vegetative pole of the egg, because a considerable portion

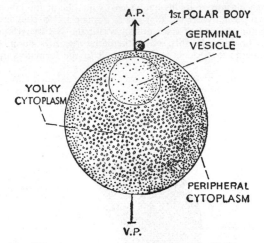

Fig. 562. AMPHIOXUS.—The egg before fertilisation.

of the animal hemisphere is occupied by the germinal vesicle (egg nucleus).   The egg is thus of the microlecithal type, the quantity of yolk present being relatively small.

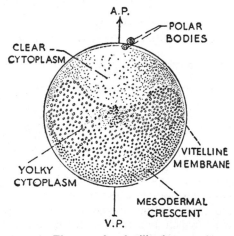

Fig. 563. AMPHIOXUS.—The egg after fertilisation seen from the dorsal side.

**Orientation of the Egg.**—From a knowledge of later development it is known that the animal pole of the egg corresponds with the future antero-ventral part, whilst the vegetative pole will give rise to the postero-dorsal part of the embryo.

**Fertilisation.**—Fertilisation takes place in the sea and the entry of the sperm near the vegetative pole provides a stimulus for the egg (secondary oocyte at this stage) to complete its maturation, for shortly after the sperm has penetrated the vitelline membrane, the second polar body is given off and comes to lie near the animal pole just within the vitelline membrane. The zygote nucleus is formed in the usual way just above the equator of the egg.

Fertilisation is also followed by a complete rearrangement of the cytoplasmic areas in the egg. After the rupture of the germinal vesicle there is left at the animal pole an area of clear cytoplasm in

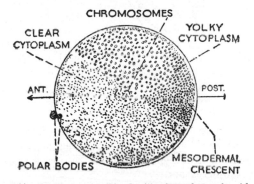

Fig. 564. AMPHIOXUS.—The fertilised egg from the side.

contact with the peripheral layer. This granular peripheral cytoplasm flows downwards towards the vegetative pole and eventually condenses to a crescent-shaped area at the future posterior end. The arms of the crescent reach round the sides of the egg and partially embrace the yolky area so that the egg is now bilaterally symmetrical with respect to the egg axis. It has been determined that the various cytoplasmic areas give rise to definite structures in the embryo and if any area is removed or damaged the embryo is lacking in some definite respect. The different types of cytoplasm are therefore termed organ-forming substances (organ specific areas), and because this regional differentiation results in a definite organ pattern being recognisable in the egg it is said to be of the **mosaic** type. During cleavage (determinate cleavage) of the zygote, the organ-forming substances become distributed in an orderly fashion to the various blastomeres and hence to the definite regions in the embryo. It has been shown that the clear cytoplasm in the animal half gives rise to the **ectoderm**, the yolky cytoplasm to the **endoderm**, whilst the crescent of granular cytoplasm will form the **mesoderm**.

**Cleavage.**—Cleavage is of the holoblastic type so that the egg is completely divided up into separate blastomeres. The first cleavage furrow is vertical and passes through both poles of the egg, and at the same time bisects the **mesodermal crescent**. The first two blastomeres thus correspond to future right and left halves of the embryo. The second cleavage plane is also vertical but at right angles to the first. The third cleavage, resulting in the formation of eight blastomeres, is at right angles to the first two and passes through just above the equator, so that four smaller cells (micromeres) lie above four larger ones (megameres). The fourth cleavage divides each of the blastomeres by meridional furrows so that eight micromeres and eight megameres are formed. The fifth cleavage planes are latitudinal and the micromeres and the megameres are each divided into upper and lower tiers. The meridional cleavages of the sixth division give a total of sixty-four cells and after this stage the cleavages become somewhat irregular.

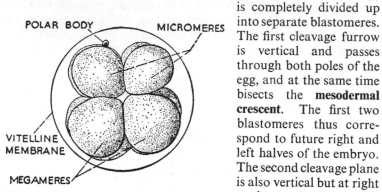

Fig. 565.   AMPHIOXUS.—The 8-celled stage from the side.

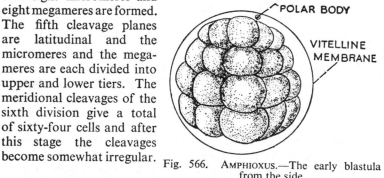

Fig. 566.   AMPHIOXUS.—The early blastula from the side.

**The Blastula.**—During the early cleavage stages a jelly-like substance accumulates between the blastomeres, and, later, by the absorption of water, it becomes fluid and fills a distinct cavity, the blastocoel, in the centre of the group of blastomeres. As cell divisions proceed, the blastocoel increases in size and in the fully-formed blastula is a spacious fluid-filled cavity. The blastula is not perfectly spherical but is pear-shaped when viewed from the side, the pointed end being posterior.

The orderly distribution of the organ-forming substances, which has taken place during cleavage, is in part responsible for areas of histologically distinct cells on the walls of the blastula. The ectodermal area consists of columnar cells occupying the bulk of the ventral half of the blastula; the larger yolky cells, which will form the endoderm, are in a plate-like area across the dorsal part of the blastula. The mesodermal crescent consists of small cells bounding the endodermal plate on its lateral and posterior borders. Immediately anterior to the **endodermal plate**, between it and the

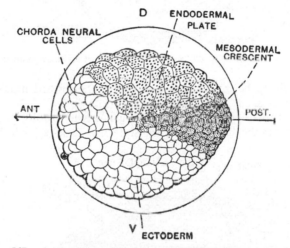

Fig. 567. AMPHIOXUS.—The fully formed blastula from the side.

ectodermal region, is a small area of slightly smaller cells (**chorda cells**) which later gives rise to the notochord, and in front of these again is a strip of specialised ectoderm destined to form the **neural plate**.

**Gastrulation.**—The first stage in gastrulation is announced by a flattening of the endodermal plate. The flattening becomes more and more pronounced, and the endodermal plate, which is roughly triangular in outline and is surrounded on its lateral and posterior faces by the mesodermal crescent, becomes concave and gradually sinks inwards into the blastocoel, which thus becomes progressively reduced in size. In this way the embryo becomes cup-shaped, the cavity of the cup being the archenteron, its opening the blastopore, and its rim the lips of the blastopore. The dorsal lip, which at this stage is really in an anterior position, contains the chorda cells, whilst its lateral lips and ventral lip are formed by the cells of the mesodermal crescent. The cells from this region also

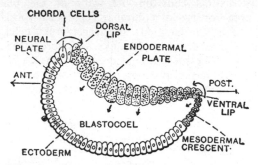

Fig. 568. Amphioxus.—Vertical section of an early gastrula.

soon begin to be invaginated and follow the inward movement of the endoderm.

The arms of the mesodermal crescent come to form groove-like areas of rapidly dividing cells (mesodermal grooves) which run in

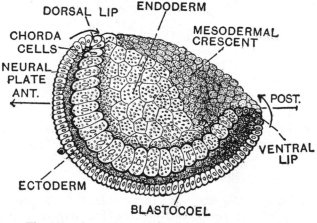

Fig. 569. Amphioxus.—Sagittal half of a late gastrula.

the lateral walls of the archenteron from the lateral lips of the blastopore. The middle of the mesodermal crescent still remains in the ventral lip of the blastopore. The cells invaginated at the dorsal lip are, of course, chorda cells and they come to lie in the mid-dorsal

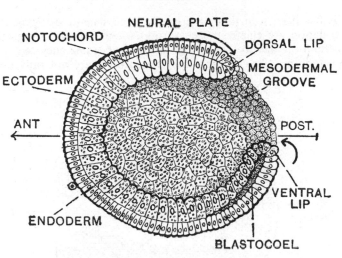

Fig. 570. AMPHIOXUS.—Sagittal half of a later gastrula.

wall of the archenteron whilst, laterally, they are in contact with the mesodermal grooves.

The embryo has, meanwhile, gradually lengthened and the blastopore has become smaller, shifting, at the same time, from a nearly dorsal to a posterior position by the backward growth of the dorsal lip.

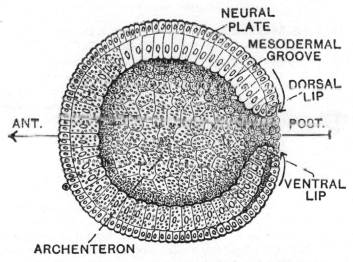

Fig. 571. AMPHIOXUS.—Sagittal half of a completed gastrula.

It has been seen that the formative movements bringing about gastrulation are chiefly those causing straightforward invagination of certain regions so that, in the end, they become enclosed within the ectoderm. The mechanics of this process are not fully under-

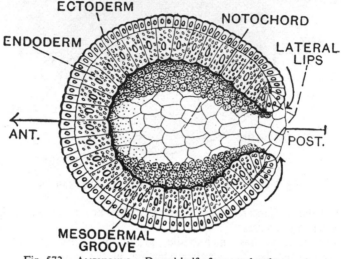

Fig. 572. AMPHIOXUS.—Dorsal half of a completed gastrula.

stood, but, clearly, gastrulation of this kind is possible only in embryos where there is little yolk.

**The Formation of the Neural Tube.**—Towards the end of gastrulation, a strip of ectoderm cells in the mid-dorsal line becomes larger to form the **neural plate**. This flattens and sinks inwards and the ectoderm at its sides rises up to form **neural folds**. These neural folds extend around the lateral lips of blastopore and, beginning at the posterior end, the folds grow towards each other over the neural plate, and finally meet in the mid-dorsal line. Meanwhile the lateral edges of the neural plate have also started to grow towards one another, a rolling up process,

Fig. 573. AMPHIOXUS.—Diagram of the changes in shape of the blastopore.

which results in the formation of the **neural tube**. It follows that, since the neural folds extend laterally to the blastopore, when they meet, the blastopore becomes roofed over by them, and this means that at this stage the neural tube communicates by a short tube (**neurenteric canal**) with the hinder end of the archenteron. The blastopore, which is now a minute orifice, soon closes up and the neurenteric canal is interrupted.

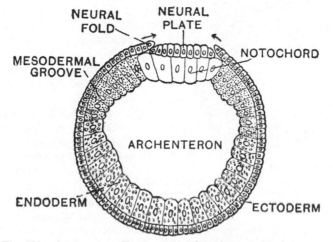

Fig. 574. AMPHIOXUS.—Transverse section of a completed gastrula.

The rudiments of the central nervous system of higher chordates are laid down in a slightly different way and the method of formation of the neural tube in amphioxus is peculiar in that the neural plate sinks inwards before

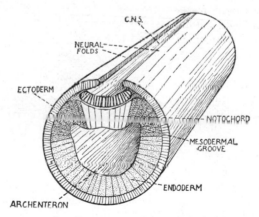

Fig. 575. AMPHIOXUS.—Stereogram of a part of a post-gastrula stage.

rolling up; the neural folds merely roof over the plate and do not contribute to the formation of the neural tube; and the neural folds make their first appearance at the posterior end.

The Further Development of the Notochord and Mesoderm.—
The chorda cells, which were invaginated as a strip forming the mid-

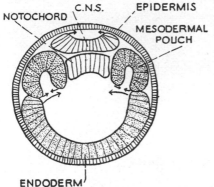

NOTOCHORD   C.N.S.   EPIDERMIS
MESODERMAL POUCH
ENDODERM

Fig. 576. AMPHIOXUS.—Transverse section to show the separation of the mesodermal pouches.

dorsal part of the arch-enteric wall, undergo a stretching and rearrangement to form a notochord consisting of a single row of disc-shaped cells, which later become vacuolated and surrounded by a notochordal sheath.

The mesodermal grooves gradually become metamerically segmented into **mesodermal pouches**\* by the development of transverse partitions. These make their appearance, first of all, at the anterior end, but the segmentation of the grooves soon extends posteriorly. Next, again beginning at the anterior end, the mesodermal pouches become separated from the wall of the archenteron and the aperture by which each communicates with the archenteric cavity closes up, so that a series of closed **mesodermal sacs**, each having a small cavity, is formed on each side of the gut-rudiment between the ectoderm and the endoderm. The chink-like cavities then enlarge and the ventral end of each sac grows downwards towards the ventral surface of the embryo, and finally the ventral wall of each member of a pair (right and left) meets in the mid-ventral line. The walls separating right from left sacs in the

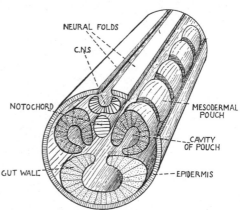

NEURAL FOLDS
C.N.S
NOTOCHORD
GUT WALL
MESODERMAL POUCH
CAVITY OF POUCH
EPIDERMIS

Fig. 577. AMPHIOXUS.—Stereogram to show the arrangement of the mesodermal pouches.

---

\* These are often termed "mesodermal somites", but each corresponds to a whole mesodermal segment in a vertebrate and not merely to the dorsal segmented part.

mid-ventral line then fuse and break down so that right and left cavities are made continuous. Further, except in the more solid, dorsal portions, the cross partitions also become lost so that metamerism is no longer evident in the ventral and lateral parts of the mesoderm, and a cavity, which is continuous throughout

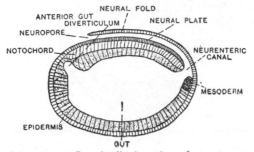

Fig. 578. AMPHIOXUS.—Longitudinal section of a post-gastrula stage.

the length of the embryo and which extends around the ventral and lateral parts of the gut, is established. This cavity is the **splanchnocoel**. It forms by far the greater part of the coelom. It is bounded laterally and ventrally by the two layers of lateral-plate mesoderm and dorsally by the segmental mesoderm. The dorsal parts of the mesoderm thus retain their metamerism and form structures which correspond to the **somites** of vertebrates, and may be so termed. Soon after their formation the members of each pair on one side of the body become arranged alternately with respect to those on the other.

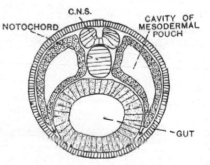

Fig. 579. AMPHIOXUS.—Transverse section of a young larva.

**Formation of the Gut.—** After separation of the notochord and mesodermal pouches from the archenteric wall the endoderm remains, for a short time, divided along its mid-dorsal line; but the edges of the endoderm soon grow towards each other and fuse below the notochord to complete the roof of the endodermal part of the gut (**mesenteron**).

**The Anterior Gut Diverticula.—** After about eight mesodermal pouches have been delimited, two diverticula arise from the dorsal

side of the anterior part of the archenteron.   The left diverticulum
remains small, and eventually, after separation from the gut, acquires

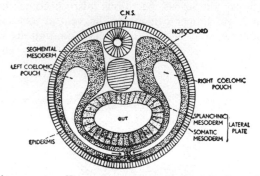

Fig. 580.   AMPHIOXUS.—Transverse section of a larva to show the formation
of the coelom.

an opening into the oral hood on the left-hand side of the larva to
form **Hatschek's pit**.   The right diverticulum also becomes separated
from the gut and enlarges and extends forward to form the cavity
of the "head".

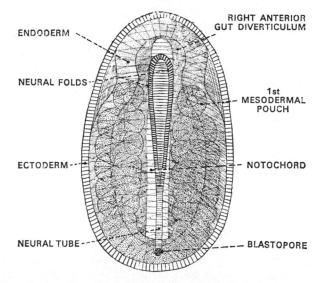

Fig. 581.   AMPHIOXUS.—Optical view of a young larva from the dorsal aspect.

**Differentiation of the Somites.**—The somites retain for some
time small cavities (myocoels) which at one time were continuous with

the splanchnocoel, that is, the somitic cavities are the dorsal remnants of the cavities of the mesodermal pouches, for which reason they are considered as coelomic spaces. The inner wall (the one nearest the mid-line) of each myocoel then thickens to form the **myotome**, the cells of which later form muscle-cells. The outer wall, which remains thin, contributes to the formation of the **dermatome**, from which the sub-epidermal connective tissue is derived. The **sclerotome** and the **ventral parts of the dermatome** originate as sheets of cells from the ventral end of the somite, from which region also the **germ-cells** are proliferated.

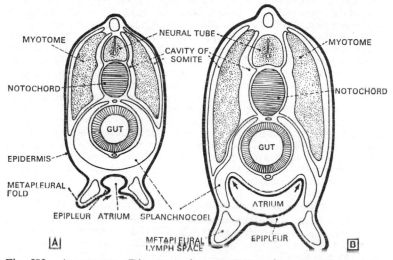

Fig. 582. AMPHIOXUS.—Diagrammatic transverse sections through the larva during the formation of the atrium.

**Hatching.**—The embryo, whose outer cells have become ciliated, breaks free from the egg membrane towards the end of gastrulation and by means of its cilia swims in the surface layer of the sea. It may now be termed a larva (**gastrula larva**) although, because the mouth and anus are not yet formed, it is incapable of feeding.

**Larval Development.**—Soon, changes occur which make feeding on the smaller plankton possible. The larva elongates rapidly, and, by means of movement set up by its myotomes, swimming becomes more active. A patch of ectoderm fuses with the endoderm of the gut on the left-hand side of the ventral surface of the head, and then perforation occurs at this point to form a **mouth**, which has ciliated borders. The **anus** is formed in a similar manner on the left of the ventral mid-line near to the hinder end of the body. The pharyngeal portion of the gut also acquires other openings to the exterior which are the **visceral clefts**. They do not, however, develop in a regular sequence from front to rear, nor are they at first arranged in anything approaching bilateral symmetry. The first cleft to be formed arises as a diverticulum from the anterior end of the pharynx on the right-hand side at a time before the mouth is opened. It eventually forms a coiled glandular tube—known as the **club-shaped gland**—and opens just ventral to the mouth. The club-shaped gland disappears at metamorphosis. Other clefts, fourteen in all, which later form the first set of gill slits of the left-hand side, then perforate in the mid-ventral line. Later on they shift so that they come to lie just to the right of the mid-line.

The first series of gill slits of the right-hand side are eight in number and are formed on the right side of the pharynx above the fourteen previously formed slits. Thus two linear series of slits are present on the same side of the body, the right. The slits of the left-hand side then move across the ventral mid-line and take up their definitive position, and, by the closure of six of them, their number is reduced to eight; that is, they now correspond in number to those of the right-hand side. The gill slits of both sides then become subdivided by tongue bars and further slits are added to the hinder end of both series.

The endostyle is at first a V-shaped strip of ciliated cells on the right-hand side at the anterior end of the pharynx, but the two limbs of the V fuse and extend back in the ventral mid-line between the ventral ends of the gill slits.

Shortly before the gill slits of the right-hand side are perforated, a groove is formed along the ventral surface of the body. It is bordered on each side by a fold of body wall—the right and left metapleural folds—which, in the pharyngeal region, curve over to the right-hand side so that one fold lies above and the other below the gill slits. A sort of shelf, the epipleur or sub-atrial fold, then grows towards the mid-line from the ventral edge of each metapleural fold, and eventually the shelves meet and join so that the groove is converted into a tube. This tube is the atrium. It becomes closed anteriorly but remains open at its hinder end by the atriopore. As the larva gradually assumes some degree of bilateral symmetry, the cavity of the atrium enlarges to its adult proportions and in so doing displaces the body wall so that the splanchnocoel around the anterior end of the intestine becomes much restricted in extent.

**Metamorphosis.**—For some considerable time the larva continues to feed on the plankton of the surface layers. It grows to a stage when it forsakes the surface for the sea-floor, where it enlarges and undergoes a slow metamorphosis into the adult.

## THE DEVELOPMENT OF THE FROG

The embryology of the frog, the example to be considered after amphioxus, provides an instance of development from a moderately telolecithal egg, the young animal hatching as a larva (tadpole) in a much more advanced stage of life than the gastrula larva of amphioxus.

The early stages of development of the frog,* and for that matter of other craniates, follow, in the main, the same general lines as those outlined for amphioxus, but there is at least one important

* Most of the recent work on the development of anuran amphibia has been carried out on continental species, and whilst there is no reason for believing that their development differs in any fundamental respect from that of the common English frog—*Rana temporaria*—the following account is largely a summary of those features which are common to all Anura. It does not refer to *R. temporaria* alone except in the description of the tadpole.

difference. The differentiation or determination of the cytoplasm of the embryonic cells does not take place until a later stage—usually towards the end of cleavage or during gastrulation—and the different types of cytoplasm, being by then confined within the cell membranes of the blastomeres, are redistributed in the embryo by streaming movements of the cells.

**The Egg.**—Each egg when released into the body cavity from the ovary is in the primary oocyte stage and is a spherical cell about 1·6 mm. in diameter. Roughly one-half of its surface is a deep blackish-brown colour owing to the presence of a superficial layer of pigment. The other half is lighter in colour and it is in this half

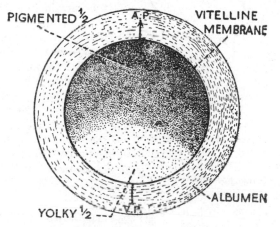

Fig. 583. FROG.—The unfertilised egg, animal pole uppermost.

that the bulk of the yolk is aggregated. The large egg nucleus lies in the pigmented half, which is therefore the animal hemisphere of the egg. The first maturation division of the egg occurs during its passage down the oviduct and the first polar body remains within the thin vitelline membrane at a point near the animal pole. The secondary oocytes receive a thick coat of albumen as they pass along the oviduct, and this swells up on contact with the water in which the eggs are laid and sticks them together, so forming the familiar frog spawn.

**Fertilisation.**—It is in this form, as secondary oocytes, that the eggs are laid, and since the male immediately deposits the spermatozoa upon them, fertilisation occurs at once, usually before the albuminous covering has had time to swell up. The entry of the spermatozoon into the egg appears to induce the completion of maturation, for the second polar body is extruded almost immediately. At the point where the head of the spermatozoon comes into contact with the vitelline membrane, the surface of the egg is raised into a small protuberance into which the spermatozoon penetrates. The passage of the sperm nucleus (male pronucleus) through the cytoplasm is marked by the carrying in with it of some of the pigment granules from the surface, and it traverses a definite path towards the nucleus (female pronucleus) of the now mature egg. This path is commonly straight, passing directly towards the female pronucleus when that nucleus is in the middle of the cytoplasm, but may, on occasion, when the female pronucleus is excentrically placed, consist of a preliminary "penetration path" and a subsequent "copulation path" inclined at an angle to one another. As will be seen later, the direction of the sperm path is important in cleavage. The penetration of the spermatozoon causes certain changes in the distribution of the external pigment in addition to the carrying in of pigment by the spermatozoon. At a point diametrically opposite the point of entry, there is an inward flux of pigment and water which results in the formation, on the surface, of a crescentic area of a greyish, instead of the normal dark, colour. This development is accompanied in some amphibian eggs by streaming movements in the **pigment**.

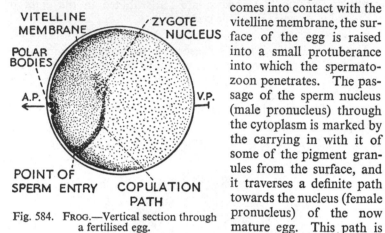

Fig. 584. Frog.—Vertical section through a fertilised egg.

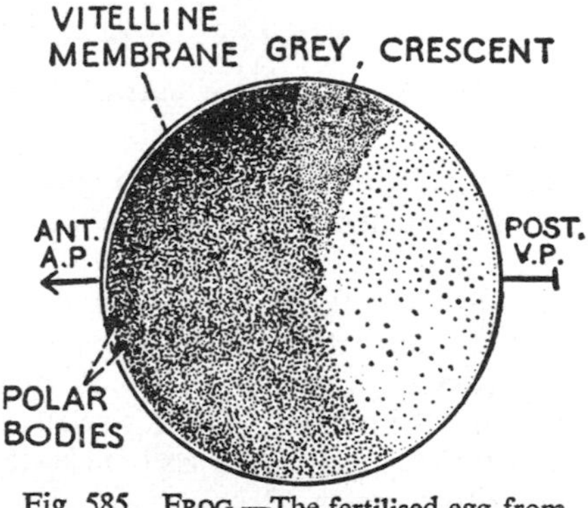

Fig. 585. Frog.—The fertilised egg from the side.

**Symmetry.**—The appearance of the grey crescent on the surface of the egg has materially altered its symmetry. Before fertilisation took place the arrangement of its parts was apparently radially symmetrical about an axis passing directly from the animal to the vegetative poles. Now, however, a definite bilateral symmetry* has been imposed about a plane passing through the middle of the grey crescent and the two poles. Normally this marks the plane of the first cleavage division so that it divides the egg into right and left halves, but actually the position of the first cleavage plane is determined by other conditions (see below).

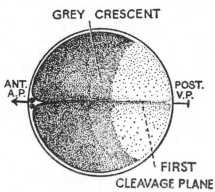

Fig. 586. FROG.—2-celled stage; dorsal view.

Another event which follows on the appearance of the grey crescent is that the orientation of the future embryo can now be established, for the grey crescent marks the position of the **dorsal lip of the blastopore** when it is first formed, and consequently of the dorsal surface of the early embryo. It is therefore possible now to indicate the main axis of the future embryo, though

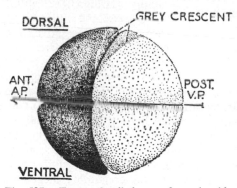

Fig. 587. FROG.—8-celled stage from the side.

for the present the egg always floats into position with the animal pole uppermost however it may be displaced. This is because the egg has become free from the vitelline membrane as a result of a slight shrinkage due, in all probability, to the extrusion of a little water.

**Cleavage.** — Cleavage is of the holoblastic, unequal type for the quantity of yolk present

---

* There is, however, much evidence in support of the view that the position of the grey crescent has already been determined before the entry of the sperm. This merely causes its presence to be revealed. The copulation path of the sperm happens to coincide, in most cases, with the plane of bilateral symmetry.

is sufficient to cause considerable hampering of cleavage in the vegetative hemisphere.

The first cleavage plane is meridional and its position is determined by the direction of the terminal portion of the sperm path. If the penetration path is in a straight line with the copulation path, then the first cleavage plane passes through the centre of the grey crescent; but if the female pronucleus is excentric and the copulation path is inclined to the penetration path, then the first cleavage plane may pass to one side of the grey crescent, since it always coincides with the plane of the copulation path or, more accurately, the equator of the first spindle. This does not, however, interfere with the symmetry of the future embryo, for the crescent will be divided at a subsequent cleavage. The first cleavage divides the fertilised egg or zygote into two blastomeres, the cleavage being indicated on the surface by a meridional furrow which is more pronounced in the pigmented hemisphere than in the yolk-containing region.

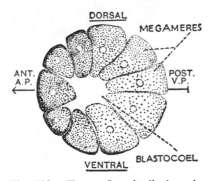

Fig. 588. FROG.—Longitudinal section through an early blastula.

Even before the furrow of the first cleavage plane has extended completely into the vegetative region, the second division has begun. This is also meridional, but at right angles to the first. It extends rapidly through the animal hemisphere and more slowly through the vegetative, producing four blastomeres. The third cleavage plane is latitudinal and is situated well above the equator of the egg. It produces eight blastomeres, four pigmented micromeres at the animal pole, and four large yolk-containing megameres at the vegetative pole. The fourth cleavage division is meridional and each of the micromeres divides into two, producing eight, and somewhat later divides the megameres in a similar way. But the fact that the third cleavage division has separated the almost yolk-less, pigmented micromeres from the yolk-containing megameres has removed from the former the hampering effect of the yolk to the passage of the cleavage planes. Thus, although for a time the cleavage planes follow an orderly sequence of alternating meridional and latitudinal direction, the division of the micromeres proceeds much more rapidly than that of the megameres with the production of a blastula, the upper hemisphere of which is composed externally of small pigmented cells and the lower of large yolk-containing cells, the blastocoel being nearer the animal pole. In all these happenings the position of the grey crescent remains indicated by the colour of the cells formed within its limits.

**The Blastula.**—In the upper or animal hemisphere, surrounding the blastocoel, the wall of the blastula consists of several layers of small cells, the outer ones of which have pigment on their external surfaces. The lower, or vegetative hemisphere, is composed of a mass of large yolk-containing cells.

From a mere inspection of the late blastula, however close and detailed, it would appear that there is little distinction between its constituent cells except degree of pigmentation, size, and yolk content. There are thus no visual clues or indications of the parts that the various cells will play in future development. Fortunately, an application of the technique of *intra vitam* staining has gone a long way to solving this difficulty of interpretation, and it is now possible to construct maps (fate-maps) of the amphibian blastula which show in a very detailed way the **prospective** (or presumptive) **fate** of different regions of the blastula wall. Briefly, small parts of the blastula are coloured with various harmless dyes so that any particular cell area can be recognised for a considerable time and thus followed into the gastrula and even into later stages of development.

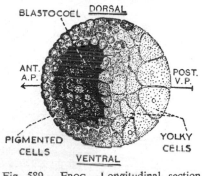

Fig. 589. FROG.—Longitudinal section through a late blastula.

It must be explained that the limits of any particular group of cells are not sharply defined, and the fate-maps show the mean results of very numerous series of marking experiments. It is also important to realise that, with the exception of the cells derived by division of the grey crescent, the fate of the different cell areas is by no means fully fixed or determined, for, as will be mentioned later, they can, under conditions which depart from the normal, show a great versatility of behaviour. The fate-maps, in fact, merely show the part which any particular group of blastula cells may be expected to play in the formation of the main organ rudiments under normal conditions. But, as is implied in the terms "prospective" or "presumptive" fate, they give no account of the full range of qualities lying latent in any group of cells and which can be revealed only by experimental analysis. This does not, however, imply that although the blastula is capable of considerable readjustment if interfered with by injury, or other means, its cells lack all traces of individual characteristics acquired during the earlier stages of development. This matter will be alluded to in a later section (p. 715).

The mapping out of the areas with different prospective fates has in many amphibia been carried out in great detail, and the following main areas may be distinguished. The **prospective ectoderm** occupies almost all of the animal (anterior) half of the blastula and is subdivisible into **prospective epidermis,** which lies

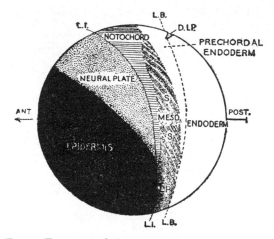

Fig. 590. FROG.—Fate-map of the prospective areas from the side (based on Vogt).

along its ventral and ventrolateral parts and **prospective neural (medullary) plate,** a broad band stretching over the antero-dorsal and lateral parts. Posterior to this (*i.e.* nearer the vegetative pole) on the dorsal surface, is a smaller area with two lateral horns passing down on each side of the blastula. This is the **prospective notochord.** The **prospective mesoderm** lies chiefly in two lateral areas behind the horns of the notochordal area. The two meso-dermal fields are separated in the mid-dorsal line by the notochordal cells, but are continuous across the mid-ventral line, so that the mesoderm lies in a crescentic area in much the same relative position that it does in the blastula of amphioxus. The remainder of the vegetative (posterior) half is occupied by the large **yolky endoderm**

**cells** while the **prechordal plate** is a small area in the mid-dorsal line from which endoderm and mesenchyme are derived.

It will be realised from this description that the future axial structures lie in broad bands transverse to the body axis. That is, they are different not only in position but in shape from the structures to which they will give rise in the later embryo. The next important step in development is the re-arrangement of these areas so that they come to occupy their definitive positions and proportions in the embryo. This process is, of course, gastrulation, but because of the appreciable quantity of yolk present, it does not take place by straightforward invagination as in amphioxus, but by more devious means.

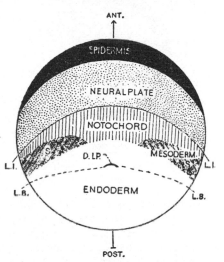

Fig. 591. FROG.—Fate-map of the prospective areas from the dorsal aspect (Vogt).

**Gastrulation.**—As seen from the outside, gastrulation appears to consist of the progressive growth of the pigmented cells over the lighter-coloured, yolk-containing cells, until all but a small circular patch, the **yolk-plug** (marking the position of the blastopore), is covered. Actually, gastrulation is far more complicated than this and is due to several types of activity going on at one and the same time; but the end result is, of course, that the prospective ectoderm comes to enclose the prospective notochord, mesoderm, and endoderm. Gastrulation is due to a remarkable series of mass migrations of cells, which may be termed formative movements, and is not caused by the localised production of new

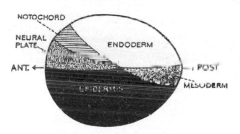

Fig. 592. AMPHIOXUS.—Map of the organ-specific areas from the side for comparison with Fig. 590.

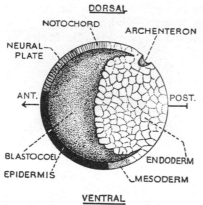

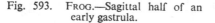

Fig. 593. FROG.—Sagittal half of an early gastrula.

embryonic material as was formerly supposed. It is essentially a rearrangement of material already present, and it has been shown that whilst the cells continue to divide actively they do so at more or less the same rate throughout the whole embryo. There is thus, *at this stage*, no special region of "proliferation" in the frog or, for that matter, in any other vertebrate embryo.

Although during normal development the various formative movements are perfectly integrated, it has been shown by various surgical experiments and by observation of isolated living portions that the effects are, to a large extent, independent of one another, and they can thus be conveniently described separately. Firstly, the smaller cells of the dorsal side of the animal (anterior) half of the blastula begin to migrate towards the posterior end and spread out over the cells of the vegetative (posterior) half. Then along a small crescentic area, on the dorsal side, some way behind the posterior margin of the prospective notochord and the prechordal-plate area, the cells start to roll over and migrate inwards beneath the outer layers. In this way there is formed a crescentic groove bounded in front by the intucking prechordal plate, and behind by the endodermal cells. This groove is the beginning of the **archenteron** and its anterior margin is the **dorsal lip of the blastopore**, soon to be occupied by the intucking notochordal cells. Later the area of intucking extends laterally so that the groove becomes a

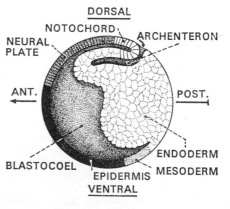

Fig. 594. FROG.—Sagittal half of a later gastrula.

wide crescent with backwardly projecting horns (lateral lips).

Meanwhile, as fresh material rolls inwards over the lips of the blastopore by the migration of cells from the animal half, the groove becomes deeper and deeper so that the archenteron soon forms an extensive cavity and, at the same time, the blastocoel becomes diminished. Eventually the blastopore becomes a complete

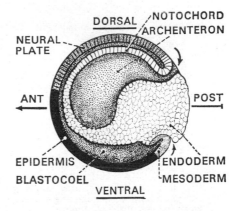

Fig. 595. FROG.—Sagittal half of a gastrula at a later stage than Fig. 594.

circle, by the meeting of the lateral lips on the ventral side, and after this it becomes progressively smaller by gradual contraction of its margin. Whilst the intucking (**invagination**) of cells over the lips of the blastopore has been taking place, the gradual spread of the small cells over the yolky cells has continued without interruption and the blastopore becomes progressively carried to the posterior end, so that by the time the ventral lip is established the blastopore is seen as a circular aperture near the original vegetative pole.

Gastrulation is, then, brought about by three *main* processes: overgrowth (epiboly) or spreading of the cells of the animal half over those of the vegetative half, intucking (invagination) of material around the margins of the blastopore, and contraction of the margins (lips) of the blastopore. These movements are, however, accompanied by others which result in a gradual extension in length and a convergence towards the midline of the areas which will form the axial

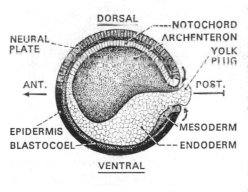

Fig. 596. FROG.—Sagittal half of a completed gastrula.

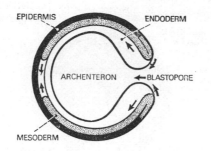

Fig. 597. FROG.—A purely diagrammatic horizontal section of a gastrula showing mesoderm and endoderm migrating inwards at the lateral lips of the blastopore.

structures of the embryo (prospective neural plate, notochord, and somites); an extension in all directions of prospective epidermis; and by expansion of the lateral-plate mesoderm when it has been enclosed. But despite this great extension of the epidermis, some of the yolky cells are not enclosed until a relatively late stage in development and protrude through the blastopore as the **yolk-plug.**

The chief result of gastrulation is that the main regions of the wall of the blastula undergo changes in shape and come to lie in their definitive positions in the embryo. Thus the material first carried in at the dorsal lip is prechordal ento-mesoblast of the prechordal plate from which cells wander away to contribute to the head mesoderm. It forms the extreme anterior end of the archenteric wall and remains continuous behind with the yolky endodermal cells, which at this stage are not yet enclosed. They later form the sides and the floor of the archenteron. The next material, in order, to pass inwards at the dorsal lip is prospective notochord. It soon forms a strip of cells in the roof and is continuous in front with the anterior (endodermal) wall of the archenteron. Mesoderm begins to be invaginated as the area of intucking extends to the lateral lips. It passes inwards beneath the prospective epidermis and comes to lie *between this and the endoderm*, which has meanwhile become enclosed to form the sides of the archenteron. However, even when fully formed, the archenteric cavity does not extend far into the ventral part

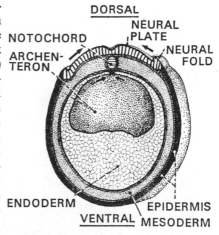

Fig. 598. FROG.—Anterior half of an early neurula.

of the gastrula because of the accumulation there of the large yolky endoderm cells, some of which (yolk-plug) also lie between the archenteric cavity and the ventral lip of the blastopore. With the completion of the blastopore the mesoderm of the ventral surface is invaginated at the ventral lip. It will, of course, be realised that the sequence given above merely indicates the order of cell migration; the process of migration being a continuous one.

Another effect of gastrulation is that owing to the migration of cells to the mid-dorsal line and the general stretching of the animal half of the embryo, the roof and sides of the blastocoel become thinner, but when gastrulation is completed the original thickness is re-established, though now the cells are arranged in two definite layers above and around the sides of the archenteric cavity, the line of demarcation between them being the remains of the now nearly obliterated blastocoel.

Gastrulation also causes a shift in the centre of gravity of the embryo which, in the blastula stage, floats with the animal pole uppermost. This is because the blastocoel, a fluid-filled cavity, lies almost entirely within the animal half and renders it lighter than the vegetative half which is occupied by cells full of heavy yolk.

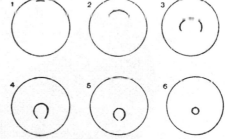

Fig. 599. FROG.—Successive stages in the formation of the blastopore as seen from the vegetative pole.

The development of the archenteron within the dorsal part of the embryo gradually occludes the blastocoel, and the embryo, which is free to rotate inside the vitelline membrane, swings round until it lies with the new cavity, the archenteron, uppermost—that is, until the embryo axis is horizontal.

**The Onset of Differentiation.**—From this account it will be understood that, although the effect of cleavage and gastrulation has been to produce an embryo with the same essential parts as in amphioxus, the influences operating to produce that end have been somewhat different, and it will be of interest at this point to examine the results of recent experimental work on the factors involved, and to contrast the methods of differentiation of the frog embryo with those noticed in amphioxus. In that example, on completion of the cytoplasmic streaming after fertilisation, the fate of the different cytoplasmic regions was fixed, and if a portion of the egg was removed, then the embryo was deficient in some particular feature. In the frog, however, the determination of the embryonic regions is not final until the completion of gastrulation. Yet there is a progressive determination and a gradual mapping out of the main regions from the time of fertilisation onwards. For example, fertilisation imposes bilaterality and the first cleavage plane normally determines the right and left halves of the embryo; but the embryo can still readjust itself to injury, as is shown by the experiment of separating the first two blastomeres. Each develops into a smaller but otherwise perfect embryo. Final determination at this stage (and even much later stages) is confined to the grey crescent, the cytoplasmic precursor of the prechordal plate and prospective

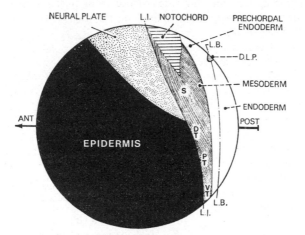

Fig. 600. Fate-map of the prospective areas in *Discoglossus*.

notochord and mesoderm. Thus, if a lateral or a dorsal half of an embryo in the two-celled, early cleavage, or blastula stage, is sliced off it will so readjust its growth that a normal embryo results, whereas the separated ventral halves either shrivel up and die or, at best, produce ventral structures only. That is, only portions of the embryo which contain some of the grey crescent material have the power of regulating their development.

The nuclei of the blastomeres apparently play but little part in the onset of determination and up to a certain stage in development they are known to be mutually equivalent, and such differentiation as exists resides in the cytoplasm of the cells. A single experiment out of many may be cited as showing this equivalence of the nuclei. A fine hair was tied around the fertilised egg of a newt, constricting it into the shape of a dumb-bell and confining the nucleus to one side of the constriction. Only the side with the nucleus underwent cleavage. Later the ligature was released and one nucleus was allowed to pass through into the enucleate half of the egg. The loop was again applied and tightened in the plane of bilateral symmetry so as to separate the egg into two halves. Both halves grew into normal embryos, provided that separation was carried out not later than the sixteen-celled stage. After this point normal embryos failed to develop from the half which had previously been without nuclei, presumably because its cytoplasm had undergone degenerative changes. Thus the presence of *any* nucleus out of sixteen is sufficient to induce normal development which its absence rendered impossible.

After the formation of the grey crescent, the next important step in differentiation is the blocking out of the main regions of the embryo on the surface of the blastula. This differentiation is quite invisible and has only been detected by experimental analysis, which has chiefly taken the form of isolation of the regions of different prospective fates and their subsequent culture in suitable salt solutions. These experiments show that although in normal development the main organs of the embryo arise from particular regions of the blastula wall, the degree of differentiation that they have attained at the beginning of gastrulation is variable. Thus, pieces of prospective epidermis if removed and cultured develop into epidermal tissue; but prospective neural plate, on the other hand, also gives rise to epidermis. Prospective endoderm, when isolated, forms endodermal structures alone and, depending on the endodermal region from which it was removed, the fragment will build definite regions of the alimentary system. Similarly, the prospective notochordal and mesodermal regions always follow their prospective fate. These experiments show that the cells of the early gastrula, although not visibly different from one another, are not indifferent, for they have a tendency to develop in a certain way; but it has also been shown that their ultimate fate is by no means completely determined or fixed at this stage, although the cells retain their individual characteristics.

**Loss of Plasticity—Organisers.**—The regions of the blastula, with the striking exception of the grey crescent cells (chorda-mesoderm), are, at the beginning of gastrulation, still plastic with regard to the part they will play in the building of the embryo. This has been shown by an extensive series of transplantation experiments (chiefly performed on newts), during which pieces of the embryo were exchanged with others in differing positions. To facilitate subsequent recognition the exchanges of material were usually made between embryos of different species of newt, for the cells retain their own characteristics as regards size, pigment, and speed of division after transplantation (thus showing that they are not completely dominated by the host tissues). If, during the early gastrula stage, a piece of prospective epidermis is grafted into the neural plate region of the same or of another embryo it will differentiate into nervous tissue and form an integral part of the central nervous system, whilst prospective neural plate develops in epidermal regions into typical epidermis. Fragments of prospective epidermis introduced into the mesodermal area are carried inwards during gastrulation and produce quite normal mesodermal structures, somitic or lateral-plate structures, according to the exact position of the graft. In a like way, prospective ectoderm can give rise to endoderm if transplanted into a suitable position.

The grafts behave, then, very differently from their prospective fates when grafted into abnormal situations in the early gastrula. Their normal tendencies, as indicated by *intra vitam* staining and by culture in indifferent media, have been overridden by other influences. Grafts of prospective mesoderm or notochord, however, behave in quite a different fashion; for no matter into what region they are transferred they always sink inwards and follow out their prospective fates. That is, they give rise to mesodermal structures and notochord alone and (with certain exceptions to be noted later) are not influenced by their surroundings. In fact, exactly the reverse is true: the grafts exert a profound effect on the surrounding tissues, causing them to produce structures which normally they would not have done. For example, a piece of tissue from the dorsal lip of the blastopore (future notochord and axial mesoderm) when grafted into the prospective epidermal region of another or of the same embryo in the early gastrula stage will invaginate and form a second set of somites and a notochord, and will induce the overlying prospective epidermis of the host to form a typical neural tube. At the same time, the endoderm below the graft will be modified to form a second gut. In successful cases a fairly complete secondary embryo, formed partly from the host tissues and partly from those of the graft, may result. From these experiments it will be seen that the dorsal lip of the blastopore has the power of forcing the cells of the early gastrula to swerve from their normal developmental paths and to organise themselves into a secondary embryo.

The dorsal lip of the blastopore is therefore termed the **organiser** (primary organiser). Other grafting experiments show that this organising property is possessed, although to a lesser degree, by the whole of the prospective mesoderm, that is, by the tissue which later lies in the lateral and ventral lips of the blastopore. From these and from other very numerous experiments it is concluded that during normal development, the tissue invaginated at the dorsal and lateral lips of the blastopore not only enters into the composition of, but plays a predominant part in, the organisation of other cell regions of the embryo. Its influence decides once and for all what part the main areas will play in development. Normally the final determination imposed by the organiser is the same as the prospective fate of the tissues, but the organiser is capable of overriding the plastic or labile determination already noticed at the beginning of gastrulation. The fate of the cells is sealed, so to speak, after they have come under the influence of the organiser. The organiser itself, however, possesses regional differentiation, as is shown by the fact that the early dorsal lip of the blastopore (which is carried during invagination to the anterior end of the embryo) when grafted into the trunk region of an embryo, will induce the epidermis to form head region structures—brain, eyes, ears, and the like—so that it may be termed the head organiser.

The cell area lying close to the dorsal lip of the blastopore acts, then, as a head organiser and comparable grafting experiments show that the later invaginated material of the lateral lips acts as a trunk and tail organiser. But, of course, there is no sharp boundary between head and trunk organisers, and both are necessary for the formation of a complete and perfect embryo.

As gastrulation proceeds, the various main organ systems of the embryo are laid down under the influence of the organiser and they themselves then acquire the power of inducing later-formed structures to develop. These, in turn, develop organising properties —and so on. It is thus possible to recognise a series of organisers of the second, third, and even fourth, degree which are arranged in a sort of hierarchy at whose apex is the primary organiser.

Even in the early stages of gastrulation the head organiser can be broken down experimentally into separate components, each of which is concerned with organising some particular portion of the head, such as fore-brain plus eyes, hind-brain, gills, auditory vesicles, etc. During later stages of gastrulation the segregation of these organisers of the second degree becomes more pronounced and is usually accompanied by the formation of distinct tissues and organs. Nevertheless, the segregation of subsidiary organisers is rarely complete; their fields of influence overlap and they may sometimes co-operate.

The classic example of an organiser taking effect after the main embryonic axial structures have been laid down under the influence of the primary organiser is the tissue of the optic cup, which causes the overlying ectoderm to differentiate into a lens. In *Rana temporaria* it was found that the optic cup will differentiate normally if grafted into unusual situations and it is said to be **self-differentiating**. If, on the other hand, the optic cup is removed, a lens fails to develop so that the lens is said to be a **dependent-differentiating** structure, for it will not arise in the absence of an organising stimulus from the optic cup.

Other comparable effects have since come to light and among these may be mentioned that the development of the nasal placodes depends on the presence of the fore-brain, the posterior lobe of the pituitary body depends on the presence of the tip of the notochord, teeth on the pharyngeal endoderm, dorsal fin on the mesenchyme derived from trunk neural crests, the cornea on the optic cup and lens, the tympanic membrane on the annular cartilage, whilst many other examples could be added to this list.

So far no mention has been made of the way in which the organisers make their effects felt, and it may be stated at the outset that the process is far from being fully understood. It is fairly certain, however, that the stimulus is due to the diffusion of some chemical substance or substances from the organising tissue. This is indicated by the fact that pieces of prospective epidermis or even pieces of agar, after being placed in contact with the dorsal lip of the blastopore for a short time, acquire the ability to induce the formation of a neural plate from prospective epidermis.

Many of the properties of the primary organiser can be imitated by a great variety of chemical substances including methylene blue, steroids, fatty acids, and practically any animal tissue in which the proteins have been denatured by heat, precipitation by alcohol or other means. Any of these substances, when placed in position in the young gastrula, induce the overlying ectoderm to differentiate into a neural plate, and in favourable circumstances into a more or less complete neural tube. It should be noted that it is only after killing that ordinary tissues acquire the special property of evocation, that is the ability of calling forth a neural plate from prospective epidermis and that the active substance concerned with evocation is released in the natural state only by the living organiser. This natural organiser, as has been mentioned, has other properties which cannot be imitated artificially, for, as shown by numerous grafting experiments, it possesses a definite regional differentiation and the final structure and nature of the graft depends partly on the exact region of the organiser selected and partly on the position into which it was grafted. Nevertheless, the organiser always tends to produce complete structures, partly out of its own and partly from the tissues of the host. This integrating effect—the tendency of host and graft to combine to produce a workable "whole"—has been termed **individuation,** and it must be supposed that in its natural state and position the normal organiser is exerting an individuating effect on the developing embryo, helping to ensure its harmonious development.

In the frog, then, the orderly sequence of changes in cell arrangement and differentiation which result in the young embryo are bound up with the formative movements of gastrulation and the effects of the primary organiser. At first the organiser is the dorsal lip of the blastopore, the position of which is revealed at fertilisation by the formation of the grey crescent. When the primary organiser has performed its office, comparable duties, but of a more restricted nature, are handed on to other regions of the embryo as they become differentiated. Regarded in another way, development is a process of gradual decentralisation accompanied by the differentiation of subsidiary fields of influence which soon become self-differentiating, as can be shown by transplantation experiments carried out before any histological differences are apparent. Thus, by suitable experimentation in the neurula stage, eye-fields, gill-fields, neural-tube fields, and limb-fields can be mapped out. They overlap to some extent and their properties are most strongly evinced in the centre of each field. Thus when a district (as subsidiary fields are sometimes called), or even a portion of a district, is transplanted it retains its individuality to some extent, but is also influenced by the host district into which it is transplanted (cf. transplants of dorsal lip). When a large number of successful transplants have been made and examined statistically it is found that they give rise to ears, eyes, nasal pits, suckers, limbs, and so on, even when separated by some distance from the actual regions which will give rise to the particular organ in the host. Yet the frequency of the complete formation of a certain organ is greatest when the transplant lies nearest to the centre of the field for that particular organ in the host, so that the intensity of the tendency to form a particular organ falls off with increasing distance from the centre of the field or district.

This field influence does not exhaust itself when, under normal conditions, it has produced its effect, *i.e.* the formation of the organ rudiment. It can repeat its performance if the rudiment is extirpated and replaced by some reactive or competent tissue. For example, if an early eye-rudiment is cut out and replaced by competent epidermis, a new eye will develop. Similarly, an optic cup can organise a succession of lenses.

In spite of the overlap of various organ districts, no confused or intermediate types of organs make their appearance in the embryo, for each field seems to claim any particular cell-group or else reject it. Any one cell-group can, it seems, fall under the influence of one field and one only.

The early development of the frog, therefore, illustrates in a very clear way the confirmation of bilateral symmetry by fertilisation; the modification of cleavage by the unequal distribution of yolk and its effect upon gastrulation; the differentiation of areas in the blastula and the mode of operation of organisers. It also serves as a contrast to the type of development such as that found in amphioxus where differentiation was visibly present in the zygote even before fertilisation.

**Organogeny.**—The later phases in development are concerned with the elaboration of the main embryonic organs already present or indicated at the end of gastrulation and with the laying down of other structures, with the result that the embryo gradually becomes adapted to lead an active free-living existence, during which it feeds and grows in preparation for metamorphosis into a miniature frog. In other words, the events which follow the completion of gastrulation lead to the production of a form which eventually hatches from the egg as the familiar tadpole of our ponds and ditches.

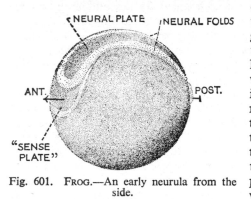

Fig. 601. FROG.—An early neurula from the side.

**The Formation of the Neural Tube.**—By the conclusion of gastrulation the prospective neural-plate area has come to occupy an elongated pear-shaped tract along the length of the mid-dorsal region. It now constitutes the **neural plate** from which the brain and spinal cord will be developed. Along the edges of this neural plate appear the **neural folds**, and between them the plate sinks downwards due to their progressive uprising and incurving towards the median line. Eventually the folds meet and fuse, with the formation of the neural tube lying beneath the re-formed ectodermal surface and above the notochord.

The closure of the neural tube begins just in front of the mid-region and proceeds both anteriorly and posteriorly. At the front end the tube remains open for a time as a **neuropore**, but posteriorly the neural folds extend laterally on each side of the blastopore which, by their closure, becomes enclosed. As a result of this enclosure of the blastopore the neural canal is in continuity with the archenteron, for by this time the yolk-plug has been withdrawn. The communicating channel thus formed is the **neurenteric canal**.

Because of the pear-like shape of the neural plate the neural tube is much wider at its anterior end. This wider end gives rise to the brain, and in it the development of internal ridges soon indicates the **three primary cerebral vesicles**. From the remainder of the neural tube the spinal cord is derived.

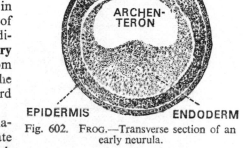

Fig. 602. FROG.—Transverse section of an early neurula.

During the differentiation of the neural plate there appears along each edge a linear band of cells as a thickening of the ectoderm, and when the neural tube is completed, they lie along the dorsolateral surface of the tube. These bands of cells are the **neural crests**, and from them are derived the **dorsal root ganglia** of the

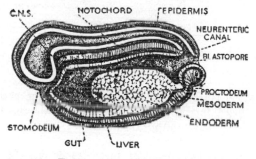

Fig. 603. FROG.—Sagittal half of a late embryo.

spinal nerves and parts of the ganglia of the sensory components of the cranial nerves. The arrangement of the ganglia is metameric and their positions correspond with those of the somites. Some of the neural crest cells give rise to mesenchyme, an important product of which is the visceral arch skeleton.

**The Separation and Development of the Notochord.**—As has been seen above, the chorda cells in the completed gastrula lie as a narrow sheet of cells in the mid-dorsal region of the roof of the archenteron. At first they are not very clearly distinguishable from the mesoderm lying on either side, but soon the mesodermal cells become separated from the chorda cells by a narrow cleft. The chorda cells then arrange themselves in the form of a cylindrical rod, and the characteristic vacuolated appearance is assumed. Around the cells the notochordal sheath is developed.

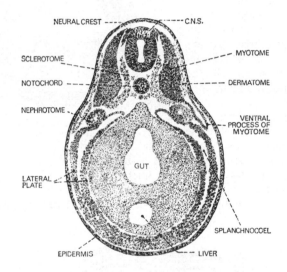

Fig. 604.    FROG.—Transverse section of a young tadpole through the trunk region.

Immediately below the notochord some of the endoderm cells become arranged in the form of a slender rod—the hypochordal rod—which eventually disappears.

**The Further Development of the Mesoderm.**—At the end of gastrulation the bulk of the mesoderm cells lie closely applied, almost like a mantle, to the endoderm cells, but usually clearly demarcated from the ectoderm. Very soon the differentiation between the endoderm and mesoderm becomes apparent also, when it is seen that the mesoderm does not extend right to the front end of the embryo. It extends as a sheet of tissue, several cells thick, from each side of the notochord almost down to the mid-ventral

line, even in the region where the large yolk-containing cells lie
piled up in the floor of the archenteron. As development proceeds
a split appears in the mesoderm, the cavity thus formed being the
coelom. The split first appears in the dorsolateral region on each
side and progresses ventralwards. Eventually, the two halves of
the mesoderm having now met in the mid-ventral line, the intervening wall thus formed breaks down so that the coelom extends,
except near the dorsal mid-line, around the developing gut. At
the same time that these developments are taking place the mesoderm is being further differentiated into its parts. That lying on

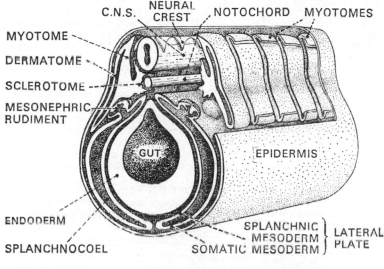

Fig. 605. Frog.—Stereogram of a portion of the trunk region of a young
tadpole.

each side of the notochord forms the **somites** and is distinguished
from the remainder—which forms the lateral-plate mesoderm by
becoming metamerically segmented. Between the somite and the
lateral-plate mesoderm lies an intermediate region which forms the
**nephrotome**.

Each somite makes three contributions to future embryonic
structures: the **myotome,** from which the body muscles are derived;
the **dermatome,** the mesodermal contribution to the skin; and the
**sclerotome,** from which the axial skeleton is developed. Within
the somite and nephrotome, portions of the coelom are present,
forming the myocoel and nephrocoel respectively, whilst that contained by the somatic and splanchnic layers of the lateral-plate

mesoderm becomes the general body cavity (perivisceral cavity or splanchnocoel).

Anteriorly, in the region of the pharynx, where the development of the somites is interfered with by the development of the sense capsules (particularly the auditory capsules) and visceral clefts, the coelom is restricted to a ventral portion which becomes modified and separated from the splanchnocoel to form the pericardial cavity. It is in this region that the rudiments of the heart are laid down.

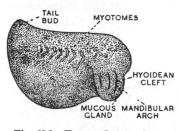
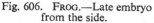

Fig. 606. FROG.—Late embryo from the side.

**The Gut.**—In the description of gastrulation it was emphasised that the invaginated prospective endodermal area forms the walls of the archenteron with the exception of a narrow cleft immediately beneath the notochordal cells. The anterior and dorsolateral walls are thin but the floor of the archenteron is occupied by the piled-up, yolk-containing cells. At the time of the differentiation of the notochord the upper edges of the endoderm meet and fuse in the mid-dorsal line so that the roof of the archenteron is completed.

Posteriorly, the blastopore is occluded by the yolk-plug, but later this is withdrawn into the floor of the archenteron and the blastopore loses its circular shape, becoming elongated dorso-ventrally and, by the meeting of the central portions of the lateral lips, 8-shaped. It is the upper portion of this 8-shaped aperture which becomes enclosed by the fusion of the neural folds to form the neurenteric canal. The lower part marks the position of a future invagination

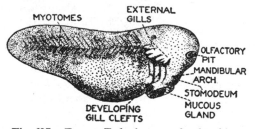

Fig. 607. FROG.—Tadpole soon after hatching.

of the ectoderm to form the **proctodeal invagination** which eventually establishes connection with the archenteric cavity.

Anteriorly, ventral to the front end of the neural tube in the middle of an elevated area of ectoderm (the so-called "sense-plate"), a similar inpushing of the ectoderm appears, giving rise to the **stomodeum** from which the **buccal cavity** will be developed. It is from this stomodeal invagination, below the forebrain, that the **hypophysis** arises to contribute to the formation of the **pituitary body**. The archenteron, which is, of course, lined by endoderm, will form the mesenteron and its derivatives. Very soon after gastrulation has been completed, differentiation of the mesenteron begins and the anterior region, which will form the **pharynx, oesophagus,** and **stomach,** is recognisable as a wider, more spacious cavity with relatively thin walls. Immediately behind it is a depression in the floor of the gut which is the so-called **liver diverticulum,** the cells of whose

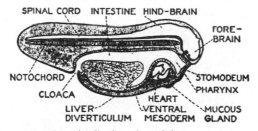

SPINAL CORD  INTESTINE  HIND-BRAIN

FORE-BRAIN

NOTOCHORD

STOMODEUM

CLOACA

PHARYNX

LIVER DIVERTICULUM  HEART  VENTRAL MESODERM  MUCOUS GLAND

Fig. 608. FROG.—Longitudinal section of the same stage as Fig. 607.

walls later proliferate to form the **liver,** whilst the lumen of the diverticulum persists as the lumen of the **bile duct.** The pharyngeal region itself is particularly wide from side to side and soon its lateral walls push outwards in three places on each side towards the ectoderm. These outpushings, which are the rudiments of the first-formed **visceral pouches (hyoidean** and **first** and **second branchial pouches),** are at first merely double folds of endoderm whose walls only later draw apart to form pouches or clefts as such. Two more pouches are formed later on, making a total of five. As they grow outwards towards the ectoderm, the mesoderm lying between them becomes displaced and concentrated in the region between successive pouches, so that a series of more or less vertical strips is formed which are the mesodermal forerunners of the **visceral arches** derived from neural crest cells of the head. Then, in the ectoderm on the sides of the head a vertical furrow appears opposite each visceral pouch; and, except in the case of the hyoidean pouch, which remains as the cavity of the **middle ear** and **Eustachian tube,** perforation occurs in these places, so that eventually the pharynx communicates with the exterior by way of a series of **visceral clefts.** This, however, does not take place until after hatching.

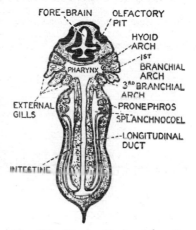

FORE-BRAIN    OLFACTORY PIT

HYOID ARCH

1ST BRANCHIAL ARCH

PHARYNX

3RD BRANCHIAL ARCH

EXTERNAL GILLS

PRONEPHROS

SPLANCHNOCOEL

LONGITUDINAL DUCT

INTESTINE

Fig. 609. FROG.—Horizontal section through a newly hatched tadpole.

### The Elongation of the Embryo.—

Up to the end of gastrulation the embryo is roughly spherical in shape, but after this, elongation in an antero-posterior direction sets in. This affects all regions of the body, but is accentuated by the formation of a tail, that is, with the extension of the body backwards beyond the proctodeal invagination. Into this caudal region, notochord, central nervous system, somitic mesoderm, and mesenchyme (but not endoderm) extend. It is now generally agreed that this embryonic post-anal extension of the body, usually termed the **tail bud**, arises chiefly by a rearrangement of the tissues present in and near the hinder end of the late gastrula or neurula, and not as a result of a rapid proliferation of indifferent cells around the margin of the nearly-closed blastopore, so that in one sense the term "tail bud" is misleading. Of particular importance in the formation of the young tail is the

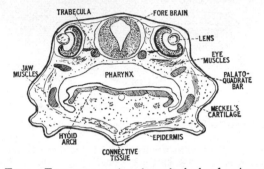

Fig. 610.  FROG.—Transverse section through the head region of a 12-mm tadpole.

stretching in the longitudinal direction of the posterior part of neural plate and neural folds, for it has been shown that in certain Amphibia (and there is no good reason for supposing that the frog behaves differently) the somitic mesoderm of the tail arises from the posterior part of what is *morphologically* the neural plate, but is, *developmentally*, prospective tail somites. The somitic mesoderm of the tail thus becomes enclosed by the hinder part of the neural folds, and these structures, stretching enormously, later form the dorsal and

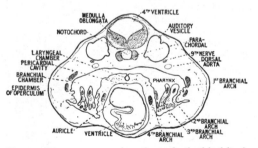

Fig. 611.  FROG.—Transverse section through the hind-brain region of a 12-mm tadpole.

ventral fins of the tail. Mesenchyme required to build other tail structures is derived, in all probability, from the hinder parts of the neural crests.

**External Gills.**—Shortly before hatching, the skin covering the dorsal part of the first and second, and later that of the third, branchial arch (*i.e.* visceral arches three, four, and five) grows out to form three pairs of feathery external gills, which are the first gills to become functional as respiratory organs.

**Mucous Glands.**—At about the same time as the external gills are forming, a deep pit lined by glandular epithelium develops in the ectoderm at the ventral end of each **mandibular arch** (*i.e.* the tissue between the hyomandibular pouch and the stomodeal invagination). These pits then coalesce to form the **mucous gland** by which the larva attaches itself to weeds and other objects after it has hatched and wriggled free from the jelly.

**Chief Features of the Tadpole just prior to Hatching.**—Under average conditions in this country the tadpole is ready to hatch from the egg membranes about fourteen days after fertilisation. It is a small, blackish, fish-like creature about seven millimetres in length. The neural tube is completed and the neurenteric canal closed, whilst the rudiments of the related structures including the organs of special sense are already laid down. The gut is a straight tube of variable dimension, and four pairs of visceral pouches are present in the pharyngeal region, although they do not as yet open to the exterior. At this time the gut has only one opening, the anus. The heart is a simple S-shaped tube, but is beginning to show divisions into chambers. Blood vessels are

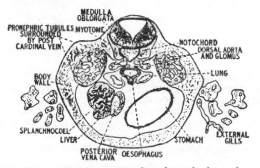

Fig. 612. FROG.—Transverse section through the regions of the pronephros of a 12-mm tadpole.

developing in the mesoderm. Only three tubules of the pronephros are as yet present, but the longitudinal duct reaches back to the cloacal (proctodeal) region.

**The Larval Life of the Tadpole.**—After about two weeks the young tadpole wriggles from the enclosing jelly and attaches itself to surrounding objects by means of its mucous glands, which produce a very sticky secretion. For some time it continues to live at the expense of the yolk, but after about a week the stomodeum becomes opened into the pharynx, and the mouth so established becomes fringed by a pair of **horny jaws** bearing papillae ("horny teeth"). These are used for rasping off small fragments of food, which is usually vegetable matter.

The gut then becomes greatly elongated and coiled, and begins to function as a digestive and absorptive organ. In the pharyngeal region the third branchial (fourth visceral) pouch develops and then all four branchial pouches perforate to the exterior, forming complete gill clefts. Direct communication with the outside world lasts for only a short time, for from the hinder border of the hyoid arch a fold of skin grows back over the gill slits. The folds, which are termed opercular folds, soon become continuous with each other on the ventral surface, and so by their continued growth the gill slits of both sides open externally into a common gill chamber.   The hinder wall of the chamber is completed by the fusion of the opercular folds with the skin behind the last (fourth) gill slit, but at one point on the left-hand side a spout-like opening— the "spiracle"—persists as the aperture by which the gill chamber opens to the exterior.   True or **internal gills** have meanwhile been gradually forming from the tissue on the branchial arches, and at the same time the external gills are diminished in size and importance.

The tadpole is then about ten millimetres in length, the body has assumed a globular shape, and the tail has been greatly lengthened so as to function as a

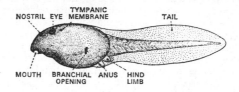

Fig. 613.   FROG.—A 14-mm tadpole.

powerful propulsive organ.   Feeding is carried on actively and growth in size is rapid.

Internal changes have proceeded apace and continue to do so without interruption.   The nervous system and organs of special sense—among which it is interesting to note the **lateral line organs**—are practically completed by the time that a length of fifteen millimetres has been attained.   The vascular system (p. 523) is fish-like in all essential features and remains like this until the lungs—which were apparent as small buds from the posterior end of the pharynx soon after hatching—become functional as respiratory organs.   This is usually

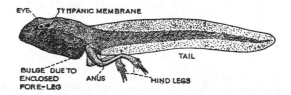

Fig. 614.   FROG.—A tadpole with the hind limbs showing.

after about three months of larval life, when the tadpoles come to the surface of the pond and gulp bubbles of air.   The pronephros, which up to a point increases in size, is gradually replaced as the larval excretory organ by the mesonephros, which begins to develop tubules when the larva is about twelve millimetres long.   The limbs appear as buds from the body wall, but at first only the hind limbs are visible from the outside because the fore limbs lie hidden beneath the operculum, through which they emerge shortly before metamorphosis.   On the right-hand side the operculum becomes thin and weak and then perforates, in response to thyroid secretions, to form a hole through which the limb pushes out, whilst the left limb emerges through the "spiracle".

In the embryo and early larva the only skeletal structure is the notochord, but with the appearance of the limb buds other parts of the skeleton are gradually laid down as a system of cartilages. The vertebrae are derived from cartilage deposited by cells from the sclerotomes, and the **urostyle** has a similar origin. The neurocranium is largely derived from two pairs of rods (trabeculae and parachordals), and to it soon fuse the sense capsules and the upper segment (plalato-quadrate bar) of the mandibular arch. This latter which forms the upper jaw, together with Meckel's cartilage (lower jaw), the hyoid arch, and the four branchial arches represent the visceral arch skeleton and are laid down in the pharyngeal wall by neural crest cells. The limb skeletons and girdles appear somewhat later, after the limbs have been further differentiated.

This cartilaginous skeleton persists throughout most of larval life, but during and after metamorphosis it is largely replaced by bone (endochondral or cartilage bone), whilst other bones (membrane bones) are added to complete the adult skeleton.

**Metamorphosis.**—Towards the end of the third month the tadpole undergoes a more or less rapid series of changes which transforms it into a young frog. This metamorphosis, involving many very striking changes during which certain larval organs undergo extensive modifications whilst others disappear, is of course a

Fig. 615. A young frog.

necessary corollary of a radical change in habits, for, as soon as it is completed, the young frogs leave the water and become to all intents and purposes land-dwellers.

The first sign that metamorphosis is imminent is that the tadpoles frequent the surface and take air into the lungs via the mouth. Soon the tadpoles cease to feed; the larval skin is cast off, and with it the frilly lips and horny jaws; the tail begins to shorten and the legs lengthen. The lungs assume more and more importance as respiratory organs; the gill slits close; and the vascular system becomes modified in a way suitable for an air-breathing creature (pp. 522-4). The mouth grows wider and the true jaws and jaw muscles begin to function; the tongue enlarges; and the eyes, now much larger, begin to bulge out like those of the adult. The stomach and liver enlarge but the intestine shortens—structural changes which are correlated with a change from a herbivorous to a carnivorous diet. The young frogs still bear a stump of a tail, but they leave the water for the land. They show a preference for damp pastures, in which they may sometimes be found in vast numbers (cf. Biblical Plague of Frogs). They continue to grow, feeding chiefly on insects, and those which survive gradually assume the adult form and colour.

## THE DEVELOPMENT OF THE CHICK

**The Egg.**—The egg of the fowl provides a good and familiar example of a large-yolk, telolecithal ovum, surrounded by tertiary membranes, and its subsequent development into the young chick has probably been the subject of more embryological investigations than any other animal type. Despite this, the various interpretations of many of the early stages are still matters of controversy and the following account cannot pretend to be anything but a brief and oversimplified summary of recent views.

The **eggs** are released from the ovary as **primary oocytes** and, owing to a large accumulation of yolk, each is an enormous cell with a diameter of something over an inch. There is a complete separation of yolk and cytoplasm, which is confined to a small area (**blastodisc**), forming a cap at the animal pole. In the centre of the blastodisc lies the **germinal vesicle** (egg nucleus). The yolk is not of uniform composition, but consists of a central mass of white yolk around which are arranged alternating concentric layers of yellow and white yolk (not to be confused with the "white of the egg"). Both types of yolk contain a high proportion of fat but only slightly more protein than is required to "build the embryo",

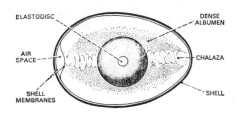

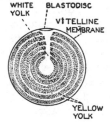

Fig. 616. A hen's egg with half of the shell removed.

Fig. 617. Vertical section of the ovum.

an arrangement which results in the production of only small amounts of nitrogenous waste. A pillar of white yolk runs from the centre of the yolk to a point just below the blastodisc.

The oocyte is invested in a transparent membrane, which may be a true vitelline membrane, but is more probably a secretion of the follicle cells of the ovary. That is, it is a **secondary membrane**. It serves to separate the yolk from the albumen and to preserve differences in water content and osmotic pressure, for the yolk contains about 50 per cent. water, while the albumen has about 85 per cent. As the egg passes down the oviduct it receives tertiary membranes which are deposited in successive layers. Firstly, the glandular walls of the upper part of the oviduct secrete a rather thin albumen. Then, the middle region adds a layer of very viscous albumen which, in turn, becomes covered by thin albumen secreted by the lowest region of the oviduct, the so-called "uterus". Part of the dense albumen forms twisted, cord-like structures, the **chalazae**, at each end of the egg but the way in which these are formed is not understood.

In the lower part of the "uterus", the tough **shell membranes,** formed of matted keratin fibres, are secreted. These are in close contact with each other except over a small area at the blunt end of the egg, where they become separated by an **air-space** shortly after the egg is laid. Osmotic absorption of water from the lower part of the oviduct takes place and has two effects. It causes the egg to swell, and the outer layers of albumen to become diluted. Still further down the "uterus" the egg becomes enclosed in the calcareous shell, the last of the tertiary membranes, which stops the imbibition of water. At first soft and pliable, the shell rapidly hardens. It is covered with a thin layer of protein which is continuous through minute pores with a similar layer on the inside of

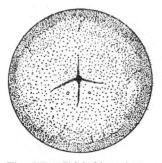

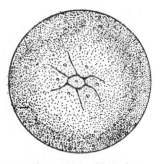

Fig. 618. Chick blastoderm —4-celled stage.

Fig. 619. Chick blastoderm showing early cleavage.

the shell. The pores, which serve to transmit gases, are particularly numerous at the blunt end of the shell over the air-space between the shell membranes.

**Fertilisation and Cleavage.**—The oocyte is fertilised in the upper part of the oviduct before the deposition of the albumen around it. Several spermatozoa enter in the region of the blastodisc, but only one plays an active part in the fertilisation processes. The rest soon die, although their nuclei may undergo a few divisions before disappearing. Almost immediately after the entry of the sperm the egg undergoes two maturation divisions, but both polar bodies soon disappear.

As is inevitable in such a heavily-yolked egg, cleavage is confined to the blastodisc, which becomes converted into a **blastoderm.** That is, cleavage is of the meroblastic type. Cleavage begins soon after the completion of fertilisation and continues without interruption as the egg passes down the oviduct. This usually takes from twelve to sixteen hours. By the time the "egg" is laid the embryo is in the blastula stage (see later) or, if laying is delayed, development may have gone much further.

The first division of the zygote nucleus is soon followed by the formation of the first cleavage furrow. This does not extend right across the blastodisc, nor does it cut completely through it in a vertical plane, so that the first two blastomeres are only incompletely delimited. The second cleavage furrow is similar to, but at right angles to, the first and is soon succeeded, in a more or less irregular manner, by others. These, also, do not involve the whole area of the blastoderm, and this results in the demarcation of central cells which are surrounded peripherally by a rim of unsegmented

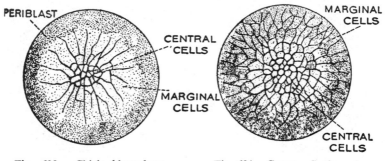

Fig. 620. Chick blastoderm showing central and marginal cells.

Fig. 621. CHICK.—Surface view of the blastoderm at a later cleavage stage.

cytoplasm (**periblast** or **marginal zone**). Later cleavages take place both in the marginal zone (future **area opaca**) and in the central area to which are added cells from the dividing marginal zone.

Soon horizontal cleavages make their appearance, at first in the central area and later in the marginal zone, and intersect with the vertical furrows. In this way the cells of the blastoderm become completely delimited. A cavity, the **sub-germinal cavity**, filled with a coagulable fluid, is then formed below the central cells, but the marginal cells remain as a layer overlapping and lying directly on the yolk. The sub-germinal cavity at this stage may be regarded as the equivalent of the blastocoel, and the embryo is now in the blastula stage. In the floor lie a few large yolky cells of uncertain origin which possibly correspond to the cells of the vegetative hemisphere of less heavily yolked eggs. The central cells which form its roof would then correspond to the smaller cells of the animal half. Whatever their homology, the large cells soon disappear. Further horizontal cleavages cause the blastoderm to become several cell-layers in thickness. It also grows larger, gradually spreading over the yolk which is engulfed and digested by the marginal cells as they encroach upon it. These marginal cells are also several layers deep, except at the extreme periphery, and the lower-most layer, which lies directly on top of the yolk, forms the

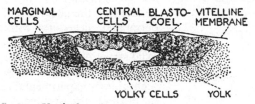

Fig. 622. CHICK.—Vertical section of a blastoderm at early blastula stage.

**yolky endoderm.** This later forms the **extra-embryonic endoderm.** Its function is to digest the yolk, and it plays no part in the formation of the endodermal structures of the embryo which arise from a different endodermal contribution.

Seen from above by transmitted light, the blastoderm at this stage shows two distinct areas—a relatively clear area, the **area pellucida,** lying centrally, and a narrower, darker zone surrounding it, the **area opaca.** The area pellucida owes its translucency to the fact that it overlies the sub-germinal cavity, whilst the area opaca corresponds to the extent of the marginal cells which, it will be remembered, lie immediately on the opaque yolk.

Fig. 623. CHICK.—Vertical section of a blastoderm at a later stage than Fig. 622.

So far, no mention has been made of the orientation of the blastoderm with respect to the embryo. This is because the first two cleavage planes, and hence the subsequent ones, bear no constant relation to the future embryo axis. The determination of the antero-posterior axis of the embryo is not possible, from a mere examination of the blastoderm, until gastrulation has begun. It has, however, been observed that the longitudinal axis of the future embryo is orientated across the longitudinal axis of the egg shell in such a way, that the future anterior end lies away from the observer if the egg is held with the blunt end of the shell to the left.

**Gastrulation.**—It will be recalled that in the previously described examples (amphioxus and frog), the formative movements of gastrulation resulted in the enclosure of the endoderm and mesoderm within the ectoderm and the laying down of the main axial structures of the embryo. These processes, although to some extent separable by experimental analysis, are, in normal circumstances, very accurately geared with one another and go on together at one and the same time.

Fig. 624. CHICK.—Surface view of a blastoderm at the end of cleavage.

Gastrulation in the chick, however, is separable into three distinct processes or phases which, although they overlap to a certain extent, occur, not at the same time, but in chronological sequence. These phases are the laying down of the embryonic endoderm, mesoderm formation, and the formation of the embryonic axial structures.

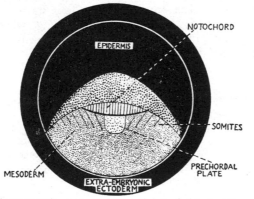
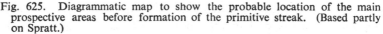

Fig. 625. Diagrammatic map to show the probable location of the main prospective areas before formation of the primitive streak. (Based partly on Spratt.)

Then again, as a direct consequence of the large amount of yolk present, the movements of gastrulation have departed considerably from the pattern found in other chordate types with less yolk. Enclosure of the deeper germ-layers by any direct sort of invagination is impossible and an archenteron as such is no longer found.

**Endoderm Formation.**—The first stage in gastrulation is marked by a rapid expansion of the blastoderm and the cells of the area

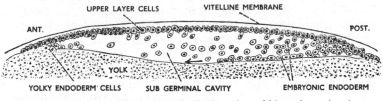

Fig. 626. CHICK.—Median longitudinal section of blastoderm showing formation of the endoderm.

pellucida become rearranged to form a single layer of cubical cells in the form of a definite epithelium. The areas which will form the main embryonic structures have been mapped out on this single-layered blastoderm, but their exact positions and limits are not so well known as are the corresponding regions in the early amphibian gastrula. Their probable arrangement is shown in Fig. 625, but it will be noticed that unlike other fate-maps which have been given (pp. 710-11) no endoderm is shown. This is because in the very earliest stages of gastrulation the prospective endoderm cells migrate rapidly downwards into the sub-germinal cavity. There, they soon reform into a sheet which spreads forwards and laterally so that soon the sub-germinal cavity acquires a complete floor of **embryonic endoderm**. Later, this sheet of embryonic endoderm becomes continuous with the **yolky (extra-embryonic) endoderm** around the margins of the blastoderm.

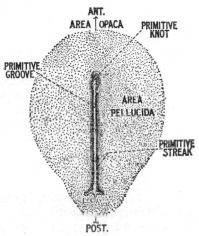

Fig. 627. CHICK.—Surface view of the area pellucida showing early primitive streak.

**Mesoderm Formation.**—The separation of the endoderm leads to the establishment of a two-layered condition in the blastoderm, the lower layer being the endoderm and the upper containing the prospective notochord, mesoderm, and ectoderm. The next phase in gastrulation is concerned with the inward migration of the mesodermal and notochordal material, so that the upper-layer cells consist of ectoderm alone when the process is completed.

After the completion of the inward migration of the endoderm certain other cells begin to leave the upper layer of the blastoderm and to migrate inwards between it and the endoderm which is, by now, recognisable as a distinct cell-layer. This second migration of upper cells at first takes place, in a somewhat irregular fashion, from a region just anterior to the area recently vacated by most of the endoderm, *i.e.* from the posterior part of the prospective mesodermal area. The inwardly migrating cells are, of course, the vanguard of the mesoderm. Soon the movement of the mesodermal cells becomes more organised and much more rapid, and is accompanied by a mass movement of the upper-layer cells towards the mid-line. This results in a heaping-up of prospective mesodermal and notochordal cells along the mid-line from a point about a quarter of the way from the anterior end of the area pellucida to a point near its posterior end. The accumulation of cells is seen on the surface of the blastoderm as a dark or opaque streak, the **primitive streak**. A distinct swelling of the anterior end of the streak is termed the **primitive knot** or **Henson's node**. It is formed of cells from the prospective notochordal area and those which will form the floor of the neural tube. When first formed the primitive streak is shorter than it is when completed and the attainment of its full length is, at least partly, due to a general stretching of the main embryonic areas in an antero-posterior direction (cf. cells of the animal hemisphere of the gastrula of the frog).

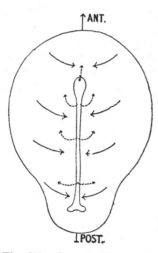

Fig. 628. CHICK.—Diagram to show cell migrations within the area pellucida.

A further result of the cell movements involved in the formation of the primitive streak is that the area pellucida loses its circular shape and acquires a pear-shaped outline, the blunt end being anterior.

The primitive streak, then, is a result of the convergence or **crowding towards the mid-line of the notochordal and mesodermal** cells, but their migrations are by no means over when they reach the streak. The mesodermal cells on arriving there pass inwards and then leave the sides of the streak and migrate out laterally to form a sheet of cells on each side between the endoderm and the upper-layer cells. Later, some of the cells take a more anterior course to form two forwardly directed horns to the sheets of mesoderm. The cells of the primitive knot (chiefly prospective notochord or "head process") migrate in a forward direction only, so as to form a column of cells connected at its hinder end with the primitive knot and lying in the mid-line below the prospective neural plate whose middle part or floor plate is formed by cells lying in the upper parts of the primitive knot.

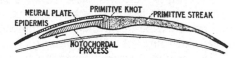

Fig. 629. CHICK.—Diagrammatic longitudinal section of the area pellucida to show the formation of the notochordal process.

At its first appearance the streak is (morphologically) merely a simple median thickening of the upper-layer cells in the posterior three-quarters (approximately) of the area pellucida, but later the rate at which the mesodermal cells leave its lower lateral borders is greater than the rate at which the streak is renewed from the

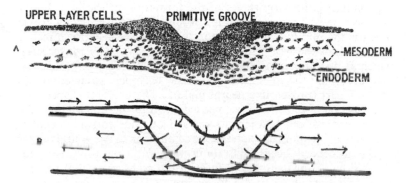

Fig. 630. CHICK.—A, transverse section through the primitive streak; B, diagram to show cell migrations through the streak.

upper-layer cells along its upper borders and consequently a groove, the **primitive groove**, appears along the length of the streak.

**The Mesoderm.**—The somitic mesoderm, which will later give rise to the somites, passes through the antero-lateral portions of the primitive streak, migrates forward, and becomes extended to flank the notochordal process. The nephrotomal and lateral plate mesoderm passes inwards mainly through the middle portion of the streak and then becomes extended both anteriorly and posteriorly, the lateral plate also diverging laterally. Extra-embryonic mesoderm goes in mainly through the posterior part of the streak, later migrating laterally and anteriorly.

The limits between these mesodermal contributions to the embryo are far less exactly known than in the frog and the above remarks must be taken merely as indicative of the most probable places and routes of cell migrations.

The primitive streak, which up to a certain stage increases in length, becomes shorter, and the primitive knot appears to migrate towards the posterior end of the area pellucida. Actually, the shortening of the streak seems to be partly due to the ectoderm (neural plate) at its sides encroaching upon it, and finally meeting in the mid-line as the mesodermal cells vacate the upper layer of the blastoderm. Thus, with the separation of the notochordal process and mesoderm, the fusion of right and left ectodermal sheets gradually extends backwards and, at the end of this process, the upper layer consists of prospective epidermis alone.

**Formation of the Central Nervous System.**—The next important step in development is the formation of the **neural plate**. This begins during the early stages of the formation of the notochordal process ("head" process), and is a result of the general convergence towards the mid-line and extension in an antero-posterior direction of the cells of the area pellucida already noticed. The convergence, which at first affects mainly the more posterior regions (prospective mesoderm) of the area pellucida, now becomes extended to the anterior part (prospective neural plate). The crescent-shaped area of neural plate cells thus becomes extended in the longitudinal axis of the embryo, but becomes narrower from side to side. That is, the arms of the crescent approach one another in the mid-line, and as the notochordal cells infiltrate inwards from the primitive knot, the two halves of the crescent meet in the mid-line so that some of their cells contribute to the formation of the primitive knot. In this way a thickened plate of cells, the neural plate, is formed above the "head" process, and soon, as a continuation of the formative movement described, neural folds rise up at its sides. Later, beginning at a point near to the anterior end of the neural plate, the neural folds approach one another and fuse in the mid-line to form the neural tube.

**Gastrulation of Chick Compared with other Types.**—At first sight, the process of gastrulation in the chick appears to be totally different from that described for amphioxus and the frog, but a closer examination reveals many important points of similarity in all three types, and many of the differences seem to be directly attributable to the extremely telolecithal condition of the egg.

Gastrulation, as in other examples, is brought about mainly by cell migrations, but the endoderm is segregated as a separate cell-layer before mesoderm production begins. The precocious formation of the endoderm is due to the necessity for providing a yolk-digesting layer, for the yolk is much too abundant to be enclosed *within* the cells of the embryo. This need does not arise in amphioxus or the frog where the yolk is enclosed during cleavage *within* the blastomeres and hence, from an early stage, in the body of the embryo.

Gastrulation in the chick begins at a stage when the blastoderm occupies only a small area on top of the unsegmented yolk, so that gastrulation by invagination is not possible and an archenteron, as such, is not formed. Instead, the endoderm is set aside early on as a separate sheet of cells and the notochordal and mesodermal cells pass inwards through the primitive streak—a structure which is not represented in lower forms. From this stage onwards, the embryo develops very rapidly at the expense of the yolk which lies below it. The primitive streak marks the place at which mesoderm and notochordal material pass inwards below the superficial layers and hence, in this respect, it is *comparable* to the dorsal and lateral lips of the amphibian blastopore. The primitive streak, however, is not in any way directly concerned with the formation of the archenteron, or its equivalent, and no true blastopore is ever developed as a consequence of its activity and it therefore seems best to regard it, not as the homologue of the "fused lips of the blastopore" (as has sometimes been done in the past), but as a new structure which has evolved to meet the needs of a gastrulation of material spread out in a single-layered blastoderm.

The general lay-out of the areas of prospective epidermis, neural plate, notochord, and mesoderm on the blastoderm is much the same as on the surface of the frog's blastula, the main point of difference being the complete girdle of ectoderm around the other embryonic areas. This may be accounted for by the necessity for the ectoderm (and, later, of the deeper-lying layers) to spread in all directions so that the yolk eventually becomes enclosed from all sides.

The main embryonic areas undergo displacements similar to those of the frog, i.e. *convergence* towards the mid-line, *migration inwards* and *forwards* (of the notochordal and mesodermal material), and *extension* in the longitudinal axis of the embryo.

To sum up, it may be said that the process of gastrulation in the chick, in the frog, and in amphioxus is fundamentally similar but, as has already been pointed out, it is difficult, if not impossible, to find a stage in any of the three types which may be termed a "gastrula" in the older sense of that term. That is to say, at no stage does the embryo consist merely of the two primary germ-layers, ectoderm and endoderm.

In the account given of gastrulation of the frog it was pointed out that the differentiation of the main organs of the embryo depends very largely on the influence of the primary organiser and that, later, other structures acquire organising properties to become organisers of the second or third degree. During recent years, the early developmental processes of the chick have also been subjected to experimental analysis, and although so far they have not been so fully investigated as those of lower vertebrates (amphibians and fishes), it has been possible to arrive at many important conclusions. These can only receive scant attention here, but it seems certain that the

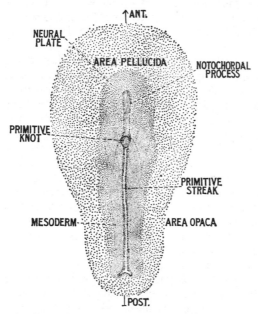

Fig. 631. CHICK.—Surface view of the area pellucida to show the early stages in the formation of the notochordal process and the mesoderm.

final determination of the main embryonic organs takes place under the influence of organisers in much the same way as in the Amphibia, but the primary organiser is at first situated in the endoderm which induces a primitive streak to appear above it in the mid-line.

Experiments in which primitive streaks were transplanted into other blastoderms show, however, that the host exercises some large measure of control over the transplant for if a streak is transplanted so that its orientation is reversed with respect to the host, when it organises a second set of embryonic structures, these follow the host orientation so that the organiser properties of the primitive streak are comparable only in certain respects with those of the amphibian organiser.

The primitive streak (together with the notochordal process) and the axial mesoderm then assume the role of an organiser, and under the influence of these the main axial structures are differentiated.

**Later Embryonic Development.**—It has already been pointed out that, as a result of mass migrations of cells on the blastoderm, the rudiments of the main parts of the young chick are laid down in a restricted part of the blastoderm in front of the primitive knot. The young chick is thus, from the start, surrounded by a wide zone of blastoderm which does not contribute to the formation of the embryo and is therefore said to be extra-embryonic. It is composed of ectodermal, endodermal, and mesodermal sheets continuous with those of the embryo.

The main features of the embryo already formed towards the close of gastrulation are: the notochord or notochordal process, which is formed of material passing forwards from the primitive knot and is flanked on each side by a sheet of mesoderm which has migrated forwards from the side near the anterior end of the primitive streak; the neural plate cells in the upper layer of the embryonic area which have converged to the mid-line to form the rudiment of the central nervous system, whilst the rest of the upper-layer cells form the epidermis. Like other parts of the blastoderm, the embryonic area is floored by a sheet of endoderm which at first rests flat on the yolk.

The next phase in the embryonic history of the young chick is chiefly concerned with foldings and differential growth of the various parts of the embryo, and with its gradual enclosure within membranes formed by the cell-layers of the extra-embryonic area. Also, with the continual spreading of the blastoderm, the yolk becomes progressively enclosed. From this point onwards it is impossible to deal with all the events as a whole, and, therefore, for the sake of clarity, one sequence of events will be followed at a time. It will be found helpful to consult the summary which is given in tabular form at the end of this section.

**Head-fold and Fore-gut.**—With the shortening of the primitive streak, the various embryonic structures become demarcated in the posterior as well as in the anterior part of the area pellucida, and practically the whole of the central area becomes occupied by the embryo. Then, in addition to the growth of its individual parts, the embryo becomes more clearly defined by the formation of folds of the blastoderm which bound its margins. The first of these limiting folds to appear is the **head-fold** for, soon after the notochordal process has been formed, a crescent-shaped fold appears just in front of the neural plate. This fold is at first composed of ectoderm and endoderm alone, for, as yet, no mesoderm has penetrated to such an anterior position. Later, however, both the ectoderm and the endoderm come to have a layer of mesoderm applied to them. The head-fold is a result of the ectodermal and endodermal layers of the blastoderm becoming bent downwards and backwards, below the anterior edge of the neural plate. It follows that the edges of the fold bound a crescentic groove, the **anterior limiting groove**, which appears as a wide depression on the surface of the blastoderm. The folding is soon accentuated by the forward growth of the neural plate and by the backward growth of the bottom of the groove below the neural plate so that the anterior limiting groove gets deeper and deeper.

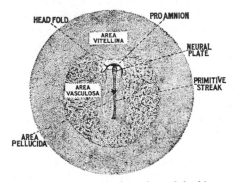

Fig. 632. CHICK.—Surface view of the blastoderm showing the main regions at a stage soon after the appearance of the head-fold.

This folding of the blastoderm has two important effects, for not only does it cause the head of the embryo to be raised off the surface of the blastoderm but it also results in the formation of and endodermal pocket below the anterior end of the central nervous system. This pocket is the first sign of the **fore-gut**.*

**Differentiation of the Mesoderm.**—When the head-fold first makes its appearance the newly separated mesoderm is arranged as two sheets of cells, one right and one left of the notochordal process and primitive streak. Each sheet extends laterally nearly to the margins of the blastoderm. Later the lateral sheets of mesoderm grow forwards and fuse in front of the notochordal

---

* The terms **fore-gut** and **hind-gut** used in the following description are the ones in common usage. It is, however, important to realise that they have quite a different meaning when applied to adult structures, where they are usually held to be synonymous with the terms stomodeum and proctodeum. In the chick embryo both fore- and hind-guts are lined by endoderm and hence form part of the mesenteron.

process, but a small area ("proamnion") below the head-fold is still, for a time, free from mesoderm.

The first parts of the mesoderm to become histologically differentiated lie between the ectoderm and the endoderm of the area opaca. Here some of the mesodermal cells group themselves into small masses and the cells occupying the inside of each mass differentiate into blood-corpuscles, while the cells on the outside become arranged as an epithelium and so form a wall around the corpuscles. The

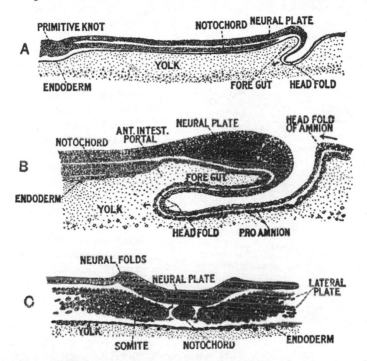

Fig. 633. CHICK.— A, longitudinal section of a young embryo to show the beginning of the head-fold, B, the anterior end at a later stage; C, transverse section through the trunk region of a young embryo to show the neural plate.

separate cell masses are called **blood-islands**, but they soon lose their identity, and run together to form a plexus of small blood vessels in the central part of the area opaca which is then known by a new name— the **area vasculosa**. The peripheral zone of the area opaca into which mesoderm has not yet penetrated is called the area vitellina.

Next, except for a short distance near the mid-line, each of the lateral sheets of mesoderm becomes split into an upper and a lower layer by the running together of small chinks and spaces between the cells. The space formed by this splitting process is coelomic and it has divided the lateral-plate mesoderm into an upper, somatic, layer (next to the ectoderm) and a lower, splanchnic, layer (next to the endoderm). The coelomic space is, of course, by no means confined to the embryonic region of the blastoderm, for it extends right to the margins of the mesoderm and hence the greater part of it is termed the **extra-embryonic coelom**, and only that portion confined between the embryonic mesoderm will form the splanchnocoel of the chick.

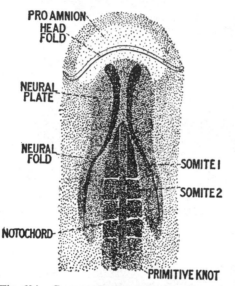

Fig. 634. CHICK.—Surface view of a 5-somite embryo.

The ectoderm, together with its somatic layer of lateral-plate mesoderm, then becomes known as the somatopleure, whilst the endoderm and the layer of splanchnic lateral-plate mesoderm together form the splanchnopleure.

Meanwhile, the undivided mesoderm on each side of the mid-line in the embryonic region becomes thickened as more and more paraxial mesodermal cells are added to it from the primitive streak. Then, shortly after the formation of the head-fold, its cells begin to be segregated into rather cube-like blocks lying right and left of the notochord. These blocks of cells are the somites. The first pair of somites* is separated off just in front of the primitive knot but soon, as the primitive streak shortens, other intersomitic furrows are formed behind the first so that a metameric series of somites

---

* The term "first pair of somites" means the first pair of somites to be differentiated in time and not the most anterior somites of the body. The first pair of somites is formed by the hinder part of the tissue cut off by the first intersomitic furrow, whilst the rest of this tissue becomes scattered and contributes to the mesoderm of the head. It is in this tissue that condensations later appear to form the head somites. They are never very clearly marked but are indicated by the large cavities (head cavities) which they contain. It will be realised that the first pair of somites are metotic somites. (See p. 581.)

is demarcated, the process beginning at the anterior and later extending to the more posterior parts of the somitic mesoderm. In the later embryo the trunk and neck somites themselves undergo differentiation and each contributes a myotome, sclerotome, and dermatome, as in all other craniate animals. The nephrotomes in the chick, although later giving rise to segmental nephric tubules, do not share in the metamerism of the somites, for they are amalgamated into a continuous sheet which joins up the lateral edges of the somites and links them to the two layers of lateral-plate mesoderm. The nephrotome region of the mesoderm thus appears similar to the lateral plate, but it is not split by a coelomic space and a series of nephrocoels is not formed until the nephric tubules make their appearance.

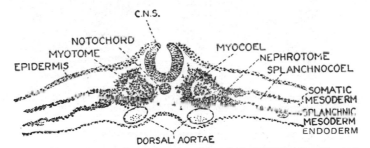

Fig. 635. CHICK. —Transverse section through the middle of a 10-somite embryo.

**The Laying Down of the Chief Embryonic Organs.**—Soon after this stage the rudiments of many of the main organ systems make their appearance.

CENTRAL NERVOUS SYSTEM.—Shortly after the formation of the head-fold the neural plate, which up to this stage was a thick, crescentic plate of cells, begins to give rise to the neural tube. Its median portion sinks downwards and its sides rise up to form neural folds. These grow towards each other in the mid-line, and by the time that four or five somites have been delimited, the folds have met and fused to form that part of the neural tube which later proves to be the mid-brain. The process of fusion then extends forwards to the front end of the neural plate and also posteriorly so that the rudiments of the fore- and hind-brains, and later of the spinal cord, are established. But, for some time, the anterior end of the neural tube remains open by a neuropore whilst the posterior end of the neural tube is, of course, not completed until the final disappearance of the primitive streak. **Neural crests**, which later become metamerically constricted to form the dorsal root ganglia, are differentiated from the lateral margins of the neural plate.*

By the time that twelve somites have been formed, the **primary optic vesicles** have already grown out from the embryonic thalamencephalon.

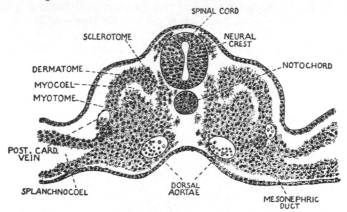

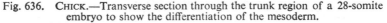

Fig. 636.   CHICK.—Transverse section through the trunk region of a 28-somite embryo to show the differentiation of the mesoderm.

THE NOTOCHORD.—The notochordal process, at first a rod of more or less undifferentiated cells, gradually assumes a more compact form, the notochord, and later it becomes surrounded by a sheath of mesenchyme cells derived from the sclerotomes.

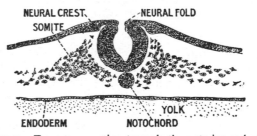

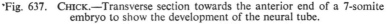

'Fig. 637.   CHICK.—Transverse section towards the anterior end of a 7-somite embryo to show the development of the neural tube.

It is from these cells and their descendants that the skeletogenous tissue, in which the vertebrae are laid down, is derived.

* The segmental arrangement is imposed on the neural crests by the disappearance of the cells in the intersegmental positions.   The cells which wander away do not, however, die, but become mesenchyme cells and seem to be able to give rise to various types of tissue (visceral arches, etc.), as in other vertebrates.

THE GUT.—The gut, which with its various outgrowths and appendages is not completed until a very late stage in development, is represented in the early embryo merely by the fore-gut. This is a blindly-ending pocket of splanchnopleure which at first opens at its sides, as well as posteriorly, on to the yolk. It is formed as a direct consequence of the folding of the blastoderm in front of

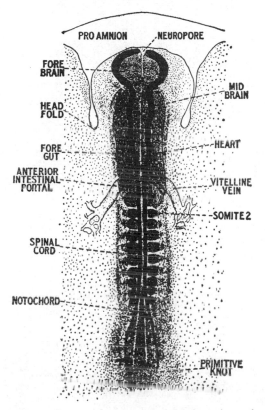

Fig. 638. CHICK.—Surface view of a 9-somite embryo.

the head. Soon, however, as the head-fold pushes backwards, the blastoderm at the sides of the embryo becomes folded to form **lateral limiting folds** and a groove, the **lateral limiting groove**, which is, of course, continuous with the anterior limiting groove. The anterior limiting groove continues to progress towards the hinder end and, as it does so, its lateral limbs merge into one another below the embryo.

It follows, therefore, that, beginning at the front end, the right and left sheets of splanchnopleure of the lateral folds meet and fuse below the embryo so that at first side walls, and later the ventral

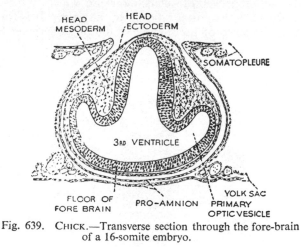

Fig. 639.   CHICK.—Transverse section through the fore-brain of a 16-somite embryo.

wall of the fore-gut, are established.   The fore-gut thus becomes a blindly-ending tube, its posterior opening on to the yolk being termed the **anterior intestinal portal**.   The fore-gut continually increases in length as the fusion of right and left sheets of splanch-

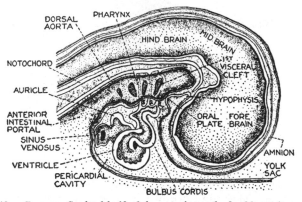

Fig. 640.   CHICK.—Sagittal half of the anterior end of a 23-somite embryo.

nopleure progresses and because the endoderm gradually with-draws from the point of fusion, the splanchnic mesoderm from each side comes into direct contact below the fore-gut.   In this way the floor of the fore-gut remains attached over a short distance to the extra-embryonic splanchnopleure by a ventral mesentery (double vertical sheet of splanchnic mesoderm). This, however, disappears almost as soon as it is formed so that right and left coelomic spaces become continuous below the gut to form the embryonic splanchnocoel.

By the time that twenty somites have been formed, the blastoderm, just behind the posterior end of the embryo, becomes folded to form the **tail-fold** and a **posterior limiting groove**. These arise in just the same way as the other limiting folds but the tail-fold faces, and grows towards, the anterior end of the embryo. The folding of the posterior splanchnopleure causes a hind-gut to be formed, which is soon converted into a blindly-ending tube by the

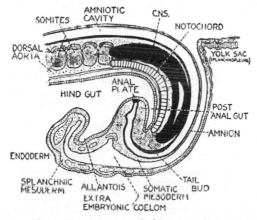

Fig. 641. CHICK.—Vertical section through the hind end of a 35 somite embryo.

fusion in the ventral mid-line of the splanchnopleure of the lateral limiting grooves. The hind-gut opens at its anterior end by the **posterior intestinal portal.** The fusion of right and left folds of splanchnopleure extends progressively forwards so that the anterior and posterior intestinal portals become closer and closer together, but their margins do not fuse, remaining separated by a short region where the lateral folds of splanchnopleure do not meet. In this region the wall of the gut remains, therefore, continuous on all sides with the extra-embryonic splanchnopleure (**yolk sac**). The gut is not completed as a continuous tube until the yolk is totally enclosed in the yolk sac, which then appears as a bag-like appendage of the gut to which it remains connected by a hollow stalk—the **yolk-sac stalk.**

By the time that about twelve somites have been delimited, the bottom (*i.e.* posterior margin) of the head-fold has grown as far back as the posterior end of the hind-brain. The part of the fore-gut formed at this stage is very broad and will give rise to the pharynx. The endoderm bounding its anterior end soon fuses near the ventral surface with the ectoderm to form the oral plate. This marks the spot at which a perforation will later appear to place the pharynx into communication with the stomodeum. Then, at four places along each side, the pharyngeal wall pushes out laterally and the tips of the pharyngeal pouches so formed fuse with the ectoderm. The first three of these pouches will, by perforation at the points of fusion, give rise to **visceral clefts**, but the fourth never actually opens out to the exterior of the embryo.

The hind-gut is placed in connection with the short proctodeum by the perforation of an anal plate, formed by the fusion of the

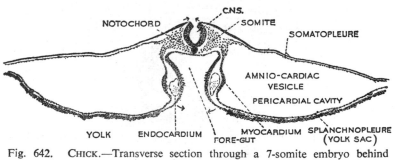

Fig. 642.   CHICK.—Transverse section through a 7-somite embryo behind the anterior intestinal portal.

wall of the hind-gut with the ectoderm on the ventral surface a short way in front of the blind, hinder end of the gut.

EMBRYONIC BLOOD VESSELS AND HEART.—It has already been mentioned that the first parts of the blood vascular system to appear are the small blood vessels in the area vasculosa. These arise precociously in the sense that a conspicuous plexus of vessels is already evident in the area vasculosa even before any somites of the embryo have been delimited. Soon, by the time the first somite has been cut off, small vessels, in continuity with those of the area vasculosa, are differentiated in the area pellucida. The appearance given is that the vascular network of the area vasculosa gradually encroaches on the area pellucida from its margins and, by the time that seven somites have been formed, vessels are appearing in the lateral-plate mesoderm of the embryo itself. At first, the embryonic vessels are all of small calibre and seem to follow no set plan but, later, those in particular regions enlarge so as to conform to a definite pattern and an orderly circulation is established in the embryo.

The differentiation of the embryonic blood vascular system is progressive, but rapid, and accompanies the formation of a definite system of vessels in the extra-embryonic area so that a complete circulation extends throughout the whole blastoderm.

In the general plan of this system, by the time that about thirty somites have been laid down, the following points may be noted. The area vasculosa is limited by a vessel, the **sinus terminalis**, which makes a complete circuit of its boundary. The sinus terminalis is connected along its whole length to the more centrally placed network by numerous small vessels and also more directly with the embryo by a pair of large veins, the **anterior vitelline veins**. Each of these veins passes backwards along the inner margin of the area

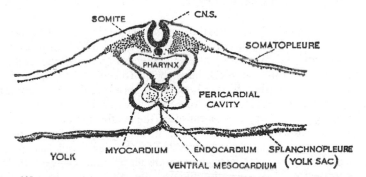

Fig. 643. CHICK.—Transverse section through a 7-somite embryo in front of the anterior intestinal portal.

vasculosa and then, at the level of the heart, is connected with the embryo by a transversely running vessel, the main **vitelline vein**, which fuses with its fellow of the opposite side to form a short, but large, vessel, the **ductus venosus**. This empties into the posterior chamber of the heart, the **sinus venosus**. The heart (to be described below), at this stage, is in effect, a large, contractile, S-shaped tube, but in which the rudiments of the main chambers are already evident. From the anterior chamber of the heart a short vessel, the **ventral aorta**, passes forwards below the pharynx. Its anterior end, which is situated at about the level of the oral plate, then bifurcates to form the first pair of **aortic arches**. These pass laterally and dorsally in the tissue (mandibular arch) immediately in front of the first (hyoidean) visceral cleft and, after encircling the pharynx in this way, each joins one of the **dorsal aortae**. These are paired longitudinal trunks running in the back of the embryo towards the posterior end. Two other aortic arches (the second and third) are also given off from the ventral aorta and each passes dorsally in the tissue between the visceral clefts to join the dorsal aortae. Later, three more arches are added to the series.

Small paired segmental arteries are given off in between the somites from the dorsal aortae before they finally diverge and leave the body near its posterior end as the right and left **vitelline arteries**. Each of these breaks up into small vessels which connect up with those of the area vasculosa and so complete the circulation in the extra-embryonic areas.

The blood supplied to the *embryo* by the branches from the dorsal aortae is returned to the sinus venosus of the heart by the right and left **ductus Cuvieri**. Each of these receives an **anterior cardinal vein**, draining the head, and a **posterior cardinal vein** bringing in blood from the trunk.

The Heart.—The formation of the heart is intimately bound up with the differentiation of the mesoderm and completion of the coelom and the fore-gut in the head region of the embryo. When about three somites have been formed, the lateral-plate mesoderm has pushed forward below the brain region of the neural plate and the right and left coelomic cavities become very large in this region. They are, because of their subsequent fate, known as the **amnio-cardiac vesicles**. The mesoderm and cavities of the amnio-cardiac vesicles soon extend into the head-fold (which still consists only of ectoderm plus endoderm) and convert it into somatopleure and splanchnopleure. In the hind-brain region, however, as the amnio-cardiac vesicles approach each other in the mid-line below the pharynx, there appears between their inner (mesial) walls and the endoderm of the pharynx on each side, a large blood vessel which connects at its hinder end with the extra-embryonic blood vessels. These large vessels, which later fuse to form a single one below the pharynx, are the rudiments of the **endocardium** or **endothelial lining** of the heart. The inner walls (splanchnic mesoderm) of the amnio-cardiac vesicles, which lie against the endocardium, will form the muscular wall of the heart and are known as the **myocardium**. The right and left parts of the myocardium then approach one another and fuse below the pharynx and also below the endocardium, so that it becomes enclosed in a complete **myocardial tube**. At the place where the right and left myocardial sheets came together there is formed a mesentery, both above and below the heart. These mesenteries are known respectively as the **dorsal** and **ventral mesocardium**, but both soon disappear except for a small portion of the dorsal mesocardium which persists at each end of the heart, and serves to suspend it in the fluid-filled space formed by the fusion of the right and left amnio-cardiac vesicles. This space is the pericardial cavity. It is, at first, continuous with the splanchnocoel, but later it becomes separated from it by a transverse septum, whilst it is limited at its front end by the tissue below the visceral clefts but remains open ventrally.

The heart when first formed is a straight tube. It soon increases rapidly in length and, because it is attached at each end, it buckles into an S-shape. Next, by the formation of transverse constrictions it becomes divided into a series of chambers separated by valves. The most posterior chamber is the sinus venosus, which receives blood from the ductus venosus (fused vitelline veins) and from the Cuvierian ducts. Anterior to the sinus venosus lies the atrium and in front of this the ventricle which opens into the most anterior

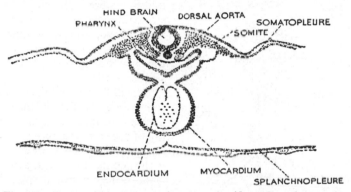

Fig. 644. CHICK.—Transverse section through a 10-somite embryo to show the completed cardiac tube.

chamber, the bulbus cordis. All the chambers have, to a varying degree, muscular walls derived from the myocardium.

The heart and pericardium, almost from the very beginning, seem disproportionately large, and as they grow still more they cause a conspicuous bulge to appear on the embryo. The relatively large size of the heart is correlated with the circulation of the blood through the extensive extra-embryonic vascular system, which is concerned with the nutrition of the embryo.

**Folding of the Embryo from off the Blastoderm and the Formation of the Body Wall.**—Up to this point, only certain features arising from the formation of the limiting folds have been alluded to, but these foldings also result in the establishment (from the somatopleure) of the body wall of the embryo and in the gradual pinching off, so to speak, of the embryo from the surface of the blastoderm.

It will be recalled that the folding of the blastoderm first takes place over a crescentic area with backwardly directed horns just in front of the embryo. The crescentic groove (anterior plus lateral limiting grooves) then gradually extends posteriorly beneath the embryo, and later another crescentic fold (tail-fold), similar to the previous one but facing in the opposite direction, appears beneath the hinder end of the embryo. This fold and its groove gradually progress in a forward direction until, finally, the two lateral arms (lateral limiting grooves and folds) of both folds meet below the embryo and surround a small area from which protrudes the stalk of the yolk sac and, later, the allantoic stalk. This folding process may perhaps be visualised in this way: it is just as if a pair of invisible fingers, working from each end towards the middle were pinching and stretching the layers beneath the embryo, and so gradually separating it from the rest of the blastoderm except for a slender stalk.

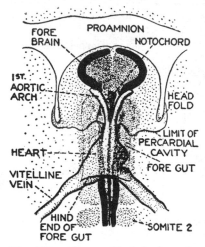

Fig. 645. CHICK.—Ventral view of the anterior end of a 9-somite embryo.

It will be realised that at the same time as the splanchnopleure is being folded, the folding of the somatopleure follows in much the same way, and it thus comes to form the outer layer to the embryonic splanchnocoel. That is, the folding of the somatopleure produces the ventral and lateral parts of the body wall, the dorsal and dorsolateral parts having already been formed by the somitic mesoderm and the ectoderm overlying it and the neural tube.

**Flexure and Rotation of the Embryo.**—For the sake of simplicity, no mention has so far been made of certain definite bendings (flexures) undergone by the embryo as its main organ systems become differentiated. These flexures, which materially alter the relation of the embryonic axis to the rest of the blastoderm, may perhaps, in part, be initiated by the heavier parts of the embryo sinking into the semi-liquid yolk for, at about the fourteen-somite stage, when the head is becoming well formed, it becomes bent downwards (ventrally) with respect to the main axis of the embryo.

This bending, which occurs in the region of the mid-brain, is the cranial flexure. Then, almost at the same time that the cranial flexure appears, the anterior end of the embryo becomes twisted about the long axis of the body in such a way that the head lies over

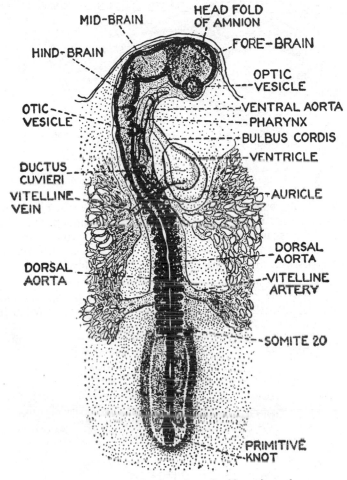

Fig. 646. CHICK.—Surface view of a 20-somite embryo.

on its left side. The more posterior parts of the embryo are, at this stage, still unaltered but the twisting gradually extends posteriorly, and eventually the whole embryo falls over, so to speak, on its left side. As this rotation of the embryo progresses,

the anterior end becomes still more bent over towards the ventral
surface by the development of a second flexure, the cervical flexure,
in the region where the hind-brain passes into the spinal cord.

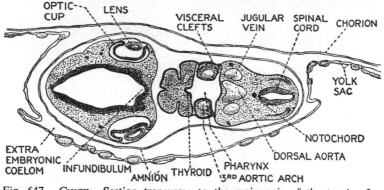

Fig. 647.   CHICK.—Section transverse to the main axis of the trunk of a
32-somite embryo through the pharyngeal region.

**The Embryonic Membranes.**—It has already been mentioned
that, in addition to the embryo proper, the blastoderm gives rise to
certain other structures which lie outside the embryo and are therefore said to be extra-embryonic.   These structures are the embryonic
membranes and are essential for the complete development of the
embryo.

THE AMNION AND THE CHORION.—The first of the embryonic
membranes to arise are the amnion and the chorion.   When fully

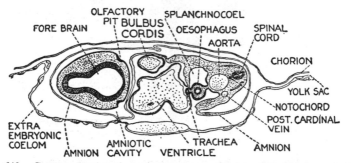

Fig. 648.   CHICK.—Section through the same embryo as Fig. 647 through the
region of the heart.

formed, the amnion is a sheet of somatopleure enclosing the entire
embryo in a fluid-filled cavity, the amniotic cavity, and the chorion,
which is also a sheet of somatopleure, forms the outer boundary for

the extra-embryonic coelom. The amnion and chorion make their first appearance at the same time as the cranial flexure when the blastoderm, immediately in front of the head of the embryo (*i.e.* the blastoderm forming the anterior edge of the anterior limiting fold), becomes elevated as a semi-circular ridge or fold. This fold

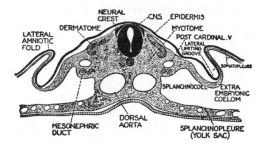

Fig. 649. CHICK.—Transverse section through the trunk region of a 29-somite embryo.

is the head-fold of the amnion. It begins to grow rapidly backwards over the head of the embryo over which it soon forms a complete arch. The head fold of the amnion is at first merely a double fold of ectoderm in front of the proamnion but, by the time its free edge has reached a little way over the embryo, it becomes invaded by the extra-embryonic mesoderm and coelom. The two layers (splanchnic and somatic) of mesoderm spread between the

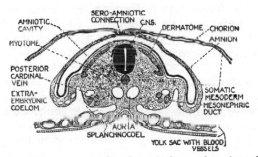

Fig. 650. CHICK.—Transverse section through the trunk region of a 32-somite embryo in the region of the sero-amniotic connection.

ectoderm and endoderm and separate them, the endoderm withdrawing from the fold and again coming to lie on the surface of the yolk. The endoderm becomes overlain by the splanchnic mesoderm and thus forms the splanchnopleure in front of the head. The somatic mesoderm applies itself to the ectoderm of the

amniotic head-fold and converts it into a double layer of somatopleure. That is to say, the head-fold of the amnion now consists
of an outer and an inner layer of somatopleure. The outer layer
(chorion) has its ectoderm outermost and its mesoderm innermost,
but, because of the doubling back of the fold, the inner layer
(amnion) has its ectoderm next to the embryo and its mesoderm,
outside this, facing the mesoderm of the chorion. The two layers
of mesoderm in the fold are separated by the extra-embryonic
coelom. The head-fold of the amnion continues its growth back
over the embryo and as it does so, the somatopleure at the sides of
the embryo rises up in continuity with the sides of the amniotic
head-fold, and in this way lateral amniotic folds are established.
They are not really independent developments, for they are continuous with the free edge of the amniotic head-fold.

In this way the free margin of the head-fold of the amnion is
being continually added to and the embryo becomes gradually
enclosed in a cavity, the amniotic cavity, which is lined by ectoderm,
since it is bounded by the amnion. Outside this membrane is an
extension of the extra-embryonic coelom which is itself limited on
the outside by the chorion. By the time that the amniotic head-
fold has grown back over the embryo as far as the eighteenth-somite,
another amniotic fold, the tail-fold of the amnion, rises up
immediately behind the posterior end of the embryo. Unlike
the amniotic head-fold it is, *from the first,* composed of a double
layer of somatopleure. It grows forwards in much the same way
as the anterior fold grows backwards and soon develops lateral
folds from the somatopleure, forming the lateral edges of the
lateral limiting grooves. Eventually, the head- and tail-folds of the
amnion meet and fuse at a point about a quarter of the way from
the posterior end of the embryo but the fusion of the layers is incomplete and involves only the ectodermal layers of the two folds.
To put it another way, unlike other regions, at this point the
chorion is not separated by the coelom from the amnion, and the
place of junction is termed the sero-amniotic connection.

When completed, the amnion and chorion are not simply
undifferentiated cell-layers, some of the mesoderm cells forming
unstriped muscle-fibres so that the membranes are endowed with
the power of contractility. The chorion continues to grow in
extent after the completion of the amnion and it keeps pace with
the extra-embryonic coelom (whose outer boundary it forms) as it
expands so as to continue to surround the still uncompleted yolk sac.

THE ALLANTOIS.—Soon after (at about the twenty-eight-
somite stage) the hind-gut has been formed by the tail-fold of the

embryo, the ventral wall of the hind-gut, just in front of the anal plate, grows out to form a sort of bud which projects out into the coelom. This bud is the **allantois** and, since it arises from the wall of the gut, it is formed of splanchnopleure, that is, an inner layer of endoderm and an outer layer of splanchnic mesoderm. The continued rapid growth of the allantois soon causes it to extend into the extra-embryonic coelom, in which it is seen as a pear-shaped vesicle between the amnion and the chorion. Later, the allantois forms a large membranous sac which fills practically the whole of the extra-embryonic coelom and its mesodermal layer fuses with

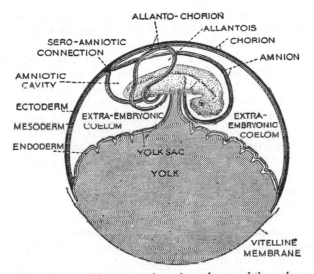

Fig. 651. CHICK.—Diagram to show the embryo and the embryonic membranes.

that of the chorion, forming a composite membrane, the **allanto-chorion**. This membrane later comes to be reflected around the albumen so as to enclose it in an albumen sac. The allantois remains, however, connected to the hind-gut by a narrow hollow stalk, the allantoic stalk or neck. When the head- and tail-folds of the embryo approach one another, the allantoic stalk and the stalk of the yolk sac become surrounded on all sides (but not below) by the somatopleure forming the ventral body wall. That is to say, the stalks of the allantois and yolk sac become enclosed in a tube of somatopleure to form the **umbilical cord**. Finally, some time before hatching, when the yolk sac and allantois have become considerably shrivelled, the umbilicus heals over.

Right from the time of its first appearance, the allantois is richly vascularised by the general system of vessels in the splanch-nopleure. Later, it receives blood by way of two large arteries, the allantoic or umbilical arteries which arise from the sciatic arteries of the chick embryo. Following a circulation in an extensive capillary network in the walls of the allantois, the blood is collected into the paired allantoic (umbilical) veins which pass forwards in the body wall to empty into the right and left ductus Cuvieri. As in the amnion and chorion, muscle-fibres are developed in the walls of the allantois.

THE YOLK SAC.—The yolk sac has already received some mention and it will be recalled that it is formed by a layer of extra-embryonic splanchnopleure. It is continuous with the embryonic splanchnopleure which later, by the formation of the head- and tail-folds of the embryo, becomes restricted to the wall of the gut and then joins the yolk sac splanchnopleure only by way of the yolk sac stalk. As the blastoderm grows in extent the yolk sac spreads over the yolk until, finally, the yolk is completely enclosed; but after this stage has been reached, as the yolk is gradually digested to provide nourishment for the embryo, it gets progressively smaller.

The endoderm lining the yolk sac is specialised both to digest the yolk and to absorb the products of digestion, and its surface is considerably increased by folds of the yolk sac wall which dip down into the yolk forming the so-called yolk sac septa. The mesoderm of the yolk sac forms mainly connective tissue, which serves to carry the extensive yolk sac circulation by means of which the embryo is supplied with nourishment. This circulation, it will be remembered, makes its appearance very early on in develop-ment as a vascular network (area vasculosa) surrounding the embryo. Later, it becomes connected to blood vessels in the embryo itself and, with the establishment of a definite circulation, certain larger vessels become defined. In this way, blood is carried to the yolk sac from the dorsal aorta by the vitelline arteries and is returned to the posterior end of the heart by a series of vitelline veins. The first vitelline veins to become evident are the anterior vitelline veins, but later, posterior vitelline veins and lateral vitelline veins are formed. Not all of these vessels persist throughout embryonic life and in the end, the yolk sac is supplied by fewer, but larger, vessels whose proximal ends enter the embryo along the yolk sac stalk.

THE ROLE OF THE MEMBRANES.—The majority of vertebrates (and, for that matter, of other animals) lay their eggs in water which, be it sea or pond or river, seems to provide a situation in which the embryos can develop with the minimum of external interference and consequently of structural adaptation. Water provides, so to speak, a friendly environment, for the embryo is surrounded by a medium from which its oxygen is readily obtainable by diffusion and into which it can pass its excretory products, which are then carried away. Water also buoys up the embryo and so reduces the stresses on the delicate young tissues whilst,

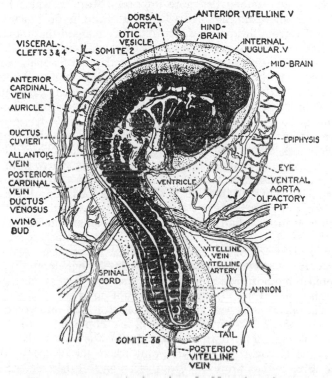

Fig. 632. CHICK.—Surface view of a 35-somite embryo.

because of the high specific heat of water, the embryo is not subjected to any sudden temperature fluctuations. The food problem is also to a large extent solved for the offspring of creatures laying their eggs in water, because food suitable for the young animal is, as a rule, abundant, whilst much of the inorganic salts required are absorbed through the egg membrane from the surrounding water. None of these favourable features are offered by conditions on dry land and moreover, there are the attendant risks of desiccation and large and sudden temperature changes. Most terrestrial animals have therefore developed special adaptive embryonic features which overcome these difficulties and so allow them to reproduce without recourse to water. Certain of them, however, like the frog, are compelled to return to the water to breed, for their spawn and the embryos hatching from it are suited only for life in the water. The reptiles, birds, and mammals, on the other hand, are completely emancipated from the consequences of their aquatic ancestry, and among the more important features which fit them for a thorough-going life on dry land are the embryonic membranes. Regarded from this point of view it will be realised that the amnion encloses the embryo in a fluid-filled cavity, which is the physiological equivalent of the water in which the embryos of aquatic animals develop. To put it another way, the amniote embryo builds and lives in a "little pond all of its own", through which a circulation of the fluid is effected by waves of contraction set up by the muscle-fibres in the amnion. The embryo is, to a large degree, protected from desiccation by the outer egg membranes—the shell, the shell membranes, and the albumen. It also seems that one of the chief functions of the albumen is actually to provide a source from which the embryo can absorb water. The shell is, however, by no means completely impermeable. It is, in fact, porous and allows of an exchange of gases between the inside of the egg and the air. Nevertheless, it forms a barrier between the egg-cell and its surroundings so that all the materials (including inorganic salts and most of the water) necessary for embryonic life must be provided by the egg contents. The energy required by the developing embryo is largely supplied by fats, which yield $CO_2$ and $H_2O$ only, during their metabolism. The $CO_2$ can readily be got rid of.

The allantois serves as a receptacle for the embryonic urine (which contains the practically insoluble compound, uric acid, a uricotelic type of nitrogen metabolism), for the neck of the allantois joins the hind-gut near to the point where the cloaca is formed. The urine is thus prevented from escaping to other regions where it might be harmful to the embryo. The allanto-chorion, which with its rich vascular supply comes into contact with the shell membrane, forms the embryonic respiratory surface and continues to function as such until the chick begins to breathe air, a short time before hatching. It also provides a means whereby calcium required for the skeleton can be withdrawn from the shell.

The amnion, chorion, and allantois can thus be collectively regarded as adaptations which allow the embryos as well as the adult to live on dry land. The yolk sac is in a different category for it is found in the eggs of fishes and it is, in fact, an inevitable development of the spreading of the blastoderm over the surface of a large-yolked, telolecithal egg. It serves as the digestive and absorptive surface by means of which the yolk is rendered available for the embryo.

**The Embryonic Stages.**—The stage to which embryonic development has proceeded by the time that the egg is laid is quite variable and depends to a large extent on the time of day at which the egg entered the oviduct. It is known that if the passage through the oviduct as far as the lower portion (vagina) has been completed by about four or five o'clock in the afternoon, the egg is laid that evening. If however, the egg reaches the vagina after about five o'clock then it is retained until the following day. During its period of retention the embryo will, of course, have undergone considerable development. From this it follows that any correlation between a certain embryonic stage in development and the age of the embryo in hours must, at the best, be approximate.

After the egg is laid, embryonic development is arrested by the drop in temperature and will continue only if the egg is incubated, that is, if the egg is kept moist and at some suitable temperature. The temperature of incubation also affects the rate of development, and within limits, the higher the temperature, the more rapid the development.

Nevertheless, it is a common practice to describe the stage of development reached by the embryo in terms of the hours during which the egg has been incubated [at a temperature around 40° C (103° F)]. A more satisfactory method is to express the embryonic stage in terms of the number of somites formed, for these are easily counted, and it is known that the rate at which organs develop relative to one another is fairly constant.

A very brief summary of changes undergone by the embryo as the egg is incubated up to seventy-two hours is given on p. 767.

## THE DEVELOPMENT OF MAMMALS

From evidence derived from many sources it may be taken as established that both mammals and birds have evolved from reptilian ancestors but along evolutionary lines which have been distinct from the beginning. The embryology of the chick may in most respects be regarded as typical of development from a large-yolked egg laid on dry land, and most of its essential features find their parallel in the development of the present-day reptiles. During the evolution of the mammals, not only the adult, but the developmental features also have become profoundly modified, for it is characteristic of most mammals that the young develop within a specially adapted part of the oviduct with which they establish intimate connection so that they are enabled to be "born alive". The development of the Mammalia is therefore of particular interest, not only because it is to this group that we ourselves belong, but because it sheds a good deal of light on the way in which the developmental features characteristic of the eggs of reptiles and birds have become modified and adapted to the needs of the embryonic life within the body of the mother. The exact stage in the evolutionary history of the mammals at which this viviparous habit was adopted is uncertain, and it is quite possible that certain of the extinct mammal-like reptiles may have been truly viviparous; nevertheless, it is only in the most highly evolved mammals (Eutheria) that viviparity reaches its maximum, and in one group (Prototheria or Monotremata) the young develop and hatch from eggs in a manner which is essentially reptilian. It is possible, therefore, that viviparity became established only after the mammals had become a separate and distinct group.

**The Eggs and Early Development.**—The eggs of the three main groups of mammals—the monotremes (Prototheria), marsupials (Metatheria), and placentals (Eutheria)—can be arranged, according to the amount of yolk present, to form a graded series. Those of the monotremes are the largest and those of the eutherians the smallest. The eggs of *Ornithorhynchus* (duck-billed platypus), although much smaller than those of any reptile, are really quite large, for they measure from 4 to 6 mm in diameter. When laid, they are enclosed in a layer of albumen and a tough calcified shell. The eggs are extremely telolecithal, the cytoplasm forming a small cap or blastodisc at the animal pole and, as would be expected, cleavage is meroblastic, resembling that in a bird or reptile. The details of gastrulation are not known, but it appears that the process resembles that in a reptile more closely than any other type. The eggs of *Echidna* (spiny anteater), which have a horny shell, are transferred to a pouch on the abdomen of the mother, where they hatch out as immature young which are then fed from a peculiar type of mammary glands.

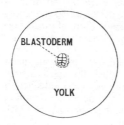

Fig. 653. Surface view of the egg of *Ornithorhynchus* (removed from shell).

The eggs of the marsupials measure from 0·15 to 0·25 mm in diameter but, although relatively small, they still contain appreciable quantities of yolky substance. Reptilian affinities are further indicated by the presence of albumen, shell membrane, and a horny shell (which disappears during early development). The yolk in some types (e.g. *Dasyurus*, native cat) is localised in a sphere lying at one end of the egg, but in others (e.g. *Didelphys*, opossum) it is found as separate small masses. Subsequent to fertilisation in the oviduct, the marsupial egg begins to divide into the first two blastomeres and, at the same time, the yolk is cast out into the fluid just within the egg membrane. The next two cleavages are

| AGE IN HOURS AFTER LAYING | NUMBER OF VISIBLE SOMITES (*i.e.* METOTIC SOMITES) | CONDITION OF EMBRYO |
|---|---|---|
| Unincubated | — | Endoderm separated from upper layer, *i.e.* blastoderm "two-layered". |
| 20 | 2 | Neural folds nearly meeting in mid-brain region—head-fold begun. |
| 24 | 6 | Area vasculosa well formed—rudiments of heart present. |
| 26 | 8 | Primary optic vesicles indicated—neural tube still open in front—fore-gut well formed. |
| 33 | 12 | Cranial flexure appearing—heart bent out to right side—auditory pit beginning to form. |
| 38 | 16 | Head beginning to turn on left side—head-fold of amnion covers fore-brain region—heart S-shaped. |
| 46 | 26 | Cervical and cranial flexures very pronounced—brain regions well differentiated—auditory vesicle practically closed—main parts of eye evident—tail-fold of amnion beginning to form. |
| 48 | 27 | Primitive streak practically gone—3rd visceral pouch formed—two complete aortic arches—hind-gut a distinct bay. |
| 72 | 36 | Oral membrane ruptured—4th visceral pouch forming—cerebral hemispheres clearly marked—limb buds present—amnion completely closed—allantois pushing out into extra-embryonic coelom—tail curved. |

also vertical and complete, and the eight blastomeres arrange themselves in the form of a ring. The sixteen-celled stage is reached by a horizontal division and thus two tiers or rings of blastomeres are formed, one above the other, the upper cells being slightly smaller than the lower ones. Subsequent divisions in both tiers result in the production of a hollow sphere of flattened cells which surround a fluid-filled cavity which also contains the yolky masses thrown out during the first cleavage of the egg. The embryo in this stage is termed a **blastocyst**. The cells forming its wall are probably the equivalent of those of the chick blastoderm at the end of cleavage, and the cavity of the blastocyst may be compared to the sub-germinal cavity; but unlike the chick, the cells of the marsupial embryo can enclose at an early stage the small amount of yolk present.

In other marsupials (e.g. *Didelphys*) a blastocyst is formed in a similar way, but the yolk is distributed throughout the cytoplasm and is cast out in the form of small particles or vesicles. During cleavage the blastomeres do not form very distinct tiers of cells, and

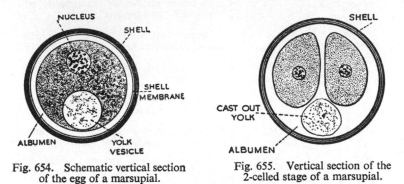

Fig. 654.   Schematic vertical section of the egg of a marsupial.

Fig. 655.   Vertical section of the 2-celled stage of a marsupial.

from the eight-celled stage onwards a more or less complete ball of cells is recognisable. There is no abrupt transition from the eggs and early developmental stages of the marsupials to those of the eutherians, which although smaller and lacking all membranes except a thin secondary membrane (and in some, a thin layer of dense albumen), still con-

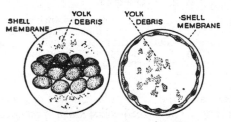

Fig. 656.   The 16-celled stage and early blastocyst of a marsupial (*Dasyurus*).

tain a small amount of yolk. In the guinea-pig and certain other types, this is accumulated chiefly to one end of the egg and is ejected immediately after fertilisation. Cleavage results in a ball of cells, the morula, which by acquiring a cavity becomes a blastocyst.

Many of the points briefly referred to above will become clearer after a more detailed description of the early development of the rabbit, which is chosen as a representative of the Eutheria.

## THE DEVELOPMENT OF THE RABBIT

**The Egg.**—The egg is released from the ovary, by rupture of the follicle, into the oviduct as a secondary oocyte. It is a small spherical cell (diameter of about 0·1 mm) and is surrounded by a thin striated membrane (**zona pellucida** or **zona radiata**) which is apparently secreted by the follicle cells of the ovary. If this is so, then a true vitelline or primary membrane is lacking. The zona itself is enclosed for a short time by a number of large follicle cells (**corona radiata**), but these soon disappear being, it seems, dispersed by the enzyme, hyaluronidase, carried in the semen, and their place is taken by a thin coat of dense albumen secreted by the oviduct. This layer becomes thicker as the egg passes down the oviduct. The small amount of yolk present is distributed fairly evenly in the cytoplasm and is only slightly more abundant in the hemisphere opposite to that from which the polar bodies are detached, so that

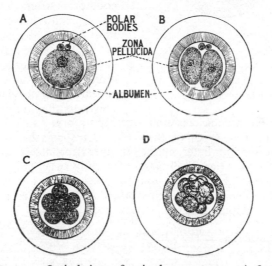

Fig. 657.   RABBIT.—Optical views of early cleavage stages.   A, fertilised egg; B, 2-celled stage;   C, early morula;   D, later morula.

the egg has a distinct polarity. This is further shown by the excentric position of the egg nucleus which lies above the equator in an area quite free from yolk. The egg is therefore only to a very slight degree, telolecithal, that is, it is *secondarily microlecithal*. The animal pole may most easily be recognised by the presence there of one or more polar bodies according to the stage of maturation.

**Fertilisation.**—The egg is fertilised in the upper part of the oviduct, and an unusual feature of mammalian development is that the whole of the spermatozoon passes into the egg cytoplasm. The tail, however, soon degenerates. Soon after the entry of the sperm the egg completes its maturation divisions, but no yolk is cast out.

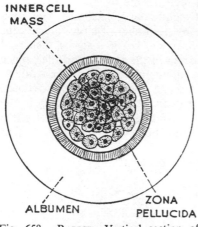

Fig. 658.    RABBIT.—Vertical section of a late morula stage.

**Cleavage.**—After a pause, the fertilised egg divides completely by a vertical furrow to form the first two blastomeres, one of which is smaller and more opaque than the other. The next cleavage is also vertical but at right angles to the first, whilst the third is horizontal and slightly above the equator. After this, cell division is no longer synchronous (it may become irregular before this stage) and the resulting blastomeres, now all more or less alike in size, appear to bear no relation whatsoever to the original polarity of the egg but become clustered together to form a solid ball of cells or morula. Two types of the cells are recognisable in the fully formed morula: an outer, epithelial layer (**trophoblast**) and a compact central mass (**inner cell mass**) of slightly larger polyhedral cells. It is in this stage that the embryo reaches the uterus, cleavage having taken place during the passage down the upper part of the oviduct or Fallopian tube.

**The Blastocyst.** — The morula, although consisting of very many cells, is not appreciably larger than the fertilised egg, but in the next phase of development growth in size is very rapid. The embryo, on reaching the uterus, comes into contact with the mucous membrane and begins to absorb the fluid secreted by it. The embryo swells rapidly as the fluid which it absorbs accumulates in a cavity which, except over a small area at one end of the

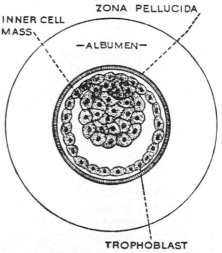

Fig. 659.    RABBIT.—Vertical section of early blastocyst stage.

blastocyst, separates the trophoblast from the inner cell mass. The embryo is now termed a **blastocyst**, and the trophoblast cells which lie immediately above the inner cell mass are called the **cells of Rauber**.

By rapid cell division and by continued imbibition of fluid, the blastocyst grows in size so that the inner cell mass appears as a knob-like thickening at one pole. It is from this thickening that all the cells of the embryo proper are derived and for this reason it is now usually termed the **embryonal knob**.

The possession of a distinct embryonal knob is one point of difference between the blastocyst of a marsupial and that of

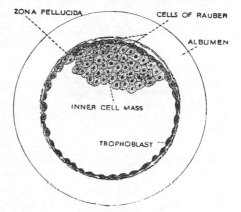

Fig. 660. RABBIT.—Vertical section of a later blastocyst stage.

a eutherian mammal, for in the marsupial the whole of the upper hemisphere of the blastocyst gives rise to the embryonic tissue and hence corresponds with the embryonal knob of higher types. The lower half of the marsupial blastocyst is a region from which extra-embryonic material is derived. In both types, the fluid in the blastocyst is coagulable (*i.e.* contains proteins) and is probably nutritive in function. The mammalian blastocyst may therefore be thought of as being the equivalent of the chick blastoderm at the

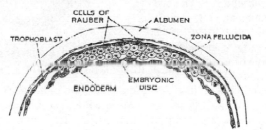

Fig. 661. RABBIT.—Vertical section through the embryonal disc.

end of cleavage, and the cavity as the equivalent of the sub-germinal cavity plus the space occupied by the yolk. As has already been mentioned, only the cells of the embryonal knob will give rise to embryonic tissue and for this reason it is comparable to the area pellucida of the chick, whilst the trophoblast is the equivalent of the extra-embryonic (chorionic) ectoderm which, because of the minuteness of the yolk, is enabled at an early stage to form the walls of a complete sac or vesicle.

These comparisons, interesting as they are, must not be taken too literally, for the blastocyst is an embryonic stage peculiar to the mammals.

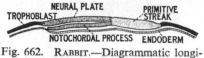

Fig. 662. RABBIT.—Diagrammatic longitudinal section through the early embryo for comparison with Fig. 629.

**Gastrulation.**—After the blastocyst has been growing for some time, certain cells, which stain more darkly begin to leave the lower surface of the embryonal knob and wander out in all directions beneath the trophoblast, thus making the upper part of the blastocyst two layers in thickness. The newly formed inner layer of cells is the endoderm, and its separation is the first step in the conversion of the blastocyst (apart from the embryonal knob) into what is really a yolk sac, although, of course, it is filled merely with fluid and, in the marsupials and certain types of eutherians (but not the rabbit), contains a few traces of the disintegrating yolk cast out during cleavage.

After separation of the endoderm the embryonal knob flattens and its area becomes correspondingly increased, its cells becoming rearranged to form a fairly regular epithelium. The cells of Rauber meanwhile disintegrate and are then either sloughed off, or perhaps are incorporated in the trophoblast, so that the cells of the embryonal knob or disc, as it

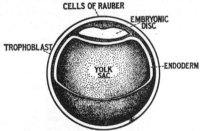

Fig. 663. Diagrammatic sagittal half of an advanced blastocyst of the rabbit type.

may now be called, are exposed at the surface of the blastocyst. Meanwhile, the blastocyst has continued to grow very rapidly and soon the zona pellucida and albuminous layers, already considerably stretched, are dissolved away.

The map of the main embryonic areas on the embryonal disc is far from being well known, but the available evidence points to it being, in broad outline at any rate, similar to that in reptiles and birds. The **primitive streak** and the formative movements which result in the separation of the **mesoderm** and **notochordal** process also seem similar to those of the chick embryo and will not be redescribed. It may, however, be mentioned that in some types, including man, the notochordal process is, for a time, a hollow structure being traversed by a chorda canal just as in a reptile. The differentiation and folding of the embryo from the embryonic area takes place in a way so like that of a chick, that some practice is required in distinguishing between chick and mammal embryos at comparable stages in development. In fact, from gastrulation onwards, the main points which require further treatment are the development and elaboration of the extra-embryonic regions of the blastocyst.

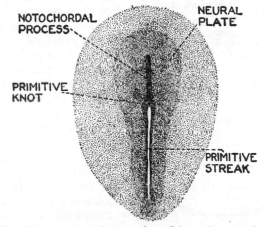

NOTOCHORDAL PROCESS

NEURAL PLATE

PRIMITIVE KNOT

PRIMITIVE STREAK

Fig. 664. RABBIT.—Surface view of the embryonal disc.

**Implantation of the Blastocyst.**—It will be recalled that the blastocyst on reaching the uterus comes into close contact with the uterine mucosa. Soon, from the trophoblast on the side opposite to the embryonic disc, small papillae (**trophoblastic villi**) grow out and penetrate into corresponding crypts in the uterine wall, so that the blastocyst becomes fixed or implanted.

**The Embryonic (Foetal) Membranes.**—As in the chick, the embryonic membranes are the amnion, chorion, allantois, and yolk sac.

AMNION AND CHORION.—After the formation of the primitive streak, the trophoblast around the margins of the embryonic disc rises up as folds, which grow over and enclose the embryo. These are the amniotic folds. The tail-fold of the amnion, which is larger and grows more rapidly than the head-fold, receives, almost from the very beginning, its complement of somatic mesoderm, but, for a time, the head-fold of the amnion lacks any mesodermal layers. Soon, however, both head- and tail-folds of the amnion come, as in the chick, to consist of an outer layer and an inner layer of somatopleure. The outer layer (**chorion**) of each amniotic fold has an outer layer of trophoblast (ectoderm) and an inner layer of somatic mesoderm, whilst the inner layer of the fold (**amnion**) has an outer layer of somatic mesoderm and an inner layer of trophoblast (ectoderm). Between the amnion and chorion lies a portion of the extra-embryonic coelom. As in the chick, the final enclosure of the embryo in an ectoderm-lined amniotic cavity is effected by the fusion of head- and tail-folds of the amnion; and the only significant point of difference between the chick and the rabbit is that in the latter it is the tail-fold which is the larger. As a result, the sero-amniotic connection lies above the anterior region of the embryo. In other eutherian mammals the amniotic cavity is formed in a variety of different ways, but when the amnion and chorion are completed they bear much the same relationship to the embryo as that described above.

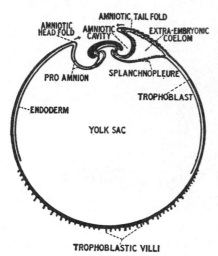

Fig. 665. RABBIT.—Diagrammatic vertical section of blastocyst to show the embryonic membranes.

After the fusion of the amniotic folds the trophoblast of the chorion develops villi, like those on the lower half of the blastocyst, which burrow into the uterine wall.

ALLANTOIS.—Even before the amniotic cavity has become closed, the endoderm beneath the hinder end of the embryo grows out as a sac or vesicle and carries with it the overlying layer of splanchnic mesoderm. This vesicle is the allantois. It grows extremely rapidly and soon comes to extend into the extra-embryonic coelom between the amnion and chorion where it enlarges and grows over the embryo outside the amnion. From its very early stages the wall of the allantois is well supplied with blood vessels and, later, certain of these enlarge to form the umbilical (allantoic) arteries and veins. With continued growth of the allantois, the splanchnic mesoderm forming its outer layer applies itself to the mesodermal layers of both the amnion and chorion and fuses with them. The outer layer of extra-embryonic tissue thus comes to consist of trophoblast, thickened mesoderm (chorionic and allantoic), and endoderm—the three layers forming collectively the **allanto-chorion.** As in the chick, the allanto-chorion plays an important part in the excretory and respiratory processes of the young animal but, in the mammal, it acquires the additional function of being nutritive for, as will be seen later, it forms the major contribution to the **placenta.**

YOLK SAC.—The **yolk sac** or **umbilical vesicle**, as it is often called in the mammal, is developed from the lower part of the

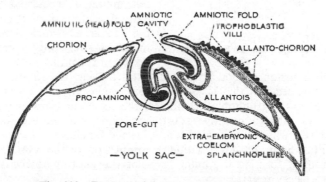

Fig. 666. RABBIT.—A later stage than Fig. 665.

blastocyst, which becomes progressively lined by endoderm. Later, the two layers of extra-embryonic lateral-plate mesoderm, together with the extra-embryonic coelom, extend downwards from the embryonic area and penetrate between the endoderm and tropho-blast of the upper part of the yolk sac. The splanchnic layer of mesoderm then applies itself to the endoderm to complete the wall of the yolk sac and in this splanchnopleure a vitelline circulation is established. The lower part of the yolk sac, however, always remains two-layered (endoderm plus trophoblast), for mesoderm does not penetrate into that region. After a time the lower wall of the blastocyst (lower wall of yolk sac) ruptures so that the cavity of the yolk sac communicates with that of the uterus, and thus becomes filled with secretions from the uterine glands. These secretions are absorbed into the vitelline circulation and are no doubt nutritive. Thus the yolk sac plays a similar, but less important, part in the nutrition of the embryo to that which it does in embryos developing from large yolked eggs. In certain mammals (carnivores, ungulates) the supply of nutritive uterine secretions (**uterine milk**) is copious and, at first, these are largely absorbed by the yolk sac whilst, in the marsupials, the wall of the yolk sac comes into close connection with the wall of the uterus to form a **yolk sac placenta**.

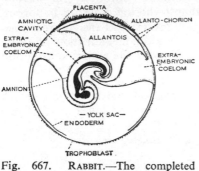

Fig. 667. RABBIT.—The completed embryonic (foetal) membranes.

**The Nutrition of the Embryo and Foetus*—The Placenta.**—The almost complete absence of yolk from the eggs of eutherian mammals necessitates the early development of structures by means of which nourishment for the embryo can be obtained from the mother. In the first instance food is provided by uterine secretions which are absorbed by the blastocyst from the cavity of the uterus but, later, the trophoblastic villi on the lower hemisphere of the blastocyst absorb nutritive substances directly from the wall of the uterus, which is to some extent digested by enzymes secreted by the villi. Soon, however, with the completion of the amnion, the trophoblast forming the outer layer of the chorion thickens and becomes syncytial whilst from it villi grow into the wall of the uterus which has also by this time undergone hypertrophy. The wall of the allantois has meanwhile fused with the chorion to form the allanto-chorion, and the mesoderm in this region soon becomes

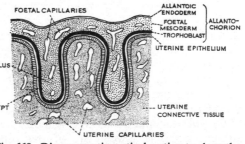

Fig. 668. Diagrammatic vertical section to show the layers entering into the composition of a placenta.

richly vascularised and grows outwards to form a core of foetal connective tissue and vessels in each of the trophoblastic villi in

* After the laying down of the main organs of the embryo and the establishment of the enclosing membranes, the young mammal is referred to as a **foetus** and the membranes as foetal membranes.

this region. The tissues forming this intimate structural connection between foetus and mother are termed a **placenta**, and from the time that it is established it provides a means of rapid but selective exchange of substances between foetal and maternal blood circulations and thus ensures that the nutritional, respiratory, and excretory requirements of the foetus are satisfied.

The ease of communication between mother and foetus is soon increased by the breaking down and disappearance of certain of the layers entering into the composition of the placenta for, almost from the beginning, the syncytial trophoblast begins to erode (by a sort of digestive process) the uterine wall. In succession, the uterine epithelium, the uterine connective tissue, and then the walls of the maternal blood-capillaries are broken down, whilst the foetal connective tissue in the villi also, to a large extent, disappears. Maternal blood comes to circulate in lacunae in the thickened trophoblast of the villi, so that in the placenta at this stage, during the exchange of substances between foetus and mother, diffusion is through the trophoblast and the walls of the foetal capillaries only.

It is only in certain mammals (including man) that a placenta of this type (**haemo-chorialis**) is found. In others (*e.g.* pig) all the layers

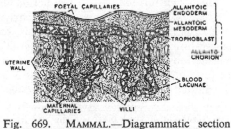

Fig. 669. MAMMAL.—Diagrammatic section through the placenta (haemo-chorialis).

of the placenta are retained (**epithelio-chorialis** type of placenta), whilst in some (*e.g.* cow and sheep) the uterine epithelium is the only layer to be eroded (**syndesmo-chorialis** type of placenta). It follows that in placentae of these types diffusion is more difficult and the placenta involves practically the whole of the trophoblast whereas, as has been seen, placentae of the haemo-chorialis type, which are more "efficient", occupy only a small area above the embryo. A more or less intermediate type of placenta is found in the carnivores where both the uterine epithelium and uterine connective tissue break down, but the walls of the maternal capillaries remain intact (**endothelio-chorialis** type of placenta). Placentae of this type are formed as a broad girdle around the foetus.

In the later stages of foetal life in the rabbit, parts of the trophoblast finally break down so that maternal blood bathes the foetal capillaries directly. The final rabbit placenta is therefore termed a **haemo-endothelial** placenta.

It must be emphasised that, no matter what the type of placenta, there is never any *direct* communication between foetal and maternal blood-streams and reasons for this are not far to seek. In the first place, the delicate foetal tissues would be quite unable to withstand the pressure developed in the maternal arteries. Secondly, substances do not pass through the placenta by simple diffusion, for the placenta exerts a selective influence and acts as a barrier to those substances which might prove harmful to the foetus. Thus, the trophoblast has the power of breaking down proteins before they enter the foetal circulation. This is necessary because some proteins are specific to individuals and their direct transference would prove injurious. Lastly it may be recalled that the secondary sexual characters are controlled by the hormones circulating in the blood, and, were foetal and maternal streams to become confluent, it is difficult to see how the secondary sexual characters of a male foetus and the mother could escape alteration.

**Birth (Parturition).**—After a **period of gestation**, varying from a few weeks in smaller mammals to nearly two years in the elephant, the foetus reaches a state at which it is, to all intents and purposes, a miniature adult. The wall of the uterus then undergoes rhythmical contractions, the connection between foetal and maternal layers in the placenta is broken, and the foetus is gradually expelled through the dilated vagina. In most mammals the foetal contributions to the placenta and the other foetal membranes are expelled some time after the actual birth of the young animal and are termed the "**after-birth**", but the only maternal tissue to be lost is a certain amount of blood, the maternal portions of the placenta being absorbed. The connection (umbilical cord) between the newly-born animal and the foetal part of the placenta is sometimes severed by the mother, although a greater or lesser portion of it always remains attached to the young animal. Later it shrivels until finally its position is indicated merely by the **umbilicus** or navel.

# CHAPTER XX

## GENETICS

**INTRODUCTION**

Every organism has a number of characteristics by which we recognise it. Some of these it shares with others, some are peculiar to itself. They may be features of the animal's morphology or, equally likely, aspects of its physiology. Many of these characteristics are passed on from generation to generation. Genetics is the study of the mechanism by which this information (see p. 28) is communicated from parent to offspring and of how these inherited characters are produced. Thus genetics is essentially the study and interpretation of the results of breeding experiments.

It is often said that "children resemble their parents" or "like begets like" but these sayings are only broadly true. If you look at the offspring of almost any animal you will be able to find differences from its parents, though in broad outline parents and offspring are alike. The sibs (sisters and brothers) will rarely be exactly alike; identical twins derived from the same zygote will be more alike than other pairs of brothers or sisters. It is plain that far from being exactly alike, organisms vary. This variation is the essential material upon which evolution can work to alter species (p. 825).

It is always important to distinguish between variation produced through the genetic mechanism and that produced by the external conditions, numerous examples of which will spring to the mind. Insects whose food supply as larvae is restricted may develop into small-size adults, yet these same dwarfs will produce offspring which are of normal size provided they, in their turn, are adequately fed as larvae. These are variations due to the environment and will not be passed on to the offspring. Such variations will not concern us here.

Sometimes characters appear in an offspring though they were absent in the parent, often they can be shown to have been present in the grandparents or even in some earlier ancestor. Or the features possessed by an organism may *appear* to be a blending of some of those of the parents. Observations like this suggest that characters may be inherited in different ways

In the lower organisms where multiplication involves some process of fission, either binary or multiple, and where the offspring are formed directly from the parental body, the information upon which the transmission of characters is based can be passed in a direct way. But in more complex animals or plants where the body is made up of a large number of cells, the only direct connection between parent and offspring is through the germ-cells which are set aside for reproductive purposes, often from an early stage in development. Though they are not isolated from the rest of the body, for hormones affect them, they play little part in the general functioning of the organism. Thus from the start one supposes that the hereditary mechanism must reside in these cells and the fertilised egg must carry the information necessary to direct the formation of the new organism. In simple organisms such as the bacteriophages (allied to viruses) and bacteria, the relationship of the hereditary mechanism with the expression of the character is closer and in consequence in recent years these organisms have proved immensely useful in research on the hereditary mechanism.

But it must be quite clearly understood that the character which is shown by an organism is not inherited as such but is the result of the interaction of the capacity to acquire that character which is handed from parent to offspring and the environment which has the power to control the form in which the capacity is expressed. Here the environment means not only the external environment but also the internal conditions within the organism against which background the character develops. Nevertheless, because they are firmly established in common speech, the term "heritable" and "inherited character" are still used. The collection of inherited information, the genes, in fact, which is responsible for the capacity is called the **genotype**, or **genome**, while the final expression in the organism of this information is the **phenotype**. The extent to which an inherited character is shown in the phenotype is known as its penetrance. While most genes studied in breeding experiments have 100 per cent. penetrance appearing in the phenotype whenever they are present, there are many which show variable penetrance. It must be stated that there is evidence of other hereditary mechanisms apart from the genes (p. 799).

The problems, then, are: what is the behaviour of the characters which are inherited as they appear in generation after generation and by what means are they passed?

## HERITABLE CHARACTERS AND THEIR BEHAVIOUR

The object of breeding experiments is the observation of a character, noting its appearance or non-appearance after crosses of animals which show it or whose ancestors showed it. The fruit fly, *Drosophila*, is very convenient material for these studies because it is easily reared in the laboratory in large numbers and has a generation time of about three weeks. This means that the results of a cross can be found out relatively quickly. An additional advantage is that a fruit fly may lay as many as 1,000 eggs so any conclusions drawn from such a large number will have statistical significance. The small number of children that human parents have in comparison mean that it is statistically unsafe to draw any conclusions from observing the offspring of one mother and father.

These fruit flies show many quite easily observable characters. Those which appear most frequently in flies caught in the wild are referred to as the wild type. Thus, they have fully developed wings. There are other flies, occurring occasionally in nature, whose wings are little more than stubs and quite useless for flying. The percentage of them vary in different populations but they are always in the minority. Differences also occur from the wild type in the shape and colour of eyes, the pattern and arrangement of bristles on the body, and so forth. Such different forms will breed true if they are inter-bred and will continue to show the character for hundreds of generations. If they are crossed with true breeding wild type flies, the offspring will always be like their wild type parents. But breeding experiments show that though the phenotypes are the same as the wild types the genotypes are different.

Now let us consider what might happen when we mate a wild type fly with one of these true-breeding vestigial winged flies. The offspring, called the $F_1$ generation, are all fully winged. The character for vestigial wings seems to have disappeared. But the disappearance is only temporary as we find when we mate one of the flies from the $F_1$ generation with a vestigial winged fly. Approximately half of the offspring are winged and the other half with vestiges of wings only. There is no statistically significant difference between the proportions of the two kinds, in other words, the offspring are equally divided into winged and vestigial individuals.

Again, if flies of the $F_1$ generation are mated together, about a quarter of their progeny, the $F_2$ generation, will have vestigial wings while the remainder will be fully winged.

These results can be explained if we suppose that the characters of the wings are controlled by a pair of genes. The wild type is usually designated + and the vestigial by initials to indicate what the character is, in this case, $vg$. Each fly contains two of these genes, the wild one $+/+$ and the vestigial one $vg/vg$. When eggs and sperm are formed meiosis takes place and one only of each pair of genes goes into each gamete. Thus, the wild fly has gametes bearing + and the vestigial fly ones bearing $vg$. When these two sorts of gametes fuse to restore the diploid condition, they yield zygotes with the genotypic constitution $+/vg$. However, in such a combination the wild type factor suppresses and over-rules the vestigial, so all the flies with this genotype show full development of their wings. But the mating of a $+/vg$ fly with a vestigial winged $vg/vg$ fly shows that though the vestigial gene has not expressed itself in the phenotype of the $F_1$ flies it is unchanged. For the $F_1$ flies will give equal numbers of gametes bearing $+$ and $vg$, while the vestigial fly will give $vg$ only. The fusion of the $+$ gametes with the $vg$ from the vestigial fly will yield zygotes with the constitution $+/vg$ in which the expression of the $vg$ will be suppressed and the flies have fully developed wings. On the other hand, an equal number of combinations of $vg/vg$ will be formed and these flies will have vestigial wings. In the cross of $F_1$ flies, since one gene must come from each parent, the possible combinations will be $+/+$, $+/vg$, $vg/+$, and $vg/vg$. Flies with any, except the last, of these will have full wings but those bearing the last combination will have vestigial ones.

The $+$ gene is said to be dominant over the $vg$ gene which is recessive to the wild type gene. Genes which bear such relationships are called alleles or allelomorphs (see p. 794). They are believed to occupy the same position on a pair of chromosomes so that when the two chromosomes pair up the alleles lie opposite each other. During meiosis one member of each pair of chromosomes will go to each end of the dividing cell and thus one will pass to each new cell (see p. 677), carrying one of the pair of alleles with it.

Supposing, however, that one of the $F_1$ flies is mated with another of the same generation, the result will be quite different. For then the number of winged flies among the offspring ($F_2$ generation) will not be statistically significantly different from three times the number of vestigial flies. This we can explain for each of the parents will now contain both wild and vestigial factors. Each will give gametes which contain the wild type gene and the vestigial gene in equal numbers. Thus a wild type gene from one parent may pair with a wild type one from the other and produce a fully winged fly. Or that same gene may pair with a vestigial one from the other parent. In this case as the wild type is dominant the fly will have fully developed wings. Or finally a vestigial gene from each parent may fuse to give flies which have reduced wings. If we tested these flies further we would find that some of the winged flies bred true but others produced vestigial offspring in later generations. All the flies with vestigial wings breed true. We show this result diagrammatically on facing page.

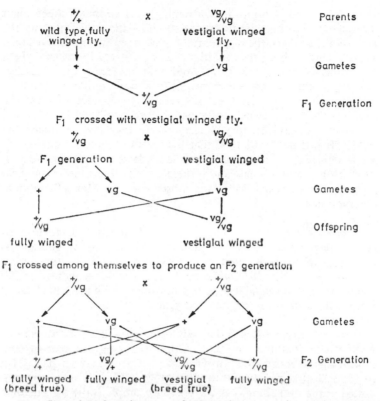

The flies that breed true, whether they contain wild type or vestigial genes, are known as homozygotes. They produce gametes which are all of the same genetic constitution. The other flies which do not breed true are heterozygotes, producing gametes of two different kinds. The genetically different gametes pair quite at random as is shown by the way the actual figures for numbers of offspring in the various classes fit in with the theoretical chance expectation based on our diagram. But the distribution of these ratios is a matter of statistics and as such, the smaller the numbers we deal with, the less likely we are to get results which immediately appear to agree closely with the theoretical ones though after the appropriate statistical test they can be shown to agree with theory.

These results are similar to the observations made by Gregor Mendel. He was an Abbot of Brunn (in what is now Czechoslovakia). His hobby was to study the inheritance of characters of garden peas such as their flower colour, the appearance (wrinkled or smooth) of their seeds, and so forth. The characters he chose always appeared in opposite contrasting pairs so that if a pea plant showed one of them it did not also show the other, just as the *Drosophila* could either have fully developed wings or vestigial wings. Having bred true-breeding lines of plants he crossed them in various combinations and found that the pairs of characters behaved in the subsequent offspring in a way similar to those in our example above. In 1866, his results which were to form the basis for the whole science of genetics were published in a journal of a local Natural History Society where they lay unnoticed for years. It was not until 1900 that they were found, and his experiments repeated. The so-called Mendel's laws were in fact formulated then by de Vries and others in the light of cytological discoveries on cell division made subsequently to Mendel's original observations.

*Mendel's First Law.*—This is the law of the segregation of factors responsible for the characters. Of a pair of characters, only one can be represented in a single gamete by its factor. Thus the gametes of the *Drosophila* could carry either a factor for full wings or one for vestigial ones but not both. What Mendel called factors are what we nowadays call genes.

*Mendel's Second Law.*—The law of independent (or free) assortment of germinal units or factors: that each one of a pair of contrasted characters may be combined with either of another pair. Let us suppose that we study the inheritance of two characters transmitted at the same time. A convenient example is the transmission of ebony ($e$), a body colour recessive gene, and vestigial wing ($vg$). If we cross a female fly with normal body colour who is vestigial winged with a male which is ebony bodied but with normal wings, the resulting $F_1$ flies will all be wild type in appearance having normal coloured bodies and being fully winged. The parents must have been homozygous for vestigial and ebony respectively for these characters are recessive and therefore can only appear when they are carried as a like pair. The $F_1$ flies must therefore be heterozygous for each of the genes. But because the wild type genes are dominant they will have the phenotype of a wild type fly.

If we now cross one of these $F_1$ flies with a fly showing both vestigial wings and ebony colour, the result is that four different kinds of flies are produced in approximately equal numbers. They are wild type, vestigial winged, wild coloured type, fully winged, ebony coloured type, and vestigial winged, ebony coloured flies. This result can be explained if we suppose that the $F_1$ parent produces equal numbers of gametes of each of these kinds: $vg+$,

$+e$, $++$, and $vg\,e$. The other parent will produce one kind of gamete only, that is $vg\,e$.

$$(\female) \qquad \frac{+}{+}\frac{vg}{vg}\ (\female) \quad \times \quad \frac{e}{e}\frac{+}{+}\ (\male) \qquad \text{Parents}$$

$$+\ vg \qquad\qquad\qquad e\ + \qquad \text{Gametes}$$

$$\frac{+}{e}\frac{vg}{+} \qquad \text{F}_1 \text{ generation}$$

$$\frac{+}{e}\frac{vg}{+} \quad \times \quad \frac{+}{e}\frac{vg}{+} \qquad \text{F}_1 \text{ generation}$$

$$+vg \ ++ \ e\,vg \ e+ \qquad +vg \ ++ \ e\,vg \ e+ \qquad \text{Gametes}$$

|  | $+vg$ | $++$ | $e\,vg$ | $e+$ |
|---|---|---|---|---|
| $+vg$ | $\dfrac{+}{+}\dfrac{vg}{vg}$ <br> wild : vestigial | $\dfrac{+}{+}\dfrac{+}{vg}$ <br> wild : wild | $\dfrac{e}{+}\dfrac{vg}{vg}$ <br> wild : vestigial | $\dfrac{e}{+}\dfrac{+}{vg}$ <br> wild : wild |
| $++$ | $\dfrac{+}{+}\dfrac{vg}{+}$ <br> wild : wild | $\dfrac{+}{+}\dfrac{+}{+}$ <br> wild : wild | $\dfrac{e}{+}\dfrac{vg}{+}$ <br> wild : wild | $\dfrac{e}{+}\dfrac{+}{+}$ <br> wild : wild |
| $e\,vg$ | $\dfrac{+}{e}\dfrac{vg}{vg}$ <br> wild : vestigial | $\dfrac{+}{e}\dfrac{+}{vg}$ <br> wild : wild | $\dfrac{e}{e}\dfrac{vg}{vg}$ <br> ebony : vestigial | $\dfrac{e}{e}\dfrac{+}{vg}$ <br> ebony : wild |
| $e+$ | $\dfrac{+}{e}\dfrac{vg}{+}$ <br> wild : wild | $\dfrac{+}{e}\dfrac{+}{+}$ <br> wild : wild | $\dfrac{e}{e}\dfrac{vg}{+}$ <br> ebony : wild | $\dfrac{e}{e}\dfrac{+}{+}$ <br> ebony : wild |

Gametes of 1st parent

Gametes of 2nd parent

Results: wild/wild 9, wild/vestigial 3, ebony/wild 3, ebony/vestigial 1.

The results of the cross can be looked at in another way. There are two classes of the flies which are like the original parents; they are the vestigial flies and the ebony flies. Two classes show combinations which did not appear in the original parents: wild type for both characters, and vestigial-ebony. These are *re-combinants* because the combinations of characters are new ones.

It is clear from these results that the genes have been assorting themselves quite independently and randomly so that chance produces equal numbers of all possible sets of genes in the gametes. The factors for vestigial have not been tied in any way with those for ebony. This demonstrates Mendel's Second Law.

At this point it will be convenient to summarise the previous pages by stating the definitions of some of the terms used by geneticists:

**Gene.** This corresponds to Mendel's germinal unit or factor and is the effective genetical cause present in the gamete which may lead to the appearance of the character in the adult.

**Dominant character.** A character due to a dominant gene, the one of a pair of contrasted characters which expresses itself in the outward appearance of a heterozygote.

**Recessive character.** A character due to a recessive gene, the one of a pair of contrasted characters which does not express itself in the outward appearance of the heterozygote, but which is transmitted to later generations without change.

**Alleles** (Allelomorphs). The members of a series of genes any two of which can behave in a Mendelian manner. They all lie at the same locus on a chromosome.

**Homozygous.** Applied to an individual receiving like alleles from its two parents.

**Heterozygous.** Applied to an individual receiving unlike alleles from its two parents.

**Phenotype.** The outward expression of the genes, thus a type of organism defined as possessing certain observable characters. This must be contrasted with:

**Genotype,** which is a type of organism defined by its genetical constitution.

**Monohybrid** inheritance. The inheritance of one pair of genes.

**Dihybrid** inheritance. The inheritance of two pairs of genes.

## The Relationship of Inheritance with the Chromosomes

The discovery of the chromosomes and the study of their behaviour in cell division eventually led to a provisional explanation of the mechanism for the transfer of genetic information from cell to cell and generation to generation. It will be recalled that during the formation of the gametes reduction division takes place so that by separation of the members of pairs of homologous chromosomes they contain the haploid number of chromosomes. This echoes the separation of the pair of factors (genes) into the constituents which is demanded by Mendel's First Law.

When the zygote is formed the diploid condition is restored and the chromosomes associate in pairs in each of which one must have come from the father and the other from the mother. The likeness of the movements of the chromosomes and the behaviour of the pairs of factors postulated by Mendel is one of the main arguments for the belief that the chromosomes carry the genetic factors. Additional support comes from the equal distribution of the chromosomes to each daughter-cell in mitosis. By this mechanism, each cell has the same material both qualitatively and quantitatively as the other. All cells will then have the same chromosome complements.

Mendel was fortunate in his choice of the characters of the garden pea for the ones he used always showed completely independent assortment. However, though many crosses show results in keeping with this law, many cases have been found of apparent exceptions to the rule. They fall into two groups: those in which a new phenotype arises from the cross due to interaction between the gene factors (incomplete dominance) and those in which the assortments are not random, so that the results of the cross do not fit in with the expected ratios. The inheritance of comb form in poultry is a good example of the first kind of exception.

### Incomplete Dominance

Two forms of comb, the rose-comb and the pea-comb, among the many varieties which occur in poultry, breed true. But when birds with these combs are bred together, a totally different form appears, that termed walnut. This is an example of incomplete

Fig. 670. Rose-comb.

Fig. 671. Pea-comb.

dominance. The relations between the genes for rose and pea are such that they are not mutually exclusive in their effect on the phenotype as they would be if they were in the normal dominance-recessive relationship. Where the genes for rose- and for pea-combs occur in the same genotype they interact to produce the new character, walnut.

If the walnut-bearing birds are inter-bred, all types of comb appear in the succeeding generation with the addition of a different one, single, which is also characteristic of the wild relatives of the domesticated birds. The results can be expressed as shown overleaf. It is plain that the single-combed birds, which breed true, must be carrying a double recessive factor. The walnut-combs are therefore due to a double dominant whereas the original comb forms, which also bred true, must be due to a pair of genes, not a single one like the characters which we have dealt with so far. This shows that there are occasions when genes interact. In fact this is far

| Adults | $pp r^+ r^+$ (Rose) | × | $p^+ p^+ rr$ (Pea) | |
|---|---|---|---|---|

Types of Gametes    $pr^+$   |   $p^+ r$

| Adults | $p^+ pr^+ r$ (Walnut) | × | $p^+ pr^+ r$ (Walnut) | F₁ |
|---|---|---|---|---|

TYPES OF GAMETES.

| ♂→ ♀↓ | $p^+ r^+$ | $pr^+$ | $p^+ r$ | $pr$ |
|---|---|---|---|---|
| $p^+ r^+$ | $p^+ r^+$ $p^+ r^+$ walnut | $pr^+$ $p^+ r^+$ walnut | $p^+ r$ $p^+ r^+$ walnut | $pr$ $p^+ r^+$ walnut |
| $pr^+$ | $p^+ r^+$ $pr^+$ walnut | $pr^+$ $pr^+$ rose | $p^+ r$ $pr^+$ walnut | $pr$ $pr^+$ rose |
| $p^+ r$ | $p^+ r^+$ $p^+ r$ walnut | $pr^+$ $p^+ r$ walnut | $p^+ r$ $p^+ r$ pea | $pr$ $p^+ r$ pea |
| $pr$ | $p^+ r^+$ $pr$ walnut | $pr^+$ $pr$ rose | $p^+ r$ $pr$ pea | $pr$ $pr$ single |

Adults. 9 walnut : 3 rose : 3 pea : 1 single.     F₂

commoner than was at first thought and the influence of the neighbouring genes on the expression of any one of them is great, as we shall see later (p. 796).

*The Determination of Sex.* In most animals sex is determined genetically. There are exceptions to this rule, but they are rare. The sex, for example, of certain insect parasites may be determined by the amount of food available to them. If the parasites are crowded the relative numbers of each sex may be altered.

In general, however, the mating of a male and a female produces equal numbers of males and females. The final ratio of males to females may be affected for other reasons, more males may die during development, for instance, but the sexes usually begin equal in numbers.

Fig. 672. Single-comb.

The chromosome set of a male and a female animal differ in appearance. *Drosophila*, for example, has four pairs in the diploid set but one of the partners of one pair is differently shaped in the male whereas in the female both are identical. This also is true of men. But in birds and butterflies, to take two examples, it is the female that has the odd chromosome.

In any case the existence of a pair of unlike chromosomes means that the bearer of them will give unlike gametes, half of them with one chromosome and the other half with the unlike partner. The sex which does this is called the heterogametic sex and is the male in *Drosophila* and man, but the female in birds and butterflies. The other sex will give gametes, all of which have the same constitution with regard to the sex chromosome. This is the homogametic sex and is the female in the fly and man, and the male in birds and butterflies.

These chromosomes, called the X and Y chromosomes, which determine the sex are differentiated as sex chromosomes from the rest of the set which are called autosomes. A male human being is XY while a female is XX. Sometimes the balance between the autosomes and the sex chromosomes is the important factor in deciding the sex. Animals which have XXX or XXXY may then be super-females showing accentuated female characters. This is the situation in some moths. Often, however, the Y chromosome is unimportant in deciding the sex. Thus it is the ratio of the X chromosomes to the autosomes which determines the sex. A *Drosophila* with X is a male despite the fact that it has no Y chromosome. In humans the Y chromosome is of the greater importance and without it maleness is not developed. Whatever the genetic mechanism may be it is only a switch device to put in train the hormonal and other differences which bring about the outward phenotypic differences between the sexes.

There are, however, occasions when the offspring of a cross are not distributed in the expected ratio and which cannot be explained by incomplete dominance. Under the conditions we have mentioned so far the products of the meiotic divisions in the formation of the sex cells in the two parents are equal in number. If, however, these products are unequal, then the final ratio will be altered. An example of this is the result of crossing wild type *Drosophila* with true breeding flies having the characters glass eye (*gl*) and ebony body colour (*e*). The result is an $F_1$ generation which are wild in appearance but among the offspring ($F_2$ generation) of a cross of these $F_1$ females with glass eyed, ebony males similar to the original parents, 92 per cent. are either glass eyed and ebony bodied or wild type, there being far less than would be expected of normal eyed, ebony bodied and glass eyed, normal bodied individuals. Thus the parental types represent a much larger proportion than the 50 per cent. that we would expect if the genes were sorting independently. This can be shown as on the following page. The fact that the genes for glass eye and ebony body tend to remain together is indicated by the horizontal lines. Such a tendency to

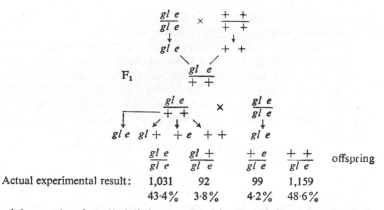

| | $\dfrac{gl\ e}{gl\ e}$ | $\dfrac{gl\ +}{gl\ e}$ | $\dfrac{+\ e}{gl\ e}$ | $\dfrac{+\ +}{gl\ e}$ | offspring |
|---|---|---|---|---|---|
| Actual experimental result: | 1,031 | 92 | 99 | 1,159 | |
| | 43·4% | 3·8% | 4·2% | 48·6% | |

stick together is called linkage. In this case linkage for the genes concerned does not occur on the sex chromosomes but on the autosomes (autosomal as opposed to sex linkage, where the linked genes are born on the sex chromosomes and which has a number of special consequences as we shall see later, p. 791).

The combinations of characters shown by the 8 per cent. of the offspring, *i.e.* those not found in the parents, are called the recombination groups. In a cross the extent to which these recombination groups are found is a measure of the amount of crossing over, and thus the reciprocal of linkage. If the proportion of recombination groups approaches the theoretical ones expected for independent assortment then clearly linkage is very weak. On the other hand, in the example quoted above the proportions are very different from the expected and indicate strong linkage.

It is found, in any one species, that certain genes normally associate together to form linkage groups, but different linkage groups sort independently of others. The number of these linkage groups is equal to the number of chromosomes characteristic of that species. This strongly suggests that linkage of two or more genes is due to their occurrence on the same chromosome and this is, indeed, the accepted explanation.

## Sex Linkage

Occasionally the inheritance of characteristics which appear in the phenotype are strongly (but not completely) linked with one or the other sex. The inheritance of haemophilia in man is a good example. Almost without exception women are never haemophiliacs though they transmit the condition; it is men who are seriously affected. The disease, a condition in which the blood does not clot, occurs only in males but must be transmitted by females who do not, however, show the condition. This is because the gene responsible is borne on the X chromosome; it is recessive to the normal gene. Thus in a man bearing an affected X chromosome there can be no normal gene to counteract the haemophilia gene and prevent its phenotypic effect. But a woman has another X chromosome on which there will very probably be a normal gene and thus she will not show the condition but only transmit it when she passes on the affected X chromosome; she is a "carrier". The likelihood of a woman having two affected X chromosomes is very small for she has to be the daughter of an affected father and a carrier mother; even though such marriages have been reported,

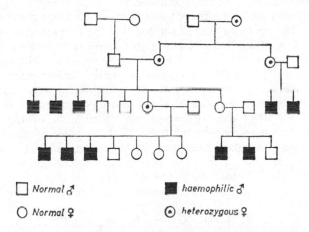

Fig. 673. A pedigree of haemophilia.

there have been no daughters resulting from them. In all probability the homozygous condition is lethal during development.

Cytological proof that linkage occurs because the characters are borne upon the same chromosome has been obtained in various ways but one of the most elegant used characters carried on the X chromosome of *Drosophila*. These concerned the dominant Bar eye and the recessive eye colour mutant carnation. One strain carried both of these genes while another carried the wild type alleles. The X chromosomes of the two strains were morphologically different from normal, in one by a chance natural occurrence, in the other after bombardment with X-rays. The result of the crosses was to show that where recombinations occurred the X chromosomes of the individuals showing the recombinations (in this case, wild eye shape and carnation eye colour, or Bar eye and wild eye colour) also had X chromosomes which clearly resulted from recombination between the X chromosomes of the two strains.

Mendel's Law of recombination probably would not have been formulated as such had he studied a greater variety of characters, or had he known of the chromosomes and their behaviour. Nowadays it is obvious to us that there are far more genes than there are chromosomes and believing as we do that genes occur in them, it seems plain that gene linkage *must* occur.

### Genetic Maps

When it became clear that the frequency of recombination depended upon the particular genes being studied, Morgan suggested that the genes must be arranged in a linear order along the chromosomes. The next step was to suggest that the frequency of recombination (*i.e.* how often linkage *breaks down*) is a measure of the distance apart of the genes. Those close together would be closely linked and show little recombination, those farther apart more recombination. From this sort of information *map distances* could be allocated to pairs of genes based on the extent of their linkage and from these distances the distribution of the genes along the chromosome found out. If three pairs of characters, for example, are taken, the genes for them can be represented by A, a, B, b, and C, c. The percentage recombination of the characters can be observed in crosses. Suppose that of A, a and B, b is 1 per cent. and that for B, b and C, c is 5 per cent., the relative positions of A, B, and C may be either

A  B  C  or  C  A B.
←1→ ←5→          ←1→
                ←— 5 —→

But if in addition the percentage for A, a and C, c is 6 per cent. then the positions must be

A  B      C.
←1→ ←—5—→
←——6——→

By slowly and patiently accumulating data of this sort of all the known characters of a species it has been possible for geneticists to produce genetic maps of whole chromosomes.

Crossing over can also be confirmed by the observation of the giant chromosomes of many Diptera (Fig. 32). These are in fact interphase chromosomes, clearly visible under the lower powers of the light microscope after staining because each consists of many copies lying parallel to each other. They are banded and these bands coincide with the position of the gene loci derived from genetic mapping. Thus changes in the gene distribution can be easily seen.

## The Mechanism of Crossing Over

Recombination occurs because of crossing over between chromosomes but how this comes about is not completely clear.

Probably the process occurs when the tetrads have been formed, that is when the two homologous chromosomes are closely applied in meiosis. Each is a bivalent, being in the form of two sister chromatids which have not yet parted. At first they are coiled around each other. Possibly at this time breaks occur in two non-sister chromatids. The broken ends attach themselves to chromatids of the other bivalent. Since all four chromatids can enter into crossing over, this seems the most likely explanation but an alternative is for the mixture of genes to occur during reduplication so that one bivalent is formed partly by its parent, as it were, and partly by the parent's partner. But there are more difficulties in accepting this explanation than the first one.

Recombination is also found in organisms like viruses where there are no chromosomes to break and rejoin. Here the effect does seem to come about during the duplication process. Perhaps work done on these organisms will lead to some addition to our ideas of how crossing over occurs in higher organisms.

## MUTATIONS

In any naturally occurring population animals with characters strikingly dissimilar to those of the rest of the population are constantly appearing. The white coloured albinos found occasionally among populations of most wild mammals are an example, as are also men suffering from haemophilia. These altered individuals are mutants. The phenotypic change which they display may arise from two sources in the genotype. First, there may be structural alteration in a chromosome, or second, a change in the structure of the gene itself, a gene mutation. The first are visible under the light microscope—under relatively low magnification where giant chromosomes can be observed—while the second are invisible even under the highest magnifications of the electron microscope; their presence is detected only by their phenotypic effects.

**Chromosome Mutation.**—A great variety of structural alterations in chromosomes can occur which change the expression of the genes borne upon them and thus give rise to altered appearance of the individuals. These occur because chromosomes break and the broken ends appear to have a certain amount of stickiness which permits them to make new connections (crossing over is an example of this).

To give a few examples, two breaks in a chromosome can lead to a section being dropped out, or *deleted*, the broken ends joining and leaving the middle piece on one side. Or, a section may be *inverted*, so that the sequence of genes becomes ABCFEDGH instead of ABCDEFGH, the segment bearing the DEF having been reversed. Or, again, a section may be duplicated, so that the sequence ABCDEFGH becomes ABCDEDEFGH. Such a duplication is the cause of the mutation Bar eye in *Drosophila*. Lastly, a section of one chromosome may become attached to another. This is *translocation*; some forms of mongolian idiocy in man are associated with a chromosome change in which chromosome 21 appears in triplicate (because of a failure to sort correctly at gamete formation), one of these may then become attached to another chromosome in the set.

Most of the chromosomal changes can be detected by their effect on pairing of homologous chromosomes at zygotene stage of meiosis. Since similar parts of each partner will tend to be opposed, it will not be possible for the chromosomes to lie parallel to each other. To take a simple example, a deletion in one partner will require that a loop be formed in the other. The loop will contain those parts which are missing from the first chromosome.

Another form of chromosome mutation is the duplication of whole sets of chromosomes in polyploidy. This is an important source of variation in plants but is probably rare in occurrence in animals. However, a number of species of earthworm are polyploids.

### Gene Mutation

Thus far we have considered the gene as a stable entity unchanged in its transmission, which occupies a particular locus on a chromosome. While this is in general true the nature of a gene can be altered so that its expression is altered as well. This is a mutation. A gene mutates to produce a certain set of variations upon itself; on any one occasion any one of these variations are produced. These are the alleles and since they are altered forms of a gene, they occupy the same locus on the chromosome.

A gene will mutate into one only of a certain number of alleles. Thus, the gene responsible for wild type wing shape and size in *Drosophila* mutates to one of a series of alleles (12) ranging from the reduced wings of the *vestigial* allele to the fully winged *wild* type. The alleles of the *Hb* locus responsible for the production of haemoglobin in human blood cause the production of different haemoglobins which differ only in the presence or absence of a few amino acids in the molecule.

Alleles in a series can be caused to mutate into each other as well as to the wild type. If the wild type is thought of as the extreme end of such a series mutation can cause changes towards or away from the wild type but always along the same line of alleles.

Mutations are naturally occurring as we have said. But they can also be produced by submitting animals to a variety of treatments; X-rays and other ionising radiations, ultra-violet light, chemical treatments such as exposure to mustard gas (nitrogen mustard), and even heat can all be used to cause mutations. Though with any one treatment there is no certainty as to which gene will be caused to mutate, or the allele which will be formed. Thus, this is a very gross process whose mode of action may differ according to the method used.

Ionising radiations may act by producing actual damage for they can cause chromosomal as well as gene mutations. But it is also clear that the rate of mutation produced is directly correlated with the dosage and, further, that the dosage is cumulative. Thus an X-ray dosage of 6,000 röntgen units produces sex-linked lethal mutations in 12 per cent. of the offspring of an irradiated population of *Drosophila*, while three doses of 2,000 units have the same effect. But it may be that the effect is indirect through the production of some substance as a result of the effect of irradiation on other chemicals in the cell. Water might be turned, for example, to hydrogen peroxide, a highly reactive compound which would then affect the genetic material itself.

Ultra-violet light may have yet other effects for it has been shown to cause closer binding of the bases in the DNA chain by the formation of more hydrogen bonds between them. This would of course affect the pairing of the bases and thence the reduplication of the DNA effecting a gene mutation.

Another class of mutagens, as substances causing mutations are called, are those like bromouracil which are chemically analogous to one of the bases in the DNA molecule. They replace the base, thymine in the case of bromouracil, and produce alterations in the sequences of base pairs during reduplication of the DNA.

**Mutation Rates.**—Mutation rates are usually expressed as the number occurring in a million gametes produced by the organism. Since in matings not all gametes take part in the production of a zygote and hence an offspring in the phenotype of which the genes carried in the gametes can be recognised, the greater the number of matings tested, the more accurate is the estimate of the mutation rate likely to be. Figures for some rates are given overleaf.

| ORGANISM | GENE | MUTATIONS PER MILLION GAMETES |
|----------|------|-------------------------------|
| Man | Haemophilia | 20—30·2 |
|  | Muscular dystrophy | 8 |
| *Drosophila* | *y* (yellow) | 29 |
|  | *w* (white) | 29 |
|  | *ct* (cut) | 150 |

Most mutation rates are extremely slow but the rates per million gametes must be viewed against the generation time of the organism. The same rate in a bacterium with a generation time of minutes will produce faster change than in a mammal with a generation time measurable in years.

On the other hand, an apparently slow rate may nevertheless be evolutionarily effective because in evolution one is dealing with change which has occurred over very long periods of time, giving the possibility of even very rare events to occur.

**The Effect of Mutations.**—The great number of mutations which are produced artificially are lethal, causing the death of the animal at some time, early or late in life. Others appear in animals which as a result are only viable because they are in laboratory culture and protected therefore against the action of what would be their normal environment. Mutations of both these kinds are also found in wild populations after series of matings by which their presence can be revealed. Their action has, however, been suppressed by mechanisms which will be considered later.

The great majority are also recessive, though exceptions are known. Whether they be recessive or dominant to begin with, this may be altered under the influence of selection. It has been shown that genes can increase their dominance or recessiveness as a result of artificial selection for those animals in which their action is either enhanced or suppressed. When this happens the gene itself is not being altered as is shown by transferring the gene by means of cross-breeding into another genotype. What is happening is that the expression of the gene is being modified by the other genes. In other words, alterations in the genic background are altering the expression of the gene.

**The Genic Background.** This demonstrates the great importance of the other genes in altering the expression of one of their number. Indeed, just as the general environment of an animal determines the expression of the genes in the phenotype, so also does the genic environment of the chromosome (p. 780).

The genotypes of many wild species seem to be self-regulating so that mutations which occur are controlled and prevented from having a harmful effect, such effects being modified by the other genes in the set.

## The Gene

So far we have used the word gene to mean three different hereditary units: as a unit controlling the appearance of a character, as a unit capable of mutation, and as a unit of recombination (between two of which breakage may occur bringing about recombination of characters but within which breakage does not occur). Breeding experiments on *Drosophila* and other animals do not, on the whole, indicate any difference between these units.

But more complete investigation of crosses may reveal a condition which indicates that the "gene" is divisible. Such a position was shown in a series of alleles, known as *lozenge*, which affect eye shape and are carried on the X chromosome of *Drosophila*, when after some 20,000 male offspring had been obtained, some five to ten were found to be wild type in eye structure, by recombination. Since the original parents bore mutants at this locus, this could only have come about by a break within what was thought to have been one locus, so that a wild type chromosome would be found. This later proved to be the case, and it was shown that within a section 0·2 map units in length three possible sites of mutation existed. Female flies who were heterozygous for alleles at two of these sites could, however, have two different phenotypes. If the alleles occurred on different chromosomes, the fly had a mutant appearance, but if both mutant alleles occurred on the same X chromosome, the female had eyes which were wild type in appearance. The first of these arrangements is known as the *trans* form and the second the *cis* form.

This suggests that the duplication of some substance, probably messenger RNA, is easier if the two wild type genes are on the same chromosome than if they are on two of them (*trans*), for then duplication would have to go along one chromosome and then along its homologue. *Cis-trans* configurations of this sort have now been found in a number of other organisms, insects, moulds, bacteria, and viruses among them.

It is work on viruses (bacteriophages) which has led to reconsideration of the idea of a gene as a particulate entity and shown how the different definitions can be reconciled. One region of the phage chromosome can produce a mutant which causes the phage to be unable to destroy the bacterium *Escherichia coli* strain K. However, this region proves to contain two segments, the activities of which are complementary unless there is a mutation in one and not the other. These units are called *cistrons* and are units of function in producing phenotypic effect. Within each cistron are

many places where mutations can occur. Recombination can
occur only between these places, that is, breakage can occur only
between them. They are *recons*, while "the smallest element,
alteration of which can be effective in causing a mutation" is a
*muton*. These last two units may, in the phage DNA, be no more
than one or two nucleotide pairs. Evidence exists for similar
units in higher organisms as well as bacteria. It means that we
cannot continue to view the gene as an ultimate particle of heredity,
as Dalton's billiard ball atom was the ultimate particle of matter.
What is the "gene" will alter according to which one of its
properties we are considering.

**Gene Action.**—There are many processes in the living body
which require the production of a series of chemical substances, as
a result of which one material is changed into another. Each of
the steps of the series is controlled by an enzyme. One such series
in man is that which results in the production of homogentisic acid
which is then converted into fumaric acid and acetoacetic acid, which
are excreted. Normally, therefore, no homogentisic acid appears in
the urine of humans, but there are individuals whose urine goes
black, the blackening being brought about by the oxidisation of
homogentisic acid. This condition is inherited in every way like a
Mendelian recessive character. This suggests that the recessive
allele does not have the ability to produce the enzyme responsible
for the breakdown of the homogentisic acid in the body as the wild
type gene has.

There are many other examples of a connection between the gene
and the presence or absence of a specific enzyme responsible for a
step in such a metabolic series. For example, the absence of the
enzyme responsible for the production of melanin in the eye of
*Drosophila* means that the flies will have pink eyes in place of the
dark wild coloured ones. In addition, there is a great deal of very
clear evidence that a single gene controls the production of a
specific enzyme in the metabolism of the mould *Neurospora*.
There remains no doubt that in many instances, and perhaps
all, the relationship can be summed up as "one gene—one
enzyme".

With the discovery of the mechanisms of control of cellular
activity by DNA in the nucleus, it is not difficult to conceive how
this relationship happens. The DNA produces messenger RNA
(p. 25) which leaves the nucleus and passes into the cytoplasm.
There it organises the production of specific proteins on the ribosomes with the aid of the shorter chained transfer RNA. The
specific protein is the enzyme responsible for the metabolic step.

**The Expression of the Gene.**—The cells of an animal's body contain identical sets of chromosomes within their nuclei (there are exceptions, for example, when one-half of an insect's body is male in character and the other half female, but this is a general rule) yet the cells are not alike either in their morphology or their function. It is clear that all the information carried by the genes is not utilised at one time; parts will be important at the stage of differentiation of the cell, others when it settles down to function as a mature cell. Some suggestive observations of the giant chromosomes of midges, like *Chironomus*, indicate that the process of control is being seen at work. At times bulges can be seen to develop on parts of the chromosomes; these are called Balbiani rings. On staining they prove to be mainly composed of RNA and by the use of tracers can be shown to be formed from substances in the cell. Thus they may represent the active part of the chromosome from which messenger RNA is being produced. Control of the activity of the genes probably comes through the cytoplasm. There is a characteristic patterning of the order in which parts of the chromosomes in midges become active during moulting of the larva into pupa and adult. These changes are known to be connected with the presence of the moulting hormone ecdysone (p. 331) in the blood. This stresses the fact that the nucleus cannot be considered in isolation from the cytoplasm for there are interactions in both directions: indeed there is very much evidence that the cell surface is implicated as well, being very potent in determining differentiation of cells; this work is not done by the genes alone.

**Cytoplasmic Inheritance.**—There is evidence that inheritance of some characters may not depend on nuclear genes but on some sort of self-reproducing gene-like factors in the cytoplasm. The suspicion arises when reciprocal crosses do not give the same results, in other words, when the numbers of offspring showing a certain character is different according to which of the parents bears the characters. Usually the mother is more influential in these cases and inheritance follows the mother's characteristics. It is important to distinguish true inheritance by cytoplasmic agents from maternal influences when this happens for both may have their effect through the cytoplasm of the egg. For example, the direction of coiling of the shell of the snail *Limnea peregra* is determined not by the animal's own genotype but by that of its female parent. And, in another case, the colour of the eyes of the offspring of the moth *Ephestia* depends upon whether the egg receives a chemical precursor of the eye pigment from the mother while the egg is in the ovariole. These effects seem to occur through the cytoplasm of the egg which is supplied mainly by the female parent. But they are not inherited, they are an effect of the environment.

The existence of true cytoplasmic inheritance does appear to be proved in one case at least to the satisfaction of critical geneticists. The existence of the kappa particles in the cytoplasm of *Paramecium* (p. 85) by which one *Paramecium* kills others of the sensitive strain depends upon nuclear genes, thus they do not show cytoplasmic inheritance. But other recently discovered particles (metagons) concerned with the formation of *mu* particles (which cause the death of the sexual partner and are therefore "mate killers") can multiply even when in the body of the ciliate *Didinium* which preys upon *Paramecium*. These are true extra-nuclear, self-reproducing particles.

# CHAPTER XXI

## THE INTERRELATIONS AND ORIGINS OF ANIMALS—
## ORGANIC EVOLUTION

### INTRODUCTION

The method of presentation which has been adopted in the preceding chapters, that of proceeding from the simpler animals to the more complex, has implied the existence of an interrelationship between the various forms, as opposed to the idea that they represent isolated and unrelated examples taken at random from the multiplicity of animals existing at the present day. This arrangement is an expression of a point of view, so generally accepted nowadays that it excites little comment, that in the natural world there has been in operation a process called **Organic Evolution.** By this process, the present-day forms have arisen by gradual changes from pre-existing forms, unlike them in many features, so that it may be said that the more complex have been derived from the simpler and that the animal kingdom can be viewed as a series of progressively increasing complexity.

This idea is really a quite ancient one for it goes back to the early Greek philosophers and has been applied to other than biological concepts. This is particularly so in philosophical and religious discussions. To the biologist, organic evolution means the gradual unfolding of the living world on this planet. There also exists an equally ancient opposing view found in many mythologies, namely, **Special Creation**, which postulates that present-day forms were created by some outside influence when life came into existence on this planet and have continued with little alteration ever since.

Both views assume the existence of life; the latter as an accompaniment of each act of creation, whilst the former, if pushed to a logical conclusion, postulates the ultimate origin of the organic world from the inorganic. There have, in fact, been several theories to account for the formation of self-replicating organic molecules from simple chemical compounds. **Spontaneous Generation**, that is, the sudden appearance of living organisms, whatever their mode of origin, is discounted as happening nowadays, for Pasteur (1822-95) showed conclusively that to-day no organism arises except as the offspring of pre-existing organisms, but this is *negative, inductive* evidence and some simple forms of life **certainly** *must* have arisen spontaneously, although perhaps under very different conditions, early in the Palaeozoic. For the present, then, biologists are content to accept the *fact* of life and are more concerned with the *way* in which the various forms of living organisms arose.

Although it is impossible, here, to give in detail the varied histories of these two schools of thought, one attempt to effect a compromise between them may be mentioned. The examination by Cuvier (1769-1832) of the fossilised remains of animals, quite unlike those found living to-day, led him to postulate—since he was a believer in special creation—that there had been in the history of the world several creations and that each, at the end of its period, had been wiped out by some cataclysmic upheaval and a fresh set of organisms created on different lines. This theory was termed **Catastrophism**, but gains no credence now.

Similarly, with the idea of evolution, various attempts have been made to explain the *way* in which it has come about, with the more important of which are associated the names of Darwin (1809-82), Wallace (1823-1913), and Lamarck (1744-1829). But before it is possible to evaluate the means by which it has been effected, it will be well to examine the evidence upon which the statement that evolution *has* taken place is based.

This evidence is, perhaps, directed towards establishing two important ideas. Firstly, the *fact of change*; that is, that it is possible for a species to alter in measurable time. This involves the admission that the species is not immutable, and if the environment changes, then the *species* may change. Secondly, an estimation of the *extent to which change has taken place* in the form of organisms in the history of this planet. This, naturally, must in a measure be indirect evidence, since no one now living was present when the bulk of the changes took place.

## THE FACT OF CHANGE

This is the most direct evidence which can be offered and comes from several sources.

In the previous pages of this book the individual members of each species of animal have been dealt with as if they were all exactly alike, and so far as their *main* features of form and function were concerned, this is undoubtedly true. For example, a particular frog shares the features of its skin, skeleton, vascular system, etc., with all other frogs of the same species, so that it is immaterial what particular individual is taken for the examination of these features. Moreover, the egg of a frog always produces a frog and no other type of animal. But if all the frogs emerging from a batch of eggs produced by a single female are examined critically, then in some small particulars they will all be found to differ from one another. This is even more true of the progeny of different parents. These differences may be due to hereditary influences (genetical), or may be due to the effect of the conditions under which the individuals developed (environmental). It is to these genetical differences that attention must be directed in considering whether they can be the source of changes in the species itself.

Mankind discovered early in history that the structure and physiology of plants and animals can be altered by careful selection and breeding. A classic example is that of the many and varied forms of pigeons produced in historic time as the result of the efforts of pigeon fanciers. There is a great deal of difference between the wild rock-pigeon, the original stock, and the fantail or pouter pigeon. Similarly, the stud-book of the Turf Club records the very obvious differences between the modern racehorse and the arab steed from which it was bred. The selection of milck cows for their milk yield has resulted in the production of strains which excel in this direction. Again, strains of cows have been selected for their flesh output; of sheep for length of wool; of rabbits for their fur. Also strains of wheat have been bred for their grain yield, rust resistance, and the like. No useful purpose would be served by the multiplying of instances, but these examples will serve to give some idea of the many variations which are possible within what is still regarded as the same species, and it is probable that if the development of such differences was continued, there would emerge forms which would be regarded as new species. (See also p. 804.)

In a few instances genetical changes have been noticed and their spread recorded in populations of animals in the wild. An interesting example is that of industrial melanism in moths, particularly in the peppered moth *Biston betularia*. Very dark melanic forms of this species were first noticed near Manchester in 1850. These were called the *carbonaria* variety. Nowadays they are by far the commonest type in all industrial areas having replaced the original stock. Moreover, it has been shown experimentally that this can be explained by differential predation by moth-eating birds which find the normal moths easier to see against soot-blackened backgrounds than the *carbonaria* variety. The reverse is true in open country in the relatively unpolluted West of England where the original light-coloured variety is still the commoner one. The caterpillars of *carbonaria* have a better survival rate than those of the original variety when food is scarce. Here then is an example of an obvious genetical change which has been favoured by selection within a short while not by man, however, but by naturally occurring stresses and hence is called Natural Selection (p. 825).

All this (and many more examples could be added) establishes that animal populations can and do change in captivity and in the wild. Where a species is distributed over a wide area in which there are variations of climate, temperature, altitude, and humidity, forms arise in these various situations which show very obvious departures from the normal, so much so that they are dubbed subspecies and a trinomial nomenclature is adopted to distinguish them. For example, the European bullfinch, *Pyrrhula pyrrhula* has a form *P.p. pyrrhula* which ranges from west Siberia through northern Russia and Scandinavia, whilst in southern and western Europe from northern Italy to the west, embracing western Germany, a different form, *P.p. minor*, occurs. Again, in North America, woodpeckers are found right across the continent. On the Pacific coast these birds are regarded as belonging to the species *Colaptes cafer*. The eastern species, though it also extends westwards into Alaska, is termed *C. auratus*. Between the two regions there exists a territory over 1,000 miles long and from 300-400 miles wide, where most of the birds have characteristics of both species. The accepted explanation is that the original population became separated into two isolated groups by the glaciation of the Rocky Mountains and subsequently, when the ice receded, met once more and interbred. But by that time the modification of the ancestral form, by adaptation to environmental conditions, had produced the two types which taxonomists have elevated to the status of distinct species.

## THE EXTENT OF CHANGE

Having once accepted as valid that organisms can alter their characters, the next problem to be explored is the extent to which such change has taken place on the earth. Here the evidence must, of necessity, be indirect.

**Evidence from Classification.**—In the account of the principles of classification (Chapter III) it is stated that a phylum is composed of animals which show a sufficient number of common features to be regarded as varieties of expression of a common ground-plan of organisation. Although it is possible to conceive that the species within a phylum have been separately created, it is much more reasonable to conclude that they have all arisen from a common stock and that each variant from it is the expression of adaptations to its particular environmental conditions. When a large phylum such as the Arthropoda which includes animals which have occupied a great variety of ecological niches is considered, the force of this contention is evident.

The problem of relating the phyla one to another is more difficult. Yet it is possible to arrange the phyla in a series of progressive complexity, beginning with the Protozoa (uni-cellular), and passing through the Coelenterata (multi-cellular diploblasts), Platyhelminthes (triploblastic acoelomates), and Annelida (triploblastic coelomates), and culminating eventually in the Chordata (mainly represented by the Vertebrata). Admittedly there are large gaps between the various phyla and the series, which is made up mostly of present-day forms, is really only one of grades of organisation, and is not phylogenetic. But all these difficulties serve to bring into prominence and place in their correct perspective, such considerations based upon *present-day* animals, about which, naturally, the greatest amount of information is available. If evolution has occurred, then the fundamental relationships between the animal groups lies not between *existing* types but between the *stocks* from which they arose. Modern types represent merely the terminal twigs of the vast phylogenetic tree, and it is to the main branches and the trunk that attention must be directed in the endeavour to discover relationships.

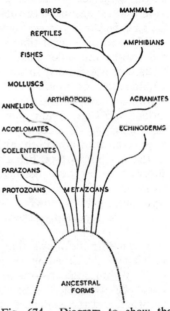

Fig. 674. Diagram to show the possible interrelationships of the main animal stocks.

However, in spite of these differences between the phyla, the *fact* that they can be arranged in a series based on their present-day characters is evidence in support of their interrelations.

**Evidence from Geology and Palaeontology.**—Most of the rocks that form the crust of the earth are arranged in layers of strata one above the other, but this formal arrangement is often obscured or upset by foldings and other irregularities. Yet by the application of Hutton's principle, that "the present is the key to the past", the strata can be arranged in a series in which the oldest lie at the base and the most recently formed at the top. Thus a study of the rocks at present exposed on the earth's crust shows that they are continually being weathered or eroded, to be broken into fragments and undergo considerable chemical alteration by solution of some of their constituents. The fragments of disintegration of the rocks are then carried away, chiefly by streams and rivers, but to a lesser extent by winds, at the same time being sorted into particles of different grades. Eventually, however, the fragments become deposited in places, such as lakes or the sea, where they are no longer subjected to erosion. They are then in the process of forming new **sedimentary rocks**. It is reasonable to conclude that the stratified rocks as we know them have been formed in a similar way in past geological time and that, therefore, the younger rocks will overlie the older ones. Certain other rocks, termed **igneous rocks**, are not formed in this way but result from the cooling and solidification of molten materials and the oldest of these may represent those formed by the original solidification of the earth's surface. Others, again, for example basalt which flows from volcanoes, may be of relatively recent origin. The various processes of erosion, transportation, and deposition, would in time have led to the earth's surface being reduced to a uniform plain but at intervals the crust has become folded so as to form mountains and hills, whilst the land masses themselves have moved radially so becoming higher or lower. Such changes have caused the erosion cycle to be repeatedly renewed and made stratification difficult to follow.

From this it follows that if animals or their remains became buried and preserved in the sedimentary rocks during the phase of deposition some record of the animal populations of the past would be available. This preservation has, in fact, repeatedly happened and the evidences of the living organisms of the past are called **fossils**. It is also clear that those fossils occurring in the deeper strata are representatives of more ancient faunas than those in the higher strata. But although the sedimentary rocks yield abundant fossils as evidence of life in past geological time, geology itself could point only in a general way to the sequence of successive faunas. Indeed, it was only by a study of the nature of the fossils that the strata in one part of the world were identified as the same age or a different age from strata in other regions. More recently, however, an assessment of the proportion of radio-active elements has made possible a more direct method of dating certain rocks, but even now, the study of fossils plays an essential part in the understanding of the nature and time sequence of sedimentary rocks.

The present estimated age of the earth is 4,500,000,000 years and the first fossils are bacteria known from South African rocks 3,000,000,000 years old but the first photosynthetic cells, *Gunflintia*, are found in 2,000,000,000 year old rocks from Ontario. However, the first metazoan animal fossils, *Spriggina*, from the Ediacara hills in Australia is a relative newcomer being only 700,000,000 years old.

Fossils may be produced in a variety of ways, from the preservation of a complete animal to the recording in receptive material of mere traces or impressions. Because they are less likely to have been broken by earth movements, those found in the more recent deposits are likely to be more complete than those from the older strata, but much depends upon the way in which the process of fossilisation occurred. Thus a complete mammoth—a form of elephant now extinct—was found in Siberia completely frozen; flies entangled in the gummy exudations from trees are preserved in amber, the form which the gum has taken after burial; the skeletons of ancient man have been recovered from burial-grounds and other situations. In all these instances either the whole animal or those parts resistant to decay have been preserved. But in many instances, particularly in the more ancient fossils, the actual animal and even the hard parts of it have completely disappeared and the fossils consist of impressions or moulds. Footprints, for example, made by an animal walking over an expanse of mud have been preserved by the subsequent hardening of the material. Again, an organism buried in a rapidly hardening deposit may decay and completely disappear, but the space which it occupied, which preserves its form and outline, may become filled up by infiltration with another kind of material forming a mould or copy which exhibits all its main external features.

From a consideration of all these kinds of fossils, a large amount of evidence has been accumulated about the types of animals (and plants) which have existed upon this earth since life first appeared, but it must be realised that the story is far from complete, for many animals are, from the nature of their bodies, unsuitable for producing fossils, even granted that they ended their existence in a place where fossilisation was possible. The Geological Record, or the Record of the Rocks, must therefore inevitably be incomplete for, as Darwin aptly described it, it is like a book from which many pages have been torn and upon those that remain but a few words are legible.

But palaeontology has greatly extended its scope in the hundred years which have elapsed since the publication of the *Origin of Species* and a great deal is known about the faunas of the various stages of geological time. It is, for example, now possible to say that most of the major phyla appeared during the Cambrian which makes it practically certain that a wealth of simpler animals must have lived in Pre-Cambrian times, although leaving scant trace, possibly because their bodies lacked hard parts. Two major phyla emerge only after the Cambrian: the Bryozoa in the later

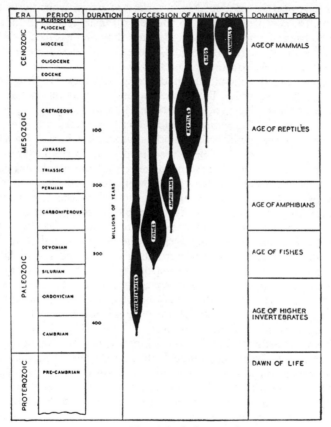

Fig. 675.   The geological succession.

part of the Ordovician, followed shortly afterwards by the early members of the phylum Chordata. Thus the palaeontological record lends little direct support to the idea that the more lowly phyla always arose before the more "advanced" ones which is, perhaps, a rather unexpected finding. They probably did but so early on that we have no record of them. It is also somewhat of a surprise to find that none of the main phyla became extinct. All have representatives living to this day and the same is, in the main, true for major groups within the phyla. But a moment's reflection tells us that there is no reason why a protozoan, perfectly adapted to its particular mode of life, *should* become extinct. Indeed, it becomes increasingly clear that major groups of animals arise and survive for a time, not because they are better suited to a particular way of life, but in response to changing environmental conditions or because they colonise new niches.

This process may take a long time and, *within each phylum* it is the rule that the more generalised groups appear before the more specialised ones. This is particularly well known for the vertebrates, for which group it is clear that there is a period of geological time in which each of the main classes reaches a maximum in numbers and diversity of types and subsequently declines (although, with the exception of the Aphetohyoidea, never to extinction). It is thus common to speak of the Mesozoic as the age of reptiles, of the Permian and Carboniferous as the age of amphibians, and of the Devonian as the age of fishes, each of the groups named being particularly abundant as fossils in rocks laid down at each of those times, so much so that they are often referred to as *dominant* types of animals. But it is possible to look at groups of animals in this way only at levels *below* the rank of phylum, as may easily be appreciated by remembering that whilst the present-day (and the periods of the Tertiary) are often called the age of mammals, by far the greatest number of animals, both absolutely and in number of species, are insects. Yet within each phylum the fossil record provides the strongest possible evidence that could be expected for evolution and the same is true for the progressive changes which take place within classes and the smaller taxonomic categories, in that what are judged on morphological grounds to be "simpler" or "less specialised" groups appear earlier in geological time, though simple types may survive to-day.

Besides this general picture of the results of evolutionary processes which the fossils provide, evidence has been obtained in a few instances where the fossilised remains present a moderately complete series of the changes during the evolution of one group of animal from another. Among these is that of the modern horse, which runs on the tips of the middle digits of both fore and hind limbs, an extreme example of the unguligrade mode of progression; has modifications of other parts of the limbs correlated with this feature; and has teeth of considerable length (*i.e.* height) with complicated configurations on the surface of the crowns suitable for chewing grass. In spite of the fact that the modern horse became extinct in North America during the Pleistocene era, an extensive series of fossilised remains found in that country shows a gradual and progressive reduction in the number of digits in the horse-like animals from the small *Eohippus* (Eocene) with five

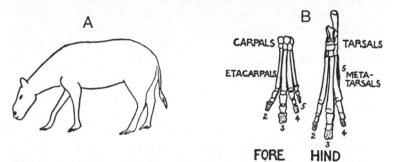

Fig. 676. EOHIPPUS.—A, restoration of the animal; B, skeleton of the limb extremities (redrawn after Lull).

digits (one vestigial) in the fore limbs and four (one vestigial) in the hind limbs to the modern horse (see Fig. 676).

The small *Eohippus*, no larger than a fox-terrier, lived on the lush vegetation of the sedgy meadows bordering the streams, where its digitigrade limbs were suited to the soft nature of the ground and its short-crowned, simple teeth to its food.   By the time of the Oligocene period the countryside had become much drier, with fewer streams, whilst forests with broader meadows and even

Fig. 677. MESOHIPPUS.—A, restoration of the animal; B, skeleton of the limb extremities (redrawn after Lull).

prairie land predominated.   Under these conditions speed became of importance in escape from enemies.   The more horse-like forms (*Mesohippus; Miohippus*) showed a further decrease in the number of digits (three in fore and hind limbs) and an increase in the length of the teeth and the complexity of their crowns. Similarly, through the Miocene and Pliocene, remains were found of a variety of forms approaching more and more closely to that of the modern horse, in the extent of the reduction of the lateral digits and the lengthening and increasing complexity of the teeth, until in the

Fig. 678. MERYCHIPPUS (redrawn after Lull).

Pleistocene, horses appeared belonging to the same genus (*Equus*) as that of the modern horse. This is by no means the whole story; for during its progress horse-like forms had migrated from the New to the Old World with the production of parallel forms and also, even in America, many off-shoots from the main stock had appeared and died out. The particular interest in the series, however, lies in the correlation of morphological change with changing conditions in the environment, the modifications adapting the organism to the new conditions.

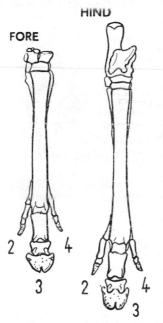

Fig. 679. HIPPARION.—Skeleton of limb extremities.

From the wider point of view the history of the horse is merely an example of the evolution of a genus, but the other fossils provide material whereby different *classes* of vertebrates may be linked up. Among the ancient reptiles was a group called the Synapsida (Theromorpha, or mammal-like reptiles) which had many features in common with the mammals. It is clearly impossible here to deal with all the features, but reference to some of the changes in the skull will suffice as an illustration. Apart from the enlargement of the cranium, the characteristic features of the mammalian skull (see p. 437) are the presence of a hard (secondary) palate; the reduction of the bones in each

Fig. 680.  MODERN HORSE.          Fig. 681.  MODERN HORSE.
                                   —Skeleton of hind limb.

half of the lower jaw to one, the dentary; the attainment of a
new articulation by the dentary with the squamosal; and the
presence of three ear ossicles.

The most primitive (in the Permian and Triassic) of the Theromorpha (Theriodontia and Dicynodontia) had no defined secondary

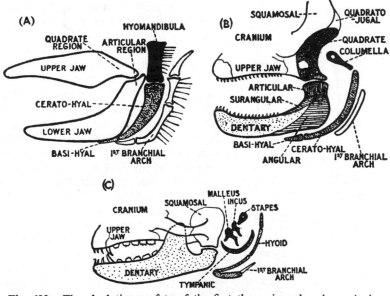

Fig. 682.  The evolutionary fate of the first three visceral arches.  A, in a
         cartilaginous fish;  B, a reptile;  C, a mammal.

palate; the ear ossicle was one slender bone; the lower jaw was largely composed of a series of membrane bones, similar to other reptiles, in which the dentary was not extensively developed; and the articulation was between the articular and the quadrate. In the more specialised Theromorpha (Cynodontia) of the later phases of the Triassic period, a progressive series of forms is known which show a gradual enlargement of the squamosal and dentary; the disappearance of the splenial and coronoid bones; and reduction of the angular, surangular, and prearticular bones. The quadrate of the upper jaw, like the articular of the lower, ceases to be of importance in jaw articulation and a backward process of the

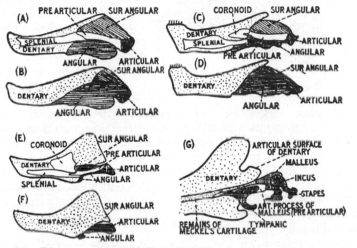

Fig. 683. Diagrams of lower jaws to show the evolution of the ear ossicles. A and B, inner and outer sides in an early theromorph; C and D, inner and outer sides in a later theromorph; E and F, inner and outer sides in a cynodont; G, in an embryonic marsupial.

dentary attains a new articulation with the squamosal. The discarded articular and quadrate become modified to form additional ear ossicles, the articular forming the malleus and the quadrate the incus, the stapes having been derived from the columella auris. The tympanic bone characteristic of the mammalian middle ear is formed from the reptilian angular.

These illustrations, meagre as they are, cannot pretend to give anything approaching a complete picture of the evolutionary examples provided by the study of fossil forms, but they do serve to show the kind of events which have been taking place in the animal world in prehistoric time.

**Evidences from Comparative Morphology and Anatomy.**—The comparative study of the form and structure of animals (such as the treatment of the vertebrates in the preceding chapters) reveals that structures of the same embryonic origin and basic structure may have different functions and, alternatively, those having the same functions may have different origins in the embryo. The limbs of a rabbit, the wings of a bird or a bat, the paddles of a

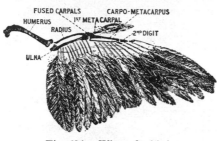

FUSED CARPALS    CARPO-METACARPUS
HUMERUS
RADIUS
    1ST METACARPAL
        2ND DIGIT
ULNA

Fig. 684.  Wing of a bird.

whale, all have a common foundation in the penta-dactyl limb skeleton, so that it can be reasonably assumed that they have been derived from a common ancestral type. They are therefore said to be **homologous structures**. On the other hand, the wings of a bird or a bat and those of an insect serve the same function—flight—but their structure and development is not in any way comparable. Such structures are said to be **analogous**. The fact that homologous structures can be clearly recognised is then evidence that they have arisen as the result of a process of modification of a basic type. Examples of homologies are easier to find within a phylum than between different phyla, emphasising that the evidence for intra-phyletic is more apparent than that for inter-phyletic evolution, a difficulty which has already been mentioned.

The homologies of limb skeleton are but one example of this kind of phenomenon, but there are many others, particularly in the vertebrates, which will recur to the memory.

Fig. 685.  Outline of a bat to show the skeleton of the wings.

The recognition of homologous structures is, then, good evidence for a fundamental similarity of design, and this can be best explained by assuming that the groups possessing it originated from a common stock. Its line of reasoning is closely allied to that drawn from schemes of classification for these rely mainly on the study of comparative morphology.

Further evidence from morphology and anatomy comes from **vestigial structures**, the best examples of which, perhaps, again come from the vertebrates. One of the main characteristics of birds, for instance, is the modification of the fore limbs to act as wings. But there are some birds, such as the kiwi, which cannot fly, yet in them

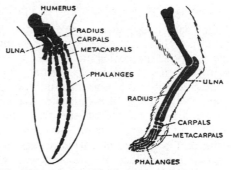

Fig. 686. The paddle of a dolphin and the fore limb of a rabbit.

the wing-plan is the same. Again, although most terrestrial vertebrates are quadrupedal, there are some, like the snakes, which are limbless, yet vestiges of the girdles and parts of the limb skeleton are still present in the boas. In man, those muscles which in other mammals move the pinna of the ear, are still represented but so feebly that they are incapable of producing appreciable movements.

**Evidences from Embryology.**—Von Baer (1792-1867), and later Ernst Haeckel (1843-1919) (in a more categorical form), were responsible for the idea known as the **Recapitulation Theory** which postulates that an organism during its development repeats its ancestral history. Put tersely, it states that "ontogeny recapitulates phylogeny". It is impossible to accept completely Haeckel's theory, yet the study of embryology, and particularly that of the higher Metazoa, provides evidence of the existence of evolutionary processes. Every chordate, for example, begins life as a single cell, which, by division, produces a multi-cellular embryo. These cells, by rearrangement into layers, by folding and other alterations, give rise to a series of stages which, in general organisation, *resemble* those of adult animals at

Fig. 687. A dragon-fly.

a lower level of organisation. The temptation to homologise these stages with adult animals proved too great for Haeckel and his followers. Thus the egg was stated to represent the protozoan stage, the gastrula, the coelenterate, and so on. But it is obvious that many features of the developmental stages are adaptations to particular conditions of the surroundings in which these stages find themselves. Thus, it would not be surprising to find that the embryological history of the frog, which is passed through in water entirely apart from the body of the parent, is different from that of a mammal which takes place in a specially modified part of the oviduct of the female. The interest lies in finding that, discounting differences which are clearly adaptational, the main features shown by the two types of embryo are very similar, and that the general sequence of events follows much the same course. An example is the occurrence of the visceral clefts in chordate embryos, even in those (*e.g.* mammals) where the majority of them never function at all. In fact, there is a remarkable similarity between all early chordate embryos, which is cogent evidence of their common evolutionary origin, but the differences become more apparent as the embryos increase in age as was pointed out by Von Baer.

Other instances of this constancy of developmental ground-plans could be cited. In some parasitic cirripedes, where the adults have lost almost all their crustacean characters, the constant appearance of characteristic larval forms (nauplius, cypris, etc.) enables the organism to be placed in its correct phylogenetic position. The caution which has to be exercised in distinguishing between basic larval or embryonic characters and those which are adaptive to the environment and life-history is emphasised when the complicated developmental stages of parasitic Platyhelminthes are considered, stages which are entirely unrepresented in the life-histories of non-parasitic forms and can have no evolutionary significance.

The fact that similarities exist between the larvae of different phyla such as Annelida and Mollusca, or Hemichordata and Echinodermata, provides evidence for affinities between phyla.

A factor which has to be taken into account in the evaluation of embryological stages is that, just as the stages themselves have been modified, so also may the sequence of the stages be altered. Phases may be telescoped, condensed, or even omitted altogether so that the general picture is materially altered, yet the cropping up of a feature here and another there may betray affinity. For instance, some Amphibia have become so terrestrial in habit that they do not return to the water to breed. Yet within the egg and before it hatches, a tadpole stage is passed through.

**Evidences from Geographical Distribution.**—From the very large amount of information now available of the occurrence of the different kinds of animals in various parts of the earth's surface, it is clear that they are by no means evenly distributed. Not only have certain zones, determined by latitude, elevation, temperature, etc., their own characteristic faunas, but places with the same climatic conditions in different regions of the world do not always possess the same animal forms. Elephants, for example, occur in Africa and India but not in Brazil. The climatic conditions of the British Isles and of New Zealand are very alike, but the familiar animals in the two places are very different.

Yet many British animals and plants introduced to New Zealand flourished to such an extent that they have become a pest. Among these may be mentioned the blackberry plant and the skylark. Much the same is true of the rabbits in Australia. The climatic and other conditions of the land near the north and south poles are nearly identical and yet each has its own totally different fauna. For example, penguins are confined to the south polar and polar bears to the north polar regions. Such examples suffice to show that the present-day distribution of animals does not depend by any means entirely on the suitability of the indigenous animals to their surroundings.

Again, representatives of particular groups of animals may occur at widely separated places but not in the regions in between them. Examples of this "discontinuous distribution", as it is called, are the lung-fishes, certain species of which are found in the rivers of South America, West Africa, and South Australia, and nowhere else. Animals with a discontinuous distribution are, as a rule, members of groups of primitive animals known from the geological record to have had a much wider distribution in past times.

It is now believed that the answer to these and related problems can be given only by considering the nature and distribution of the animal populations of past epochs in relation to the geological history of the ancient seas and land masses. These populations, like those of the present day, undertook migrations, sometimes of a rapid and extensive nature but also a slow and steady spreading from their places of origin, often undergoing evolutionary changes which would better fit them to any changing environments that they encountered. Thus the present-day distribution of any group of animals depends partly on its centre of origin, partly on the extent to which its migrations were facilitated or hindered in the past, and partly on its ability to survive in regions which it reached.

In such migrations limits are set by natural barriers. High mountain ranges, stretches of water, varying topographical and climatic features with correlated differences in vegetation and potential food supply, prevent the spread of purely land forms; differences in salinity and temperature, and currents and tides influence the distribution of aquatic forms. Aerial forms are less affected, their range corresponding to their powers of sustained flight. Barriers have by no means remained constant throughout geological time, for the configuration of the earth's surface has been altered by subsidences and upheavals; land masses now completely separated by water were at one time linked up, enabling migration by land animals to take place.* In this way it has frequently happened that representatives of once widely distributed groups have become isolated in situations free from competition with more recent and highly evolved forms and have there survived but have died out elsewhere. On the other hand, by the formation of new routes for migration, groups originating in a restricted area have spread to distant situations. Many animals (*e.g.* house-fly, earthworm, and whale) have become practically cosmopolitan.

Some of the most cogent evidence in support of evolution is obtained from the study of island faunas. Where islands have arisen by separation from a land mass, their faunas can be reasonably expected to resemble that of the neighbouring mainland and often do so, but almost invariably differences arise which can only be explained by the operation of evolutionary processes. Thus there is evidence that the vast island continent of Australia was cut off before the eutherian mammals had achieved their present dominant position, and it is therefore most significant that almost all the mammals of that great continent are either monotremes or marsupials. Such native eutherian mammals as are represented, like the dingo and rat, etc., were introduced by man when he invaded these lands.

Again, the fauna of the British Isles affords an interesting example, because of its familiarity, of an island fauna of the so-called continental type. On the whole it may be said that the fauna of those regions nearest to the mainland approximates more closely to the continental than does that of districts lying farther away. In these more remote regions, too, the fauna is, on the whole, poorer in numbers of species. Thus snakes, common in eastern counties, are rarer in the west and are absent from Ireland. Many butterflies, *e.g.* clouded yellow and painted lady which do not breed in

* Wegener's theory of continental drift, which postulates that the present continents were originally a vast land mass in the southern hemisphere which has become split up and the constituent parts have drifted apart, has much to support it.

England, are common on the Continent, but reach our shores as summer migrants. The small liver-fluke (*Dicrocoelum lanceolatum*) is fortunately absent from the British Isles, though common in Europe. On the other hand, some species (birds, fresh-water fishes, and insects), though related to continental ones, are peculiar to these islands. Many other similar examples could be given, but these will serve to illustrate the differences.

It is also to be noted that the fauna and flora of the more easterly parts of the British Isles more closely resemble that of Western Europe mainland, while that of south-west England and of southern Ireland show resemblances to that of Mediterranean countries.

These facts can be explained when it is realised that during the glacial periods of the Pleistocene the greater part of the British Isles was covered by ice sheets which destroyed the animals and plants, which at that time were similar to the continental ones. During the warm interglacial periods, the surviving forms on the Continent (to which England was still joined) migrated here, but this migration was still incomplete as the glaciers receded for the last time, when subsidence of the land took place with the separation, first of Ireland and then of Great Britain from the main land mass. From this now isolated fauna the present one has been derived with its marked similarities and differences from the continental one.

It is assumed, therefore, that the few species which are peculiar to these islands must have *evolved* from the original fauna after separation from the mainland took place.

Instances of island faunas of a quite different kind are those of islands of volcanic origin which, when they first appeared, were devoid of land animal life. The Galapagos islands situated on the equator, six hundred miles west of South America, originated in this way. Since these islands first arose above the surface of the sea as the result of some subterranean volcanic disturbance they had no indigenous land animals then, those now found there must have come from the mainland either by flying, swimming, being carried by wind, ocean currents, or on driftwood or floating vegetation, or introduced by man on his rare visits. Yet an examination of the animals has shown that although they bear some resemblances to those of the mainland, they clearly differ from them. Here the giant tortoise, *Testudo ephippium*, resembles other species of that genus but is characterised by its enormous size. Also, more than one distinct species of these giant forms is found on the islands. But some of the birds are even more illustrative of this point. Five well-marked genera of ground-finches, *Geospizidae*, occur on the islands, and as many as ten distinct

species may be found on one island. Whatever may have been the
original form which invaded these originally sterile islands when
they appeared above the surface of the sea, with the passage of
time, all these new species have arisen, protected from competition
by their isolation. Also, although the islands in the group are not
separated from one another by very great distances, each individual
island has its own particular species even among the birds. This is
explicable by supposing that after the species from the mainland
had arrived, they evolved along lines quite different from those of
their progenitors who remained behind, with the production of
entirely new species. It was the peculiarities of the fauna of these
islands noticed by Darwin when he visited them in 1835 which
contributed to his formulation of his theory of evolution.

All these facts point to the conclusion that the present-day
distribution of animals cannot be explained completely on the
hypothesis of "suitability to environment". It is understandable,
however, if two conditions are accepted: (1) that existing animals
are the descendants of extinct populations which were of a more
generalised type, and that these wandered out from their places of
origin, becoming modified to meet the new conditions they encountered, intermediate populations possibly dying out (this is called
**adaptive radiation**); (2) that with the changes in the configuration
of the land during geological time, certain groups became isolated
due to the erection of barriers of one sort or another and were then
free to evolve along different lines but were prevented from interbreeding with other populations, eventually becoming so different
that they could not do so even if brought together again.

**The Evidence from Comparative Physiology and Genetics.—**
All the facts so far brought forward in support of evolution have
dealt with structural features; but it is clear that changes in structure
*must* have been accompanied by corresponding physiological
changes, and thus there must have been a physiological as well as a
morphological side to evolution. All available evidence shows
that, viewed as a whole, the physiology of the structurally simpler
is usually less complex than that of the higher animals, and that
animals said to be related to one another on structural grounds
usually show physiological similarities. Direct evidence for an
evolution of physiological processes from a simpler to a more
complex condition is obviously unobtainable, but many interesting
inferences (of which only a few can be given) can be made from
present-day animals.

Evidence from many sources points to the view that the primitive
members of most of the main phyla and classes lived in the Palaeozoic

seas or estuaries, and that only later did their land-living relatives become adapted to their new environment. The gradual transition, and some of the structural alterations which accompanied it in the vertebrates, have already been mentioned, but it may now be said that in their physiology even the most fully adapted of terrestrial vertebrates retain unmistakable traces of their descent from aquatic ancestors.

The analysis of the body fluids of both aquatic and terrestrial animals shows that they are fundamentally similar in composition and, furthermore, that in their ionic composition they resemble sea water. Now the sea seems to provide a particularly suitable environment for most types of living organisms and also has the merit of maintaining, within the rather narrow limits within which cells can live, conditions which are fairly constant. A point of interest in this connection is that the body fluids of most animals contain less magnesium but more potassium than does the water of our present-day oceans, but there is no good reason for the belief that the sea of past geological time differed from modern sea water in a similar way.

In most of the simpler marine organisms the body is freely permeable and the body fluids automatically vary with changes in the surrounding sea water; but all vertebrates (and many other animals) are able to maintain the composition of their body fluids constant. This is a relatively easy matter for marine animals, but presents a greater problem for fresh-water and terrestrial animals. Nevertheless, the problem has been solved by the development of excretory organs which, be they nephridia, Malpighian tubules or kidneys, as well as of salt absorbing organs such as gills, may be regarded from this point of view as having evolved to meet the need for maintaining a constancy of internal environment.

But true excretory organs have another important function; that of eliminating waste nitrogenous compounds from the body. It is interesting to find that it is characteristic of most aquatic animals that the bulk of the nitrogenous waste is in the form of ammonia, excess protein being merely deaminated and the products not synthesised to form urea or uric acid. This is particularly true of the simpler invertebrates, but in amphibians urea is the chief excretory product. This difference is attributed to the fact that the very soluble but highly toxic substance, ammonia, must be rapidly eliminated from the body in dilute solution, and this is only possible in aquatic organisms. The elaboration of the innocuous substance, urea, from the ammonia released during deamination, is a necessary complication of the excretory processes of land animals where the intake of water is of necessity restricted. In animals most fully adapted to life on dry land, *e.g.* the insects, reptiles, and birds, the bulk of the excess nitrogen is excreted as uric acid or urates. These are practically insoluble substances, a property which favours resorption of water from the contents of the excretory tubules with the result that the "urine" is so concentrated as to be semi-solid, and a considerable economy in water is effected.

Thus it will be seen that an advancing complication in the physiology of excretion accompanies the structural complexity by which terrestrial ("higher") vertebrates are distinguished from the aquatic ("lower") vertebrates.

A similar correspondence between increasing morphological complexity and physiology has been found in the developing hen's "egg". The young embryo up to the fifth day excretes chiefly ammonia; from the fifth to ninth day, chiefly urea; and from the ninth day onwards chiefly uric acid. Sometimes cited as chemical evidence for the theory of recapitulation, this instance is better regarded as an example of developmental adaptation. The ammonia is produced in relatively small amounts by the early embryo and to a large extent probably diffuses out through the shell. As time goes on the larger embryo, still confined within the shell, produces greater quantities of nitrogenous waste and this is elaborated into substances which are less harmful. These can be passed out of the body, and because of their relative insolubility, stored in the allantois "out of harm's way". In any event, the urea is produced by a series of reactions quite different from those in fishes and Amphibia. At the most then, there is a parallelism between evolutionary and developmental changes in the nature of the substance excreted.

From what has been mentioned above it will have been gathered that the idea of homology between structures and organs in different kinds of animals implies that the evolution of new types has taken place by a series of small steps, the various parts of the animal body being altered almost imperceptibly in succeeding generations. The ground-plan of the organ is always retained for long periods of time and evolutionary changes affect chiefly the details of structure. An organ or structure is not quickly discarded to be replaced by an entirely different one. Furthermore, structures must (with certain exceptions) at each stage of evolution have been of positive benefit to their possessors or they would have been eliminated. In somewhat the same way animals have evolved physiologically. Innumerable examples of obvious similarities between the physiology of (structurally) related animals spring to the mind and all of these add weight to the view that animals took origin in common stocks.

The probably universal occurrence of cytochrome in the cytoplasm of all aerobic organisms is a striking instance of chemical similarity between living matter and on a more restricted plane, but equally interesting, is the presence of the respiratory pigment, haemoglobin, in the red corpuscles of all vertebrates. The haemoglobin of different animals is not, however, identical for the protein component differs and endows the pigment with different properties which can be seen to be adaptations to the respiratory needs of the body. Indeed, the probability is that haemoglobin has been evolved many times, quite independently.

In various animals which have been studied, both in the laboratory and in nature, it has been found that occasionally a character is altered and is inherited in an altered form as a mutation (p. 793). Most of those which have been studied have proved to be recessive and their effects are only evident in the homozygous condition, that is, when a recessive gene from both parents is present in the zygote. Mutations which behave in a dominant fashion are much rarer. Now it is clear that, owing to the rareness of a mutation in any animal population, the chances are strongly in favour of an animal possessing the mutant gene to mate with one not possessing it, that is, with one of the normal type, so that if the mutation is truly recessive, the offspring, which are heterozygous for that character, will not show any change in body form. Hence selection will not begin to operate on the heterozygotes carrying the recessive mutant gene. On the other hand, dominant mutations, whether favourable or unfavourable to the species, immediately come under the action of selection, and if favourable the animals and their offspring showing the change tend to persist at the expense of the normal type.

In other instances where the mutant gene is not completely recessive but produces some effect in the heterozygous condition, it is found that if the mutation produces an unfavourable effect, then, by selection, those heterozygotes which, owing to differences in the gene complex manifest the least effect, tend to persist, and after many generations the mutation therefore becomes completely recessive. The reverse effect takes place when favourable mutations occur. They tend to become dominant. Thus it is only to be expected that the bulk of the mutations which crop up in animals studied in the laboratory are both recessive and disadvantageous. They show themselves as departures from the normal or wild type of animal in which favourable mutations have been incorporated in the normal gene complex as dominant genes. There is, however, strong evidence for the belief that unfavourable recessive mutations identical with those observed in the laboratory occur also in nature.

On the other hand, the effect of a gene is affected by whatever genes are next to it on the chromosome (the genic background, p. 796). So the lethal effects of a harmful mutation may be suppressed by the genes among which it finds itself.

It must be emphasised that selection in almost all instances acts on the effects produced by the gene complex as a whole, and only rarely on the effects produced by a single gene. Conversely, it is known that a single gene may influence more than one character. Certain of its effects may control definitely beneficial characters, but others may result in the production of characters which are apparently of no benefit nor positive harm to an animal. It is in this way that many so-called non-adaptive characters may have arisen during evolution.

The actual cause of the mutational changes in the genes in nature is quite obscure, but the behaviour of induced mutations is in every way comparable to those which occur without any apparent stimulus. Whether or not these experiments have any real bearing on evolutionary processes occurring naturally is still uncertain; but the objection that all induced mutations are pathological has been answered by an experiment in which the subjection of mutant flies to the action of X-rays had the result of producing mutations in the reverse direction. That is, mutations were produced which caused the mutant characters to be replaced in the next generation by those of the normal or wild type.

A few instances are known of mutations of considerable magnitude by means of which a complete alteration of a species took place in a single generation. Many of the known instances were the result of a doubling or even of a trebling or quadrupling of the somatic chromosome number. Thus triploid, tetraploid, and polyploid organisms are sometimes formed in one bound. This type of genetic change is really quite different from true mutations and clear-cut examples of it are known only among the plants. It is doubtful if it has played any significant part in animal evolution, though it is suspected that some earthworm species are polyploids.

## THE WAY IN WHICH EVOLUTION HAS COME ABOUT

The acceptance of the fact of evolution, however, leaves unanswered the important question of the way in which it has been effected in the natural world. Of the various solutions which have been put forward from time to time, those associated with the names of Lamarck (1744-1829) and Darwin (1809-82) may alone be considered.

**Lamarckism.**—Essentially, Lamarck's solution depends upon the acceptance of two postulates: that of the **use and disuse** and the **transmission of acquired characters**, and at the same time endows the organism with an inherent capacity to increase in size and complexity. Lamarck postulated that if, to meet the needs of the environment, certain structures were called into unusual activity or subject to undue strain, these structures became modified— were strengthened, elongated, and the like and that these features were handed on to the progeny. Similarly, structures which were not used tended to be less developed and this condition was equally heritable. To use the classic example, the giraffe obtained its long neck by stretching upwards to reach the food available only from the tall trees, and the long neck thus produced was transmitted to the offspring with the result that there arose a race of long-necked animals, to wit, the modern giraffe. In the same way, the wings of flightless birds have arrived at their present ill-developed condition because they are no longer used for flight.

Although it is plain that structures and organs can be much altered by use and disuse during an individual's lifetime, there is no proof that such modifications will generally be transmitted to the offspring. Thus it is difficult to conceive this as a general mechanism for evolution. But Lamarck seems to have understood the role of environmental needs in producing changes, which was so strongly stressed by Darwin.

**Darwinism.**—Darwin and Wallace were responsible for a solution on quite different lines which became known as the **Theory of Natural Selection.** This can be expressed briefly as a logical sequence combining observed facts with the inferences dependent thereon.

(*a*) Most animals produce large numbers of offspring and, with even the slow-breeding animals, if all their progeny survived and reproduced, the resulting numbers would be very large. Actually, the numbers of adults of any particular species remain, over a reasonable period, relatively constant.

*From this it is evident that among the offspring there is considerable competition for the maintenance of life, from which it is deduced that a* **struggle for existence** *operates in Nature.*

(*b*) The individuals of a species (and even the progeny of the same parents) differ from one another in small features (continuous variation).

*Such differences may confer upon certain individuals an advantage over their fellows in the struggle for existence, so that they have a greater chance of survival and consequently of procreation, resulting in the* **survival of the fittest.**    *This implies the adaptation of the organism to its environment.*

(*c*) The inheritance of characters.    In general, the progeny have the characters of their parents.

*Since the possession of the advantageous variation has enabled the variant to survive and reproduce, the offspring should also possess this advantage and chance of survival.*

In this way new forms will arise, each fitting into its particular niche in the environment.    In fact, the individuals close to the norm of the population often prove better fitted than those which represent the extremes.    But environmental conditions are liable to, and do, change.    Hence there is in operation a perpetual process

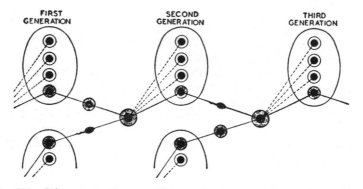

Fig. 688.    Diagram to illustrate Weismann's theory of the continuity of the germ-plasm.

favouring the evolution of new forms, or as Darwin expressed it, the **origin of species.**

On this hypothesis, to use the same example as before, the long neck of the giraffe has arisen by selection since those individuals whose necks were long enough to reach the vegetation on the tall trees would have a better chance to survive and reproduce, resulting with the progress of time, in a long-necked race.

It can scarcely be said that adherents of either of these schools of thought would accept them in the bald form in which they are stated above.    In fact, Darwin himself was fully aware of difficulties in the way of acceptance of his views, nor did he think that Natural Selection provided a complete answer to all the problems of evolution.    Lamarckism received a shrewd blow when Weismann (1834-1914) attacked its basic thesis of the transmission of acquired characters, and showed that alterations in the body (soma) were not inherited and postulated that effective changes could only take place through the germ-cells (germ-plasm). In fact, he emphasised the point of view that the body of an organism with all its characteristics is the product of the germ-cells and not *vice versa*, so that the only real connection between successive generations is through the germ-cells, the body or soma being merely their custodian.

Nowadays, Lamarckism has few adherents, for it is realised that a (body) character is often not a single attribute, but is the result of the interaction of heritable factors (genes) and environmental conditions. Therefore, there is no such thing as an "acquired character" in the Lamarckian sense. This, however, does not preclude the possibility of environmental conditions influencing the combination of genes, which, by their interaction with environmental conditions, result in the production of new characters.

For the Natural Selectionists, an apparent solution of some of their difficulties came from the discovery by De Vries (1848-1935) of large mutations, which were not only heritable but caused wide divergences from the original character complex. This seemed to provide a means whereby advantageous variations could arise which might, in one step, have a survival value and come to be handed on to the next and subsequent generations, but it is known that such large mutations only rarely lead to new species and examples of such are known only in plants. Moreover, when it was shown mathematically, notably by Fisher and Haldane, that small mutations due to minor alterations in the gene complex are adequate to explain evolution in the time since life appeared on this earth, then selection acting in conjunction with these provides in broad outline a mechanism adequate to account for evolution.

Indeed, the available evidence tends to prove that selection does operate in the natural world and the fact that different forms can be produced by selection is evident from the results obtained by breeders and fanciers by artificial selection, a point which was emphasised by Darwin himself. Equally, too, the fact that there is a struggle for existence seems indisputable, but it is highly doubtful if this, as commonly understood, plays an important part in evolution. From an evolutionary point of view what matters is not the survival nor even the existence of individual animals, but the rate at which particular groups reproduce and multiply. Those reproducing most rapidly will be those most likely to survive. The common-sense meaning of the word "struggle" is some form of combat between animals of the same or different species and struggles undoubtedly do occur, but whether or not they have any important effect on rates of reproduction is quite a different matter. Indeed, the destruction of herbivores by carnivores is apt to be exaggerated and it is obvious if killing was indiscriminate the carnivores would quickly suffer extinction by the falling off of their food supplies. Under natural conditions parasites rarely do much harm to their hosts and a "balance" is arrived at in which state the parasites survive and the host shows little or no symptoms of disease. Many groups of animals have developed social and other devices to minimise "struggles" and the phrase "Nature red in tooth and claw" is merely a dramatic phrase for a myth.

Observers of the fossil record have claimed that there are trends in certain groups to evolve in a particular direction, often apparently to the disadvantage of the animals concerned. This has been called **Directive Evolution, Orthogenesis,** or **Orthoselection,** and would appear to be contrary to our ideas that natural selection encourages the evolution of more efficient organisms. However, on close examination of the record these trends are often found not to be as continuous and persistent as they were once thought to be. The trend of increase in size did not continue in the evolution of the horses; for example, some lines were characterised by a reduction in size. None of the so-called trends among horses lasted longer than 15-20 million years, a short time by palaeontological standards.

Nevertheless, it is true that once a mutation which has survival value has become established and results in specialisation in any direction, the occurrence of further changes in the same direction are extremely likely and will tend to preserve the established line of specialisation and enhance it, so long as it continues to be advantageous.

There are trends of a more general nature; as an example of a widely spread evolutionary trend the gradual lightening of the bony skeleton of all the vertebrate classes may be considered. The earlier known members of each class are almost invariably more heavily ossified than their present-day relatives and, furthermore, their fossil history shows that this reduction has gone on steadily with time. The tendency to increase to a vast size which was common to so many groups of Mesozoic reptiles is another example of a trend, a trend which is said by many to have contributed to their extinction. Large horns are in deer associated with a large size of the body. That is to say, larger species of deer have *relatively* larger antlers than do small species. This results from different relative rates of growth, and it may well be that the advantage of a large body overrides what seems to us to be disproportionately large antlers. Probably, if more knowledge was available, all "trends" could be shown to result in better adaptation and it is likely that extinction is a result of changing environmental conditions and not because a group of animals became ill-adapted to static conditions. From an examination of living species, particularly those groups of animals which are verging upon becoming true species, we visualise the origin of species as being due in the first place to a geographical barrier, such as a river or a mountain range, isolating a part of a uniform population of animals from the rest. In their isolation natural selection functions to produce morphological changes as well as biological characteristics such as different feeding habits or courtship behaviour. Should the geographical barrier be removed, the river dry up, for example, the changed group may be unable to mate with their cousins when they meet because of the differences which have arisen. The isolated group is now a true species of its own, separate from the original stock (which may itself have altered considerably since the split first took place).

# CHAPTER XXII

## ECOLOGY

Ecology or environmental biology can be broadly defined as scientific natural history. It pays particular attention to the attributes of groups of organisms ranging in scope from populations of one species (autecology) to all the populations of diverse species (synecology) within an area and their relations with each other and with the physical environment. At all levels ecology may require the application of other branches of biology and indeed, of other sciences—physics, chemistry, meteorology, statistics, and so on—and it is in this respect that it has grown away from the older-fashioned type of natural history.

A basic concept of ecology is that of the ecosystem, that is the totality of the physical environment of any area together with all the organisms living in it for, as will be more fully realised when examples have been given, no one population (assemblage of a single species) can exist in isolation nor apart from a particular habitat. Always species occur together to form communities whose nature depends largely on that of the physical surroundings which in turn may be profoundly modified by the communities it supports. Naturally, the closeness of association and the degree of dependence between the various species making up a community is variable. It ranges from the close dependence of parasites on their hosts to mere chance association between species able to survive in the same physical environment. Nevertheless, certain features are common to all ecosystems be they lakes, shallow sea areas, rocky shores, grassland, deciduous forest, and so on. Each offers a particular type of physical environment which supports populations falling into two broad categories: (1) autotrophic (holophytic) organisms (of which green plants are usually the most important), and (2) heterotrophic organisms of two main kinds, (a) animals which feed directly on the plants (herbivores) or on other animals (carnivores), and (b) heterotrophic micro-organisms, such as bacteria and fungi which utilise dead organic matter as a source of nutrition, breaking it down to its inorganic constituents and so playing an essential part in the return of plant nutrients to the environment after they have been temporarily locked up in the bodies of organisms. The autotrophs, on land mainly the higher plants and in the sea and lakes mainly the small floating plants or phytoplankton, are the primary producers of organic material, utilising for this purpose the energy of sunlight (photosynthesis). It follows that most primary production is carried out on the surface of the land and in the well-lit surface layers of seas and lakes. In certain situations, however, chemosynthetic bacteria may be important autotrophs, as in soil and in the depths of the sea where light is not available. Obviously, the productivity of any ecosystem, that is the total amount of organic material produced in any unit of time, say hours or years, depends on the efficiency of its primary producers and this in turn on the amount of plant nutrients available, on the duration and intensity of light, on temperature and other features of the environment as well as upon complex interactions between its various populations. The biomass of any particular population at any particular moment of time is called the standing crop and may vary from season to season.

The economy of an ecosystem can be expressed as a series of food chains which show the relations of the carnivores to the herbivores and of the herbivores to the plants. At the base of the food chains are the herbivores which are said to form the key-industry animals of the community. These are preyed on by carnivores which in turn may be eaten by larger carnivores, but eventually a climax is reached and animals are found which are not preyed on by others. The following food chains will serve to illustrate these points:

Oak tree→aphids→spiders→small birds→hawks.
Acorns→mice→owls.
Clover→bees (honey)→mice→cats or owls.
Phytoplankton→copepods→small fishes→dogfishes.

Usually food chains do not occur as isolated series of events but are cross-linked to form a food web as in this example:

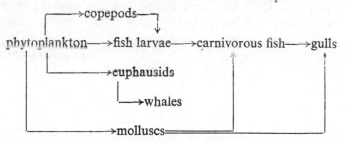

A consideration of numerous examples shows that there are rarely more than five links in any food chain and more usually only three. It is to be noticed that, as a rule, the animals nearer the end of a food chain are larger than those at the beginning. This is not invariably so, but it is clear that unless doomed to extinction, the lower members of a food chain must be more numerous than the carnivores that prey on them since at each link there is a loss of potential energy. This is the so-called pyramid of numbers (Fig. 689). A similar pyramid could be drawn to represent the number of species present at each trophic level in any one ecosystem. Numerous species of key industry animals usually occur, fewer of small carnivores and fewer still of large carnivores—perhaps one species only of climax animals which are preyed on by no others and hence are placed at the apex of the pyramid. Thus one lion kills about fifty zebras in a year. Also, in all food chains there is an upper and a lower size limit for the food of any given species of animal. A small carnivore like the weasel could not tackle a large herbivore like the cow. Conversely, larger carnivores do not prey on very small animals, for it would involve an expenditure of energy not commensurate with the returns.

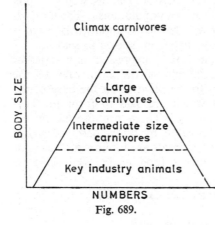

Fig. 689.

It will be realised from what has gone before that the examples of food chains and food webs given represent only some of the series of steps essential for the maintenance of an ecosystem and that their completion should include the small heterotrophs, mainly bacteria, which break down the excreta and dead bodies of organisms to inorganic materials which are then available as plant nutrients, the sum total of the events making up the food cycle of the ecosystem, thus:

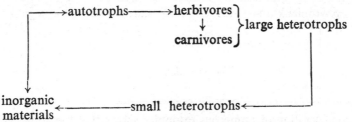

It should also be remembered that the carbon dioxide in the atmosphere is maintained at a constant level by the interactions of the plants and animals in the various ecosystems of the earth, the animals giving out carbon dioxide during respiration and the plants on balance withdrawing it when photosynthesising.

Total productivity and the nature of the various plant and animal populations are largely governed by the physical conditions in the external environment but in a previous section it was pointed out that the great diversity of animal life in different parts of the globe cannot be explained by the supposition that the nature of the animal population is entirely determined by the geographical and physical features of the surroundings. It does not necessarily follow that two regions with apparently identical climatic and topographical conditions will have even remotely similar animal populations. To choose but two examples, the lands of the Arctic and especially the shores of the Polar Seas, have a distinctive mammalian fauna of which the following animals are important representatives:—Seals, the Arctic Fox, the Polar Bear, Wolves, Lemmings, and Reindeer. None of these animals are found in the Antarctic. On the other hand, Penguins are abundant there but are absent from all other regions of the earth except the Galapagos Islands. Eutherian mammals were (with the exception of the Dingo Dog and certain Rats), until recent years, entirely unknown in Australia, whose native mammals are the specialised surviving members of a more archaic group— the Marsupials (Metatheria). Yet,

Fig. 690. A Marsupial.
[*Fox Photo.*]

when Rabbits (and many other animals and plants not native to that region) were introduced into Australia they did not die out, nor merely survive, but thrived to such an extent that they now are pests. It may be concluded, therefore, that Australia is in every way suitable to support at least certain types of Eutherian mammals, and that its faunistic peculiarities are solely a result of evolutionary processes in a region which had received its fauna and then in past time become isolated from other land masses.

These, and many other instances of anomalous distribution of animals each require some particular application of the theory of Organic Evolution. Nevertheless, it is quite clear that the main factors which influence the distribution of animal life as we know it to-day are environmental ones and that the struggle for existence, usually rigorous, ensures that only those animals which are suited to their surroundings, survive. It is of interest, therefore, to review, quite briefly, the way in which the physical, climatic, and other features, which collectively make up the external environment, affect the lives of the animals living in it.

TEMPERATURE.—Probably no other single factor has such a far-reaching effect on the fauna as temperature. In many instances, for example the reef-forming corals, it decides the limits of their distribution so that coral reefs are not found even on the west coast of tropical Africa which is swept by cold oceanic currents. Reptiles are practically confined to the warmer regions of the earth and attain their maximum size and activity only in the tropics. The same is true for many other animals such as amphibians and moths and butterflies. Even more striking and widespread are the effects which temperature exerts indirectly on the animal populations through its influence on the vegetation. This is clearly brought out by a consideration of the changes met with in the flora and fauna from the Arctic to the Tropics. All gradations from the cold desert-like tundra to luxuriant rain-forest are met with, and each region has its typical fauna. A somewhat similar gradation of vegetation may be noticed in the ascent of a mountain in the tropics. The effects of temperature on aquatic animals makes itself felt chiefly in the surface layers, for in the depths the water is at the practically uniform temperature of 2-3° C., and, on the whole, most creatures of the deeper waters have an extremely wide distribution. Ocean currents, themselves in part due to temperature variations, play a large part in the distribution of marine organisms and in some ways are analogous to the effects of wind on the smaller land animals and on seeds. Temperature frequently acts as a limiting factor, but its direct influence is often difficult to distinguish from that of salinity, for salinity is, as a rule, higher in the warmer than in the colder seas.

LIGHT.—The amount of light, that is, its duration and intensity, influences the fauna chiefly by its effects on the vegetation, for, no less than between animals, there is a struggle for existence among plants, the dominant type of vegetation flourishing to the detriment of other less successful kinds. In this struggle light is frequently the deciding factor.

In the sea, no less than on the land, variation in the light is important because of its effects on the metabolism of the plants. The most vital of these are the myriads of minute algae (phytoplankton) which float in the surface waters and form the bulk of the basic food supply of marine animals. It is the duration rather than the intensity of the light which has the most far-reaching effects and it is noticed that in the springtime of temperate latitudes, as the days become longer, the phytoplankton increases enormously, giving the so-called spring outburst. In late spring and summer the surface layers warm up and become less dense than the deeper layers, warm water floating, so to speak, on cold dense water, there being little mixing between the two because the conditions are stable. The upper layers are separated from the lower layers by a thin layer in which the temperature drops very quickly with depth. This is the discontinuity layer or thermocline, an example of which as it occurs in the English Channel in June is shown in Fig. 691.

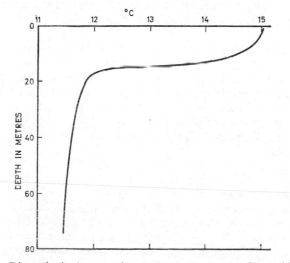

Fig. 691. Discontinuity layer or thermocline in the English Channel in June.

The spring outburst of phytoplankton gradually exhausts the supply of plant nutrients (mainly the phosphates and nitrates) in the surface layers and the phytoplankton can then no longer reproduce itself fast enough to keep pace with the rate at which it is eaten by the animals of the plankton and so declines to a low level during the summer. In the autumn the surface layers cool and become denser than the deeper layers so that a state of instability ensues and the first equinoctial gales cause a mixing of upper and lower layers. Because of this, plant nutrients, which have accumulated in the deeper layers during the summer owing to the bacterial decomposition of dead plants and animals which have sunk to the sea floor, are renewed at the surface. The intensity and duration of the light suffices for a second minor, autumnal, outburst of phyto-plankton but this is of brief duration and soon declines to the low winter level.

Light also plays an important part in regulating the seasonal activities and day to day activities of plants and animals. Often it is of major importance in determining the season at which plants flower and fruit, some flowering when days are lengthening; others when it is shortening—with various repercussions on the animals which feed on them. Many animals use the direction and intensity of the incident light as a cue to which they can orientate. This is so, for example, of many insects like bees. Many, perhaps most animals have a preferred light intensity and move so as to keep within it. This is true, for example, of zooplankton which moves upwards at night and downwards during the day.

SALINITY.—Broadly speaking, two main types of watery environ-ments are recognised—fresh water and marine. The distinction between them is not an absolute one and regions of brackish water, where the salinity is greater than fresh, but less than sea water, are common, especially in the estuaries of rivers. The animal populations of each of these regions of varying salinity are, viewed as a whole, distinctive. This is particularly true of certain groups of animals such as the echinoderms, which are entirely marine. Many smaller groups of animals are also distributed throughout the seas but absent from fresh waters. Of these, the polychaete worms and cephalopod molluscs may be mentioned. On the other hand, many fishes and all aquatic amphibia are restricted to inland waters. A closer examination of aquatic environments reveals, however, that certain animals are tolerant of wide changes in salinity. The common rag worm (*Nereis diversicolor*) and the flounder are found in both the sea and in brackish waters, whilst the salmon undertakes yearly migrations from the sea to the rivers. Eels and certain other fishes migrate in the reverse direction—from ponds and rivers to the sea.

It seems clear that the effects of differences of salinity on animals are almost entirely due to disturbances that they set up in the osmotic pressure and ionic composition of the body fluids, for part at least of the body wall of aquatic animals behaves as a permeable membrane. An exchange of water or salts between the fluids of the body and the surrounding water is, indeed, often essential for excretion, numerous examples of which can be found in preceding chapters, whilst a permeable membrane is a prerequisite of gaseous interchange. It is found that the bodies of most marine invertebrate animals contain fluids which approximate in their osmotic pressure to that of the water in which they live and vary with changes in salinity of the environment. Creatures of this type, which cannot tolerate any change in the osmotic pressure of their body fluids, are therefore restricted to the sea. Others, which can exist when the osmotic pressure of their fluids is decreased, may be found in brackish water. Yet again, others can, by means of their excretory organs, eliminate excess water from the body and so have some measure of control over their internal environment, and thus may also range over areas of differing salinity. The blood of vertebrate animals is, as a rule, kept at constant osmotic pressure, but there is a limit to changes in salinity which are tolerated and each type of animal is specialised to regulate its osmotic pressure in a particular direction. The osmotic pressure of the blood of marine Teleost fishes is about half that of sea water, so that water continually tends to leave the blood and pass to the surrounding medium. To counterbalance this loss these fishes drink and swallow sea water and get rid of the excess salts through the gills. Thus energy is expended to maintain the osmotic balance. Selachian fishes maintain the balance in a different way, for they keep up the osmotic pressure of the blood by retaining urea in it, thus preventing any osmotic loss of water. In both types of fishes the urine is relatively concentrated, so that little water is lost in this way. Fresh-water animals are faced with quite a different problem, for their body fluids must always be kept at a much higher osmotic pressure than that of the water. Water, therefore, continually diffuses inwards and is as continually excreted whilst salts are absorbed through gills or other sites. The means adopted for the elimination of excess water from the body vary from the contractile vacuoles of the Protozoa to the kidneys of the vertebrate animals.

OTHER PHYSICAL FACTORS.—Many other physical factors play a part in controlling animal distribution. Of these, many have an indirect effect only, making their influence felt at first on the plant life. Others act mainly on the plants but may, on occasions, have a direct limiting action on the distribution of certain animals. In this category would come any chemical peculiarities of the water or soil. Crayfish, for example, are found only in streams where the calcium content of the water reaches a certain minimum which is required in the preparation of the exoskeleton. The influence of the concentration of salts, such as phosphates, on the phytoplankton has already been mentioned.

A complete list of the physical factors concerned with environment would include rainfall, humidity, the nature of the winds, the acidity of the soil, and many others, but it is not possible to deal with these except to mention them by name.

BIOTIC FACTORS.—THE INTERDEPENDENCE OF ANIMALS AND PLANTS.—Acting in conjunction with the various physical factors mentioned above are others, which may be termed biological or biotic factors, for animals live, not in "splendid isolation" but in communities which, as a rule, are composed of diverse species and types of species.

Each of these, no less than the physical factors, is an integral part of the environment, and each plays its part in the maintenance of the "balance of nature". Of prime importance is the nature of the vegetation, for it provides a relatively stable basis upon which the varied activities of the animal population rests. This has been emphasised in a previous section. On the whole it may be said that the type of vegetation depends on the interplay of physical factors, but complexities due to reactions of the vegetation on the climate are often met with. For example, trees tend to increase the available water by retaining the rainfall, and in general tend to tone down extremes of climate.

Fig. 692. ". . . other ducks, and other water birds came there to nest."

Again, animals often produce profound effects on the plant life. Rabbits by their burrowing often destroy trees and larger plants, and may be responsible for the reversion of forests to grassland.

The history of a Scottish heather moor is interesting as showing the interrelation of plants and animals. In 1892 gulls were introduced there and protected from molestation. They multiplied, and by their waddling and excreta the heather was gradually exterminated. The area became swampy and grasses, rushes, and docks grew up in its place. Grouse, formerly plentiful, disappeared, whilst teal, other ducks, and other water birds came there to nest. After 1900 the gulls were no longer protected and decreased in numbers. The area then slowly reverted to its former condition and by 1917 the grouse had returned and the whole region had practically regained its nature of a true heather moor.

In certain instances plant distribution may act as a definite limiting factor on the range of an animal. The northern limit of the beaver, for example, coincides with the northern limit of the aspen tree upon which the animal is, to a large extent, dependent for food.

In a similar way certain animals, and this is particularly true of parasites, symbionts, and commensals, are utterly dependent on the presence of particular kinds of other animals. The distribution of the liver fluke coincides with that of its secondary host, *Limnaea*; the malaria parasite is found only in those regions populated by an appropriate species of mosquito; *Monocystis* is found only in the seminal vesicles of the earthworm. It must not, however, be imagined that all animals are so completely hemmed in by their environment, and many show a considerable elasticity with respect to certain variations in their surroundings.

The tiger, normally an inhabitant of the jungle, can flourish well above the snow line when food is plentiful. Again, some animals, although showing a preference for a particular diet, will change it when occasion demands. This elasticity to habitat makes the variety of animal and plant communities almost infinite and complicates an already extremely complex study. Nevertheless, when a broad view is taken, it appears that each type of habitat has its own general *type* of animal population, although in comparable regions in different part of the earth the species may be quite unrelated. Thus, the

Fig. 693. "The koala is bear-like . . ." [*London Zoological Society*.]

rain-forests of both the New and Old Worlds abound in monkeys. Superficially, they are very alike and both show marked adaptations to arboreal life. Yet really, they are of quite distinct types as is shown by important variations in their anatomy. Perhaps even more striking are similarities between the marsupial mammals and eutherian mammals which live under comparable environmental conditions. The marsupial mole resembles in appearance and skeleton not only our own European mole but the golden mole of South Africa; the Tasmanian wolf, as its name implies, is wolf-like not only in appearance but in its behaviour. The koala is bear-like and there are even marsupials (phalangers) which have a patagium between their limbs which enables them to parallel the "flying squirrels".

Very often, then, in different habitats of any given type, there are found animals which show comparable structural adaptations.

They are examples of convergent evolution. Even if such as these cannot be found, it is always possible to distinguish certain animals which play comparable roles in the general life of the respective communities. Animals which do this are said to occupy a particular niche. The Arctic fox scavenges for the remains of seals killed by polar bears and also eats guillemot eggs. It fills the same niche as the hyaena, which feeds on the remains of zebra killed by lions and also eats ostrich eggs. In South America the rabbit niche is filled by another rodent—the agouti—whilst on coral islands, the earthworm niche is filled by land crabs.

**The Dynamic Nature of the Environment.**—From the considerations set out above it is evident that the environment, made up as it is by the interaction of numerous factors, many of which are themselves really expressions of the activity of living organisms, does not provide a constant or unchanging background. It is essentially dynamic and the phrase "the balance of nature" is quite misleading if taken in its literal sense. A variation in one environmental feature has repercussions on others and such equilibrium as exists is, as a rule, quite unstable. Thus the numbers of each kind of animal in a given habitat do not remain even approximately constant but fluctuate about a mean which, as a rule, represents the optimum density for the species. In many species the fluctuations in numbers are not sufficiently striking to obtrude themselves except in certain seasons or years when such violent fluctuations occur that they are bound to excite comment. In certain instances these fluctuations are to some extent predictable, for it has been noticed that for most species the numbers increase up to a maximum over a definite period of years and then drop, rapidly, as a rule, back to a minimum. In some animals the cycle takes eleven years but cycles of ten, seven, three, and other numbers of years are known. The cycles can be shortened or lengthened by environmental changes, but the fundamental causes of the phenomenon are unknown. It is possible that there exists a correlation with cyclical climatic changes. It is only to be expected that the most extreme instances of fluctuations, resulting from a reinforcing of the cyclical variations in numbers by environmental variations, are the best known. Of these, the lemming story provides a striking and clear-cut example. Lemmings are small, rather rat-like creatures, inhabiting the cold desert (Tundra) regions of the northern countries. Favourable conditions lead to such a rapid increase in the reproductive rate of these small animals, that after two consecutive good years the lemmings are found in vast swarms. The owls and Arctic foxes which prey on the lemmings also increase in numbers because of the abundance of food, but these predators lag behind the faster-breeding lemmings which, despite the heavy toll taken on their numbers, continue to multiply until at last they over-populate their surroundings. The overcrowding soon becomes severe and although there may be ample supplies of food, the lemming population undergoes a "crash"—not because its members migrate out on "suicide marches" as was for a long time believed— but because, it seems, that crowding alters the time at which they attain sexual maturity, the rate at which they breed and die, and so on. During the phase of increase lemmings become sexually mature when relatively young and breed in the winter as well as the summer. Crowding leads to increased fighting, to "stress", and early mortality. Reverse changes occur when the population level is low. Here then is an example of a population regulating its numbers by behavioural means. It is but one of many that could be cited. Well known is the territorial behaviour of birds and carnivorous mammals which have the effect of spacing out the members of a population. At certain times of the year it may, however, be advantageous for members of the population to aggregate. This is true, for example, of fishes which shoal for breeding purposes but afterwards disperse.

Fig. 694. "Lemmings are small, rather rat-like creatures..."

After a population crash of the lemmings the owls and foxes are deprived of sufficient food for their increased numbers and they in turn decrease in numbers. The lemming case is typical in its general features of many others such as voles, field-mice, rabbits, rats, and many insects, including locusts. As a rule, animal populations which fluctuate in this extreme manner lie near the base of the food chains, and certain of them are economically important for their effects on agriculture. They are nearly always small animals with a rapid reproduction rate. If injurious to the interests of man they are said to form plagues. In a few instances the fact that the natural predators of a species can act as a curb to rapid increase in numbers of animals (or plants) harmful to agriculture has been made use of to effect a control. This is termed biological control, a well-known example of which is the case of the fluted scale insect, a pest parasitic on citrus fruits in North America. A very great measure of control has been gained by the introduction of a coccinellid beetle from Australia which preys on the scale insects. It is often found that when a species is introduced into a new region where its natural predators are absent, there is a

vast and sometimes permanent increase in numbers far above that
possible in the original habitat. The instance of the rabbits in
Australia has already been quoted. There are many other comparable examples and they serve as warnings to the unwary for, as a
rule, little or no control is possible except at great expense. In the
present state of knowledge the results which will follow the artificial
shifting of animal populations are practically unpredictable. The
species introduced to new surroundings may die out or may run
riot and the simultaneous introduction of natural predators may
not have the expected results at all. It is said that an attempt to
control rabbits in New Zealand by bringing weasels to the country
failed because weasles showed a preference for poultry. In general,
under natural conditions, regulatory forces of one sort or another
operate to keep the various populations of an ecosystem fairly
constant in number over a period of years. But these regulatory
forces are always extremely complex and rarely are they fully
understood, the stability of any ecosystem depending on the interaction of all its various communities and having been arrived at by
evolutionary process over long periods of time—many millions of
years in certain instances.

**Migrations.**—The nature and density of animal populations are
often strongly affected by wanderings from one situation to another.
These vary enormously in kind and extent but in the first place there
are those slow migrations which have taken place throughout the
ages and which affect all the animals in the community. These
movements may be termed radiations, for it is thought that all
animals tend to radiate outwards in search of food and other
habitats from the centre in which they originally evolved. During
their wanderings the various species are spread over wide areas but
also, as new conditions are encountered, the animals become modified
in an adaptive way to meet them. Only such as are adapted survive,
so that in the course of time new species arise. This is called
adaptive radiation. In contrast to this slow radiation, but differing
from it chiefly in degree, are other more violent and erratic migrations which frequently occur as expressions of extreme over-popula-
tion. Many of the migrating animals meet with disaster for one
reason or another as in locust plagues.

In quite another category are the regular seasonal migrations
undertaken by many classes of animals in connection with their
breeding habits. As is well known, at certain seasons of the year
many birds migrate for long distances either to, or from, their
nesting grounds. Some of these seasonal migrations are of extraordinary distances and are carried out with great precision. For

example, the Arctic tern migrates from the Canadian Arctic to the Antarctic pack ice, taking in the west coast of Africa as part of its route. The great shearwater ranges over the whole of the Atlantic yet returns to breed in the tiny islands of the Tristan da Cunha group in the southern Atlantic. Many of our better-known birds, among which may be mentioned swallows, swifts, cuckoos, and martins, migrate far to the south after breeding here. Others, such as the purple sandpiper, pass the winter in the British Isles and migrate northwards in the spring. In general, it may be said of the migratory birds in the northern hemisphere that they fly northwards to breed in the springtime and return to more southerly countries in the autumn. In many instances the advantage to the species of these travels is obscure, but in others it seems that the comparative solitude of the northern districts is essential for mating and nesting, whilst the long hours of daylight allow of longer times for feeding. It is unfortunate that from many points of view the migrations of birds are but little understood. This is partly because birds fly at such great heights when migrating that only very rarely have they been seen directly. Moreover, many migrations occur mainly at night. It is only recently that, by the use of radar, these high altitude flights have been plotted and they give a much more complex picture than was formerly thought to be true, many of the seasonal migrants temporarily reversing their direction from time to time. At present it is impossible to say exactly how the migrating birds find their way to their destination, but recent work on the homing of pigeons has thrown a great deal of light on the way in which some species of birds navigate. It is believed that pigeons find their way home by sun-navigation, using methods which are basically the same as those used by early mariners. That is to say, pigeons react as if they are able to compare the apparent height of the sun at its zenith above the horizon, and its height at any given time of day with the position of the sun as seen at home.

This was suggested by finding that pigeons "home" much more slowly when the weather is cloudy; that they become confused if they are prevented from seeing the sun for a week or so (during which time it changes its apparent position in the sky); by disrupting their daily rhythms by subjecting them to unnatural hours of artificial daylight, so destroying any "time sense", and so on, when again the ability to "home" is diminished. In order to navigate by the sun two essential conditions must be fulfilled. A bird, it must be assumed, can remember the apparent arc of the sun across the sky as seen from its home and must be able to compare it with that at the release point. By comparing the apparent height of the sun at its zenith at the release point with that at its home, it could gain information as to its latitude, for if the sun's altitude was higher at the release point than at home, then the bird is south of home and *vice versa*. But pigeons are not, of course, always released at noon (local time) and it must be assumed that they can, after observation of the movement of the sun for a short time, extrapolate its apparent position to find the zenith. But this would give information only about the latitude of the release point. In order to appreciate the longitude, *i.e.* the course to be set east or west, a pigeon must have some "sense of time" in order that it can compare the apparent altitude of the sun east or west of the zenith at any particular time of the day. There is nothing intrinsically difficult in believing that birds do have some sort of "internal clock". We, ourselves, have and so have many animals, as has been shown by critical experiments. Sun navigation leads pigeons near enough to home for them to pick up remembered landmarks.

Perhaps the seasonal migrations of some other birds depend partly on a similar mechanism. But the new broods will migrate to the usual destinations even if kept back behind the migrations of the parent flocks. This shows that behaviour learned during their life time is playing no part in directing their migrations. Nor, can they be guided by their parents. Rather it would seem that their reactions are innate or inborn and are common to each member of the species. The cues to which they react and orientate may in some instances be features of the geographical landscape, coastlines, and so on but the height of the migrations and the fact that they often take place at night suggest that sometimes birds orientate to the pattern of stars in the sky.

The stimulus to begin migrations seems to be connected with the state of the endocrine balance of the body, and this in turn is affected by seasonal climatic changes in the external environment. It has been shown, for example, that in certain birds increased light has a stimulating effect *via* the neuro-sensory system on the anterior lobe of the pituitary, one of whose hormones activates and causes a seasonal enlargement of the gonads preparatory to breeding.

It is probable that a similar mechanism is at work in fishes, some of which migrate in a most astonishing way for thousands of miles. Best known, but by no means the most striking, of the migratory fishes is the Atlantic salmon, which returns in the autumn and winter to breed in rivers and streams, often the same ones which they left in the previous year. The eggs hatch in the upper parts of the rivers and the larvae live and grow there until they reach a size of several inches in length. Then they swim downstream and forsake the rivers for the sea, mainly off Greenland, where they feed, grow, and when sexually mature (after about three or four years), return once again to breed in their turn. The Pacific salmon apparently returns to breed but once, for the effort of migrating up the often densely-crowded rivers proves too much for it and death, as a rule, follows spawning. Many other fishes behave in a comparable, but usually less extreme, way. They are said to be anadromous, that is, they move into or towards fresh water to spawn.

Other fishes migrate in the reverse direction and are catadromous. Most famous of all these are the eels, but even now details of their migrations are in dispute. The ordinary European eel lives the greater part of its life in river sand ponds, where it grows to a length of about two feet and is then ready to assume sexual maturity. It is greenish in colour and is called a green eel. Towards the end of the summer the green eels make their way to the sea, often crossing the land for quite considerable distances. Some weeks are then spent in the shallow coastal waters and during this period the colour changes to a silvery hue (silver eels) and the gonads begin to mature. The eyes gradually enlarge and the purplish visual pigment is replaced by a yellowish one which absorbs preferentially the blue-green rays forming most of the light penetrating the surface layers of the sea. The silver eels then set out on the last

Fig. 695. The Salmon.

stage of their journey from which they never return. Heading westward, they swim across the Atlantic and eventually, after a period of several months, they are believed to reach the Sargasso Sea. Meanwhile the gonads have become mature and the eyes still further enlarged (an adaptation to the conditions of dim light found in the deeper waters in which the eels now swim). Breeding takes place well below the surface of the sea, but the eggs and the larvae are pelagic and are found in abundance at the surface. The larvae are minute creatures and are flattened from side to side. They are called **leptocephali**, for having an appearance so unlike that of an eel, they were at one time mistakenly referred to a separate genus—

*Leptocephalus.* The American eels migrate eastward and also breed in the Sargasso. The leptocephali of both species then start to travel outwards from the Sargasso in all directions and eventually they reach the shores when they metamorphose into tiny eels (elvers). The sorting of the European from the American species is apparently automatic because, whereas the European leptocephali take about three years to complete their eastward migration and metamorphosis into elvers, the America leptocephali take only one year. It follows that any European leptocephali which pass westward reach the shores of America well before they have metamorphosed or have become adapted to life on the shores and fresh water, and they consequently perish. On the contrary, American leptocephali migrating eastward change into elvers in mid-Atlantic and die because they are not suited to life in the ocean.

This account is based on the work of Schmidt but recently evidence has been brought forward which casts doubt on part of this story. It is pointed out that European eels have never been caught far out from our coasts—certainly not in mid-Atlantic—so there is no real proof that they ever reach the Sargasso. Perhaps they all die without breeding! If this is true it means that all the eels in Europe are the offspring of American eels whose leptocephali have drifted to our shores. It is also necessary to assume that the different lengths of larval life, as well as the slight morphological differences which distinguish American from European eels, are due to differences in the temperature of the water east and west of the Sargasso Sea. But serological studies show that European and American eels are distinct, which does not suggest they have a common origin. More information is needed before the remarkable story of the eel can be told in full, but if Schmidt's theory is true then it represents an extreme example of a type of migration undergone by other kinds of fishes, the adults of which swim against the current to reach their spawning grounds and the young fry being drifted back passively to situations where they feed and grow.

# CHAPTER XXIII

## THE MAIN GROUPS OF ANIMALS

### PHYLUM PROTOZOA

Protozoa form a collection of microscopic (with few exceptions) organisms with such a wide range of structural and physiological features that they are difficult to define briefly. Its various groups have little in common other than that the whole organism is comparable with a single cell. There are exceptions even to this rule but they are rare enough for it to be meaningful to speak of Protozoa as unicellular organisms since the vast majority function as independent cells. There is no single character or collection of characters which will serve sharply to delineate Protozoa from simple plants. True it is that most Protozoa are heterotrophic and all unicellular plants are autotrophic but the mode of nutrition may change according to the food available. Moreover, organisms clearly akin as judged by structure and life-history may be either plants or animals as judged by their nutrition.

Protozoa is, then, a Phylum containing a great diversity of minute organisms which, for convenience, are lumped together by systematists until more information is available which will enable their relationships to be assessed with more certainty. The same attitude should be adopted to the subdivisions of the Phylum into classes, orders, and so on.

**Class: Rhizopoda:** Protozoa which are usually free-living and which move and ingest their food by means of pseudopodia, at least during the predominant phase of the life-history. Differentiation of the body is rarely carried to an advanced degree.

ORDER: AMOEBINA.—Amoeba and its relatives.—The body consists of a naked mass of protoplasm which is differentiated into ectoplasm and endoplasm. —Pseudopodia, blunt or lobose in type, are used for feeding and locomotion.— A few are parasitic.—Reproduction is usually by binary fission.—Encystment is a common means of dispersal.

Example: *Amoeba.*

ORDER: RHIZOMASTIGINA.—Small amoeboid forms which also have a flagellum —Fresh-water.

*Mastigamoeba.*

ORDER: TESTACEA.—Amoebae with shells formed of protein impregnated with silica on to which sand grains may be attached, with a single aperture through which blunt pseudopodia are protruded Mostly fresh-water.

*Arcella.*

ORDER: THALAMMOPHORA.—Foraminiferans.—Amoeboid forms protected by a shell or test in which are single or many apertures through which the pseudopodia protrude and join to form a network.—Reproduction is by fission and sexually, by the fusion of gametes, which often bear flagella.—Marine.

*Elphidium, Globigerina.*

ORDER: HELIOZOA.—The Sun Animalcules and their allies.—The body is radially symmetrical.—The pseudopodia are slender and stiff.—Vacuolated ectoplasm.—Many nuclei usually present.—Reproduction is by fission and, sexually, by the fusion of gametes.—A skeleton of spicules is frequently present.—Mostly fresh-water.

*Actinophrys, Actinosphaerium, Acanthocystis.*

ORDER: RADIOLARIA.—Radially symmetrical rhizopods living in the surface layers of the sea.—The cytoplasm is in two distinct layers separated by a membranous capsule. The nucleus lies in the cytoplasm inside the capsule.—Radial symmetry is lost in some genera.—A skeleton, usually of spicules, is always present.—The pseudopodia are stiff and radiating.—Marine, planktonic.

*Thalassicolla, Sphaerozoum.*

ORDER: ACANTHARIA.—Formerly included with Radiolaria but probably not closely related to them.—Differing from Radiolaria in that spicules are not siliceous (strontium sulphate?) and that there is no membrane between central and peripheral cytoplasm.

*Acanthometra.*

ORDER: MYCETOZOA.—The Slime Fungi.—Semi-terrestrial rhizopods (or fungi according to the botanists) which form encrusting masses on rotten wood and the like.—The masses are really colonies of amoeba-like rhizopods, but the colony has the power to move as a unit.—Reproduction takes place by fission and by the formation of spores in sporangia like those of fungi. The spores hatch out little amoebae which may acquire flagella.—Dispersal may take place by the formation of multi-nucleate cysts.

**Class: Mastigophora.**—The Flagellates.—All are small Protozoa which in the adult stage are motile, swimming by means of flagella.—Very diverse forms may be met with in the life-history of a single species.—The nutrition is variable.—A pellicle is present.—Reproduction is by longitudinal fission. Flagellates are usually grouped into two main subdivisions, often called sub-classes, the Phytomastigina, which are plant-like in their nutrition and the Zoomastigina, which are heterotrophic.

**Subclass: Phytomastigina.**—Plant-like flagellates with not more than eight flagella.—Chloroplasts present.—The first three orders are related to brown algae; the last three to green algae.

ORDER: CHRYSOMONADINA.—Small, simple in structure with one or two flagella.—Reserve food usually fatty—not starch.—Cyst wall contains silica.—Yellowish brown in colour.—The marine silicoflagellates and coccoliths are the best known. Other minute ($\mu$ flagellates) species, important in plankton, belong here.

ORDER: CRYPTOMONADINA.—Ovoid, flattened with a pronounced gullet from whose wall arise two unequal flagella.—Cellulose cyst wall.—Many are the symbiotic algae known as zooxanthellae.

*Cryptomonas.*

ORDER: DINOFLAGELLATA.—Two flagella, one pointing forward, the other forming a girdle around the cell.—Some are naked, others have a covering of cellulose plates.—Most are autotrophic and many are important members of the phytoplankton.—Others are heterotrophic and some have both methods of nutrition.

*Ceratium, Noctiluca.*

ORDER: EUGLENOIDINA.—The Euglenas and their allies.—Elongated, spindle-shaped body, whose shape is capable of some degree of alteration by means of myonemes.—Two flagella arising from a crypt into which opens the main contractile vacuole.—Reserve food is paramylum.—An "eye spot" is usually present.—The creatures swim in a spiral path.

*Euglena, Peranema.*

ORDER: PHYTOMONADINA.—Classified by botanists as unicellular green algae in a group, Volvocales.—Plant-like flagellates with 2-8 (but usually 2) flagella.—Usually with a chloroplast and a cellulose wall.—Colourless species feed saprozoically. Many are colonial and approach the multicellular condition.

*Chlamydomonas, Polytoma, Pandorina, Volvox.*

ORDER: CHLOROMONADINA.—Two unequal flagella.—Minute.—Food reserve is oil. —Not well-known.

*Chloramoeba.*

**Subclass: Zoomastigina.**—Flagellates with a definitely holozoic nutrition and sometimes with more than four flagella.—Some are well-known parasites.

ORDER. PROTOMONADINA.—Small, colourless flagellates with rarely more than two flagella.—The flagella may be of unequal size. —In the Trypanosomes there is a single flagellum which is continuous with an undulating membrane along the body.

*Bodo, Trypanosoma.*

ORDER: POLYMASTIGINA.—Flagellates with four or more flagella.—An appreciable degree of differentiation of the body can usually be noticed.

*Giardia.*

ORDER: HYPERMASTIGINA.—Small flagellates, with numerous flagella, living as symbionts in the gut of insects.—They assist in the digestion of wood in the gut of the White Ants (Termites).

*Trichonympha.*

Polymastigina and Hypermastigina are often put together in one order, Metamonadina.

**Class: Sporozoa.**—Parasitic Protozoa which are propagated by spores,—The spores usually have a resistant envelope.—Reproduction may be sexual, asexual, or both may take place in a single life-history.—Trophic organelles are lacking.

**Subclass: Telosporidia.**—The trophic phase (trophozoite) is quite distinct from the reproductive phase.—The spores usually hatch out as young trophozoites.

ORDER: GREGARINIDA.—The Gregarines.—Parasites of invertebrates.—They may be found in the gut, the coelom, or in the tissues.—One phase of the life-history is intra-cellular, but for the rest it is outside the cells of the host.—They are not, as a rule, very dangerous to their hosts.—The trophozoite may have three distinct regions.—One group of gregarines has a phase of asexual reproduction which results in a multiplication of the trophozoites by schizogony.

*Monocystis, Gregarina, Porospora, Selenidium.*

### ORDER: COCCIDIOMORPHA.

SUBORDER: COCCIDIA VERA.—The Coccidians.—Parasites chiefly in invertebrates but also common in vertebrates.—Typically, the trophic phase is intra-cellular.—Sporogony alternates with schizogony in the life-history.—The sporoblasts are produced by division of the zygote.—The sporocysts themselves contain numerous sporozoites.—The gametes are not all alike (anisogamy).—There are microgametes and megagametes.

*Eimeria, Merocystis.*

SUBORDER: HAEMOSPORIDIA.—Coccidiomorph parasites of warm-blooded animals.—They produce various types of malaria.

*Plasmodium.*

**Subclass: Neosporidia.**—Rather obscure sporozoans.—Composite spores.—The various types are probably not related to one another.

*Sarcocystis.*

**Class: Ciliophora.**—Protozoa with relatively simple life-histories, but with bodies which often reach a high degree of differentiation.—The locomotor structures are cilia arising from grooves or pits in the pellicle arranged in orderly tracts.   There are two nuclei—a highly polyploid meganucleus concerned with trophic activities and one or more micronuclei mainly, if not solely, concerned in sexual reproduction.—Asexual reproduction is by transverse binary fission, and there is nearly always a sexual process involving the fusion of gametic nuclei.—In certain ciliates true unlike gametes are formed.

**Subclass: Holotricha.**—Ciliates with not very obvious zones of composite cilia around the mouth, the general cilia over the body forming the main swimming organelles.

ORDER: GYMNOSTOMATIDA.—Very generalised ciliates with uniform ciliation in longitudinal tracts.—No modified cilia around the mouth.—Fresh-water mainly.

*Holophrya.*

ORDER: SUCTORIDA.—Epizoic as adults and with no cilia and no regular arrangement of kinetosomes.—A few to many straight, hollow, sticky tentacles used for catching and digesting prey.—No mouth.—Ciliated young resembling gymnostomes.

*Tokophrya, Acineta, Dendrocometes.*

ORDER: TRICHOSTOMATIDA.—Mouth at base of funnel-like depression.—Otherwise resembling gymnostomes.   Some are endozoic.

*Colpoda, Balantidium.*

ORDER: HYMENOSTOMATIDA.—Mouth depression long and bearing an undulating membrane with three or more parallel rows of fused cilia.

**Paramecium, Tetrahymena.**

ORDER: PERITRICHIDA.—Bell-shaped ciliates often with a stalk and an adhesive organelle by which they are attached to the substratum or onto other organisms.—A double or triple row of strong cilia around the mouth depression, the rest of the body being unciliated.—True gametes.—Marine and fresh-water.

*Vorticella, Zoothamnion.*

ORDER: ASTOMATIDA.—Endozoic, mouthless ciliates mainly living in oligochaete worms.

*Anoplophrya.*

ORDER: CHONOTRICHA.—Epizoic on aquatic crustacea.—Body is vase-like.

Subclass: Spirotricha.—Prominent adoral membranes which usually form the main locomotory as well as feeding apparatus.

*Spirochona*

ORDER: HETEROTRICHIDA.—Body bears normal ciliation as well as special adoral cilia.

*Stentor, Spirostomum, Nyctotherus.*

ORDER: OLIGOTRICHIDA.—Spirotrichs with cilia restricted to certain areas of the body.

*Diplodinium.*

ORDER: ENTODINIOMORPHIDA.—Ciliates endozoic in herbivorous mammals.— The body is very complex but the body ciliature is sparse.

*Ophryoscolex.*

ORDER: TINTINNIDA.—Mainly marine and planktonic.—Body cilia, sparse.— Body contained in a vase-like case or lorica.—Sometimes placed with oligotrichs.

*Tintinnopsis.*

ORDER: HYPOTRICHIDA.—Few, if any, simple cilia; instead there are compound cilia called cirri.—Body flattened devoid of cilia and cirri. Adoral membranelles prominent.—Creepers on substrates often moving like "clockwork mice".

*Euplotes, Stylonichia.*

Class: Opalinuta.—Superficially resembling holotrichous ciliates but differing from them in having numerous nuclei all alike and in not conjugating.— Mouthless.—Endozoic, mainly in rectum of Amphibia.—Flattened body.— Considered by some as flagellates for division takes place along the kineties.

*Opalina.*

## BRANCH PARAZOA—PHYLUM PORIFERA

The Sponges.—Aquatic (most are marine) animals having a multi-cellular body with the cells arranged in two layers around a central cavity (gastric cavity). The body is thus sac-like, but it invariably becomes much complicated by foldings of its walls.—Many sponges are colonial and reproduce asexually by budding.—The body cells are at a lower level of organisation and differentiation even than in coelenterates, and are capable under certain conditions of reverting to a generalised type of cell (de-differentiation) which may then differentiate again into any type of cell found in the sponge-body.—Co-ordination of bodily functions is by direct mechanical action or by chemical activators, for no trace of a nervous system has been detected.—The type of embryonic development is unique.—Sponges are for the above (and many other) reasons, therefore, regarded as something rather apart from other Metazoa, and to avoid confusion the structures of the body receive names different from those applied to analogous parts in coelenterates.—The outer cell-layer is termed the pinnacoderm; the inner, the choanoderm.—The aperture opening out from the gastral cavity is the osculum.—The pinnacoderm consists of flattened cells, certain of which (pore cells) are perforated by a pore through which water enters.—The choanoderm consists of peculiar cells (collar cells or choanocytes) which bear a single flagellum, which is surrounded at the base by a delicate protoplasmic collar.—It is the action of the collar cells which maintains a current of water in through the pores and out through the osculum.  The pinnacoderm is separated from the choanoderm by a layer of jelly (mesogloea) into which wander cells of various kinds.  Certain of them secrete spicules which are often of a complicated structure and may join up to form a basket-like skeleton.—In a few sponges (e.g. the Bath Sponge) a spicular skeleton is absent and is functionally replaced by a network of horny fibres (spongin fibres), whilst, in others again, even these may be lacking.—Many sponges have a ciliated larva.

**Class: Calcarea.**—Sponges having a spicular skeleton of calcium carbonate. *Grantia, Sycon.*

**Class: Hexactinellida.**—Sponges with a skeleton built of six-rayed spicules made of silica.—Some of the deep-sea types are well known as curios, for their glistening delicate skeleton is extremely ornamental, e.g. Venus's Flower-basket and the Glass-rope Sponge.
*Euplectella, Hyalonema.*

**Class: Demospongia.**—Sponges with a skeleton of four-rayed siliceous spicules; of spicules together with spongin fibres; of fibres alone or with no skeleton at all.—The Bath Sponge (order Euceratosa) belongs here.
*Halichondria, Spongilla, Euspongia.*

## BRANCH EUMETAZOA—PHYLUM COELENTERATA

Hydroids, Jelly-fishes, Anemones, Corals, etc.—Mainly marine.—Diploblastic, radially symmetrical animals in which the two-layered (ectoderm and endoderm) body wall encloses a single cavity, the enteron.—Communication between the enteron and the exterior is by *one* aperture only, the mouth.—There are two basic forms of individual, the hydroid and the medusoid, but polymorphism is frequent.—The characteristic weapons of offence and defence are nematocysts.  Both solitary and colonial forms are common.

## SUB-PHYLUM CNIDARIA

Coelenterates having nematocysts and almost perfect radial symmetry.

**Class: Hydrozoa.**—The Hydroid Zoophytes and Velum-bearing Medusae.—Mostly marine colonial forms and during the life-history both hydroid and medusoid stages are usually represented.

ORDER: GYMNOBLASTEA.—Mostly marine colonial forms in which the coenosarc is enclosed within a perisarc which, however, is not extended to form hydrothecae around the hydranths.—Bell-shaped medusae usually produced from blastostyles, but not always set free.—Gonads located in the walls of the manubrium.—Statocysts not usually present.
**Hydra, Coryne, Tubularia.**

ORDER: CALYPTOBLASTEA.—Marine colonial forms in which the perisarc is extended to form hydrothecae around the hydranths and gonothecae around the blastostyles.—Medusae usually saucer-shaped.—Gonads located on the radial canals.—Statocysts usually present.

*Obelia, Sertularia.*

ORDER: TRACHYMEDUSAE (= Trachylina).—Mainly marine forms in which, usually, only the medusoid stage is represented, the hydroid being reduced or absent.—A prominent velum.

*Liriope, Geryonia (= Carmarina), Aglantha.*

ORDER: NARCOMEDUSAE.—Marine medusae with scalloped margin to bell and no manubrium.—No hydroid stage.

*Cunina.*

ORDER: LIMNOMEDUSAE.—The hydroid stage is minute.—Most are fresh-water.—The medusa may bear gonads either on the manubrium or on the walls of the radial canals or on both.

*Gossea, Craspedacusta.*

ORDER: MILLEPORINA.—Marine Hydrozoa with the hydroid colony on the surface of a massive calcareous corallum perforated by pores into which the polyps can be retracted.—Two types of polyps, dactylozooids and gastero-zooids.—Dactylozooids in a circle around the gasterozooids.

*Millepora.*

ORDER: STYLASTERINA.—Similar to the Milleporina but with branched corallum.

*Stylaster.*

ORDER: SIPHONOPHORA.—Pelagic, colonial animals which show poly-morphism (pneumatophores, nectocalices, gasterozooids, tentaculozooids, hydrophyllia, gonophores, etc.) in a very marked way.—Both hydroid and medusoid forms are modified to subserve particular functions.—Individuals may be arranged in groups (cormidia) along a long central axis or collected beneath a large float (pneumatophore).

*Halistemma, Physalia, Velella.*

Class: Scyphozoa.—Jelly-fishes.—Marine animals in which the adult is always medusoid.—Hydroid stage may be represented in the life-history.—Medusae arise from the hydroid by strobilisation producing ephyrae.—Enteron divided into gastric pouches by ridges.—Extensive system of radial canals.—Sense organs, tentaculocysts.

*Aurelia.*

Class: Anthozoa.—Sea Anemones, Sea Fans, Sea Pens, Stony Corals, etc.—Solitary or colonial, marine forms in which only hydroid individuals are represented.—Enteron divided by mesenteries.—Entrance to enteron by stomodeum.—Gonads endodermal in origin and position.

ORDER: ALCYONARIA.—Anthozoa with eight pinnate tentacles and eight mesenteries.—One ciliated groove (siphonoglyph) in stomodeum.—Muscle bands on mesenteries face the siphonoglyph.—Corallum, when present, internal and usually consists of spicules.

*Alcyonium, Gorgonia, Corallium, Pennatula.*

ORDER: ZOANTHARIA.—Tentacles usually simple and number six or a multiple of six.—Mesenteries varying in number.—Stomodeum generally with two siphonoglyphs.—Mesenteries in relation with the siphonoglyphs (directive mesenteries) have the muscle bands facing outwards; other mesenteries, both primary and secondary, arranged in pairs with the muscle bands facing one another.—Mesenteric filaments trilobed in section.—Both solitary (sea anemones) and colonial forms (madreporarian corals) are represented.—In the corals, a massive external calcareous corallum secreted by the ectoderm is present.

*Actinia, Caryophyllia, Madrepora.*

## SUB-PHYLUM CTENOPHORA

The Sea Gooseberries, etc.—Solitary, freely swimming forms which show many differences from the coelenterates.—A considerable measure of bilateral symmetry.—Characteristic locomotor structures ctenidia ("combs"), formed of rows of fused cilia.—Nematocysts absent, but special "lasso cells" are present.—Extensive internal system of canals opening out from the enteric cavity.

*Hormiphora, Pleurobrachia.*

## PHYLUM PLATYHELMINTHES

The Flatworms.—Triploblastic acoelomate animals usually of small size and leaf-like shape.—The alimentary canal has a single aperture, the mouth.—Nitrogenous excretion and osmoregulation are carried out by a flame-bulb system.—The reproductive system is complicated and usually hermaphrodite.—The flatworms have some affinities with the ctenophores.

**Class: Turbellaria.**—Free-living flatworms found living in water or in damp situations.—The epidermis is glandular, ciliated, and bears only a thin cuticle.

ORDER: ACOELA.—Small turbellarians with no gut and simple gonads and ducts.

*Convoluta.*

ORDER: RHABDOCOELA.—Turbellarians having a fairly simple sac-like intestine or no intestine.

*Mesostoma.*

ORDER: TRICLADIDA.—The Planarians.—The intestine has three main branches.—Sense organs are frequently well developed.

*Planaria, Dendrocoelum.*

ORDER: POLYCLADIDA.—The Polyclads.—Marine turbellarians.—The gut has very numerous caeca which ramify all through the body.—Numerous simple eyes are usually present.

*Leptoplana, Thysanozoon.*

**Class: Monogenoidea** (= Monogena).—This new class is now separated from the Trematoda by the structure of the larvae. Ectoparasites with relatively simple life-histories, no intermediate hosts. Mesenchyme secretes a tough cuticle and cilia are lacking in adults.   Hooks and suckers are present and enable the parasites to cling to their hosts which are often aquatic animals.

*Polystoma.*

**Class: Cestoda.**—The Tapeworms.—Internal parasites.   No alimentary canal or obvious sense organs, nervous system is poorly developed.   There is a thick and many layered "cuticle".   Either, hooks, or, suckers or, both, are present at one end, they have great powers of both sexual and asexual reproduction.   Most undergo strobilisation, producing a chain of proglottides.   The larvae are hexacanths and the life cycle requires either one or, two, intermediate hosts.   The adult and one of the larval stages often parasitise vertebrates.

*Diphyllobothrium, Taenia, Echinococcus.*

**Class: Trematoda** (= Aspidogastrea + Digenea).—The Flukes.—The adults usually live in the vertebrate alimentary canal or associated organs.   They have complex life cycles involving one or two intermediate hosts, the first is often a mollusc.   Larvae undergo polyembryony.   This group contains some of the most economically important flatworm parasites.

*Fasciola, Schistosoma.*

### PHYLUM NEMERTINI (NEMERTEA)

The Proboscis Worms.—Triploblastic acoelomate animals with a body so extensible that it may become thread-like.—Most are marine.—The body is not truly metamerically segmented although the gonads are serially repeated, an arrangement that perhaps foreshadows the metamerism of annelids.— Both mouth and anus are present.—Above the stomodeum, but quite separate from it, is an eversible proboscis which may be exceedingly long.   When retracted it lies in a fluid-filled cavity lined by mesoderm.—The epidermis is glandular and ciliated.—The nervous system is arranged on a plan similar to that of the flatworms.—A simple blood vascular system is present.—There is a flame-bulb excretory system.—Simple sense organs are usually to be found near the anterior end.—The musculature of the body wall consists of several layers of both circular and longitudinal muscles.—The arrangements of the muscle layers, and their relation to the nerve cords, is used in classifying the group into orders.

*Lineus.*

### PHYLUM NEMATODA

The Roundworms.—Triploblastic acoelomate animals with elongated spindle-shaped bodies.—There is no trace of metamerism.—Respiratory organs and a blood vascular system are lacking.—The organs in the young worms, like those of most other acoelomate animals, are packed in parenchymatous tissue but, in the adult, this almost entirely disappears, so that the viscera lie freely in a fluid-filled cavity.—The epidermis secretes a tough cuticle and then may, to a large extent, disappear as a distinct cell-layer.—Cilia are completely lacking even from the developmental stages.—The nervous system consists of a simple collar around the anterior end of the gut and from six to eight longitudinal cords.—Apart from papillae, sense organs are lacking.—A system of large cells, which appear to be glandular and excretory, is usually present.   The cells sometimes communicate with the exterior by a pair of lateral canals.— The contractile elements of the body consists of a layer of large cells which have contractile processes next to the epidermis.—Both free-living and parasitic nematodes are known. They seem to have mastered practically every situation that can support life and yet, despite their very wide dispersal, their anatomy is remarkable for its uniformity.—The life-histories of the parasitic forms are frequently complicated.

*Ditylenchus, Heterodera, Rhabditis, Ascaris, Ancylostoma, Filaria.*

## PHYLUM ROTIFERA

The Rotifers or Wheel-animals.—Minute triploblastic animals lacking a true coelom.—The body is of variable shape and protected by a cuticle.— There is a somewhat complicated gut.—A crown of cilia and a spiral, ciliated tract lies in front of the mouth.—The nervous system is seldom more than a ganglion connected to a few nerves.—The excretory organs are canals connected to solenocytes.—The sexes are separate.—The body cavity consists of spaces in the mesoderm traversed by muscles and connective tissue.—Most rotifers are fresh-water animals, but a few are marine.—They show great powers of resisting desiccation both in the adult stage and in the egg.

*Rotifer, Hydatina.*

## PHYLUM CHAETOGNATHA

The Arrow Worms.—Relatively simple triploblastic coelomate animals.— The body is elongated and transparent.—The gut is a straight tube.—The body has three regions; a head, trunk, and post anal region (tail).—The coelom is similarly divided into three compartments.—The head bears a hood and rows of chitinous hooks.—A blood vascular system and nephridia are absent.— The nervous system is simple and consists of a circum-oesophageal ganglionated nerve ring connected with longitudinal nerves.—Eyes are usually present.— Cilia are present in the gut walls.—Male gonads lie in the tail; ovaries in the trunk.—Arrow worms are pelagic and are frequently found in swarms in the plankton.

*Sagitta.*

## PHYLUM POLYZOA (BRYOZOA)

Small colonial animals usually superficially resembling sea-weeds.—Most are marine.—The individual animals of the colony are termed zooids and bear a crown of ciliated tentacles.—Two very distinct classes are included in the phylum, and it is unlikely that they are closely related. Indeed, by many, they are rightly regarded as separate phyla.

**Class: Entoprocta.**—The Endoproct Polyzoans.—Triploblastic acoelomate animals.—The gut is U-shaped and the anus opens within the circlet of ciliated tentacles.—The excretory organs are tubules and flame-bulbs.

*Pedicellina.*

**Class: Ectoprocta.**—The Ectoproct Polyzoans.—The vast majority of polyzoans belong to this group.—Triploblastic *coelomate* animals.—The gut is U-shaped but the anus opens outside the circlet of tentacles.—The gonads develop from the walls of the coelom.—The epidermis usually secretes a strong covering which may be strengthened by calcium salts.—There is no flame-bulb system.—Polymorphism may be very marked, some of the zooids being modified to form avicularia and vibracula, curious structures looking respectively like a bird's head and a whip. They are chiefly of use in keeping the colony clear of sediment.

*Flustra, Bugula.*

## PHYLUM PHORONIDA

Small marine gregarious zooids each enclosed in a membranous tube.—Triploblastic coelomate animals.—The gut is U-shaped and the mouth and anus are surrounded by a horseshoe-shaped ridge bearing very numerous tentacles.—The excretory organs are a pair of nephromyxia.—The animals are hermaphrodite.—The larva is the actinotrocha.

*Phoronis.*

## PHYLUM BRACHIOPODA

The Lamp Shells.—Triploblastic coelomate animals enclosed in a dorsal and a ventral shell.—The anterior part of the body bears a horseshoe-shaped ridge from which spring ciliated tentacles.—There is a pair of coelomoducts but no nephridia.—The coelom is extensive and is continued into the folds of body wall (mantle) below the shell and into the tentacles.—A simple vascular system is present.—The central nervous system is a ganglionated nerve ring around the anterior end of the gut.—The body is usually attached to the substratum by a stalk.

*Lingula, Terebratula.*

## PHYLUM ANNELIDA

The Segmented Worms.—Triploblastic, metamerically segmented, coelomate Metazoa.—Body wall covered by a glandular epidermis.—Longitudinal and circular muscles present.—Cuticle is thin and made of collagen plus mucopolysaccharides.—Perivisceral cavity coelomic.—Central nervous system consisting of dorsally situated cerebral ganglia and ventral nerve cord.—Nerve-cells distributed along the length of the cord and not confined to the ganglionic swellings.—Peripheral nerves segmentally arranged.—Excretory organs are nephridia, though they may be combined with coelomoducts.—Chaetae typically present.

**Class: Polychaeta.**—The Bristle Worms.—Marine annelids with obvious segmentation and cephalisation.—Head usually [] tentacles and palps and frequently eyes.—Parapodia with numerous chaetae arise from the post-cephalic segments.—Sexes usually separate with gonads which may extend throughout the length of the body, and are never confined to a few "genital segments".—Larva occasionally a free-swimming trochophore.

*Nereis, Arenicola, Sabella.*

**Class: Oligochaeta.**—The Earthworms and their allies.—Mainly terrestrial and fresh-water annelids with marked segmentation but reduced cephalisation.—Parapodia absent and chaetae few in number and arranged singly.—Hermaphrodite with gonads confined to a few segments.—Gonoducts are coelomoducts.—Eggs laid in cocoons produced by a modified region of the epidermis called the clitellum.—Development direct without larval form.

*Lumbricus, Eisenia, Tubifex.*

**Class: Hirudinea.**—The Leeches.—Ectoparasitic annelids with few true segments, but each segment showing numerous annuli.—Chaetae are, as a rule, absent.—Suckers present at both ends of the body.—Perivisceral cavity filled with parenchymatous tissue and coelom reduced to longitudinal tubes.—Hermaphrodite.—Eggs laid in cocoons secreted by the clitellum.—Development direct.

*Hirudo, Clepsine.*

**Class: Archiannelida.**—Small marine annelids related to the Polychaeta, from which certain of them may have been derived, but some are deficient in such characters as parapodia and chaetae.

*Polygordius, Dinophilus.*

In addition to the preceding four groups, there are also sometimes included in the Annelida the **Echiuroidea** and the **Sipunculoidea**. These are rather sedentary aberrant forms which, although possessing many of the chief annelid characters, such as nephromyxia, trochophore larva, annelid type of central nervous system, and body wall, yet show little or no traces of segmentation. The preoral part of the body is large and usually forms a ciliated proboscis. In the Echiuroidea chaetae are present, but not in the Sipunculoidea. These groups are sometimes regarded as separate phyla, but are frequently placed as Classes in the Annelida.

## PHYLUM ECHINODERMATA

The Starfishes, Sea Urchins, Brittle Stars, and Sea Cucumbers.—Triploblastic coelomate animals.—They are entirely marine.—The larvae are bilaterally symmetrical, but a large measure of radial symmetry (they are usually five-rayed) has been acquired by the adults.—A calcareous skeleton is developed in the mesoderm.—The coelom is divided into several compartments, each more or less separate from the other and carrying out a different function in the life of the animal. One portion of the embryonic coelom becomes constricted off to form the perivisceral coelom; a second forms the water vascular system and retains a connection with the exterior by a pore or pores (the madreporite); a third part of the coelom, which is much less extensive, is the "perihaemal system", whilst the cavities from whose walls the gonads are developed are also coelomic.—There is no definite excretory system or blood vascular system.—The water vascular system probably assists in respiration and is linked up with the cavities of special outpushings from the body wall termed tube feet.—The nervous system is, for such highly organised animals, in a relatively diffuse state. Typically, the central nervous system consists of a ring of nervous tissue around the oral part of the gut and strands of nervous tissue from the nerve ring along the arms or radii. This part of the nervous system and the nerves connected with it are closely associated with the ectoderm. In addition, in many echinoderms there is a deeper-lying nervous system, largely motor in function, lying in the mesoderm and connected to the radial nerves by strands running through the connective tissue.—A superficial nerve plexus is found all over the body beneath the epidermis.—The larvae are pelagic and swim by means of ciliated bands which traverse processes jutting out from the body.

**Class: Stelleroidea.**—The Starfishes and Brittle Stars.—Active, predaceous free-living echinoderms with a five-rayed symmetry clearly indicated by the "arms" which radiate out from the central portion of the body.—The body is flattened.—The mouth lies on the lower surface.—Spines are borne on the body.

**Subclass: Asteroidea.**—The Sea Stars or Starfishes.—Free-living echinoderms with a flattened star-shaped body. The anus (if present) opens on the
upper (aboral surface); the mouth, on the lower (adoral surface).—Tube
feet spring from a groove (the ambulacral groove) along the under surface of
each of the rays (arms).—The tube feet can be retracted and withdrawn and
can adhere to the substratum. They are the locomotor structures of the
animal.—The aboral surface of the body is beset by spines which are movable
about their bases.—Other structures, which are modified spines and are termed
pedicellariae, are also present on the outside of the body. They are of a
variable nature but consist essentially of minute jaw-like structures borne on
the apex of a muscular mobile stalk. They are used for removing foreign
particles from the surface.—Starfishes are carnivorous animals feeding largely
on molluscs, but ciliary tracts assist in feeding in some forms.—The larva is
termed a Bipinnaria.

*Asterias, Solaster.*

**Subclass: Ophiuroidea.**—The Brittle Stars.—Free-living echinoderms with
a flattened body clearly marked into two regions—the disc and the arms.—
The tube feet protrude from the lower surface of the arms but the ambulacral
groove is closed to form a tube (epineural canal).—Spines are present.—The
tube feet assist in locomotion, but this is usually effected chiefly by muscular
movements of the arms.—Gut caeca and gonads do not extend into the arms.—
The larva is called an Ophiopluteus.—Feeding is by means of tube feet which
convey food to the mouth, but some ophiuroids have ciliary feeding tracts.

*Ophiothrix, Ophiura.*

**Class: Echinoidea.**—The Sea Urchins.—Active free-living echinoderms
with a body which, typically, is globular, but which in some forms is flattened
and acquires a large measure of bilateral symmetry.—The skeletal plates fit
closely together to form a rigid "test".—Spines are present and are movably
articulated with knobs on the skeletal plates.—The tube feet are arranged in
ten meridional rows.—The mouth lies at the lower pole; the anus at the upper
pole of the body.—Various types of pedicellariae are present between the
spines.—A complicated jaw apparatus (Aristotle's Lantern) is found in connection with the mouth.—The larva is termed an Echinopluteus.—Some
echinoids are ciliary feeders; most use the jaw apparatus; others depend
chiefly on the tube feet to bring food to the mouth.

*Echinus, Echinocardium.*

**Class: Holothuroidea.**—The Sea Cucumbers.—The body is elongated so as
to be sausage or worm-like in appearance.—The skeletal plates are minute
and embedded in the fleshy body wall.—There are no spines.—The mouth
lies at one end of the body; the anus at the other.—A pair of very much
branched diverticula from the hinder end of the intestine assist in respiration.—
Special organs which can discharge sticky threads are often present near the
hinder end of the gut.—Certain of the tube feet around the mouth are enlarged
to form tentacles; other tube feet are reduced.—The larva is termed an
Auricularia.—Feeding is carried out by the tube feet.

*Holothuria, Synapta.*

**Class: Crinoidea.**—The Sea Lilies.—Echinoderms with a cup-shaped body
attached to the substratum by a stalk, at least during the young stages.—Both
the mouth and the anus lie on the side away from the stalk attachment.—

The five main arms bifurcate, and on the branches are small side branches (pinnules).—The skeleton consists of calcareous plates jointed together and in the arms the plates resemble vertebrae and are provided with a powerful musculature.—The water vascular system does not open to the exterior but into the perivisceral coelom.—The tube feet are small and sensory in function.—Crinoids feed by ciliary currents maintained by ciliated tracts on the arms.—The larva, after a free-swimming stage, becomes attached to the substratum, but the adult may again become free-living.

*Antedon, Rhizocrinus.*

## PHYLUM ARTHROPODA

The Crayfishes, Crabs, Lobsters, Centipedes, Insects, Spiders, Scorpions, Mites, etc.—Metamerically segmented, bilaterally symmetrical animals.— Typically, each segment has a pair of jointed appendages and at least one pair is modified as jaws.—Cephalisation is pronounced, and the number of segments included in the head is variable.—The exoskeleton is strongly developed.—The central nervous system consists of cerebral ganglia and a ventral nerve cord made up of separate paired ganglia connected by commissures.—The body cavity is haemocoelic.—A contractile heart with ostia lying in a haemocoelic pericardial cavity is present.—Cilia are absent.

**Class: Crustacea.\***—The Crayfishes, Crabs, Lobsters, Shrimps, etc.— Mainly aquatic arthropods breathing by gills.—The exoskeleton is frequently massive.—There are two pairs of antennae and three pairs of head appendages act as jaws.—Some of the anterior thoracic appendages may also serve as mouth parts.

**Subclass: Cephalocarida.**—A recently discovered group of minute crustacea living in sand just below tidemarks but including fossil representatives.— Very generalised appendages inviting comparison with those of Trilobites.— Little division of body into distinct taginate.—Several pairs of excretory organs.

*Hutchinsoniella.*

**Subclass: Branchiopoda.**—The Fairy Shrimps, Brine Shrimps, etc.—Free-living crustacea in which the trunk appendages are broad, lobed, and fringed with hairs.

*Triops, Chirocephalus, Daphnia.*

**Subclass: Ostracoda.**—Small crustacea with a bivalved carapace.—The trunk and abdominal limbs are reduced.

*Cypris.*

**Subclass: Mystacarida.**—A recently discovered group of small cructacea living in marine sands, possibly allied to Copepods.—No carapace.—Simple appendages.—No clear divisions of the body.

*Derocheilocaris.*

---

\* Crustacea is such a large class and contains such a diversity of animals that it presents special problems for taxonomists who find it convenient to use rather unusual categories such as series, super orders, and so on. For simplicity these have been omitted.

**Subclass: Copepoda.**— Free-living or parasitic crustacea without a carapace.—The antennules are frequently enlarged and used for swimming.—The thoracic appendages are biramous.

*Cyclops, Calanus.*

**Subclass: Cirripedia.**—The Barnacles.—Sedentary animals, either sessile or stalked.—The head and abdomen are much reduced.—The body is enclosed in a fold of the body wall termed the mantle which bears calcareous plates.— The thoracic appendages are plumose and are used for feeding.—Some forms are parasitic.—There are free swimming larval stages in the life-history.

*Balanus, Lepas, Sacculina.*

**Subclass: Branchiura.**—Fish lice. Ectoparasites on fresh-water fishes and Amphibia.—Flattened bodies.—Head appendages modified for adhesion and sucking.

*Argulus.*

**Subclass: Malacostraca.**—The Krill, Woodlice, Lobsters, Crabs, Shrimps, etc.—The eyes are usually stalked.—Typically, the carapace covers the thorax, with consists of eight segments.—The abdomen is made up of six appendage-bearing segments.—The exopodite of the antenna is scale-like.—The uropods and the telson form a tail fan.

ORDER: LEPTOSTRACA.—The carapace takes the form of a bivalved shell.— In the front of the head region is a movable head plate.—There are eight thoracic and eight abdominal segments.—All are marine.

*Nebalia.*

ORDER: SYNCARIDA.—The carapace is absent.—The thoracic appendages have exopodites.—Fresh-water forms.

*Anaspides*

ORDER: THERMOSBAENACEA.—Inhabitants of hot springs.—Rather like Isopods but having a dorsal brood pouch.

*Thermosbaena*

ORDER: MYSIDACEA.—The thorax is covered by the carapace.—The exopodite of the antenna is scale-like.—The thoracic appendages have exopodites.—The abdomen terminates in a tail fan. There is a statocyst in the endopodites of the last pair of abdominal appendages (uropods).—Mainly pelagic, marine animals.

*Mysis.*

ORDER: CUMACEA.—The carapace is small, four or five of the thoracic segments being exposed.—The abdomen is slender.—The uropods are slender and do not form a tail fan with the telson.—Marine

*Diastylis.*

ORDER: TANAIDACEA.—The carapace is reduced.—The abdomen short.— There is a resemblance in bodily form to an Isopod.—Marine.

*Apseudes.*

ORDER: ISOPODA.—The carapace is absent.—The body is flattened dorso-ventrally.—The abdomen is often reduced.—The endopodites of the

abdominal appendages function as gills.—Fresh-water, marine, and terrestrial.
*Ligia, Asellus, Oniscus.*

ORDER: AMPHIPODA.—The carapace is absent and the body is laterally compressed.—The abdomen is elongated.—Fresh-water and marine.
*Gammarus, Orchestia, Caprella.*

ORDER: EUPHAUSIACEA.—The carapace encloses the thorax.—None of the anterior thoracic appendages (which are biramous) is modified as maxillipeds.—There is only one series of gills.
*Nyctiphanes.*

ORDER: DECAPODA.—The carapace completely covers the thorax.—The abdomen may be large (Macrura) or reduced and flexed beneath the cephalothorax (Brachyura).—The exopodite (scaphognathite) of the maxilla is large. —There are three pairs of maxillipeds.—The thoracic limbs are uniramous.— Generally more than one series of gills is present.—Fresh-water, marine, and terrestrial.

*Astacus, Cancer.*

Class: Onychophora.—Tracheate arthropods with a thin cuticle.—The head is composed of three segments and has only one pair of jaws.—The body segments are similar and bear appendages.—Cilia are present in the reproductive system.—Terrestrial.
*Peripatus.*

Class: Myriapoda.—The Centipedes and Millipedes.—Tracheate arthropods in which the head region is clearly demarcated.—There is one pair of antennae. —The leg-bearing body segments are similar.

ORDER: CHILOPODA.—The Centipedes.—The body segments are similar.— The appendages of the first body segment are modified as poison jaws.—The genital aperture is posterior.
*Lithobius, Scolopendra.*

ORDER: DIPLOPODA.—The Millipedes.—The body segments are similar and each apparent segment bears two pairs of appendages.—The antennae are club-shaped.—The genital aperture is anterior.
*Julus.*

Class: Insecta.—The Insects.—Tracheate arthropods in which the body is divided into head, thorax, and abdomen.—Wings, when present, arise from the second and third thoracic segments.—There are three pairs of walking legs, one pair to each thoracic segment.—The abdomen of the adult usually bears no ambulatory appendages.—The life-history usually shows a metamorphosis.

Subclass: Apterygota.*—Wingless insects, which change little during metamorphosis.—Abdominal appendages in adult in addition to genitalia and cerci.

| ORDER: THYSANURA. | Bristle tails. |
| ORDER: COLLEMBOLA. | Spring tails. |

* The number of orders of insects is so large that full details of their characteristics cannot be given, but only an indication of familiar examples.

**Subclass: Pterygota.**—Insects which usually have wings.—The abdomen bears no appendages for locomotion.—Malpighian tubules present.—Mouth-parts usually hanging free of head.—Metamorphosis complete or incomplete.

**Section I: Palaeoptera.**—Wings do not fold over body when in repose.—Metamorphosis incomplete, the wings developing outside the body (exopterygote).—Numerous Malpighian tubules.—Nymphs aquatic.

| | |
|---|---|
| ORDER: EPHEMEROPTERA. | May-flies. |
| ORDER: ODONATA. | Dragon-flies. |

**Section II: Polyneuroptera.**—Wings fold over body when in repose.—Metamorphosis incomplete.—Numerous Malpighian tubules.

| | |
|---|---|
| ORDER: DICTYOPTERA. | Cockroaches and Mantids. |
| ORDER: ISOPTERA. | Termites. |
| ORDER: PLECOPTERA. | Stone-flies. |
| ORDER: CHELEUTOPTERA. | Stick- and Leaf-insects. |
| ORDER: ORTHOPTERA. | Locusts, Grasshoppers, and Crickets. |
| ORDER: EMBIOPTERA. | Embiids. |
| ORDER: DERMAPTERA. | Earwigs. |

**Section III: Oligoneuroptera.**—Wings folded in repose.—Metamorphosis complete, the wings developing inside the body (endopterygote).—Few Malpighian tubules.

| | |
|---|---|
| ORDER: COLEOPTERA. | Beetles. |
| ORDER: MEGALOPTERA. | Alder-flies. |
| ORDER: RAPHIDIOPTERA. | Snake-flies. |
| ORDER: PLANIPENNIA. | Lace-wings and Ant-lions. |
| ORDER: MECOPTERA. | Scorpion-flies. |
| ORDER: TRICHOPTERA. | Caddis-flies. |
| ORDER: LEPIDOPTERA. | Butterflies and Moths. |
| ORDER: DIPTERA. | True flies. |
| ORDER: SIPHONAPTERA. | Fleas. |
| ORDER: HYMENOPTERA. | Ants, Bees, and Wasps. |
| ORDER: STREPSIPTERA. | |

**Section IV.: Paraneuroptera.**—Wings folded in repose.—Metamorphosis incomplete, the wings developing outside the body (exopterygote).—Few Malpighian tubules.

| | |
|---|---|
| ORDER: PSOCOPTERA. | Book-lice and Bark-lice. |
| ORDER: MALLOPHAGA. | Bird-lice. |
| ORDER: ANOPLURA. | Body-lice. |
| ORDER: THYSANOPTERA. | Thrips. |
| ORDER: HOMOPTERA. | Cicadas, Aphids, and Scale-insects. |
| ORDER: HETEROPTERA. | Plant-, Water-, and Bed-bugs. |

**Class: Arachnida.**—Scorpions, Spiders, King-crabs, Mites.—The body is divided into an anterior prosoma and a posterior mesosoma and metasoma (sometimes fused to form an opisthosoma).—The head appendages are peculiar (prehensile chelicerae, sensory or prehensile pedipalpi).—There are four pairs of walking legs on the mesosoma.—The respiratory organs are gill-books, lung-books, or tracheae.

ORDER: SCORPIONIDEA.—The Scorpions.—The body is divided into pro-soma, mesosoma, and metasoma.—The pedipalps are prehensile.—There are

four pair of lung-books in the mesosoma.—The post-anal telson forms a sting.
*Scorpio.*

ORDER: EURYPTERIDA.—Extinct aquatic forms known only from fossils.—
They resemble scorpions.

ORDER: XIPHOSURA.—The King-crabs.—Aquatic arachnids in which the
body is enclosed in a carapace.—The respiratory organs are gill-books.
*Limulus.*

ORDER: ARANEIDA.—Spiders.—The body is divided into prosoma and
opisthosoma.—The pedipalps are sensory; they are modified in the male for
reproductive purposes.—Spinnerets are present on the abdomen for spinning
the web.
*Epeira.*

ORDER: ACARINA.—The Mites and Ticks.—The body is rounded with no
demarcation between the prosoma and opisthosoma.
*Tyroglyphus, Argas.*

## PHYLUM MOLLUSCA

The Snails, Slugs, Bivalves, Octopuses.—Coelomate animals which do
not show segmentation.—The body consists of the head-foot and visceral
mass.—The skin of the visceral mass is extended into soft folds which form
the mantle which often secretes a shell.—The respiratory organs are usually
a pair of ctenidia (gills).—The pericardial cavity is coelomic in origin.—The
nervous system is made up of ganglia connected by commissures.—There is
sometimes a larval form which resembles an annelid trochophore.

**Class: Monoplacophora.**—Mainly fossil molluscs with a limpet-like
appearance. One genus, *Neopilina*, lives in the depths of the ocean at the
present day.—The shell of *Neopilina* is uncoiled.—There is a simple foot used
for creeping.—Gills, kidneys, coelomic sacs, and certain muscles are serially
repeated.

**Class: Amphineura.**—The Chitons.—The body is bilaterally symmetrical.—
The mouth and anus are at opposite ends of the body.—The foot is flattened
and the mantle bears calcareous plates.
*Lepidochitona.*

**Class: Gastropoda.**—The Slugs, Snails, Limpets, etc.—The tentacle-bearing
head is distinct.—The foot is flattened.—The visceral hump is often twisted so
that the anus is anterior.—The mantle secretes a shell.
*Littorina, Patella, Helix, Archidoris.*

**Class: Scaphopoda.**—The Elephant-tooth Shell.—The foot is reduced and
the shell is tubular.
*Dentalium.*

**Class: Lamellibranchiata.**—The Mussels, Oysters, Scallops, etc.—The
laterally compressed body is bilaterally symmetrical.—The head is reduced.—
The mantle encloses the body and secretes a bivalved shell.—The foot is wedge-
shaped.—There are usually two ctenidia in the mantle cavity.
*Mytilus, Anodonta.*

**Class: Cephalopoda.**—The Squids, Octopuses, etc.—The head is well developed and surrounded by prehensile tentacles (probably derived from part of the foot).—The mantle cavity is capacious and communicates with the exterior by a siphon (developed from the foot) through which water is expelled.—The shell is often reduced and internal.—The eyes are large and complex in structure.

*Sepia, Octopus.*

## PHYLUM CHORDATA

The characters which apply to most members of this phylum have been explained in a previous section (pp. 362-5). Of these features the following are the most important and are possessed by all true chordates.—A notochord is present at least during some stage of the life-history.—The pharynx has visceral clefts.—The central nervous system is tubular and dorsally situated in the body.—There is a blood vascular system in which the circulation is backwards in the dorsal vessel (or vessels) and forwards in the main ventral vessel.—The dorsal and ventral vessels are connected in the region of the pharynx by a series of commissural vessels which run in the tissue between successive visceral clefts.—There is a post-anal tail.

It is now accepted practice to recognise four main grades of organisation within the Chordata, each of which receives the rank sub-phylum. Sometimes the first three are spoken of as protochordates—a useful common name which serves to contrast them with vertebrates (craniates). There can be little doubt that hemichordates represent a distinct evolutionary line and that they should be excluded from Chordata. They are kept in this category merely for the sake of orthodoxy. Yet urochordates and cephalochordates share so many features that they must at some remote time have had a common ancestry.

**Sub-phylum: Hemichordata.**—Peculiar, marine, worm-like creatures having only very doubtful affinities with other chordates from which they differ in several respects. The body is not metamerically segmented but is divisible into three main regions—preoral lobe (proboscis), collar, and trunk.—There is no post-anal tail, for the anus is terminal.—The central nervous system is not typically present, but a cord of nervous material in the dorsal part of the collar receives this name although it is not definitely tubular and contains only irregular spaces. The collar cord (central nervous system) is connected by a commissure to a ventral nerve cord, which, like the commissures, is merely a specialisation of the sub-epidermal plexus found throughout the body.—There is a simple blood system in which the direction of flow is the reverse of that in true chordates.—The structure which is termed the notochord is certainly not homologous with the notochord of higher forms, for it is restricted to the proboscis region and, developing rather late in the embryo, it is not derived from tissue at the dorsal lip of the blastopore.—An endostyle is absent.—The embryonic development is quite unlike that in any true chordate but is comparable in many ways with that of echinoderms.—Some hemichordates have indirect development from a larva (Tornaria) very like some types of echinoderm larvae.

**ORDER: ENTEROPNEUSTA.**—Elongated worm-like, burrowing animals with numerous gill slits.—Ciliated glandular epidermis.

*Saccoglossus, Balanoglossus, Ptychodera.*

**ORDER: PTEROBRANCHIA.**—Sedentary, colonial animals.—The preoral lobe bears ciliated tentacles which are concerned with ciliary feeding currents of water.—Gill slits two in number or absent.—Gut is U-shaped. *Cephalodiscus, Rhabdopleura.*

**Sub-phylum: Urochordata (Tunicata).**—Remarkable, ciliary-feeding, marine animals—the Sea Squirts and their allies.—Most are sedentary.—Except in one group the bulk of the chordate characters are lost in the adult, but the affinities of the urochordates are clearly shown by the number of important chordate characters possessed by the free-swimming larvae—these have a simple but hollow dorsal central nervous system; pharyngeal clefts—a tail to which the notochord is restricted (hence Urochordata)—a well-developed endostyle—and the embryonic development shows a close parallel with that of amphioxus. At metamorphosis the larvae settle on the sea-floor to which they adhere by means of glandular papillae on the head. Growth in size is rapid, but many of the organs undergo great modification, or else reduction, so that many of the chordate characters are lost. The tail is cast off. The larval nervous system is lost completely but becomes replaced by a simple ganglion situated dorsal to the anterior end of the pharynx.—The body becomes surrounded by a gelatinous or horny coat (the test or tunic).—The pharynx enlarges to form an enormous sac perforated by numerous apertures (stigmata) formed by subdivision of the larval gill slits and by the addition of new ones.—The stigmata do not open directly to the exterior, but into an ectoderm-lined cavity (the atrium).—Other peculiar features of the Tunicates are the absence of a coelom and the lack of metamerism.

The structural changes of metamorphosis may be regarded as specialisation to a sedentary mode of life in which the ciliary feeding habit is perfected and renders superfluous the search for food. Locomotory and neuro-sensory systems therefore become simplified and reduced.

ORDER: LARVACEA.—Free-swimming, pelagic forms which retain the larval characters into the adult stage.—Two gill slits only.—No atrium.—The test is thin and periodically cast off and then renewed.—The test has a meshed inhalant aperture which serves as the sieve in ciliary feeding. *Oikopleura, Fritillaria.*

ORDER: ASCIDIACEA.—The Sea Squirts.—Usually fixed to the substratum.—Often colonial.—Reproduction by sexual methods and by budding.—Metamorphosis is complete. *Ascidia, Clavelina, Botryllus, Pyrosoma.*

ORDER: THALIACEA.—The Salps.—Large free-swimming forms.—Transparent gelatinous test.—Body wall muscles in distinct bands, wholly or partially encircling the body.—Two distinct phases in the life-history which differ from one another structurally (polymorphism), and by the fact that one reproduces asexually; the other sexually. *Salpa, Doliolum.*

**Sub-phylum: Cephalochordata.**—The Lancelets.—Small fish-like animals possessing most of the chief chordate characters.—The notochord extends the whole length of the body even to the tip of the snout (hence Cephalochordate).—Pharynx large and sac-like, perforated by numerous gill slits which are the equivalents of the stigmata of the ascidians.—There is a prominent endostyle and definite ciliated tracts which, as in the ascidians, are concerned with the ciliary-feeding habit.—The gill slits open into an atrium.—There is a definite coelom.—Organs of nitrogenous excretion are nephridia.—Metamerism is very well marked.

*Branchiostoma* (amphioxus), *Asymmetron.*

## SUB-PHYLUM CRANIATA

The Vertebrates.—Chordates which show a very definite advance in complexity of structure and activity on the Acrania.—There is a very high degree of cephalisation, so that a proper head region can be recognised.—Organs of special sense are prominently developed.—There is a definite brain which is enclosed in a cranium (hence Craniata).—The endoskeleton, formed of cartilage or of cartilage and bone, is a conspicuous feature and is associated with the large and complex musculature.—The organs of nitrogenous excretion are kidneys, which require for their functioning a high blood pressure.—A true muscular heart provided with valves is differentiated from part of the ventral vessel.—The visceral clefts are restricted in number and are primarily associated with respiration and not with feeding.—Whilst an endostyle is to be found in the larval stages of certain of the more primitive craniates it becomes replaced in the adults by the thyroid gland.—A definite system of endocrine organs is present.—There is one or more portal systems of veins.

**Superclass: Agnatha. Class: Agnatha (Cyclostomata).**—The Lampreys, Hagfishes, and their extinct relatives.—The most primitive of all craniates.—Cephalisation is less complete than in higher forms.—The mouth is round (hence Cyclostomata) and is not bounded by jaws (hence Agnatha).—The skull is unique and its parts are difficult to homologise with those of other craniates.—A premandibular gill slit and arch are present at least in a rudimentary form.—The profundus nerve has a separate ganglion and is not fused to the trigeminal.—The brain is primitive and parts of its roof are composed of non-nervous material.—The pineal body is eye-like and is sensitive to light.—The pituitary body is peculiar in that the hypophysis retains its opening with the outside world and has a common opening (nasohypophysial aperture) with the olfactory organ which is a median and not a paired structure.—The vertebral column is but feebly developed, for there are no centra and the notochord persists throughout life.—The larva, which is called an Ammocoete, has an endostyle and other ciliated tracts in the pharynx. It is a ciliary feeder.—The adult kidney (a mesonephros) is not a compact organ, but consists of segmentally arranged tubules.—Gills are sac-like and formed of endoderm.—Only one or else two semi-circular canals in the ear.

**Class: Ostracodermi.**—Palaeozoic Cyclostomes.—Endoskeleton very extensive and solid and composed of bone.—Head and anterior end of the body and most of the important organs are encased in bone.—Exoskeleton of placoid scales.—As a rule ten sac-like or pouched gills between which is ossified tissue representing the visceral arches.—At least one gill pouch lies in front of the mouth.—Mouth on the ventral surface.—Pectoral appendages of a fin-like form sometimes present.—Ostracoderms were very heavy creatures and were probably relatively inactive, feeding passively on debris or small invertebrates on or near the substratum. This class is a highly artificial grouping of unlike forms and is often split into separate classes.

*Cephalaspis, Pteraspis.*

**Class: Cyclostomata.**—The modern Lampreys and Hagfishes. Eel-like creatures predatory on fishes.—The endoskeleton is cartilaginous and rudimentary and exoskeleton is lacking (secondary features).—All traces of fins and girdles are absent.—Skin is slimy because of the abundance of mucous glands.—The first gill sac is supplied by the glossopharyngeal nerve indicating that anterior gills have been lost.—The respiratory water current can enter and leave the gill sacs when the mouth is occupied with sucking on to the prey.— Mouth round and suctorial.—Buccal cavity provided with a muscular, piston-like tongue bearing horny (epidermal) teeth. This arrangement allows the cyclostomes to adhere to and rasp the flesh from fishes.—Visceral arch skeleton and gills are placed in a relatively posterior position and the heart is enclosed in a cartilage which is really derived from the hinder visceral arch.—The tail is practically diphycercal.

*Petromyzon, Myxine.*

**Superclass: Gnathostomata.**—All the more familiar vertebrates.—Craniates with a greater degree of cephalisation than cyclostomes.—The mouth is bounded by jaws which are formed from the mandibular visceral arch.—The premandibular visceral arch is probably represented by the trabeculae.—The profundus nerve is fused to the trigeminal.—The brain is more advanced in structure than in the cyclostomes.—The olfactory organs and nostrils are always paired structures.—The hypophysis never opens to the exterior in the adult.—There is never an endostyle in the larvae, which are rarely ciliary feeders.—The greater part of the respiratory epithelium of the gills is of ectodermal origin and the nerves and blood vessels supplying the gills (or visceral clefts if gills are absent) pass down on the outside of the visceral arches.—Paired limbs, both pectoral and pelvic, are almost invariably present but are secondarily lost in a few instances.—There are always three semi-circular canals in the ear.—The glands associated with the alimentary canal are more compact than in cyclostomes and are arranged on a common basis in all gnathostomes.

The lower gnathostomes, all of which are aquatic, were formerly classed together as Pisces (Fishes). This class has, largely as a result of new knowledge of fossil forms, been split into four classes. Nevertheless, we can still usefully employ the term fishes to mean creatures more completely adapted to aquatic life than any other craniates.—The body is streamlined to facilitate passage through the water.—The tail is muscular and forms a powerful propulsive organ.—The paired limbs are the pectoral and pelvic fins. They are concerned with regulating the swimming level.—Median fins (including the tail fin) are always present. They act as "keels" and maintain stability.—Fins of all types are not just folds of skin, but are provided with muscles, endoskeletal rods, and dermal fin rays.—The respiratory organs are gills borne on the branchial arches.—Accessory breathing organs may be present.—Except in a few instances the heart pumps deoxygenated blood only, and shows no division into right and left halves.—The lateral line organs are prominently developed.— There is no middle ear.—There is an exoskeleton of scales.

**Class: Aphetohyoidea.**—This includes several orders of Palaeozoic fishes which, although at first sight very different from one another, yet have certain interesting and important features in common. They represent a lower level of organisation than the fishes as we know them to-day.—The endoskeleton is bony.—Jaw apparatus in a very primitive state.—Both upper and lower jaws consist of two pieces, and from the hinder end of the jaws there projects back a series of bony rays which forms a covering (operculum) over the succeeding visceral arches. The jaws thus bear a very real resemblance to a branchial arch.—Hyoid arch is complete and the hyoidean cleft is respiratory and not modified to form a spiracle.—Hyoid arch plays no part in suspending the jaws to the cranium (hence Aphetohyoidea = free hyoid).—Fins preceded by a spine.—Some forms have paired fins in addition to pectorals and pelvics.— Eyes are large and well forward on the head.—Tail is heterocercal and pectoral fins broad, features correlated with a body which is heavier than the surrounding water.

*Acanthodes, Coccosteus, Pterichthys.*

**Class: Chondrichthyes (Elasmobranchii).**—The Sharks, Dogfishes, Skates, Rays, Chimaeras, and many extinct cartilaginous fishes.—Fishes with an endoskeleton composed entirely of cartilage which, however, is usually impregnated with calcium salts.   The cartilaginous skeleton is probably derived by reduction from an aphetohyoidean-like ancestral stock.—Exoskeleton is of placoid scales.—Jaw structure shows an advance on that of the Aphetohyoidea.— Jaws are well formed and show little or no trace of branchial arch plan.— Upper jaw (palato-quadrate bar) and the lower jaw (Meckel's cartilage) are each single cartilages.—Hyoid arch (except in one small group) always plays an important part in suspending the jaws to the cranium.—Jaw suspension is amphistylic or hyostylic except in the Chimaeras.—Hyomandibular cleft reduced to a spiracle.—Gills (except in the Chimaeras) are not covered by an operculum. —External nares and mouth on ventral surface of the head.—No internal nares. — Tail is heterocercal and paired fins are broad.—The flexible body is heavier than sea water.—The heart has a well-formed conus arteriosus.—The brain has large olfactory lobes and a large cerebellum.—There is a spiral valve in the intestine.—A lung (or its homologue, the swim-bladder) is lacking.—Fertilisation is internal and the male has an intromittent organ (claspers).—Eggs are large-yolked and are enclosed in horny capsules.

**Subclass: Selachii.**—The Sharks, the Rays, and their allies both fossil and living.   Amphistylic or hyostylic jaw suspension.—No operculum.—Teeth are numerous and replaced in rapid succession throughout life.

*Cladoselache, Notidanus, Carcharias, Selache, Scyliorhinus, Raja, Torpedo.*

**Subclass: Bradyodonti.**—The Chimaeras and their fossil relatives.—Palato-quadrate bar is fused to the cranium, giving a peculiar type of autostyly.— Gills covered by an operculum, which is a fold of skin projecting back from the hyoid arch.—Spiracle is absent.  Eyes are very large and high up on the head.—Teeth are few in number and form crushing plates.

*Cochliodus, Chimaera.*

**Class: Osteichthyes.**—The Bony Fishes.—The cartilaginous endoskeleton of the embryo is, in the adult, more or less completely replaced by bone, and to this primary skeleton membrane bones are added.—The exoskeleton consists of bony scales.—The mouth is terminal and the external nostrils lie on the dorsal surface of the snout.—An air-bladder, which primitively functions as a lung, is present except in rare instances, when it is known to be secondarily absent.— Olfactory lobes and cerebellum are relatively small.—The jaw suspension is of a variable type.—Fins are variable in structure.

Bony fishes are first found as fossils in the Devonian and from that time onwards to the present day they are the dominant type of fishes.   They probably arose from the Aphetohyoidean stock.   Several interesting evolutionary trends are to be noticed.—Both the endoskeleton and exoskeleton undergo a progressive  reduction so that the body becomes lighter.— The tail, which is heterocercal in most of the primitive members of the group, gradually acquires some measure of symmetry, and in the highest bony fishes (teleosts) and in certain others it is homocercal.—The paired fins, as a rule, do not act as elevators, for the air-bladder renders the head buoyant. The pelvic fins tend to become anterior in position.—The endoskeletal elements of the fins become reduced in number.

**Subclass: Crossopterygii.**—The Lung Fishes, certain extinct groups, and the Coelacanths (until recently believed to be extinct).—The paired fins are very narrow at the base and, as a rule, articulate with the girdles by a single basal element. The fins are thus freely movable about their axis.—A lobe of body wall projects on to the fin (hence Crossopterygii = tassel fin). Relatively large cerebral hemispheres.—A single or double lung is present.—The heart and vascular system are modified in connection with the air-breathing habit.—Internal nostrils often present.—The Crossopterygii are adapted to life in shallow fresh water in which the oxygen content is low and in which gills do not provide a surface adequate to supply the oxygen requirements of the body.

**Order: Osteolepidoti.**—Palaeozoic Crossopterygians related to the stock from which the Amphibia evolved.—Fusiform body covered with thick bony scales which have a shiny outer layer (cosmin).—Tail is heterocercal.—All parts of the skeleton including the teeth show great resemblances to the earliest known Amphibia (Labyrinthodontia); the skeleton of the paired fins of certain osteolepids even gives a hint as to how the pentadactyl limb may have arisen.
*Osteolepis.*

**Order: Coelacanthini.**—Fishes related to the osteolepids, which they resemble in general form and structure.—The skeleton is somewhat reduced.—There are no internal nares.—The tail is diphycercal.

The peculiar interest attaching to the group is that until a specimen (referred to a new genus *Latimeria*) was caught off the east coast of Africa in 1939, it was thought to have become extinct in the Cretaceous.
*Undino, Macropoma, Latimeria.*

**Order: Dipnoi.**—The living and extinct Lung Fishes.—On the whole, the Devonian forms show great resemblances to the osteolepids.—Jaws are short and certain bones are lacking.—Teeth form crushing plates.—The skeleton of the paired fins consists of pre- and post-axial jointed radials articulating with a median jointed axis.—Internal nostrils are present.—The modern Dipnoi have undergone considerable specialisation.—The skeleton (including the scales) is much reduced and the tail is diphycercal.—The lung is a very efficient respiratory organ and the vascular system and heart are accordingly adapted to deal with pulmonary respiration. They resemble those of Amphibia.
*Dipterus, Ceratodus, Protopterus, Lepidosiren.*

**Subclass: Actinopterygii.**—Many different groups of extinct bony fishes as well as the vast majority of bony fishes living at the present day.—The paired fins, especially in the more primitive members, have broad bases and lack fleshy lobes.—External nares are double, but internal nares are absent.—The cosmin layer is lacking from the scales, which are of the ganoid type.—The jaw suspension is always hyostylic.—The air-bladder becomes, during the course of evolution, modified to form a hydrostatic organ, so that the fishes are enabled to adjust their specific gravity to that of the surrounding water.

ORDER: PALAEONISCOIDEI.—Actinopterygians ranging from the Devonian to the present day, but most are extinct.—Carnivorous fishes with sharp and often large teeth.—A single dorsal fin.—Tail is usually very obviously heterocercal.—Epichordal lobe usually represented merely by a row of ridge scales.— The preopercular bone is usually arched over the hinder end of the maxilla.— Paired fins fan-like.—Scales are ganoid.—Eyes are large.—Extinct "Ganoids". —Numerous families.

The Birchirs of the Nile and its tributaries.

*Cheirolepis, Polypterus.*

ORDER: CHONDROSTEI.—The Sturgeons and their allies.

*Acipenser.*

ORDER: HOLOSTEI.—Extinct "Ganoids" and also the modern Bow Fin and Bony Pike.

*Amia, Lepisosteus.*

SUBORDER: TELEOSTEI.—The vast majority of modern Bony Fishes.—The skeleton has undergone considerable reduction during the evolution of this group.—The scales are thin overlapping bony plates, completely covered by the epidermis.—The hinder ends of the maxilla and premaxilla are not fixed, so that they can swing forward, allowing a wide gape to the mouth.—The airbladder has a hydrostatic function.—The specific gravity approximates to the surrounding water and the tail is homocercal.—There is no spiracle.—There are only four gill slits.—The urinary and genital ducts have separate openings.— A conus arteriosus is lacking from the heart.—There is no spiral valve in the intestine.—Jurassic—recent.

The teleosts represent the highest types of actinopterygians and have become adapted to practically all types of aquatic habit. They therefore show considerable variation in form, and the group includes such diverse types as the Carps, Trout, Eels, Flying Fishes, Coffer Fishes, Sea Horses, and Flat Fishes.

Class: Amphibia.—Craniates, which during their evolution have only partially succeeded in overcoming the difficulties attending life on the land.— The adults haunt the damper places or else have become secondarily adapted to a thorough-going, aquatic life.—Except for some which are viviparous, or have special breeding habits they breed in water, for their eggs are not able to survive in arid conditions.—The skin is permeable, and amphibians cannot tolerate salt water.—The earliest known amphibians differ greatly from the modern Newts, Salamanders, Frogs, and Toads, and, in fact, in most ways they bear obvious resemblances to the osteolepid fishes. They differ from these chiefly in having pentadactyl limbs instead of fins and in having a middle ear apparatus.—Modern amphibians are characterised by having a smooth, moist skin rich in gland cells.—The skin serves as a respiratory surface and is plentifully supplied with blood vessels.—Scales (with the exception of the minute scales of the Apoda) are absent.—Gills are always present in the larvae and may persist into adult life.—Lungs, which open into the buccopharyngeal cavity by way of a glottis are found in the adults.—The heart is better adapted to deal with oxygenated as well as deoxygenated blood than that of any fish. There are two auricles, a single ventricle and a complicated mechanism which ensures that the two different kinds of blood are distributed in an appropriate way.—The kidney is a mesonephros.—There is a cloaca which has a voluminous diverticulum, the bladder. This is said to be homologous with the allantois of the amniotes.—The skull bears two (exoccipital) condyles.—The middle ear apparatus has a single auditory ossicle.—There is no external ear.—Lateral line organs are present in the larvae and in the adults of aquatic forms.—Fins, if present, are not supported by skeletal structures.

**Subclass: Stegocephalia.**—Extinct Amphibia first appearing in the Carboniferous.—The endoskeleton is very massive, and there is an exoskeleton of bony scales.—The skull has a complete casing of membrane bones which are deeply sculptured.

Five separate orders are recognised in this group, but none appears to be a likely ancestor of modern Amphibia.

**Subclass** (and ORDER): **Anura.**—The Frogs and the Toads.—Amphibians which lose the tail at metamorphosis.—The gill slits close in the adult.—The vertebral column is remarkable for its shortness. There are only five vertebrae in some forms.—Ribs are only rarely present.—The limbs are adapted both for swimming and for leaping. The fore limbs are short; the hind limbs are long, powerful, and have a web between the digits.

*Rana, Bufo, Discoglossus, Xenopus.*

**Subclass** (and ORDER): **Urodela.**—The Newts and the Salamanders.—Amphibia in which the tail persists throughout life and assists in progression, both on the land and in swimming.—The fore and hind limbs are relatively feebly developed and are about equal in size. They serve to keep the belly clear of the ground when the creature is moving on land.—The vertebral column is long and there may be as many as ninety or so vertebrae. These are usually of the opisthocoelous type.—Both external and internal gills may persist in the adult and if they do, then the lungs tend to atrophy.—The skull is, as a rule, poorly ossified.

*Cryptobranchus, Proteus, Triton, Ambystoma, Salamandra.*

**Subclass** (and ORDER): **Apoda (Gymnophiona).**—Limbless burrowing Amphibia.—Girdles are absent.—The tail is practically absent, so that the anus becomes sub-terminal.—The eyes are covered with opaque skin, are small, and practically functionless.—The middle ear apparatus is lacking.—The body is marked externally by rings, which add to the worm-like appearance.—The head bears sensory tentacles.—Minute calcified scales are present, sunk deeply into the skin.—The larval stage may be omitted when the young hatch direct from large-yolked eggs, which are laid in moist earth.

**Class: Reptilia.**—The Lizards, Crocodiles, Snakes, Turtles, and their allies, both living and extinct.—The reptiles living at the present day are but the insignificant survivors of a group which evolved towards the end of the Palaeozoic and, conquering every type of habitat, held sway throughout the Mesozoic period.

Reptiles are craniates which are completely adapted to life on dry land both as adults and in their embryonic development.—The skin is dry and bears horny epidermal scales.—Bony plates may also be present.—The lungs form the sole respiratory organs.—The division of the heart into right and left halves (in connection with pulmonary respiration) is complete or nearly so.—The aortic arches spring directly from the ventricles.—The visceral clefts never develop gills.—The kidney is a metanephros and, associated with the need for water conservation, the nitrogenous excretory matter is largely uric acid.—There is a cloaca.—There is no external ear, but the lagena is often large and coiled so that it may be dignified with the name of a cochlea.—The eggs are large-yolked and enclosed in a shell. They are laid and developed on dry land, for a larval stage is omitted.—The embryo is enclosed in embryonic membranes.—The skull is longer than in Amphibia, so that twelve nerves are intra-cranial.—The skull articulates with the vertebral column by a single condyle.—The vertebral column is usually very long and flexible.—A distinct neck region is present.—In most reptiles the outer casing of dermal bones over the cranium in incomplete owing to the presence of one or more windows (temporal vacuities) which develop to allow a greater space for the housing of the jaw muscles. The number of temporal vacuities and their relations to the dermal bones of the skull provide a basis for the classification of the group.—The lower jaw consists of several bones.—There is a single auditory ossicle.

**Subclass: Anapsida.**—Reptiles in which there are no temporal vacuities, although the bones may be emarginated from the hinder end of the skull.

ORDER: COTYLOSAURIA.—A group of small primitive reptiles which evolved in the Carboniferous and died out in the Triassic.—The cotylosaurs were probably a heterogeneous assemblage which included the ancestors of later groups.—In many ways they resembled the Stegocephalia, from which they inherited their heavily ossified skull.—The jaws and teeth were poorly developed. —The limbs were also relatively feeble structures and projected out laterally from the body, the upper arm and the thigh being practically parallel to the ground.—Much muscular energy is expended in resisting gravitational forces by animals having this type of limbs.

*Diadectes.*

ORDER: CHELONIA (TESTUDINATA).—The Turtles and Tortoises.—The trunk is relatively short and broad, and protected by a dorsal shield (carapace) and a ventral shield (plastron).—Typically, each shield is composed of bony plates overlain by large horny epidermal plates (tortoise shell).—The expanded and flattened neural spines and ribs contribute to the bones of the carapace, whilst the clavicles and interclavicle become separated from the rest of the pectoral girdle to form three of the bones of the plastron.—The head, neck, tail, and limbs can be withdrawn for protection beneath the shields.— There are no true temporal vacuities in the skull, but the dermal bones may be partially reduced from the rear (emargination).—The jaws are strong but lacking in teeth, which are functionally replaced by a horny bill or beak.—The quadrate is immovably fixed to the cranium.—A short bony secondary palate is developed so that the internal nares open into the posterior part of the buccal cavity.— There is no sternum.— The limbs of aquatic forms are modified to form paddles.

*Chelone, Dermochelys, Testudo.*

**Subclass: Synapsida.**—The Mammal-like Reptiles.—Also termed Therapsida or Theromorpha.—They arose in the Permian from an unknown group of cotylosaurs and after undergoing a remarkable series of evolutionary changes they eventually, in the Upper Triassic, gave origin to the Mammalia.—Very many groups of mammal-like reptiles are known, but all seem to experience similar evolutionary trends which result in the evolution of creatures bearing resemblances to the mammals themselves (parallel evolution). It is therefore difficult to name many characters which apply to all the evolutionary stages of the group.

The skull has a single temporal vacuity which is bounded by the post-orbital, jugal, squamosal, and parietal bones.—The cranial cavity is relatively large.— The auditory capsule lies towards the floor of the cranium.—The palate is high and vaulted.—There are two coracoid elements in the pectoral girdle, a precoracoid and a coracoid.

The most primitive members of the subclass in general form resemble the cotylosaurs from which they arose but, with time, others arise which show an advance in structure towards mammal-like animals.—The body becomes more compact and the limbs rotate beneath the body so that they carry its weight vertically. The progression may then be described as running or striding.— The cranial cavity increases in size to accommodate the larger brain, an enlargement made possible by the ventral position of the ear.—The teeth enlarge and certain become specialised so that canines, carnassials, and the like become recognisable.—The jaws become deeper and stronger.—A secondary palate develops and allows of retention of food in the mouth where it is dealt with by the heterodont dentition.—The dentary bone increases in size and importance, whilst the bones at the rear of the lower jaw diminish in size and may become loosened from it. In some forms the dentary articulates with the squamosal. Thus the mammalian arrangement is foreshadowed.—It is possible that the higher, mammal-like reptiles had hair and were warm-blooded.

*Dimetrodon, Cynognathus.*

**Subclass: Parapsida.**—Other extinct reptiles which are worthy of mention because they show such perfect adaptation to aquatic life are the Ichthyosaurs and the Plesiosaurs. They are difficult to classify, for their anatomy shows many specialisations and their affinities do not seem to lie with any known group. They may provisionally be placed in two separate orders.

ORDER: ICHTHYOPTERYGIA.—The Ichthyosaurs.—Reptiles adapted to life on the high seas.—The body is streamlined.—The facial region is elongated and the jaws are armed with very numerous teeth.—The limbs are paddle-like.— There is a large vertical tail fin.—The skull has a single temporal vacuity.

ORDER: SAUROPTERYGIA.—The Plesiosaurs.—Like the Ichthyosaurs they are adapted to marine or estuarine life.—The small head is borne on a very long flexible neck.—The fore limbs form large paddles used for swimming.— The hind limbs are similar but smaller.—The tail is short and bears no fin.— The skull has a single temporal vacuity.

**Subclass: Diapsida.**—All the familiar reptiles are included in this subclass, together with practically all the extinct groups.—The skull has two temporal vacuities, an upper and a lower, but in some groups the lower margin of the lower vacuity is incomplete.—The pectoral girdle has a single coracoid element.— The ear lies fairly high up on the cranium.—Diapsid reptiles have had such a remarkable evolutionary history and have become adapted to so many different ways of life that it is difficult to generalise about their characters. The features of the various subsidiary groups will therefore be dealt with separately.

SUPER-ORDER: LEPIDOSAURIA.

ORDER: THECODONTIA.—Primitive and little-understood extinct diapsids.

ORDER: RHYNCHOCEPHALIA.—Extinct Lizard-like Diapsids and the modern Tuatara of New Zealand.—The skull retains the boundaries of both upper and

The upper vacuity is bounded by the ⸺
⸺ones; the lower, by the post-orbital, squa⸺
⸺d to the jaws.—The pineal eye is a photore⸺
*Sphenodon.*

⸺—The Lizards and Snakes.—The lower border of t⸺
⸺vacuity is incomplete owing to a reduction of the processes o⸺
⸺nd squamosal.—The body is covered by horny scales.—Animals⸺
⸺ted to life on dry land in warmer climates.

SUBORDER: LACERTILIA.—Lizards and their relatives.—The tail is very long as a rule.—The rami of the lower jaw, although usually slender, are united by suture so that the gape of the mouth is not excessively large.—Although limbless lizards are known, they always show some trace of girdles.
*Lacerta, Iguana, Anguis.*

SUBORDER: OPHIDIA.—The Snakes.—The long slender shape is due, not to the presence of a long tail, but to an elongation of the body.—Limbs and (with the exception of a rudiment of a pelvic girdle in some, *e.g.* Pythons) limb girdles are absent.—The mouth is very extensible and even the upper jaw can move relative to the cranium.—The middle ear apparatus is lacking.—The tongue is sensory.—Some snakes have poison fangs.—The vertebrae are procoelous and bear extra articulating facets.—The lower border of both temporal vacuities has vanished.
*Tropidonotus, Pelias, Boa.*

SUPER-ORDER: ARCHOSAURIA.

ORDER: SAURISCHIA.—Dinosaurs.—Mostly very large animals, now extinct. —Carnivorous and herbivorous types are known.—Limbs and feet and girdles of the reptilian type.
*Brontosaurus, Diplodocus.*

ORDER: ORNITHISCHIA.—Dinosaurs.—Often very large extinct animals.— All are herbivorous.—Limbs and girdles showing certain affinities with those of Birds.—Fore limbs short; hind limbs long and powerful.
*Iguanodon, Stegosaurus, Triceratops.*

ORDER: PTEROSAURIA.—Extinct flying reptiles, the Pterodactyls.—Wings supported by the fore limbs.—The fourth digit is greatly elongated; the fifth is absent.
*Pteranodon, Rhamphorhynchus.*

ORDER: PSEUDOSUCHIA.—Extinct, lizard-like creatures with a two-arched skull.

ORDER: PHYTOSAURIA.—Extinct, lizard-like forms.

ORDER: CROCODILIA.—The Alligators and Crocodiles.—Large reptiles adapted to life in rivers and lakes.—The skin is armoured with horny and bony scales.—The tail is flattened from side to side and is a powerful swimming organ.—The skull has two temporal vacuities and is very heavily ossified.—The jaws are beset with numerous strong teeth.—There is an extensive secondary

~~ich is formed almost entirely of bone.—The internal nares open
~ of the pharynx and the opening into the oesophagus is guarded by
~rial epiglottis.—The external nares are on the dorsal surface of the
~ated snout.—The heart shows a complete division of the ventricular
~n into right and left halves.—There is a muscular but incomplete diaphragm
~tially separating the abdominal from the thoracic cavity.

The Crocodiles and Alligators are, because of the position of their internal
and external nares, able to swim practically submerged in the water, only the
tip of the snout projecting above the surface of the water.

*Crocodilus, Alligator, Gavialis.*

**Class: Aves.**—The Birds.—Craniates with such marked affinities to the
Reptilia that they have been aptly described as "glorified reptiles".—They
probably evolved from some group of arboreal reptiles related to the primitive
ornithischian dinosaurs, but they have become so adapted to flying that in
many ways they are now the most specialised of all craniate classes.—The
body is provided with a covering of feathers which are not only concerned
with flight, but also make possible the maintenance of the high body temperature.—The fore limbs are modified to form the wings. There are only three
digits: the first, second, and third. The second is the longest. The three
metacarpals are fused into a single mass. There are only two carpals remaining
separate; the rest fuse to the metacarpals.—The pectoral girdle is strong and the
sternum large and provided (except in some flightless birds) with a keel for
the attachment of the large pectoral muscles which are concerned with moving
the wings.—The hind limbs are large, strong, and in them there has been considerable fusion of bones.—The pelvic girdle is large and attached to several
sacral vertebrae which are fused together. Thus provision is made for the
stresses resulting from the bipedal method of walking and running.—Horny
epidermal scales are found on the legs and feet.—The skull has many reptilian
features, but the cranial cavity is large and the component bones fuse up very
early on in life.—The jaws are toothless in all modern birds and the jaws are
ensheathed in a horny covering—the beak.—The lower jaw, which is composed
of several bones, articulates with the quadrate.—There is a single auditory
ossicle.—The skull is really of the diapsid type, but the lower border of the upper
temporal vacuity has vanished so that the two vacuities have become confluent.—
All the bones are very light but strong, and many contain spaces into which
project air-sacs—thin-walled expansions from the lungs. The air-sacs assist
in maintaining a constant draught of oxygenated air through the lungs.—The
heart and the rest of the vascular system follow a modification of the reptilian
plan, but there is a single carotico-systematic arch—the right one.—The brain
is large, especially certain regions of it. The so-called cerebral hemispheres
are not really the equivalent of those structures in the mammal, for they derived
from the corpus striatum (sides and floor of the telencephalon). The two optic
lobes are enormous, for the sense of sight in birds is very acute.—Birds not only
make provision for their young by laying large-yolked eggs protected by a shell,
but both parents show a complicated behaviour pattern designed to ensure
the safety of the eggs and young after they have hatched. This "parental
care" exceeds in degree that of many mammals.

**Subclass: Archaeornithes.**—Extinct birds from the Upper Jurassic.—Among
the many primitive features may be mentioned the presence of teeth on the
jaws; the presence of claws on the wing digits; the tail, which had many separate
vertebrae; the metacarpals were not fused together.

*Archaeopteryx.*

**Subclass: Neornithes.**—Birds which have departed further from the reptilian stock than the Archaeornithes.—The metacarpals are fused.—The tail is short.—Teeth are present only in a few extinct forms.—Claws are only rarely present on the wing digits.

ORDER: ODONTOGNATHAE.—Extinct (Cretaceous) birds adapted for life on or near the water.—The jaws bore numerous sharp teeth.—Wings were poorly developed, but the legs were strong and used for swimming.—Most of the vertebrae were biconcave.

*Hesperornis.*

ORDER: RATITAE (PALAEOGNATHAE).—The large Running-birds, the Ostrich, the Emu, and their relatives.—Most are very large and heavy—a fact that may account for their inability to fly. The wings are small and the sternum is not provided with a keel.—The legs are extremely powerful.—Clavicles are small or absent.—The feathers lack barbs.

*Struthio, Dromaeus, Rhea, Apteryx.*

ORDER: CARINATAE (NEOGNATHAE).—All the more familiar birds whose chief features have been set out as being typical of the class. They differ from the Ratitae chiefly by those features which fit them for flight, but there are other differences including certain features of the skull.

**Class: Mammalia.**—The Mammals.—Craniates which have attained the most complete structural, developmental, and physiological adaptation to terrestrial life.—A few have successfully reconquered the sea and fresh water and have become modified accordingly.—Mammals are characterised by their great activity, high rate of metabolism, "intelligence", and by their display of parental care, features in which they are rivalled only by the birds.—The young are nourished by milk, a secretion of modified skin glands of the mother. —Typically, the skin is covered by a coat of hair which acts as a heat insulating layer and allows the high body temperature to be maintained.—In many aquatic mammals, where hair may be practically absent, heat insulation is provided by layers of subcutaneous fat (blubber).—The skin has two kinds of glands: sweat glands and sebaceous glands. The former produce a watery fluid by evaporation of which a control is effected over the body temperature. The latter secrete a greasy substance which helps to maintain the hair in a water-resisting condition.—The brain is remarkable for the large cerebral hemispheres and the large and complicated cerebellum. There are four optic lobes.—The senses of sight and hearing are acute.—The membranous labyrinth has a complex cochlea.—There are three auditory ossicles.—There is an external ear.—The skull has many interesting features, some of which are shared by the higher mammal-like reptiles. The cranial cavity is large.—The lower jaw consists of one bone only, the dentary, which articulates with the squamosal.—Turbinal bones covered by mucous membrane (some of which is olfactory) are found in the nasal cavity.—There is an extensive secondary palate which is largely bony. It shuts off a dorsal air passage from the rest of the buccal cavity—an arrangement which allows of the retention of food in the mouth. The internal nares open at the extreme hinder end of the pharyngeal region.—An epiglottis and a larynx are present.—The dentition is heterodont and diphyodont. The teeth are firmly set in sockets by cement.

The heart has two auricles and two ventricles, but the sinus venosus is reduced to a mere vestige, the sinuauricular node.—There is only one caROTICO-systemic arch, the left—an interesting point which is confirmatory evidence for the view that mammals arose from a reptilian stock quite distinct from that which gave origin to all the other reptiles and the birds.—A muscular diaphragm completely separates a thoracic from an abdominal cavity.—The ribs assist in respiratory movements.—The long bones and the vertebrae ossify from three centres; an arrangement allowing for bone growth over long periods of time. The ossifications at the ends of the bones are termed epiphyses.— The cloacal region of the embryo becomes (except in the Monotremes) divided into two chambers by a partition, the perineum, so that the urinogenital aperture is separated from the anus.—As in the mammal-like reptiles there is a precoracoid as well as a coracoid, but in the higher mammals both of these bones become reduced to mere projections on the scapula.—Reptilian-like scales are sometimes found; on the rat's tail, for example.

Fossil mammal material is remarkably abundant but, except for certain parts of the body, the animals are very badly preserved and the various groups are distinguished chiefly on the nature of the limb bones and the teeth. In general, it may be said that the tooth pattern was simpler in the more primitive than in the later evolved mammals.

The extinct mammalian orders will not be dealt with, for they form a very specialised branch of palaeontology. Suffice it to say that any comprehensive scheme of classification would, of course, take them into account.

The surviving members of the class are numerous and are the dominant land craniates of to-day. They fall into three distinct subclasses.

**Subclass: Prototheria.**—The Duckmole (Duck-billed Platypus) and the Spiny Anteater.

ORDER: MONOTREMATA.—Mammals which show greater evidence of their reptilian descent than any other.—The pectoral girdle is of the type found in all mammal-like reptiles. A precoracoid and a coracoid are present as separate bones. There is a large T-shaped interclavicle behind the clavicles.—The rectum and urinogenital ducts open into a cloaca.—The brain is relatively simple and there is no corpus callosum.—Large-yolked eggs are laid.—There are no mammae, but the young are nourished after hatching by a secretion which is poured into a depression on the abdomen. The secretion may be termed milk, but it is secreted not by modified sebaceous, but by modified sudorific glands.— The body temperature is comparatively low (25° to 28° C. as compared with 35° to 40° C. in the higher mammals).—True teeth are absent in the adults.

*Ornithorhynchus, Echidna.*

**Subclass: Metatheria.**—The Marsupials.

ORDER: MARSUPIALIA—Mammals, which, although more advanced than the monotremes, yet are primitive in most respects when compared to the eutherian mammals.—All are viviparous, but the young are born in a very immature state and migrate into a pouch (marsupium) on the abdomen of the mother, where they are nourished by milk.—The milk glands are modified sebaceous glands and the milk is passed into the mouth of the young through mammae.—The placenta is only rarely an allantoic placenta, and then only of a simple type.—On the whole their skeleton is like that of higher mammals, but there are certain peculiarities, notably in the skull and dentition. The cranial cavity is relatively small for a mammal. The palate often has vacuities. The angle of the lower jaw is bent inwards towards the mid-line. The tympanic bulla, if present, is formed by the alisphenoid. The canal for the external carotid pierces the basisphenoid and not the alisphenoid. More than $\frac{3}{3}$ incisor teeth may be present.—A corpus callosum is present only in a few species.

Marsupials flourish only in the Australasian regions, but Opossums and a small rat-like marsupial are found in South America. Elsewhere they are absent. The degree of adaptive radiation shown by the group rivals that of the higher mammals, for they have become adapted to fill practically all the niches available to a mammal.

The fact that marsupials originated so long ago and soon underwent an extensive radiation makes it difficult to subdivide them into "natural" categories. It is sometimes convenient to separate the carnivorous ones, polyprotodonts (with more than three incisors in each jaw, above and below) from the herbivorous ones, diprotodonts (with never more than three incisors in each jaw, above and below, and often having the second and third digits bound together by tissue). This character gives the following grouping but nowadays marsupials are arranged merely into super-families.

DIDACTYLA.—The Opossums, Native Cats, Tasmanian Wolf, and Marsupial Mole.—Carnivorous and insectivorous marsupials.—All the digits are free.— There are usually more than $\frac{3}{3}$ incisors (polyprotodont dentition).—The pouch, if present, opens ventrally or posteriorly.

*Didelphys, Dasyurus, Thylacinus, Notoryctes.*

SYNDACTYLA.—The second and third toes are bound together in a common mass of tissue.—The pouch opens forwards, as a rule.—A few (Peramelidae— the Bandicoots) are carnivorous and polyprotodont. The remainder— Kangaroos, Native Bears, Phalangers, and the like are herbivorous and have fewer than $\frac{3}{3}$ incisors (diprotodont condition).

*Perameles, Macropus, Phascolarctos, Phalanger.*

Subclass: Eutheria.—The "higher mammals".—All the better-known mammals belong here.—The young are born as miniature adults after a prolonged period of gestation.—The allantoic placenta reaches a very high degree of complication.—Mammae and mammary glands are present.—The brain is highly organised and there is always a corpus callosum.—The skull and dentition have certain features which are peculiar to the group. Only rarely are there palatal vacuities. The angle of the lower jaw is not inflected. The tympanic bulla, if present, is formed by the tympanic bone. The carotid canal pierces the alisphenoid.—There are never more than $\frac{3}{3}$ incisors.

Including fossil forms, over twenty orders of eutherian mammals are usually recognised. Only certain of these will be mentioned. As far as possible the genera mentioned are British and are living to-day, since a proper treatment would occupy too much space.

ORDER: INSECTIVORA.—Small mammals whose diet is typically insectivorous and whose teeth are adapted accordingly.—Familiar British examples are the Mole, the Hedgehog, and the Shrews.—The teeth are small and have pointed cusps.—Clavicles are usually present.—The zygomatic arch is weak and the skull rarely bears pronounced ridges for muscle attachments.—The intestine is short; the caecum, small.

*Talpa, Erinaceus, Sorex.*

ORDER: RODENTIA.—The Rodents or gnawing animals: Rats and Mice, Porcupines and Beavers.—The incisors are chisel-like and used for gnawing, and grow from persistent pulps.—Canines are absent.—The cheek teeth are adapted for grinding.—The diet is herbivorous or omnivorous.—The intestine is long and the caecum large.—Clavicles slender, but only rarely absent.

*Mus, Hystrix, Castor.*

ORDER: LAGOMORPHA.—The Hares and Rabbits, distinguished from the rodents chiefly by the possession of a second pair of upper incisors immediately behind the first pair. This order was distinct from the Rodentia as far back as the basal Eocene.

*Oryctolagus, Lepus.*

ORDER: CARNIVORA.—Carnivores: the Cats and the Dogs; the Bears, the Otters, and Martens; the Seals.—The diet is almost always the flesh of other craniate animals.—The teeth are strong and deep-rooted.—The canines and carnassials are prominent.—The zygomatic arch is strong and stands well out from the cranium to allow for the powerful jaw muscles.—The cranium usually bears ridges.—The clavicles are usually absent.—The gut is short and the caecum small or absent.

*Felis, Canis, Ursus, Lutra, Mustela, Phoca.*

ORDER: ARTIODACTYLA.—The Even-toed Ungulates: Cattle and Sheep; Pigs, Deer, and Camels.—Herbivorous mammals usually of large size.—They are specialised to deal with large quantities of vegetable food.—The incisors and canines are of variable nature.—The cheek teeth are adapted for grinding and are either bunodont (hummocky cusps) or else selenodont (having ridge-like cusps parallel to the long axis of the jaw).—The intestine is fairly short, but the stomach is complicated and often has several chambers.—The caecum is small.—Some Artiodactyla chew the cud.—The limbs, adapted for running, show a considerable fusion of bones and the digits are reduced in number.— The axis of symmetry of the feet passes between the third and fourth digits, which are equal in size and bear hooves (cloven hoof).

*Bos, Ovis, Sus, Cervus, Camelus, Hippopotamus.*

ORDER: CETACEA.—The Whales and the Dolphins.—Aquatic mammals having a fish-like general appearance.—The body is streamlined and devoid of hair.—There is a thick layer of blubber.—The cervical vertebrae are reduced to thin plates so that a distinct neck region is not formed.—The hind limbs are absent and the fore limbs modified to form paddles.—The end of the body is expanded to form a flattened tail with horizontal "flukes".—There is an internarial epiglottis.—Networks of small blood vessels serve to store oxygenated blood when the creatures dive from the surface.—In the Toothed Whales the teeth are very numerous and all more or less alike, but in the Whalebone Whales all teeth are lost in the adult.—The nostrils are dorsal.

*Balaena, Physeter.*

ORDER: PERISSODACTYLA.—The Odd-toed Ungulates.—Horses, Tapirs, and Rhinoceroses.—Herbivorous, usually of considerable size.—The intestine is long and there is a large caecum.—The stomach remains in a simple state.— The teeth are adapted for grinding, and in most, the cusps of the cheek teeth are united to form ridges or lophs, which often give a complicated pattern to the crown of the tooth. The premolars resemble the molars. The ridges run transverse to the long axis of the jaws.—The limbs show a considerable fusion of bones and the digits are usually reduced in number.—The axis of

symmetry of the feet is through the third digit, which is always the largest.—
The digits bear hooves.

*Equus, Tapirus, Rhinoceros.*

ORDER: HYRACOIDEA.—The biblical Conies.—Small rabbit-like creatures
having certain affinities with the ungulates.

*Hyrax.*

ORDER: PROBOSCIDEA.—The Elephants.—Large animals of the ungulate
type.—The skull is very specialised and the dentition unique.—The two upper
incisors are of great size and form tusks. Canines and premolars are lacking.
Only two or three molars find room enough in the jaws at one time, but as they
wear out others move along from rear to front to take their place. The tooth
pattern is complicated and is made up of ridges transverse to the long axis of
the jaw (lophodont teeth).—The brain cavity is relatively small but the skull
bones remarkably thick.—The stomach is simple, the intestine is long, and the
caecum large.—The limbs are massive and the limb bones but little modified.

*Elephas.*

ORDER: SIRENIA.—The Sea Cows.—Large herbivorous animals adapted to
life in estuarine waters or shallow sea.—They show certain resemblances to the
Elephants.

*Halicore, Manatus.*

ORDER: CHIROPTERA.—The Bats.—Small mammals with strong powers of
flight.—The fore limbs are modified to form wings. The limb bones are long
and slender and, except for the first, the digits are also elongated. The web of
the wing (patagium) is formed of skin.—The hind limbs are small and are
sometimes partly attached to the patagium.—The teeth bear pointed cusps.—
Some bats are frugivorous; others are insectivorous.

*Pteropus, Vespertilio.*

ORDER: DERMOPTERA.—The "flying lemur".—Related to insectivores but
with a patagium, for gliding.—Fingers not elongated.—Pectinate lower incisors.

ORDER: PRIMATES.—Lemurs and their relatives, Monkeys and Apes.

SUBORDER: LEMUROIDEA.—The Lemurs.—Animals related to the Monkeys,
but they are more primitive in most respects.—They are adapted for arboreal
life.—The general appearance is fox-like.

*Cheiromys, Lemur.*

SUBORDER: ANTHROPOIDEA.—The Monkeys and the Apes.—The brain is the
most highly developed of any mammals.—The cranial cavity is large.—The eyes
face forwards and the orbits are completely separated from the temporal vacuity
by a bony partition.—The limbs remain in a primitive condition, there being
but little fusion of bones.—The digits bear nails.—The cheek teeth are bunodont.

The Anthropoidea fall into two well-defined groups—the Platyrrhina and
the Catarrhina. It is to the latter that Man himself is related.

INFRAORDER: PLATYRRHINA.—The Monkeys of the New World.—The nostrils
face forwards.—The internasal septum is broad.—There is no bony external
auditory meatus.—The tail is often prehensile.—The jugal meets the parietal.—
There are three premolars.

*Hapale, Mycetes.*

INFRAORDER: CATARRHINA.—The Old World Monkeys and Apes.—The nostrils face downwards.—The internasal septum is narrow.—There is a bony external auditory meatus but no tympanic bulla.—The tail is never prehensile.—The alisphenoid separates the jugal from the parietal.—Cheek pouches and ischial callosities are often present.

*Macacus, Simia, Pan, Gorilla, Australopithecus, Pithecanthropus, Homo.*

# SELECTED FURTHER READING

CHAPTER I

J. A. Ramsay, 1965. *The Experimental Basis of Modern Biology.* Cambridge University Press.

J. Riegel, 1965. *Energy, Life, and Animal Organisation.* English Universities Press.

CHAPTER II

E. H. Mercer, 1961. *Cells and Cell Structure.* Hutchinson Educational.

A. B. Novikoff and E. Holtzman, 1970. *Cells and Organelles.* Holt, Rinehart, and Winston.

CHAPTER III

A. J. Cain, 1966. *Animal Species and Evolution.* Hutchinson University Library.

G. G. Simpson, 1961. *The Principles of Taxonomy.* Oxford University Press.

CHAPTER IV

G. H. Beale, 1954. *The Genetics of Paremecium Aurelia.* Cambridge University Press.

D. L. Mackinnon and R. S. J. Hawes, 1961. *An Introduction to the Study of Protozoa.* Oxford, Clarendon Press.

D. R. Pitelka, 1963. *Electron Microscopic Structure of Protozoa.* Pergamon.

H. Sandon, 1963. *Essays on Protozoology.* Hutchinson Educational.

K. Vickerman and F. E. G. Cox, 1967. *The Protozoa.* John Murray.

CHAPTER VI

L. H. Hyman, 1940. *The Invertebrates. Vol. I. Protozoa through Ctenophora.* McGraw-Hill.

H. M. Lenhoff and W. F. Loomis, 1961. *The Biology of Hydra and of Some Other Coelenterates.* University of Miami Press.

CHAPTER VII

R. B. Clark, 1964. *Dynamics in Metazoan Evolution.* Oxford, Clarendon Press.

CHAPTER VIII

L. H. Hyman, 1951. *The Invertebrates. Vol. II. Platyhelminthes and Rhyncocoela. The Acoelomate Bilateria.* McGraw-Hill.

883

CHAPTER IX
L. H. Hyman, 1951. *The Invertebrates. Vol. III. Acantho-cephala, Ascelminthes and Entoprocta. The Pseudocoleomate Bilateria.* McGraw-Hill.
D. L. Lee, 1965. *The Physiology of Nematodes.* Oliver and Boyd: University Reviews in Biology.

CHAPTER X
M. Rothschild and T. Clay, 1952. *Fleas, Flukes, and Cuckoos.* Collins: New Naturalist.
J. D. Smyth, 1962. *Introduction to Animal Parasitology.* English Universities Press.

CHAPTER XI
R. B. Clark, 1964. *Dynamics in Metazoan Evolution.* Oxford, Clarendon Press.

CHAPTER XII
R. P. Dales, 1967. *Annelids.* Hutchinson University Library.

CHAPTER XIII
L. A. Borradaile, F. A. Potts, L. E. S. Eastham, and J. T. Saunders (rev. G. A. Kerkut), 1961. *The Invertebrata.* Cambridge University Press.
J. D. Carthy, 1965. *The Behaviour of Arthropods.* Oliver and Boyd: University Reviews in Biology.
T. H. Waterman, 1960 and 1961. *The Physiology of Crustacea.* Vols. I and II. Academic Press.
V. B. Wigglesworth, 1964. *The Natural History of Insects.* Weidenfeld and Nicholson.
V. B. Wigglesworth, 1965. *Principles of Insect Physiology.* Methuen.

CHAPTER XIV
V. Fretter and A. Graham, 1963. *British Prosobranch Molluscs.* Ray Society.
E. Step (rev. by A. L. Wells), 1945. *Shell Life.* Warne & Co.

CHAPTER XV
J. Z. Young, 1962. *The Life of Vertebrates.* Oxford University Press.

CHAPTER XVI
E. J. W. Barrington, 1965. *The Biology of Hemichordata and Protochordata.* Oliver and Boyd: University Reviews in Biology.

CHAPTER XVII

E. J. W. Barrington, 1963. *An Introduction to General and Comparative Endocrinology.* Oxford University Press.

J. D. Carthy, 1965. *Animal Behaviour.* Aldus Books.

H. S. D. Garven, 1957. *A Student's Histology.* Livingston.

G. M. Hughes, 1963. *Comparative Physiology of Vertebrate Respiration.* Heinemann.

J. Gray, 1959. *Animal Locomotion.* Pelican Books.

F. R. Jevons, 1965. *The Biochemical Approach to Life.* Unwin University Books.

A. M. Lockwood, 1964. *Animal Body Fluids and their Regulation.* Heinemann.

K. Schmidt-Nielsen, 1970. *Animal Physiology.* Prentice Hall Foundations of Modern Biology Series.

W. B. Yapp, 1960. *An Introduction to Animal Physiology.* Oxford University Press.

J. Z. Young, 1957. *The Life of Mammals.* Oxford University Press.

CHAPTERS XVIII AND XIX

W. H. Freeman and B. Bracegirdle, 1966. *An Atlas of Embryology.* Heinemann.

O. E. Nelson, 1953. *Comparative Embryology of the Vertebrates.* Blakiston.

P. Weiss, 1939. *The Principles of Development.* Holt.

CHAPTER XX

R. P. Levine, 1962. *Genetics.* Holt, Rinehart, and Winston, Modern Biology Series.

CHAPTER XXI

A. J. Cain, 1966. *Animal Species and their Evolution.* Hutchinson University Library.

P. M. Sheppard, 1967. *Natural Selection and Heredity.* Hutchinson.

G. G. Simpson, 1949. *The Meaning of Evolution.* Yale University Press.

J. Maynard Smith, 1958. *The Theory of Evolution.* Pelican Books.

CHAPTER XXII

J. Hillaby, 1960. *Nature and Man.* Phoenix House.

A. Macfadyen, 1970. *Animal Ecology.* Pitman.

E. P. Odum, 1963. *Ecology.* Holt, Rinehart, and Winston, Modern Biology Series.

J. D. Carthy, 1963. *Animal Navigation.* Unwin Books.

The Institute of Biology is sponsoring a series of booklets, "Studies in Biology", published in collaboration with Edward Arnold.   New ones are continually being added.

# INDEX

ABDOMEN, 272, 315, 403
— rabbit, 403
Abdominal appendages, 282
— ganglia, 327, 355
Abducens, 583, 586
Abductor muscle, 283
Absorption, 2, 298, 507
Accelerometer, 627
Acceptor cell, 34
Accessory heart, 292
— glands, 484
Accommodation, 610, 611, 615
Acetabularia, 36
Acetabulum, 450, 456, 457, 458
Acetylcholine, 472, 477, 601, 603
Aciculum, 224
Acidophils, 536
Acini, 498
Acoelomata, 158-61, 215
Acoelous, 423
Acorn barnacles, 312
Acoustico-lateralis system, 574, 619
Acrania, 366-89, 865
Acromion, 457
— process, 457
Acrosome, 673, 682
Actin, 474, 476
Actinopterygii, 870
Action potential, 477, 601
Actomyosin, 476
Adaptive feature, 387
— radiation, 820, 842
Addison's disease, 665
Adductor femoris, 466, 467
— muscle, 283
Adenine, 16, 27
Adenohypophysis, 657, 658
— hormones, 658
Adenosine diphosphate, 476, 548, 549
— triphosphate (ATP), 42, 476, 547, 549
Adipose tissue, 399
A-disc, 473
Ad-radial tentacles, 151
Adrenal body—anatomy and development, 663

Adrenal body—physiology, 664
— cortex, 663
— medulla, 663
Adrenalin, 472, 602, 664, 671
Adrenocortical tissue, 663
Adrenocorticotropic hormone (ACTH), 658
Aestivation, 356
Afferent arterioles, 652
— branchial arteries, 516, 519
— — vessel, 376, 551, 552
— cell, 232
— fibres, 124, 232, 592
— nephridial vessels, 244
— nerve-fibres, 160
— nerve-root, 379
— neurons, 477, 572
After-birth, 778
Agglutination, 538
Agglutinin, 538
Agnatha, 390, 867
Air-bladder, 628
Air-sac, 555, 558
Air-space, 733
Alae cordis, 291
Alanine, 21
Alary muscles, 323
Albumen, 255, 265, 354, 641, 645, 690, 705, 732, 769
— gland, 354, 357
— sac, 761
Albuminous gland cell, 239
Albumins, 535
Aldosterone, 652, 665
Alimentary canal, 215, 482
— system, 482
— — amphioxus, 369
— — Ascaris, 198
— — cockroach, 321
— — crayfish, 284-7
— — dogfish, 484
— — Fasciola, 183
— — frog, 486
— — Helix, 358
— — Lumbricus, 240
— — Nereis, 228
— — rabbit, 488
— — turbellaria, 163
Alisphenoid, 439
All or none rule, 478, 479

All-or-nothing response, 125
Allanto-chorion, 761, 775
Allantoic artery, 762
— bladder, 643
— stalk, 756
— vein, 762
Allantois, 643, 761, 764
— chick, 760
Alleles, 782, 786, 794, 797
Allelomorphs, 782
Alternation of generations, 57, 155
Alveolar membrane, 443
Alveoli, 203, 305, 443, 558
Amines, 296
Amino acids, 21, 36, 77, 499, 506
Amino acid code, 23
— — derivatives, 655
— — peptidases, 507
Amitotic division, 91
Ammocoete, 387
— larva, 371, 383
Ammonia, 260
Ammonium carbonate, 296
Amnio-cardiac vesicle, 754
Amnion, 758, 764, 774
— chick, 758
— rabbit, 773
Amniota, 621, 635
Amniotic cavity, 758, 760
— folds, 773
Amoeba, behaviour, 116
— digestion, 74
— encystment, 76
— form, 71
— growth and reproduction, 76
— locomotion, 72
— nutrition, 74
— osmoregulation, 75
— respiration, 75
Amoebic dysentery, 105
Amoeboid movement, 73
Amoebulae, 76, 211
Amphibia, 871
— ear, 623
Amphicoelous, 420, 423
Amphids, 201
Amphineura, 357, 864
Amphioxus, 220
— alimentary canal, 369

Amphioxus, anterior gut diverticula, 701
— atrium, 373
— blastula, 694
— blood vascular system, 375
— body cavity, 373
— — wall, 367
— cleavage, 694
— digestion and absorption, 386
— egg, 691
— excretion, 387
— excretory system, 377
— external features, 366
— feeding, 383
— fertilisation, 693
— gaseous interchange, 386
— gastrulation, 695
— habits, 383
— hatching, 703
— larval development, 703
— nervous system, 378
— neural tube, 698
— notochord and mesoderm, 700
— orientation of the egg, 692
— reproduction, 387
— reproductive system, 381
— sense organs, 380
— skeletal system, 368
Ampulla, 622
Ampullary sense organs, 621
Amylase, 287, 505
Amylolytic enzymes, 503
Anadromous forms, 845
Anaerobic glycolysis, 548
— respiration, 203
Anal cirri, 225
— plate, 761
— pore, 87
— styles, 316
Analogous structures, 814
Anaphase, 679
Anapophysis, 425
Anapsida, 873
Anatomy, 1
Androgens, 668
Angiotensin, 652
Angiotensinogen, 652
Angstrom, 15
Angulosplenial, 437
Animal hemisphere, 705, 709, 738
— pole, 687, 766
Anion, 476

Anisogamy, 104
Ankle, 449
Annelida, 220, 223, 267, 857
Anopheles, 99
Anopheles gambiae, 103
Antenna, 276, 277, 299, 315, 317, 339
Antennal socket, 317
Antennary gland, 296
— nerve, 300
Antennule, 276, 277, 299, 303, 309, 310
Anterior abdominal vein, 526
— aorta, 292, 352
— cardinal sinus, 521
— — vein, 515, 524, 754
— cervical sympathetic ganglion, 598
— choroid plexus, 562, 564, 566, 568
— commissure, 563
— cornua, 437
— dorsal fin, 401
— frontal lobe, 568
— intercostal vein, 532
— intestinal portal, 750
— limiting groove, 744, 749
— mesenteric artery, 519, 532
— — vein, 532
— peduncle, 572
— rectus, 582
— renal tubules, 636
— sucker, 183
— testis sac, 253
— vena cava, 524, 526, 529, 532
— vitelline vein, 753
Anthozoa, 853
Antibodies, 210, 538, 540
Anti-diuretic hormone, (ADH), 652, 660
Anti-enzyme, 210, 505
Antigen, 538
Anti-thrombin, 537
Antrycide, 110, 111
Anura, 401, 872
Anus, 199, 215, 228, 229, 322, 347, 350, 352, 357, 361, 372, 403, 509, 686, 703
Aortic arches, 390, 514, 522, 528, 753
Aphetohyoidea, 868
Aphids, 335
Apical sense organ, 235
Apis mellifera—see Honey bee

Apoda, 872
Apodemes, 269, 283, 317
Appendages, 271
— dogfish, 401
— frog, 402
— rabbit, 403
Appendicular skeleton, 417
— — dogfish, 446
Appendix, 490
Apposition image, 302, 328
Apterygota, 862
Aqueduct of Sylvius, 463, 567
Aqueous humour, 607
Arachnida, 863
Arachnoid layer, 563
Arbor vitae, 572
Archaeornithes, 876
Archenteron, 559, 685, 695, 697, 712, 713, 726
Archiannelida, 858
Archipallium, 576
Arciform muscles, 263
Area opaca, 734, 735
— pellucida, 735, 739, 752
— vasculosa, 745, 752
— vitellina, 745
Areolar tissue, 398
Armadillidium, 313
Arolium, 318
Arteries, 510
Arteriole, 510
Arthrobranchs, 294
Arthrodial membrane, 268, 273, 290, 294
Arthropoda, 268, 860
— appendages, 271
— haemocoel, 271
— metamerism, 271
— nervous system, 272
Articular, 438
— bone, 632
— membranes, 316
Arytenoid, 442
— cartilage, 554, 556
Ascaris, alimentary canal, 198
— body cavity, 197
— — wall, 196
— excretory system, 199
— form of body, 195
— life-history, 202
— nervous system, 200
— nutrition, 203
— reproductive system, 201
— respiration, 203
— sense organs, 200
Asexual multiplication, 90
Assimilation, 2, 507

Association centre, 574
— cyst, 94
*Astacus*, alimentary canal, 284
— appendages, 274
— autotomy, 290
— blood vascular system, 291
— cuticle, 273
— digestion and absorption, 287
— endophragmal skeleton, 283
— excretion and osmoregulation, 296
— feeding, 287
— form of the body, 272
— growth and moulting, 289
— life-history, 305
— locomotion, 283
— metamerism, 274
— nervous system, 299
— reproductive system, 305
— respiratory system, 294
— sense organs, 301
Asteroidea, 859
Asters, 50, 684
Astragalus, 453, 455
Atlas vertebra, 423, 427
Atrial cavity, 367
Atrichous isorhizas, *see* Small glutinants, 136
Atrio-coelomic canal, 373
Atriopore, 373, 382, 387, 691
Atrium, 325, 367, 373, 513, 518, 523, 528, 755
Auditory capsule, 431, 435, 438, 439, 624, 626, 632
— nerve, 585, 588, 622, 630
— ossicles, 440, 627, 628
— placode, 631
— vesicle, 606, 631
Auricular appendices, 530
Auricle, 347, 350, 352, 518, 524, 529, 530
Auriculo-ventricular valves, 518, 524
Autoecology, 830
Autogamy, 88, 90, 91
Autonomic nervous system, 559, 597
— system, 580
Autosome, 790
Autostylic, 446
— suspension, 435
Autotrophic, *see* Holophytic

Autotrophic organisms, 830
Aves, 876
Avoiding reaction, 86
Axial skeleton, 417
— — dogfish, 419
— — frog, 421
— — rabbit, 424
Axillary sclerites, 319
Axis cylinder, 591
— vertebra, 427
Axon, 51, 591
Azygos vein, 529, 532

BACTERIA, 32
Bacteriophages, 34
Balance of nature, 840
*Balanus*, 312
Balbiani rings, 799
Baler, *see* Scaphognathite, 279
Basal body, 49, 50
— cartilage, 447
— disc, 139
— granule, 81
— — (kinetosome), 85
— metabolism, 661
— plate, 430
Basement membrane, 392
Bases cranii, 438
Basibranchial cartilage, 433
Basi-dorsals, 419
Basihyal, 432
Basilar membrane, 624, 630
Basioccipital, 438, 439
Basipodite, 276, 277, 278
Basi-pterygium, 448
Basisphenoid, 438, 439
Basi-ventrals, 419
Basophilia, 36
Basophils, 536
Behaviour, 216
— of protozoa, 116
Bell-rudiment, 154
Benign tertian fever, 101
Biceps, 473
— femoris, 467
Bicuspid valve, 531
Bilateral symmetry, 129, 163
Bile, 484, 505
— canaliculi, 497
— duct, 486, 488, 490, 497, 727
— ductule, 497
Binary fission, 89, 105, 108
Binominal nomenclatures, 67

Biological inheritance, 25
Biotic factors, 838
Bipolar nerve-cells, 608
— neuron, 590
Birth, 533
— pore, 189
Bladder, 296, 643
Bladder-worm, *see* Cysticercus
Blastocoel, 147, 235, 684, 685, 688, 694, 709, 734
Blastocyst, 767, 771
Blastoderm, 306, 689, 733, 737
Blastodisc, 689, 732, 766
Blastomere, 684, 688, 693, 694, 708, 767, 770
Blastopore, 306, 685, 691, 696, 707, 711, 713, 726
Blastostyle, 148, 154, 155
Blastula, 147, 154, 684, 695, 734
Blastulation, 62
*Blatta orientalis*, 315
Blepharoplast, 112
Blind spot, 609
Blood, 215, 326, 398
— agglutination, 538
— chloride shift, 545
— clotting, 537
— corpuscles, 534
— group, 538
— histology and physiology, 533
— island, 745
— loading tension, 543
— platelets, 534, 537
— proteins, 535
— vascular system, 122, 215, 510
— vessels, 510
Body, 278
— cavity, 481
— of the hyoid, 437, 442
— shape, 542
— wall muscles, 217
Bohr effect, 543
Bolus, 503
Bone, 470
— histology, 460
— lamellae, 462
Botany, 1
Bowman's capsule, 220, 394, 633
Brachial plexus, 580
— vein, 526
Brachiopoda, 362, 857

Bradyodonti, 869
Brain, 160, 222, 390, 559, 560
— cerebral membranes, 563
— commissures, 563
— dogfish, 564
— evolution, 572
— flexure, 563
— frog, 566
— rabbit, 567
— ventricles, 562
Branchial arch, 432, 433, 549
— chamber, 274
— cleft, 364, 432, 484, 542, 549
— pouch, 727
— sac, 371
Branchialis branch, 588
Branchio-cardiac groove, 274, 291
Branchiopoda, 309, 860
Branchiostegite, 274, 294
*Branchiostoma lanceolatum, see* Amphioxus, 366
Branchiura, 861
Breaking plane, 291
Bristle-worm, 223
Bronchi, 395, 557
Bronchioles, 558
Brown bodies, 260
— funnel, 373
— tube, 260
Brunner's glands, 490, 495
Brush-border, 199
Buccal branch, 586
— cavity, 228, 241, 252, 345, 350, 482, 484, 726
— cirri, 385
— mass, 350, 360
Bucco-pharyngeal, 488
— cavity, 484, 486, 551, 554
— region, 482
Bulbus cordis, 513, 518, 528, 755

CAECA, 163
Caecum, 490
*Calanus*, 311
Calcaneum, 453, 455
Calcarea, 852
Calcarious node, 823
Calciferous glands, 242, 261
Cambrian, 808

Canal system, 199
Canaliculi, 409, 462
Cancellous bone, 464
Canines, 444, 445
Capillaries, 511
Capillary, 230, 244
Capitulum, 426
Capsule, 133
Caput epididymis, 646
Carapace, 274
Carbohydrates, 499
Carbon dioxide, 260
Carbonic acid, 544
— anhydrase, 544
Carboniferous, 809
Carboxypeptidases, 507
Cardia, 483, 489
Cardiac limb, 485
— muscle, 465, 472, 479, 514
— ossicle, 284
— tube, 528
Cardinal vein, 376
Cardio-pyloric muscles, 285
— valve, 286, 287
Cardo, 317
Carnassials, 445
Carnivores, 830
Carnivorous, 499
Carotico-systemic aorta, 528
Carotid arch, 523, 525, 527, 529
— artery, 377
— labyrinth, 525
Carpal bones, 449
Carpale, 449, 454
Carpals, 454
Carpus, 449, 454
Cartilage, 470
— bones, 434, 435, 438, 446, 462
— histology, 459
Catadromous forms, 845
Catastrophism, 802
Catecholamines, 664
Cation, 476
Cauda equina, 580
Caudal artery, 376, 532
— fin, 401
— sympathetic system, 328
— vertebra, 424, 429
Caudofemoralis, 468
Cell-body, 137, 579, 590, 593
Cell cycle, 50
— membranes, 31, 479
— migration, 715
— structure, 29

Cells, 7, 8, 41
— of Rauber, 771
— of Sertoli, 674
Cellular organelles, 41
— transformation, 34
Cellulases, 499
Celluloses, 64
Cement, 443
— layer, 270
Central canal, 378
— cell, 245
— nervous system, 124, 160, 216, 247, 364, 378, 390, 479, 559, 604, 686
Centrale, 449, 454, 455
Centriole, 50, 53
Centrolecithal, 305
Centromeres, 677, 678
Centrosome, 673, 678, 683
Centrum, 418, 419, 422, 426
*Cepaea nemoralis*, 348
Cephalisation, 222, 232, 272
Cephalocarida, 860
Cephalochordata, 366, 866
— affinities, 387
Cephalopoda, 361, 865
Cephalothorax, 272
Ceratobranchial cartilage, 433
Ceratohyal, 432, 442
Cercaria, 183, 189, 191, 211
Cerci, 316, 318
Cerebellum, 562, 564, 566, 567, 569
Cerebral ganglia, 165, 175, 231, 235, 247, 272, 299, 327, 355
— hemispheres, 560, 566, 567, 574, 575
— vesicle, 379
Cerebro-spinal fluid, 562, 563, 593
Cerebrum, 568
Cervical flexure, 758
— groove, 273
— plexus, 580
— rib, 426, 427
— sympathetic nerve, 598
— — system, 670
— vertebra, 424, 426
Cervicum, 315
Cervix, 670
Cestoda, 169, 172, 855
— pathology, 180
Chaetae, 224, 237, 267

Chaetigerous sac, 224, 227, 237
Chaetognatha, 856
Chalazae, 732
Cheek teeth, 445
Chelae, 277
Cheliped, 280
Chemo-receptors, 166, 201, 209, 277, 304, 328, 355, 491, 605, 616
— contact, 328
— distant, 328
Chiasmata, 678
Chick, central nervous system, 747
— development, 731
— egg, 731
— embryonic blood vessels, 752
— — membranes, 758
— endoderm formation, 737
— fertilisation and cleavage, 733
— flexure, 756
— gastrulation, 736, 740
— gut, 749
— head-fold, 743
— heart, 752, 754
— mesoderm, 744
— — formation, 737, 739
— neural plate, 740
— notochord, 748
— role of membranes, 762
— yolk sac, 762
Chirocephalus, 311
Chitin, 64, 237, 269
Chitons, see Amphineura
Chlamydomonas, 113
Chloragogenous cells, 242, 260
— tissue, 240
Chlorocruorin, 230
Chlorohydra, 137
Chlorophyll, 49
Chloroplast, 49, 77, 111
Choline esterase, 477
Choloroquine, 102
Chondrichthyes, 869
Chondrin, 459
Chondroblasts, 418, 419, 459
Chondrocranium, 431, 438
Chorda cells, 695, 700, 724
— mesoderm, 717
Chordae tendinae, 524, 531
Chordal centrum, 418
Choroid, 613
— plexus, 593

Chordata, 69, 362, 865
Chordates, 214
Chordotonal organ, 329
Chords, 196
Chorion, 758, 764, 774
— chick, 758
— rabbit, 773
Choroid, 606
Chromaffin tissue, 664
Chromatic aberration, 614
Chromatids, 54, 678, 679, 793
Chromatin tissue, 663
Chromatophores, 410
Chromomeres, 677
Chromosome, 12, 25, 40, 51, 678, 788
— mutation, 793
Chronaxie, 478
Chrysalis, 332
Chrysopsin, 615
Chyle, 506
Chyme, 505
Chymotrypsin, 505, 507
Cilia, 50, 81, 83, 234, 245, 346, 395
Ciliary body, 606, 607
— feeders, 383, 388
— muscles, 613
— structure, 83
Ciliated cells, 163, 371
— epithelium, 163, 395, 593
Ciliophora, 850
Circular canal, 128, 151
— fibre, 607
— muscles, 163, 239, 258
Circulus cephalicus, 523
Circum-oesophageal connectives, 299, 300, 327
Circum-pharyngeal connectives, 247
Cirri, 224, 225
Cirripedia, 861
Cis form, 797
Cis-trans configuration, 797
Cistrons, 797
Citric acid cycle, see Krebs cycle
Clasper, 401, 448, 639
Class, 68
Clavicle, 450, 456, 458
Claw, 65, 412, 415
Cleavage, 147, 682, 684, 688
— furrow, 734
— holoblastic, 689
— meroblastic, 689

Cleavage, plane, 707, 708, 735
Clitellata, 267
Clitellum, 237, 238, 252, 253, 255, 263, 264
Clitoris, 648
Cloaca, 195, 199, 488, 633, 638, 642, 644
Cloacal aperture, 401
Clone, 88, 91
Close-packed tissues, 63, 392
Club-shaped gland, 704
Clypeus, 316
Cnemial crest, 451
Cnidaria, 852
Cnidocil, 134, 136
Cochlea, 624, 630, 632
Cochlear canal, 624
— duct, 624
Cockroach, 272
— alimentary canal, 321
— appendages, 317
— blood vascular system, 323
— cuticle, 316
— excretory system, 326
— form of the body, 315
— habitat, 315
— internal skeleton, 317
— life-history, 330
— metamorphosis, 331
— neuro-sensory system, 327
— nutrition, 322
— reproductive system, 329
— respiratory system, 324
— wings, 318
Cocoon, 264, 334
— membrane, 264
Coelenterata, 126, 127, 852
— nutrition, 143
Coeliac artery, 519, 532
Coeliaco-Mesenteric artery, 525
Coelom, 158, 214, 217, 267, 271, 701, 725
Coelomata, 158
Coelomates, general, 214, 246
Coelomic corpuscle, 239, 246
— epithelium, 217, 227, 240, 245, 481
— fluid, 217, 260
— funnel, 636
— sacs, 220, 348
Coelomoducts, 218, 253
Coelomostome, 633
Coenosarc, 148

Co-enzymes, 31
Coiled filament, 133
Collagen, 47, 65, 196, 204, 398, 418
Collar, 349
Collateral ganglia, 598
Collecting tubule, 649
Colleterial glands, 330
Colloid, 660
Colon, 490, 509
Colour vision, 612
Columella, 349, 435
— auris, 437, 624, 632
Columellar muscle, 350
Columnae carneae, 531
Columnar epithelium, 392, 394
Commensal, 839
Commissural vessel, 217, 230
Commissure, 200, 299, 566, 567
Common carotid, 531
Communities, 830
Comparative physiology, 820
Complete metamorphosis, 332, 338
Complex cells, 39
Compound epithelia, 393, 395
— eye, 301, 315
Compression, 470
— member, 470
Conchiolin, 349
Conditioned reflex, 120, 628
Cones, 608, 611, 613
Conglobate gland, 329
Conjugants, 89
Conjugation, 88
Conjunctiva, 609, 613
Connectives, 272, 580
Connective tissue, 391, 392, 397, 653
Continental drift, 818
Contractile vacuole, 75, 87, 114
Control of cell division, 58
— systems in bacteria, 56
— — complex cells, 59
Conus arteriosus, 518
— terminalis, 579
Convergent evolution, 840
Convoluta, 137
Convoluted tubule, 651
Copepoda, 310, 861
Coprophagy, 509
Copulation, 261, 357
— path, 706, 708

Copulatory spicules, 195
Coraco-branchial muscles, 448, 551
Coraco-hyoid muscle, 448, 551
Coracoid, 450, 455, 456
— region, 447
— spur, 457
Coraco-mandibular muscle, 448
Cornea, 361, 606, 613
Corona radiata, 769
Corpora allata, 328, 331, 671
— cavernosa, 647
Corpora lutea, 658
Corpora pedunculata, 231, 299
— quadrigemina, 569, 572, 578
— striata, 560, 571, 575, 576
Corpus albicans, 568
— callosum, 563, 571
— luteum, 655
— mammillare, 568
— spongiosum, 647
— striatum, 561
Corpuscle, 230, 510
Cortex, 414
Cortical layer, 196
Cortico-spinal tracts, 595
Corticosteroids, 664
Corticosterone, 664
Cortisol, 665
Cortisone, 665
Costal cartilages, 459
Cotyloid, 458
Coughing reflex, 553
Counter current exchange system, 553, 652
Coxa, 318
Coxopodite, 276, 277, 291, 303
Cranial flexure, 563, 757, 759
— ganglia, 585
— nerves, 439, 559, 581, 585
— — dogfish, 585
— — frog, 588
— — mammal, 589
Craniata, 67, 390, 867
Cranium, 390, 430, 438
— dogfish, 431
Crayfish, see Astacus
Crescentic groove, 756
Cretinism, 661
Cribriform plate, 439
Cricoid, 442
— cartilage, 554, 556

Crinoidea, 859
Cristae, 622, 628
Crithridia, 105, 107
Crop, 242, 321, 351
Cross bridges, 476
Crossing over, 678, 790, 792, 793
Crossopterygii, 870
Crown, 443
Crura cerebri, 562, 565, 567, 569, 572
Crustacea, 272, 308, 860
Crypts of Lieberkühn, 490
Crystalline cone, 301
— tract, 301
Ctenidium, 346, 358, 360, 361
Ctenophora, 129, 854
Cubical epithelium, 394, 607
Cupula, 622
Cutaneous artery, 526
— sense organs, 604
Cuticle, 132, 170, 182, 194, 195, 196, 225, 228, 242, 252, 268, 269, 289, 381
Cuticular lens, 301
Cuvierian sinus, 521
— vein, 515, 524
Cyclops, 310
Cyclosis, 87
Cyclostomata, 390, 867
Cypris, 311
Cypris lava, 313
Cysticercus, 178
Cysts, 76, 92, 94, 100, 104, 115, 147, 191, 211
Cytase, 351, 499
Cytochrome, 548, 549
Cytomembranes, 32
β-cytomembranes, 45
Cytopharynx, 79, 86, 111
Cytoplasm, 29, 39, 41, 72
Cytoplasmic inheritance, 49, 62, 799
Cytosine, 17
Cytostome, 86, 92

DAPHNIA, 309
Darting, 283
Dart sac, 351, 354
Darwinism, 825
Daylight vision, 612
D.D.T., 102
Deamination, 260, 508
Decapoda, 308
Dehydrogenases, 547
Demi-facet, 426

Demospongia, 852
Dendrite, 51, 591
Dense bone, 461
Dental formula, 445
— groove, 443
— lamina, 443
— papilla, 410, 443
Dentalium, 360
Dentary, 437, 442
Dentine, 409, 443
Dependent differentiating, 720
Depolarisation, 600
Derbyshire neck, 661
Dermal bone, 434, 436, 465
— denticle, 409, 410
— papilla, 410
Dermatome, 382, 416, 703, 725, 747
Dermis, 407, 413
Dermotrichia, 410, 448
Descending artery, 292
Desmonemes, see Volvents, 135
Desmosome, 63, 132, 393, 474, 479
Desoxyribose, 15
Determinate cleavage, 685, 693
Determination, 716
— of sex, 788
Deutoplasm, 673
Development, amphioxus, 691
— frog, 704
— mammals, 765
Devonian, 809
Diabetes insipidus, 660
— mellitus, 666
Diakinesis, 678
Diaphragm, 403, 481, 556, 558
Diaphysis, 464
Diapsida, 874
Diastatic enzymes, 503
Diastole, 293
Differentiation, 58, 61, 705, 717
Digenea, 169
Digestion, 2, 215, 503
Digestive cells, 137
— diverticula, 287, 307
— enzymes, 142, 287
— gland, 190, 347, 351, 352, 354, 373
Digitigrade, 452
Digits, 402, 448
Dihybrid, 786
Dipeptidases, 507

Diphosphopyridine Nucleotide, 548
Diphyodont, 444
Diploblastic, 157
Diploblastica, 126
Diploid, 155, 675
— condition, 786
— number, 677, 684
Diplospondyly, 421
Diptera, 319
Direct muscles, 319
Disaccharides, 499
Disc, 151
Discontinuous distribution, 817
Dispersal, 212
Distal pigment, 302
Ditremata, 645
DNA, 14, 24, 33, 52, 798
— and control of protein synthesis, 35
— and proteins, 23
— cycle, 28
— replication, 18
Dogfish, alimentary system, 484
— appendages, 401
— appendicular skeleton, 446
— axial skeleton, 419
— brain, 564
— cranial nerves, 585
— cranium, 431
— external features, 400
— head, 400
— locomotion, 403
— respiration, 549
— skin, 408
— trunk and tail, 401
— Urino-genital system, 637
— vascular system, 518
Dolichosaccus, 191
Dominance, 787
Dominant character, 786
— gene, 791
Donnan effect, 545
Dorsal abdominal artery, 292
— aorta, 323, 376, 390, 514, 519, 523, 525, 531, 753
— ciliated organ, 233
— commissure, 563
— diaphragm, 324
— diverticulum, 241
— horn, 579
— lip, 707
— — of the blastopore, 712, 718

Dorsal lobe, 486
— longitudinal canals, 374
— — muscles, 319
— mesocardium, 513
— nerve cord, 200
— pores, 237
— root, 579, 593
— — ganglion, 579, 581, 593, 723, 748
— surface, 162
— vessel, 217, 230, 243, 245, 375
Dorso-intestinal vessel, 244
Dorsolateral channels, 288
— placodes, 620
Dorso-lumbar vein, 526
Dorso-subneural vessel, 244
Dorso-ventral muscles, 163
Double circulation, 518
Drain-pipe cell, 246
Drone, 341
Drosophila, 52, 780
— chromosomes, 55
Dryopteris, 155
Ductus arteriosus, 523, 529, 533
— botalli, 523
— caroticus, 523, 529
— cuvieri, 376, 515, 754
— ejaculatorius, 329
— endolymphaticus, 622, 623, 632
— venosus, 753, 755
Duodenal vein, 532
Duodenum, 483, 484, 486, 488, 490
— histology, 495
Dura mater, 563, 593
Dysentery, 337

EAR, 621
— amphibians, 623
— development, 631
— mammals, 624
— physiology, 627
— ossicles, 438, 630
Ecdysis, 270, 289
Eedysone, 799
Echinodermata, 362, 858
Echinoidea, 859
Echiuroidea, 858
Ecology, 1, 830 46
Ecosystem, 830
Ectoderm, 127, 147, 148, 157, 215, 306, 685, 691, 694, 740, 744
Ectoparasites, 169, 208
Ectoplasm, 72, 79, 93, 113

Ectoprocta, 856
Effector neuron, 300
— organs, 5
Efferent branchial arteries, 519
— — loop, 551
— — vessel, 376, 516
— fibres, 572, 591, 592
— nephridial vessel, 244
— nerve-cells, 160
— nerve-root, 379
— neuron, 124, 477
— Pseudobranchial artery, 520
Egestion, 2, 143
Egg, 211, 354, 387, 641
— -case, 641
— -cells, 654
— membrane, 202, 689
— orientation, 690
Eisenia foetida, 262
Elastica externa, 418
Elastic cartilage, 460, 632
Elastin, 398
Elephantiasis, 206
Elytra, 319
Embryo, 265, 687
Embryology, 1, 682, 815
— chordate, 682
Embryonal disc, 772
— knob, 771
Embryonic cell, 188
— endoderm, 737
— field, 721
— membranes, 758
— period, 687
— vessels, 752
Embryophore, 177
Enamel, 409, 443
— organ, 410, 443
Endite, 276, 278, 279
Endocardiac tube, 513
Endocardium, 754
Endochondral ossification, 463
Endocrine organs, 655, 667
— system, vertebrates, 655
Endocuticle, 269, 290
Endoderm, 127, 157, 306, 685, 691, 694, 711, 744
Endodermal cells, 137, 143
— lamella, 151
— plate, 695
Endolymph, 622, 624, 630
Endoparasites, 169, 208, 210
Endophragmal skeleton, 269
Endoplasm, 72, 79, 93

Endoplasmic reticulum (ER), 47, 182
Endopleurites, 284
Endopodite, 276, 279
Endopterygota, 333
Endoskeleton, 417
Endospores, 33
Endosternites, 284
Endosteum, 460
Endostylar cilia, 384
Endostyle, 371, 384, 388
Endothelia, 392
Endothelial lining, 754
Endothelio-chorialis placenta, 777
Endothelium, 513
End plate potential, 477
— sac, 296
Entamoeba histolytica, 104
Enteric disease, 337
— plexus, 249
Enteroceptors, 605
Enterokinase, 507
Enteron, 127, 143, 147, 151, 163
Ento-mesoblast, 714
Entoprocta, 856
Entropy, 9, 31
Enzyme induction, 56, 57
— repression, 56
Enzymes, 22, 33, 476, 503, 798
— as catalysts, 22
Eohippus, 810
Ephemeroptera, 183
Ephippium, 310
Epiboly, 713
Epibranchial artery, 519
— cartilage, 433
Epicardium, 513
Epicoracoid, 456
Epicranial plate, 316
Epicuticle, 269
Epidermal sense organs, 252
Epidermis, 196, 225, 269, 407, 413
Epididymis, 636, 646
Epigenetic sequential, 60
Epiglottis, 489, 557, 618
Epihyal, 442
Epimeron, 273
Epimysium, 473
Epineurium, 249, 591
Epipharyngeal groove, 372, 385
Epipharynx, 318, 321
Epiphragm, 356
Epiphysis, 424, 461, 464
Epipleur, 367, 373

Epipodite, 276, 295
Episomes, 49
Episternum, 456
Epithelia, 392
Epithelio-chorialis placenta, 777
Epithelium, 391, 392
Equus, 811
Erector muscle, 414
Erythroblasts, 535
Erythrocytes, 98, 535
Erythropioetin, 59
Escape movement, 259, 283
Escherichia coli, 20
Ethmoid, 439, 440
Ethmo-palatine ligament, 433
Euglena, behaviour, 119
— form and habitat, 110
— locomotion, 112
— nutrition, 113
— osmoregulation, 114
— reproduction, 115
— respiration and excretion, 114
Euglenoid movement, 110
Euplotes, 82
Eustachian tube, 483, 487, 489, 589, 623, 626, 629, 632, 727
— valve, 530
Eutheria, 766, 879
Evocation, 720
Execretion, 1, 4
Excretory organs, 215, 821
— pore, 195, 199
Exites, 276
Exoccipital, 435, 439
Exocrine glands, 655
Exocuticle, 269
Exo-erythrocytic phase, 97
Exopeptidases, 507
Exopodite, 276, 280
Exopterygota, 333
Exoskeleton, 64, 269, 273
Expiration, 559
Exponential growth, 58
Extension, 713
Extensor, 271
— cruris, 467
— muscles, 405, 466
External auditory meatus, 402, 440. 626
— carotid, 529
— — artery, 525, 531
— ear, 621, 624
— gills, 728
— iliac vein, 532
— jugular vein, 524, 526, 532

External mandibular branch, 586
— nares, 554
— nostrils, 401
— respiration, 3, 216, 542
Exteroceptors, 605
Extrabranchial cartilages, 433
Extra-cellular, 247
— digestion, 143
— fluids, 31
Extra-embryonic coelom, 746, 759, 760, 761, 774, 775
— endoderm, 735, 737
— mesoderm, 740, 775
Extra-oesophageal vessel, 243
Extrinsic eye muscles, 613
— muscles, 285, 606
Eye, 166, 171, 225, 355, 380, 401
— development, 612
— physiology, 610
— structure, 606
Eyeball, 606
Eyelids, 609

FACIAL nerve, 583, 585, 586
Facultative parasite, 210
Faeces, 2, 212, 326, 509
Fallopian tube, 648, 655
False ribs, 429
Family, 68
Fasciculi, 473
Fasciola, alimentary canal, 183
— excretory system, 184
— form and structure, 181
— life-history, 186
— nervous system, 184
— pathology, 192
— reproductive system, 184
Fat cells, 399
Fate-map, 709
Fats, 499
Fatty acids, 203
Feather, 65
Feed-back mechanism, 160
— principle, 479
Female pronucleus, 706
Femoral artery, 532
— vein, 526
Femur, 318, 449, 451, 453, 455, 467
Fenestra ovalis, 624, 626, 630

Fenestra rotunda, 625, 626, 630
Fertilisation, 357, 680, 682, 693
— membrane, 683
Fibril, 153, 473
Fibrin, 537
Fibrinogen, 294, 535, 537
Fibro-cartilage, 424, 459, 460
Fibrocytes, 47, 63, 398
Fibrous sheath, 418
— zonule, 615
Fibula, 449, 455
Fibulare, 449, 453, 455
Filaroidea, 206
Filter, 286
Filter-chamber, 284, 286, 287
Filum terminale, 579, 580
Final host, 212
Fine axon, 232
— fibres, 196
Fin ray, 369
First polar body, 677, 691
Fishes, membranous labyrinth, 622
Flagella, 83, 106, 111, 137, 166, 317, 354, 673
Flame-bulbs, 161, 166, 190, 218
Flatworms, 208
Flavin adenine dinucleotide, 549
Flea, 108
Flexor, 271
muscle, 405, 466
Flexures, 756
Floating ribs, 429
Flocculus, 569
Flukes, 169
Foetus, 533, 776
Follicle, 676
— stimulaying hormone, 658
Follicular epithelium, 383
Fontanelles, 430
Food, 499
— chains, 78, 831
Food-cup, 73
Food-vacuole, 73, 87
Foot, 449
Foramen, 442
— magnum, 431, 435, 439
— of Monro, 561, 565, 567
— ovale, 530, 533
Foraminifera, 360
Force pump, 552, 554
Fore arm, 448,
— brain, 560

Fore gut, 744, 749
Formative cell, 224
— movements, 698
Fossa ovalis, 530
Fossils, 806
Fourth ventricle, 562, 567
Fovea centralis, 608
Frog, alimentary system, 486
— appendages, 402
— axial skeleton, 421
— brain, 566
— blastula, 709
— cleavage, 707
— cranial nerves, 588
— development of gut, 726
— — — mesoderm, 724
— — — notochord, 724
— differentiation, 715
— egg, 705
— fertilisation, 705
— gastrulation, 711
— girdles, 455
— head, 401
— limbs, 453
— locomotion, 405
— metamorphosis, 731
— neural tube, 722
— organisers, 717
— organogeny, 722
— respiration, 554
— skin, 411
— skull, 434
— spawn, 645, 705
— tadpole, 729
— teeth, 444
— trunk, 402
— urino-genital system, 642
— vascular system, 522
Frons, 316
Frontal, 436, 439
— cilia, 384
— ring, 439
— segment, 439
Frontoparietal, 436
Fusion nucleus, 89

GALEA, 317, 341
Gall bladder, 488, 491
Gametes, 90, 233
Gametic nuclei, 89, 90
Gametocyst, see Association cyst
Gametocyte, 94, 98
Gametogenesis, 673, 674
Gametophytes, 155
Gamont, see Gametocyte

Ganglia, 200, 232, 299, 591
Ganglion cells, 249, 608
Gasserian ganglion, 585
Gastric caeca, 334
— glands, 395, 483, 494
— juice, 483, 494
— mill, 284, 287
— vein, 526
Gastrin, 504
Gastrolith, 285
Gastropoda, 347, 358, 864
Gastrula, 147
— larva, 703
Gastrulation, 62, 147, 306, 682, 685, 686, 691, 698
Gel, 73
Gelatinous zonule, 615
Genae, 316
Gene, 12, 24, 60, 781, 782, 784, 785, 792, 797, 827
— action, 798
— complex, 91
— distribution, 792
— mutation, 793, 794
Generic name, 67
Genetical changes, 803
Genetic code, 24
— map, 792
— molecule, 25, 29
— — structural changes, 26
— substance, 11
— units, 49
Genetics, 1, 779
Geniculate ganglion, 585
Genital armature, 318, 329
— artery, 532
— atrium, 164, 171, 176, 186, 355
— ducts, 636
— pore, 355
— pouch, 330
— ridge, 635
— rudiments, 306
— vein, 532
Genitalia, 164, 175
Genito-intestinal canal, 172
Genome, 12, 25, 28, 33, 48, 60, 780
Genotype, 780, 786, 796
Genus, 67, 68
Geographical distribution, 817
Geological records, 807
— succession, 808
Geotaxis, 146

Germarium, 164
Germ-cells, 131, 154, 253 674, 703
Germinal cells, 57, 191
— epithelium, 653, 654, 673, 674
— unit, 785
— vesicle, 692, 732
Germinative layer, 396
Germ-layer, 157, 682, 685
— theory, 158
Germ mother-cells, 146
Germplasm, 126
Gestation, 778
Giant axon, 232, 251
— cell, 197, 251
— fibre, 249
— — system, 251, 283, 300
Gill, 216, 294, 390, 516, 522, 542, 546
— bar, 369, 388
— cleft, 400
— lamellae, 551
— ray, 551
— slit, 369, 388
Girder, 470
Girdles, frog, 455
— mammal, 457
— tetrapods, 450
Gizzard, 242, 256, 284, 321, 322
Gland-cell, 131, 137, 163, 371
Glandular cell, 225
— epithelium, 326, 392
Glans penis, 647
Glenoid cavity, 457
— fossa, 442
Globulin, 535
Glomerular filtrate, 650
Glomerulus, 633, 650
Glomus, 634
Glossa, 317
Glossina, see Tsetse flies
Glossopharyngeal ganglion, 585
— nerve, 584, 588
Glottis, 487, 489, 542, 554, 557
Glucagon, 498, 665, 666
Glucocorticoid hormone, 664
Glucose, 203
Glutamic acid, 27
Glutathione, 139
Gluteus, 468
Glutinants, 136
Glycine, 21
Glycogen, 111, 187, 260, 476

Glycylalanine, 21
Gnathobase, 317
Gnathostomata, 390, 868
Goblet cells, 239, 241, 255
Goitre, 661
Golgi body, 44
Gonad, 137, 233, 252, 390, 667
Gonadial artery, 519, 526
— hormones, 667
— vein, 526
Gonadotropic hormones, 658
Gonads, vertebrates, 635
Gonapophysis, 318
Gonocoel, 382
Gonoducts, 220, 252, 305
Gonotheca, 150
Gonotome, 382
Graafian follicle, 655
Gracilis, 467, 468
Granular leucocytes, 536
Granulocytes, 536
Green gland, 278, 297, see also Antennary gland
Gregarine movement, 94
Gregarinida, 94
Grey crescent, 683, 707, 709, 716, 717
— matter, 593
Ground substance, 398
Group, 67
Growth, 1, 2, 58, 674 curve, 58
— differentiation hormone, 331
— hormone, 659
Guanin, 615
Guanine, 17
Gubernaculum, 646
Guinea worm, 206
Gullet, 111
Gut, 122, 215
Gutter cell, 245
Gyrodactyloids, 172

HAEM, 543
Haemal arch, 421
Haemo-chorialis placenta, 777
Haemocoel, 100, 218, 268, 291, 323, 347
Haemocyanin, 294, 352
Haemo-endothelial placenta, 777
Haemoglobin, 26, 203, 230, 309, 536, 543
Haemolymph, 271, 290, 291, 323

Haemophilia, 537, 790
Hair, 65, 413, 414
— follicle, 414
— papilla, 414
— shaft, 414
Hallation, 302
Hallux, 449
Hand, 449
Haploid, 90, 104, 155, 675
— number, 88, 677, 684, 786
Hatschek's groove, 367
— pit, 702
Haversian canals, 461
— system, 462, 465
Head, 272, 315, 403, 450, 451, 673, 683
— capsule, 316, 327
— fold, 744, 754, 759, 773
— foot, 345
— organiser, 717
— segments, 582
Hearing, 603, 627
Heart, 323, 347, 350, 352, 364, 390, 510, 512, 514, 528, 726
— muscle, 465
Helicotrema, 624, 630
Helix aspersa, alimentary canal, 350
— — blood vascular system, 352
— — excretory system, 352
— — external features, 348
— — life-history, 357
— — locomotion and behaviour, 356
— — mantle and lung, 350
— — nervous system, 355
— — reproductive system, 353
— — sense organs, 355
— — shell, 349
— pomatia, 356
Hemibranch, 551
Hemichordata, 366, 865
Hemimetabolous metamorphosis, 332
Henson's node, 738
Heparin, 537
Hepatic artery, 292, 497
caeca, 321
— portal system, 521, 666
— — vein, 203, 376, 496, 508, 516, 521, 526
— sinus, 521
— vein, 376, 496, 508, 516, 526, 532
Hepato-pancreatic duct, 488

Herbivores, 830
Herbivorous, 499
Hermaphrodite, 103, 164, 205, 252
— duct, 354
Heterocercal tail, 405
Heterodont, 444
Heteronereis, 234
Heterophils, 536
Heterotrophic, see Holozoic
— micro-organisms, 830
— organisms, 830
Heterozygotes, 783
Heterozygous, 786
Hexacanth, 178, 856
— embryo, 177
Hexactinellida, 852
Hibernation, 356
Hind-brain, 560, 562
Hind-gut, 372, 744, 752
Hippocampal lobe, 568, 576
Hippocampus, 571
Hirudinea, 267, 857
Hirudo medicinalis, 267
Histology, 1, 391
— blood, 533
— bone, 460
— cartilage, 459
— duodenum, 495
— liver, 495
— muscles, 472
— nervous tissue, 590
— oesophagus, 493
— ovary, 654
— pancreas, 498
— spinal cord, 592
— stomach, 493
— sub-mandibular gland, 491
— tongue, 491
Histones, 62
Holoblastic cleavage, 707
Holobranch, 551
Holometabolous metamorphosis, 332
Holophytic, 77, 113, 209
Holothuroidea, 859
Holotricha, 850
Holotrichous isorhizas, see Large glutinants, 135
Holozoic, 77, 113
Homeostatic, 31
Homocercal tail, 405
Homodont, 444
Homogentistic acid, 798
Homoiothermic, 412
Homologous structures, 814

Homozygotes, 783
Homozygous, 786
Honey bee, behaviour, 343
— structure and life-history, 340
— spoon, 342
Hooks, 170
Hookworm, 194, 205
Hooves, 415
Hopper, 334
Hormones, 59, 512, 535, 562, 653, 655
Horny jaws, 729
Host, 212, 213
Housefly, see Musca domestica
Humerus, 449, 450, 454
Humus, 256
Hyaline cartilage, see Cartilage
Hyaloid canal, 607
Hydatid cyst, 180
Hydra, asexual reproduction, 145
— behaviour, 140
— digestion, 142
— excretion, 145
— feeding, 138
— gaseous interchange, 145
— locomotion, 139
— movements, 138
— osmoregulation, 145
— regeneration, 146
— sexual reproduction, 146
— structure, 130
Hydranths, 148, 155
Hydrocaulus, 148
Hydroid, 127
— zooids, 128
Hydrorhiza, 148
Hydrostatic pressure, 198, 204, 293
— skeleton, 270
Hydrotheca, 148
5-Hydroxytryptamine, 133
Hydrozoa, 852
Hygroreceptors, 605
Hymen, 641
Hymenoptera, 340
Hyoid arch, 432, 437, 632
Hyoidean artery, 520, 521
— branch, 586
— pouch, 727
Hyomandibular branch, 586
— cartilage, 432, 435
— cleft, 583, 588

Hyostylic suspension, 432, 435
Hypapophysis, 425
Hypertonic, 261
Hypobranchial cartilage, 433
— plexus, 584
Hypochordal rod, 724
Hypodermis, 331
Hypoglossal nerve, 580, 584, 589
Hypopharynx, 109, 318, 321
Hypophysis, 561, 726, 564, 657
Hypostome, 137, 148
Hypothalamus, 657, 659
Hypotonic fluid, 75
H-zone, 474

ICHNEUMON fly, 213
I-discs, 473
Igneous rocks, 806
Ileo-caecal valve, 490
Ileum, 486, 488, 490, 506
Iliac artery, 526, 532
Iliacus, 469
Ilo-colonic ring, 372, 386
Iliolumbar vein, 532
Ilium, 450, 457, 458, 467
Image formation, 610
Imaginal buds, 332
Imago, 333
Incisor process, 278
— teeth, 402
Incisors, 444, 445
Incomplete dominance, 787
— metamorphosis, 332
Incubation period, 101
Incus, 446, 627, 632
Independent effectors, 142
Indeterminate cleavage, 685
Indirect muscles, 319
Individuation, 721
Induction, 61
Inductive cue, 60
Infective egg, 203
— stage, 212
Inferior jugular sinus, 521
— — vein, 524
— oblique muscle, 582
— rectus, 582
Infundibular organ, 380
Infundibulum, 561, 564, 565, 657
Ingestion, 2, 215, 501
Inguinal canal, 646

Inheritance, 24, 786
Inhibitor genes, 57
Inner cell mass, 770
— ear, 619, 621
Innominate artery, 519, 529, 531
— bone, 458
— vein, 526
Insecta, 862
Insect flight, 319
Inspiration, 558
Insulin, 498, 665, 666
Interauricular septum, 517, 523, 528, 530
Interbranchial septum, 550
Intercalated discs, 479
Intercellular bridge, 393
— cement, 392
— gap, 392
Intercostal arteries, 532
Inter-dorsals, 419
Interferon, 540
Intermediary neuron, 300
Intermedin, 659
Intermedium, 449, 454
Intermoult period, 289
Intermuscular plexus, 248
Internal carotid, 525, 529, 531
— — artery, 520
— fertilisation, 353
— gills, 730
— iliac vein, 532
— jugular vein, 524, 526, 532
— nares, 554, 556
Inter-nasal septum, 440
Inter-neural facilitation, 140
Internuncial nerve-cells, 160
— neuron, 124, 232, 249, 251, 252, 594
Interphase, 679
Inter-radial tentacles, 151
Inter-radii, 151
Interrenal bodies, 663
Intersomitic furrow, 746
Interstitial cell, 131, 133, 135, 137, 146, 149, 653
— — stimulating hormone (ICSH), 658
— fluid, 511
— matrix, 397
— tissue, 668
Inter-ventrals, 419
Interventricular septum, 528
Intervertebral disc, 424, 460

Intervertebral neural plate, 420
Intestinal glands, 395
— vein, 526
Intestine, 190, 229, 242, 287, 352, 483, 485, 488, 490
Intra-cellular, 246
— digestion, 143, 203
— duct, 218
— fluids, 30
Invagination, 713
Inverted retina, 609, 615
Iris, 606, 613
Ischiopodite, 291
Ischiopubic bar, 448
— symphysis, 467
Ischium, 450, 457, 458
Island faunas, 819
Islets of Langerhans, 498, 665
Isogametes, 95
Isometric contraction, 478
Isopoda, 313
Isoptera, 340
Isotonic contraction, 478
Iter, 563, 565

JACOBSON'S Organ, 619
Janus green, 41
— — reaction, 41
Jaw, 228, 350
Jejunum, 490, 506
Joints, 450, 452, 458
Jugal bone, 441
Juvenile, 205, 206
— hormone, 234, 331
Juxta-glomerular complex, 652

KAPPA particles, 92
Karyosome, 111
Keratin, 65, 196
Kidney, 347, 350, 352, 361, 390, 633
— ducts, 634
— histology and physiology, 648
— tubule, 633
Killer strains, 92
Kinase, 505
Kinetodesma, 50, 82
Kinetodesmata, 85
Kinetoplast, 106
Kinetosome, see Basal body
— see Basal granule
Kinety, 82

Kingdom, 68
Kölliker's pit, 380
Krebs cycle, 548

LABIAL cartilages, 433
Labium, 315, 317, 321
Labrum, 284, 316, 318, 321
Labyrinth, 296
Lachrymal duct, 609
— gland, 609
Lacinia, 317
Lacteal, 508
Lactic acid, 203, 548
Lactogenic hormone (LTH), 659
Lacunae, 459, 462, 495
Lagena, 623, 624, 632
Lamarckism, 825
Lamellae, 239, 242
Lamellibranchiata, 360, 864
Lamina terminalis, 560
Laminaria, 148
Large glutinants, 135
— intestine, 322
Larva, 148, 332, 687
Laryngeal cartilages, 437, 442
— chamber, 487, 554
Larynx, 460, 489, 556
Latent period, 478
Lateral amniotic folds, 760
— cephalic artery, 292
— chords, 196
— cilia, 384, 385
— ganglia, 598
— limiting folds, 749
— — groove, 749, 760
— line canal, 400
— — organs, 730
— — system, 620
— lip, 713
— lobe, 569
— nerve cord, 200
— neural vessel, 243
— plate mesoderm, 365, 416, 472, 581, 633, 701, 725, 740
— pouches, 161
— teeth, 284
— temporal lobe, 568
— ventricle, 561, 562, 567, 568
Lateralis branch, 588
Laurer's canal, 176
Left subclavian artery, 531
Leishmania, 105, 107
Lens, 225, 361
— capsule, 610

Lens rudiments, 613
Lepas, 312
Lepidochitona, 357
Lepidoptera, 338
Leptocephali, 845
Leptomonas, 105, 107
Leptotene, 677
Leucocytes, 535
Lienogastric artery, 519
— vein, 532
Life, 9
Life-cycle, 28
Ligaments, 399, 470
Ligamentum nuchae, 399
Light, 834
Ligia, 313
Lignins, 64
Limb bud, 731
— girdle, 390
— skeleton, 390
Limbs, 390
— frog, 453
— rabbit, 454
Limiting folds, 755
Limnaea, 188, 191, 348
Lingua, 318
Lingual artery, 526
— vein, 526
Lining epithelium, 242
Linkage groups, 790
Lipase, 287, 505, 507
Lipid membrane, 7
Lipids, 31
Lipoid layer, 270, 536
Lipolytic enzymes, 504
Lips of the blastopore, 685
Liver, 191, 394, 483, 484, 486, 490, 727
— diverticulum, 727
— fluke, see Fasciola
— histology, 495
— lobes, 490
— lobule, 497
— rot, 192
— sinusoids, 495
Living organisms, 6
Lobi inferiores, 564
Locomotion, 1, 2, 270
— dogfish, 403
— frog, 405
— rabbit, 406
Locomotor organs, 217
Locusta, external features, 333
— internal features, 334
— life-history, 334
Loligo, 361
Longitudinal channels, 242
— duct, 633, 634
— excretory canal, 176

Longitudinal fission, 115
— muscles, 163, 196, 204, 239, 258
— nerve, 176
— — cord, 347
Loop of Henle, 652
Lorenzini's ampullae, 621
Loudness, 630
Lower jaw, 442
Lozenge, 797
Limbar vertebra, 424
Lumbricus, alimentary canal, 240
— blood vascular system, 243
— body wall, 238
— burrowing, 259
— coelom, 239
— excretory system, 245
— external features, 236
— gaseous interchange, 259
— habitat and nutrition, 256
— life-history, 261
— locomotion, 257
— nervous system, 247
— nitrogenous excretion, 260
— oogenesis, 256
— reproductive system, 252
— sense organs, 251
— spermatogenesis, 255
Lung, 216, 349, 390, 516, 542, 546, 555, 557
Luteinising hormone (LH), 658
Luteotropic hormones, (LTH), 668
Lymph, 260, 539
— heart, 527
— spaces, 411
— vessels, 511
Lymphatic system, 206, 521, 526, 532
Lymphocytes, 536, 662
Lyra, 571

MACROPHAGES, 399
Macrophagous feeders, 274
Maculae, 622, 628
Maggots, 337
Main aorta, 352
Malacostraca, 276, 308, 861
Malaria, 96
— control, 101

Malaria parasite, 210, see also *Plasmodium*
Male pronucleus, 706
Malignant tertain fever, 101
Malleus, 446, 627, 632
Malpighian body, 633, 634
— layer, 407, 411
— tubule, 321, 326, 334
Maltase, 507
Mammal, cranial nerves, 589
— ear, 624
— eggs, 766
— girdles, 457
— respiration, 555
— skull, 437
— urino-genital system, 645
Mammalia, 877
Mammalian brain, 570
Mammary glands, 415
Mammotropic hormone (MH), 659
Mandibles, 276, 278, 315, 317, 321
Mandibular arch, 432, 583, 729
— stylets, 206
— vein, 526
Mantle, 312, 350
— cavity, 346
Manubrium, 150, 429, 458
Manus, 449
Marginal cells, 245
— vesicle, 152
— zone, 734
Marrow, 460
Massa intermedia, 563, 572
Mast cells, 399
Mastigophora, 105, 848
Matrix, 397
Maturation, 674
Maxillae, 276, 279, 315, 317, 335, 341, 436, 440
Maxillary stylets, 206
Maxillipeds, 277, 279
Maxillule, 276, 278, 279
Measly pork, 178
Mechanical stimulation, 142
Mechanoreceptors, 304, 605
Meckel's cartilage, 432, 436
Median eminence, 658, 659
— fissure, 568
— passage, 288
— tooth, 284

Mediastinal cavity, 481
Medulla, 414, 652
— oblongata, 562, 565, 567, 569, 578, 619
Medullated fibres, 591
Medusa, 127, 128, 150, 153, 154, 155
— development, 154
— reproductive organs, 154
— swimming, 153
Megagametes, 99
Megagametocyte, 98
Megakaryocytes, 534
Megamere, 177, 694, 708
Meganucleus, 79, 88, 89, 91
Mehlis gland, 164, *also see* Shell gland
Meiosis, 54, 675, 677, 680, 793
Melanin, 607
Melanism, 803
Melanophore-stimulating hormone (MSH), 659
Melatorin, 670
Membrane bone, 434, 436, 446, 462, 465
Membranous labyrinth 439, 619, 621, 622, 624, 631, 632
— — fishes, 622
Mendel, 783
Mendel's First Law, 784
— Laws, 784
— Second Law, 784, 785
Meninges, 563, 593
Mentomeckelian bones, 436
Mentum, 317
Mepacrine, 102
Meroblastic cleavage, 733, 766
Merozoite, 97, 98, 102, see *also* Schizozoites
Mesenchymal cells, 418
Mesenchyme, 158, 163, 198, 235, 267, 306, 463, 534, 711, 723, 727, 748
— cell, 197
Mesenteron, 229, 307, 321, 701, 727, 744
Mesentery, 143, 230, 374, 481
Mesethmoid, 440
Mesocardium, 754
Mesoderm, 157, 217, 271, 272, 306, 685, 691, 694, 714, 725, 744, 772

Mesodermal crescent, 694, 695
— field, 710
— grooves, 696, 700
— pouches, 218, 700
— sacs, 700
— somites, 415
Mesogloea, 127, 132
*Mesohippus*, 810
Mesomeres, 188
Mesonephric duct, 638, 640, 643, 646
— kidney, 642
— tubule, 634
Mesonephros, 633, 639
Meso-pterygium, 447
Mesorchium, 638
Mesosome, 33
Mesosternum, 456
Mesothoracic ganglia, 327
Mesothorax, 315, 318
Mesovarium, 640
Mesozoic, 809
Messenger RNA, 47, 797, 798
Metabolism, 1
Metabolites, 60
Metacarpals, 449, 454, 455
Metacercariae, 183
Metachronal rhythm, 85
Metacromion process, 457
Metagons, 800
Metameres, 272
Metamerism, 172, 221, 272, 348, 365, 415
— vertebrate head, 580
Metamorphosis, 234, 687
Metanephric duct, 635, 645
— kidney, 645
— tubule, 635
Metanephros, 633
Metaphase, 52, 678, 679
— plate, 52
Metapleural folds, 367, 373
Metapophysis, 425, 471
Meta-pterygium, 447
Metatarsals, 449, 453, 455
Metatheria, 766, 878
Metathoracic ganglia, 327
Metathorax, 315, 318
Metazoa, 121, 125, 157
Microfilaria, 206
Microgametes, 99
Microgametocyte, 98
Microlecithal, 688
— egg, 692
Micromere, 694, 708
Micronucleus, 79, 88, 89, 91
Microphagous feeders, 274
Microsomes, 36

Microvilli, 176, 199, 326, 394
Mid-brain, 560, 562, 567, 747
Middle ear, 440, 621, 623, 632, 727
— peduncle, 572
— piece, 673, 683
— tube, 245, 260
Mid-gut, 287, 334, 372
— caecum, 287
— diverticulum, 373, 386
Migration, 842
Migratory nucleus, 89
Milk, 415
Mineralocorticoid, 665
Mineral salts, 499, 500
Miocene, 811
Miohippus, 810
Miracidia, 188, 191
Miracidium larva, 209
Mitochondria, 41, 182, 673
— reproduction, 43
Mitochondrion, 36
Mitosis, 51, 675, 677
Mitotic protein, 53
Mitral valve, 531
Mixed nerve, 595
Modulations, 61
Molar, 444, 445
— process, 278
Molecular phylogeny, 63
Mollusca, 864
— structure, 345
Monoculture, 193
Monocystis, 92, 103
— form and habitat, 93
— locomotion, 93
— mode of infection, 95
— nutrition, 94
— reproduction, 94
— respiration and excretion, 94
Monocytes, 536
Monogenoidea, 170
— excretory system, 171
— life-history, 172
— nervous system, 171
— reproductive system, 171
Monohybrid, 786
Monomeric units, 49
Monophyodont, 444
Monoplacophora, 346, 864
Monosacchrides, 499
Monotonous structure, 193
Monotremata, 645, 688, 766
Morphogenesis, 58

Morphogenetic events, 58
— potential, 63
Morphological differentiation, 121
Morphology, 1
Morsitans group, 110
Morulae, 255
Mosaic development, 63
— eggs, 684
— image, see Apposition image
— type, 693
Motor cell, 249
— end plate, 477, 603
— fibre, 251, 300, 581
— impulse, 251, 564
— nerve, 479
— neuron, 124, 232, 574
Moulting, 289
— fluid, 289
— hormone, 290
Mouth, 151, 215, 228, 240, 284, 349, 357, 369, 484, 486, 703
— cavity, 321
— parts, 277
Movement, 541
Muco-protein, 261
Mucosa, 482
Mucous cone, 229
— glands, 354, 484, 729
Mucus, 210, 346, 385
— secreting glands, 410
Müllerian duct, 637, 639, 640, 644, 647
Müller's fibres, 609
Multienzyme system, 42
Multinucleate, 30
Multiple fission, 109
Multiplication, 674
Multiplicative phase, 211
Multipolar neuron, 590
Musca domestica, 336
Muscle, 465, 470
— cell, 197
— contraction, 474
— electrical properties, 476
— fibre, 163, 473, 479
— histology and physiology, 472
— insertion, 466
— origin, 466
— spindle, 477, 479
— tails, 132
— twitch, 478
Muscular bulb, 198
— collar, 189
— system, 217
— tissue, 391
— tube, 245, 247,

Musculi papillares, 531
Musculo-cutaneous vein, 526, 527
— -epithelial cells, 131, 137, 140, 149, 152
— -gland cells, 137
— -glandular organ, 164
Mushroom-shaped gland, 329
Mutagens, 795
Mutant gene, 823
Mutations, 13, 21, 28, 793, 794, 824
— effects, 796
— rates, 795
Muton, 798
Mycetozoa, 78
Myelin sheath, 591, 593
Myelocytes, 536
Myocardial tube, 754
Myocardium, 513, 754, 755
Myocoel, 702, 725
Myocommata, 368, 469
Myocytes, 43, 46
Myogenic beat, 480
Myomere, 368
Myoneme, 82, 93, 94, 112, Myosin, 476
Myotome, 365, 368, 382, 416, 469, 472, 703, 725, 747
Myriapoda, 862
Mystacarida, 860
Mytilus edulis, 83
Myxoedema, 661

NAGANA, 109
Nares, 487, 618
Narrow tube, 245
Nasal, 436, 440
— cavity, 395
— sinus, 609
Natural classification, 66, 69, 89
— selection, 5, 13, 804, 825
Naturally acquired immunity, 97
Nauplius, 310
— eye, 311
Neck, 402, 443, 451
Nectochaete, 234
Negentropy, 10
Nematoblast, 131, 133, 136, 149
Nematocyst, 133, 149
— discharge, 133, 142
Nematoda, 193, 204, 855
— abundance, 193
— distribution, 193

Nematoda, host parasite relations, 204
— life-cycles, 204
— locomotion, 204
Nematode, 208, 211
Nematomorpha, 194
Nemertine worms, 215
Nemertini, 855
Neopallium, 571, 575
Neopilina, 347
Neornithes, 877
Neosporidia, 850
Nephric tubule, 747
Nephridia, 161, 218, 244, 245, 260, 377, 387
Nephridiopore, 237, 245, 247, 261
Nephridiostome, 219, 245
Nephridium of Hatschek, 378
Nephrocoel, 633, 634, 725, 747
Nephrocoelar chamber, 634
Nephromyxia, 220
Nephrostome, 245
Nephrotomal mesoderm, 740
Nephrotome, 416, 633, 725, 747
Nephthyids, 227
Nereis, alimentary canal, 228
— blood vascular system, 230
— body wall, 225
— coelom, 227
— diversicolor, external features, 223
— excretory system, 231
— feeding, 229
— life-history, 234
— locomotion, 228
— neuro-sensory system, 231
— reproductive system, 233
Nerve, 124
— -cells, 131, 136, 231, 378, 590
— collar, 231, 247
— components, 596
— cord, 227, 258
— ending, 477, 602
— -fibre, 123, 137, 160, 231, 249, 272, 303, 380, 476, 479, 560, 593, 599, 604, 617, 618
— impulse, 5, 123, 125, 152, 217, 320, 476, 591, 604

Nerve net, 124, 140, 149, 160
— physiology, 599
— ring, 152, 184, 200, 248, 347
— root, 581
— tissue, 391
— — histology, 590
Neural arch, 418, 422, 426
— crest, 579, 581, 723, 747
— folds, 698, 722, 740
— plate, 559, 695, 698, 714, 722, 744, 747
— spine, 420, 427, 471
— tube, 378, 379, 559, 612, 698, 738, 740, 747
Neurenteric canal, 559, 699, 723, 726
Neurilemma, 591, 602
Neurocranium, 430, 431, 731
Neurofibrillae, 252, 591
Neuroglia, 249, 590
Neuroglial cells, 609, 658
Neurohypophysis, 652, 657, 658
— hormones, 659
Neuromast organs, 619, 620
— system, 619
Neuromuscular junction, 301, 477
Neuron, 123, 200, 216, 590, 604
Neuropodium, 224
Neuropore, 381, 722, 747
Neuro-secretory cells, 234, 290, 331, 656, 659, 664, 671
— system, 572
Neuro-sensory system, 159, 209, 214, 343
— — vertebrates, 559
Neutrophils, 536
Niche, 840
Nictitating membrane, 609
Nissl granules, 591
Nitrogenous excretion, 4, 298
— waste, 144
Nodes of Ranvier, 591
Non-granular leucocytes, 536
Non-medullated fibres, 591
Noradrenaline, 656, 664
Nostrils, 402
Notochord, 363, 366, 368, 388, 417, 686, 714

Notochordal process, 740, 772
— sheath, 368, 418, 724
Notopodium, 224
Notum, 316, 319
Nuchal organs, 225
Nuclear membrane, 29, 39, 673, 684
— spindle, 678, 679
Nucleic acid, 13, 14, 92
Nucleoli, 40, 94, 673
Nucleotide triphosphates, 38
Nucleotides, 17
Nucleus, 29, 39, 72, 94, 106, 111
— pulposus, 424
Nuptial pad, 412, 645
Nutrition, 1, 2, 499
Nymphs, 331

OBELIA, blastostyle, 150
— hydranths, 148
— life-history, 154
— medusa, 150
— nervous system, 152
— sense organs, 152
Oblique muscle, 227
Obturator foramen, 458
— internus, 469
Occipital condyle, 427, 439
— ring, 439
— segment, 439
Occipito-vertebral artery, 525
Ocelli, 328
Octopus, 361
Oculomotor, 582
— nerve, 586
Odonata, 183
Odontoblasts, 409, 410, 443
Odontoid fossa, 428
— process, 427
Odontophore, 345, 350, 351
Oesophageal artery, 525
— glands, 242
— pouches, 242, 261
— sympathetic system, 328
Oesophagus, 229, 241, 284, 321, 347, 351, 372, 385, 483, 484, 487, 489, 727
— histology, 493
Oestrogens, 668, 670
Oestrous cycle, 655
Olecranon process, 451
Olfactory capsule, 431

Olfactory cells, 560, 618
— lobe, 560, 564, 566, 567
— nerve, 585, 618
— organ, 487, 564, 616, 618
— pallium, 575
— pit, 380
— sac, 606, 618
— tract, 568
Oligocene, 810
Oligochaeta, 236, 267, 857
Oligoneuroptera, 863
Ommatidia, 301, 328
Omnivorous, 499
Omosternum, 456
Onchosphere, 177
Onchomiracidia, 172
Onychophora, 862
Oocyte, 144, 146, 202, 254, 256, 264, 654
Oogenesis, 676
Oogonia, 233, 256, 676
Öoplasm, 62
Ootheca, 330
Opalinata, 851
Open systems, 10
— tissues, 63
Operand, 57
Operator gene, 57
Operculum, 133
Ophiuroidea, 859
Ophthalmic branch, 586
Ophthalmicus profundus, 582
Optic capsule, 431
— chiasma, 564, 566, 568, 614, 657
— cup, 612, 719
— ganglia, 290
— lobe, 562, 564, 565, 566, 567, 578
— nerve, 585, 607
— organelle, 252
— stalk, 290, 612
— tectum, 578
— thalami, 567, 568, 572, 578
— vesicle, 606
Optocoel, 565, 567
Oral cirri, 367, 369
— groove, 79, 86, 92
— hood, 367, 368, 385
Orbit, 431, 606
Orbitosphenoid, 439
Order, 68
Ordovician, 808
Organelles, 92
Organ-forming substances, 684, 693, 695
Organic evolution, 801, 833

Organic evolution, evidence, 804-24
Organiser, 718, 742
Organ of Corti, 625, 630
— specific areas, 684, 693
Organogeny, 682, 686
Organs of special sense, 604
Orientation, 627
Origin of species, 826
Oro-nasal grooves, 400
Orthogenesis, 828
Orthoselection, see Orthogenesis, 828
Osmoregulation, 211, 261
Ossification, 462
Osteichthyes, 869
Osteoblasts, 463
Osteoclasts, 463, 464
Osteocytes, 462, 463
Ostia, 271, 292, 323
Ostracodermi, 867
Ostracoda, 312, 860
Otocith, 622
Otocysts, 356
Ovarian funnel, 255
— tubule, 330
Ovary, 146, 164, 171, 176, 186, 201, 254, 305, 330, 381, 635, 640, 644, 648, 668, 673
— histology, 654
Oviducal aperture, 282, 641
— gland, 641
Oviduct, 177, 186, 201, 238, 330, 354, 636, 644, 648
Ovipositor, 334
Ovisacs, 254, 644
Ovotestis, 354
Ovoviviparity, 689
Ovum, 103, 146, 673, 676, 677
Oxidative phosphorylation, 548
Oxygen, 259
— debt, 548
Oxyhaemoglobin, 259, 543
Oxyntic cells, 495
Oxytocin, 660

PACEMAKER, 85
Pachytene, 678
Palaeontology, 1, 805, 807
Palaeopallium, 576
Palaeoptera, 863
Palate, 440, 488

Palatine, 436, 437, 440
— branch, 586
Palato-pterygo quadrate bar, 432, 434
Pallial groove, 357
Pallium, 560, 564, 575
Palp, 225, 278
Palpalis group, 110
Palpifer, 317
Paludrine, 102
Pancreas, 483, 484, 486, 488, 490
— anatomy and development, 665
— histology, 498
— physiology, 666
Pancreatic duct, 486, 490, 498
— juices, 484, 498, 505
Papilla, 188, 200, 386
Parabasal body, 106, 107
Parachordal plates, 430
Paraglossa, 317
Paragnaths, 228
Paramecium, 103
— behaviour, 118
— dispersal, 92
— form, 79
— locomotion, 85
— nutrition, 86
— osmoregulation, 87
— reproduction, 88
— respiration and excretion, 87
— structure, 81
Paramylum, 111
— granules, 93
Paraneuroptera, 863
Parapodia, 217, 224
Parapodial muscle, 227
Parapsida, 874
Parasite, 92, 209, 274, 839
Parasitic ciliates, 209
Parasitism, general, 208, 209
Parasphenoid, 436
Parasympathetic nervous system, 472, 480
— system, 598
Parathormone, 662
Parathyroids, anatomy and development, 661
— physiology, 662
Parazoa, 851
Parenchyma, 158, 182
Parietal, 439
— ganglia, 355
— ring, 439
— segment, 439
Parotid glands, 488

Pars distalis, 658
— intermedia, 659
Pars nervosa, 658
Pars tuberalis, 659
Parthenogenesis, 309, 335
Particle-covered membranes, 47
Parturition, 670
Patella, 451
Patella groove, 451
Pathetic nerve, 586
Patheticus, 583
Pectineus, 468
Pectoral fin, 401, 446
— girdle, 446, 447, 450
Pedal artery, 352
— cord, 347
— ganglia, 355
Pedicel, 317
Peduncle, 572
Pellicle, 78, 79, 92, 106, 110
Pelvic fin, 401, 446
— girdle, 424, 428, 446, 448, 450, 456, 458
— vein, 526
Pelvis, 649
Penetrants, 133, 138, 142
Penetration gland, 188
— path, 706, 708
Penis, 164, 171, 176, 354, 647
Pentadactyl limb, 448
— — modifications, 451
Pepsin, 504
Peptic cells, 495
Peptide bond, 21
— link, 506
Peranema, 113
Periblast, 734
Pericardial cavity, 218, 271, 291, 364, 390, 726, 755
— depression, 447
Pericardio-peritoneal septum, 515
Pericardium, 291, 347, 481, 518, 530
Perichondral bone, 463
Perichondrium, 459, 462
Perichordal centrum, 419
Peri-enteric plexus, 244
Perigonadial cavity, 382
Perilymph, 622, 625, 630
Perilymphatic space, 622
Perimysium 473
Perineum, 645
Perineurium, 591
Peridontal membrane, 443
Periosteum, 399, 460, 463, 625

Periostracum, 349
Periotic, 439
Peripharyngeal band, 372, 386
Peripheral nervous system, 124, 160, 247, 559
— neuron, 300
Periplaneta, see Cockroach, 314
— americana, 315
Perisarc, 148
Peristalsis, 484, 506
Peristomial cirri, 232
— nerves, 247
Peristomium, 225, 236, 241
Peritoneal funnel, 633, 634
Peritoneum, 225, 240, 246, 368, 394, 481, 482, 638
Peritrophic membrane, 242, 334
Perivisceral cavity, 217, 220, 726
Permian, 809
Per-radii, 151
Pes, 449
Peyer's patches, 490, 663
Phage DNA, 798
Phages, see Bacteriophage, 34
Phagocyte, 98, 210, 399, 535
Phagocytic cells, 386
— corpuscle, 294
Phalanges, 449, 453, 455
Pharyngeal epithelium, 241
— glands, 241
— pouch, 752
Pharyngobranchial cartilage, 433
Pharynx, 163, 168, 198, 208, 228, 241, 321, 364, 369, 374, 385, 387, 482, 484, 489, 727, 752
Phase of growth, 96
Phenotype, 12, 780, 786, 796
Phoronida, 857
Phospholipid, 31, 48
— inclusions, 44
Phosphorylation, 548
Photoreceptor, 111, 252, 380, 608
Photosynthesis, 77, 144
Phototactic, 144
Phrenic nerve, 580
Phyllopod limbs, 311
Phyllopodia, 275, 279, 309
Phylogenetic relationships, 66

Phylogenies, 70
Phylum, 68
Physiological division of labour, 122
Physiology, 1
Phytomonadina, 113
Phytomastigina, 848
Phytoplankton, 834
Pia mater, 563, 592, 593
Pigment cell, 411
— cup, 380
— layer, 612
Pineal body, 564, 568, 670
— organ, 561
— stalk, 564, 565
Pinna, 402, 627, 632
Pinocytosis, 74, 176
Pinocytotic vesicles, 182
Pitch, 630
Pituicytes, 658
Pituitary, 652
— body, 561, 565, 566, 567, 568, 726
— — anatomy and development, 657
Placenta, 533, 775, 777
Placode, 605
Placoid scale, 409
Plankton, 308
Plantigrade, 452
Plantulae, 318
Planula, 154
Plasma, 510, 534
— membrane, 29, 33, 50, 392
— — pores, 31
— — structure, 31
Plasmalemma, 72, 76
Plasmodium, 100, 103
— life-history, 96
— falciparum, 101
— malariae, 101
— vivax, 101
Plasticity, 122, 126
Plastid, 49
Platyhelminthes, 854
Pleistocene, 810
Pleopods, 277, 282
Pleura, 481
Pleural cavities, 481, 557
— ganglia, 355
Pleurites, 316
Pleurobranchs, 295
Pleuron, 273, 276, 316
Plexus, 216, 416, 580
Pliocene, 811
Pneumostome, 349
Podobranchs, 294
Podomere, 271, 275

Poikilothermic, 412
Polar bodies, 254, 705, 769
— molecules, 31
Polarised light, 344
Polarity, 687
Pollen-basket, 342
Pollex, 449
Polychaeta, 223, 857
Polyclads, 163
Polyembryony, 191
Polyglucosamine, 269
Polymorphic, 105
Polymorphism, 129, 155
Polymorphous form, 536
Polyneuronal fibre, 301
Polyneuroptera, 863
Polynucleotides, 21
Polyp, 127
Polypeptide chain, 47
Polypeptides, 21, 24
Polyplacophora, 346
Polyploid, 30
— nucleus, 91
Polyploidy, 55, 794
Polysaccharides, 499
Polysome, 38
Polyspermy, 683
Polystoma, excretory system, 171
— form and structure, 170
— life-history, 171
— nervous system, 171
— reproductive system, 171
Polyzoa, 856
Pomatias elegans, 348
Pons varolii, 569, 572
Population crash, 841
Porifera, 851
Porphyrin, 543
Portal canal, 497
— system, 515, 516, 521, 526, 529
— vein, 532
Post-axial, 450
Posterior aorta, 352
— cardinal sinus, 521
— — vein, 515, 754
— cervical sympathetic ganglion, 598
— choroid plexus, 562, 565, 567, 569
— commissure, 563, 572
— cornua, 437
— dorsal fin, 401
— intestinal portal, 751
— limiting groove, 751
— mesenteric artery, 519, 532

Posterior mesenteric vein, 532
— peduncle, 572
— processes, 189
— rectus, 583
— testis sac, 254
— vena cava, 524, 526, 529, 532
Post-ganglionic fibre, 591
Post-spiracular branch, 586
— nerve, 583
Post-synaptic membrane, 603
Post-trematic branch, 584, 588
Post-zygapophysis, 423, 425, 426
Potassium, 476
Pre-axial, 449
Pre-cambrian, 808
Prechordal plate, 711, 712, 716
Precoracoid, 450, 456
Precoxa, 276, 278
Predator, 209, 274, 339
Pre-erythrocytic phase, 97
Preganglionic fibre, 300, 380, 591, 597
Pregancy, 670
Prehallux, 454
Prehension, 215, 275
Premaxilla, 436, 440, 554
Premolars, 444, 445
Premunition, 180
Prepuce, 647
Prepyloric ossicle, 284
Presphenoid, 438, 439
Pre-spiracular branch, 586
— nerve, 583
Press, 286
Pre-synaptic membrane, 603
Pretarsus, 318
Pre-trematic branch, 584, 588
Pre-zygapophysis, 423, 425, 426
Primaquin, 102
Primary cerebral vesicles, 560, 723
— egg membrane, 689
— filament, 474
— follicles, 654
— germ-layer, 157
— gill bar, 369, 374, 376
— — slits, 370
— gonadial cavity, 382
— host, 206
— metamerism, 272

Primary oocyte, 676, 677, 705, 732
— optic vesicle, 562, 612, 748
— organiser, 718, 720, 742
— spermatocytes, 255, 675
Primitive groove, 739
— knot, 738, 746
— streak, 738, 739, 746, 772
Primordial germ-cells, 674
Proamnion, 745
Proboscis, 336, 338
Procoelous, 422
Proctodeal invagination, 726
Proctodeum, 215, 229, 268, 287, 321, 482, 744, 752
Productivity, 831
Profundus ganglion, 585
Progesterone, 670
Progestin, 668, 670
Proglottide, 211
Proglottis, 173
Proleptus, 194
Prone, 454
Pronephric duct, 634
— tubule, 634
Pronephros, 633, 730
Pro-otic, 435
Propagative cell, 188
Prophase, 51, 677, 678, 679
Prophylaxis, 101
Proprioceptive fibres, 477
Proprioceptors, 160, 252, 258, 321, 477, 479, 564, 605
Pro-pterygium, 447
Proscolex, 178
Prosobranchia, 358
Prospective ectoderm, 710
— endoderm, 737
— epidermis, 710, 718
— fate, 709, 717
— mesoderm, 710, 740
— neural plate, 710, 718, 740
— notochord, 710, 716
Prostheca, 317
Prosthetic group, 543
Prostomial tentacles, 231
Prostomium, 225, 231, 236, 247, 252, 259
Protandric hermaphrodite, 201
Protease, 287
Protein synthesis, 35, 36, 45, 77
Proteins, 14, 21, 22, 52, 499
— as enzymes, 22

Proteins, structural, 22
Proteolytic enzyme, 241, 504
Prothalli, 155
Prothoracic glands, 331
Prothorax, 315
Prothrombin, 537
Protochordates, 366
Protonephridia, 235, 377
Protoplasm, 7
Protopodite, 276, 296
Prototheria, 766, 878
Protozoa, 71, 847
— sexual processes, 103
Protractor lentis muscle, 202, 406, 615
Protractors, 227
Protrusion, 227
Proventriculus, 321
Proximal pigment, 302
Psalterium, 571
Pseudobranch, 551
Pseudocoel, 198, 572
Pseudoheart, 243
Pseudopodia, 72, 92, 137, 209
Pseudostratified epithelium, 396
Pseudotrachea, 336
Psoas, 468
Pterocardiac ossicle, 284
Pterygoid, 436, 437, 440
Pterygota, 863
Ptilinum, 337
Ptyalin, 504
Pubic symphysis, 458, 670
Pubis, 450, 457, 458
Pulmo-cutaneous arch, 523, 526
Pulmonary aorta, 528, 531
— arch, 529
— artery, 523, 526
— chamber, 190
— vein, 350, 352, 523, 526, 530
Pulmonata, 360
Pulmonate, 348
Pulp cavity, 409, 443
Pupa, 332, 337
Puparium, 337
Pupil, 606, 613
Purine, 18
Pus, 540
Pygidium, 225, 229, 234, 236, 272
Pyloric limb, 485
— ossicle, 284
— region, 490
— sphincter, 483, 485, 505
Pyramid, 650

Pyramidines, 17
Pyrenoids, 111
Pyriform lobe, 576
Pyruvic acid, 548

QUADRATE, 438
— bone, 632
Quadratojugal, 436
Quartan fever, 101
Queen, 341

RABBIT, alimentary system, 488
— allantois, 774
— appendages, 403
— axial skeleton, 424
— blastocyst, 770, 773
— brain, 567
— cleavage, 770
— embryonic membranes, 773
— fertilisation, 769
— gastrulation, 772
— head and neck, 402
— limbs, 454
— locomotion, 406
— ovum, 768
— parturition, 778
— placenta, 776
— ribs and sternum, 429
— skin, 412
— skull, 429
— trunk and tail, 403
— vascular system, 528
— yolk sac, 775
Radial canals, 128, 143, 151
— cartilage, 446, 448
— fibre, 607
— muscle, 446
— symmetry, 129, 163
Radiale, 449, 454
Radioreceptors, 605
Radio-ulna, 454
Radius, 449, 454
Radula, 350, 356, 360
— sac, 351
Rami, 583
Ramus communicans, 580, 597
— lateralis, 585
— mandibularis, 583, 586
— maxillaris, 583, 586
— ophthalmicus, 583, 586
— visceralis, 585
Rana, see Frog
Ranidae, 68
"Reaction Profile", 47

Recapitulation theory, 685, 815
Receptaculum seminis, 354
Receptor, 200, 217, 300
— cell, 123, 604, 626
— organ, 380, 592, 622
Recessive character, 786
— gene, 791, 823
Re-combinants, 785
Recons, 798
Rectal gland, 486
— papilla, 334
Rectum, 195, 322, 347, 350, 352, 483, 488, 490
Recurrent laryngeal nerve, 589
Red bone marrow, 460
— corpuscles, 535
Redia, 188, 211
Reflex arc, 124, 300, 594
Refractive system, 610
Refractory period, 320, 479, 480, 600
Regulation eggs, 685
Reissner's membrane, 624, 630
Relaxin, 668, 670
Renal artery, 519, 525, 532, 650
— portal system, 521, 526
— — vein, 515, 521, 526
— system, vertebrates, 633
— vein, 515, 521, 526, 532, 650
Rennin, 504, 652
Repression, 60
Repressive cue, 60
Repressor genes, 57
Reproduction, 1, 4
Reproductive cells, 137
Reptilia, 872
Residual cytoplasm, 95
Resorption, 290
Respiration, 1, 3, 42
— dogfish, 549
— frog, 554
— mammal, 555
Respiratory pigment, 215
— quotient, 546
— surface, 215, 216
Resting potential, 476, 599
Retaining cells, 46, 47
Reticulocyte, 46
Reticulum, 44, 47
Retina, 361, 607, 609, 613
Retinulae, 301
Retractor femoris, 468, 469
— muscle, 202, 406
Retractors, 227

Rhabdites, 163
Rhabditis, 194
Rhabdom, 301, 302
Rhabdomeres, 301
Rheobase, 478
Rhinal fissure, 571
Rhinocoel, 563, 565, 567
Rhizopoda, 847
Rhodopsin, 608, 611
Rib, 403, 419, 420, 558
Ribonucleic acid (RNA), 33
Ribose, 16
— nucleic acid (RNA), 16
Ribosomes, 34, 36, 45
Right lymphatic duct, 532
— subclavian artery, 531
RNA, 56
RNA (m-RNA), 37, 38, 60
RNA, polymerases, 38
RNA (s-RNA), 36
RNA, types and properties, 37
Rods, 608, 611, 613
Rolling, 321, 405
Root, 443
Rostral cartilages, 433
— subneural artery, 292
Rostrum, 274, 335
Rotifera, 856
Royal jelly, 343

SACCI vasculosi, 564, 565
Sacculus, 622, 623, 624, 632
— rotundus, 490
Saccus endolymphaticus, 622
Sacral plexus, 580
— vertebra, 423, 424, 428, 456
Sacrum, 428
Salinity, 836
Saliva, 484
Salivary glands, 101, 322, 351, 484, 488
Saprophytic, 209
Saprozoic, 113, 194
— feeding, 205
Sarcolemma, 474, 476, 479, 603
Sarcomere, 474
Sartorius, 467
Scala media, 624
— tympani, 625, 630
— vestibuli, 624, 630
Scale, 278, 408
Scape, 317
Scaphognathite, 279, 295

Scaphopoda, 360, 864
Scapula, 450, 456
Scapular region, 447
Scapulo-coracoid, 458
— bone, 450
— cartilage, 447
Scavengers, 274
Schizogony, 96, 97, 98, 101
Schizont, 97, 98
Schizonticides, 102
Schizozoites, 98, 102
Schwann cells, 591
Sciatic artery, 526
— vein, 526
Sclerites, 268, 273, 316
Sclerotic, 431, 606, 613
Sclerotisation, 269
Sclerotome, 382, 416, 703, 725, 747
Scolex, 173
Scroll bones, 440
Scrotal sac, 646
Scrotum, 646
Scyphozoa, 853
Sebaceous glands, 414, 415
Second Law of Thermodynamics, 9
— polar body, 683
Secondary cercariae, 191
— filament, 474
— gill slits, 370
— host, 106, 178, 206, 212
— membrane, 689, 732
— microlecithal, 769
— oocyte, 655, 677, 691, 705, 768
— optic vesicle, 612
— palate, 438, 556
— polar body, 677
— rediac, 191
— spermatocytes, 675
Secretin, 505
— cells, 47
Secretory cells, 44, 46
Sedimentary rocks, 806
Segmental artery, 519
— ganglia, 232, 248
— mesoderm, 701
— vessel, 230
Segmentation, 172, 221
Selachii, 421, 462, 869
Self-differentiating, 720
Semen, 674
Seminal fluid, 262, 305, 645, 674
— groove, 238, 263
— vesicles, 96
Semi-circular canals, 622, 632

Semi-lunar valve, 292, 518
Semimembranosus, 467
Seminiferous tubule, 636, 653, 675
Semitendinosus, 468
Sense capsules, 430, 431, 726
— organs, 360
— plate, 726
Sensillae, 329
Sensitives, 92
Sensitivity, 1, 4
Sensory appendages, 225
— cell, 131, 137, 225, 239, 249, 381, 617
— fibre, 124, 252, 579, 581
— hairs, 328, 618, 630
— lines, 619
— nerve, 416, 479
— papillae, 234
— process, 381
— system, 604
— tube, 620
Septa, 227
Septum pellucidum, 571
Seroamniotic connection, 760, 774
Sertoli cells, 653, 668
Serum, 539
Sesamoid bone, 451
Setae, 276
Setobranchs, 295
Sex, 103
— linkage, 790
Sexual cycle, 96
— dimorphism, 103
— reproduction, 680
— transformation, 34
Shaft, 133
Shank, 449
Shearing, 470
Sheath nuclei, 591
Shell, 186, 349
— gland, 171, 176, 186
— membrane, 733
Shock, reaction, 119
Sickle cell anaemia, 26
Sight, 603, 610
Signal molecules, 59
Signet ring stage, 98
Simple epithelia, 393
— reflex arc, 124
Single circulation, 516
Sinu-auricular node, 529
— valves, 518, 524, 530
Sinus glands, 290, 671
— terminalis, 753
— venosus, 513, 518, 523, 524, 528, 753, 755

Sinuses, 520
Sinusoids, 496, 497
Siphon, 360, 639
Sipunculoidea, 858
Skeletal muscle, 465
— plate, 371
— tissue, 398
Skeleton, 417
— functions, 470
Skin, 407
— dogfish, 408
— frog, 411
— rabbit, 412
Skull, 417
— frog, 434
— mammal, 437
Sleeping sickness, 105
Small glutinants, 136
— intestine, 322
Smell, 345, 603, 616, 618
Smooth membranes, 44
Snail, see *Helix aspersa*
Sodium, 476
Sol, 73
Solar plexus, 598
Sole, 349
Solenocytes, 218, 220, 378
Soma, 90, 126
Somatic cells, 58, 188, 191
— layer, 158, 217, 746
— mesoderm, 416
— motor fibre, 581, 589, 596
— — system, 479
— reflex arc, 597
— sensory fibre, 581, 596
Somatopleure, 746, 754, 756, 760, 774
Somatropic hormone (STH), 659
Somite, 365, 579, 701, 702, 714, 725, 746
Somitic mesoderm, 739
Sound receptor, 328, 627
Special creation, 801
Species, 67, 68
Specific name, 67
Specificity, 212
Sperm, 176, 354
— duct, 354
— funnels, 254
— morula, 95
— sacs, 638
Spermatic artery, 647
— cord, 647
— nerve, 647
— vein, 647
Spermatid, 256, 675
Spermatogenesis, 674

Spermatogonia, 146, 233, 255, 674
Spermatophore, 334, 355, 357
Spermatozoa, 146, 256 305, 387, 636, 639, 643, 653
Spermatozoon, 103, 673, 676
Spermatheca, 238, 253, 255, 263, 330, 354, 357
Sphenethmoid, 435
Spherical aberration, 614
Sphincter muscle, 240, 258
Spicules, 202
Spigelian lobe, 490
Spinal accessory nerve, 589
— cord, 390, 559, 567
— — histology, 592
— nerve, 559, 579, 585, 593
Spindle, 53, 679, 684
— attachments, 678
Spinneret, 133
Spino-thalamic tracts, 595
Spiracles, 324, 400, 432, 484, 550, 730
Spiracular ligament, 433
Spiral septum, 527
— valve, 486, 525
Spirotricha, 851
Splanchnic layer, 158, 217, 240, 746
— mesoderm, 416, 480, 512
Splanchnocoel, 218, 481, 513, 633, 634, 635, 701, 726, 756
Splanchnopleure, 746, 751, 754, 756, 759, 775
Spleen, 488
Splenic vein, 526
Spongy bone, 461, 464
— protein, 196
Spontaneous generation, 801
Spore, 95
Sporoblasts, 95
Sporocyst, 95, 96, 188, 211
Sporogony, 96, 101
Sporophyte, 155
Sporozoa, 209, 210, 849
Sporozoites, 95, 97, 100, 101, 211
Spring outburst, 835
Squamosal 437, 439, 441, 442
Squamous epithelium, 393
Stapes, 435, 624, 627, 632

Statocysts, 152, 166, 277, 299, 303
Statolith, 152, 303
— hairs, 303
Stegocephalia, 872
Stellate ganglion, 598
Stelleroidea, 858
Stenopodium, 275, 280
Stenoteles, 133
*Stentor*, 82
Stereoscopic vision, 614
Sternal plate, 274
— sinus, 292
Sternebrae, 429
Sternites, 316
Sternopleural muscles, 320
Sternum, 273, 316, 403, 429, 456
Steroid hormones, 668
Steroids, 655
Stigma, 111
Stigmata, *see* Spiracles
Stipes, 317
Stomach, 347, 351, 483, 487, 489, 727
— histology, 493
Stomatogastric nervous system, 300
Stomodeum 215, 229, 268, 284, 321, 482, 561, 726, 744, 752
Stratified epithelium, 396, 489, 491
Stratum corneum, 413
— granulosum, 413
— malpighii, 396
Striated border, 394
Stridulatory crest, 333
Striped muscle, 465, 472, 473
Strobila, 173, 221
Strobilisation, 211
Stroma, 654, 676
Strngyloidea, 205
Struggle for existence, 825
Stylets, 133, 335
Stylohyal, 442
Subclacian artery, 519, 525, 529
— vein, 526, 527, 532
Sub-endostylar coelom, 374
Sub-epidermal plexus, 239, 248, 250
Sub-epithelial plexus, 232
Sub-germinal cavity, 689, 734, 737
Sub-intestinal vessel, 376
Sub-kingdom, 68
Subliminal stimuli, 478

Sublingual glands, 488
Sub-mandibular gland, histology, 491
Submaxillary glands, 488
Submentum, 317
Subneural vessel, 243
Sub-oesophageal ganglia, 299, 327
Sub-pharyngeal ganglion, 232, 247, 251
Sub-phylum, 68
Subscapular vein, 526
Subsidiary organiser, 719
Substrate, 503, 546
Sub-umbrella, 154
Succus entericus, 505, 507
Suckers, 170, 208
Suction pump, 552
Suctoria, 209
Suctorial bulb, 171
— pharynx, 190, 209
Sudorific glands, 413
Sulci, 570
Sulcus limitans, 572
Summation of stimuli, 478
Superficial ophthalmis, 583
Superior rectus, 582
Superposition image, 302
Supporting cell, 225, 239
Supra-dorsals, 419
Supraoccipital, 439
Suprarenal body, anatomy and development, 663
Supra-scapula, 456
Survival of the fittest, 826
Suspensory ligament, 607, 610, 615
Suture, 316
— line, 290
Sweat glands, 394, 413
Swimmerets, see Pleopods
Sylvian fissure, 568
Symbionts, 144, 499, 509, 839
Symbiosis, 92, 144
Sympathetic autonomic ganglia, 598
— nervous system, 300, 472, 480
— neuron, 597
— system, 580, 598
Sympathin, 602
Synapse, 123, 140, 251, 252, 300, 572, 591, 603
Synapsida, 873
Synapticula, 370, 376
Syndesmo-chorialis placenta, 777

Synecology, 830
Syngamy, 91, 104
Synovial cavity, 458
— fluid, 458
— membrane, 459
Systematics, 66
Systemic arch, 523, 525, 527, 529
Systole, 292, 293, 527

TADPOLE, 170, 212, 522
Taenia solium, excretory system, 174
— — form, 173
— — life-history, 177
— — nervous system, 174
— — proglottides, 175
— — reproductive system, 176
Tail, 402
— bud, 728
— fan, 277
— fold, 751, 760, 773
— organiser, 719
Tanned protein, 314
Tanning, 269
Tapetal layer, 302
Tapetum, 615
Tapeworm, 169, 172, 209, 210
Tarsale, 449
Tarsal podomeros, 318
Tarus, 318, 455
Taste, 603, 616
— bud, 491, 616, 617
Taxonomy, 1, 66
Tectorial membrane, 626, 630
Tectum opticum, 562
Teeth, 409, 442
Tegument, 175, 182
Telencephalon, 560, 564, 568, 575
Teleosts, 408
Telolecithal, 687
— egg, 704, 706
— ovum, 731
Telophase, 54, 679
Telosporidia, 849
Telson, 273, 277
Temperature, 834
Tendon, 399, 466
Tension, 470
— member, 470
Tensor tympani, 627
Tentacles, 127, 128, 151, 225, 349
Tentorium, 317
Teredo, 352

Tergites, 316
Tergum, 273, 316
Terminal bars, 393
Tertiary membrane, 690, 731, 732
Tessellated epithelium, 394
Testis, 146, 164, 171, 176, 184, 201, 253, 305, 329, 381, 635, 636, 638, 642, 646, 668, 673
— histology, 653
— sac, 253
Testosterone, 668
Tetanus, 479, 662
Tetrads, 793
Tetrapoda, 448
Thalamencephalon, 60, 564, 567, 568, 657, 748
Thalami, 562
Thalamo-cortical tracts, 595
Thalamus, 578
Thebesian valve, 530
Thermocline, 835
Theromorpha, 812
Thigh, 449
Third commissure, 563
— ventricle, 562, 564, 565, 567, 568
Thoracic duct, 532
— ganglia, 327
— limbs, 280
— vertebra, 424
Thorax, 272, 315, 403
Threshold stimulus, 478, 479
Thrombin, 537
Thrombocytes, 534
Thrombokinase, 537
Through-conduction paths, 141
Thymidine, 38
Thymine, 17
Thymus, anatomy, and development, 662
— physiology, 662
Tyroglobulin, 661
Thyrohyal, 442
Thyroid, anatomy and development, 660
— cartilage, 442, 556
— gland, 371, 394
— hormone, 371
— physiology, 661
— stimulating hormone (TSH), 658
Thyroxin, 658, 661
Tibia, 318, 449, 455

Tibiale, 449, 453
Tibio-fibula, 453, 467
Timbre, 630
Tissue respiration, 3, 542
Tongue, 488
— bar, 370
— histology, 491
Torsion, 358
Touch, 603
Toxins, 204
Trabeculae, 430, 431, 463
Trachea, 395, 460, 482, 556, 557
Tracheoles, 325
Transducer, 124
Transduction, 604
Transfer RNA, 798
— — (S-RNA), 36
Transferred genes, 34
Trans-form, 797
Transitional epithelium, 397
Translocation, 794
Transplantation, 717
Transport mechanisms, 542
Transverse excretory canal, 176
— fission, 87
— ligament, 428
— process, 419, 420, 422, 425, 428, 471
— septum, 515
Trematoda, 169, 854
Trematodes, 209
Tri-carboxylic cycle, 548
Triceps femoris, 467
Trichocysts, 81
Trichoptera, 183
Tricuspid valve, 531
Trigeminal nerve, 582, 583, 586
Tri-iodothyronine, 661
Triploblastica, 261, 157
Triploblastic system, 60
Trochanter, 318, 451
Trochlea, 450
Trochlear nerve, 586
Trochophore, 234
Trophi, 277
Trophic level, 832
— organ, 208
Trophoblast, 770, 772
Trophoblastic villi, 773, 776
Trophozoite, 93, 94, 95, 97
Tropocollagen, 47, 63, 398
Truncus, arteriosus, 523, 524

Trunk, 402
— organiser, 719
Trypanosoma, 105
— control, 109
— reproduction, 107
Trypanosome, 108
Trypsin, 507
Tsetse flies, 109
Tube-feet, 217
Tuberculum, 426
Tuberosities, 451
Tunica albuginea, 653
Turbatrix, 194
Turbellaria, 854
— alimentary canal, 163
— asexual reproduction and regeneration, 169
— excretory system, 166
— feeding, 168
— form, 162
— gaseous interchange, 169
— life-history, 169
— movement, 167
— neuro-sensory system, 164
— reproductive system, 163
— structure, 163
Turbinals, 440
Twilight vision, 612
Tympanal organ, 334
Tympanic, 440
— bone, 626, 632
— chamber, 483, 623, 630
— membrane, 401, 623, 626, 629
Tympanohyal, 442
Tympanum, 334, 630
Typhlosole, 242

ULNA, 449, 454
Ulnare, 454
Ultra-filtration, 199
Umbilical artery, 774
— cord, 761, 778
— vein, 775
— vesicle, 775
Umbilicus, 778
Undulating membrane, 86, 106
Ungulate, 213
Unguligrade, 453
Unipolar neuron, 590
Universal donor, 538
— recipient, 538

Unstriped muscle, 465, 472
Upper arm, 448
— jaw, 441
Uracil, 27, 38
Urates, 327
Urea, 144, 260, 296, 508, 535
Ureter, 350, 352, 635, 639, 643, 645, 647
Urethra, 646, 647, 648
Uric acid, 296, 326, 327, 352
Uricotelic excretion, 327, 353
Urinary bladder, 645
— duct, 636, 637
— papilla, 640, 641, 644
— sinus, 640
Urine, 260, 296, 650
Uriniferous tubule, 394, 633, 649
Urino-genital aperture, 403
— duct, 636, 643
— papilla, 638, 642
— sinus, 638
— system, dogfish, 637
— — frog, 642
— — mammal, 645
Urocardiac ossicle, 284
Urochordata, 366, 866
Urodela, 872
Uropods, 282
Urostyle, 422, 424, 731
Uterine milk, 776
Uterus, 171, 177, 186, 201, 637, 648, 655
— masculinus, 647
Utriculus, 622, 628, 632

VAGINA, 171, 177, 202, 330, 355, 648
Vagus, 585
— ganglion, 585
— nerve, 584, 588
Valerianic acid, 203
Valine, 27
Vascular system, 307
— — dogfish, 518
— — frog, 522
— — rabbit, 528
Vas deferens, 164, 171, 176, 185, 201, 238, 254, 262, 282, 305, 329, 636, 638, 646
— efferens, 254, 636, 638, 643
Vector, 100. 101, 102, 109, 206, 212

Vegetative hemisphere, 709
— pole, 687, 713
Vein, 319, 510, 515
Velar tentacles, 369
Velum, 154, 369
— interpositum, 572
Vena cava, 530
Venous system, 520, 526
Ventral aorta, 376, 518, 753
— diaphragm, 324
— fin, 401
— fissure, 592
— horn, 579
— lobe, 486
— nerve cord, 200, 231, 247, 249, 253, 272
— pocket, 241
— root, 579, 581, 593
— sucker, 182
— surface, 162
— sympathetic system, 328
— thoracic sinus, 292
— vessel, 217, 230, 243, 245, 375
Ventricle, 347, 352, 513, 518, 523, 524, 528, 529, 530, 562, 569, 755
Ventro-intestinal vessel, 244
— -parietal vessel, 244
Venule, 511
Vermis, 569
Vertebra, 471
Vertebral column, 364, 366, 390, 417, 421
— neural plate, 420
Vertebraterial canal, 420, 427
Vertebrates, 390
— cranial nerves, 580
— endocrine system, 655
— gonads, 635
— metamerism in the head, 580
— peripheral nervous system, 579
— renal system, 633
— sensory system, 603
— spinal cord, 578
— — nerves, 579
Vertex, 316
Vertical muscles, 319
Vesicle, 606
Vesiculae seminales, 95, 185, 201, 254, 638, 643

Vestibule, 648
Vestigial structures, 815
Vibrissae, 402, 414
Villus, 490, 507
Virus, 6, 29
Viscera, 481
Visceral arch, 390, 431, 445, 727
— — skeleton, 431
— clefts, 364, 390, 416, 431, 513, 514, 528, 582, 703, 726, 727, 752
— loop, 358
— mass, 345, 349
— motor fibre, 582, 589, 596
— muscle, 465
— pouches, 727
— sensory fibre, 581, 596
— skeleton, 430, 437, 438
Visceralis branch, 588
Vision, 610
Visual acuity, 302
— purple, 608
Vitamin A, 608
Vitamins, 499, 500, 501
Vitellarium, 164
Vitelline artery, 754
— duct, 177, 186
— gland, 164, 171, 176, 186
— membrane, 682, 689, 691
— vein, 753, 762
Vitrellae, 301
Vitreous body, 607
— humour, 607
Viviparity, 689, 766
Vocal cords, 556
Voluntary muscle, 582
Volvents, 135, 138, 142
Vomer, 436, 440
Vorticella, 82
Vulva, 648

WALKING, 283
— legs, 277, 280, 315
— pattern, 451
Water, 499, 500, 509
— vascular system, 214
Wac, 341
Weberian ossicles, 623, 628
Wheel organ, 367, 386
White corpuscles, 294, 399, 535

White fibres, 398, 461
— fibrous tissue, 399
— matter, 593
— tube, 296
Wide tube, 245
Wing, 315
— pad, 331
Wolffian body, 634
— duct, 634, 638, 640, 642, 643, 644
Workers, 340
Wrist, 448

XENOPUS laevis, 527
Xiphisternal cartilage, 429, 456
Xiphisternum, 429
Xiphoid cartilage, 456
X-organs, 290
X-ray diffraction, 41

YAWING, 321, 405
Yellow elastic tissue, 399
— fibres, 398, 460
Yolk, 236, 306, 673, 687, 741
— and development, 688
— cells, 186
— gradient, 688
— plug, 711, 714
— sac, 306, 751, 760
— — placenta, 776
— — septa, 762
— — stalk, 751
Yolky endoderm, 735
— — cells, 710
Y-organ, 290

Z-LINE, 473
Zona pellucida, 768, 772
— radiata, 768
Zoochlorellae, 137, 144
Zooid, 127
Zoology, 1
Zoomastigina, 849
Zoophytes, 127
Zooxanthellae, 144
Zygapophysis, 421, 422, 429
Zygomatic arch, 441
— glands, 488
Zygote, 90, 95, 100, 680, 682, 684
Zygotene, 677
Zygotic nucleus, 89, 90